Decision Engineering

Series Editor

Professor Rajkumar Roy
Department of Enterprise Integration
School of Industrial and Manufacturing Science
Cranfield University
Cranfield
Bedford
MK43 0AL
UK

Other titles published in this series

Cost Engineering in Practice
John McIlwraith

IPA – Concepts and Applications in Engineering
Jerzy Pokojski

Strategic Decision Making
Navneet Bhushan and Kanwal Rai

Product Lifecycle Management
John Stark

From Product Description to Cost: A Practical Approach
Volume 1: The Parametric Approach
Pierre Foussier

From Product Description to Cost: A Practical Approach
Volume 2: Building a Specific Model
Pierre Foussier

Decision-Making in Engineering Design
Yotaro Hatamura

Composite Systems Decisions
Mark Sh. Levin

Intelligent Decision-making Support Systems
Jatinder N.D. Gupta, Guisseppi A. Forgionne and Manuel Mora T.

Knowledge Acquisition in Practice
N.R. Milton

Global Product: Strategy, Product Lifecycle Management and the Billion Customer Question
John Stark

Enabling a Simulation Capability in the Organisation
Andrew Greasley

Mitsuo Gen • Runwei Cheng • Lin Lin

Network Models and Optimization

Multiobjective Genetic Algorithm Approach

Springer

Mitsuo Gen, PhD
Lin Lin, PhD
Graduate School of Information,
Production & Systems (IPS)
Waseda University
Kitakyushu 808-0135
Japan

Runwei Cheng, PhD
JANA Solutions, Inc.
Shiba 1-15-13
Minato-ku
Tokyo 105-0013
Japan

ISBN 978-1-84996-746-4 e-ISBN 978-1-84800-181-7

DOI 10.1007/978-1-84800-181-7

Decision Engineering Series ISSN 1619-5736

British Library Cataloguing in Publication Data
Gen, Mitsuo, 1944-
Network models and optimization : multiobjective genetic algorithm approach. - (Decision engineering)
1. Industrial engineering - Mathematical models 2. Genetic algorithms
I. Title II. Cheng, Runwei III. Lin, Lin
670.1'5196

Cover design: eStudio Calamar S.L., Girona, Spain

Printed on acid-free paper

9 8 7 6 5 4 3 2 1

springer.com

Preface

Network design optimization is basically a fundamental issue in various fields, including applied mathematics, computer science, engineering, management, and operations research. Network models provide a useful way for modeling various real world problems and are extensively used in many different types of systems: communications, mechanical, electronic, manufacturing and logistics. However, many practical applications impose on more complex issues, such as complex structure, complex constraints, and multiple objectives to be handled simultaneously and make the problem intractable to the traditional approaches.

Recent advances in evolutionary algorithms (EAs) focus on how to solve such practical network optimization problems. EAs are stochastic algorithms whose search strategies model the natural evolutionary phenomena; genetic inheritance and Darwinian strife for survival. Usually it is necessary to design a problem-oriented algorithm for the different types of network optimization problems according to the characteristics of the problem to be treated. Therefore, how to design efficient algorithms suitable for complex nature of network optimization problems is the major focus of this research work. Generally, EAs involve following metaheuristic optimization algorithms, such as genetic algorithm (GA), evolutionary programming (EP), evolution strategy (ES), genetic programming (GP), learning classifier systems (LCS), and swarm intelligence (comprising ant colony optimization: ACO and particle swarm optimization: PSO). Among them, genetic algorithms are perhaps the most widely known type of evolutionary algorithms today.

In the past few decades, the study on how to apply genetic algorithms to problems in the industrial engineering world has aroused a great deal of curiosity of many researchers and practitioners in the area of management science, operations research and industrial and systems engineering. One major reason is that genetic algorithms are powerful and broadly applicable stochastic search and optimization techniques, and work well for many complex problems which are very difficult to solve by conventional techniques. Many engineering problems can be regarded as a kind of network type based optimization problems subject to complex constraints. Basic genetic algorithms usually fail to produce successful applications for these thorny engineering optimization problems. Therefore, how to tailor genetic algorithms to meet the nature of these problems is a major focus in this research. This book is intended to cover major topics of the research on application of the multiobjective genetic algorithms to various network optimization problems, including basic network models, logistics network models, communication network models, advanced planning and scheduling models, project scheduling models, assembly line balancing models, tasks scheduling models, and advanced network models.

Since early 1993, we have devoted our efforts to the research of genetic algorithms and its application to optimization problems in the fields of industrial engineering (IE) and operational research (OR). We have summarized our research results in our two early books entitled *Genetic Algorithms and Engineering Design*, by John Wiley & Sons, 1997 and *Genetic Algorithms and Engineering Optimization*, by John Wiley & Sons, 2000, covering the following major topics: constrained optimization problems, combinatorial optimization problems, reliability optimization problems, flowshop scheduling problems, job-shop scheduling problems, machine scheduling problems, transportation problems, facility layout design problems, and other topics in engineering design.

We try to summarize our recent studies in this new volume with the same style and the same approach as our previous books with the following contents: multiobjective genetic algorithms, basic network models, logistics network models, communication network models, advanced planning and scheduling models, project scheduling models, assembly line balancing models, tasks scheduling models and advanced network models. This book is suitable for a course in the network model with applications at the upper-level undergraduate or beginning graduate level in computer science, industrial and systems engineering, management science, operations research, and related areas. The book is also useful as a comprehensive reference text for system analysts, operations researchers, management scientists, engineers, and other specialists who face challenging hard-to-solve optimization problems inherent in industrial engineering/operations research.

During the course of this research, one of the most important things is to exchange new ideas on the latest developments in the related research fields and to seek opportunities for collaboration among the participants and researchers at conferences or workshops. One of the authors, Mitsuo Gen, has organized and founded several conferences/workshops/committees on evolutionary computation and its application fields such as C&IE (International Conference on Computers and Industrial Engineering) in Japan 1994 and in Korea 2004, IMS (International Conference on Information and Management Science) in 2001 with Dr. Baoding Liu, IES (Japan-Australia Joint Workshop on Intelligent and Evolutionary Systems) in 1997 with Dr. Xin Yao, Dr. Bob McKay and Dr. Osamu Katai, APIEMS (Asia Pacific Industrial Engineering & Management Systems) Conference with Dr. Hark Hwang and Dr. Weixian Xu in 1998, IML (Japan-Korea Workshop on Intelligent Manufacturing and Logistics Systems) in 2003, ILS (International Conference on Intelligent Logistics Systems) in 2005 with Dr. Kap Hwan Kim and Dr. Erhan Kosan and ETA (Evolutionary Technology and Applications) Special Interest Committee of IEE Japan in 2005 with Dr. Osamu Katai, Dr. Hiroshi Kawakami and Dr. Yasuhiro Tsujimura. All of these conferences/workshops/committees are continuing right now to develop our research topics with face to face contact. Dr. Gen called on additional conference, ANNIE2008 (Artificial Neural Networks in Engineering; Intelligent Systems Design: Neural Networks, Fuzzy Logic, Evolutionary Computation, Swarm Intelligence, and Complex Systems) organized by Dr. Cihan Dagli from 1990, as one of Co-chairs. He gave tutorials on genetic algorithms twice and a plenary talk this year.

It is our great pleasure to acknowledge the cooperation and assistance of our colleagues. Thanks are due to many coauthors who worked with us at different stages of this project: Dr. Myungryun Yoo, Dr. Seren Ozmehmet Tasan and Dr. Azuma Okamoto. We would like to thank all of our graduate students, especially, Dr. Gengui Zhou, Dr. Chiung Moon, Dr. Yinzhen Li, Dr. Jong Ryul Kim, Dr. YongSu Yun, Dr. Admi Syarif, Dr. Jie Gao, Dr. Haipeng Zhang and Mr. Wenqiang Zhang and Ms. Jeong-Eun Lee at Soft Computing Lab. at Waseda University, Kyushu campus and Ashikaga Institute of Technology, Japan.

It was our real pleasure working with Springer UK publisher's professional editorial and production staff, especially Mr. Anthony Doyle, Senior Editor and Mr. Simon Rees, Editorial Assistant.

This project was supported by the Grant-in-Aid for Scientific Research (No. 10044173: 1998.4-2001.3, No. 17510138: 2005.4-2008.3, No. 19700071: 2007. 4-2010.3) by the Ministry of Education, Science and Culture, the Japanese Government. We thank our wives, Eiko Gen, Liying Zhang and Xiaodong Wang and children for their love, encouragement, understanding, and support during the preparation of this book.

Waseda University, Kyushu campus
January 2008

Mitsuo Gen
Runwei Cheng
Lin Lin

Contents

Chapter 1
Multiobjective Genetic Algorithms

Many real-world problems from operations research (OR) / management science (MS) are very complex in nature and quite hard to solve by conventional optimization techniques. Since the 1960s there has been being an increasing interest in imitating living beings to solve such kinds of hard optimization problems. Simulating natural evolutionary processes of human beings results in stochastic optimization techniques called evolutionary algorithms (EAs) that can often outperform conventional optimization methods when applied to difficult real-world problems. EAs mostly involve metaheuristic optimization algorithms such as genetic algorithms (GA) [1, 2], evolutionary programming (EP) [3], evolution strategys (ES) [4, 5], genetic programming (GP) [6, 7], learning classifier systems (LCS) [8], swarm intelligence (comprising ant colony optimization (ACO) [9] and particle swarm optimization (PSO) [10, 11]). Among them, genetic algorithms are perhaps the most widely known type of evolutionary algorithms used today.

1.1 Introduction

The usual form of genetic algorithms (GA) is described by Goldberg [2]. GAs are stochastic search algorithms based on the mechanism of natural selection and natural genetics. GA, differing from conventional search techniques, start with an initial set of random solutions called *population* satisfying boundary and/or system constraints to the problem. Each individual in the population is called a *chromosome* (or *individual*), representing a solution to the problem at hand. Chromosome is a string of symbols usually, but not necessarily, a binary bit string. The chromosomes *evolve* through successive iterations called *generations*. During each generation, the chromosomes are *evaluated*, using some measures of *fitness*. To create the next generation, new chromosomes, called *offspring*, are formed by either merging two chromosomes from current generation using a *crossover* operator or modifying a chromosome using a *mutation* operator. A new generation is formed by *selection*, according to the fitness values, some of the parents and offspring, and rejecting others so as to keep the *population size* constant. Fitter chromosomes have higher probabilities of being selected. After several generations, the algorithms converge to the best chromosome, which hopefully represents the optimum or suboptimal solution to the problem.

1.1.1 General Structure of a Genetic Algorithm

In general, a GA has five basic components, as summarized by Michalewicz [12]:

1. A genetic representation of potential solutions to the problem.
2. A way to create a population (an initial set of potential solutions).
3. An evaluation function rating solutions in terms of their fitness.
4. Genetic operators that alter the genetic composition of offspring (crossover, mutation, selection, *etc.*).
5. Parameter values that genetic algorithm uses (population size, probabilities of applying genetic operators, *etc.*).

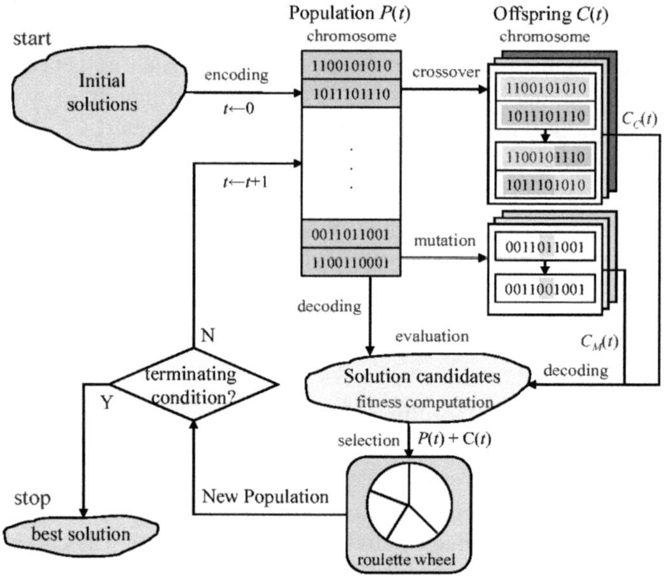

Fig. 1.1 The general structure of genetic algorithms

Figure 1.1 shows a general structure of GA. Let $P(t)$ and $C(t)$ be parents and offspring in current generation t, respectively and the general implementation structure of GA is described as follows:

procedure: basic GA
input: problem data, GA parameters
output: the best solution
begin
  $t \leftarrow 0$;
  initialize $P(t)$ by encoding routine;
  evaluate $P(t)$ by decoding routine;
  while (**not** terminating condition) **do**
    create $C(t)$ from $P(t)$ by crossover routine;
    create $C(t)$ from $P(t)$ by mutation routine;
    evaluate $C(t)$ by decoding routine;
    select $P(t+1)$ from $P(t)$ and $C(t)$ by selection routine;
    $t \leftarrow t+1$;
  end
  output the best solution
end

1.1.2 Exploitation and Exploration

Search is one of the more universal problem solving methods for such problems one cannot determine a prior sequence of steps leading to a solution. Search can be performed with either *blind strategies* or *heuristic strategies* [13]. Blind search strategies do not use information about the problem domain. Heuristic search strategies use additional information to guide search move along with the best search directions. There are two important issues in search strategies: exploiting the best solution and exploring the search space [14]. Michalewicz gave a comparison on hillclimbing search, random search and genetic search [12]. Hillclimbing is an example of a strategy which exploits the best solution for possible improvement, ignoring the exploration of the search space. Random search is an example of a strategy which explores the search space, ignoring the exploitation of the promising regions of the search space.

GA is a class of general purpose search methods combining elements of directed and stochastic search which can produce a remarkable balance between exploration and exploitation of the search space. At the beginning of genetic search, there is a widely random and diverse population and crossover operator tends to perform wide-spread search for exploring all solution space. As the high fitness solutions develop, the crossover operator provides exploration in the neighborhood of each of them. In other words, what kinds of searches (exploitation or exploration) a crossover performs would be determined by the environment of genetic system (the diversity of population) but not by the operator itself. In addition, simple genetic operators are designed as general purpose search methods (the domain-independent search methods) they perform essentially a blind search and could not guarantee to yield an improved offspring.

1.1.3 Population-based Search

Generally, an algorithm for solving optimization problems is a sequence of computational steps which asymptotically converge to optimalsolution. Most classical optimization methods generate a deterministic sequence of computation based on the gradient or higher order derivatives of objective function. The methods are applied to a single point in the search space. The point is then improved along the deepest descending direction gradually through iterations as shown in Fig. 1.2. This point-to-point approach embraces the danger of failing in local optima. GA performs a multi-directional search by maintaining a population of potential solutions. The population-to-population approach is hopeful to make the search escape from local optima. Population undergoes a simulated evolution: at each generation the relatively good solutions are reproduced, while the relatively bad solutions die. GA uses probabilistic transition rules to select someone to be reproduced and someone to die so as to guide their search toward regions of the search space with likely improvement.

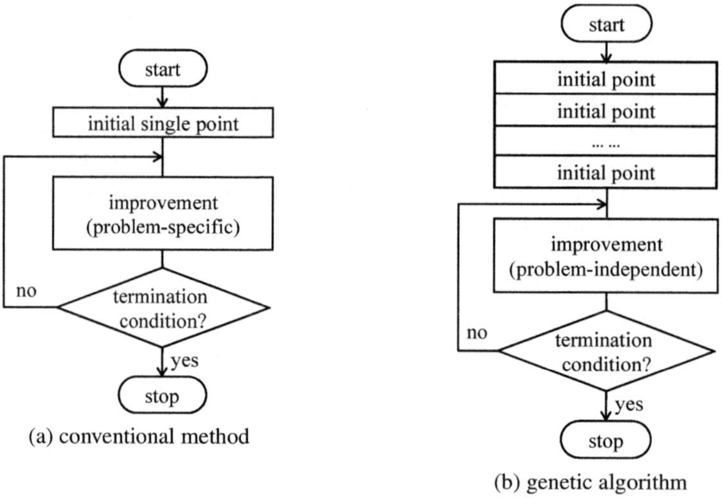

Fig. 1.2 Comparison of conventional and genetic approaches

1.1.4 Major Advantages

GA have received considerable attention regarding their potential as a novel optimization technique. There are three major advantages when applying GA to optimization problems:

1. *Adaptability*: GA does not have much mathematical requirement regarding about the optimization problems. Due to the evolutionary nature, GA will search for solutions without regard to the specific inner workings of the problem. GA can handle any kind of objective functions and any kind of constraints, *i.e.*, linear or nonlinear, defined on discrete, continuous or mixed search spaces.
2. *Robustness*: The use of evolution operators makes GA very effective in performing a global search (in probability), while most conventional heuristics usually perform a local search. It has been proved by many studies that GA is more efficient and more robust in locating optimal solution and reducing computational effort than other conventional heuristics.
3. *Flexibility*: GA provides us great flexibility to hybridize with domain-dependent heuristics to make an efficient implementation for a specific problem.

1.2 Implementation of Genetic Algorithms

In the implementation of GA, several components should be considered. First, a genetic representation of solutions should be decided (*i.e.*, *encoding*); second, a fitness function for evaluating solutions should be given. (*i.e.*, *decoding*); third, genetic operators such as crossover operator, mutation operator and selection methods should be designed; last, a necessary component for applying GA to the constrained optimization is how to handle constraints because genetic operators used to manipulate the chromosomes often yield infeasible offspring.

1.2.1 GA Vocabulary

Because GA is rooted in both natural genetics and computer science, the terminologies used in GA literatures are a mixture of the natural and the artificial.

In a biological organism, the structure that encodes the prescription that specifies how the organism is to be constructed is called a *chromosome*. One or more chromosomes may be required to specify the complete organism. The complete set of chromosomes is called a *genotype*, and the resulting organism is called a *phenotype*. Each chromosome comprises a number of individual structures called *genes*. Each gene encodes a particular feature of the organism, and the location, or *locus*, of the gene within the chromosome structure, determines what particular characteristic the gene represents. At a particular locus, a gene may encode one of several different values of the particular characteristic it represents. The different values of a gene are called *alleles*. The correspondence of GA terms and optimization terms is summarized in Table 1.1.

Table 1.1 Explanation of GA terms

Genetic algorithms	Explanation
Chromosome (string, individual)	Solution (coding)
Genes (bits)	Part of solution
Locus	Position of gene
Alleles	Values of gene
Phenotype	Decoded solution
Genotype	Encoded solution

1.2.2 Encoding Issue

How to encode a solution of a given problem into a chromosome is a key issue for the GA. This issue has been investigated from many aspects, such as mapping characters from a genotype space to a phenotype space when individuals are decoded into solutions, and the metamorphosis properties when individuals are manipulated by genetic operators

1.2.2.1 Classification of Encodings

In Holland's work, encoding is carried out using binary strings [1]. The binary encoding for function optimization problems is known to have severe drawbacks due to the existence of Hamming cliffs, which describes the phenomenon that a pair of encodings with a large Hamming distance belongs to points with minimal distances in the phenotype space. For example, the pair 01111111111 and 10000000000 belong to neighboring points in the phenotype space (points of the minimal Euclidean distances) but have the maximum Hamming distance in the genotype space. To cross the Hamming cliff, all bits have to be changed at once. The probability that crossover and mutation will occur to cross it can be very small. In this sense, the binary code does not preserve locality of points in the phenotype space.

For many real-world applications, it is nearly impossible to represent their solutions with the binary encoding. Various encoding methods have been created for particular problems in order to have an effective implementation of the GA. According to what kind of symbols is used as the alleles of a gene, the encoding methods can be classified as follows:

- Binary encoding
- Real number encoding
- Integer/literal permutation encoding
- a general data structure encoding

The real number encoding is best for function optimization problems. It has been widely confirmed that the real number encoding has higher performance than the binary or Gray encoding for function optimizations and constrained optimizations. Since the topological structure of the genotype space for the real number encoding method is identical to that of the phenotype space, it is easy for us to create some effective genetic operators by borrowing some useful techniques from conventional methods. The integer or literal permutation encoding is suitable for combinatorial optimization problems. Since the essence of combinatorial optimization problems is to search for a best permutation or combination of some items subject to some constraints, the literal permutation encoding may be the most reasonable way to deal with this kind of issue. For more complex real-world problems, an appropriate data structure is suggested as the allele of a gene in order to capture the nature of the problem. In such cases, a gene may be an n-ary or a more complex data structure.

According to the structure of encodings, the encoding methods also can be classified into the following two types:

- One-dimensional encoding
- Multi-dimensional encoding

In most practices, the one-dimensional encoding method is adopted. However, many real-world problems have solutions of multi-dimensional structures. It is natural to adopt a multi-dimensional encoding method to represent those solutions.

According to what kind of contents are encoded into the encodings, the encoding methods can also be divided as follows:

- Solution only
- Solution + parameters

In the GA practice, the first way is widely adopted to conceive a suitable encoding to a given problem. The second way is used in the evolution strategies by Rechenberg [4] and Schwefel [5]. An individual consists of two parts: the first part is the solution to a given problem and the second part, called strategy parameters, contains variances and covariances of the normal distribution for mutation. The purpose for incorporating the strategy parameters into the representation of individuals is to facilitate the evolutionary self-adaptation of these parameters by applying evolutionary operators to them. Then the search will be performed in the space of solutions and the strategy parameters together. In this way a suitable adjustment and diversity of mutation parameters should be provided under arbitrary circumstances.

1.2.2.2 Infeasibility and Illegality

The GA works on two kinds of spaces alternatively: the encoding space and the solution space, or in the other words, the genotype space and the phenotype space. The genetic operators work on the genotype space while evaluation and selection work on the phenotype space. Natural selection is the link between chromosomes and the performance of the decoded solutions. The mapping from the genotype space to the phenotype space has a considerable influence on the performance of the GA. The most prominent problem associated with mapping is that some individuals correspond to infeasible solutions to a given problem. This problem may become very severe for constrained optimization problems and combinatorial optimization problems.

We need to distinguish between two basic concepts: *infeasibility* and *illegality*, as shown in Fig. 1.3. They are often misused in the literature. Infeasibility refers to the phenomenon that a solution decoded from a chromosome lies outside the feasible region of a given problem, while illegality refers to the phenomenon that a chromosome does not represent a solution to a given problem.

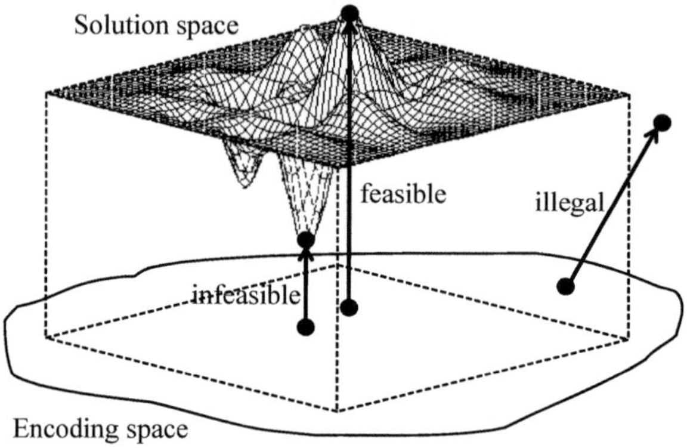

Fig. 1.3 Infeasibility and illegality

The *infeasibility* of chromosomes originates from the nature of the constrained optimization problem. Whatever methods are used, conventional ones or genetic algorithms, they must handle the constraints. For many optimization problems, the feasible region can be represented as a system of equalities or inequalities. For such cases, penalty methods can be used to handle infeasible chromosomes [15]. In constrained optimization problems, the optimum typically occurs at the boundary between feasible and infeasible areas. The penalty approach will force the genetic search to approach the optimum from both sides of the feasible and infeasible regions.

The *illegality* of chromosomes originates from the nature of encoding techniques. For many combinatorial optimization problems, problem-specific encodings are used and such encodings usually yield illegal offspring by a simple one-cut-point crossover operation. Because an illegal chromosome cannot be decoded to a solution, the penalty techniques are inapplicable to this situation. Repairing techniques are usually adopted to convert an illegal chromosome to a legal one. For example, the well-known PMX operator is essentially a kind of two-cut-point crossover for permutation representation together with a repairing procedure to resolve the illegitimacy caused by the simple two-cut-point crossover. Orvosh and Davis [29] have shown that, for many combinatorial optimization problems, it is relatively easy to repair an infeasible or illegal chromosome, and the repair strategy did indeed surpass other strategies such as the rejecting strategy or the penalizing strategy.

1.2.2.3 Properties of Encodings

Given a new encoding method, it is usually necessary to examine whether we can build an effective genetic search with the encoding. Several principles have been proposed to evaluate an encoding [15, 25]:

Property 1 (*Space*): Chromosomes should not require extravagant amounts of memory.
Property 2 (*Time*): The time complexity of executing evaluation, recombination and mutation on chromosomes should not be a higher order.
Property 3 (*Feasibility*): A chromosome corresponds to a feasible solution.
Property 4 (*Legality*): Any permutation of a chromosome corresponds to a solution.
Property 5 (*Completeness*): Any solution has a corresponding chromosome.
Property 6 (*Uniqueness*): The mapping from chromosomes to solutions (decoding) may belong to one of the following three cases (Fig. 1.4): 1-to-1 *mapping*, n-to-1 *mapping* and 1-to-n *mapping*. The 1-to-1 mapping is the best among three cases and 1-to-n mapping is the most undesir one.
Property 7 (*Heritability*): Offspring of simple crossover (*i.e.*, one-cut point crossover) should correspond to solutions which combine the basic feature of their parents.
Property 8 (*Locality*): A small change in chromosome should imply a small change in its corresponding solution.

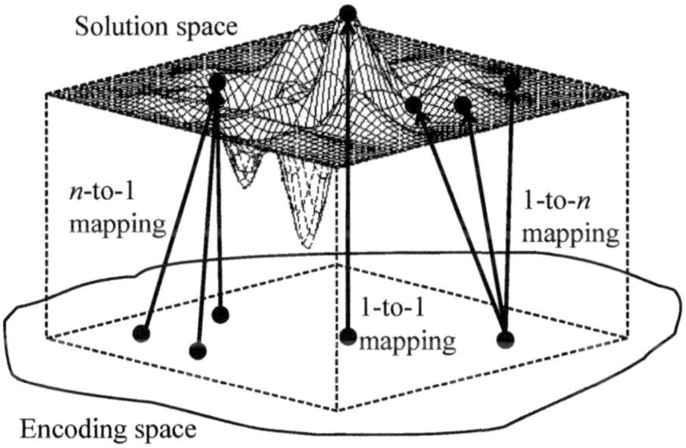

Fig. 1.4 Mapping from chromosomes to solutions

1.2.2.4 Initialization

In general, there are two ways to generate the initial population, *i.e.,* the heuristic initialization and random initialization while satisfying the boundary and/or system constraints to the problem. Although the mean fitness of the heuristic initialization is relatively high so that it may help the GA to find solutions faster, in most large scale problems, for example, network design problems, the heuristic approach may just explore a small part of the solution space and it is difficult to find global optimal solutions because of the lack of diversity in the population. Usually we have to design an *encoding procedure* depending on the chromosome for generating the initial population.

1.2.3 Fitness Evaluation

Fitness evaluation is to check the solution value of the objective function subject to the problem constraints. In general, the objective function provides the mechanism evaluating each individual. However, its range of values varies from problem to problem. To maintain uniformity over various problem domains, we may use the fitness function to normalize the objective function to a range of 0 to 1. The normalized value of the objective function is the fitness of the individual, and the selection mechanism is used to evaluate the individuals of the population.

When the search of GA proceeds, the population undergoes evolution with fitness, forming thus new a population. At that time, in each generation, relatively good solutions are reproduced and relatively bad solutions die in order that the offspring composed of the good solutions are reproduced. To distinguish between the solutions, an evaluation function (also called fitness function) plays an important role in the environment, and scaling mechanisms are also necessary to be applied in objective function for fitness functions. When evaluating the fitness function of some chromosome, we have to design a *decoding procedure* depending on the chromosome.

1.2.4 Genetic Operators

When GA proceeds, both the search direction to optimal solution and the search speed should be considered as important factors, in order to keep a balance between exploration and exploitation in search space. In general, the *exploitation* of the accumulated information resulting from GA search is done by the selection mechanism, while the *exploration* to new regions of the search space is accounted for by genetic operators.

The *genetic operators* mimic the process of heredity of genes to create new offspring at each generation. The operators are used to alter the genetic composition of individuals during representation. In essence, the operators perform a random search, and cannot guarantee to yield an improved offspring. There are three common genetic operators: crossover, mutation and selection.

1.2.4.1 Crossover

Crossover is the main genetic operator. It operates on two chromosomes at a time and generates offspring by combining both chromosomes' features. A simple way to achieve crossover would be to choose a random cut-point and generate the offspring by combining the segment of one parent to the left of the cut-point with the segment of the other parent to the right of the cut-point. This method works well with bit string representation. The performance of GA depends to a great extent, on the performance of the crossover operator used.

The *crossover probability* (denoted by p_C) is defined as the probability of the number of offspring produced in each generation to the population size (usually denoted by *popSize*). This probability controls the expected number $p_C \times popSize$ of chromosomes to undergo the crossover operation. A higher crossover probability allows exploration of more of the solution space, and reduces the chances of settling for a false optimum; but if this probability is too high, it results in the wastage of a lot of computation time in exploring unpromising regions of the solution space.

Up to now, several crossover operators have been proposed for the real numbers encoding, which can roughly be put into four classes: conventional, arithmetical, direction-based, and stochastic.

The conventional operators are made by extending the operators for binary representation into the real-coding case. The conventional crossover operators can be broadly divided by two kinds of crossover:

- Simple crossover: one-cut point, two-cut point, multi-cut point or uniform
- Random crossover: flat crossover, blend crossover

The *arithmetical operators* are constructed by borrowing the concept of linear combination of vectors from the area of convex set theory. Operated on the floating point genetic representation, the arithmetical crossover operators, such as convex, affine, linear, average, intermediate, extended intermediate crossover, are usually adopted.

The *direction-based operators* are formed by introducing the approximate gradient direction into genetic operators. The direction-based crossover operator uses the value of objective function in determining the direction of genetic search.

The *stochastic operators* give offspring by altering parents by random numbers with some distribution.

1.2.4.2 Mutation

Mutation is a background operator which produces spontaneous random changes in various chromosomes. A simple way to achieve mutation would be to alter one or more genes. In GA, mutation serves the crucial role of either (a) replacing the genes lost from the population during the selection process so that they can be tried in a new context or (b) providing the genes that were not present in the initial population.

The *mutation probability* (denoted by p_M) is defined as the percentage of the total number of genes in the population. The mutation probability controls the probability with which new genes are introduced into the population for trial. If it is too low, many genes that would have been useful are never tried out, while if it is too high, there will be much random perturbation, the offspring will start losing their resemblance to the parents, and the algorithm will lose the ability to learn from the history of the search.

Up to now, several mutation operators have been proposed for real numbers encoding, which can roughly be put into four classes as crossover can be classified. *Random mutation* operators such as uniform mutation, boundary mutation, and plain mutation, belong to the conventional mutation operators, which simply replace a gene with a randomly selected real number with a specified range. *Dynamic mutation* (non-uniform mutation) is designed for fine-tuning capabilities aimed at achieving high precision, which is classified as the arithmetical mutation operator. *Directional mutation* operator is a kind of direction-based mutation, which uses the gradient expansion of objective function. The direction can be given randomly as a free direction to avoid the chromosomes jamming into a corner. If the chromosome is near the boundary, the mutation direction given by some criteria might point toward the close boundary, and then jamming could occur. Several mutation operators for integer encoding have been proposed.

- *Inversion mutation* selects two positions within a chromosome at random and then inverts the substring between these two positions.
- *Insertion mutation* selects a gene at random and inserts it in a random position.
- *Displacement mutation* selects a substring of genes at random and inserts it in a random position. Therefore, insertion can be viewed as a special case of displacement. *Reciprocal exchange mutation* selects two positions random and then swaps the genes on the positions.

1.2.4.3 Selection

Selection provides the driving force in a GA. With too much force, a genetic search will be slower than necessary. Typically, a lower selection pressure is indicated at the start of a genetic search in favor of a wide exploration of the search space, while a higher selection pressure is recommended at the end to narrow the search space. The selection directs the genetic search toward promising regions in the search space. During the past two decades, many selection methods have been proposed, examined, and compared. Common selection methods are as follows:

- Roulette wheel selection
- $(\mu+\lambda)$-selection
- Tournament selection
- Truncation selection
- Elitist selection
- Ranking and scaling
- Sharing

Roulette wheel selection, proposed by Holland, is the best known selection type. The basic idea is to determine selection probability or survival probability for each chromosome proportional to the fitness value. Then a model roulette wheel can be made displaying these probabilities. The selection process is based on spinning the wheel the number of times equal to population size, each selecting a single chromosome for the new procedure.

In contrast with proportional selection, $(\mu + \lambda)$-*selection* are deterministic procedures that select the best chromosomes from parents and offspring. Note that both methods prohibit selection of duplicate chromosomes from the population. So many researchers prefer to use this method to deal with combinatorial optimization problem.

Tournament selection runs a "tournament" among a few individuals chosen at random from the population and selects the winner (the one with the best fitness). Selection pressure can be easily adjusted by changing the tournament size. If the tournament size is larger, weak individuals have a smaller chance to be selected. *Truncation selection* is also a deterministic procedure that ranks all individuals according to their fitness and selects the best as parents. *Elitist selection* is generally used as supplementary to the proportional selection process.

The *ranking and scaling mechanisms* are proposed to mitigate these problems. The scaling method maps raw objective function values to positive real values, and the survival probability for each chromosome is determined according to these values. Fitness scaling has a twofold intention: (1) to maintain a reasonable differential between relative fitness ratings of chromosomes, and (2) to prevent too-rapid takeover by some super-chromosomes to meet the requirement to limit competition early but to stimulate it later.

Sharing selection is used to maintain the diversity of population for multi-model function optimization. A sharing function optimization is used to maintain the diversity of population. A sharing function is a way of determining the degradation of an individual's fitness due to a neighbor at some distance. With the degradation, the reproduction probability of individuals in a crowd peak is restrained while other individuals are encouraged to give offspring.

1.2.5 Handling Constraints

A necessary component for applying GA to constrained optimization is how to handle constraints because genetic operators used to manipulate the chromosomes often yield infeasible offspring. There are several techniques proposed to handle constraints with GA [26]-[29]. Michalewicz gave a very good survey of this problem[30, 31]. The existing techniques can be roughly classified as follows:

- Rejecting strategy
- Repairing strategy
- Modifying genetic operators strategy
- Penalizing strategy

Each of these strategies have advantages and disadvantages.

1.2.5.1 Rejecting Strategy

Rejecting strategy discards all infeasible chromosomes created throughout an evolutionary process. This is a popular option in many GA. The method may work reasonably well when the feasible search space is convex and it constitutes a reasonable part of the whole search space. However, such an approach has serious limitations. For example, for many constrained optimization problems where the initial population consists of infeasible chromosomes only, it might be essential to improve them. Moreover, quite often the system can reach the optimum easier if it is possible to "cross" an infeasible region (especially in non-convex feasible search spaces).

1.2.5.2 Repairing Strategy

Repairing a chromosome means to take an infeasible chromosome and generate a feasible one through some repairing procedure. For many combinatorial optimization problems, it is relatively easy to create a repairing procedure. Liepins and his collaborators have shown, through empirical test of GA performance on a divers set of constrained combinatorial optimization problems, that the repair strategy did indeed surpass other strategies in both speed and performance [32, 33].

Repairing strategy depends on the existence of a deterministic repair procedure to converting an infeasible offspring into a feasible one. The weakness of the method is in its problem dependence. For each particular problem, a specific repair algorithm should be designed. Also, for some problems, the process of repairing infeasible chromosomes might be as complex as solving the original problem.

The repaired chromosome can be used either for evaluation only, or it can replace the original one in the population. Liepins *et al.* took the *never replacing* approach, that is, the repaired version is never returned to the population; while Nakano and Yamada took the *always replacing* approach [34]. Orvosh and Davis reported a so-called *5%rule*: this heuristic rule states that in many combinatorial optimization problems, GA with a repairing procedure provide the best result when 5% of repaired chromosomes replaces their infeasible originals [29]. Michalewicz *et al.* reported that the 15% replacement rule is a clear winner for numerical optimization problems with nonlinear constraints [31].

1.2.5.3 Modifying Genetic Operators Strategy

One reasonable approach for dealing with the issue of feasibility is to invent problem-specific representation and specialized genetic operators to maintain the feasibility of chromosomes. Michalewicz *et al.* pointed out that often such systems are much more reliable than any other genetic algorithms based on the penalty approach. This is a quite popular trend: many practitioners use problem-specific representation and specialized operators in building very successful genetic algorithms in many areas [31]. However, the genetic search of this approach is confined within a feasible region.

1.2.5.4 Penalizing Strategy

These strategies above have the advantage that they never generate infeasible solutions but have the disadvantage that they consider no points outside the feasible regions. For highly constrained problem, infeasible solution may take a relatively big portion in population. In such a case, feasible solutions may be difficult to be found if we just confine genetic search within feasible regions. Glover and Greenberg have suggested that constraint management techniques allowing movement through infeasible regions of the search space tend to yield more rapid optimization and produce better final solutions than do approaches limiting search trajectories only to feasible regions of the search space [35]. The *penalizing strategy* is such a kind of techniques proposed to consider infeasible solutions in a genetic search. Gen and Cheng gave a very good survey on penalty function [15].

1.3 Hybrid Genetic Algorithms

GA have proved to be a versatile and effective approach for solving optimization problems. Nevertheless, there are many situations in which the simple GA does not perform particularly well, and various methods of *hybridization* have been proposed. One of most common forms of *hybrid genetic algorithm* (hGA) is to incorporate local optimization as an add-on extra to the canonical GA loop of recombination and selection. With the hybrid approach, *local optimization* is applied to each newly generated offspring to move it to a local optimum before injecting it into the population. GA is used to perform *global exploration* among a population while heuristic methods are used to perform *local exploitation* around chromosomes. Because of the complementary properties of GA and conventional heuristics, the hybrid approach often outperforms either method operating alone. Another common form is to incorporate GA parameters adaptation. The behaviors of GA are characterized by the balance between exploitation and exploration in the search space. The balance is strongly affected by the strategy parameters such as *population size, maximum generation, crossover probability*, and *mutation probability*. How to choose a value to each of the parameters and how to find the values efficiently are very important and promising areas of research on the GA.

1.3.1 Genetic Local Search

The idea of combining GA and local search heuristics for solving optimization problems has been extensively investigated and various methods of hybridization have been proposed. There are two common form of genetic local search. One featurs *Lamarckian evolution* and the other featurs the *Baldwin effect* [21]. Both approaches use the metaphor that an individual learns (hillclimbs) during its lifetime (generation). In the Lamarckian case, the resulting individual (after hillclimbing) is put back into the population. In the Baldwinian case, only the fitness is changed and the genotype remains unchanged. According to Whitley, Gordon and Mathias' experiences on some test problems, the Baldwinian search strategy can sometimes converge to a global optimum when the Lamarckian strategy converges to a local optimum using the same form of local search. However, in all of the cases they examined, the Baldwinian strategy is much slower than the Lamarckian strategy.

The early works which linked genetic and Lamarkian evolution theory included those of Grefenstette who introduced Lamarckian operators into GA [17], Davidor who defined *Lamarkian probability* for mutations in order to enable a mutation operator to be more controlled and to introduce some qualities of a local hill-climbing operator [18], and Shaefer who added an intermediate mapping between the chromosome space and solution space into a simple GA, which is Lamarkian in nature [19].

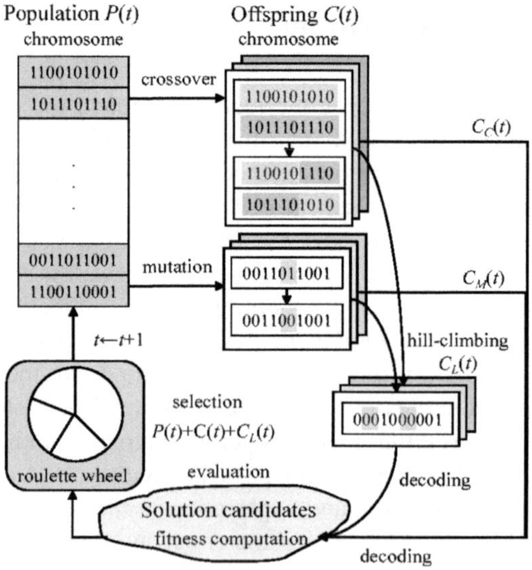

Fig. 1.5 The general structure of hybrid genetic algorithms

Kennedy gave an explanation of hGA with Lamarckian evolution theory [20, 21]. The simple GA of Holland were inspired by Darwin's theory of natural selection. In the nineteenth century, Darwin's theory was challenged by Lamarck, who proposed that environmental changes throughout an organism's life cause structural changes that are transmitted to offspring. This theory lets organisms pass along the knowledge and experience they acquire in their lifetime. While no biologist today believes that traits acquired in the natural world can be inherited, the power of Lamarckian theory is illustrated by the evolution of our society. Ideas and knowledge are passed from generation to generation through structured language and culture. GA, the artificial organisms, can benefit from the advantages of Lamarckian theory. By letting some of the individuals' experiences be passed along to future individuals, we can improve the GA's ability to focus on the most promising areas. Following a more Lamarckian approach, first a traditional hill-climbing routine could use the offspring as a starting point and perform quick and localized optimization. After it has *learned* to climb the local landscape, we can put the offspring through the evaluation and selection phases. An offspring has a chance to pass its experience to future offspring through common crossover.

Let $P(t)$ and $C(t)$ be parents and offspring in current generation t. The general structure of hGA is described as follows:

---

procedure: hybrid GA
input: problem data, GA parameters
output: the best solution
begin
    $t \leftarrow 0$;
    initialize $P(t)$ by encoding routine;
    evaluate $P(t)$ by decoding routine;
    while (**not** terminating condition) **do**
        create $C(t)$ from $P(t)$ by crossover routine;
        create $C(t)$ from $P(t)$ by mutation routine;
        climb $C(t)$ by local search routine;
        evaluate $C(t)$ by decoding routine;
        select $P(t+1)$ from $P(t)$ and $C(t)$ by selection routine;
        $t \leftarrow t+1$;
    end
    output the best solution
end

In the hybrid approach, artificial organisms first pass through Darwin's biological evolution and then pass through Lamarckian's intelligence evolution (Fig. 1.5). A traditional hill-climbing routine is used as Lamarckian's evolution to try to inject some "smarts" into the offspring organism before returning it to be evaluated.

Moscato and Norman have introduced the term *memetic algorithm* to describe the GA in which local search plays a significant part [22]. The term is motivated by

Dawkins's notion of a *meme* as a unit of information that reproduces itself as people exchange ideas [23]. A key difference exists between genes and memes. Before a meme is passed on, it is typically adapted by the person who transmits it as that person thinks, understands and processes the meme, whereas genes get passed on whole. Moscato and Norman linked this thinking to local refinement, and therefore promoted the term memetic algorithm to describe genetic algorithms that use local search heavily.

Radcliffe and Surry gave a formal description of *memetic algorithms* [24], which provided a homogeneous formal framework for considering memetic and GA. According to Radcliffe and Surry, if a local optimizer is added to a GA and applied to every child before it is inserted into the population, then a memetic algorithm can be thought of simply as a special kind of genetic search over the subspace of local optima. Recombination and mutation will usually produce solutions that are outside this space of local optima, but a local optimizer can then repair such solutions to produce final children that lie within this subspace, yielding a memetic algorithm as shown in Fig. 1.6.

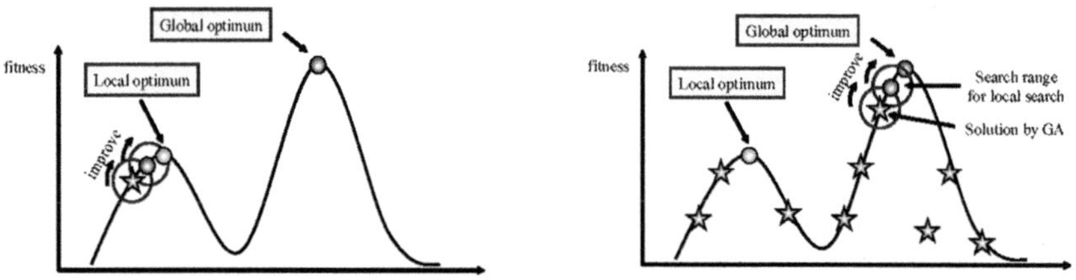

Fig. 1.6 Applying a local search technique to a GA loop

1.3.2 Parameter Adaptation

Since GA are inspired from the idea of evolution, it is natural to expect that the *adaptation* is used not only for finding solutions to a given problem, but also for tuning the GA to the particular problem. During the past few years, many adaptation techniques have been suggested and tested in order to obtain an effective implementation of the GA to real-world problems. In general, there are two kinds of adaptations:

- Adaptation to problems
- Adaptation to evolutionary processes

The difference between these two adaptations is that the first advocates modifying some components of GA, such as representation, crossover, mutation and selection, in order to choose an appropriate form of the algorithm to meet the nature of a given problem. The second suggests a way to tune the parameters of the changing configurations of GA while solving the problem. According to Herrera and Lozano, the later type of adaption can be further divided into the following classes [36]:

- Adaptive parameter settings
- Adaptive genetic operators
- Adaptive selection
- Adaptive representation
- Adaptive fitness function

Among these classes, parameter adaptation has been extensively studied in the past ten years because the strategy parameters such as mutation probability, crossover probability, and population size are key factors in the determination of the exploitation vs exploration tradeoff. The behaviors of GA are characterized by the balance between exploitation and exploration in the search space. The balance is strongly affected by the strategy parameters such as *population size, maximum generation, crossover probability*, and *mutation probability*. How to choose a value for each of the parameters and how to find the values efficiently are very important and promising areas of research of the GA.

Usually, fixed parameters are used in most applications of the GA. The values for the parameters are determined with a *set-and-test approach*. Since GA is an intrinsically dynamic and adaptive process, the use of constant parameters is in contrast to the general evolutionary spirit. Therefore, it is a natural idea to try to modify the values of strategy parameters during the run of the algorithm. It is possible to do this in various ways:

- By using some rule
- By taking feedback information from the current state of search
- By employing some self-adaptive mechanism

1.3.2.1 Classification of Parameter Adaptation

A recent survey on adaptation techniques is given by Herrera and Lozano [36], and Hinterding *et al.* [37]. According to these classifications of adaptation, there are three main categories.

1. **Deterministic adaptation:** Deterministic adaptation takes place if the value of a strategy parameter is altered by some deterministic rule. Usually, a time-varying approach is used, measured by the number of generations. For example, the mutation ratio is decreased gradually along with the elapse of generation by using the following equation:

$$p_{\mathrm{M}} = 0.5 - 0.3\frac{t}{maxGen}$$

   where t is the current generation number and *maxGen* is the maximum generation. Hence, the mutation ratio will decrease from 0.5 to 0.2 as the number of generations increase to *maxGen*.

2. **Adaptive adaptation:** Adaptive adaptation takes place if there is some form of feedback from the evolutionary process, which is used to determine the direction and/or magnitude of the change to the strategy parameter. Early examples of this kind of adaptation include Rechenberg's "1/5 success rule" in volution strategies, which was used to vary the step size of mutation [38]. The rule states that the ratio of successful mutations to all mutations should be 1/5, hence, if the ratio is greater than 1/5 then increase the step size, and if the ratio is less than 1/5 then decrease the step size. Davis's "adaptive operator fitness" utilizes feedback on the success of a larger number of reproduction operators to adjust the ratio being used [39]. Julstrom's adaptive mechanism regulates the ratio between crossovers and mutations based on their performance [40].
3. **Self-adaptive adaptation:** Self-adaptive adaptation enables strategy parameters to evolve along with the evolutionary process. The parameters are encoded onto the chromosomes of the individuals and undergo mutation and recombination. The encoded parameters do not affect the fitness of individuals directly, but better values will lead to better individuals and these individuals will be more likely to survive and produce offspring, hence propagating these better parameter values. The parameters to self-adapt can be ones that control the operation of evolutionary algorithms, ones that control the operation of reproduction or other operators, or probabilities of using alternative processes. Schwefel developed this method to self-adapt the mutation step size and the mutation rotation angles in evolution strategies [5]. Hinterding used a multi-chromosome to implement the self-adaptation in the cutting stock problem with contiguity, where self-adaptation is used to adapt the probability of using one of the two available mutation operators, and the strength of the group mutation operator.

1.3.2.2 Auto-tuning Probabilities of Crossover and Mutation

Gen and Cheng [16], in their book, surveyed various adaptive methods using several *fuzzy logic control* (FLC). Subbu *et al.* [41] suggested a *fuzzy logic controlled genetic algorithm* (flcGA), and the flcGA uses a fuzzy knowledge-based development. This scheme is able to adaptively adjust the rates of crossover and mutation operators. Song *et al.* [42] used two FLCs; one for the crossover rate and the other for the mutation rate. These parameters are considered as the input variables of GA and are also taken as the output variables of the FLC. For successfully applying an FLC to a GA in [41, 42], the key is to produce well-formed fuzzy sets and rules. Recently, Cheong and Lai [**?**] suggested an optimization scheme for the sets and rules. The GA controlled by these FLCs were more efficient in terms of search speed and search quality of a GA than the GA without them.

Yun and Gen [44] proposed *adaptive genetic algorithm* (aGA) using FLC. To adaptively regulate GA operators using FLC, they use the scheme of Song *et al.* [42] as a basic concept and then improve it. Its main scheme is to use two FLCs: the crossover FLC and mutation FLC are implemented independently to regulate the rates of crossover and mutation operators adaptively during the genetic search process. The heuristic updating strategy for the crossover and mutation rates is to consider the change of the average fitness which is occurred at each generation.

By using this basic heuristic updating strategy, they can build up a detailed scheme for its implementation. For the detailed scheme, they use the changes of the average fitness which occur at parent and offspring populations during two continuous generations of GA: it increases the occurrence rates of p_C and p_M, if it consistently produces a better offspring during the two continuous generations; however, it also reduces the occurrence rates if it consistently produces a poorer offspring during the generations. This scheme is based on the fact that it encourages the well-performing operators to produce much better offspring, while also reducing the chance for poorly performing operators to destroy the potential individuals during a genetic search process.

1.3.2.3 Auto-tuning Population Size

As discussed in [45], varying the population size between two successive generations affects only the selection operator of the GA. Let n_t and n_{t+1} denote the population size of the current and the subsequent generation, respectively. The selection of the individuals can be considered as a repetitive process of n_{t+1} selection operations, with p_j being the probability of selection of the $j-$th individual. For most of the selection operators, such as fitness proportionate selection and tournament selection with replacement, the selection probability p_j remains constant for the n_{t+1} selection operations. The expected number of copies of the $j-$th individual after selection can be expressed as

$$c(j,t+1) = p_j n_{t+1} c(j,t)$$

where $c(j,t)$ is the number of copies of the $j-$th individual at generation t. The expected number of the $j-$th individual is directly proportional to the population size of the subsequent generation. Therefore, the portion of the population related to the $j-$th individual after selection can be expressed as

$$\rho(j,t+1) = \frac{c(j,t+1)}{n_{t+1}} = \frac{p_j n_{t+1} c(j,t)}{n_{t+1}} = p_j c(j,t)$$

which is independent of the population size of the subsequent generation, provided that the variation of the population size is not significant enough to modify the probability p_j. However, a GA with decreasing population size has bigger initial population size and smaller final population size, as compared to a constant population size GA with the same computing cost (*i.e.*, equal average population size). This is expected to be beneficial, because a bigger population size at the beginning provides a better initial signal for the GA evolution process whereas, a smaller population size is adequate at the end of the run, where the GA converges to the optimum. The above ascertainment motivated this paper, offering good arguments for a successful statistical outcome.

Based on previous research, Koumousis and Katsaras proposed *saw-tooth GA* (stGA). A variable population size scheme combined with population reinitialization can be introduced to the GA to improve its performance further. The mean population size $\bar{n}$ of the periodic scheme corresponds to the constant population size GA having the same computing cost. Moreover, the scheme is characterized by the amplitude D and the period of variation T. Thus, at a specific generation $n(t)$, the population size is determined as

$$n(t) = \textbf{int}\left(\bar{n}+D-\frac{2D}{T-1}\left(t-T\cdot\text{int}\left(\frac{t-1}{T}\right)-1\right)\right)$$

The Koumousis and Katsaras proposed stGA utilizes a variable population size following the periodic scheme presented in Fig. 1.7 in the form of a saw-tooth function. A detailed explanation is in [45].

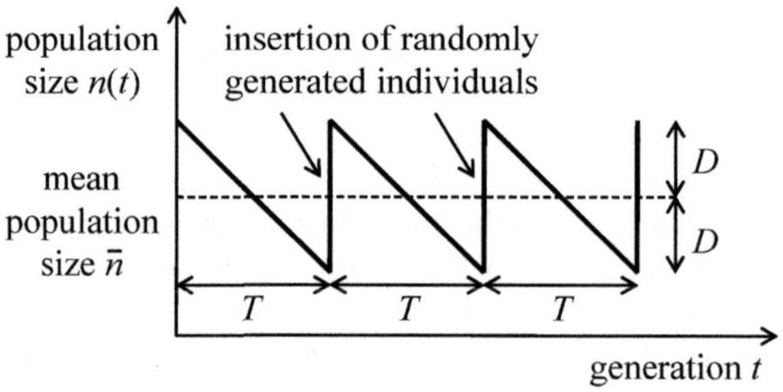

Fig. 1.7 Population variation scheme of saw-tooth GA [45]

1.3.2.4 Auto-tuning Strategy by FLC

GA has been known to offer significant advantages against conventional methods by using simultaneously several search principles and heuristics. However, despite the successful application of GA to numerous optimization problems, it is not easy to deaide if the identification of the correct settings of genetic parameters for the problems is effective or not. These genetic parameters determine the exploitation–exploration trade-off on the search space. All of genetic parameters should be auto-tuning depending on the convergence situation of current generation. If most chromosomes of current generation fasten on the same area of the search space, exploitation should be increased. Contrarily, exploration should be increased. Recently, Lin and Gen [46] proposed a new *auto-tuning strategy* that fuzzy logic control (FLC) has been adopted for auto-tuning the balance between exploitation and exploration based on change of the average fitness of the current and last generations . Different to conventional auto-tuning strategy, this auto-tuning based GA (atGA) comprises the following two components:

- There are three genetic operators: crossover (same with a simple GA), immigration (a kind of random search), heuristic mutation (a kind of heuristic search).
- Auto-tuning strategy turns the probabilities of above three genetic operators. (p_C: crossover probability, p_I: immigration probability, p_M: heuristic mutation probability; $p_C + p_I + p_M \equiv 1$)

Heuristic Mutation: This kind of heuristic search is a same techniques with genetic local search discussed above. An LS technique is applied to generate some new offspring from selected parents. This heuristic mutation can move the offspring to a local optimum. The number of offspring is dependent on the probability of mutation p_M. In other words, the probability of mutation determines the weight of exploration for the search space.

Immigration Operator: Moed *et al.* [47] proposed an immigration operator which, for certain types of functions, allows increased exploration while maintaining nearly the same level of exploitation for the given population size. It is an example of a random strategy which explores the search space ignoring the exploitation of the promising regions of the search space.

The algorithm is modified to (1) include immigration routine, in each generation, (2) generate and (3) evaluate $popSize \cdot p_I$ random members, and (4) replace the $popSize \cdot p_I$ worst members of the population with the $popSize \cdot p_I$ random members (p_I, called the immigration probability).

The probabilities of immigration and crossover determine the weight of exploitation for the search space. The main scheme is to use two FLCs: auto-tuning for exploration and exploitation $T\left[p_M \wedge (p_C \vee p_I)\right]$ and auto-tuning for genetic exploitation and random exploitation ($T[p_C \wedge p_I]$) are implemented independently to regulate adaptively the genetic parameters during the genetic search process. For the detailed scheme, we use the changes of the average fitness which occur in parents and offspring populations during continuous u generations of GA: it increases the occurrence probability of p_M and decreases the occurrence probability of p_C and p_I if it consistently produces better offspring; otherwise, it decreases the occurrence probability of p_M and increases the occurrence probability of p_C and p_I, if it consistently produces poorer offspring during the generations. For example, in minimization problems, we can set the change of the average fitness at generation t, $\triangle f_{avg}(t)$ as follows:

$$\triangle f_{avg}(t) = \overline{f_{parSize}}(t) - \overline{f_{offSize}}(t) \tag{1.1}$$

$$= \frac{1}{parSize}\sum_{k=1}^{parSize} f_k(t) - \frac{1}{offSize}\sum_{k=1}^{offSize} f_k(t) \tag{1.2}$$

where $parSize$ and $offSize$ are the parent and offspring population sizes satisfying constraints, respectively. We define the regulation of p_M, by using the values of

$\triangle f_{avg}(t-i)$, i=1, 2, $\cdots$, u, and we define p_{C} and p_{I} by using correlation coefficient of the individuals in the current generation. They are shown as follows:

$$\begin{aligned} p_{\mathrm{M}} &= \textbf{regulation 1}(\triangle f_{avg}(t-i), \quad i=1,2,\cdots,u) \\ p_{\mathrm{C}} &= \textbf{regulation 2}(\triangle f_{avg}(t-i), \quad i=1,2,\cdots,u) \\ p_{\mathrm{I}} &= 1-(p_{\mathrm{M}}+p_{\mathrm{C}}) \end{aligned}$$

where the routines of "**regulation 1**" and "**regulation 2**" are shown in Figures 1.8 and 1.9; the membership function for p_{M} and p_{C} are shown in Figures 1.10 and 1.11, respectively. $\triangle f$ and μ_M, μ_C are

$$\triangle f = \sum_{i=1}^{u} 2^{t-i} \cdot \lambda \left(\triangle f_{avg}(t-i)\right), \quad t \geq u \tag{1.3}$$

$$\mu_M = \frac{0.8-0.2}{2^u-2\alpha}, \quad 0 \leq \alpha \leq \frac{2^u}{4} \tag{1.4}$$

$$\mu_C = \frac{0.8(1-p_{\mathrm{M}})-0.1}{2^u-2\alpha}, \quad 0 \leq \alpha \leq \frac{2^u}{4} \tag{1.5}$$

$$\lambda(x) = \begin{cases} 1, & \text{if } x \geq 0 \\ 0 & \text{otherwise} \end{cases} \tag{1.6}$$

Let $P(t)$ and $C(t)$ be parents and offspring in current generation t, respectively. The general structure of auto-tuning GA is described as follows:

---

procedure: auto-tuning GA
input: problem data, GA parameters
output: the best solution
begin
  $t \leftarrow 0$;
  initialize $P(t)$ by encoding routine;
  evaluate $P(t)$ by decoding routine;
  while (**not** terminating condition) **do**
    create $C(t)$ from $P(t)$ by crossover routine;
    create $C(t)$ from $P(t)$ by mutation routine;
    create $C(t)$ from $P(t)$ by immigration routine;
    evaluate $C(t)$ by decoding routine;
    if $t > u$ **then**
      auto-tuning p_{M}, p_{C} and p_{I} by *FLC*;
    select $P(t+1)$ from $P(t)$ and $C(t)$ by selection routine;
    $t \leftarrow t+1$;
  end
  output the best solution
end

```
procedure: Regulation 1 for p_M
begin
    if Δf ≤ α then
        p_M = 0.2;
    if α ≤ Δf ≤ 2^u − α then
        p_M = 0.2 + μ_M · (Δf − α);
    if Δf ≥ 2^u − α then
        p_M = 0.8;
    end
end
```

Fig. 1.8 Pseudocode of regulation for p_M

```
procedure: Regulation 2 for p_C
begin
    if Δf ≤ α then
        p_C = 0.8;
    if α ≤ Δf ≤ 2^u − α then
        p_C = 0.8 · (1 − p_M) − 0.1 − μ_C · (Δf − α);
    if Δf ≥ 2^u − α then
        p_C = 0.1;
    end
end
```

Fig. 1.9 Pseudocode of regulation for p_C

1.4 Multiobjective Genetic Algorithms

Optimization deals with the problems of seeking solutions over a set of possible choices to optimize certain criteria. If there is only one criterion to be taken into consideration, they become single objective optimization problems, which have been extensively studied for the past 50 years. If there are more than one criterion which must be treated simultaneously, we have *multiple objective optimization problems* [48, 49]. Multiple objective problems arise in the design, modeling, and planning

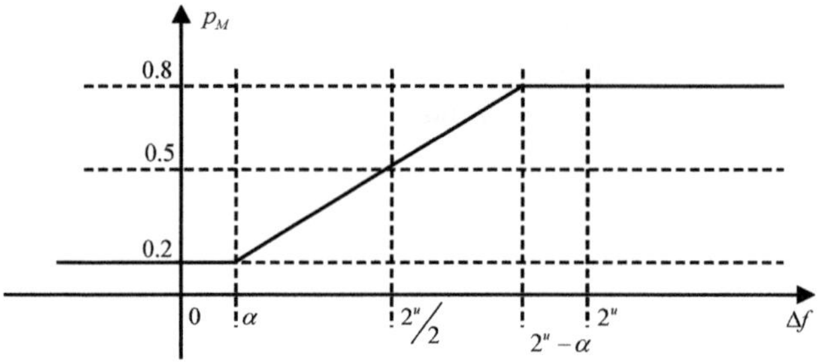

Fig. 1.10 Membership function for p_{M}

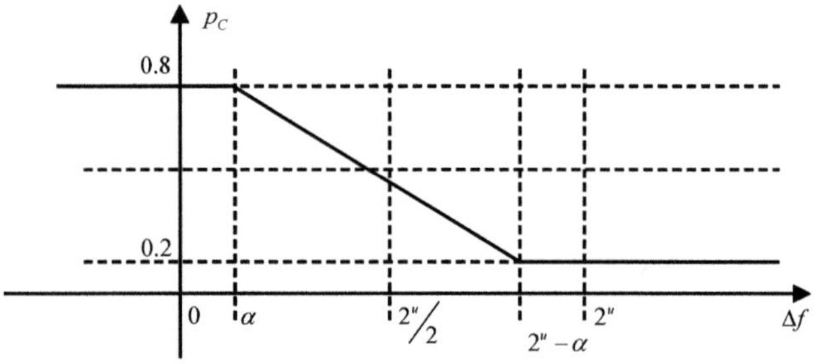

Fig. 1.11 Membership function for p_{C}

of many complex real systems in the areas of industrial production, urban transportation, capital budgeting, forest management, reservoir management, layout and landscaping of new cities, energy distribution, *etc*. It is easy to see that almost every important real-world decision problem involves multiple and conflicting objectives which need to be tackled while respecting various constraints, leading to overwhelming problem complexity. The multiple objective optimization problems have been receiving growing interest from researchers with various background since early 1960 [50]. There are a number of scholars who have made significant contributions to the problem. Among them, Pareto is perhaps one of the most recognized pioneers in the field [51]. Recently, GA has received considerable attention as a novel approach to multiobjective optimization problems, resulting in a fresh body of research and applications known as *evolutionary multiobjective optimization* (EMO).

1.4.1 Basic Concepts of Multiobjective Optimizations

A single objective optimization problem is usually given in the following form:

$$\max\ z = f(\boldsymbol{x}) \tag{1.7}$$
$$\text{s. t. } g_i(\boldsymbol{x}) \le 0, \quad i = 1, 2, \ldots, m \tag{1.8}$$
$$\boldsymbol{x} \ge 0 \tag{1.9}$$

where $\boldsymbol{x} \in \mathrm{R}^n$ is a vector of n decision variables, $f(\boldsymbol{x})$ is the objective function, and $g_i(\boldsymbol{x})$ are inequality constraint m functions, which form the area of feasible solutions. We usually denote the feasible area in decision space with the set S as follows:

$$S = \{\boldsymbol{x} \in \mathrm{R}^n | g_i(\boldsymbol{x}) \le 0, \quad i = 1, 2, \ldots, m,\ \boldsymbol{x} \ge 0\} \tag{1.10}$$

Without loss of generality, a multiple objective optimization problem can be formally represented as follows:

$$\max\ \{z_1 = f_1(\boldsymbol{x}), z_2 = f_2(\boldsymbol{x}), \cdots, z_q = f_q(\boldsymbol{x})\} \tag{1.11}$$
$$\text{s. t. } g_i(\boldsymbol{x}) \le 0, \quad i = 1, 2, \ldots, m \tag{1.12}$$
$$\boldsymbol{x} \ge 0 \tag{1.13}$$

We sometimes graph the multiple objective problem in both decision space and criterion space. S is used to denote the feasible region in the *decision space* and Z is used to denote the feasible region in the *criterion space*

$$Z = \{\boldsymbol{z} \in \mathrm{R}^q | z_1 = f_1(\boldsymbol{x}), z_2 = f_2(\boldsymbol{x}), \cdots, z_q = f_q(\boldsymbol{x}), \quad \boldsymbol{x} \in S\} \tag{1.14}$$

where $\boldsymbol{z} \in \mathrm{R}^k$ is a vector of values of q objective functions. In the other words, Z is the set of images of all points in S. Although S is confined to the nonnegative orthant of R^n, Z is not necessarily confined to the nonnegative orthant of R^q.

1.4.1.1 Nondominated Solutions

In principle, multiple objective optimization problems are very different from single objective optimization problems. For the single objective case, one attempts to obtain the best solution, which is absolutely superior to all other alternatives. In the case of multiple objectives, there does not necessarily exist such a solution that is the best with respect to all objectives because of *incommensurability* and *conflict* among objectives. A solution may be best in one objective but worst in other objectives. Therefore, there usually exists a set of solutions for the multiple objective case which cannot simply be compared with each other. Such kinds of solutions are called *nondominated* solutions or *Pareto optimal solutions*, for which no improvement in any objective function is possible without sacrificing at least one of the other objective functions. For a given nondominated point in the criterion space Z, its image point in the decision space S is called *efficient* or *noninferior*. A point in S is efficient if and only if its image in Z is nondominated. The concept of Pareto optimal solutions and dominated solutions is shown in Fig. 1.12

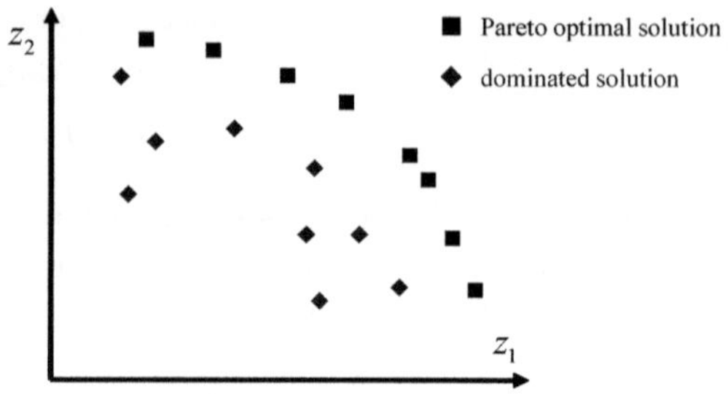

Fig. 1.12 The concept of Pareto optimal solutions (maximization case)

Definition 1.1. For a given point $\boldsymbol{z}^0 \in Z$, it is *nondominated* if and only if there does not exist another point $\boldsymbol{z} \in Z$ such that, for the maximization case,

$$z_k > z_k^0, \quad \text{for some } k \in \{1,2,\cdots,q\} \text{ and} \tag{1.15}$$
$$z_l \geq z_l^0, \quad \text{for all } l \neq k \tag{1.16}$$

where, $\boldsymbol{z}^0$ is a *dominated* point in the criterion space Z.

Definition 1.2. For a given point $\boldsymbol{x}^0 \in S$, it is *efficient* if and only if there does not exist another point $\boldsymbol{x} \in S$ such that, for the maximization case,

$$f_k(\boldsymbol{x}) > f_k(\boldsymbol{x}^0), \quad \text{for some } k \in \{1,2,\cdots,q\} \text{ and} \tag{1.17}$$
$$f_l(\boldsymbol{x}) \geq f_l(\boldsymbol{x}^0), \quad \text{for all } l \neq k \tag{1.18}$$

where $\boldsymbol{x}^0$ is *inefficient*.

A point in the decision space is *efficient* if and only if its image is a nondominated point in the criterion space Z.

To illustrate the above definitions, consider the following linear programming problem:

$$\max\ f_1(x_1,x_2) = -x_1 + 3x_2 \tag{1.19}$$
$$\max\ f_2(x_1,x_2) = 3x_1 + x_2 \tag{1.20}$$
$$\text{s. t. } g_1(x_1,x_2) = x_1 + 2x_2 - 2 \leq 0 \tag{1.21}$$
$$g_2(x_1,x_2) = 2x_1 + x_2 - 2 \leq 0 \tag{1.22}$$
$$x_1, x_2 \geq 0 \tag{1.23}$$

The feasible region S in the decision space is shown in Fig. 1.13a. The extreme points in the feasible region are $x^1(0,0)$, $x^2(1,0)$, $x^3(2/3,2/3)$, and $x^4(0,1)$. The feasible region Z in the criterion space is obtained by mapping set S by using two objectives (Eqs. 1.19 and 1.20), as shown in Fig. 1.13b. The corresponding extreme points are $z^1(0,0)$, $z^2(-1,3)$, $z^3(4/3,8/3)$ and, $z^4(3,1)$. From the figures we observe that both region are convex and the extreme points of Z are the images of extreme points of S.

Looking at the points between z^4 and z^3, it is noted that as $f_2(x_1,x_2)$ increases from 1 to 8/3, $f_1(x_1,x_2)$ decreases from 3 to 4/3, and accordingly, all points between z^4 and z^3 are nondominated points. In the same way, all points between z^3 and z^2 are also nondominated points. The corresponding efficient points in the decision space are in the segments between point x^2 and x^3, and between point x^3 and x^4, respectively.

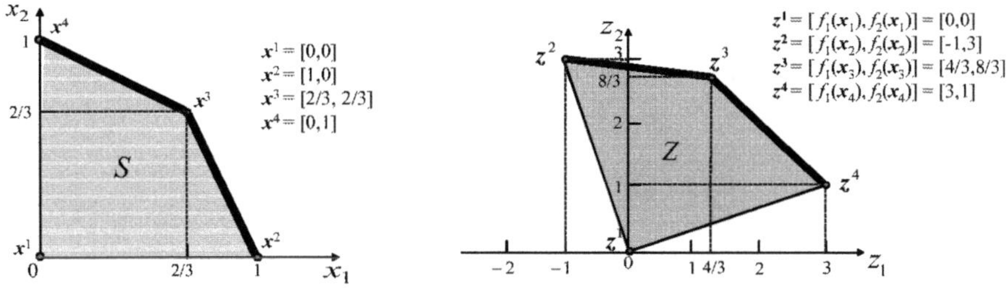

(a) Decision space S (b) Criterion space Z

Fig. 1.13 Feasible region and efficient solutions in decision space S and criterion space Z

There is a special point in the criterion space Z called an *ideal point* (or a *positive ideal solution*), denoted by $z^* = (z_1^*, z_2^*, \ldots, z_q^*)$ where $z_k^* = \sup\{f_k(x)|x \in S\}$, $k = 1,2,\ldots,q$. The point z* is called an ideal point because usually it is not attainable. Note that, individually, z_k^* may be attainable. But to find a point z^* which can simultaneously maximize each $f_k(\cdot)$, $k = 1,2,\ldots,q$ is usually very difficult.

1.4.2 Features and Implementation of Multiobjective GA

The inherent characteristics of the GA demonstrate why genetic search is possibly well suited to multiple objective optimization problems. The basic feature of the GA is the multiple directional and global search by maintaining a population of potential solutions from generation to generation. The *population-to-population* approach is useful to explore all Pareto solutions.

The GA does not have much mathematical requirement regarding the problems and can handle any kind of objective functions and constraints. Due to their evolutionary nature, the GA can search for solutions without regard to the specific inner workings of the problem. Therefore, it is applicable to solving much more complex problems beyond the scope of conventional methods' interesting by using the GA.

Because the GA, as a kind of meta-heuristics, provides us a great flexibility to hybridize with conventional methods into their main framework, we can take with advantage of both the GA and the conventional methods to make much more efficient implementations for the problems. The ongoing research on applying the GA to the multiple objective optimization problems presents a formidable theoretical and practical challenge to the mathematical community [16].

Let $P(t)$ and $C(t)$ be parents and offspring in current generation t. The general structure of *multiobjective genetic algorithm* (moGA) is described as follows:

---

procedure: moGA
input: problem data, GA parameters
output: Pareto optimal solutions E
begin
  $t \leftarrow 0$;
  initialize $P(t)$ by encoding routine;
  calculate objectives $f_i(P), i = 1, \cdots, q$ by decoding routine;
  create Pareto $E(P)$;
  evaluate $eval(P)$ by *fitness assignment routine*;
  while (**not** terminating condition) **do**
    create $C(t)$ from $P(t)$ by crossover routine;
    create $C(t)$ from $P(t)$ by mutation routine;
    calculate objectives $f_i(C), i = 1, \cdots, q$ by decoding routine;
    update Pareto $E(P,C)$;
    evaluate $eval(P,C)$ by *fitness assignment routine*;
    select $P(t+1)$ from $P(t)$ and $C(t)$ by selection routine;
    $t \leftarrow t+1$;
  end
  output Pareto optimal solutions $E(P,C)$
end

1.4.3 Fitness Assignment Mechanism

GA is essentially a kind of meta-strategy method. When applying the GA to solve a given problem, it is necessary to refine upon each of the major components of GA, such as *encoding methods, recombination operators, fitness assignment, selection operators, constraints handling,* and so on, in order to obtain a best solution to the given problem. Because the multiobjective optimization problems are the natural extensions of constrained and combinatorial optimization problems, so many useful methods based on GA have been developed during the past two decades. One of special issues in multiobjective optimization problems is the *fitness assignment mechanism*. Since the 1980s, several fitness assignment mechanisms have been proposed and applied in multiobjective optimization problems [16, 58]. Although most fitness assignment mechanisms are just different approaches and suitable for different cases of multiobjective optimization problems, in order to understanding the development of moGA, we classify algorithms according to proposed years of different approaches:

Generation 1 **Vector Evaluation Approach:**
vector evaluated GA (veGA), Schaffer [52]

Generation 2 **Pareto Ranking + Diversity:**
multiobjective GA (moGA), Fonseca and Fleming [53]
non-dominated sorting GA (nsGA), Srinivas and Deb [54]

Generation 3 **Weighted Sum + Elitist Preserve:**
random weight GA (rwGA), Ishibuchi and Murata [55]
adaptive weight GA (awGA), Gen and Cheng [16]
strength Pareto Evolutionary Algorithm II (spEA II), Zitzler *et al.* [56, 57]
non-dominated sorting GA II (nsGA II), Deb *et al.* [59]
interactive adaptive-weight GA (i-awGA), Lin and Gen [46]

1.4.3.1 Vector Evaluation Approach: Generation 1

Vector Evaluated Genetic Algorithm (veGA: Schaffer [52]): veGA is the first notable work to solve the multiobjective problems. Instead of using a scalar fitness measure to evaluate each chromosome, it uses a vector fitness measure to create the next generation. In the veGA approach, the selection step in each generation becomes a loop; each time through the loop the appropriate fraction of the next generation, or subpopulations, is selected on the basis of each of the objectives. Then the entire population is thoroughly shuffled to apply crossover and mutation operators. This is performed to achieve the mating of individuals of different subpopulations. Nondominated individuals are identified by monitoring the population as it evolves, but this information is not used by the veGA itself. This approach, illustrated in Fig. 1.14, protects the survival of the best individuals on one of the objectives and, simultaneously, provides the appropriate probabilities for multiple selection of individuals which are better than average on more than one objective.

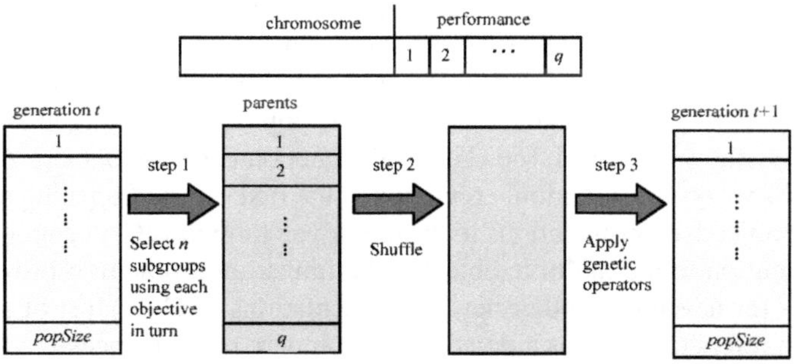

Fig. 1.14 Illustration of veGA selection

A simple two objective problem with one variable is used to test the properties of veGA by Schaffer:

$$\min f_1(\boldsymbol{x}) = x^2 \tag{1.24}$$

$$\min f_2(\boldsymbol{x}) = (x-2)^2 \tag{1.25}$$

$$\text{s. t. } x \in \mathrm{R}^1 \tag{1.26}$$

Figure 1.15a plots the problem by restricting variable x within the region from –2 to 4. It is clear that the Pareto solutions constitute all x values varying from 0 to 2. Figure 1.15b plots the problem in the criterion space.

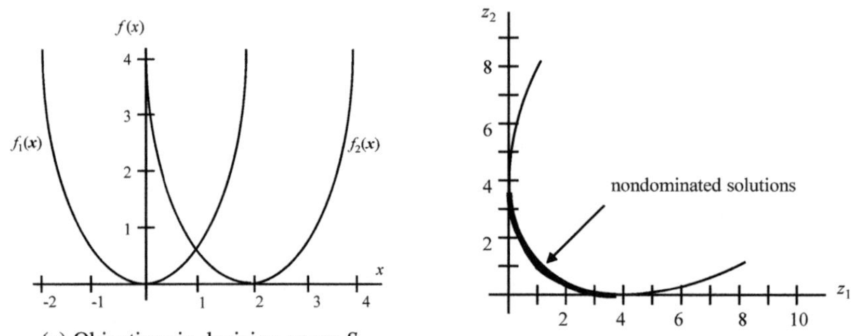

(a) Objectives in decision space S

(b) Objectives in criterion space Z

Fig. 1.15 Objectives in decision space S and criterion space Z

The following is the simulation results of veGA given by Srinivas and Deb [54]. The parameters of genetic algorithms are set as follows: maximum generation is 500, population size 100, string length (binary code) 32, the probability of crossover 1.0, and the probability of mutation 0. Initial range for the variable x is (–10.10). The initial population is shown in Fig. 1.16a. At generation 10, population converged towards to the nondominated region as shown in Fig. 1.16b. At generation 100, almost the whole population fell in the region of nondominated solutions as shown in Fig. 1.16c. Observe that at generation 500, the population converged to only three subregions as shown in Fig. 1.16d. This is the phenomenon of *speciation.*

Schaffer's works are commonly regarded as the first effort of opening the domain of multiple objective optimizations to the GA. Even though veGA cannot give a satisfactory solution to the multiple objective optimization problem, it provides some useful hints for developing other new implementations of GA. Most of subsequent works often refers to veGA as a paradigm to compare performances.

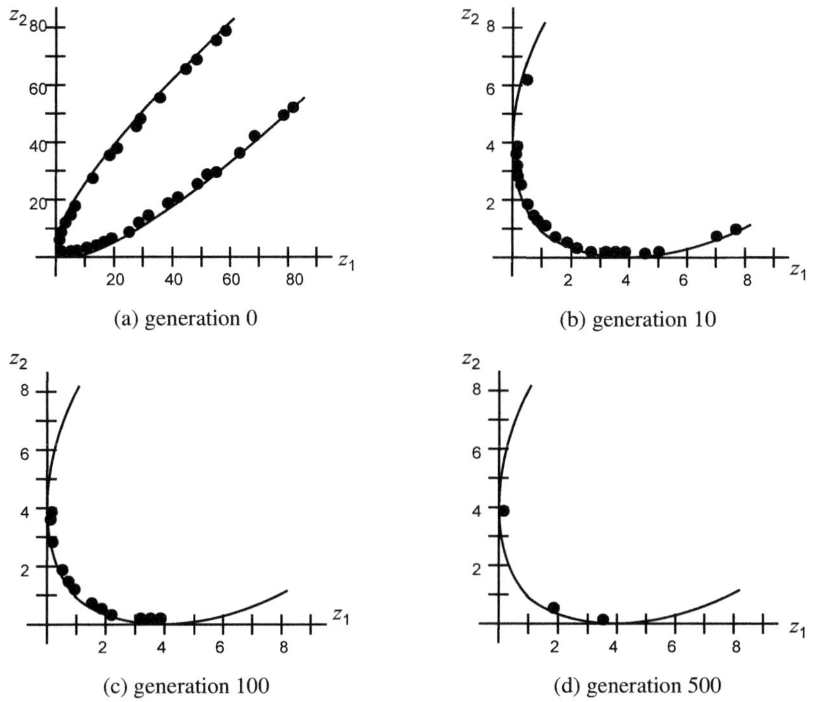

Fig. 1.16 Population in different generations obtained using veGA

1.4.3.2 Pareto Ranking + Diversity: Generation 2

Pareto ranking based fitness assignment method was first suggested by Goldberg [2]. The ranking procedure is as follows: giving rank 1 to the nondominated individuals and removing them from contention, then finding the next set of nondominated individuals and assigning rank 2 to them. The process continues until the entire population is ranked. This approach is illustrated in Fig. 1.17 for a simple case with two objectives to be simultaneously minimized.

Multiobjective Genetic Algorithm (moGA: Fonseca and Fleming, [53]): Fonseca and Fleming proposed a multiobjective genetic algorithm in which the rank of a certain individual corresponds to the number of individuals in the current population by which it is dominated. Based on this scheme, all the nondominated individuals are assigned rank 1, while dominated ones are penalized according to the population density of the corresponding region of the tradeoff surface. Figure 1.18 illustrate a simple case with two objectives to be simultaneously minimized.

Non-dominated Sorting Genetic Algorithm (nsGA: Srinivas and Deb [54]): Srinivas and Deb also developed a Pareto ranking-based fitness assignment and it called nondominated sorting Genetic Algorithm (nsGA). In their method, the nondominated solutions constituting a nondominated front are assigned the same

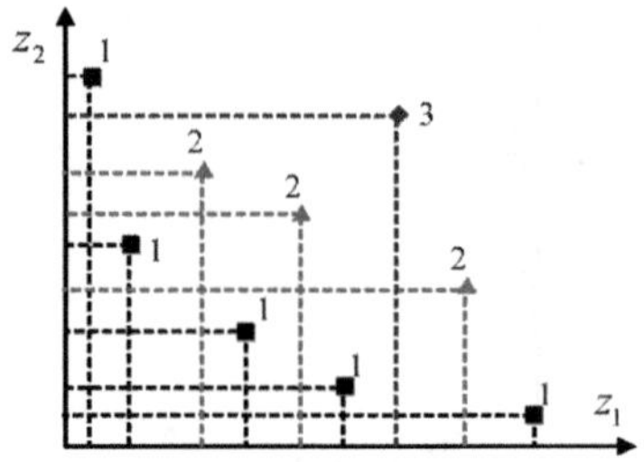

Fig. 1.17 Illustration of Goldberg's ranking (minimization case)

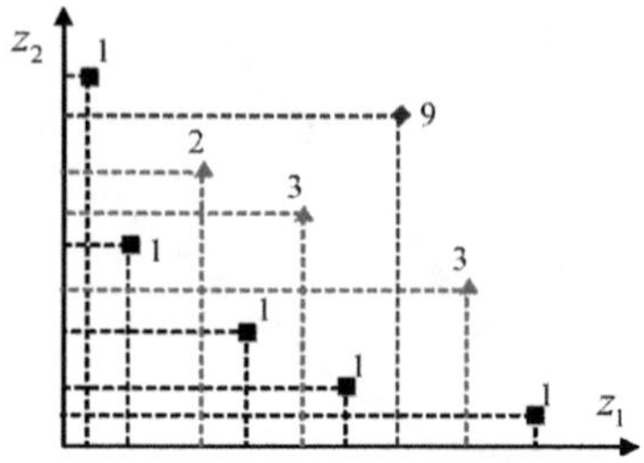

Fig. 1.18 Illustration of moGA ranking (minimization case)

dummy fitness value. These solutions are shared with their dummy fitness values (phenotypic sharing on the decision vectors) and ignored in the further classification process. Finally, the dummy fitness is set to a value less than the smallest shared fitness value in the current nondominated front. Then the next front is extracted. This procedure is repeated until all individuals in the population are classified.

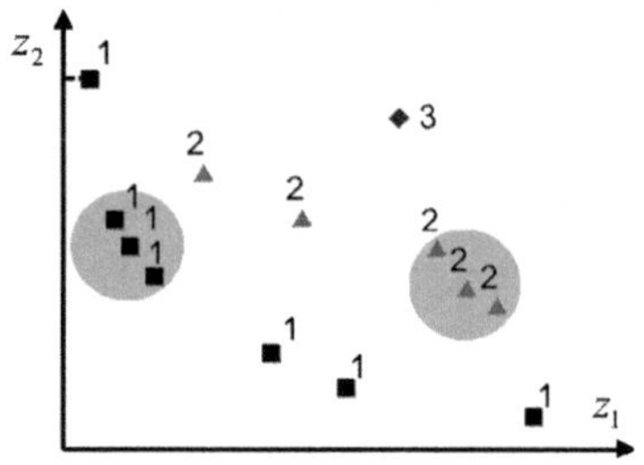

Fig. 1.19 Illustration of nsGA ranking (minimization case)

1.4.3.3 Weighted Sum + Elitist Preserve: Generation 3

Random-weight Genetic Algorithm (rwGA: Ishibuchi and Murata [55]): Ishibuchi and Murata proposed a weighted-sum based fitness assignment method, called random-weight Genetic Algorithm (rwGA), to obtain a variable search direction toward the Pareto frontier. The weighted-sum approach can be viewed as an extension of methods used in the conventional approach for the multiobjective optimizations to the GA. It assigns weights to each objective function and combines the weighted objectives into a single objective function. Typically, there are two kinds of search behaviors in the objective space: fixed direction search and multiple direction search as demonstrated in Fig. 1.20. The fixed weights approach makes the genetic algorithms display a tendency to sample the area towards a fixed point in the criterion space, while the random weights approach makes the genetic algorithms display a tendency of variable search direction, therefore being able to to sample uniformly the area towards the whole frontier. In rwGA, each objective $f_k(x)$ is assigned a weight $w_k = r_k / \sum_{j=1}^{q} r_j$, where r_j are nonnegative random number between [0, 1] with q objective functions. And the scalar fitness value is calculated by summing up the weighted objective value $w_k \cdot f_k(x)$. To search for multiple solutions in parallel, the weights are not fixed so as to enable genetic search to the sample uniformly from the whole area towards the whole frontier. The procedure of rwGA is shown in Fig. 1.21

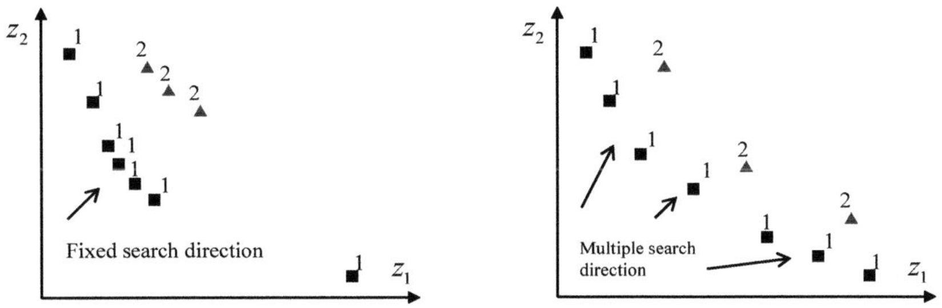

(a) Fixed search direction (b) Multiple search direction

Fig. 1.20 Illustration of fixed direction search and multiple direction search in criterion space (minimization case)

---

procedure: rwGA
input: the objective $f_k(v_i)$ of each chromosome v_i, $k = 1, 2, ..., q$, $\forall i \in popSize$
output: fitness value $eval(v_i), \forall i \in popSize$
begin
  $r_j \leftarrow$ **random**[0,1], $j = 1, 2, ..., q$; // nonnegative random number
  $w_k \leftarrow r_k \Big/ \sum_{j=1}^{q} r_j$, $k = 1, 2, ..., q$;
  $eval(v_i) \leftarrow \sum_{k=1}^{q} w_k \left(f_k(v_i) - z_k^{\min}\right)$, $\forall i$;
  output $eval(v_i)$, $\forall i$;
end

Fig. 1.21 Pseudocode of rwGA

Strength Pareto Evolutionary Algorithm (spEA: Zitzler and Thiele [56]): Zitzler and Thiele proposed strength Pareto evolutionary algorithm (spEA) that combines several features of previous multiobjective Genetic Algorithms (moGA) in a unique manner. The fitness assignment procedure is a two-stage process. First, the individuals in the external nondominated set P' are ranked. Each solution $i \in P'$ is assigned a real value $s_i \in [0,1)$, called *strength*; s_i is proportional to the number of population members $j \in P$ for which $i \succ j$. Let n denote the number of individuals in P that are covered by i and assume N is the size of P. Then s_i is defined as $s_i = n/(N+1)$. The fitness f_i of an objective i is equal to its strength: $f_i = s_i$. Afterwards, the individuals in the population P are evaluated. The fitness of an individual $j \in P$ is calculated by summing the strengths of all external nondominated solutions $i \in P'$ that cover j. The fitness is $f_j = 1 + \sum_{i \in (i \succ j)} s_i$, where $f_j \in [1, N)$. Figure 1.22 shows an illustration of spEA for a maximization problem with two objectives. The procedure of spEA is shown in Fig. 1.23

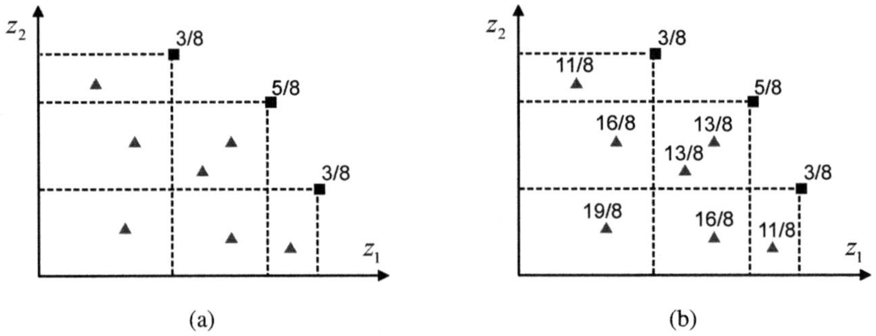

Fig. 1.22 Illustration of spEA (maximization case)

Adaptive-weight Genetic Algorithm (awGA: Gen and Cheng [16]): Gen and Cheng proposed an *adaptive weights approach* which utilizes some useful information from current population to readjust weights in order to obtain a search pressure towards to positive ideal point. For the examined solutions at each generation, we define two extreme points: the *maximum extreme point* z^+ and the *minimum extreme point* z^- in the criteria space as the follows:

$$z^+ = \left\{ z_1^{\max}, z_2^{\max}, \cdots, z_q^{\max} \right\} \tag{1.27}$$

$$z^- = \left\{ z_1^{\min}, z_2^{\min}, \cdots, z_q^{\min} \right\} \tag{1.28}$$

where $z_k^{\min}$ and $z_k^{\max}$ are the maximal value and minimal value for objective k in the current population. Let P denote the set of current population. For a given individual $\boldsymbol{x}$, the maximal value and minimal value for each objective are defined as the follows:

$$z_k^{\max} = \max\{f_k(\boldsymbol{x}) | \boldsymbol{x} \in P\}, \quad k = 1, 2, \cdots, q \tag{1.29}$$

$$z_k^{\min} = \min\{f_k(\boldsymbol{x}) | \boldsymbol{x} \in P\}, \quad k = 1, 2, \cdots, q \tag{1.30}$$

The hyper parallelogram defined by the two extreme points is a minimal hyper parallelogram containing all current solutions. The two extreme points are renewed at each generation. The maximum extreme point will gradually approximate to the positive ideal point. The adaptive weight for objective k is calculated by the follow-

```
procedure : spEA
input: the objective f_k(v_i) of each chromosome v_i, k = 1,2,...,q, ∀i ∈ popSize
output: fitness value eval(v_i), ∀i ∈ popSize
begin
    nondominated set P' ← φ;
    dominated set P ← φ;
    for i = 1 to popSize
        if v_i is nondominated solution then
            P' ← P' ∪ {v_i};
        else
            P ← P ∪ {v_i};
    N ← |P|
    n_i ← |Q_i|,  Q_i = {j | i ≻ j, j ∈ P},  ∀i ∈ P';
    s_i ← n_i/N + 1,  ∀i ∈ P';
    eval(v_i) ← s_i,  ∀i ∈ P';
    eval(v_j) ← 1 + Σ_{i∈(i≻j)} s_i,  ∀j ∈ P;
    output eval(v_i),  ∀i;
end
```

Fig. 1.23 Pseudocode of spEA

ing equation:

$$w_k = \frac{1}{z_k^{\max} - z_k^{\min}}, \quad k = 1, 2, \cdots, q \tag{1.31}$$

For a given individual $\boldsymbol{x}$, the weighted-sum objective function is given by the following equation:

$$z(\boldsymbol{x}) = \sum_{k=1}^{q} w_k (f_k(x) - z_k^{\min}) \tag{1.32}$$

$$= \sum_{k=1}^{q} \frac{f_k(\boldsymbol{x}) - z_k^{\min}}{z_k^{\max} - z_k^{\min}} \tag{1.33}$$

As the extreme points are renewed at each generation, the weights are renewed accordingly. The above equation is a *hyperplane* defined by the following extreme point in current solutions:

$$\left(z_1^{\max}, z_2^{\min}, \cdots, z_k^{\min}, \cdots, z_q^{\min}\right)$$

$$\cdots$$

$$\left(z_1^{\min}, z_2^{\min}, \cdots, z_k^{\max}, \cdots, z_q^{\min}\right)$$

$$\cdots$$

$$\left(z_1^{\min}, z_2^{\min}, \cdots, z_k^{\min}, \cdots, z_q^{\max}\right)$$

It is an adaptive moving line defined by extreme points $(z_1^{\max}, z_2^{\min})$ and $(z_1^{\min}, z_2^{\max})$, as shown in Fig. 1.24. The rectangle defined by the extreme points $(z_1^{\max}, z_2^{\min})$ and $(z_1^{\min}, z_2^{\max})$ is the minimal rectangle containing all current solutions. As show in Fig.

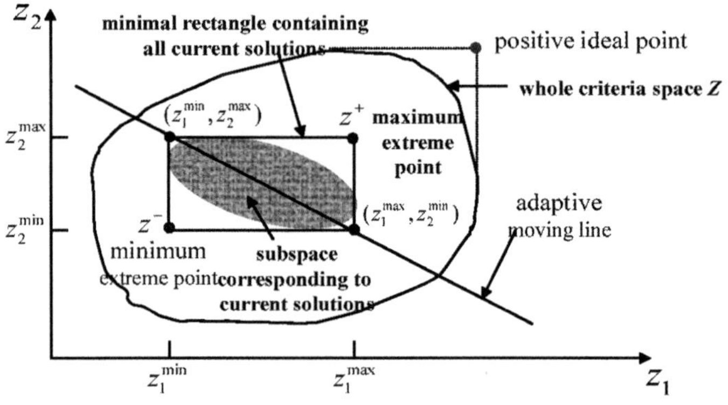

Fig. 1.24 Adaptive-weights and adaptive hyperplane (maximization case)

1.24, the hyperplane divides the criteria space Z into two half spaces: one half space contains the positive ideal point, denoted as Z^+ and the other half space contains the negative ideal point, denoted as Z^-. All examined Pareto solutions lie in the space Z^+ and all points lying in the space Z^+ have larger fitness values than the points in space Z^-. As the maximum extreme point approximates to the positive ideal point along with the evolutionary progress, the hyperplane will gradually approach the positive ideal point. Therefore, the awGA can readjust its weights according to the current population in order to obtain a search pressure towards to the positive ideal point. The procedure of awGA is shown in Fig. 1.25

Non-dominated Sorting Genetic Algorithm II (nsGA II: Deb [58]): Deb *et al.* [58] suggested a nondominated sorting-based approach, called non-dominated sorting Genetic Algorithm II (nsGA II), which alleviates the three difficulties: computational complexity, nonelitism approach, and the need for specifying a sharing parameter [59]. The nsGA II was advanced from its origin, nsGA. In nsGA II as shown in Fig. 1.26, a nondominated sorting approach is used for each individual to create Pareto rank, and a crowding distance assignment method is applied to implement density estimation. In a fitness assignment between two individuals, nsGA II prefers the point with a lower rank value, or the point located in a region with fewer points if both of the points belong to the same front. Therefore, by combining a fast nondominated sorting approach, an elitism scheme and a parameterless sharing method with the original nsGA, nsGA II claims to produce a better spread of solutions in some testing problems. After calculating the fitness of each individual i

---

procedure : awGA
input: the objective $f_k(v_i)$ of each chromosome v_i, $k = 1, 2, ..., q$, $\forall i \in popSize$
output: fitness value $eval(v_i), \forall i \in popSize$
begin

$\{z_k^{\max}\} \leftarrow \max_i\{f_k(v_i)\}$, $k = 1, 2, ..., q$; // maximum extreme point z^+

$\{z_k^{\min}\} \leftarrow \min_i\{f_k(v_i)\}$, $k = 1, 2, ..., q$; // minimum extreme point z^-

$w_k \leftarrow \dfrac{1}{z_k^{\max} - z_k^{\min}}$, $k = 1, 2, ..., q$;

$eval(v_i) \leftarrow \sum_{k=1}^{q} w_k \left(f_k(v_i) - z_k^{\min} \right)$, $\forall i$;

output $eval(v_i)$, $\forall i$;

end

Fig. 1.25 Pseudocode of awGA

by the procedure showed in Fig. 1.25, the nsGA II adopt a special selection process. That is, between two individuals with nondomination rank r_i and r_j, and crowding distance d_i and d_j, we prefer the solution with the lower (better) rank. Otherwise we select the better one depending on the following conditions. And an example of simple case with two objectives to be minimized using nsGA II is shown in Fig. 1.27:

$$i \prec j \ \textbf{ if } \ (r_i < r_j) \ \textbf{ or } \ ((r_i = r_j) \ \textbf{ and } \ (d_i > d_j))$$

Interactive Adaptive-weight GA (i-awGA: Lin and Gen [46]): Generally, the main idea of the Pareto ranking-based approach is a clear classification between nondominated solution and dominated solution for each chromosome. However, it is difficult to clear the difference among non-dominated solutions (or dominated solutions). This is illustrated in Fig. 1.28a, although there are distinct differences between the dominated solutions (2, 2) and (8, 8), but there are no distinct differences between their fitness values 13/5 and 11/5 by strength Pareto evolutionary algorithm.

Different from Pareto ranking-based fitness assignment, weighted-sum based fitness assignment assigns weights to each objective function and combines the weighted objectives into a single objective function. It is easier to calculate the weight-sum based fitness and the sorting process becomes unnecessary; thus it is effective for considering the computation time to solve the problems. In addition, another characteristic of the weighted-sum approach is used to adjust genetic search toward the Pareto frontier. For combining weighted objectives into a single objective function, the good fitness values are assigned with solutions near the Pareto frontier. However, shown in Fig. 1.28c, the fitness values (12/11, 14/11, 13/11) of

procedure: nsGA II
input: the objective $f_k(v_i)$ of each chromosome v_i, $k = 1, 2, ..., q$, $\forall i \in popSize$
output: fitness value $eval(v_i), \forall i \in popSize$
begin
    set $P \leftarrow \{v_i\}$; // nondominated-sort (r)
    $rank \leftarrow 1$;
    while $P \neq \phi$ **do**
        for $i = 1$ **to** $popSize$
            if v_i is nondominated solution **then**
                $r_i \leftarrow rank$;
        $P \leftarrow P \setminus \{v_i | r_i = rank\}$;
        $rank \leftarrow rank + 1$;
    $r_i \leftarrow 0, \quad \forall i \in popSize$ // distance-assignment (d)
    for $k = 1$ **to** q
        $\{j\} \leftarrow \textbf{sort}\left\{P \middle| \max(z_k^i), v_i \in P\right\}$;
        $d_k^1 = d_k^{popSize} \leftarrow \infty$;
        $d_k^j \leftarrow \dfrac{d_k^{j+1} - d_k^{j-1}}{\max\limits_i\{z_k^i\} - \min\limits_i\{z_k^i\}}, \quad j = 2, ..., popSize - 1$;
    $d_i \leftarrow \sum_{k=1}^{q} d_k^i, \quad \forall i \in popSize$;
    output $eval(v_i) = \{r_i, d_i\}, \quad \forall i$;
end

Fig. 1.26 Pseudocode of nsGA II

most dominated solutions ((10, 4), (8, 8), (6, 9)) are greater than the fitness values (11/11, 11/11) of some non-dominated solutions ((12, 1), (1, 12)) by using the adaptive-weight based fitness assignment approach.

Lin and Gen proposed an *interactive adaptive-weight fitness assignment approach* as shown in Fig. 1.29, which is an improved adaptive-weight fitness assignment approach with the consideration of the disadvantages of weighted-sum approach and Pareto ranking-based approach. First, there are two extreme points defined as the maximum extreme point $z^+ = \{z_1^{\max}, z_2^{\max}, \cdots, z_q^{\max}\}$, and minimum extreme point $z^- = \{z_1^{\min}, z_2^{\min}, \cdots, z_q^{\min}\}$, where $z_k^{\min}$ and $z_k^{\max}$ are the maximal value and minimal value for objective k in the current population. Afterwards, the adaptive weight for objective k is calculated by the following equation:

$$w_k = \frac{1}{z_k^{\max} - z_k^{\min}}, \quad k = 1, 2, \cdots, q \tag{1.34}$$

Then, calculate the penalty term $p(v_i)$=0, if v_i is a nondominated solution in the nondominated set P. Otherwise $p(v_{i'})$=1 for dominated solution $v_{i'}$. Last, calculate

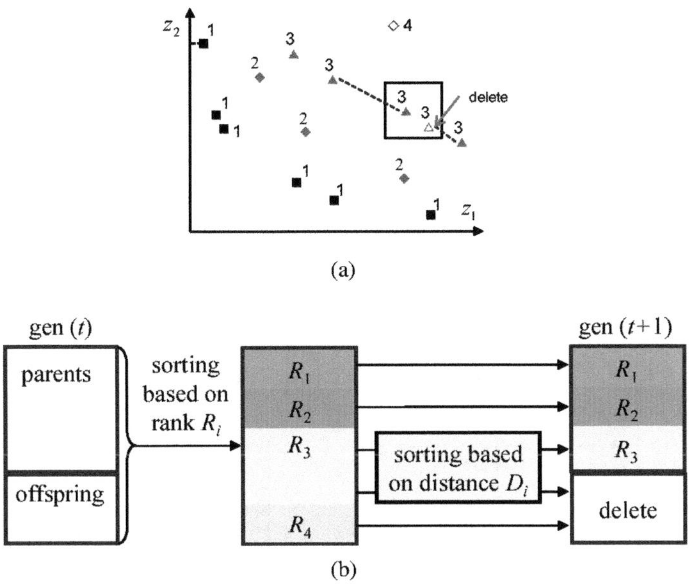

Fig. 1.27 Illustration of nsGA II (minimization case)

the fitness value of each chromosome by combining the method as follows:

$$eval(v_i) = \sum_{k=1}^{q} w_k(f_k(v_i) - z_k^{\min}) + p(v_k), \quad \forall\, i \in \; popSize$$

1.4.4 Performance Measures

Let S_j be a solution set (j=1, 2, $\cdots$, J). In order to evaluate the efficiency of the different fitness assignment approaches, we have to define explicitly measures evaluating closeness of S_j from a known set of the Pareto-optimal set S^*. In this subsection, the following three measures are considered that are already used in different moGA studies. They provide a good estimate of convergence if a reference set for S^* (*i.e.*, the Pareto optimal solution set or a near Pareto optimal solution set) is chosen as shown in Fig. 1.30.

Number of Obtained Solutions $|S_j|$

Evaluate each solution set depending on the number of obtained solutions.

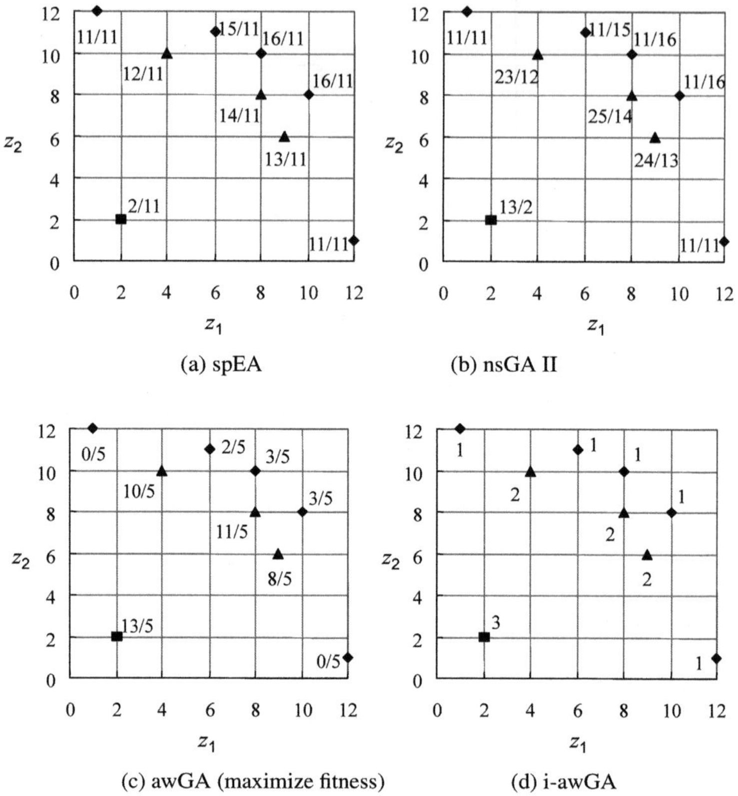

Fig. 1.28 Illustration of fitness values by different fitness assignment mechanisms. (maximization case)

Ratio of Nondominated Solutions $R_{NDS}(S_j)$

This measure simply counts the number of solutions which are members of the Pareto-optimal set S^*. The $R_{NDS}(S_j)$ measure can be written as follows:

$$R_{NDS}(S_j) = \frac{|S_j - \{x \in S_j | \exists r \in S^* : r \prec x\}|}{|S_j|} \tag{1.35}$$

where $r \prec x$ means that the solution x is dominated by the solution r. An $R_{NDS}(S_j)$=1 means all solutions are members of the Pareto-optimal set S^*, and an $R_{NDS}(S_j)$=0 means no solution is a member of the S^*. It is an important measure that, although the number of obtained solutions —S_j— is large, if the ratio of nondominated solutions $R_{NDS}(S_j)$ is 0, it may be a worse result. The difficulty with the above measures is that, although a member of S_j is Pareto-optimal, if that solution does not exist in

```
procedure: i-awGA
input: the objective f_k(v_i) of each chromosome v_i, k = 1,2,...,q, ∀i ∈ popSize
output: fitness value eval(v_i), ∀i ∈ popSize
begin
    {z_k^max} ← max_i{f_k(v_i)}, k = 1,2,...,q;   // maximum extreme point z+
    {z_k^min} ← min_i{f_k(v_i)}, k = 1,2,...,q;   // minimum extreme point z−
    w_k ← 1 / (z_k^max − z_k^min), k = 1,2,...,q;
    p(v_i) ← 0, ∀i;
    if v_i is dominated solution then
        p(v_i) ← 1;
    eval(v_i) ← Σ_{k=1}^{q} w_k (f_k(v_i) − z_k^min) + p(v_i), ∀i;
    output eval(v_i), ∀i;
end
```

Fig. 1.29 Pseudocode of i-awGA

S^*, it may be not counted in $R_{NDS}(S_j)$ as a non-Pareto-optimal solution. Thus, it is essential that a large set for S^* is necessary in the above equations.

Average Distance $D1_R(S_j)$

Instead of finding whether a solution of S_j belongs to the set S^* or not, this measure finds an average distance of the solutions of S_j from S^*, as follows:

$$D1_R(S_j) = \frac{1}{|S^*|}\sum_{\boldsymbol{r}\in S^*} \min\left\{d_{\boldsymbol{rx}} | \boldsymbol{x} \in S_j\right\} \tag{1.36}$$

where $d_{\boldsymbol{rx}}$ is the distance between a current solution $\boldsymbol{x}$ and a reference solution $\boldsymbol{r}$ in the two-dimensional normalized objective space

$$d_{\boldsymbol{rx}} = \sqrt{(f_1(\boldsymbol{r}) - f_1(\boldsymbol{x}))^2 + (f_2(\boldsymbol{r}) - f_2(\boldsymbol{x}))^2} \tag{1.37}$$

The smaller the value of $D1_R(S_j)$, the better the solution set S_j. This measure explicitly computes a measure of the closeness of a solution set S_j from the set S^*.

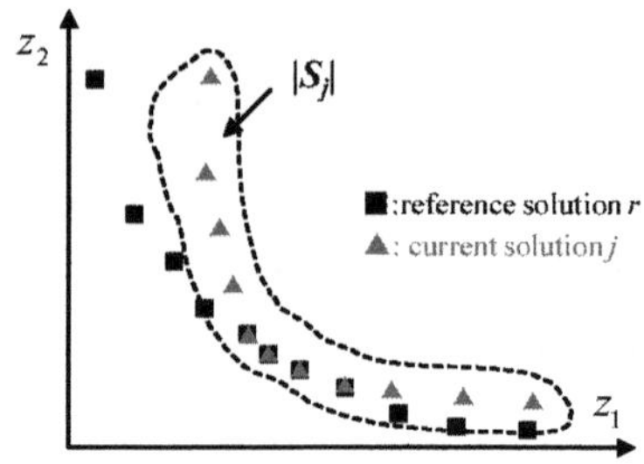

(a) Number of obtained solutions $|S_j|$

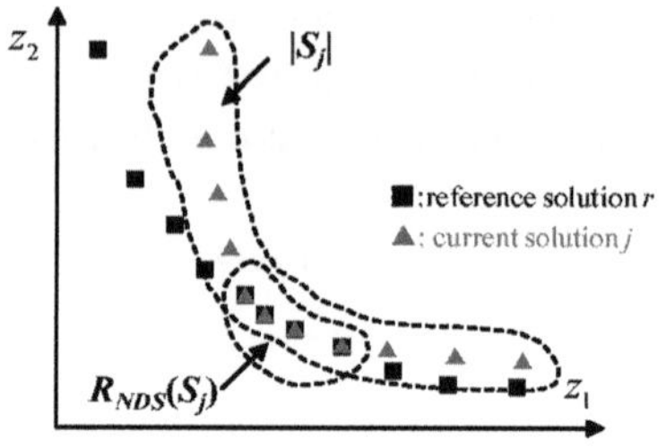

(b) Ratio of nondominated solutions $R_{NDS}(S_j)$

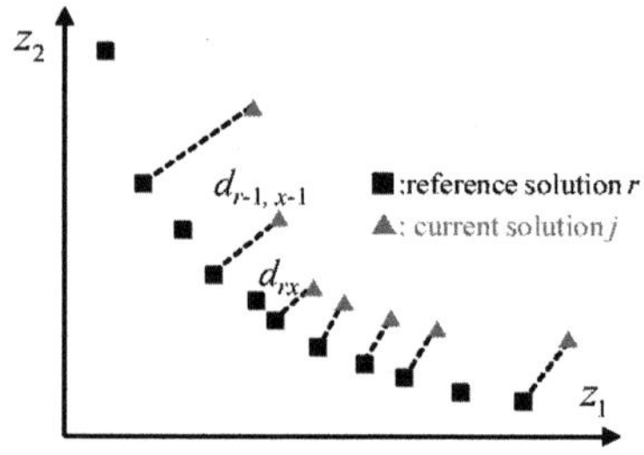

(c) Average distance $D1_R(S_j)$

Fig. 1.30 Illustration of performance measures (maximization case)

1.4.4.1 Reference Set S^*

In order to make a large number of solutions in the reference set S^*, first we calculate the solution sets with special GA parameter settings and much long computation time by each approach which is used in comparison experiments, then combine these solution sets to calculate the reference set S^*. Furthermore, we will combine small but reasonable GA parameter settings for comparison experiments. Thus ensure the effectiveness of the reference set S^*.

References

1. Holland, J. (1992). *Adaptation in Natural and Artificial System*, Ann Arbor: University of Michigan Press; 1975, MA: MIT Press.
2. Goldberg, D. (1989). *Genetic Algorithms in Search, Optimization and Machine Learning*, Reading, MA: Addison-Wesley.
3. Fogel, L., Owens, A. & Walsh, M. (1966). *Artificial Intelligence through Simulated Evolution*, New York: John Wiley & Sons.
4. Rechenberg, I. (1973). *Optimieriung technischer Systeme nach Prinzipien der biologischen Evolution*, Stuttgart: Frommann-Holzboog.
5. Schwefel, H. (1995). *Evolution and Optimum Seeking*, New York: Wiley & Sons.
6. Koza, J. R. (1992). *Genetic Programming*, Cambridge: MIT Press.

7. Koza, J. R. (1994). *Genetic Programming II*, Cambridge: MIT Press.
8. Holland, J. H. (1976). Adaptation. In R. Rosen & F. M. Snell, (eds) *Progress in Theoretical Biology IV*, 263–293. New York: Academic Press.
9. Dorigo, M. (1992) *Optimization, Learning and Natural Algorithms*, PhD thesis, Politecnico di Milano, Italy.
10. Kennedy, J. & Eberhart, R. (1995). Particle swarm optimization, *Proceeding of the IEEE International Conference on Neural Networks*, Piscataway, NJ, 1942–1948,
11. Kennedy, J. & Eberhart, R. C. (2001). *Swarm Intelligence*. Morgan Kaufmann.
12. Michalewicz, Z. (1994). *Genetic Algorithm + Data Structures = Evolution Programs*. New York: Springer-Verlag.
13. Bolc, L. & Cytowski, J. (1992). *Search Methods for Artificial Intelligence*, London: Academic Press.
14. Booker, L. (1987) Improving search in genetic algorithms, in Davis, L, ed. *Genetic Algorithms and Simulated Annealing*, Morgan Kaufmann Publishers, Los Altos, CA.
15. Gen, M. & Cheng, R. (1997). *Genetic Algorithms and Engineering Design*. New York: John Wiley & Sons.
16. Gen, M. & Cheng, R. (2000). *Genetic Algorithms and Engineering Optimization*. New York: John Wiley & Sons.
17. Grefenstette, J. J. (1991). Lamarkian learning in multi-agent environment, *Proceedings of the 4th International Conference on Genetic Algorithms*, San Francisco, Morgan Kaufman Publishers.
18. Davidor, Y. (1991). A genetic algorithm applied to robot trajectory generation, Davis, L., editor, *Handbook of Genetic Algorithms*, NewYork, Van Nostrand Reinhold.
19. Shaefer, C. (1987). The ARGOT strategy: adaptive representation genetic optimizer technique, *Proceedings of the 2nd International conference on Genetic Algorithms*, Lawrence Erlbaum Associates, Hillsdale, N, J.
20. Kennedy, S. (1993). Five ways to a smarter genetic algorithm, *AI Expert*, 35–38.
21. Whitley, D., Gordan, V. & Mathias, K. (1994). Lamarckian evolution, the Baldwin effect & function optimization, in Davidor, Y., Schwefel, H. & Männer, R., Editors, Parallel Problem Solving from Nature:PPSN III, Berlin, *Springer-Verlag*, 6–15.
22. Moscato, P. & Norman, M. (1992). A memetic approach for the traveling saleman problem: implementation of a computational ecology for combinatorial optimization on message-passing systems, *in Proceedings of the international conference on parallel computing & transputer applications*, Amsterdam.
23. Dawkins, R. (1976). *The Selfish Gene*, Oxford, Oxford University Press.
24. Radcliffe, N. & Surry, P. (1994). *Formal memetic algorithms*, Fogarty, T., editor , Berlin, Evolutionary Computing, 1–16.
25. Raidl, G. R. & Julstrom, B. A. (2003). Edge Sets: An Effective Evolutionary Coding of Spanning Trees, *IEEE Transactions on Evolutionary Computation*, 7(3), 225–239.
26. Kim, J. H. & Myung, H. (1996). A two-phase evolutionary programming for general constrained optimization problem, *Proceeding of the 1st Annual Conference on Evolutionary Programming*.
27. Michalewicz, Z. (1995). Genetic algorithms, numerical optimization, and constraints, in Eshelman, L. J. ed. *Proceeding of the 6th International Conference on Genetic Algorithms*, 135–155.
28. Myung, H. & Kim, J. H. (1996). Hybrid evolutionary programming for heavily constrained problems, *Bio-Systems*, 38, 29–43.
29. Orvosh, D. & Davis, L. (1994). Using a genetic algorithm to optimize problems with feasibility constraints, *Proceeding of the 1st IEEE Conference on Evolutionary Computation*, 548–552.
30. Michalewicz, Z. (1995). A survey of constraint handling techniques in evolutionary computation methods, in McDonnell *et al.* eds. *Evolutionary Programming IV*, MA: MIT Press.
31. Michalewicz, Z., Dasgupta, D., Riche, R. G. L. & Schoenauer, M. (1996). Evolutionary algorithms for constrained engineering problems, *Computers and Industrial Engineering*, 30(4), 851–870.

32. Liepins, G., Hilliard, M., Richardson, J. & Pallmer, M. (1990). Genetic algorithm application to set covering and traveling salesman problems, in Brown ed. *OR/AI: The Integration of Problem Solving Strategies*.
33. Liepins, G. & Potter, W. (1991). A genetic algorithm approach to multiple faultdiagnosis, Davis ed. *Handbook of Genetic Algorithms*, New York: Van Nostrand Reinhold, 237–250.
34. Nakano, R. & Yamada, T. (1991). Conventional genetic algorithms for jo-shop problems, *Proceeding of the 4th International Conference on Genetic Algorithms*, 477–479.
35. Glover, F. & Greenberg, H. (1989). New approaches for heuristic search: a bilateral linkage with artificial intelligence, *European Journal of Operational Research*, 39, 19–130.
36. Herrera, F. & Lozano, M. (1996). Adaptation of Genetic Algorithm Parameters Based On Fuzzy Logic Controllers, in F. Herrera and J. Verdegay, editors, *Genetic Algorithms and Soft Computing*, Physica-Verlag, 95–125.
37. Hinterding, R., Michalewicz, Z. & Eiben, A. (1997). Adaptation in Evolutionary Computation: A Survey, *Proceeding of IEEE International Conference on Evolutionary Computation*, Piscataway, NJ, 65–69.
38. Rechenberg, I. (1973). *Optimieriung technischer Systeme nach Prinzipien der biologischen Evolution*, Stuttgart: Frommann-Holzboog.
39. Davis, L. editor, (1991). *Handbook of Genetic Algorithms*, New York: Van Nostrand Reinhold.
40. Julstrom, B. (1995). What Have You Done For Me Lately Adapting Operator Probabilities in A Steady-state Genetic Algorithm, *Proceeding 6th International Conference on GAs*, San Francisco, 81–87.
41. Subbu, R., Sanderson, A. C. & Bonissone, P. P. (1999). Fuzzy Logic Controlled Genetic Algorithms versus Tuned Genetic Algorithms: an Agile Manufacturing Application, *Proceeding of IEEE International Symposium on Intelligent Control (ISIC)*, 434–440.
42. Song, Y. H., Wang, G. S., Wang, P. T. & Johns, A. T. (1997). Environmental/Economic dispatch using fuzzy logic controlled genetic algorithms, *IEEE Proceeding on Generation, Transmission and Distribution*, 144(4), 377–382.
43. Cheong, F. & Lai, R. (2000). Constraining the optimization of a fuzzy logic controller using an enhanced genetic algorithm, *IEEE Transactions on Systems, Man, and Cybernetics-Part B: Cybernetics*, 30(1), 31–46.
44. Yun, Y. S. & Gen, M. (2003). Performance analysis of adaptive genetic algorithms with fuzzy logic and heuristics, *Fuzzy Optimization and Decision Making*, 2(2), 161–175.
45. Koumousis, V. K. & Katsaras, C. P. (2006). A saw-tooth genetic algorithm combining the effects of variable population size and reinitialization to enhance performance, *IEEE Transactions on Evolutionary Computation*, 10(1), 19–28.
46. Lin, L. & Gen, M. Auto-tuning strategy for evolutionary algorithms: balancing between exploration and exploitation, *Soft Computing*, In press.
47. Moed, M. C., Stewart, C. V. & Kelly, R. B. (1991). Peducing the search time of a steady state genetic algorithm using the immigration operator, *Proceeding IEEE International Conference Tools for AI*, 500–501.
48. Dev, K. (1995). *Optimization for Engineering Design: Algorithms and Examples*, New Delhi: Prentice-Hall.
49. Steuer, R. E. (1986). *Multiple Criteria Optimization: Theory, Computation, and Application*,New York: John Wiley & Sons.
50. Hwang, C. & Yoon, K. (1981). *Multiple Attribute Decision Making: Methods and Applications*, Berlin: Springer-Verlage.
51. Pareto, V. (1971). *Manuale di Econòmica Polìttica, Società Editrice Libràia*, Milan, Italy, 1906; translated into English by A. S. Schwier, as *Manual of Political Economy*, New York: Macmillan.
52. Schaffer, J. D. (1985). Multiple objective optimization with vector evaluated genetic algorithms, *Proceeding 1st International Conference on GAs*, 93–100.
53. Fonseca, C. & Fleming, P. (1993). Genetic Algorithms for Multiobjective Optimization: Formulation, Discussion and Generalization, *Proceeding 5th International Conference on GAs*, 416–423.

54. Srinivas, N. & Deb, K. (1995). Multiobjective Function Optimization Using Nondominated Sorting Genetic Algorithms, *Evolutionary Computation*, 3, 221–248.
55. Ishibuchi, H. & Murata, T. (1998). A multiobjective genetic local search algorithm and its application to flowshop scheduling, *IEEE Transactions on Systems, Man, and Cybernetics*, 28(3), 392–403.
56. Zitzler, E. & Thiele, L. (1999). Multiobjective evolutionary algorithms: a comparative case study and the strength Pareto approach, *IEEE Transactions on Evolutionary Computation*, 3(4), 257–271.
57. Zitzler, E., Laumanns, M. & Thiele, L. (2002) SPEA2: Improving the Strength Pareto Evolutionary Algorithm for Multiobjective Optimization. *Proceedings of the EUROGEN Conference*, 95–100.
58. Deb, K. (2001). *Multiobjective Optimization Using Evolutionary Algorithms*, Chichester: Wiley.
59. Deb, K., Pratap, A., Agarwal, S . & Meyarivan, T. (2002). A fast and elitist multiobjective genetic algorithm: NSGA-II, *IEEE Transactions on Evolutionary Computation*, 6(2), 182–197.

Chapter 2
Basic Network Models

Network design is one of the most important and most frequently encountered classes of optimization problems [1]. It is a combinatory field in graph theory and combinatorial optimization. A lot of optimization problems in network design arose directly from everyday practice in engineering and management: determining shortest or most reliable paths in traffic or communication networks, maximal or compatible flows, or shortest tours; planning connections in traffic networks; coordinating projects; and solving supply and demand problems.

Furthermore, network design is also important for complexity theory, an area in the common intersection of mathematics and theoretical computer science which deals with the analysis of algorithms. However, there is a large class of network optimization problems for which no reasonable fast algorithms have been developed. And many of these network optimization problems arise frequently in applications. Given such a hard network optimization problem, it is often possible to find an efficient algorithm whose solution is approximately optimal. Among such techniques, the genetic algorithm (GA) is one of the most powerful and broadly applicable stochastic search and optimization techniques based on principles from evolution theory.

2.1 Introduction

Network design is used extensively in practice in an ever expanding spectrum of applications. Network optimization models such as shortest path, assignment, max-flow, transportation, transhipment, spanning tree, matching, traveling salesman, generalized assignment, vehicle routing, and multi-commodity flow constitute the most common class of practical network optimization problems.

In this chapter, we introduce some of the core models of network design (Fig. 2.1), such as shortest path models (*i.e.,* node selection and sequencing), spanning tree models (*i.e.,* arc selection) and maximum flow models (*i.e.,* arc selection and flow assignment) *etc.* These network models are used most extensively in applica-

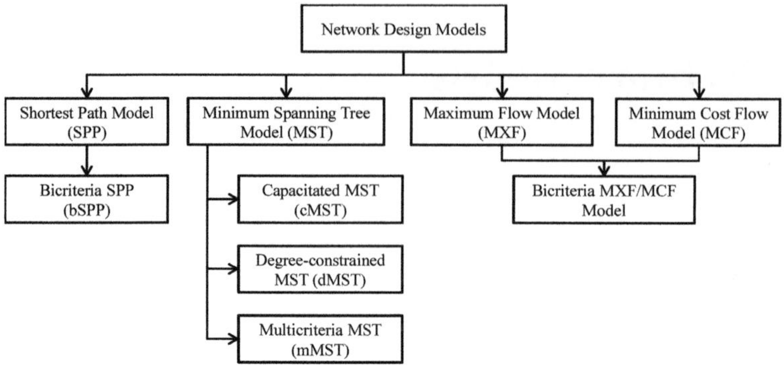

Fig. 2.1 The core models of network design introduced in this chapter

tions and differentiated by their structural characteristics. We assume that a connected graph $G = (N, A)$ is given, and it is a structure consisting of a finite set N of elements called nodes and a set A of unordered pairs of nodes called arcs. The descriptions of the models are as follows.

2.1.1 Shortest Path Model: Node Selection and Sequencing

It is desired to find a set of nodes. What is the shortest path between two specified nodes of G?

The *shortest path model* is the heart of network design optimization. They are alluring to both researchers and practitioners for several reasons: (1) they arise frequently in practice since in a wide variety of application settings we wish to send some material (*e.g.*, a computer data packet, a telephone call, a vehicle) between two specified points in a network as quickly, as cheaply, or as reliably as possible; (2) as the simplest network models, they capture many of the most salient core ingredients of network design problems and so they provide both a benchmark and a point of departure for studying more complex network models; and (3) they arise frequently as subproblems when solving many combinatorial and network optimization problems [1]. Even though shortest path problems are relatively easy to solve, the design and analysis of most efficient algorithms for solving them requires considerable ingenuity. Consequently, the study of shortest path problems is a natural starting point for introducing many key ideas from network design problems, including the use of clever data structures and ideas such as data scaling to improve the worst case algorithmic performance. Therefore, in this chapter, we begin our discussion of network design optimization by studying shortest path models.

Applications

Shortest path problems arise in a wide variety of practical problem settings, both as stand-alone models and as subproblems in more complex problem settings.

1. *Transportation planning*: How to determine the route road that have prohibitive weight restriction so that the driver can reach the destination within the shortest possible time.
2. *Salesperson routing*: Suppose that a sales person want to go to Los Angeles from Boston and stop over in several city to get some commission. How can she determine the route?
3. *Investment planning*: How to determine the invest strategy to get an optimal investment plan.
4. *Message routing in communication systems*: The routing algorithm computes the shortest (least cost) path between the router and all the networks of the internetwork. It is one of the most important issues that has a significant impact on the network's performance.

2.1.2 Spanning Tree Model: Arc Selection

> *It is desired to find a subset of arcs such that when these arcs are removed from G, the graph remains connected. For what subset of arcs is the sum of the arc lengths minimized?*

Spanning tree models play a central role within the field of network design. It generally arises in one of two ways, directly or indirectly. In some direct applications, we wish to connect a set of points using the least cost or least length collection of arcs. Frequently, the points represent physical entities such as components of a computer chip, or users of a system who need to be connected to each other or to a central service such as a central processor in a computer system [1]. In indirect applications, we either (1) wish to connect some set of points using a measure of performance that on the surface bears little resemblance to the minimum spanning tree objective (sum of arc costs), or (2) the problem itself bears little resemblance to an "optimal tree" problem – in these instances, we often need to be creative in modeling the problem so that it becomes a minimum spanning tree problem.

Applications

1. *Designing physical systems*: Connect terminals in cabling the panels of electrical equipment. How should we wire the terminals to use the least possible length of wire? Constructing a pipeline network to connect a number of towns, how should we using the smallest possible total length of pipeline.

2. *Reducing data storage*: We wish to store data specified in the form of a two-dimensional array more efficiently than storing all the elements of the array (to save memory space). The total storage requirement for a particular storage scheme will be the length of the reference row (which we can take as the row with the least amount of data) plus the sum of the differences between the rows. Therefore, a minimum spanning tree would provide the least cost storage scheme.
3. *Cluster analysis*: The essential issue in cluster analysis is to partition a set of data into "natural groups"; the data points within a particular group of data, or a cluster, should be more "closely related" to each other than the data points not in the cluster. We can obtain n partitions by starting with a minimum spanning tree and deleting tree arcs one by one in nonincreasing order of their lengths. Cluster analysis is important in a variety of disciplines that rely on empirical investigations.

2.1.3 Maximum Flow Model: Arc Selection and Flow Assignment

It is desired to send as much flow as possible between two special nodes of G, without exceeding the capacity of any arc.

The *maximum flow model* and the shortest path model are complementary. They are similar because they are both pervasive in practice and because they both arise as subproblems in algorithms for the minimum cost flow problem. The two problems differ because they capture different aspects. Shortest path problems model arc costs but not arc capacities; maximum flow problems model capacities but not costs. Taken together, the shortest path problem and the maximum flow problem combine all the basic ingredients of network design optimization. As such, they have become the cores of network optimization [1].

Applications

The maximum flow problems arise in a wide variety of situations and in several forms. For example, sometimes the maximum flow problem occurs as a subproblem in the solution of more difficult network problems, such as the minimum cost flow problems or the generalized flow problem. The problem also arises directly in problems as far reaching as machine scheduling, the assignment of computer modules to computer processors, the rounding of census data to retain the confidentiality of individual households, and tanker scheduling.

1. *Scheduling on uniform parallel machines*: The feasible scheduling problem, described in the preceding paragraph, is a fundamental problem in this situation and can be used as a subroutine for more general scheduling problems, such as the maximum lateness problem, the (weighted) minimum completion time problem, and the (weighted) maximum utilization problem.

2. *Distributed computing on a two-processor computer*: Distributed computing on a two-processor computer concerns assigning different modules (subroutines) of a program to two processors in a way that minimizes the collective costs of interprocessor communication and computation.
3. *Tanker scheduling problem*: A steamship company has contracted to deliver perishable goods between several different origin-destination pairs. Since the cargo is perishable, the customers have specified precise dates (*i.e.,* delivery dates) when the shipments must reach their destinations.

2.1.4 Representing Networks

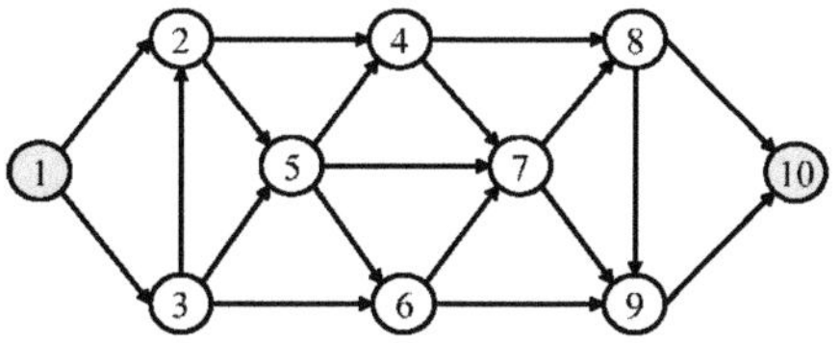

Fig. 2.2 A digraph G

Definition 2.1. Edge Lists

A *directed graph* (or *digraph*) G on the nodes set $N = \{1, 2, \cdots, n\}$ is specified by:

1. Its number of nodes n;
2. The list of its arcs, given as a sequence of ordered m pairs A=$\{(i, j)$, (k, l), $\cdots$, $(s, t)\}$.

The digraph G of Fig. 2.2 may then be given as follows:

1. Its number of nodes $n = 10$;
2. The list of its arcs $A = \{$(1, 2), (1, 3), (2, 4), (2, 5), (3, 2), (3, 5), (3, 6), (4, 7), (4, 8), (5, 4), (5, 6), (5, 7), (6, 7), (6, 9), (7, 8), (7, 9), (8, 9), (8, 10), (9, 10)$\}$.

Definition 2.2. Adjacency Lists

A digraph with node set $N = \{1, 2, \cdots, n\}$ is specified by:

1. The number of nodes n;
2. n lists $S_1, \cdots, S_i, \cdots, S_n$, where S_i contains all nodes j for which G contains an arc (i, j).

The digraph G of Fig. 2.2 may be represented by *adjacency lists* as follows:

1. Its number of nodes $n = 10$;
2. The adjacency lists: $S_1 = \{2, 3\}$, $S_2 = \{4, 5\}$, $S_3 = \{2, 5, 6\}$, $S_4 = \{7, 8\}$, $S_5 = \{4, 6, 7\}$, $S_6 = \{7, 9\}$, $S_7 = \{8, 9\}$, $S_8 = \{9, 10\}$, $S_9 = \{10\}$ and $S_{10} = \phi$.

Definition 2.3. Adjacency Matrices

A digraph with node set $N = \{1,2,\cdots,n\}$ is specified by an $(n \times n)$-matrix $A = (a_{ij})$, where $a_{ij} = 1$ if and only if (i,j) is an arc of G, and $a_{ij} = 0$ otherwise. A is called the *adjacency matrix* of G. The digraph G of Fig. 2.2 may then be given as follows:

$$
A = \begin{array}{c} \\ 1 \\ 2 \\ 3 \\ 4 \\ 5 \\ 6 \\ 7 \\ 8 \\ 9 \\ 10 \end{array}
\begin{array}{c}
\begin{array}{cccccccccc} 1 & 2 & 3 & 4 & 5 & 6 & 7 & 8 & 9 & 10 \end{array} \\
\begin{bmatrix}
0 & 1 & 1 & 0 & 0 & 0 & 0 & 0 & 0 & 0 \\
0 & 0 & 0 & 1 & 1 & 0 & 0 & 0 & 0 & 0 \\
0 & 1 & 0 & 0 & 1 & 1 & 0 & 0 & 0 & 0 \\
0 & 0 & 0 & 0 & 0 & 0 & 1 & 1 & 0 & 0 \\
0 & 0 & 0 & 1 & 0 & 1 & 1 & 0 & 0 & 0 \\
0 & 0 & 0 & 0 & 0 & 0 & 1 & 0 & 1 & 0 \\
0 & 0 & 0 & 0 & 0 & 0 & 0 & 1 & 1 & 0 \\
0 & 0 & 0 & 0 & 0 & 0 & 0 & 0 & 1 & 1 \\
0 & 0 & 0 & 0 & 0 & 0 & 0 & 0 & 0 & 1 \\
0 & 0 & 0 & 0 & 0 & 0 & 0 & 0 & 0 & 0
\end{bmatrix}
\end{array}
$$

Adjacency matrices can be implemented simply as an array $[1, \cdots, n, 1, \cdots, n, \cdots]$. As they need a lot of space in memory (n^2 places), they should only be used (if at all) to represent digraphs having many arcs. Though adjacency matrices are of little practical interest, they are an important theoretical tool for studying digraphs.

2.1.5 Algorithms and Complexity

2.1.5.1 Algorithms

Algorithms are techniques for solving problems.

Here the term problem is used in a very general sense: a problem class comprises infinitely many instances having a common structure. In general, an algorithm is a technique which can be used to solve each instance of a given problem class. An algorithm should have the following properties [3]:

1. *Finiteness of description*: The technique can be described by a finite text.
2. *Effectiveness*: Each step of the technique has to be feasible (mechanically) in practice.
3. *Termination*: The technique has to stop for each instance after a finite number of steps.
4. *Determinism*: The sequence of steps has to be uniquely determined for each instance.

Of course, an algorithm should also be correct, that is, it should indeed solve the problem correctly for each instance. Moreover, *an algorithm should be efficient, which means it should work as fast and economically as possible* [3].

2.1.5.2 Complexity

Computational complexity theory is the study of the complexity of problems – that is, the difficulty of solving them. Problems can be classified by *complexity class* according to the time it takes for an algorithm to solve them as function of the *problem size*. For example, the travelling salesman problem can be solved in time $O(n^2 2^n)$ (where n is the size of the network to visit).

Even though a problem may be solvable computationally in principle, in actual practice it may not be that simple. These problems might require large amounts of time or an inordinate amount of *space*. Computational complexity may be approached from many different aspects. Computational complexity can be investigated on the basis of *time*, memory or other resources used to solve the problem. Time and space are two of the most important and popular considerations when problems of complexity are analyzed. The *time complexity* of a problem is the number of steps that it takes to solve an instance of the problem as a function of the size of the input (usually measured in bits), using the most efficient algorithm. To understand this intuitively, consider the example of an instance that is n bits long that can be solved in n^2 steps. In this example we say the problem has a time complexity of n^2. Of course, the exact number of steps will depend on exactly what machine or language is being used. To avoid that problem, the Big O notation is generally used. If a problem has time complexity $O(n^2)$ on one typical computer, then it will also have complexity $O(n^2)$ on most other computers, so this notation allows us to generalize away from the details of a particular computer.

2.1.6 NP-Complete

In complexity theory, the *NP-complete* problems are the most difficult problems in NP ("non-deterministic polynomial time") in the sense that they are the smallest subclass of NP that could conceivably remain outside of P, the class of deterministic polynomial-time problems. The reason is that a deterministic, polynomial-time solution to any NP-complete problem would also be a solution to every other problem in NP. The complexity class consisting of all NP-complete problems is sometimes referred to as NP-C. A more formal definition is given below.
A decision problem C is NP-complete if it is complete for NP, meaning that:

1. It is in NP and
2. It is NP-hard, *i.e.* every other problem in NP is reducible to it.

"Reducible" here means that for every problem L, there is a polynomial-time many-one reduction, a deterministic algorithm which transforms instances $l \in L$ into instances $c \in C$, such that the answer to c is Yes if and only if the answer to l is Yes. To prove that an NP problem A is in fact an NP-complete problem it is sufficient to show that an already known NP-complete problem reduces to A. A consequence of this definition is that if we had a polynomial time algorithm for C, we could solve all problems in NP in polynomial time.

This definition was given by Cook [5], though the term "NP-complete" did not appear anywhere in his paper. At that computer science conference, there was a fierce debate among the computer scientists about whether NP-complete problems could be solved in polynomial time on a deterministic Turing machine. John Hopcroft brought everyone at the conference to a consensus that the question of whether NP-complete problems are solvable in polynomial time should be put off to be solved at some later date, since nobody had any formal proofs for their claims one way or the other. This is known as the question of whether P=NP.

2.1.7 List of NP-complete Problems in Network Design

Routing Problems

Bottleneck traveling salesman, Chinese postman for mixed graphs, Euclidean traveling salesman, K most vital arcs, K-th shortest path, Metric traveling salesman, Longest circuit, Longest path, Prize Collecting Traveling Salesman, Rural Postman, Shortest path in general networks, Shortest weight-constrained path, Stacker-crane, Time constrained traveling salesman feasibility, Traveling salesman problem, Vehicle Routing

Spanning Trees

Degree constrained spanning tree, Maximum leaf spanning tree, Shortest total path length spanning tree, Bounded diameter spanning tree, Capacitated spanning tree, Geometric capacitated spanning tree, Optimum communication spanning tree, Isomorphic spanning tree, K-th best spanning tree, Bounded component spanning forest, Multiple choice branching, Steiner tree, Geometric Steiner tree, Cable Trench Problem

Flow Problems

Minimum edge-cost flow, Integral flow with multipliers, Path constrained network flow, Integral flow with homologous arcs, Integral flow with bundles, Undirected flow with lower bounds, Directed two-commodity integral flow, Undirected two-commodity integral flow, Disjoint connecting paths, Maximum length-bounded disjoint paths, Maximum fixed-length disjoint paths, Unsplittable multicommodity flow

Cuts and Connectivity

Graph partitioning, Acyclic partition, Max weight cut, Minimum cut into bounded sets, Biconnectivity augmentation, Strong connectivity augmentation, Network reliability, Network survivability, Multiway Cut, Minimum k-cut

2.2 Shortest Path Model

Shortest path problem (SPP) is at the heart of network optimization problems. Being the simplest network model, shortest path can capture most significant ingredients of network optimization problem. Even though it is relatively easy to solve a shortest path problem, the analysis and design of efficient algorithms requires considerable ingenuity. Consequently, the study of models for shortest path is a natural beginning of network models for introducing innovative ideas, including the use of appropriate data structures and data scaling to improve algorithmic performance in the worst cases.

Furthermore, the traditional routing problem has been a single objective problem having as a goal the minimization of either the total distance or travel time. However, in many applications dealing with the design and efficient use of networks, the complexity of the social and economic environment requires the explicit consideration of objective functions other than cost or travel time. In other words, it is necessary to take into account that many real world problems are multi-criteria in nature. The objective functions related to cost, time, accessibility, environmental impact, reliability and risk are appropriated for selecting the most satisfactory (*i.e.,* "textitbest compromise") route in many network optimization problems [6].

Example 2.1: A Simple Example of SPP

A simple example of shortest path problem is shown in Fig. 2.3, and Table 2.1 gives the data set of the example.

Table 2.1 The data set of illustrative shortest path problem

i	1	1	1	2	2	2	3	3	3	4	5	6	6	6	6	7	7	8	8	9	9	10
j	2	3	4	3	5	6	4	6	7	7	8	5	8	9	10	6	10	9	11	10	11	11
c_{ij}	18	19	17	13	16	14	15	16	17	19	19	15	16	15	18	15	13	17	18	14	19	17

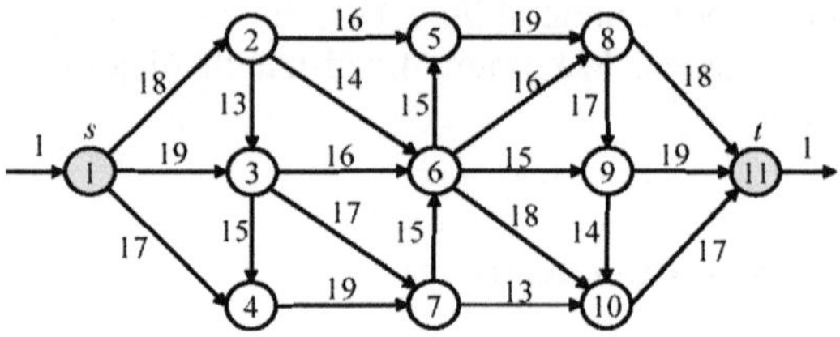

Fig. 2.3 A simple example of shortest path problem

Traditional Methods

A method to solve SPP is sometimes called a *routing algorithm*. The most important algorithms for solving this problem are:

Dijkstra's algorithm: Solves single source problem if all edge weights are greater than or equal to zero. Without worsening the run time, this algorithm can in fact compute the shortest paths from a given start point s to all other nodes.
Bellman-Ford algorithm: Computes single-source shortest paths in a weighted digraph (where some of the edge weights may be negative). Dijkstra's algorithm accomplishes the same problem with a lower running time, but requires edge weights to be nonnegative. Thus, Bellman-Ford is usually used only when there are negative edge weights.
Floyd-Warshall algorithm: An algorithm to solve the all pairs shortest path problem in a weighted, directed graph by multiplying an adjacency-matrix representation of the graph multiple times.

Recently, to address SPP, neural networks (NNs) and GAs (and other evolutionary algorithms) received a great deal of attention regarding their potential as optimization techniques for network optimization problems [7]–[9]. They are often used to solve optimization problems in communication network optimization problems, including the effective approaches on the *shortest path routing* (SPR) problem [10]–[14], *multicasting routing* problem [11], *ATM bandwidth allocation* problem [15], *capacity and flow assignment* (CFA) problem [16], and the *dynamic routing* problem [17]. It is noted that all these problems can be formulated as some sort of a combinatorial optimization problem.

2.2.1 Mathematical Formulation of the SPP Models

Let $G = (N,A)$ be a *directed network*, which consists of a finite set of nodes $N = \{1, 2, \cdots, n\}$ and a set of directed arcs $A = \{(i, j), (k,l), \cdots, (s,t)\}$ connecting m pairs of nodes in N. Arc (i, j) is said to be incident with nodes i and j, and is directed from node i to node j. Suppose that each arc (i, j) has been assigned to a nonnegative value c_{ij}, the cost of (i, j). The SPP can be defined by the following assumptions:

A1. The network is directed. We can fulfil this assumption by transforming any undirected network into a directed one.
A2. All transmission delay and all arc costs are nonnegative.
A3. The network does not contain parallel arcs (*i.e.*, two or more arcs with the same tail and head nodes). This assumption is essentially for notational convenience.

Indices

i, j, k : index of node $(1, 2, \cdots, n)$

Parameters

n : number of nodes
c_{ij}: transmission cost of arc (i, j)
d_{ij} : transmission delay of arc (i, j)

Decision variables

x_{ij} : the link on an arc $(i, j) \in A$.

2.2.1.1 Shortest Path Model

The SPP is to find the minimum cost z from a specified *source node* 1 to another specified *sink node* n, which can be formulated as follows in the form of integer programming:

$$\min z = \sum_{i=1}^{n} \sum_{j=1}^{n} c_{ij} x_{ij} \tag{2.1}$$

$$\text{s. t. } \sum_{j=1}^{n} x_{ij} - \sum_{k=1}^{n} x_{ki} = \begin{cases} 1 & (i = 1) \\ 0 & (i = 2, 3, \cdots, n-1) \\ -1 & (i = n) \end{cases} \tag{2.2}$$

$$x_{ij} = 0 \text{ or } 1 \quad \forall\, i,\ j \tag{2.3}$$

2.2.1.2 Bicriteria Shortest Path Model

The *bicriteria shortest path problem* (bSPP) is formulated as follows, in which the objectives are minimizing cost function z_1 and minimizing delay function z_2 from source node 1 to sink node n:

$$\min z_1 = \sum_{i=1}^{n} \sum_{j=1}^{n} c_{ij} x_{ij} \tag{2.4}$$

$$\min z_2 = \sum_{i=1}^{n} \sum_{j=1}^{n} d_{ij} x_{ij} \tag{2.5}$$

$$\text{s. t. } \sum_{j=1}^{n} x_{ij} - \sum_{k=1}^{n} x_{ki} = \begin{cases} 1 & (i = 1) \\ 0 & (i = 2, 3, \cdots, n-1) \\ -1 & (i = n) \end{cases} \tag{2.6}$$

$$x_{ij} = 0 \text{ or } 1 \quad \forall\, i,\, j \tag{2.7}$$

with constraints at Eqs. 2.2 and 2.6, a flow conservation law is observed at each of the nodes other than s or t. That is, what goes out of node i, $\sum_{j=1}^{n} x_{ij}$ must be equal to what comes in, $\sum_{k=1}^{n} x_{ki}$.

2.2.2 Priority-based GA for SPP Models

Let $P(t)$ and $C(t)$ be parents and offspring in current generation t, respectively. The overall procedure of *priority-based GA* (priGA) for solving SPP models is outlined as follows:

---

procedure: priGA for SPP models
input: network data (N, A, $\boldsymbol{c}$, $\boldsymbol{d}$),
  GA parameters (*popSize*, *maxGen*, p_M, p_C, p_I)
output: Pareto optimal solutions E
begin
  $t \leftarrow 0$;
  initialize $P(t)$ by priority-based encoding routine;
  calculate objectives $z_i(P), i = 1, \cdots, q$ by priority-based decoding routine;
  create Pareto $E(P)$;
  evaluate $eval(P)$ by *interactive adaptive-weight fitness assignment routine*;
  while (**not** terminating condition) **do**
    create $C(t)$ from $P(t)$ by weight mapping crossover routine;
    create $C(t)$ from $P(t)$ by insertion mutation routine;
    create $C(t)$ from $P(t)$ by immigration routine;
    calculate objectives $z_i(C), i = 1, \cdots, q$ by priority-based decoding routine;
    update Pareto $E(P, C)$;
    evaluate $eval(P, C)$ by interactive adaptive-weight fitness assignment routine;
    select $P(t+1)$ from $P(t)$ and $C(t)$ by roulette wheel selection routine;
    $t \leftarrow t + 1$;
  end
  output Pareto optimal solutions $E(P, C)$
end

2.2.2.1 Genetic Representation

How to encode a solution of the network design problem into a chromosome is a key issue for GAs. In Holland's work, encoding is carried out using binary strings. For many GA applications, especially for the problems from network design problems, the simple approach of GA was difficult to apply directly. During the 10 years, various nonstring encoding techniques have been created for network routing problems [8].

We need to consider these critical issues carefully when designing a new non-binary application string coding so as to build an effective GA chromosome. How to encode a path in a network is also critical for developing a GA application to network design problems, it is not easy to find out a nature representation. Special difficulty arises from (1) a path contains variable number of nodes and the maximal number is n-1 for an n node network, and (2) a random sequence of edges usually does not correspond to a path.

Variable-length Encoding

Recently, to encode a diameter-constrained path into a chromosome, various common encoding techniques have been created. Munemoto *et. al.* proposed a *variable-length encoding* method for network routing problems in a wired or wireless environment [10]. Ahn and Ramakrishna developed this variable-length representation and proposed a new crossover operator for solving the shortest path routing (SPR) problem [12]. The variable-length encoding method consists of sequences of positive integers that represent the IDs of nodes through which a path passes. Each locus of the chromosome represents an order of a node (indicated by the gene of the locus) in a path. The length of the chromosome is variable, but it should not exceed the maximum length n, where n is the total number of nodes in the network. The advantage of variable-length encoding is the mapping from any chromosome to solution (decoding) belongs to *1-to-1 mapping* (uniqueness). The disadvantages are: (1) in general, the genetic operators may generate infeasible chromosomes (illegality) that violate the constraints, generating loops in the paths; (2) repairing techniques are usually adopted to convert an illegal chromosome to a legal one.

An example of the variable-length chromosome and its decoded path are shown in Fig. 2.4, in terms of the directed network in Fig. 2.3.

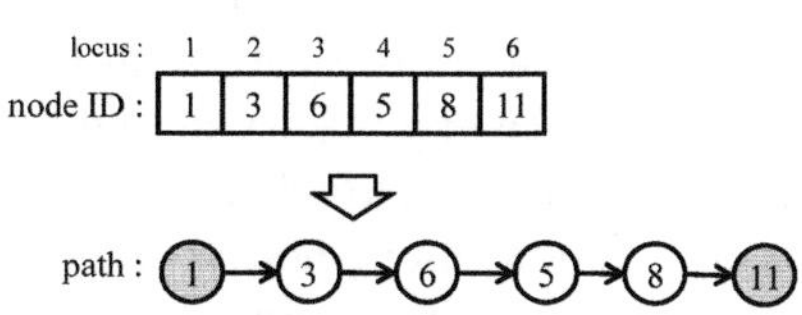

Fig. 2.4 An example of variable-length chromosome and its decoded path

Fixed-length Encoding

Inagaki *et al.* proposed a *fixed-length encoding* method for multiple routing problems [11]. The proposed method are sequences of integers and each gene represents the node ID through which it passes. To encode an arc from node i to node j, put j in the i-th locus of the chromosome. This process is reiterated from the source node 1 and terminating at the sink node n. If the route does not pass through a node x, select one node randomly from the set of nodes which are connected with node x, and put it in the x-th locus. The advantages of fixed-length encoding are: (1) any path has a corresponding encoding (completeness); (2) any point in solution space is accessible for genetic search; (3) any permutation of the encoding corresponds to a path (legality) using the special genetic operators. The disadvantages are: (1) in some cases, n-to-1 mapping may occur for the encoding; (2) in general, the genetic operators may generate infeasible chromosomes (illegality), and special genetic operator phase is required. Therefore we lose feasibility and heritability. An example of fixed-length chromosome and its decoded path is shown in Fig. 2.5, for the undirected network shown in Fig. 2.3.

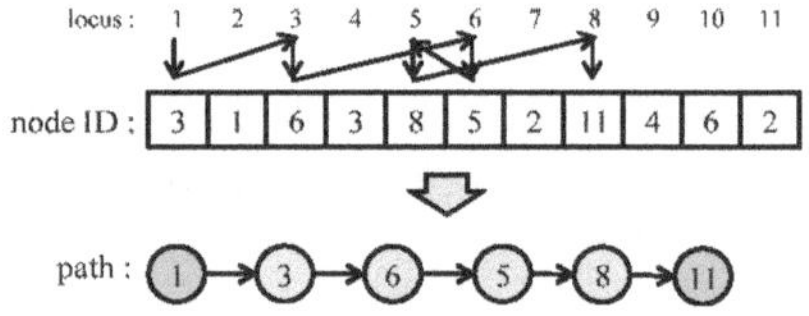

Fig. 2.5 An example of fixed-length chromosome and its decoded path

2.2.2.2 Priority-based Encoding and Decoding

Cheng and Gen proposed a priority-based encoding method firstly for solving Resource-constrained Project Scheduling Problem (rcPSP) [22]. Gen *et al.* also adopted priority-based encoding for the solving bSPP problem [23]. Recently, Lin and Gen proposed a *priority-based encoding* method [24]. As is known, a gene in a chromosome is characterized by two factors: locus, *i.e.*, the position of the gene located within the structure of chromosome, and allele, *i.e.*, the value the gene takes. In this encoding method, the position of a gene is used to represent node ID and its value is used to represent the priority of the node for constructing a path among candidates. A path can be uniquely determined from this encoding.

Illustration of priority-based chromosome and its decoded path are shown in Fig. 2.6. At the beginning, we try to find a node for the position next to source node 1. Nodes 2, 3 and 4 are eligible for the position, which can be easily fixed according to adjacent relation among nodes. The priorities of them are 1, 10 and 3, respectively. Node 3 has the highest priority and is put into the path. The possible nodes next to node 3 are nodes 4, 6 and 7. Because node 6 has the largest priority value, it is put into the path. Then we form the set of nodes available for next position and select the one with the highest priority among them. Repeat these steps until we obtain a complete path, (1-3-6-5-8-11).

The advantages of the priority-based encoding method are: (1) any permutation of the encoding corresponds to a path (feasibility); (2) most existing genetic operators can be easily applied to the encoding; (3) any path has a corresponding encoding (legality); (4) any point in solution space is accessible for genetic search. However, there is a disadvantage as that n-to-1 mapping (uniqueness) may occur for the encoding at some case. For example, we can obtain the same path, (1-3-6-5-8-11) by different chromosomes, (v_1 = [11, $\underline{1}$, 10, 3, 8, 9, $\underline{5}$, 7, 4, 2, 6] and v_2 = [11, $\underline{5}$, 10, 3, 8, 9, $\underline{1}$, 7, 4, 2, 6]).

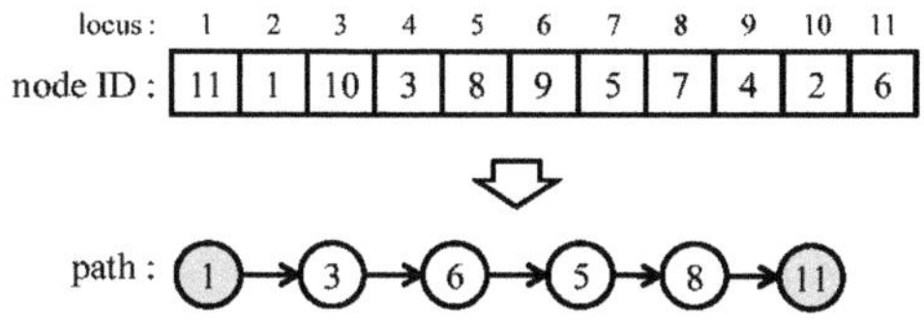

Fig. 2.6 A example of priority-based chromosome and its decoded path

In general, there are two ways to generate the initial population, *heuristic initialization* and *random initialization*. However, the mean fitness of the heuristic initialization is already high so it may help the GAs to find solutions faster. Unfortunately, in most large scale problems, it may just explore a small part of the solution space and it is difficult to find global optimal solutions because of the lack of diversity in the population [12]. Therefore, random initialization is effected in this research so that the initial population is generated with the priority-based encoding method. The encoding process and decoding process of the priority-based GA (priGA) are shown in Figures 2.7 and 2.8, respectively.

Depending on the properties of encodings (introduced on page 9), we summarize the performance of the priority-based encoding method and other introduced encoding methods in Table 2.2.

Table 2.2 Summarizes the performance of encoding methods

Chromosome Design		Space	Time	Feasibility	Uniqueness	Locality	Heritability
Variable length-based GA	Ahn and Ramakrishna [13]	m	O($m$log$m$)	Poor	1-to-1 mapping	Worst	Worst
Fixed length-based GA	Inagaki *et al.* [12]	n	O($n$log$n$)	Worst	n-to-1 mapping	Worst	Worst
Priority-based GA	Gen *et al.* [14]	n	O($n$log$n$)	Good	n-to-1 mapping	Good	Good

Fig. 2.7 Pseudocode of priority-based encoding method

```
procedure: priority-based encoding
input: number of nodes n
output: kth initial chromosome v_k
begin
    for i = 1 to n
        v_k[i] ← i;
    for i = 1 to ⌈n/2⌉
        repeat
            j ← random[1, n];
            l ← random[1, n];
        until(j ≠ l);
        swap(v_k[j], v_k[l]);
    output v_k;
end
```

```
procedure: priority-based decoding
input: number of nodes n, chromosome v_k,
       the set of nodes S_i with all nodes adjacent to node i
output: path P_k
begin
    initialize i ← 1, l ← 1, P_k[l] ← i;  // i: source node, l: length of path P_k.
    while S_1 ≠ φ do
        j' ← argmax{v_k[j], j ∈ S_i};
                                  // j': the node with highest priority among S_i.
        if v_k[j'] ≠ 0 then
            P_k[l] ← j';         // chosen node j' to construct path P_k.
            l ← l + 1;
            v_k[j'] ← 0;
            i ← j';
        else
            S_i ← S_i \ j';      // delete the node l adjacent to node i.
            v_k[i] ← 0;
            l ← l − 1;
            if l ≤ 1 then l ← 1, break;
            i ← P_k[l];
    output path P_k
end
```

Fig. 2.8 Pseudocode of priority-based decoding method

2.2.2.3 Genetic Operators

Genetic operators mimic the process of heredity of genes to create new offspring at each generation. Using the different genetic operators has very large influence on GA performance [17]. Therefore it is important to examined different genetic operators.

For priority-based representation as a permutation representation , several crossover operators have been proposed, such as *partial-mapped crossover* (PMX), *order crossover* (OX), *cycle crossover* (CX), *position-based crossover* (PX), *heuristic crossover*, *etc*. Roughly, these operators can be classified into two classes:

- Canonical approach
- Heuristic approach

The canonical approach can be viewed as an extension of two-point or multi-point crossover of binary strings to permutation representation. Generally, permutation representation will yield illegal offspring by two-point or multi-point crossover in the sense of that some priority values may be missed while some priority values may be duplicated in the offspring. *Repairing procedure* is embedded in this approach to resolve the illegitimacy of offspring.

The essence of the canonical approach is the blind random mechanism. There is no guarantee that an offspring produced by this kind of crossover is better than their parents. The application of heuristics in crossover intends to generate an improved offspring.

Partial-mapped Crossover (PMX)

PMX was proposed by Goldberg and Lingle [19]. PMX can be viewed as an extension of two-point crossover for binary string to permutation representation. It uses a special repairing procedure to resolve the illegitimacy caused by the simple two-point crossover. So the essentials of PMX are a simple two-point crossover plus a repairing procedure. The procedure is illustrated in Fig. 2.9. PMX works as follows:

Procedure PMX
Input: two parents
Output: offspring
Step 1: Select two positions along the string uniformly at random.
The substrings defined by the two positions are called the *mapping sections*.
Step 2: Exchange two substrings between parents to produce proto-children.
Step 3: Determine the *mapping relationship* between two mapping sections.
Step 4: Legalize offspring with the *mapping relationship*.

Priority values 3, 11 and 2 are duplicated in *proto-child* 1 while values 4, 9 and 6 are missing from it. According to the mapping relationship determined in step 3,

Fig. 2.9 Illustration of the PMX operator

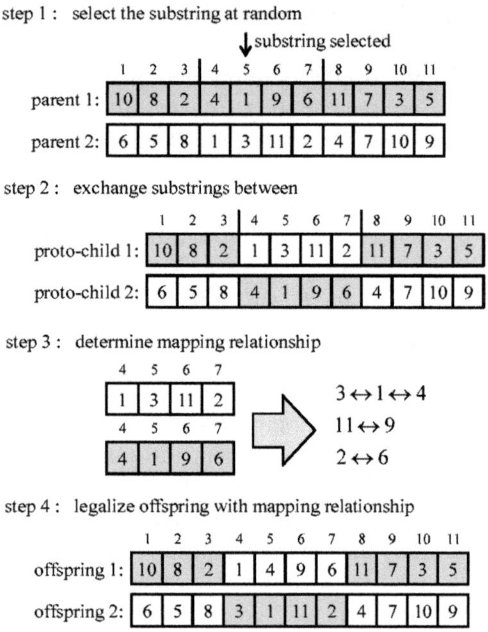

the excessive values 3, 11 and 2 should be replaced by the missing values 4, 9 and 6, respectively, while keeping the swapped substring unchanged.

Order Crossover (OX)

OX was proposed by Davis [20]. It can be viewed as a kind of variation of PMX with a different repairing procedure. The procedure is illustrated in Fig. 2.10. OX works as follows:

Procedure OX
Input: two parents
Output: offspring
Step 1: Select a substring from one parent at random.
Step 2: Produce a proto-child by copying the substring into the corresponding positions of it.
Step 3: Delete the nodes which are already in the substring from the second parent.
The resulted sequence of nodes contains the nodes that the proto-child needs.
Step 4: Place the nodes into the unfixed positions of the proto-child from left to right according to the order of the sequence to produce an offspring.

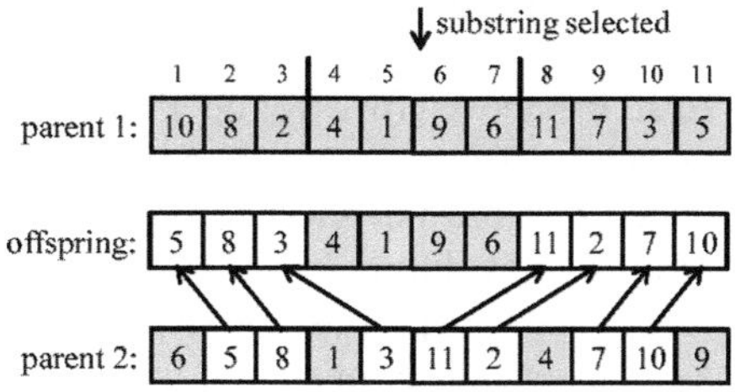

Fig. 2.10 Illustration of the OX operator

2.2.2.4 Position-based Crossover (PX)

Position-based crossover was proposed by Syswerda [21]. It is essentially a kind of uniform crossover for permutation representation together with a repairing procedure. It can also be viewed as a kind of variation of OX where the nodes are selected inconsecutively. The procedure is illustrated in Fig. 2.11. PX works as follows:

Procedure PX
Input: two parents
Output: offspring
Step 1: Select a set of positions from one parent at random.
Step 2: Produce a proto-child by copying the nodes on these positions into the corresponding positions of it.
Step 3: Delete the nodes which are already selected from the second parent. The resulted sequence of nodes contains the nodes the proto-child needs
Step 4: Place the nodes into the unfixed positions of the proto-child from left to right according to the order of the sequence to produce one offspring.

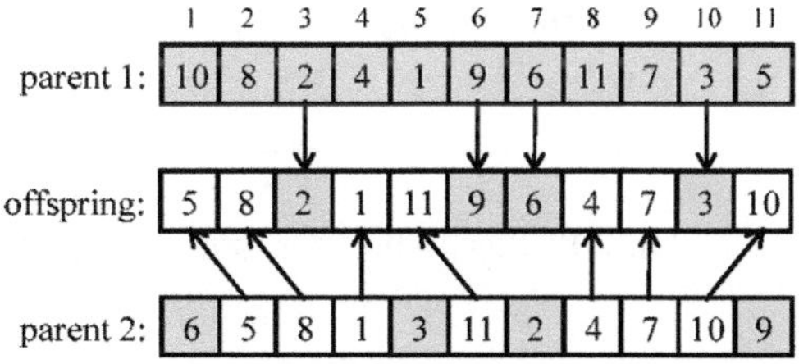

Fig. 2.11 Illustration of the PX operator

Weight Mapping Crossover (WMX):

In all the above crossover operators, the mechanism of the crossover is not the same as that of the conventional one-cut point crossover. Some offspring may be generated that did not succeed on the character of the parents, thereby retarding the process of evolution.

Lin and Gen proposed a *weight mapping crossover* (WMX) [24]; it can be viewed as an extension of one-cut point crossover for permutation representation. At one-cut point crossover, two chromosomes (parents) would choose a random-cut point and generate the offspring by using a segment of its own parent to the left of the cut point, then remap the right segment based on the weight of other parent of right segment. Fig. 2.12 shows the crossover process of WMX, and an example of the WMX is given in Fig. 2.13.

```
procedure: weight mapping crossover (WMX)
input: two parents v1[·], v2[·], the length of chromosome n
output: offspring v1'[·], v2'[·]
begin
   p ← random[1, n];        // p: a random cut-point
   l ← n − p;               // l: the length of right segments of chromosomes
   v1' ← v1[1 : p] // v2[p+1 : n];
   v2' ← v2[1 : p] // v1[p+1 : n];   // exchange substrings between parents.
   s1[·] ← sorting(v1[p+1 : n]);
   s2[·] ← sorting(v2[p+1 : n]);   // sorting the weight of the right segments.
   for i = 1 to l
      for j = 1 to l
         if v1'[p+i] = s2[j] then v1'[p+i] ← s1[j];
      for j = 1 to l
         if v2'[p+i] = s1[j] then v2'[p+i] ← s2[j];
   output offspring v1'[·], v2'[·];
end
```

Fig. 2.12 Pseudocode of weight mapping crossover

As showed in Fig. 2.12, first we choose a random cut point p; and calculate l that is the length of right segments of chromosomes, where n is the number of nodes in the network. Then we get a mapping relationship by sorting the weight of the right segments $s_1[\cdot]$ and $s_2[\cdot]$. As a one-cut point crossover, it generates the offspring $v_1'[\cdot]$, $v_2'[\cdot]$ by exchange substrings between parents $v_1[\cdot]$, $v_2[\cdot]$; legalized offspring with mapping relationship and then two new chromosomes are produced eventually. In

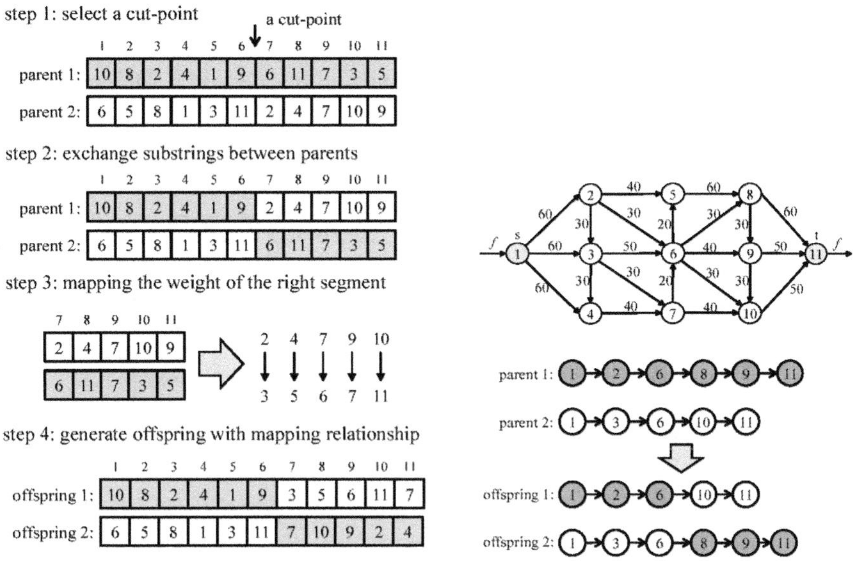

(a) Illustration of WMX

(b) Illustration of the generated paths for the directed network shown in Fig. 2.2

Fig. 2.13 Example of WMX procedure

some cases, WMX is the same as that of the conventional one-cut point crossover and can generate two new paths that exchanged sub-route from two parents.

For permutation representation, it is relatively easy to produce some mutation operators. Several mutation operators have been proposed for permutation representation, such as swap mutation, inversion mutation, insertion mutation, *etc.*

Inversion Mutation

Inversion mutation selects two positions within a chromosome at random and then inverts the substring between these two positions. It is illustrated in Fig. 2.14.

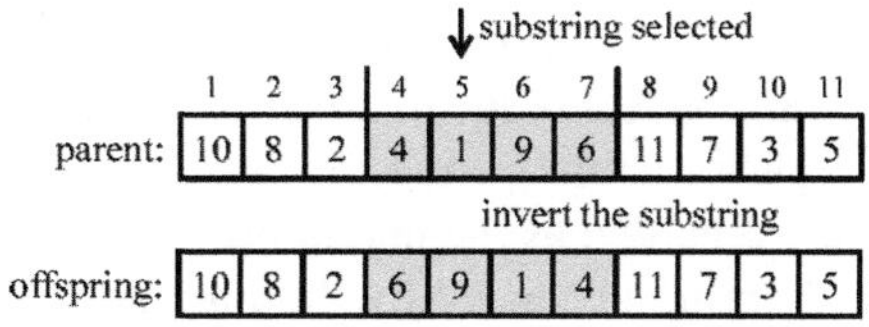

Fig. 2.14 Illustration of the inversion mutation

Insertion Mutation

Insertion mutation selects an element at random and inserts it in a random position as illustrated in Fig. 2.15.

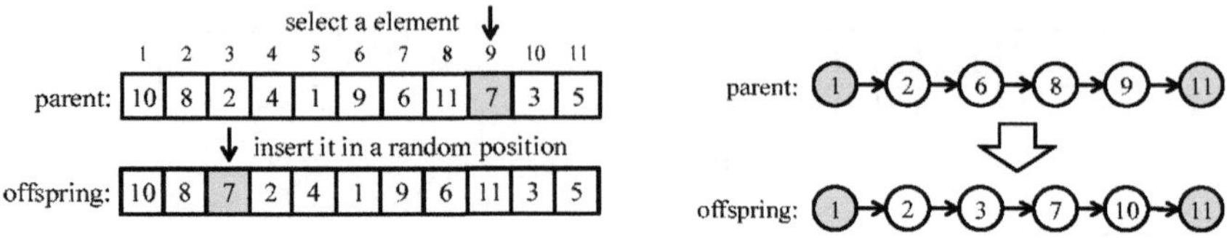

(a) Illustration of WMX

(b) Illustration of the generated paths for the directed network shown in Fig. 2.2

Fig. 2.15 Example of the insertion mutation

Swap Mutation

Swap mutation selects two elements at random and then swaps the elements on these positions as illustrated in Fig. 2.16.

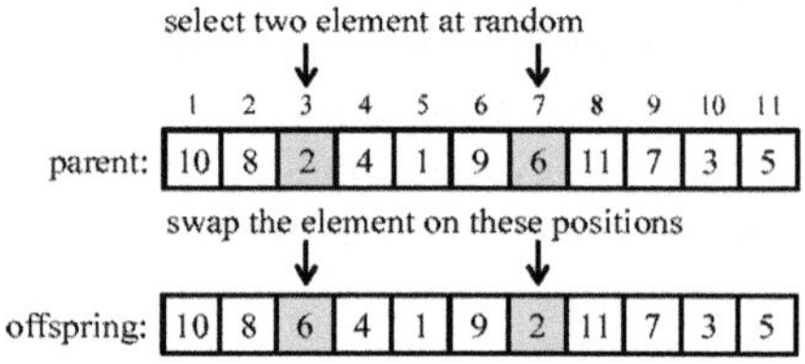

Fig. 2.16 Illustration of the swap mutation

2.2.2.5 Heuristic Mutation

Heuristic mutation was proposed by Gen and Cheng [7]. It is designed with a neighborhood technique in order to produce an improved offspring. A set of chromosomes transformable from a given chromosome by exchanging no more than λ elements are regarded as the neighborhood. The best one among the neighborhood is used as offspring produced by the mutation. The procedure is illustrated in Fig. 2.17. The mutation operator works as follows:

Procedure Heuristic Mutation
Input: parent
Output: offspring
Step 1: Pick up λ genes at random.
Step 2: Generate neighbors according to all possible permutation of the selected genes.
Step 3: Evaluate all neighbors and select the best one as offspring.
Step 4: Fulfill the child with the remaining nodes.

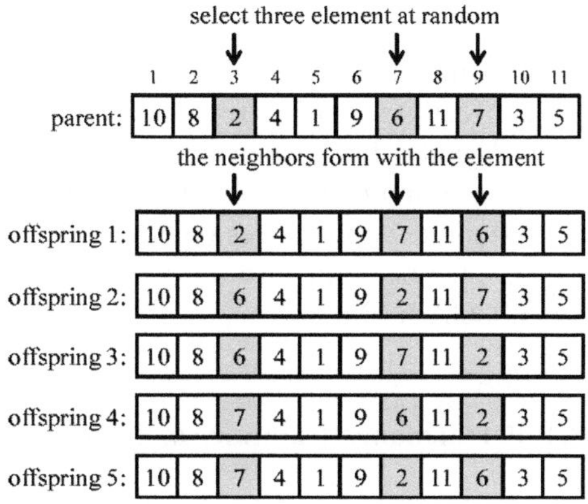

Fig. 2.17 Illustration of the heuristic mutation

Immigration Operator

The *immigration operator* is modified to (1) include immigration routine, in each generation, (2) generate and (3) evaluate $popSize \cdot p_I$ random chromosomes, and (4) replace the $popSize \cdot p_I$ worst chromosomes of the current population (p_I, called the immigration probability). The detailed introduction is showed on page 23.

Interactive Adaptive-weight Fitness Assignment

The *interactive adaptive-weight fitness assignment mechanism* assigns weights to each objective and combines the weighted objectives into a single objective function. It is an improved adaptive-weight GA [8] with the consideration of the disadvantages of weighted-sum approach and Pareto ranking-based approach. The detailed introduction is showed on page 39.

Selection Operator

We adopt the roulette wheel selection (RWS). It is to determine selection probability or survival probability for each chromosome proportional to the fitness value. A model of the roulette wheel can be made displaying these probabilities. The detailed introduction is showed on page 13.

2.2.3 Computational Experiments and Discussions

Usually during the GA design phase, we are only concerned with the design of genetic representations, neglecting the design of more effective genetic operators with depend on the characteristics of the genetic representations. In the first experiment, the effectiveness of different genetic operators will be demonstrated. Then to validate the effectiveness of different genetic representations, priority-based GA with Ahn and Ramakrishna's algorithm [12] are compared. Finally, to validate the multiobjective GA approaches for solving bSPR problems, i-awGA with three different fitness assignment approaches (spEA, nsGA II and rwGA) are compared. For each algorithm, 50 runs with Java are performed on a Pentium 4 processor (3.40-GHz clock), 3.00GA RAM.

2.2.3.1 Performance Comparisons with Different Genetic Operators

In this experiment, the different genetic operators for priority-based genetic representation are combined; there are partial-mapped crossover (PMX), order crossover (OX), position-based crossover (PX), swap mutation, insertion mutation and immigration operator. We used six shortest path problems [12, 61], and Dijkstra's algorithm used to calculate the optimal solution to each problem. GA parameter settings were taken as follows: population size, *popSize* =10; crossover probability, p_C=0.30; mutation probability, p_M =0.30; immigration rate, p_I = 0.30. In addition, there are two different terminating criteria employed. One of them is the number of maximum generations, *maxGen* =1000. Another stopping criteria is the terminating condition T=200. That is, if the best solution is not improved until successive 200 generations, the algorithm will be stopped. In order to evaluate the results of each test, the following four performance measures are adopted: average of the best solutions (ABS), standard deviation (SD), percent deviation from optimal solution (PD) and a statistical analysis by analysis of variance (ANOVA) [25].

There are seven kinds of combinations of genetic operators: PMX + Swap (Alg. 1), PMX + Insertion (Alg. 2), OX + Swap (Alg. 3), OX + Insertion (Alg. 4), PX + Swap (Alg. 5), PX + Insertion (Alg. 6) and WMX + Insertion + Immigration (Alg.7). Figure 2.18 showed the ABS on the six test problems. The values of PD, SD and ANOVA analysis with 50 runs of 7 algorithms on 6 test problems are showed in Table 2.3.

As depicted in Fig. 2.18, all PD with Alg. 7 are better than each of the other combinations of genetic operators. Also all of the SD of 50 runs with Alg. 7 are smaller than each of the other combinations of genetic operators. In addition, the SD of 50 runs with Alg. 7 were 0 in test problem #1 – #3 and #5; that means we can obtain the

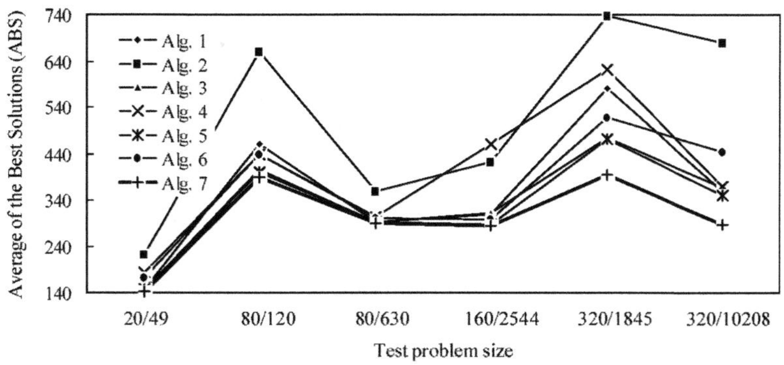

Fig. 2.18 Effectiveness comparing of ABS with seven kinds of combinations of genetic operators

Table 2.3 The Average of PD, SD for combinations of genetic operators and ANOVA analysis

Test problems	Alg. 1	Alg. 2	Alg. 3	Alg. 4	Alg. 5	Alg. 6	Alg. 7
20 / 49	4.10	55.87	7.27	28.69	8.92	21.63	0.00
80 / 120	18.12	69.98	8.96	12.91	8.50	12.72	0.00
80 / 632	1.07	23.32	0.76	4.53	2.53	3.79	0.00
160 / 2544	8.80	49.08	19.28	62.58	12.49	5.58	0.07
320 / 1845	47.20	86.82	24.95	57.70	24.52	31.24	0.00
320 /10208	24.26	136.34	43.37	28.12	37.82	53.99	0.04
Mean	17.26	70.24	17.43	32.42	15.80	21.49	0.02
SD	34.59	147.45	60.38	111.68	54.17	124.50	0.12
Variance	1196.14	21740.92	3645.65	12471.66	2934.87	15500.29	0.02
Sum of squares	358841.24	6522274.63	1093695.70	3741497.46	880462.44	4650087.65	4.55
Factors	Sum of squares	Freedom degree	Mean square	F		F ($\alpha = 0.05$)	t ($\alpha = 0.05$)
Between groups	881925.72	6	146987.62	17.84		2.10	1.96
Within-groups	17246863.66	2093	8240.26				
Total	18128789.38	2099					
LSD	14.53						
Mean Difference with Alg. 7	17.24	70.22	17.41	32.40	15.78	21.47	

global optimum for each run. In the other two test problems, #4 and #6, the SD were close to 0. In other words, Alg. 7 has good stability. In Table 2.3, ANOVA analysis depended on the average of 50 PDs with 6 test problems and are shown to analyze the differences between the quality of solutions obtained by various combinations of genetic operators. At the significant level of $\alpha = 0.05$, $F=17.84$ is greater than the reference value of ($F=2.10$). The difference between Alg. 7 and each of the other combinations is greater than LSD=14.53 (Least Significant Difference). The Alg. 7 (WMX + Insertion + Immigration) is indeed statistically better than the others.

2.2.3.2 Comparisons with Different Encoding Methods

How to encode a solution of the problem into a chromosome is a key issue in GAs. The different chromosome representations will have a very big impact on GA designs. In the second experiment, the performance comparisons between priority-based GA approach (priGA) and ahnGA are showed. In priGA, WMX + Insertion + Immigration are used as genetic operators. The six shortest path problems, as with the first experiment, are combined. In order to evaluate the results of each test, four performance measures are adopted: average of the best solutions (ABS), standard deviation (SD), percent deviation from optimal solution (PD) and a statistical analysis by analysis of variance (ANOVA) [25].

In addition, as we all know, in general the result of GA decrease with increasing the problem size means that we must increase the GA parameter settings. Therefore if the parameter setting of a GA approach does not increase the problem size, then we can say this GA approach has a very good search capability of obtaining optimal results. The effectiveness comparing with six kinds of different GA parameter settings are shown as follows:

Para 1: $maxGen$=1000, $popSize$=10, p_C=0.3, p_M=0.3, p_I = 0.3, T=200.
Para 2: $maxGen$=1000, $popSize$=10, p_C=0.5, p_M=0.5, p_I = 0.3, T=200.
Para 3: $maxGen$=1000, $popSize$=10, p_C=0.7, p_M=0.7, p_I = 0.3, T=200.
Para 4: $maxGen$=1000, $popSize$=20, p_C=0.3, p_M=0.3, p_I = 0.3, T=200.
Para 5: $maxGen$=1000, $popSize$=20, p_C=0.5, p_M=0.5, p_I = 0.3, T=200.
Para 6: $maxGen$=1000, $popSize$=20, p_C=0.7, p_M=0.7, p_I = 0.3, T=200.

To see clearly the difference between priGA and ahnGA with the different GA parameter settings, Fig. 2.19 shows the ABS of six test problems. The values of PD, SD and ANOVA analysis are showed in Table 2.4. Because of an memory error, the last two large-scale test problems with large GA parameter setting were not solved by ahnGA. As depicted in Fig. 2.19a and Table 2.4, ahnGA can solve the first four test problems successfully, but for solving last two test problems, GA parameter setting affected the efficiency of ahnGA. In addition, the memory requirements are too high and even impossible to use with the higher GA parameter settings. Figure 2.19b and Table 2.4 show the efficiency of priGA with all the test problems successfully solved without any effect grow GA parameter settings by priGA.

In Table 2.4, ANOVA analysis depends on the average of 50 PD with 6 test problems to analyze the differences between the quality of solutions obtained by two kinds of chromosome representations with six kinds of different GA parameter settings. At the significant level of $\alpha = 0.05$, F=16.27 is greater than the reference value of (F=1.79). The difference between the lower GA parameter setting (Para 1, Para 2 or Para 3) and higher GA parameter setting (Para 4, Para 5 or Para 6) is greater than the LSD=0.48 by ahnGA. The GA parameter settings indeed statistically affect the efficiency of ahnGA. However, all of the difference between the GA parameter settings is smaller than LSD=0.48 by priGA, meaning that the GA parameter settings did not affect the efficiency of priGA. Futhermore, the difference between ahnGA and priGA is greater than LSD=0.48, combining the lower GA parameter setting

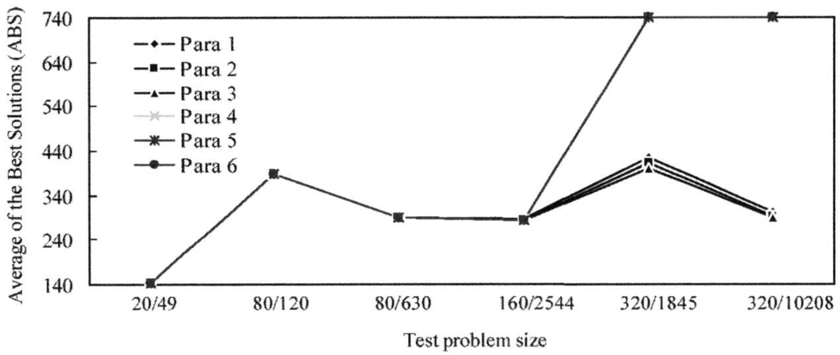

(a) Combine with ahnGA

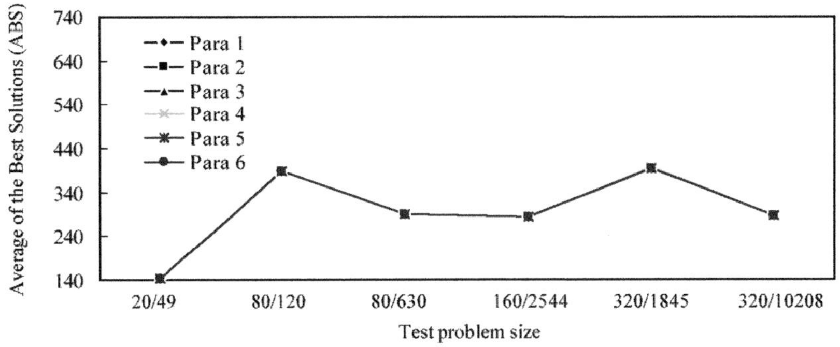

(b) Combine with priGA

Fig. 2.19 Effectiveness comparing of ABS with six kinds of GA parameter settings

Table 2.4 The average of PD, SD for different chromosome representations and ANOVA analysis

Test problems	ahnGA						priGA					
	Para 1	Para 2	Para 3	Para 4	Para 5	Para 6	Para 1	Para 2	Para 3	Para 4	Para 5	Para 6
20 / 49	0.00	0.00	0.00	0.00	0.00	0.00	0.00	0.00	0.00	0.00	0.00	0.00
80 / 120	0.00	0.00	0.00	0.00	0.00	0.00	0.00	0.00	0.00	0.00	0.00	0.00
80 / 632	0.08	0.01	0.01	0.00	0.00	0.00	0.00	0.00	0.00	0.00	0.00	0.00
160 / 2544	1.04	0.14	0.24	0.39	0.04	0.00	0.07	0.00	0.00	0.03	0.00	0.00
320 / 1845	7.44	4.87	3.13	-	-	-	0.00	0.00	0.00	0.54	0.00	0.00
320 /10208	5.33	1.27	0.83	-	-	-	0.04	0.00	0.01	0.01	0.01	0.01
Mean	2.31	1.05	0.70	0.10	0.01	0.00	0.02	0.00	0.00	0.10	0.00	0.00
SD	7.76	4.77	3.47	0.53	0.09	0.00	0.12	0.00	0.03	1.57	0.03	0.02
Variance	60.25	22.72	12.01	0.28	0.01	0.00	0.02	0.00	0.00	2.45	0.00	0.00
Sum of squares	18075.87	6815.52	3603.02	55.78	1.47	0.00	4.55	0.00	0.24	735.90	0.24	0.12
Factors	Sum of squares	Freedom degree	Mean square	F						F ($\alpha = 0.05$)	t ($\alpha = 0.05$)	LSD
Between groups	1594.87	11	144.99	16.27						1.79	1.96	0.48
Within-groups	29292.69	3288	8.91									
Total	30887.56	3299										

"-" means out of memory error

(Para 1 Para 2 or Para 3). priGA was indeed statistically better than ahnGA with lower GA parameter settings.

2.2.3.3 Performance Validations of Multiobjective GAs

In the third experimental study, the performance comparisons of different multi-objective GA (moGA) approaches are shown for solving bSPR problems by different fitness assignment approaches; there are strength Pareto evolutionary algorithm (spEA), non-dominated sorting genetic algorithm II (nsGA II), random-weight genetic algorithm (rwGA) and interactive adaptive-weight genetic algorithm (i-awGA). In each GA approach, priority-based encoding was used, WMX, insertion, immigration and auto-tuning operators were used as genetic operators. All test network topologies were constructed by Beasley and Christofides [61, 62]. The two objectives, costs and delay, are represented as random variables depending on the distribution functions. The network characteristics for test networks are shown in Table 2.5.

Table 2.5 Network characteristics # of nodes n, # of arcs m, cost $\boldsymbol{c}$ and delay $\boldsymbol{d}$ for the test networks

ID	n	m	Cost $\boldsymbol{c}$	Delay $\boldsymbol{d}$
1	100	955	runif(m, 10, 50)	runif(m, 5, 100)
2		959	rnorm(m, 50, 10)	rnorm(m, 50, 5)
3		990	rlnorm(m, 1, 1)	rlnorm(m, 2, 1)
4		999	10*rexp(m, 0.5)	10*rexp(m, 2)
5	200	2040	runif(m, 10, 50)	runif(m, 5, 100)
6		1971	rnorm(m, 50, 10)	rnorm(m, 50, 5)
7		2080	rlnorm(m, 1, 1)	rlnorm(m, 2, 1)
8		1960	10*rexp(m, 0.5)	10*rexp(m, 2)
9	500	4858	runif(m, 10, 50)	runif(m, 5, 100)
10		4978	rnorm(m, 50, 10)	rnorm(m, 50, 5)
11		4847	rlnorm(m, 1, 1)	rlnorm(m, 2, 1)
12		4868	10*rexp(m, 0.5)	10*rexp(m, 2)

Uniform Distribution: runif(m, min, max)
Normal Distribution: rnorm(m, mean, sd)
Lognormal Distribution: rlnorm(m, meanlog, sdlog)
Exponential Distribution: rexp(m, rate)

Performance Measures

In these experiments, the following performance measures are considered to evaluate the efficiency of the different fitness assignment approaches. The detailed explanations have been introduced on page 41.

Number of obtained solutions $|S_j|$: Evaluate each solution set depend on the number of obtained solutions.

Ratio of nondominated solutions $R_{NDS}(S_j)$: This measure simply counts the number of solutions which are members of the Pareto-optimal set S^*.

Average distance $D1_R(S_j)$: Instead of finding whether a solution of S_j belongs to the set S^* or not, this measure finds an average distance of the solutions of S_j from S^*.

For finding the reference solution set S^*, we used the following GA parameter settings in all of four fitness assignment approaches (spEA, nsGA II, rwGA and our i-awGA) : population size, *popSize*=50; crossover probability, p_C=0.90; mutation probability, p_M=0.90; immigration rate, p_I=0.60; stopping criteria: evaluation of 100,000 generations.

Table 2.6 The ABS of 50 runs by different fitness assignment approaches

ID	$\|S_j\|$				$R_{NDS}(S_j)$				$D1_R(S_j)$			
	spEA	nsGA	rwGA	i-awGA	spEA	nsGA	rwGA	i-awGA	spEA	nsGA	rwGA	i-awGA
1	1.64	1.70	1.64	1.84	1.00	1.00	1.00	1.00	0.00	0.00	0.00	0.00
2	5.00	5.08	4.98	5.64	0.18	0.16	0.22	0.38	0.18	0.23	0.17	0.10
3	3.30	3.04	3.22	3.48	0.91	0.93	0.92	0.91	0.00	0.00	0.00	0.00
4	7.36	7.40	7.12	7.46	0.04	0.02	0.04	0.04	0.06	0.06	0.05	0.05
5	3.26	3.22	3.12	3.46	1.00	1.00	1.00	1.00	0.00	0.00	0.00	0.00
6	1.74	2.40	2.20	1.54	0.28	0.14	0.18	0.30	0.17	0.24	0.22	0.15
7	4.16	3.96	3.66	3.70	0.52	0.59	0.66	0.68	0.40	0.42	0.40	0.05
8	5.90	4.80	5.30	5.16	0.05	0.13	0.07	0.10	1.10	0.89	0.96	0.86
9	1.16	1.24	1.28	1.36	0.99	0.96	0.91	0.99	0.00	0.01	0.01	0.00
10	2.60	2.42	2.62	2.30	0.11	0.18	0.16	0.33	1.17	0.76	0.99	0.59
11	2.86	2.90	2.70	3.22	0.31	0.30	0.30	0.43	0.01	0.01	0.01	0.00
12	5.82	6.02	6.14	6.20	0.03	0.03	0.04	0.05	0.19	0.19	0.20	0.19

As depicted in Table 2.6, most results of ABS of 50 runs by i-awGA are better than each of the other fitness assignment approach. In addition, it is difficult to say the efficiency of the approach only depends on the performance measure $|S_j|$ or $R_{NDS}(S_j)$. It is necessary to demonstrate the efficiency both of the performance measure $|S_j|$ and $R_{NDS}(S_j)$.

In Tables 2.7 and 2.8, ANOVA analysis depended on the $|S_j|$ and $R_{NDS}(S_j)$ in 50 with test problem 11 runs to analyze the difference and was shown the quality of solutions obtained by four kinds of different fitness assignment approaches. At the significance level of $\alpha = 0.05$, the F=3.56 and 3.12 is greater than the reference value of F=2.84, respectively. The difference between i-awGA and each of

Table 2.7 ANOVA analysis with $|S_j|$ in test problem 11

	spEA	nsGA II	rwGA	i-awGA
# of data	50	50	50	50
Mean	2.86	2.90	2.70	3.22
SD	0.92	0.83	0.78	0.64
Variance	0.84	0.69	0.61	0.41
Sum of squares	42.02	34.50	30.50	20.58
Factors	Sum of squares	Freedom degree	Mean square	F
Between groups	7.12	3	2.37	3.65
Within-groups	127.60	196	0.65	
Total	134.72	199		
F (α = 0.05)	2.68			
t (α = 0.05)	1.98			
LSD	0.31			
Mean Difference with i-awGA	0.36	0.32	0.52	

Table 2.8 ANOVA analysis with $R_{NDS}(S_j)$ in test problem 11.

	spEA	nsGA II	rwGA	i-awGA
# of data	50	50	50	50
Mean	0.31	0.30	0.30	0.43
SD	0.27	0.22	0.26	0.23
Variance	0.07	0.05	0.07	0.05
Sum of squares	3.62	2.43	3.33	2.62
Factors	Sum of squares	Freedom degree	Mean square	F
Between groups	0.57	3	0.19	3.12
Within-groups	12.01	196	0.06	
Total	12.58	199		
F (α = 0.05)	2.68			
t (α = 0.05)	1.98			
LSD	0.10			
Mean Difference with i-awGA	0.11	0.13	0.13	

the other approaches (spEA, nsGA II or rwGA) is greater than the LSD=0.31 and 0.10, respectively. That means i-awGA is indeed statistically better than the other approaches.

2.3 Minimum Spanning Tree Models

Given a connected, undirected graph, a *spanning tree* of that graph is a subgraph which is a tree and connects all the nodes together. A single graph can have many different spanning trees. We can also assign a weight to each edge, which is a number representing how unfavorable it is, and use this to assign a weight to a spanning tree by computing the sum of the weights of the edges in that spanning tree. A *minimum spanning tree* (MST) is a spanning tree with weight less than or equal to the weight of every other spanning tree [4].

Example 2.2: A Simple Example of MST

A simple example of minimum spanning tree problem is shown in Fig. 2.20, and Table 2.9 gives the data set of the example.

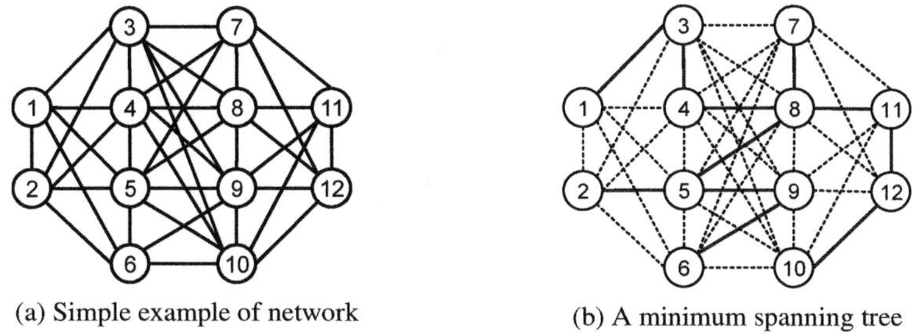

Fig. 2.20 A simple example of minimum spanning tree problem

Table 2.9 The data set of illustrative minimu spanning tree problem

k	edge (i,j)	weight w_{ij}	k	edge (i,j)	weight w_{ij}	k	edge (i,j)	weight w_{ij}	k	edge (i,j)	weight w_{ij}
1	(1, 2)	35	11	(3, 7)	51	21	(5, 7)	26	31	(7, 12)	41
2	(1, 3)	23	12	(3, 8)	23	22	(5, 8)	35	32	(8, 9)	62
3	(1, 4)	26	13	(3, 9)	64	23	(5, 9)	63	33	(8, 11)	26
4	(1, 5)	29	14	(3, 10)	28	24	(5, 10)	23	34	(8, 12)	30
5	(1, 6)	52	15	(4, 5)	54	25	(6, 7)	27	35	(9, 10)	47
6	(2, 3)	34	16	(4, 7)	24	26	(6, 8)	29	36	(9, 11)	68
7	(2, 4)	23	17	(4, 8)	47	27	(6, 9)	65	37	(9, 12)	33
8	(2, 5)	68	18	(4, 9)	53	28	(6, 10)	24	38	(10, 11)	42
9	(2, 6)	42	19	(4, 10)	24	29	(7, 8)	38	39	(10, 12)	26
10	(3, 4)	23	20	(5, 6)	56	30	(7, 11)	52	40	(11, 12)	51

Traditional Methods

Kruskal's Algorithm: Examines edges in nondecreasing order of their lengths and include them in MST if the added edge does not form a cycle with the edges already chosen. The proof of the algorithm uses the path optimality conditions. Attractive algorithm if the edges are already sorted in increasing order of their lengths [26]. Table 2.10 presents the trace table for the generating process of MST for solving a simple network example in Fig. 2.21. The procedure of Kruskal's algorithm is shown in Fig. 2.22.

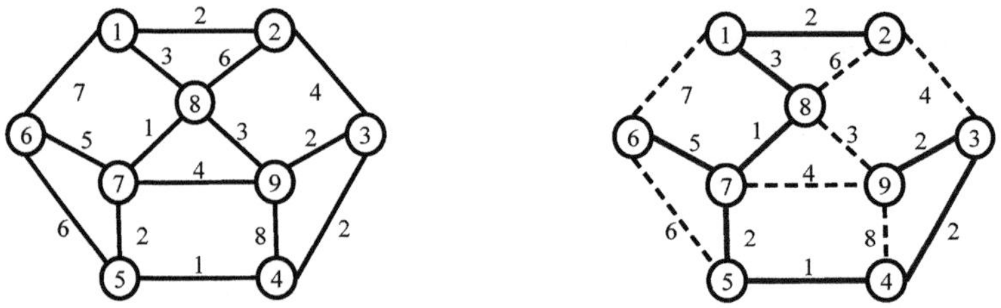

(a) a grap $G = (V, E)$ and weight $w_{ij}, \forall\ (i, j) \in V$ (b) MST

Fig. 2.21 A small-scale example of MST

```
procedure: Kruskal's Algorithm
input: graph G = (V, E), weight w_ij, ∀(i, j) ∈ V
output: spanning tree T
begin
  T ← φ;
  A ← E;      // A:eligible edges
  while |T| < |V| − 1 do
    choose an edge (u, v) ← argmin{w_ij | (i, j) ∈ A};
    A ← A \ {(u, v)};
    if u and v are not yet connected in T then
      T ← T ∪ {(u, v)};
  output spanning tree T;
end
```

Fig. 2.22 Pseudocode of Kruskal's algorithm

Prim's Algorithm: Maintains a tree spanning a subset S of node and adds minimum cost edges in the cut $[S, S]$. The proof of the algorithm uses the cut optimality condi-

Table 2.10 Trace table for Kruskal's algorithm

Step	Eligible edges A	Weight $\{w_{ij}\}$	$\|T\|$	Edge (u, v)	Spanning tree T	Fitness z
1	{(1, 2), (1, 6), (1, 8), (2, 3), (2, 8), (3, 4), (3, 9), (4, 5), (4, 9), (5, 6), (5, 7), (6, 7), (7, 8), (7, 9), (8, 9)}	{2, 7, 3, 4, 6, 2, 2, 1, 8, 6, 2, 5, 1, 4, 3}	0	(4, 5)	{(4, 5)}	1
2	{(1, 2), (1, 6), (1, 8), (2, 3), (2, 8), (3, 4), (3, 9), (4, 9), (5, 6), (5, 7), (6, 7), (7, 8), (7, 9), (8, 9)}	{2, 7, 3, 4, 6, 2, 2, 8, 6, 2, 5, 1, 4, 3}	1	(7, 8)	{(4, 5), (7, 8)}	2
3	{(1, 2), (1, 6), (1, 8), (2, 3), (2, 8), (3, 4), (3, 9), (4, 9), (5, 6), (5, 7), (6, 7), (7, 9), (8, 9)}	{2, 7, 3, 4, 6, 2, 2, 8, 6, 2, 5, 4, 3}	2	(1, 2)	{(4, 5), (7, 8), (1, 2)}	4
4	{(1, 6), (1, 8), (2, 3), (2, 8), (3, 4), (3, 9), (4, 9), (5, 6), (5, 7), (6, 7), (7, 9), (8, 9)}	{7, 3, 4, 6, 2, 2, 8, 6, 2, 5, 4, 3}	3	(3, 4)	{(4, 5), (7, 8), (1, 2), (3, 4)}	6
5	{(1, 6), (1, 8), (2, 3), (2, 8), (3, 9), (4, 9), (5, 6), (5, 7), (6, 7), (7, 9), (8, 9)}	{7, 3, 4, 6, 2, 8, 6, 2, 5, 4, 3}	4	(3, 9)	{(4, 5), (7, 8), (1, 2), (3, 4), (3, 9)}	8
6	{(1, 6), (1, 8), (2, 3), (2, 8), (4, 9), (5, 6), (5, 7), (6, 7), (7, 9), (8, 9)}	{7, 3, 4, 6, 8, 6, 2, 5, 4, 3}	5	(5, 7)	{(4, 5), (7, 8), (1, 2), (3, 4), (3, 9), (5, 7)}	10
7	{(1, 6), (1, 8), (2, 3), (2, 8), (4, 9), (5, 6), (6, 7), (7, 9), (8, 9)}	{7, 3, 4, 6, 8, 6, 5, 4, 3}	6	(1, 8)	{(4, 5), (7, 8), (1, 2), (3, 4), (3, 9), (5, 7), (1, 8)}	13
8	{(1, 6), (2, 3), (2, 8), (4, 9), (5, 6), (6, 7), (7, 9), (8, 9)}	{7, 4, 6, 8, 6, 5, 4, 3}	7	(6, 7)	{(4, 5), (7, 8), (1, 2), (3, 4), (3, 9), (5, 7), (1, 8), (6, 7)}	18

tions. Can be implemented using a variety of heaps structures; the stated time bound is for the Fibonacci heap data structure [26]. The procedure of Kruskal's algorithm is shown in Fig. 2.23. Table 2.11 presents the trace table for the generating process of MST for solving a simple network example in Fig. 2.21.

Table 2.11 Trace table for Prim's algorithm

Step	Set of connected nodes C	Eligible edges A	Weight $\{w_{ij}\}$	Node u	Node v	Spanning tree T	Fitness z
1	{5}	{(5, 4), (5, 6), (5, 7)}	{1, 6, 2}	5	6	{(5, 4)}	1
2	{5, 4}	{(5, 6), (5, 7), (4, 3), (4, 9)}	{6, 2, 2, 8}	4	3	{(5, 4), (4, 3)}	3
3	{5, 4, 3}	{(5, 6), (5, 7), (4, 9), (3, 9), (2, 3)}	{6, 2, 8, 2, 4}	3	9	{(5,4), (4, 3), (3, 9)}	5
4	{5, 4, 3, 9}	{(5, 6), (5, 7), (4, 9), (2, 3), (7, 9), (8, 9)}	{6, 2, 8, 4, 4, 3}	5	7	{(5,4), (4, 3), (3, 9), (5, 7)}	7
5	{5, 4, 3, 9, 7}	{(5, 6), (4, 9), (2, 3), (7, 9), (8, 9), (6, 7), (7, 8)}	{6, 8, 4, 4, 3, 5, 1}	7	8	{(5,4), (4, 3), (3, 9), (5, 7), (7, 8)}	8
6	{5, 4, 3, 9, 7, 8}	{(5, 6), (4, 9), (2, 3), (7, 9), (8, 9), (6, 7), (1, 8), (2, 8)}	{6, 8, 4, 4, 3, 5, 3, 6}	8	1	{(5,4), (4, 3), (3, 9), (5, 7), (7, 8), (1 8)}	11
7	{5, 4, 3, 9, 7, 8, 1}	{(5, 6), (4, 9), (2, 3), (7, 9), (8, 9), (6, 7), (2, 8), (1, 2), (1, 6)}	{6, 8, 4, 4, 3, 5, 6, 2, 7}	1	2	{(5,4), (4, 3), (3, 9), (5, 7),(7, 8), (1 8), (1, 2)}	13
8	{5, 4, 3, 9, 7, 8, 1, 2}	{(5, 6), (4, 9), (2, 3), (7, 9), (8, 9), (6, 7), (2, 8), (1, 6)}	{6, 8, 4, 4, 3, 5, 6, 7}	7	6	{(5,4), (4, 3), (3, 9), (5, 7),(7, 8), (1, 8), (1, 2), (6, 7)}	18

Sollin's Algorithm: Maintains a collection of node-disjoint trees: in each iteration, adds the minimum cost edge emanating from each such tree. The proof of the algorithm uses the cut optimality conditions [1].

Recently, the genetic algorithm (GA) and other evolutionary algorithms (EAs) have been successfully applied to solve constrained spanning tree problems of real-life instances and have also been used extensively in a wide variety of communica-

```
procedure: Prim's Algorithm
input: graph G = (V, E), weight w_ij, ∀(i, j) ∈ V
output: spanning tree T
begin
  T ← φ;
  choose a random starting node s ∈ V;
  C ← C ∪ {s};   //C: set of connected nodes
  A ← A ∪ {(s,v), ∀v ∈ V};   //A: eligible edges
  while C ≠ V do
    choose an edge (u,v) ← argmin{w_ij | (i, j) ∈ A};
    A ← A \ {(u,v)};
    if v ∉ C then
      T ← T ∪ {(u,v)};
      C ← C ∪ {v};
      A ← A ∪ {(v,w) | (v,w) ∧ w ∉ C};
  output spanning tree T;
end
```

Fig. 2.23 Pseudocode of Prim's algorithm

tion network design problems [27, 28, 29]. For example, Abuali *et al.* [30] developed a new encoding scheme for *probabilistic MST* problem. Zhou and Gen [31] gave an effective GA approach to the *quadratic MST* problem. Zhou and Gen [32] and Fernades and Gouveia [33] investigated the *leaf-constrained MST* problem with GA approaches. Raidl and Drexe [34] and Ahuja and Orlin [35] gave the EAs for the *capacitated MST* problem occurring in telecommunication applications. Zhou and Gen [36, 37], Raidl [38, 39], Knowles and Corne [40], Chou *et al.* [41], and Raidl and Julstrom [42] investigated the different encoding methods and gave the performance analyzes for the *degree-constraint MST* problem, respectively. Raidl and Julstrom [43] and Julstrom [44] developed EAs for the bounded-diameter MST problem. Zhou and Gen [45], Knowles and Corne [40], Lo and Chang [46], and Neumann [47] investigated the *multicriteria MST* problem with GA approaches and other EAs. EAs were also applied to solve other communication network design problem such as *two graph problem, one-max tree problem* and communication spanning tree problem [48]–[51].

2.3.1 Mathematical Formulation of the MST Models

The MST model attempts to find a minimum cost tree that connects all the nodes of the network. The links or edges have associated costs that could be based on their distance, capacity, and quality of line. Let $G = (V, E, W)$ be a connected weighted undirected graph with $n = |V|$ nodes and $m = |E|$ edges, and $w_{ij} \in W$ represent the weight or cost of each edge $(i, j) \in E$ where the weight is restricted to be a nonnegative real number.

Indices

$i, j = 1, 2, \cdots, n$, index of node

Parameters

$n = |V|$: number of nodes
$m = |E|$: number of edges
$w_{ij} \in W$: weight of each edge $(i, j) \in E$
$c_{ij} \in C$: cost of each edge $(i, j) \in E$
u_i: weight capacity of each node i
$d_{ij} \in D$: delay of each edge $(i, j) \in E$
d_i: degree constraint on node i

Decision Variables

x_{ij}: the 0,1 decision variable; 1, if edge $(i, j) \in E$ is selected, and 0, otherwise
y_{ij}: degree value on edge (i, j)

2.3.1.1 Minimum Spanning Tree Model

Let A_S denote the set of edges contained in the subgraph of G induced by the node set S (*i.e.*, A_S is the set of edges of E with both endpoints in S). The *integer programming* formulation of the MST problem can be formulated as follows:

$$\min z = \sum_{(i,j)\in E} w_{ij} x_{ij} \tag{2.8}$$

$$\text{s. t. } \sum_{(i,j)\in E} x_{ij} = n - 1 \tag{2.9}$$

$$\sum_{(i,j)\in A_S} x_{ij} \le |S| - 1 \quad \text{for any set } S \text{ of nodes} \tag{2.10}$$

$$x_{ij} \in \{0,1\}, \quad \forall\,(i,j) \in E \tag{2.11}$$

In this formulation, the 0-1 variable x_{ij} indicates whether we select edge (i,j) as part of the chosen spanning tree (note that the second set of constraints with $|S| = 2$ implies that each $x_{ij} \leq 1$). The constraints at Eq. 2.9 is a cardinality constraint implying that we choose exactly $n-1$ edges, and the packing constraint at Eq. 2.10 implies that the set of chosen edges contain no cycles (if the chosen solution contained a cycle, and S were the set of nodes on a chosen cycle, the solution would violate this constraint).

2.3.1.2 Capacitated Minimum Spanning Tree Model

Capacitated minimum spanning tree (cMST) problem is a extended case of MST. Give a finite connected network, the problem is to find a MST, where the capacity of nodes are satisfied. Mathematically, the problem is reformulated as follows:

$$\min z = \sum_{(i,j)\in E} c_{ij}x_{ij} \tag{2.12}$$

$$\text{s. t. } \sum_{(i,j)\in E} x_{ij} = n-1 \tag{2.13}$$

$$\sum_{(i,j)\in A_S} x_{ij} \leq |S| - 1 \quad \text{for any set } S \text{ of nodes} \tag{2.14}$$

$$\sum_{j=1}^{n} w_{ij}x_{ij} \leq u_i, \quad \forall\, i \tag{2.15}$$

$$x_{ij} \in \{0,1\}, \quad \forall\,(i,j) \in E \tag{2.16}$$

In this formulation, the constraint at Eq. 2.15 guarantees that the total link weight of each node i does not exceed the upper limit u_i.

2.3.1.3 Degree-constrained Minimum Spanning Tree Model

Degree-constrained minimum spanning tree (dMST) is a special case of MST. Given a finite connected network, the problem is to find an MST where the upper bounds of the number of edges to a node is satisfied. The dMST problem is NP-hard and traditional heuristics have had only limited success in solving small to midsize problems. Mathematically, the dMST problem is reformulated as follows:

$$\min z = \sum_{(i,j)\in E} w_{ij}x_{ij} \tag{2.17}$$

$$\text{s. t. } \sum_{(i,j)\in E} x_{ij} = n-1 \tag{2.18}$$

$$\sum_{(i,j)\in A_S} x_{ij} \le |S|-1 \quad \text{for any set } S \text{ of nodes} \tag{2.19}$$

$$\sum_{j=1}^{n} y_{ij} \le d_i, \quad \forall\, i \tag{2.20}$$

$$x_{ij}, y_{ij} \in \{0,1\}, \quad \forall\,(i,j) \in E \tag{2.21}$$

In this formulation, the constraint at Eq. 2.20 guarantees that the total edges to a node i does not exceed the degree d_i.

2.3.1.4 Multicriteria Minimum Spanning Tree Model

In the real world, there are usually such cases that one has to consider simultaneously multicriteria in determining an MST because there are multiple attributes defined on each edge, and this has become subject to considerable attention. Each edge has q associated positive real numbers, representing q attributes defined on it and denoted by w_{ij}^k. In practice w_{ij}^k may represent the weight, distance, cost and so on. The *multicriteria minimum spanning tree* (mMST) model can be formulated as follows:

$$\min z_k = \sum_{(i,j)\in E} w_{ij}^k x_{ij} \quad k = 1,2,\cdots,q \tag{2.22}$$

$$\text{s. t.} \sum_{(i,j)\in E} x_{ij} = n-1 \tag{2.23}$$

$$\sum_{(i,j)\in A_S} x_{ij} \le |S|-1 \quad \text{for any set } S \text{ of nodes} \tag{2.24}$$

$$x_{ij} \in \{0,1\}, \quad \forall\,(i,j) \in E \tag{2.25}$$

2.3.2 *PrimPred-based GA for MST Models*

Let $P(t)$ and $C(t)$ be parents and offspring in current generation t, respectively. The overall procedure of PrimPred-based GA for solving minimum spanning tree models is outlined as follows:

2.3.2.1 Genetic Representation

In GAs literature, whereas several kinds of encoding methods were used to obtain MSTs, most of them cannot effectuality encode or decode between chromosomes and legality spanning trees. Special difficulty arises from (1) a cardinality constraint implying that we choose exactly $n-1$ edges, and (2) implying any set of chosen

---

procedure: PrimPred-based GA for MST models
input: network data (V, E, W),
  GA parameters $(popSize, maxGen, p_M, p_C)$
output: a minimum spanning tree
begin
  $t \leftarrow 0$;
  initialize $P(t)$ by *PrimPred-based encoding routine*;
  evaluate $P(t)$ by *PrimPred-based decoding routine*;
  while (**not** terminating condition) **do**
    create $C(t)$ from $P(t)$ by *Prim-based crossover routine*;
    create $C(t)$ from $P(t)$ by *LowestCost mutation routine*;
    evaluate $C(t)$ by *PrimPred-based decoding routine*;
    select $P(t+1)$ from $P(t)$ and $C(t)$ by roulette wheel selection routine;
    $t \leftarrow t+1$;
  end
  output a minimum spanning tree
end

---

edges containing no cycles. We need to consider these critical issues carefully when designing an appropriate encoding method so as to build an effective GA. How to encode a spanning tree T in a graph G is critical for developing a GA to network design problem, it is not easy to find out a nature representation. We summarized the several kinds of classification of encoding methods as follows:

1. Characteristic vectors-based encoding
2. Edge-based encoding
3. Node-based encoding

Considering this classification, the following subsections review several recent encoding methods concerning how to represent a spanning tree. Figure 2.24 presents a simple network with 12 nodes and 40 arcs.

(a) Example with 12 nodes and 40 edges [42] (b) A generated example of spanning tree

Fig. 2.24 A simple undirected graph

Characteristic Vectors-based Encoding

Davis *et al.* [53] and Piggott and Suraweera [54] have used a *binary-based encoding method* to represent spanning trees in GAs. A binary-based encoding requires space proportional to m and the time complexities of binary-based encoding is $O(m)$. The mapping from chromosomes to solutions (decoding) may be *1-to-1 mapping*. In a complete graph, $m = n(n-1)/2$ and the size of the search space is $2^{n(n-1)/2}$. However, only a tiny fraction of these chromosomes represent feasible solutions, since a complete graph G has only n^{n-2} distinct spanning trees. Repairing strategies have been more successful, but they require additional computation and weaken the encoding heritability. An example of the binary-based encoding is shown in Fig. 2.25

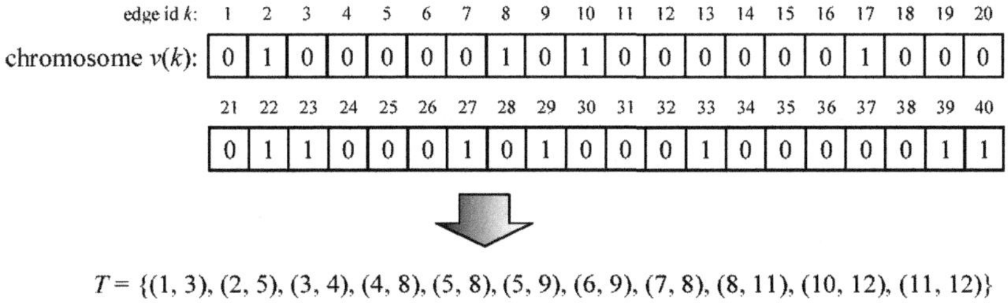

Fig. 2.25 An example of the binary-based encoding

Bean [55] described a *random keys-based encoding method* for encoding ordering and scheduling problems. Schindler *et al.* [49] and Rothlauf *et al.* [51] further investigated network random keys in an evolution strategy framework. In this encoding, a chromosome is a string of real-valued weights, one for each edge. To decode a spanning tree, the edges are sorted by their weights, and Kruskal's algorithm considers the edges are sorted order. As for binary-based encoding, random keys-based encoding requires space proportional to m and the time complexities is $O(m)$. Whereas all chromosomes represent feasible solutions, the uniqueness of the mapping from chromosomes to solutions may be n-to-1 mapping. In a complete graph with only n^{n-2} distinct spanning trees, consider random keys-based encoding, the size of the search space is $(n(n-1)/2)!$ For this reason, most different chromosomes represent the same spanning tree (*i.e.*, n is a very large number with n-to-1 mapping from chromosomes to solutions) and it weaken the encoding heritability. An example of the random keys-based encoding is shown in Fig. 2.26

Edge-based Encoding

Edge-based encoding is an intuitive representation of a tree. A general edge-based encoding requires space proportional to n–1 and the time complexities is $O(m)$. The

edge id k:	1	2	3	4	5	6	7	8	9	10	11	12	13	14	15	16	17	18	19	20
chromosome $v(k)$:	0.01	0.99	0.03	0.24	0.02	0.35	0.28	0.95	0.45	0.90	0.15	0.57	0.43	0.52	0.84	0.09	0.85	0.54	0.13	0.42

	21	22	23	24	25	26	27	28	29	30	31	32	33	34	35	36	37	38	39	40
	0.53	0.82	0.81	0.48	0.07	0.51	0.79	0.17	0.75	0.41	0.29	0.21	0.70	0.32	0.55	0.19	0.59	0.08	0.68	0.62

$T = \{(1, 3), (2, 5), (3, 4), (4, 8), (5, 8), (5, 9), (6, 9), (7, 8), (8, 11), (10, 12), (11, 12)\}$

Fig. 2.26 An example of the random keys-based encoding

mapping from chromosomes to solutions (decoding) may be *1-to-1 mapping*. In a complete graph, $m = n(n-1)/2$ and the size of the search space is $2^{n(n-1)/2}$. Edge-based encoding and binary-based encoding have very similar performance in theory. However, there are $2^{n(n-1)/2}$ possible values for a tree, and only a tiny fraction of these chromosomes represent feasible solutions, and weaken the encoding heritability. Recently, researchers investigated and developed edge-based representations in GAs for designing several MST-related problems. Considering the disadvantage of edge-based encoding, the heuristic method is adopted in the chromosome generating procedure. These methods can ensure feasibility; all chromosomes can be represented by feasible solutions.

Knowles and Corne [40] proposed a method which improves edge-based encoding. The basis of this encoding is a spanning-tree construction algorithm which is *randomized primal method* (RPM), based on the Prim's algorithm. Raidl and Julstrom [43] gave the method depending on an underlying random spanning-tree algorithm. Three choices for this algorithm, based on Prim's algorithm, Kruskal's algorithm and on random walks, respectively, are examined analytically and empirically. Raidl [39], Chou *et al.* [41], Raidl and Julstrom [43] and Julstrom [44] described the other similar heuristic methods which ensure feasibility of the chromosomes. All of this heuristic method based encoding requires space proportional to $n-1$ and the time complexities is $O(n)$. The mapping from chromosomes to solutions (decoding) may be 1-to-1 mapping. In a complete graph, $m = n(n-1)/2$ and the size of the search space is n^{n-1}. These encoding methods offer efficiency of time complexity, feasibility and uniqueness. However, offspring of simple crossover and mutation should represent infeasible solutions. Several special genetic operator and repair strategies have been successful, but their limitations weaken the encoding heritability. An example of the edge-based encoding is shown in Fig. 2.27

Node-based Encoding

Prüfer number-based encoding: Cayley [56] proved the following formula: the number of spanning trees in a complete graph of n nodes is equal to n^{n-2}. Prüfer [57] presented the simplest proof of Cayley's formula by establishing a *1-to-1 correspon-*

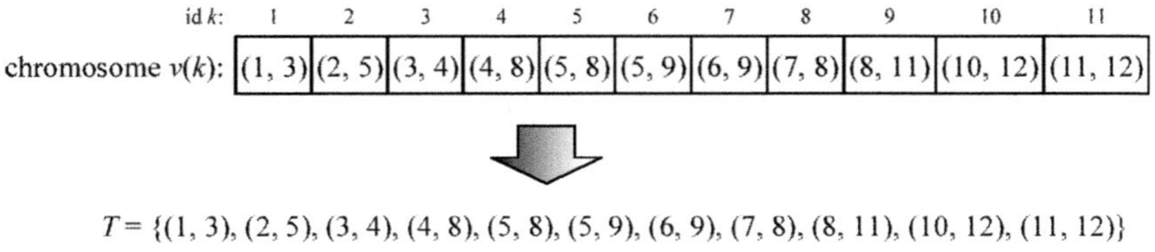

Fig. 2.27 An example of the edge-based encoding

dence between the set of spanning trees and a set of sequences of $n-2$ integers, with each integer between 1 and n inclusive. The sequence of n integers for encoding a tree is known as the Prüfer number. An example of the Prüfer number-based encoding is shown in Fig. 2.28. The trace table of Prüfer number encoding process and decoding process are shown in Tables 2.12 and 2.13, respectively.

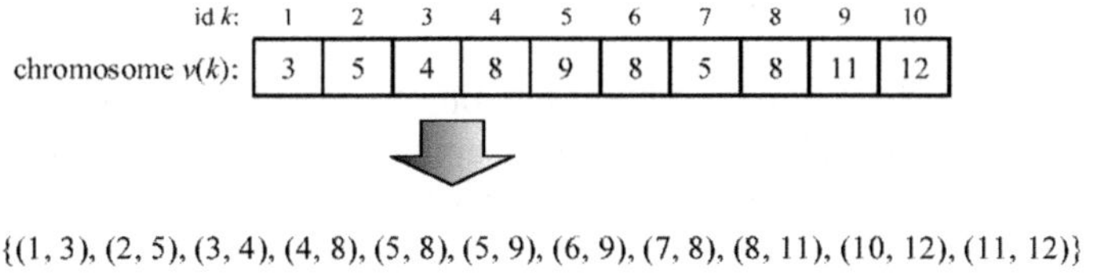

Fig. 2.28 An example of the Prüfer number-based encoding

Table 2.12 The trace table of Prüfer encoding process considering the example graph in Fig. 2.24

A tree T	i	j	Prüfer number P
{(1, 3), (3, 4), (4, 8), (7, 8), (8, 11), (11, 12), (12, 10), (8, 5), (5, 2), (5, 9), (9, 6)}	1	3	[3]
{(3, 4), (4, 8), (7, 8), (8, 11), (11, 12), (12, 10), (8, 5), (5, 2), (5, 9), (9, 6)}	2	5	[3, 5]
{(3, 4), (4, 8), (7, 8), (8, 11), (11, 12), (12, 10), (8, 5), (5, 9), (9, 6)}	3	4	[3, 5, 4]
{(4, 8), (7, 8), (8, 11), (11, 12), (12, 10), (8, 5), (5, 9), (9, 6)}	4	8	[3, 5, 4, 8]
{(7, 8), (8, 11), (11, 12), (12, 10), (8, 5), (5, 9), (9, 6)}	6	9	[3, 5, 4, 8, 9]
{(7, 8), (8, 11), (11, 12), (12, 10), (8, 5), (5, 9)}	7	8	[3, 5, 4, 8, 9, 8]
{(8, 11), (11, 12), (12, 10), (8, 5), (5, 9)}	9	5	[3, 5, 4, 8, 9, 8, 5]
{(8, 11), (11, 12), (12, 10), (8, 5)}	5	8	[3, 5, 4, 8, 9, 8, 5, 8]
{(8, 11), (11, 12), (12, 10)}	8	11	[3, 5, 4, 8, 9, 8, 5, 8, 11]
{(11, 12), (12, 10)}	10	12	[3, 5, 4, 8, 9, 8, 5, 8, 11, 12]
{(11, 12)}			

i: lowest labeled leaf node; j: predecessor node of leaf node i

Table 2.13 Trace table of Prüfer decoding process considering the example graph in Fig. 2.24

Prüfer number P	Set of residual nodes P'	Spanning tree T
[3, 5, 4, 8, 9, 8, 5, 8, 11, 12]	[1, 2, 6, 7, 10]	{(1,3)}
[5, 4, 8, 9, 8, 5, 8, 11, 12]	[2, 3, 6, 7, 10]	{(1,3), (2, 5)}
[4, 8, 9, 8, 5, 8, 11, 12]	[3, 6, 7, 10]	{(1,3), (2, 5), (3, 4)}
[8, 9, 8, 5, 8, 11, 12]	[4, 6, 7, 10]	{(1,3), (2, 5), (3, 4), (4, 8)}
[9, 8, 5, 8, 11, 12]	[6, 7, 10]	{(1,3), (2, 5), (3, 4), (4, 8), (6, 9)}
[8, 5, 8, 11, 12]	[7, 9, 10]	{(1,3), (2, 5), (3, 4), (4, 8), (6, 9), (7, 8)}
[5, 8, 11, 12]	[9, 10]	{(1,3), (2, 5), (3, 4), (4, 8), (6, 9), (7, 8), (5, 9)}
[8, 11, 12]	[5, 10]	{(1,3), (2, 5), (3, 4), (4, 8), (6, 9), (7, 8), (5, 9), (5, 8)}
[11, 12]	[8, 10]	{(1,3), (2, 5), (3, 4), (4, 8), (6, 9), (7, 8), (5, 9), (5, 8), (8, 11)}
[12]	[10, 11]	{(1,3), (2, 5), (3, 4), (4, 8), (6, 9), (7, 8), (5, 9), (5, 8), (8, 11), (10, 12)}
ϕ	[11, 12]	{(1,3), (2, 5), (3, 4), (4, 8), (6, 9), (7, 8), (5, 9), (5, 8), (8, 11), (10, 12), (11, 12)}

Many researchers have encoded spanning trees as Prüfer numbers in GAs for a variety of problems. These include the degree-constrained MST problems, multicriteria MST problems, and leaf-constrained MST problems, *etc*. However, researchers have pointed out that Prüfer number is a poor representation of spanning trees for evolutionary search, and their major disadvantages as follows:

1. It needs a complex encoding process and decoding process between chromosomes and solutions (computational cost);
2. It is difficult to consist mostly of substructures of their parents' phenotypes (poor heritability);
3. It contains no useful information such as degree, connection, *etc*., about a tree;
4. It also represents infeasible solutions if the graph G is incomplete graph.

Predecessor-based Encoding: A more compact representation of spanning trees is the predecessor or determinant encoding, in which an arbitrary node in G is designated the root, and a chromosome lists each other node's predecessor in the path from the node to the root in the represented spanning tree: if pred(i) is j, then node j is adjacent to node i and nearer the root. Thus, a chromosome is string of length $n-1$ over 1, 2, ..., n, and when such a chromosome decodes a spanning tree, its edges can be made explicit in time that is $O(n\log n)$.

Applied to such chromosomes, positional crossover and mutation operators will generate infeasible solutions, requiring again penalization or repairing. As shown in [42], Abuali *et al.* described a repairing mechanism that applies to spanning trees on sparse, as well as complete graphs. However, these strategies require additional computation and weaken the encoding heritability, and these repairing process have to spend much computational cost.

PrimPred-based Encoding

Take a predecessor-based encoding string $v[]$, with length of $n-1$, where n represents number of nodes. Assume $v(i)$ represents the allele of the i-th fixed position in chromosome $v[]$, where i starts from 2 to n. As discussed in Chou *et al.*'s research paper [41], predecessor-based encoding generates some chromosomes that are illegal (*i.e.*, not a spanning tree). Combining the simple random initialization, most of the chromosomes will be illegal for three reasons: missing node i, self-loop, or cycles. They gave the repairing function for illegal chromosomes. However, the complex repairing process will be used at each generation (computational cost), and after repairing, the offspring of the crossover and mutation are difficult to represent solutions that combine substructures of their parental solutions (worst heritability and locality).

```
procedure: PrimPred-based encoding
input: number of nodes n,
       node set S_i with all nodes adjacent to node i
output: chromosome v
begin
   node i ← 0;
   assigned node set C ← {i};
   t ← 1;
   while (t ≤ n)
      eligible edge set E ← {(i, j) | j ∈ S_i}
      choose edge (i, j) ∈ E at random;
      v(j) ← i;
      C ← C ∪ {j};
      i ← j;
      E ← E ∪ {(i, j) | j ∈ S_i};
      E ← E \ {(i, j) | i ∈ C & j ∈ C};
      t ← t + 1;
   end
   output chromosome v;
end
```

Fig. 2.29 Pseudocode of PrimPred-based encoding

```
procedure: PrimPred-based decoding
input: chromosome v, length of chromosome l
output: spanning tree T
begin
   spanning tree T ← φ;
   for i = 1 to l
      v(j) ← i;
      T ← T ∪ {i, v(i)};
   end
   output spanning tree T;
end
```

Fig. 2.30 Pseudocode of predecessor-based decoding

When a GA searches a space of spanning trees, its initial population consists of chromosomes that represent random trees. It is not as simple as it might seem to choose spanning trees of a graph so that all are equal by likely. Prim's algorithm greedily builds a minimum spanning tree from a start node by repeatedly appending the lowest cost edge that joins a new node to the growing tree. Applying Prim's algorithm to the GAs chooses each new edge at random rather than according to its cost [42]. The encoding procedure and decoding procedure are shown in Figures

2.29 and 2.30, respectively. An example of generated chromosome and its decoded spanning tree is shown in Fig. 2.31 for the undirected network shown in Fig. 2.24.

In general, there are two ways to generate the initial population, heuristic initialization and random initialization. However, the mean fitness of the heuristic initialization is already high so that it may help the GAs to find solutions faster. Unfortunately, in most large scale problems, for example network communication designing, it may just explore a small part of the solution space and it is difficult to find global optimal solutions because of the lack of diversity in the population. Therefore, random initialization is effected in this research so that the initial population is generated with the PrimPred-based encoding method.

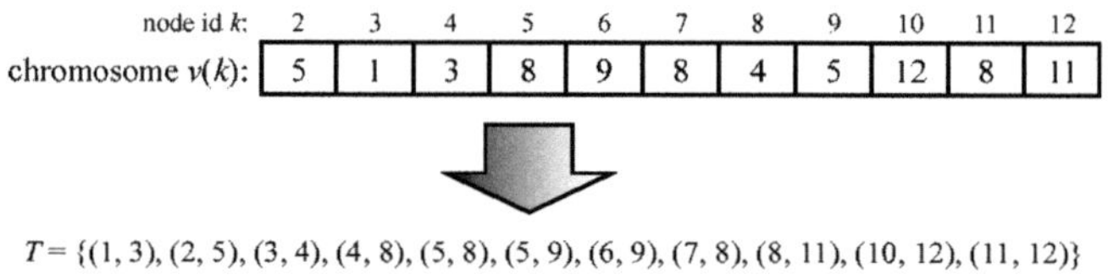

Fig. 2.31 An example of the PrimPred-based encoding

Depending on the properties of encodings (introduced on page 9), we summarize the performance of proposed PrimPred-based encoding method and other introduced encoding methods in Table 2.14.

Table 2.14 Summary of the performance of encoding methods

Representation		Space	Time	Feasibility	Uniqueness	Locality	Heritability
Characteristic Vectors-based	Binary-based encoding	m	$O(m)$	Worst	1-to-1 mapping	Worst	Worst
	Random keys-based encoding	m	$O(m)$	Good	n-to-1 mapping	Worst	Worst
Edge-based	General edge-based encoding	n	$O(m)$	Worst	1-to-1 mapping	Worst	Worst
	Heuristic edge-based encoding	n	$O(n)$	Good	1-to-1 mapping	Poor	Poor
Node-based	Prüfer number-based encoding	n	$O(n\log n)$	Good	1-to-1 mapping	Worst	Worst
	Predecessor-based Encoding	n	$O(n\log n)$	Poor	1-to-1 mapping	Worst	Worst
	PrimPred-based Encoding	n	$O(n\log n)$	Good	1-to-1 mapping	Poor	Poor

2.3.2.2 Genetic Operators

Genetic operators mimic the process of heredity of genes to create new offspring at each generation. Using the different genetic operators has very large influence on GA performance [8]. Therefore it is important to examine different genetic operators.

Prim-based Crossover

Crossover is the main genetic operator. It operates on two parents (chromosomes) at a time and generates offspring by combining both chromosomes' features. In network design problems, crossover plays the role of exchanging each partial route of two chosen parents in such a manner that the offspring produced by the crossover still represents a legal network.

To provide good heritability, a crossover operator must build an offspring spanning tree that consists mostly or entirely of edges found in the parents. In this subsection, we also apply Prim's algorithm described in the previous subsection to the graph $G' = (V, T_1 \cup T_2)$, where T_1 and T_2 are the edge sets of the parental trees, respectively. Figure 2.32 gave the pseudocode of crossover process. An example of Prim-based crossover is shown in Fig. 2.34

```
procedure: Prim-based crossover
input: parent v_1[], v_2[]
output: offspring v'[]
begin
    spanning tree T_1 ← decode(v_1);
        // decode the chromosomes to spanning tree
    spanning tree T_2 ← decode(v_1);
    edge set E ← T_1 ∪ T_2;
    subgraph G ← (V, E);
    offspring v' ← PrimPredEncode(G);
        // routine 1: PrimPred-based encoding
    output offspring v'[];
end
```

Fig. 2.32 Pseudocode of Prim-based crossover

```
procedure: LowestCost mutation
input: parent v[], node set V = {1, 2, ..., n}, cost w_ij of each edge(i, j) ∈ E
output: offspring v'[]
begin
    spanning tree T ← decode(v);
    T ← T / {(u, v) | (u, v) ∈ T} at random;
    allocate node A(k) ← 0, ∀k ∈ V;
    while (T ≠ φ)
        choose edge (i, j) ∈ T;
        T ← T / {(i, j)};
        if A(i) = φ & A(j) = φ then
            l ← min{i, j};
            A(i) ← l;
            A(j) ← l;
        else if A(i) = φ & A(j) ≠ φ then
            A(i) ← A(j);
        else if A(i) ≠ φ & A(j) = φ then
            A(j) ← A(i);
        else if A(i) ≠ φ & A(j) ≠ φ then
            if A(i) < A(j) then
                A(k) ← A(i), for all k with A(k) = A(j);
            if A(i) > A(j) then
                A(k) ← A(j), for all k with A(k) = A(i);
        end
    end
    node set C_1 ← {k | A(k) = A(u)};
    node set C_2 ← {k | A(k) = A(v)};
    edge (x, y) ← arg min{w_ij | i ∈ C_1, j ∈ C_2};
    T ← T ∪ {(x, y)};
    offspring v' ← encode(T);
    output offspring v'[];
end
```

Fig. 2.33 Pseudocode of LowestCost mutation

LowestCost Mutation

Mutation is a background operator which produces spontaneous random changes in various chromosomes. A simple way to achieve mutation would be to alter one or more genes. In GAs, mutation serves the crucial role of either replacing the genes lost from the population during the selection process so that they can be tried in a new context or providing the genes that were not present in the initial population. In this subsection, it is relatively easy to produce some mutation operators for permutation representation. Several mutation operators have been proposed for permutation representation, such as *swap mutation*, *inversion mutation*, and *insertion mutation*, and so on [7]. Swap mutation selects two positions at random. For example, to represent a feasible solutions after mutation, and to improve the heritability of offspring, a new mutation operator is proposed , first removing a randomly chosen edge, determining the separated components, then reconnecting them by the lowest cost edge. An example of LowestCost mutation is shown in Fig. 2.35. Figure 2.33 is the pseudocode of mutation process.

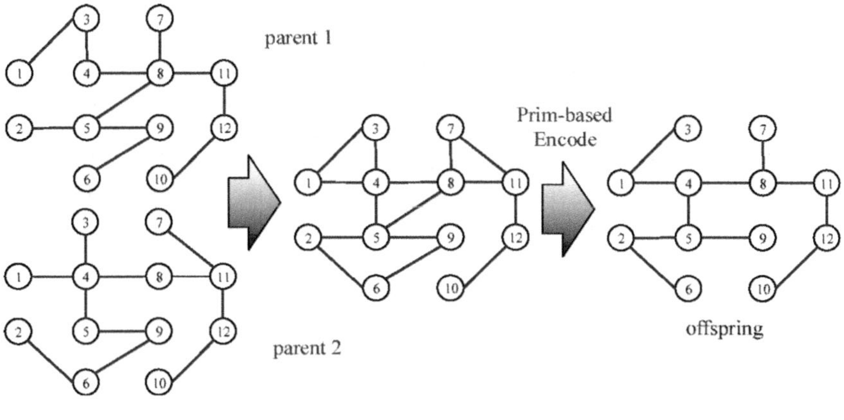

Fig. 2.34 An example of Prim-based crossover

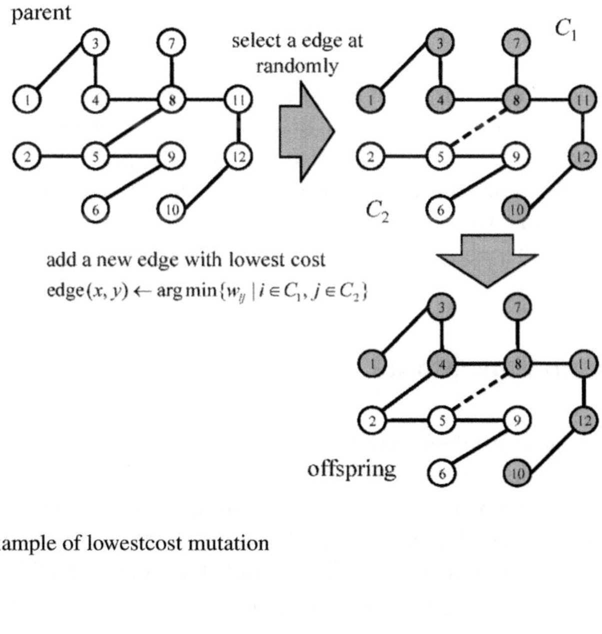

Fig. 2.35 An example of lowestcost mutation

2.3.3 Computational Experiments and Discussions

In this section, PrimPred-based GA is compared with Zhou and Gen [36] and Raidl and Julstrom [42] for solving several *large scale minimum spanning tree* (MST) problems. All the simulations were performed with Java on a Pentium 4 processor (2.6-GHz clock), 3.00GA RAM.

For examining the effectiveness of different encoding methods, PrimPred-based GA, Zhou and Gen's Prüfer number-based encoding method and Raidl and Julstrom's edge-based encoding method are applied to six test problems [61]. Prüfer number-based encoding with one-cut point crossover and swap mutation is combined, and edge-based encoding using two kinds of mutation operators is combined which is included in [42], and for initializing the chromosomes based on the edge set, Raidl and Julstrom's PrimRST (Prim random spanning tree) is combined. Each algorithm was run 20 times using different initial seeds for each test problems. And Prim's algorithm has been used to obtain optimum solutions for the problems. The GA parameter is setting as follows:

Population size: $popSize = 10$;
Crossover probability: $p_C = 0.30$, 0.50 or 0.70;
Mutation probability: $p_M = 0.30$, 0.50 or 0.70;
Maximum generation: $maxGen = 1000$;

The experimental study was realized to investigate the effectiveness of the different encoding method; the interaction of the encoding with the crossover operators and mutation operators, and the parameter settings affect its performance. Table 2.15 gives computational results for four different encoding methods on six test problems by three kinds of parameter settings. In the columns of the best cost of four encoding methods, it is possible to see that whereas the Prüfer number-based approach is faster than the others, it is difficult to build from the substructures of their parents' phenotypes (poor heritability), and the result is very far from the best one. Two kinds of mutation are used in edge-based encoding, the second one (depends on the cost) giving better performance than the first. For considering the computational cost (CPU time), because of the LowestCost mutation in the proposed approach, spending a greater CPU time to find the edge with the lowest cost they always longer than other algorithms. However, PrimPred-based GA developed in this study gives a better cost than other algorithms.

2.4 Maximum Flow Model

The maximum flow problem (MXF) is to find a feasible flow through a single-source, single-sink flow network that is maximum. The MXF model and the shortest path model are complementary. The two problems are different because they capture different aspects: in the shortest path problem model all arcs are costs but not arc capacities; in the maximum flow problem model all arcs are capacities but not costs. Taken together, the shortest path problem and the maximum flow problem combine

Table 2.15 Performance comparisons with different GA approaches: Prüfer number-based encoding (Zhou and Gen [36]), edge-based 1 and edge-based 2 with different mutation operators (Raidl and Julstrom [42]) and PrimPred-based GA

Test Problem n/m	Optimal Solutions	p_C	p_M	Prüfer Num-based		Edge-based 1		Edge-based 2		PrimPred-based	
				avg.	time	avg.	time	avg.	time	avg.	time
40/780	470	0.30	0.30	1622.20	72.20	1491.80	1075.20	495.60	1081.40	470.00	1100.20
		0.50	0.50	1624.40	87.60	1355.80	2184.40	505.80	2175.00	470.00	2256.40
		0.70	0.70	1652.60	134.80	1255.20	3287.40	497.60	3281.40	470.00	3316.00
40/780	450	0.30	0.30	1536.60	74.80	1458.20	1118.60	471.60	1093.80	450.00	1106.20
		0.50	0.50	1549.20	78.20	1311.40	2190.80	480.20	2175.00	450.00	2200.20
		0.70	0.70	1564.40	122.00	1184.40	3287.60	466.40	3262.40	450.00	3275.00
80/3160	820	0.30	0.30	3880.40	150.00	3760.20	5037.80	923.20	5059.60	820.00	5072.00
		0.50	0.50	3830.00	184.40	3692.00	10381.20	871.00	10494.20	820.00	10440.60
		0.70	0.70	3858.20	231.20	3483.80	16034.80	899.20	15871.80	820.00	15984.60
80/3160	802	0.30	0.30	3900.60	131.40	3853.00	5125.00	894.60	4934.20	802.00	5071.80
		0.50	0.50	3849.60	206.20	3515.20	10325.20	863.00	10268.80	802.00	10365.60
		0.70	0.70	3818.40	222.00	3287.20	16003.00	868.00	15965.40	802.00	15947.20
120/7140	712	0.30	0.30	5819.40	187.40	5536.60	15372.00	871.80	15306.40	712.00	15790.40
		0.50	0.50	5717.20	293.80	5141.00	31324.80	805.40	30781.40	712.00	31503.20
		0.70	0.70	5801.40	316.00	5035.20	47519.00	804.20	47047.20	712.00	47865.80
160/12720	793	0.30	0.30	7434.80	284.40	7050.40	41993.60	1353.60	42418.60	809.60	42628.20
		0.50	0.50	7361.00	421.80	7111.60	87118.80	1061.60	86987.40	793.00	86828.40
		0.70	0.70	7517.00	403.20	6735.00	163025.00	955.40	161862.40	793.00	154731.20

avg.: average solution of 20 runs; time: average computation time in millisecond (ms).

all the basic ingredients of network models [1]. As such, they have become the nuclei of network optimization models.

Example 2.3: A Simple Example of MXF

A simple example of maximum flow model is shown in Fig. 2.36, and Table 2.16 gives the data set of the example.

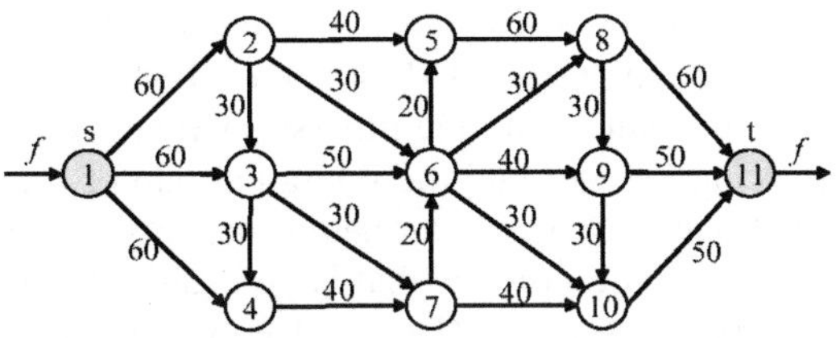

Fig. 2.36 A simple example of maximum flow model

Table 2.16 The data set of illustrative maximum flow model

i	j	u_{ij}	i	j	u_{ij}
1	2	60	6	5	20
1	3	60	6	8	30
1	4	60	6	9	40
2	3	30	6	10	30
2	5	40	7	6	20
2	6	30	7	10	40
3	4	30	8	9	30
3	6	50	8	11	60
3	7	30	9	10	30
4	7	40	9	11	50
5	8	60	10	11	50

Traditional Methods

If we interpret u_{ij} as the maximum flow rate of arc (i, j), MXF identifies the maximum steady-state flow that the network can send from node s to node t per unit time. There are several ways of solving this problem:

Ford-Fulkerson algorithm: The idea behind the algorithm is very simple. As long as there is a path from the source (start node s) to the sink (end node t), with available capacity on all edges in the path, we send flow along one of these paths. Then we find another path, and so on. A path with available capacity is called an *augmenting path*.

Edmonds-Karp algorithm: The algorithm is identical to the Ford-Fulkerson algorithm, except that the search order when finding the augmenting path is defined. The path found must be the shortest path which has available capacity. This can be found by a *breadth-first search*, as we let edges have unit length. The running time is $O(nm^2)$.

Relabel-to-front algorithm: It is in the class of *push-relabel algorithms* for maximum flow which run in $O(n^2m)$. For dense graphs it is more efficient than the Edmonds-Karp algorithm.

The MXF appears to be more challenging in applying GAs and other evolutionary algorithms than many other common network optimization problems because of its several unique characteristics. For example, a flow at each edge can be anywhere between zero and its flow capacity. Munakata and Hashier [58] give a GA approach to applied the MXF problems. In this approach, each solution is represented by a flow matrix, and the fitness function is defined to reflect two characteristics–balancing vertices and the saturation rate of the flow. Akiyama *et al.* [59] gave a mixed approach based on neural network (NN) and GA for solving a *transport safety planning problem* that is an extended version of the MXF model. Ericsson *et al.* [60] presented a GA to solve the *open shortest path first* (OSPF) weight setting problem that is also an extended version of MXF model which seeks a set of weights.

2.4.1 Mathematical Formulation

Let $G = (N,A)$ be a directed network, consisting of a finite set of nodes $N = 1,2,\cdots,n$ and a set of directed arcs $A = \{(i,j),(k,l),\cdots,(s,t)\}$ joining m pairs of nodes in N. Arc (i,j) is said to be incident with nodes i and j, and is directed from node i to node j. Suppose that each arc (i,j) has assigned to it nonnegative numbers u_{ij}, the capacity of (i,j). This capacity can be thought of as representing the maximum amount of some commodity that can "flow" through the arc per unit time in a steady-state situation. Such a flow is permitted only in the indicated direction of the arc, *i.e.*, from i to j.

Consider the problem of finding maximal flow from a source node s (node 1) to a sink node t (node n), which can be formulated as follows: Let x_{ij} = the amount of flow through arc (i,j). The assumptions are given as follows:

A1. The network is directed. We can fulfil this assumption by transforming any undirected network into a directed network.
A2. All capacities are nonnegative.
A3. The network does not contain a directed path from node s to node t composed only of infinite capacity arcs. Whenever every arc on a directed path P from note s to note t has infinite capacity, we can send an infinite amount of flow along this path, and therefore the maximum flow value is unbounded.
A4. The network does not contain parallel arcs (*i.e.*, two or more arcs with the same tail and head nodes).

This assumption is essentially a notational convenience.

Indices

$i,j,k = 1, 2, \cdots, n$ index of node

Parameters

n number of nodes
u_{ij} capacity of arc (i,j)

Decision Variables

x_{ij} the amount of flow through arc $(i,j) \in A$.

Maximum Flow Model

The mathematical model of the MXF problem is formulated as a *liner programming*, in which the objective is to maximize total flow z from source node 1 to sink node n as follows:.

$$\max\ z = v \tag{2.26}$$

$$\text{s. t.}\ \sum_{j=1}^{n} x_{ij} - \sum_{k=1}^{n} x_{ki} = \begin{cases} v & (i=1) \\ 0 & (i=2,3,\cdots,n-1) \\ -v & (i=n) \end{cases} \tag{2.27}$$

$$0 \le x_{ij} \le u_{ij} \quad (i,j=1,2,\cdots,n) \tag{2.28}$$

where the constraint at Eq. 2.27, a conservation law, is observed at each of the nodes other than s or t. That is, what goes out of node i, $\sum_{j=1}^{n} x_{ij}$ must be equal to what comes in, $\sum_{k=1}^{n} x_{ki}$. The constraint at Eq. 2.28 is flow capacity. We call any set of numbers $\boldsymbol{x}=(x_{ij})$ which satisfy Eqs. 2.27 and 2.28 a feasible flow, and v is its value.

2.4.2 Priority-based GA for MXF Model

Let $P(t)$ and $C(t)$ be parents and offspring in current generation t, respectively. The overall procedure of priority-based GA (priGA) for solving MXF model is outlined as follows:

2.4.2.1 Genetic Representation

We have summarized the performance of the priority-based encoding method and other encoding methods introduced for solving the shortest path model in a previous section (page 62).

In this section, the special difficulty of the MXF is that a solution of the MXF is presented by various numbers of paths. Table 2.17 shows two examples of the solutions of the MXF problem with various paths, for solving a simple network with 11 nodes and 22 arcs as shown in Fig. 2.36.

Until now, for presenting a solution of MXF with various paths, the general idea of chromosome design is adding several shortest path based-encoding to one chromosome. The length of these representations is variable depending on various paths, and most offspring are infeasible after crossover and mutation operations. Therefore

```
procedure: priGA for MXF model
input: network data (N, A, U),
    GA parameters (popSize, maxGen, p_M, p_C, p_I)
output: the best solution
begin
  t ← 0;
  initialize P(t) by priority-based encoding routine;
  evaluate P(t) by overall-path growth decoding routine;
  while (not terminating condition) do
    create C(t) from P(t) by WMX routine;
    create C(t) from P(t) by insertion mutation routine;
    evaluate C(t) by overall-path growth decoding routine;
    select P(t+1) from P(t) and C(t) by roulette wheel selection routine;
    t ← t+1;
  end
  output the best solution
end
```

Table 2.17 Examples of the solutions with various paths

Solution	# of paths K	Path P	Flow f	Total flow z
1	3	1-3-6-5-8-11	20	
		1-3-6-8-11	30	
		1-3-7-6-9-11	10	60
2	8	1-3-6-5-8-11	20	
		1-3-6-8-11	30	
		1-3-7-6-9-11	10	
		1-4-7-6-9-11	10	
		1-4-7-10-11	30	
		1-2-5-8-11	10	
		1-2-5-8-9-11	30	
		1-2-6-9-10-11	20	160

these kinds of representations are lost regarding the feasibility and legality of chromosomes.

In this section, the priority-based chromosome is adopted; it is an effective representation to present a solution with various paths. For the decoding process, first, a *one-path growth procedure* is introduced that obtains a path based on the chromosome generated with given network in Fig. 2.37.

Then an *overall-path growth procedure* is presented in Fig. 2.38 that removes the flow used from each arc and deletes the arcs where its capacity is 0. Based on updated network, we obtain a new path by one-path growth procedure, and repeat these steps until we obtain the overall possible path.

```
procedure: one-path growth
input: chromosome v, number of nodes n, the set of nodes S_i with all nodes adjacent to node i
output: a path P_k
begin
    initialize i ← 1, l_k ← 1, P_k[l] ← i;   // i : source node, l_k : length of path P_k.
    while S_1 ≠ φ do
        j' ← argmax{v[j], j ∈ S_i};   // j': the node with highest priority among S_i.
        if v[j'] ≠ 0 then
            P_k[l_k] ← j';   // chosen node j' to construct path P_k.
            l_k ← l_k + 1;
            v[j'] ← 0;
            i ← j';
        else
            S_i ← S_i \ j';   // delete the node j' adjacent to node i.
            v[i] ← 0;
            l_k ← l_k − 1;
            if l_k ≤ 1 then l_k ← 1, break;
            i ← P_k[l_k];
    output a path P_k[]
end
```

Fig. 2.37 Pseudocode of one-path growth

By using the priority-based chromosome which is shown in Fig. 2.39, a trace table of decoding process is shown in Table 2.18 for a simple network (shown in Fig. 2.36). The final result is shown in Fig. 2.40. The objective function value is 160

2.4.2.2 Genetic Operations

Crossover Operator

In this section, we adopt the *weight mapping crossover* (WMX) proposed on page 68. WMX can be viewed as an extension of one-cut point crossover for permutation representation. As a one-cut point crossover, two chromosomes (parents) would choose a random-cut point and generate the offspring by using segment of own par-

---

procedure: overall-path growth for MXF
input: network data (N, A, U), chromosome $v_k[\cdot]$, the set of nodes S_i with all nodes adjacent to node i
output: no. of paths L_k and the flow f_i^k of each path, $i \in L_k$
begin
 $l \leftarrow 0$; // l : number of paths.
 while $S_1 \neq \phi$ **do**
  $l \leftarrow l + 1$;
  generate a path $P_l^k[\cdot]$ by **procedure** (one-path growth);
  select the sink node a of path P_l^k;
  if $a = n$ **then**
   $f_l^k \leftarrow f_{l-1}^k + \min\{u_{ij} \mid (i,j) \in P_l^k\}$; // calculate the flow f_l^k of the path P_l^k.
   $\tilde{u}_{ij} \leftarrow u_{ij} - (f_l^k - f_{l-1}^k)$; // make a new flow capacity $\tilde{u}_{ij}$.
   $S_i \leftarrow S_i \setminus j, \quad (i,j) \in P_l^k \;\&\; \tilde{u}_{ij} = 0$; // delete the arcs which its capacity is 0.
  else
   $S_i \leftarrow S_i \setminus a, \quad \forall i$; // update the set of nodes S_i.
 output no. of paths $L_k \leftarrow l - 1$ and the flow $f_i^k, i \in L_k$.
end

Fig. 2.38 Pseudocode of overall-path growth for FMX

node ID :	1	2	3	4	5	6	7	8	9	10	11
priority :	2	1	6	4	11	9	8	10	5	3	7

Fig. 2.39 The best chromosome for MXF example

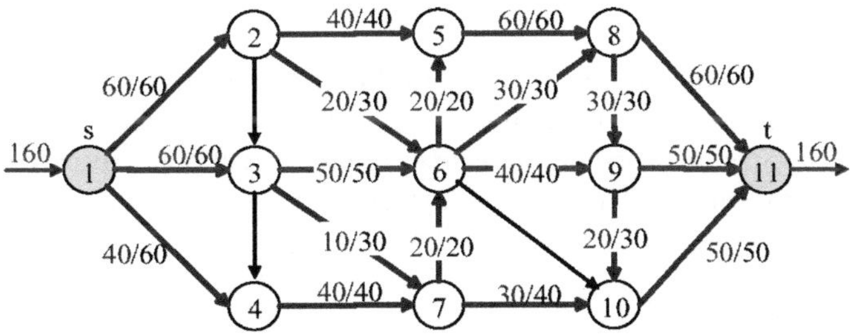

Fig. 2.40 Final result of the MXF example

Table 2.18 A trace table of decoding process for the MXF problem

k	i	S_i	j'	P_k	S_1	f_i^k
1	0			1		
	1	{2, 3, 4}	3	1-3		
	2	{4, 6, 7}	6	1-3-6		
	3	{5, 8, 9, 10}	5	1-3-6-5		
	4	{8}	8	1-3-6-5-8		
	5	{9, 11}	11	1-3-6-5-8-11	{2, 3, 4}	20
2	0			1		
	1	{2, 3, 4}	3	1-3		
	2	{4, 6, 7}	6	1-3-6		
	3	{8, 9, 10}	8	1-3-6-8		
	4	{9, 11}	11	1-3-6-8-11	{2, 3, 4}	50
3	...	...	...	1-3-7-6-9-11	{2, 4}	60
4	...	...	...	1-4-7-6-9-11	{2, 4}	70
5	...	...	...	1-4-7-10-11	{2}	100
6	...	...	...	1-2-5-8-11	{2}	110
7	...	...	...	1-2-5-8-9-11	{2}	140
8	...	...	...	1-2-6-9-10-11	φ	160

ent to the left of the cut point, then remap the right segment based on the weight of other parent of right segment.

Mutation Operator:

As introduced on page 70, insertion mutation selects a gene at random and inserts it in another random position. We adopt this insertion mutation to generate offspring in this section.

The immigration operator is modified to (1) include immigration routine, in each generation, (2) generate and (3) evaluate $popSize \cdot p_{\mathrm{I}}$ random chromosomes, and (4) replace the $popSize \cdot p_{\mathrm{I}}$ worst chromosomes of the current population (p_{I}, called the immigration probability). The detailed introduction is shown on page 23.

Selection Operator:

We adopt the roulette wheel selection (RWS). It is to determine selection probability or survival probability for each chromosome proportional to the fitness value. A model of the roulette wheel can be made displaying these probabilities. The detailed introduction is shown on page 13.

2.4.3 Experiments

The numerical examples, presented by Munakata and Hashier, were adopted [58]. All the simulations were performed with Java on a Pentium 4 processor (1.5-GHz clock). The parameter specifications used are as follows:

Population size: *popSize* =10
Crossover probability: p_C =0.50
Mutation probability: p_M =0.50
Maximum generation: *maxGen* =1000
Terminating condition: 100 generations with same fitness.

2.4.3.1 Test Problem 1

The first test problem is given in Fig. 2.41. This network contains 25 nodes and 49 arcs with no bottlenecks or loops. The maximum flow is 90.

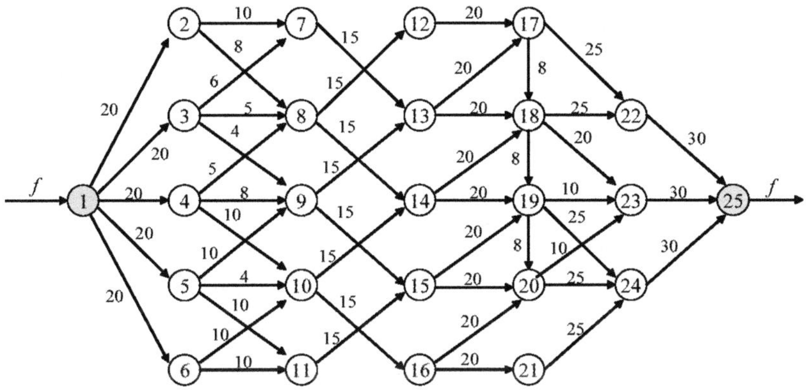

Fig. 2.41 First example of maximum flow model with 25 nodes and 49 arcs

The best chromosome using priority-based encoding is shown in Fig. 2.42 and the final result by overall-path growth decoding is shown in Fig. 2.43.

node ID :	1	2	3	4	5	6	7	8	9	10	11	12	13
priority :	8	25	2	12	15	20	1	16	21	14	7	6	18

node ID :	14	15	16	17	18	19	20	21	22	23	24	25
priority :	23	3	13	4	17	5	11	9	24	19	10	22

Fig. 2.42 The best priority-based chromosome

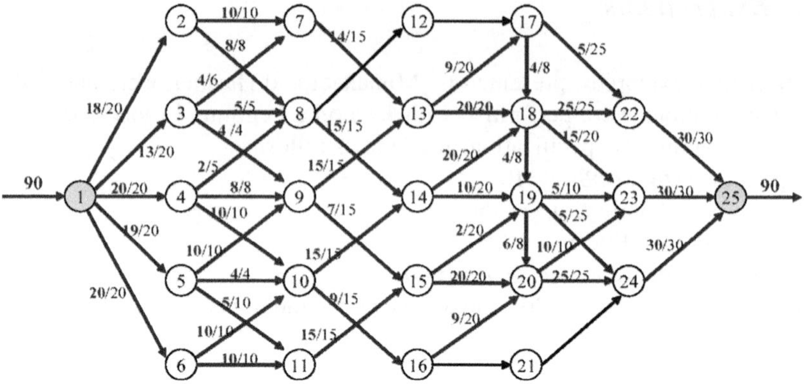

Fig. 2.43 Final result of maximum flow model with 25 nodes and 49 arcs

2.4.3.2 Test Problem 2

The second test problem is given in Fig. 2.44. This network contains 25 nodes and 56 arcs with loops but no bottlenecks. The maximum flow is 91.

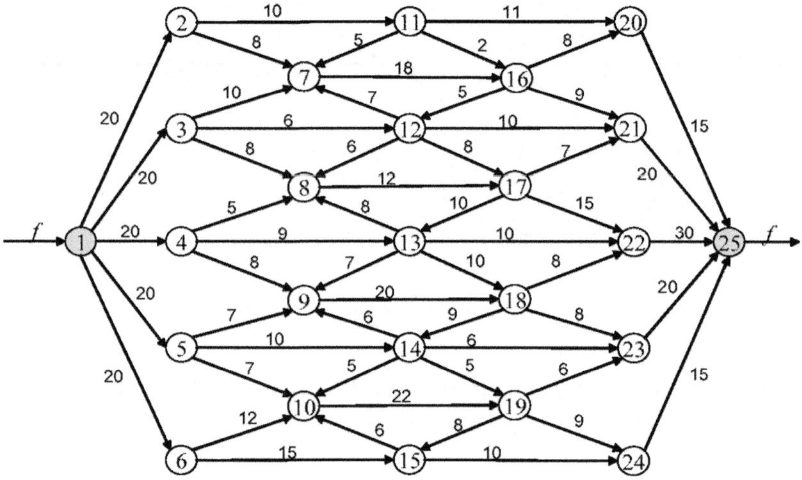

Fig. 2.44 Second example of maximum flow model with 25 nodes and 56 arcs

The best chromosome using priority-based encoding is shown in Fig. 2.45 and the final result by overall-path growth decoding is shown in Fig. 2.46.

node ID :	1	2	3	4	5	6	7	8	9	10	11	12	13
priority :	20	10	5	22	25	23	11	6	18	1	16	12	3

	14	15	16	17	18	19	20	21	22	23	24	25
	2	7	14	15	19	4	13	17	8	21	24	9

Fig. 2.45 The best priority-based chromosome

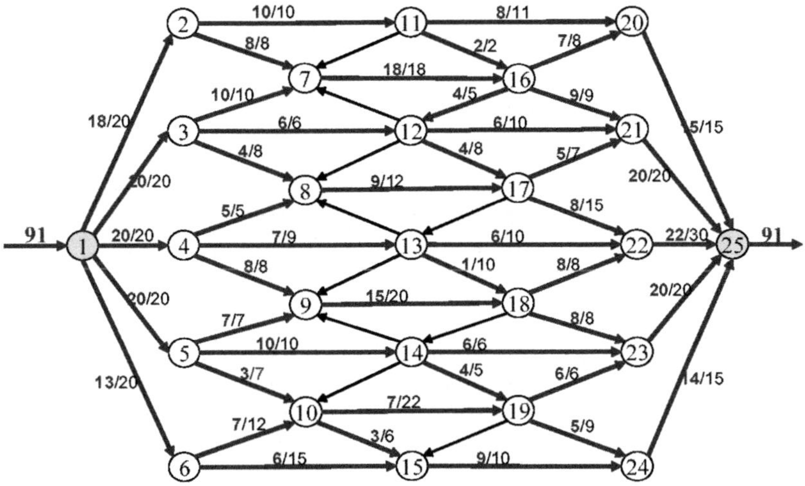

Fig. 2.46 Final result of maximum flow model with 25 nodes and 56 arcs

2.5 Minimum Cost Flow Model

The *minimum cost flow* (MCF) model is the most fundamental one of all network optimization problems. The *shortest path problem* (SPP) and *maximum flow problem* (MXF) can be formulated as two special cases of MCF: SPP considers arc flow costs but not flow capacities; MXF considers capacities but only the simplest cost structure. The MCF is finding the cheapest possible way of sending a certain amount of flow through a network.

Example 2.4: A Simple Example of MCF

A simple example of minimum cost flow problem is shown in Fig. 2.47 and Table 2.19 gives the data set of the example.

Traditional Methods

Successive shortest path algorithm: The successive shortest path algorithm maintains optimality of the solution at every step and strives to attain feasibility.

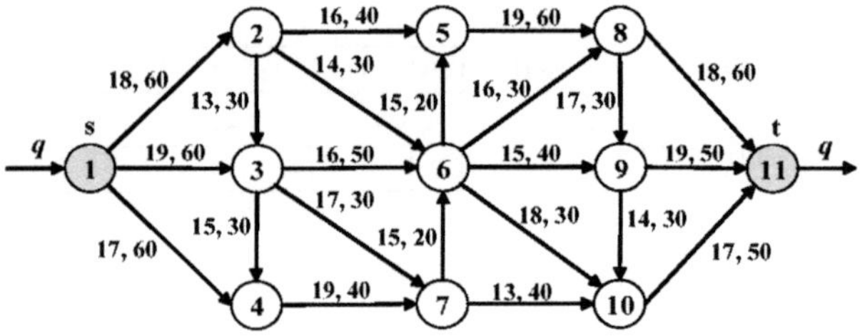

Fig. 2.47 A simple example of minimum cost flow model

Table 2.19 The data set of illustrative minimum cost network model

i	j	c_{ij}	u_{ij}	i	j	c_{ij}	u_{ij}
1	2	18	60	6	5	15	20
1	3	19	60	6	8	16	30
1	4	17	60	6	9	15	40
2	3	13	30	6	10	18	30
2	5	16	40	7	6	15	20
2	6	14	30	7	10	13	40
3	4	15	30	8	9	17	30
3	6	16	50	8	11	18	60
3	7	17	30	9	10	14	30
4	7	19	40	9	11	19	50
5	8	19	60	10	11	17	50

Primal-dual algorithm: The primal-dual algorithm for the minimum cost flow problem is similar to the successive shortest path algorithm in the sense that it also maintains a pseudo flow that satisfies the reduced cost optimality conditions and gradually converts it into a flow by augmenting flows along shortest paths.

Out-of-kilter algorithm: The out-of-kilter algorithm satisfies only the mass balance constraints, so intermediate solutions might violate both the *optimality conditions* and the *flow bound restrictions*.

2.5.1 Mathematical Formulation

Let G=(N,A) be a directed network, consisting of a finite set of nodes $N = \{1,2,\cdots,n\}$ and a set of directed arcs $A = \{(i,j),(k,l),\cdots,(s,t)\}$ joining m pairs of nodes in N. Arc (i,j) is said to be incident with nodes i and j, and is directed from node i to node j. Suppose that each arc (i,j) has assigned to it nonnegative numbers c_{ij}, the cost of (i,j) and u_{ij}, the capacity of (i,j). Associate this with an integer number q representing the supply/demand. The assumptions are given as follows:

A1. The network is directed. We can fulfil this assumption by transforming any undirected network into a directed network.

A2. All capacities, all arc costs and supply/demand q are nonnegative.
A3. The network does not contain a directed path from node s to node t composed only of infinite capacity arcs. Whenever every arc on a directed path P from node s to node t has infinite capacity, we can send an infinite amount of flow along this path, and therefore the maximum flow value is unbounded.
A4. The network does not contain parallel arcs (*i.e.*, two or more arcs with the same tail and head nodes).

Indices

$i, j, k = 1, 2, \cdots, n$ index of node

Parameters

n: number of nodes
q: supply/demand requirement
c_{ij}: unit cost of arc (i, j)
u_{ij}: capacity of arc (i, j)

Decision Variables

x_{ij}: the amount of flow through arc (i, j)

Minimum Cost Flow Model:

The minimum cost flow model is an optimization model formulated as follows:

$$\min z = \sum_{i=1}^{n} \sum_{j=1}^{n} c_{ij} x_{ij} \tag{2.29}$$

$$\text{s. t.} \quad \sum_{j=1}^{n} x_{ij} - \sum_{k=1}^{n} x_{ki} = \begin{cases} q & (i = 1) \\ 0 & (i = 2, 3, \cdots, n-1) \\ -q & (i = n) \end{cases} \tag{2.30}$$

$$0 \le x_{ij} \le u_{ij} \quad (i, j = 1, 2, \cdots, n) \tag{2.31}$$

where the constraint at Eq. 2.30, a conservation law is observed at each of the nodes other than s or t. That is, what goes out of node i, $\sum_{j=1}^{n} x_{ij}$must be equal to what comes in, $\sum_{k=1}^{n} x_{ki}$. Constraint at Eq. 2.31 is flow capacity. We call any set of numbers $\boldsymbol{x}=(x_{ij})$ which satisfy Eqs. 2.30 and 2.31 a feasible flow, and q is its value.

2.5.2 Priority-based GA for MCF Model

2.5.2.1 Genetic Representation

We have summarized the performance of the priority-based encoding method and other encoding methods for solving shortest path model in a previous section (page 62). In this section, MCF is the same special difficult with MXF that a solution of the MCF is presented by various numbers of paths. We adopt the priority-based chromosome; it is an effective representation for presenting a solution with various paths. The initialization process is same pseudocode of Fig. 2.7.

---

procedure 6 : overall - path growth
input : network data (V, A, C, U, q), chromosome v_k, the set of nodes S_i with all nodes adjacent to node i
output : number of paths L_k and the cost c_i^k of each path, $i \in L_k$
begin
  initialize $l \leftarrow 0$; // l : number of paths.
  while $f_l^k \geq q$ **do**
    $l \leftarrow l+1$;
    generate a path P_l^k by **procedure 4** (one - path growth);
    select the sink node a of path P_l^k;
    if $a = n$ **then**
      $f_l^k \leftarrow f_{l-1}^k + \min\{u_{ij} | (i,j) \in P_l^k\}$; // calculate the flow f_l^k of the path P_l^k.
      $c_l^k \leftarrow c_{l-1}^k + \sum_{i=1}^{n}\sum_{j=1}^{n} c_{ij}(f_l^k - f_{l-1}^k), \quad \forall (i,j) \in P_l^k$; // calculate the cost c_l^k of the path P_l^k.
      $\tilde{u}_{ij} \leftarrow u_{ij} - (f_l^k - f_{l-1}^k)$; // make a new flow capacity $\tilde{u}_{ij}$.
      $S_i \leftarrow S_i \setminus j, \quad (i,j) \in P_l^k \;\&\; \tilde{u}_{ij} = 0$; // delete the arcs which its capacity is 0.
    else
      $S_i \leftarrow S_i \setminus a, \quad \forall i$; // update the set of nodes S_i.
  output number of paths $L_k \leftarrow l-1$ and the cost $c_i^k, i \in L_k$.
end

Fig. 2.48 Pseudocode of overall-path growth for MCF

For the decoding process, firstly, we adopt a *one-path growth procedure* that presents same the priority-based decoding as MXF. It is given in a previous section (on page 103). Then we present an extended version of *overall-path growth procedure* that removes the used flow from each arc and deletes the arcs when its capacity is 0. Based on an updated network, we obtain a new path by one-path growth procedure, and repeat these steps until obtaining the feasible flow equal to q. The pseudocode of overall-path growth is shown in Fig. 2.48. By using the best chromosome for the MCF example which is shown in Fig. 2.49, we show a trace table of decoding process which is shown in Table 2.20 for a simple network (shown in Fig. 2.47). Where the supply/demand requirement is 60. The final result is shown in Fig. 2.50. The objective function value is

$$z = 20 \times 87 + 30 \times 69 + 10 \times 85 = 4660$$

node ID :	1	2	3	4	5	6	7	8	9	10	11
priority :	2	1	6	4	11	9	8	10	5	3	7

Fig. 2.49 The best chromosome for MCF example

Table 2.20 A trace table of decoding process for the MCF example

k	i	S_i	l	P_k	S_1	f_l^k	c_l^k
1	0			1			
	1	2, 3, 4	3	1, 3			
	2	4, 6, 7	6	1, 3, 6			
	3	5, 8, 9, 10	5	1, 3, 6, 5			
	4	8	8	1, 3, 6, 5, 8			
	5	9, 11	11	1, 3, 6, 5, 8, 11	2, 3, 4	20	87
2	0			1			
	1	2, 3, 4	3	1, 3			
	2	4, 6, 7	6	1, 3, 6			
	3	8, 9, 10	8	1, 3, 6, 8			
	4	9, 11	11	1, 3, 6, 8, 11	2, 3, 4	30	69
3	0			1			
	1	2, 3, 4	3	1, 3			
	2	4, 7	7	1, 3, 7			
	3	6, 10	6	1, 3, 7, 6			
	4	9, 10	9	1, 3, 7, 6, 9			
	5	10, 11	11	1, 3, 7, 6, 9, 11	2, 4	10	85

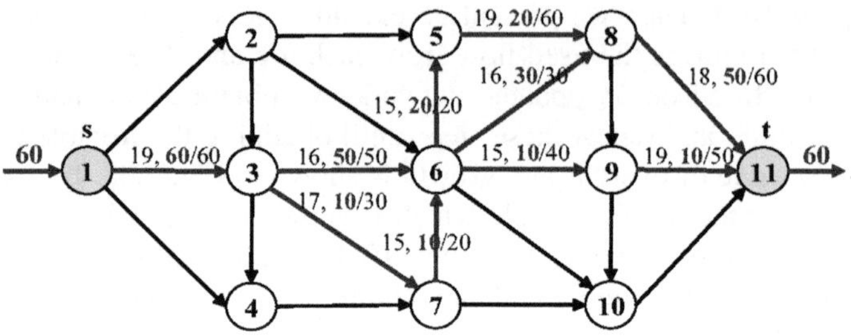

Fig. 2.50 Final result of the simple MCF example

2.5.2.2 Genetic Operators

Crossover Operator

In this section, we adopt the *weight mapping crossover* (WMX) proposed on page 68. WMX can be viewed as an extension of one-cut point crossover for permutation representation. As one-cut point crossover, two chromosomes (parents) would choose a random-cut point and generate the offspring by using segment of own parent to the left of the cut point, then remapping the right segment based on the weight of other parent of right segment.

Mutation Operator

As introduced on page 70, insertion mutation selects a gene at random and inserts it in another random position. We adopt this insertion mutation to generate offspring in this section.

The immigration operator is modified to (1) include the immigration routine, in each generation, (2) generate and (3) evaluate $popSize \cdot p_I$ random chromosomes, and (4) replace the $popSize \cdot p_I$ worst chromosomes of the current population (p_I, called the immigration probability). The detailed introduction is showed on page 23.

Selection Operator

We adopt the roulette wheel selection (RWS). It is to determine selection probability or survival probability for each chromosome proportional to the fitness value. A model of the roulette wheel can be made displaying these probabilities. The detailed introduction is showed on page 13.

2.5.3 Experiments

Two MCF test problems are generated based on the MXF network structure presented by [19], and randomly assigned unit shipping costs on each arc. All the simulations were performed with Java on a Pentium 4 processor (1.5-GHz clock). The parameter specifications are used as follows:

Population size: *popSize* =10
Crossover probability: p_C =0.50
Mutation probability: p_M =0.50
Maximum generation: *maxGen* =1000
Terminating condition: 100 generations with same fitness.

2.5.3.1 Test Problem 1

The first test problem is given in Fig. 2.51. This network contains 25 nodes and 49 arcs with no bottlenecks or loops. The supply/demand requirement is 70.

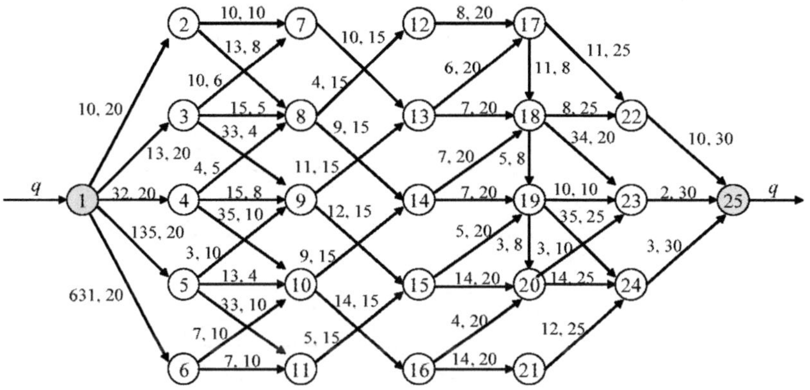

Fig. 2.51 First example of MCF model with 25 nodes and 49 arcs

The best chromosome by the priority-based encoding is shown in Fig. 2.52, the best result is 6969 and the final result by using overall-path growth decoding is shown in Fig. 2.53.

node ID :	1	2	3	4	5	6	7	8	9	10	11	12	13
priority :	1	16	11	9	6	5	7	8	15	10	3	12	13

node ID :	14	15	16	17	18	19	20	21	22	23	24	25
priority :	21	4	22	14	18	20	24	17	25	23	2	19

Fig. 2.52 The best priority-based chromosome

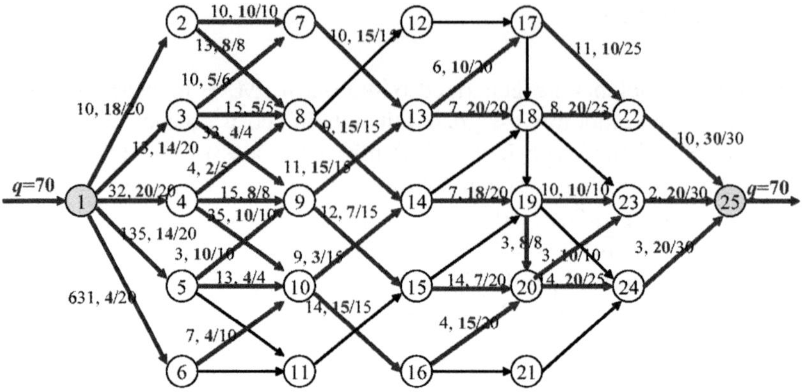

Fig. 2.53 Final result of first MCF example with 25 nodes and 49 arcs

2.5.3.2 Test Problem 2

The second test problem is given in Fig. 2.54. This network containeds 25 nodes and 56 arcs with loops but no bottlenecks. The supply/demand requirement is 72.

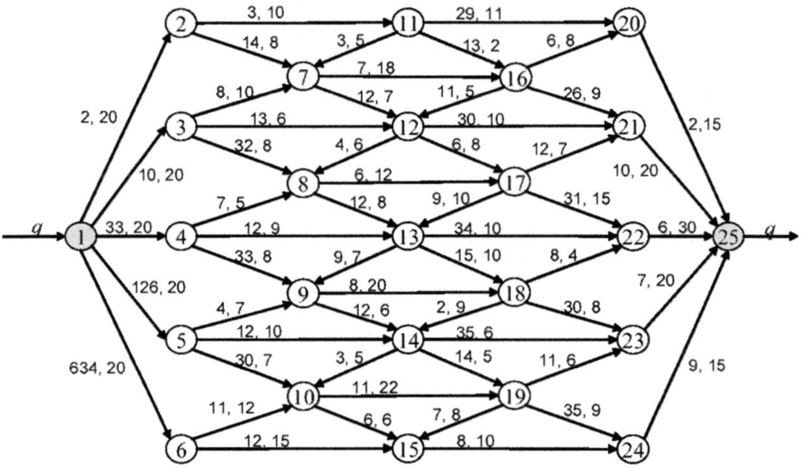

Fig. 2.54 Second example of minimum cost flow model with 25 nodes and 56 arcs

The best chromosome by the priority-based encoding is shown in Fig. 2.55, the best result is 5986 and the final result by using overall-path growth decoding is shown in Fig. 2.56.

Fig. 2.55 The best priority-based chromosome

node ID :	1	2	3	4	5	6	7	8	9	10	11	12	13
priority :	10	25	15	24	11	4	7	8	12	6	5	9	13

	14	15	16	17	18	19	20	21	22	23	24	25
	14	3	16	17	18	1	20	22	19	23	21	2

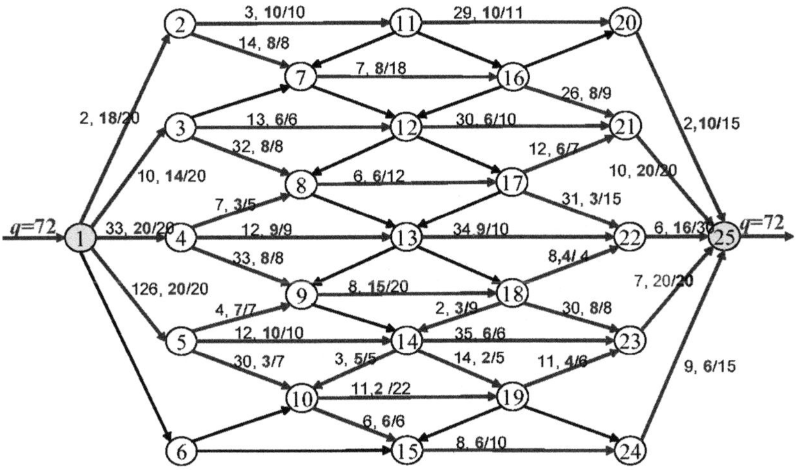

Fig. 2.56 Final result of second MCF example with 25 nodes and 56 arcs

2.6 Bicriteria MXF/MCF Model

Network design problems where even one cost measure must be minimized are often NP-hard [63]. However, in practical applications it is often the case that the network to be built is required to multiobjective. In the following, we introduce three core bicriteria network design models. (1) *Bicriteria shortest path* (bSP) model is one of the basic multi-criteria network design problems. It is desired to find a diameter-constrained path between two specified nodes with minimizing two cost functions. Hansen presented the first bSP model [64]. Recently, Skriver and Andersen examined the correlative algorithms for the bSP problems [65]; Azaron presented a new methodology to find the bicriteria shortest path under the steady-state condition [66]. (2) *Bicriteria spanning tree* (bST) model play a central role within the field of multi-criteria network modes. It is desired to find a subset of arcs which is a tree and connects all the nodes together with minimizing two cost functions. Marathe et al. presented a general class of bST model [67], and Balint proposed a non-approximation algorithm to minimize the diameter of a spanning sub-graph subject to the constraint that the total cost of the arcs does not exceed a given budget [68]. (3) *Bicriteria network flow* (bNF) model and bSP model are mutual complementary topics. It is desired to send as much flow as possible between two special nodes without exceeding the capacity of any arc. Lee and Pulat presented algorithm to solve a bNF problem with continuous variables [69].

As we know, the shortest path problem (SPP) considers arc flow costs but not flow capacities; the maximum flow (MXF) problem considers capacities but only the simplest cost structure. SPP and MXF combine all the basic ingredients of network design problems. Bicriteria MXF/MCF model is an *integrated bicriteria network design* (bND) model integrating these nuclear ingredients of SPP and MXF. This bND model considers the flow costs, flow capacities and multiobjective optimization. This bND model provide a useful way to model real world problems, which are extensively used in many different types of complex systems such as communication networks, manufacturing systems and logistics systems. For example, in a communication network, we want to find a set of links which consider the connecting cost (or delay) and the high throughput (or reliability) for increasing the network performance [70, 71]; as an example in the manufacturing application described in [72], the two criteria under consideration are cost, that we wish to minimize, and manufacturing yield, that we wish to maximize; in a logistics system, the main drive to improve logistics productivity is the enhancement of customer services and asset utilization through a significant reduction in order cycle time (lead time) and logistics costs [82].

Review of Related Work

This bND model is the most complex case of the multi-criteria network design models. The traditional bSP model and bST model only consider the node selection for a diameter-constrained path or spanning tree, and the traditional bNF model only considers the flow assignment for maximizing the total flows. But the bND model considers not only the node selection but also flow assignment for two conflicting objectives, maximizing the total flow and minimizing the total cost simultaneously. In network design models, even one measure of flow maximization is often NP-hard problem [58]. For example, a flow at each edge can be anywhere between zero and its flow capacity, *i.e.*, it has more "freedom" to choose. In many other problems, selecting an edge may mean simply to add fixed distance. It has been well studied using a variety of methods by parallel algorithm with a worst case time of $O(n^2 logn)$ (Shiloach and Vishkin), distributed algorithms with a worst case time of $O(n^2 logn)$ to $O(n^3)$ (Yeh and Munakata), and sequential algorithms *etc.*, with n nodes [58]. In addition, as one of multiobjective optimization problems, this bND problem is not simply an extension from single objective to two objectives. In general, we cannot get the optimal solution of the problem because these objectives conflict with each other in practice. The real solutions to the problem are a set of Pareto optimal solutions (Chankong and Haimes [52]). To solve this bND problem, the set of efficient solutions may be very large and possibly exponential in size. Thus the computational effort required to solve it can increase exponentially with the problem size.

Recently, genetic algorithm (GA) and other evolutionary algorithms (EAs) have been successfully applied in a wide variety of network design problems [73]. For example, Ahn and Ramakrishna developed a variable-length chromosomes and a new crossover operator for shortest path routing problem [12], Wu and Ruan proposed a gene-constrained GA for solving shortest path problem [74], Li *et al.* proposed a specific GA for *optimum path planning* in *intelligent transportation system* (ITS) [75], Kim *et al.* proposed a new path selection scheme which uses GA along with the modified roulette wheel selection method for *MultiProtocol label switching* (MPLS) network [76], Hasan *et al.* proposed a *novel heuristic GA* to solve the *single source shortest path* (ssSP) problem [77], Ji *et al.* developed a *simulation-based GA* to find multi-objective paths with minimizing both expected travel time and travel time variability in ITS [78], Chakraborty *et al.* developed *multiobjective genetic algorithm* (moGA) to find out simultaneously several alternate routes depending on distance, contains minimum number of turns, path passing through mountains [79], Garrozi and Araujo presented a moGA to solve the *multicast routing problem* with maximizing the common links in source-destination routes and minimizing the route sizes [80], and Kleeman *et al.* proposed a *modified nondominated sorting genetic algorithm II* (nsGA II) for the *multicommodity capacitated network design problem* (mcNDP), the multiple objectives including costs, delays, robustness, vulnerability, and reliability [81]. Network design problems where even one cost measure must be minimized are often NP-hard [63]. However, in practical applications, it is often the case that the network to be built is required to multiobjective. In the following, we introduce three core bicriteria network design models. (1) *Bicriteria shortest path* (bSP) model is one of the basic multi-criteria network design problems. It is desired to find a diameter-constrained path between two specified nodes with minimizing two cost functions. Hansen presented the first bSP model [64]. Recently, Skriver and Andersen examined the correlative algorithms for the bSP problems [65]; Azaron presented a new methodology to find the bicriteria shortest path under the steady-state condition [66]. (2) *Bicriteria spanning tree* (bST) model play a central role within the field of multi-criteria network modes. It is desired to find a subset of arcs which is a tree and connects all the nodes together with minimizing two cost functions. Marathe *et al.* presented a general class of bST model [67], and Balint proposed a non-approximation algorithm to minimize the diameter of a spanning sub-graph subject to the constraint that the total cost of the arcs does not exceed a given budget [68]. (3) *Bicriteria network flow* (bNF) model and bSP model are mutual complementary topics. It is desired to send as much flow as possible between two special nodes without exceeding the capacity of any arc. Lee and Pulat presented algorithm to solve a bNF problem with continuous variables [69].

As we know, shortest path problem (SPP) considers arc flow costs but not flow capacities; maximum flow (MXF) problem considers capacities but only the simplest cost structure. SPP and MXF combine all the basic ingredients of network design problems. In this chapter, we formulate an *integrated bicriteria network design* (bND) model with integrating these nuclear ingredients of SPP and MXF. This bND model considers the flow costs, flow capacities and multiobjective optimization. This bND model provide a useful way to modeling real world problems, which are extensively used in many different types of complex systems such as communication networks, manufacturing systems and logistics systems. For example, in a communication network, we want to find a set of links which consider the connecting cost (or delay) and the high throughput (or reliability) for increasing the network performance [70, 71]; as an example in the manufacturing application described in [72], the two criteria under consideration are cost, that we wish to minimize, and manufacturing yield, that we wish to maximize; in a logistics system, the main drive to improve logistics productivity is the enhancement of customer services and asset utilization through a significant reduction in order cycle time (lead time) and logistics costs [82].

2.6.1 Mathematical Formulations

Let G=(N,A) be a directed network, consisting of a finite set of nodes $N = 1, 2, \cdots, n$ and a set of directed arcs $A = (i, j), (k,l), \cdots, (s,t)$ joining m pairs of nodes in N. Arc (i, j) is said to be incident with nodes i and j, and is directed from node i to node j. Suppose that each arc (i, j) has assigned to it nonnegative numbers c_{ij}, the cost of (i, j) and u_{ij}, the capacity of (i, j). This capacity can be thought of as representing the maximum amount of some commodity that can "flow" through the arc per unit time in a steady-state situation. Such a flow is permitted only in the indicated direction of the arc, *i.e.*, from i to j.

Consider the problem of finding maximal flow and minimal cost from a source node s (node 1) to a sink node t (node n), which can be formulated as follows. Let x_{ij} = the amount of flow through arc (i, j). The assumptions are given as following:

A1. The network is directed. We can fulfil this assumption by transforming any undirected network into a directed network.
A2. All capacities and all arc costs are nonnegative.
A3. The network does not contain a directed path from node s to node t composed only of infinite capacity arcs. Whenever every arc on a directed path P from note s to note t has infinite capacity, we can send an infinite amount of flow along this path, and therefore the maximum flow value is unbounded.
A4. The network does not contain parallel arcs (*i.e.*, two or more arcs with the same tail and head nodes). This assumption is essentially a notational convenience.

Indices

$i, j, k = 1, 2, \cdots, n$ index of node

Parameters

c_{ij}: unit cost of arc (i, j)
u_{ij}: capacity of arc (i, j)

Decision Variables

x_{ij}: the amount of flow through arc (i, j)
f: the total flow

The bND problem is formulated as a mixed integer programming model, in which the objectives are maximizing total flow z_1 and minimizing total cost z_2 from source node 1 to sink node n as follows:

$$\max z_1 = f \tag{2.32}$$

$$\min z_2 = \sum_{i=1}^{n}\sum_{j=1}^{n} c_{ij}x_{ij} \tag{2.33}$$

$$\text{s. t. } \sum_{j=1}^{n} x_{ij} - \sum_{k=1}^{n} x_{ki} = \begin{cases} f \ (i=1) \\ 0 \ (i=2,3,\cdots,n-1) \\ -f \ (i=n) \end{cases} \tag{2.34}$$

$$0 \leq x_{ij} \leq u_{ij} \ \ \forall \ (i, j) \tag{2.35}$$

$$x_{ij} = 0 \text{ or } 1 \ \ \forall \ i, \ j \tag{2.36}$$

where the constraint at Eq. 2.34, a conservation law is observed at each of the nodes other than s or t. That is, what goes out of node i, $\sum_{j=1}^{n} x_{ij}$must be equal to what comes in, $\sum_{k=1}^{n} x_{ki}$. Constraint at Eq. 2.35 is flow capacity. We call any set of numbers $\boldsymbol{x}$=(x_{ij}) which satisfy Eqs. 2.34 and 2.35 a feasible flow, or simply a flow, and f is its value.

2.6.2 Priority-based GA for bMXF/MCF Model

The overall procedure of priority-based GA for solving bicriteria MXF/MCF model is outlined as follows.

2.6.2.1 Genetic Representation

To solve this bND problem, the special difficulties of the bND problem are (1) a solution is presented by various numbers of paths, (2) a path contains various numbers of nodes and the maximal number is n–1 for an n node network, and (3) a random sequence of edges usually does not correspond to a path. Thus a nature genetic representation is very difficult to use to represent a solution of bND problem.

We have summarized the performance of the priority-based encoding method and also introduced encoding methods for solving shortest path routing problems on page 62.

In this section, we consider the special difficulty of the bND problem in that a solution of the bND problem is presented by various numbers of paths. Figure

procedure: priGA for bMXF/MCF model
input: network data (N, A, C, U), GA parameters $(popSize, maxGen, p_M, p_C, p_I)$
output: Pareto optimal solutions E
begin
  $t \leftarrow 0$;
  initialize $P(t)$ by priority-based encoding routine;
  calculate objectives $z_i(P), i = 1, \cdots, q$ by priority-based decoding routine;
  create Pareto $E(P)$;
  evaluate $eval(P)$ by *interactive adaptive-weight fitness assignment routine*;
  while (**not** terminating condition) **do**
    create $C(t)$ from $P(t)$ by weight mapping crossover routine;
    create $C(t)$ from $P(t)$ by insertion mutation routine;
    create $C(t)$ from $P(t)$ by immigration routine;
    calculate objectives $z_i(C), i = 1, \cdots, q$ by priority-based decoding routine;
    update Pareto $E(P, C)$;
    evaluate $eval(P, C)$ by interactive adaptive-weight fitness assignment routine;
    select $P(t+1)$ from $P(t)$ and $C(t)$ by roulette wheel selection routine;
    $t \leftarrow t+1$;
  end
  output Pareto optimal solutions $E(P, C)$
end

2.57 presents a simple network with 11 nodes and 22 arcs and Table 2.21 shows 2 examples of the solutions of bND with various paths.

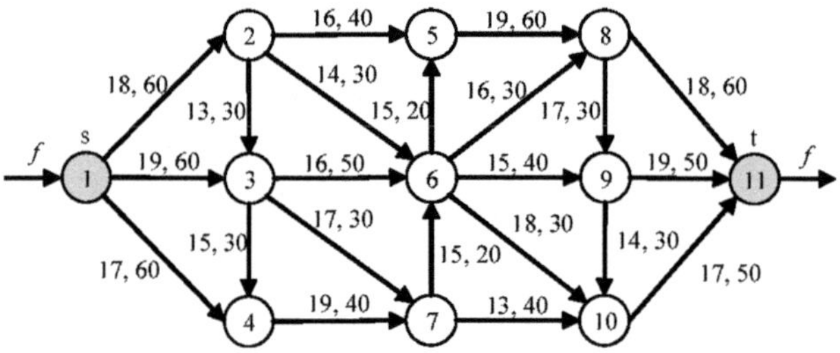

Fig. 2.57 A simple network with 11 nodes and 22 edges

Until now, to present a solution of bND problem with various paths, the general idea of chromosome design is adding several shortest path based-encodings to one chromosome. The length of these representations is variable depending on various paths, and most offspring are infeasible after crossover and mutation. Therefore these kinds of representations lose the feasibility and legality of chromosomes.

In this research, the priority-based chromosome is adopted; it is an effective representation for present a solution with various paths. For the decoding process, first,

Table 2.21 Examples of the solutions with various paths

Solution	# of paths K	Path $P(\cdot)$	Total flow z_1	Total cost z_2
1	3	(1-3-6-5-8-11), (1-3-6-8-11), (1-3-7-6-9-11)	60	4660
2	8	(1-3-6-5-8-11), (1-3-6-8-11), (1-3-7-6-9-11), (1-4-7-6-9-11), (1-4-7-10-11), (1-2-5-8-11), (1-2-5-8-9-11), (1-2-6-9-10-11)	160	12450

a *one-path growth procedure* is combined that obtains a path base on the generated chromosome with given network in Fig. 2.58.

Then present an *overall-path growth procedure* in Fig. 2.59 that removes the used flow from each arc and deletes the arcs where its capacity is 0. Based on updated network, a new path can be obtained by the one-path growth procedure, and repeat these steps until we obtain overall possible path. A trace table of the decoding process is shown in Table 2.22, for a simple network (shown in Fig. 2.57).

---

procedure: one-path growth
input: chromosome v, number of nodes n, the set of nodes S_i with all nodes adjacent to node i
output: path P_k, length of path P_k l_k
begin
  initialize $i \leftarrow 1, l_k \leftarrow 1, P_k[l] \leftarrow i$; // i: source node, l_k : length of path P_k.
  while $S_1 \neq \phi$ **do**
    $j' \leftarrow$ **argmax**$\{v[j],\ j \in S_i\}$; // j': the node with highest priority among S_i.
    if $v[j'] \neq 0$ **then**
      $P_k[l_k] \leftarrow j'$; // chosen node j' to construct path P_k.
      $l_k \leftarrow l_k + 1$;
      $v[j'] \leftarrow 0$;
      $i \leftarrow j'$;
    else
      $S_i \leftarrow S_i \setminus j'$; // delete the node j' adjacent to node i.
      $v[i] \leftarrow 0$;
      $l_k \leftarrow l_k - 1$;
      if $l_k \leq 1$ **then** $l_k \leftarrow 1$, **break**;
      $i \leftarrow P_k[l_k]$;
  output path P_k and path length l_k
end

Fig. 2.58 Pseudocode of one-path growth

```
procedure: overall-path growth for bicriteria MXF/MCF
input: network data (V, A, C, U), chromosome v_k[·],
       set of nodes S_i with all nodes adjacent to node i
output: no. of paths L_k, flow f_i^k and cost c_i^k of each path, i ∈ L_k
begin
  l ← 0;  // l: number of paths.
  while S_1 ≠ φ do
    l ← l + 1;
    generate a path P_l^k by procedure (one-path growth);
    select the sink node a of path P_l^k;
    if a = n then
      f_l^k ← f_{l-1}^k + min{u_ij | (i, j) ∈ P_l^k};  // calculate the flow f_l^k of the path P_l^k.
      c_l^k ← c_{l-1}^k + Σ_{i=1}^{n} Σ_{j=1}^{n} c_ij (f_l^k − f_{l-1}^k),  ∀(i, j) ∈ P_l^k;  // calculate the cost c_l^k of the path P_l^k.
      ũ_ij ← u_ij − (f_l^k − f_{l-1}^k);  // make a new flow capacity ũ_ij.
      S_i ← S_i \ j,  (i, j) ∈ P_l^k & ũ_ij = 0;  // delete the arcs which its capacity is 0.
    else
      S_i ← S_i \ a,  ∀i;  // update the set of nodes S_i.
  output no. of paths L_k ← l − 1, the flow f_i^k and the cost c_i^k, i ∈ L_k.
end
```

Fig. 2.59 Pseudocode of overall-path growth

2.6.2.2 Genetic Operators

Crossover Operator

In this section, we adopt the weight mapping crossover (WMX) proposed on page 68. WMX can be viewed as an extension of one-cut point crossover for permutation representation. With one-cut point crossover, two chromosomes (parents) would be chose a random-cut point and generate the offspring by using segment of one parent to the left of the cut point, then remapping the right segment based on the weight of the other parent of right segment.

Mutation Operator

As introduced on page 70, insertion mutation selects a gene at random and inserts it in another random position. We adopt this insertion mutation to generate offspring in this section.

The immigration operator is modified to (1) include immigration routine, in each generation, (2) generate and (3) evaluate $popSize \cdot p_{\mathrm{I}}$ random chromosomes, and (4)

Table 2.22 Examples of the solutions with various paths

k	i	S_i	j'	P_k	S_1	z_1^k	z_2^k
1	0			1			
	1	{2, 3, 4}	3	1-3			
	2	{4, 6, 7}	6	1-3-6			
	3	{5, 8, 9, 10}	5	1-3-6-5			
	4	{8}	8	1-3-6-5-8			
	5	{9, 11}	11	1-3-6-5-8-11	{2, 3, 4}	20	1740
2	0			1			
	1	{2, 3, 4}	3	1-3			
	2	{4, 6, 7}	6	1-3-6			
	3	{8, 9, 10}	8	1-3-6-8			
	4	{9, 11}	11	1-3-6-8-11	{2, 3, 4}	50	3810
3	...	...	...	1-3-7-6-9-11	{2, 4}	60	4660
4	...	...	...	1-4-7-6-9-11	{2, 4}	70	5510
5	...	...	...	1-4-7-10-11	{2}	100	7490
6	...	...	...	1-2-5-8-11	{2}	110	8200
7	...	...	...	1-2-5-8-9-11	{2}	140	10870
8	...	...	...	1-2-6-9-10-11	φ	160	12430

replace the $popSize \cdot p_{\mathrm{I}}$ worst chromosomes of the current population (p_{I}, called the immigration probability). The detailed introduction is shown on page 23.

Selection Operator

We adopt the roulette wheel selection (RWS). It is to determine selection probability or survival probability for each chromosome proportional to the fitness value. A model of the roulette wheel can be made displaying these probabilities. The detailed introduction is shown in page 13.

2.6.3 i-awGA for bMXF/MCF Model

As described above, a characteristic of the priority-based decoding method is its ability to generate various paths by one chromosome. The several non-dominated solutions can be combined by these paths. *i.e.*, one chromosome mapping various fitnesses. However, most of the fitness assignment methods can only be applied the case of one chromosome mapping one fitness assignment as the fitness assignment methods above (rwGA, awGA, spEA, nsGAII, *etc.*) cannot be adopted successfully to solve bND problems by using the priority-based encoding method.

To develop Pareto ranking-based fitness assignment for bND problems, two special difficulties are encountered. (1) For creating Pareto rank of each solution set,

Pareto ranking-based fitness assignment needs the process which sorts popSize (number of chromosomes) solution sets. It is a disadvantage for considering computation time. In this research, for solving bND problems, we need a more complex Pareto rank sorting process. Because of there are $\sum_{k=1}^{popSize} L_k$ solution sets in each generation, where L_k is number of solution set in k-th chromosome. The computational effort required to solve bND problem can increase with the problem size. (2) After created Pareto rank of each solution set, there are L_k Pareto ranks in k-th chromosome, and how to assign the fitness of each chromosome is another difficulty.

Different from Pareto ranking-based fitness assignment, weighted-sum based fitness assignment is to assign weights to each objective function and combine the weighted objectives into a single objective function. It is easier to calculate the weight-sum based fitness and the sorting process becomes unnecessary; thus it is effective for considering the computation time to solve the problems. In addition, another characteristic of weighted-sum approach is its use to adjust genetic search toward the Pareto frontier.

In this research, we propose an interactive adaptive-weight fitness assignment approach, which is an improved adaptive-weight fitness assignment approach with the consideration of the disadvantages of weighted-sum approach and Pareto ranking-based approach. We combine a penalty term to the fitness value for all dominated solutions. To solve bND problems, we first define two extreme points: the maximum extreme point $z^+ \leftarrow \{z_1^{\max}, z_2^{\max}\}$ and the minimum extreme point $z^- \leftarrow \{z_1^{\min}, z_2^{\min}\}$ in criteria space, where $z_1^{\max}$, $z_2^{\max}$, $z_1^{\min}$ and $z_1^{\min}$ are the maximum value and the minimum value for objective 1 (maximum total flow) and the reciprocal of objective 2 (minimum total cost) in the current population. Calculate the adaptive weight $w_1 = 1/(z_1^{\max} - z_1^{\min})$ for objective 1 and the adaptive weight $w_2 = 1/(z_2^{\max} - z_2^{\min})$ for objective 2. Afterwards, calculate the penalty term $p(v_k)$=0, if v_k is a nondominated solution in the nondominated set P. Otherwise $p(v_{k'})$=1 for dominated solution $v_{k'}$. Last, calculate the fitness value of each chromosome by combining the method as follows adopt roulette wheel selection as supplementary to the interactive adaptive-weight genetic algorithm (i-awGA):

$$eval(v_k) = 1/\left(\sum_{i=1}^{L_k} 1/\left(w_1\left(f_i^k - z_1^{\min}\right) + w_2\left(1/c_i^k - z_2^{\min}\right)\right) + p(x_i^k)\right)/L_k \qquad \forall\, k \in popSize$$

where x_i^k is the i-th solution set in chromosome v_k, f_i^k is the flow of x_i^k and c_i^k is the cost of x_i^k. We adopted the roulette wheel selection as the supplementary operator to this interactive adaptive-weight assignment approach.

2.6.4 Experiments and Discussion

In the experimental study, the performance comparisons of multiobjective GAs is demonstrated for solving bND problems by different fitness assignment approaches; there are Strength Pareto Evolutionary Algorithm (spEA), non-dominated sorting Genetic Algorithm II (nsGA II), random-weight Genetic Algorithm (rwGA) and interactive adaptive-weight Genetic Algorithm (i-awGA) proposed. Two maximum flow test problems are presented by [58], and randomly assigned unit shipping costs along each arc. In each GA approach, priority-based encoding was used, and WMX + Insertion + Immigration were used as genetic operators.

2.6.4.1 Performance Measures

In these experiments, the following performance measures are considered to evaluate the efficiency of the different fitness assignment approaches. The detailed explanations were introduced on page 41.

Number of obtained solutions $|S_j|$: Evaluate each solution set depending on the number of obtained solutions.
Ratio of nondominated solutions $R_{NDS}(S_j)$: This measure simply counts the number of solutions which are members of the Pareto-optimal set S^*.
Average distance $D1_R(S_j)$: Instead of finding whether a solution of S_j belongs to the set S^* or not, this measure finds an average distance of the solutions of S_j from S^*.

In order to make a large number of solutions in the reference set S^*, first we calculate the solution sets with special GA parameter settings and very long computation times for each approach used in comparison experiments; then we combine these solution sets to calculate the reference set S^*. Furthermore, we will combine small but reasonable GA parameter settings for comparison experiments. This ensures the effectiveness of the reference set S^*.

In this research, the following GA parameter settings are used in all of 4 fitness assignment approaches (spEA, nsGA II, rwGA and our i-awGA) for finding the reference solution set S^*: population size, *popSize* =50; crossover probability, p_C=0.90; mutation probability, p_M =0.90; immigration rate, p_I = 0.60; stopping criteria: evaluation of 100000 generations.

The number of the obtained reference solutions for two test problems is summarized in Table 2.23. We chose nondominated solutions as reference solutions from four solution sets of the four algorithms for each test problem. We show the obtained reference solution sets for the 25-node/49-arc test problem in Fig. 2.60a, 25-node/56-arc test problem in Fig. 2.60b, respectively. We can observe the existence of a clear tradeoff between the two objectives in each Figure We can also see that the obtained reference solution set for each test problem has a good distribution on the tradeoff front in the objective space.

Table 2.23 Number of obtained reference solutions and the their range width for each objective

Test Problems (# of nodes / # of arcs)	# of obtained solutions $\|S_j\|$	range width $W_{f_i}(S^*)$	
		$f_1(r)$	$f_2(r)$
25 / 49	69	85	19337
25 / 56	77	89	16048

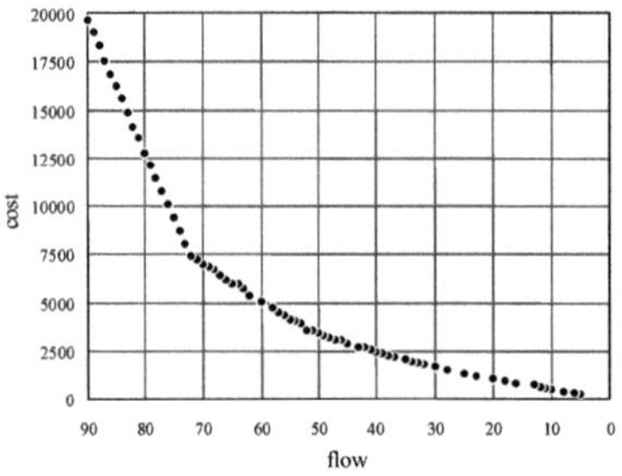

(a) 25-node/49-arc test problem

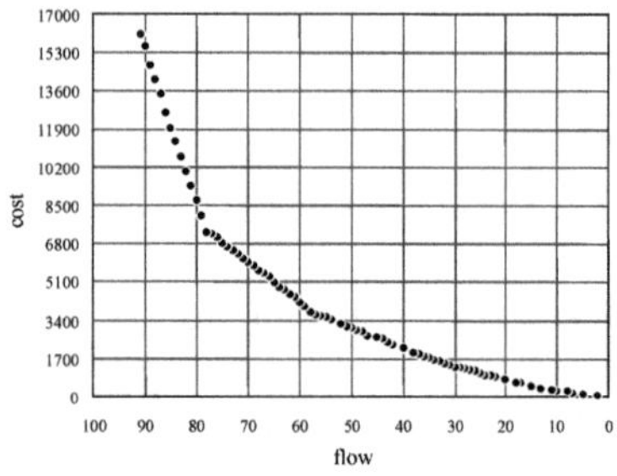

(b) 25-node/56 arc test problem

Fig. 2.60 Reference solutions obtained from the four GA approaches

The range width of the i-th objective over the reference solution set S^* is defined as

$$W_{f_i}(S^*) = \max\{f_i(r)|r \in S^*\} - \min\{f_i(r)|r \in S^*\}$$

2.6.4.2 Discussion of the Results

The i-awGA with spEA, nsGA II and rwGA are combined through computational experiments on the 25-node/49-arc and 25-node/56-arc test problems under the same GA parameter settings: population size, $popSize$ =20; crossover probability, p_C=0.70; mutation probability, p_M =0.70; immigration rate, p_I = 0.30; stopping condition, evaluation of 5000 solutions. Each simulation was run 50 times. In order to evaluate the results of each test, we consider the average results and standard deviation (SD) for three performance measures ($|S_j|$, $R_{NDS}(S_j)$ and $D1_R(S_j)$), and average of computation times (ACT) in millisecond (ms). Finally we give statistical analysis with the results of $R_{NDS}(S_j)$ and $D1_R(S_j)$ by ANOVA.

As depicted in Figures 2.61 and 2.62, the best results of $|S_j|$, $R_{NDS}(S_j)$ and $D1_R(S_j)$ are obtained by our i-awGA. As described above, the difference is in their fitness calculation mechanisms and the generation update mechanisms. While the spEA and nsGA II used fitness calculation mechanisms based on the Pareto-dominance relation and the concept of crowding, we need a more complex Pareto rank sorting process for solving bND problem; rwGA and i-awGA used a simple scalar fitness function. Figures 2.61 and 2.62 show the results of comparing computation time; i-awGA and rwGA are always faster than spEA and nsGA II. The selection process of rwGA is simplest; therefore the computation time is also fastest. However for combining weighted-sum based fitness assignment, assigning the effective weight for each objective function is difficult. As a result of computational experiments, the rwGA cannot obtain a better result than the three performance measures.

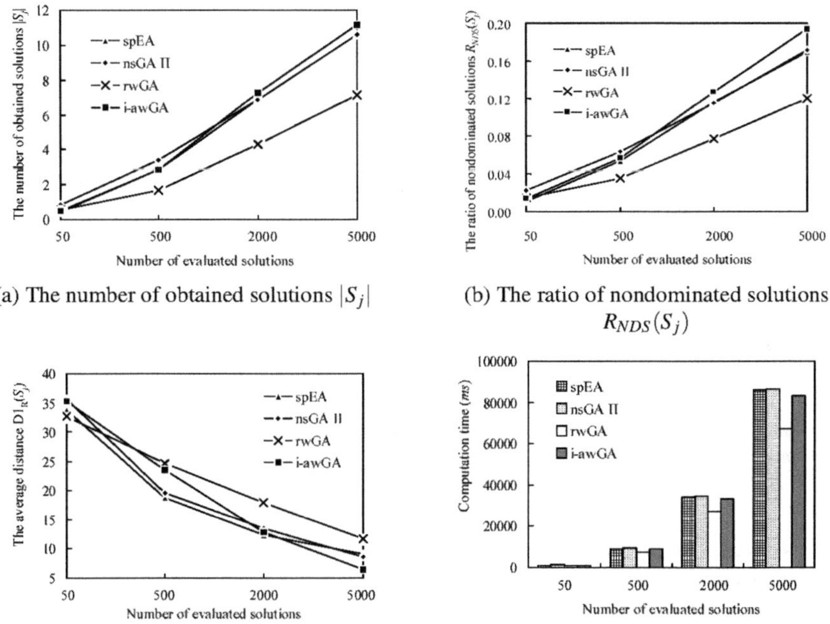

(a) The number of obtained solutions $|S_j|$ (b) The ratio of nondominated solutions $R_{NDS}(S_j)$

(c) The average distance $D1_R(S_j)$ (d) Computation time (ms)

Fig. 2.61 Performance evaluation of fitness assignment approaches using the performance measures for the 25-node/49-arc test problem

In Tables 2.24 and 2.25, ANOVA analysis is used depending on the average of $R_{NDS}(S_j)$ for 50 times with 2 test problems to analyze the difference among the quality of solutions obtained by various 4 kinds of different fitness assignment approaches. At the significant level of α=0.05, the F=21.99 and 16.59 is grater than the reference value of (F=2.68), respectively. The difference between our i-awGA and each of the other approaches (spEA, nsGA II or rwGA) is greater than the LSD=0.019 and 0.02, respectively. Similarly with ANOVA analysis depending on the average of $R_{NDS}(S_j)$, as depicted in Tables 2.26 and 2.27, we use ANOVA analysis depended on the average of $D1_R(S_j)$ for 50 times with 2 test problems to analyze the difference among the quality of solutions obtained by four kinds of different fitness assignment approaches. At the significant level of α=0.05, the F=11.22 and

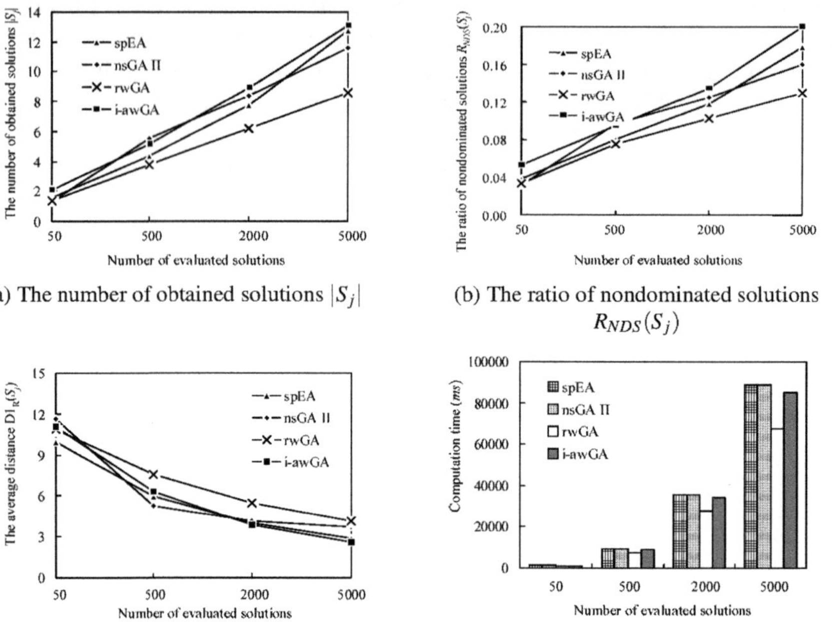

(a) The number of obtained solutions $|S_j|$

(b) The ratio of nondominated solutions $R_{NDS}(S_j)$

(c) The average distance $D1_R(S_j)$

(d) Computation time (ms)

Fig. 2.62 Performance evaluation of fitness assignment approaches using the performance measures for the 25-node/56-arc test problem

8.38 is greater than the reference value of (F=2.68), respectively. The difference between i-awGA and each of the other approaches (spEA, nsGA II or rwGA) is greater than the *LSD*=1.18 and 0.70, respectively. It is shown that i-awGA is indeed statistically better than the other approaches.

2.7 Summary

In this chapter, we introduced some of the core models of network design, such as shortest path models (*i.e.*, node selection and sequencing), spanning tree models (*i.e.*, arc selection) and maximum flow models (*i.e.*, arc selection and flow assignment) etc. These network models are used most extensively in applications and differentiated between with their structural characteristics.

To solve the different type models of network design, we introduced the performance analysis of different encoding methods in GAs. Furthermore we gave a priority-based encoding method with special decoding method for different models, and a predecessor-based encoding depended on an underlying random spanning-tree algorithm for different types of minimum spanning tree models. The effectiveness

Table 2.24 ANOVA analysis with the average of $R_{NDS}(S_j)$ in 25-node/49-arc test problem

	spEA	nsGA II	rwGA	i-awGA
# of data	50	50	50	50
Mean	0.17	0.17	0.12	0.19
SD	0.05	0.05	0.04	0.05
Variance	0.00	0.00	0.00	0.00
Sum of squares	0.10	0.13	0.08	0.13
Factors	Sum of squares	Freedom degree	Mean square	F
Between groups	0.15	3	0.05	21.99
Within-groups	0.43	196	0.00	
Total	0.58	199		
F ($\alpha = 0.05$)	2.68			
t ($\alpha = 0.05$)	1.98			
LSD	0.019			
Mean Difference with i-awGA	0.02	0.02	0.07	

Table 2.25 ANOVA analysis with the average of $R_{NDS}(S_j)$ in 25-node/56-arc test problem

	spEA	nsGA II	rwGA	i-awGA
# of data	50	50	50	50
Mean	0.18	0.16	0.13	0.20
SD	0.05	0.05	0.04	0.06
Variance	0.00	0.00	0.00	0.00
Sum of squares	0.13	0.14	0.09	0.16
Factors	Sum of squares	Freedom degree	Mean square	F
Between groups	0.13	3	0.04	16.59
Within-groups	0.52	196	0.00	
Total	0.66	199		
F ($\alpha = 0.05$)	2.68			
t ($\alpha = 0.05$)	1.98			
LSD	0.02			
Mean Difference with i-awGA	0.02	0.04	0.07	

and efficiency of GA approaches were investigated with various scales of network problems by comparing with recent related researches.

Table 2.26 ANOVA analysis with the average of $D1_R(S_j)$ in 25-node/49-arc test problem

	spEA	nsGA II	rwGA	i-awGA
# of data	50	50	50	50
Mean	9.03	8.59	11.67	6.48
SD	5.17	4.04	4.94	3.43
Variance	26.70	16.33	24.37	11.79
Sum of squares	1335.01	816.55	1218.41	589.42
Factors	Sum of squares	Freedom degree	Mean square	F
Between groups	679.97	3	226.66	11.22
Within-groups	3959.39	196	20.20	
Total	4639.36	199		
F ($\alpha = 0.05$)	2.68			
t ($\alpha = 0.05$)	1.98			
LSD	1.78			
Mean Difference with i-awGA	2.55	2.10	5.18	

Table 2.27 ANOVA analysis with the average of $D1_R(S_j)$ in 25-node/56-arc test problem

	spEA	nsGA II	rwGA	i-awGA
# of data	50	50	50	50
Mean	2.84	3.73	4.16	2.63
SD	1.78	1.98	1.90	1.28
Variance	3.17	3.91	3.60	1.64
Sum of squares	158.42	195.28	179.78	82.25
Factors	Sum of squares	Freedom degree	Mean square	F
Between groups	79.02	3	26.34	8.38
Within-groups	615.73	196	3.14	
Total	694.75	199		
F ($\alpha = 0.05$)	2.68			
t ($\alpha = 0.05$)	1.98			
LSD	0.70			
Mean Difference with i-awGA	0.22	1.11	1.53	

References

1. Ahuj, R. K., Magnanti, T. L. & Orlin, J. B. (1993). *Network Flows*, New Jersey: Prentice Hall.
2. Bertsekas, D. P. (1998). *Network Optimization: Continuous and Discrete Models*, Massachusetts: Athena Scientific.
3. Bauer, F.L. & Wossner, H. (1982). *Algorithmic Language and Program Development*,Berlin: Springer.
4. Wikipedia [Online] http://en.wikipedia.org/wiki/Main_Page

5. Cook, S. (1971). The complexity of theorem-proving procedures, *Proceeding of the 3rd Annual ACM Symposium on Theory of Computing*, 151–158.
6. Climaco, J. C. N., Jose M. F. Craveirinha. & Pascoal, M. B. (2003). A bicriterion approach for routing problems in multimedia networks, *Networks*, 41(4), 206–220.
7. Gen, M. & Cheng, R. (1997). *Genetic Algorithms and Engineering Design*. New York, John Wiley & Sons.
8. Gen, M. & Cheng, R. (2000). *Genetic Algorithms and Engineering Optimization*. New York, John Wiley & Sons.
9. Gen, M. & Cheng, R. (2003). Evolutionary network design: hybrid genetic algorithms approach. *International Journal of Computational Intelligence and Applications*, 3(4), 357–380.
10. Munemoto, M., Takai, Y. & Sate, Y. (1997). An adaptive network routing algorithm employing path genetic operators, *Proceeding of the 7th International Conferece on Genetic Algorithms*, 643–649.
11. Inagaki, J., Haseyama, M. & Kitajim, H. (1999). A genetic algorithm for determining multiple routes and its applications. *Proceeding of IEEE International Symposium. Circuits and Systems*, 137–140.
12. Ahn, C. W. & Ramakrishna, R. (2002). A genetic algorithm for shortest path routing problem and the sizing of populations. *IEEE Transacion on Evolutionary Computation.*, 6(6), 566–579.
13. Gen, M., Cheng, R. & Wand,D. (1997). Genetic algorithms for solving shortest path problems. *Proceeding of IEEE International Conference on Evolutionary Computer*, 401–406.
14. Gen, M. & Lin, L. (2005). Priority-based genetic algorithm for shortest path routing problem in OSPF. *Proceeding of Genetic & Evolutionary Computation Conference.*
15. Bazaraa, M., Jarvis, J. & Sherali, H. (1990). *Linear Programming and Network Flows*, 2nd ed., New York, John Wiley & Sons.
16. Mostafa, M. E. & Eid, S. M. A. (2000). A genetic algorithm for joint optimization of capacity and flow assignment in Packet Switched Networks. *Proceeding of 17th National Radio Science Conference*, C5–1–C5–6.
17. Shimamoto, N., Hiramatsu, A. & Yamasaki, K. (1993). A Dynamic Routing Control Based On a Genetic Algorithm. *Proceeding of IEEE International Conference Neural Networks*, 1123–1128.
18. Michael, C. M., Stewart, C. V. & Kelly, R. B. (1991). Reducing the Search Time of A Steady State Genetic Algorithm Using the Immigration Operator. *Proceeding IEEE International Conference Tools for AI*, 500–501.
19. Goldberg, D. & Lingle, R., Alleles. (1985). logic and the traveling salesman problem. *Proceeding of the 1st Internatinal Conference on GA*, 154–159.
20. Davis, L. (1995). Applying adaptive algorithms to domains. *Proceeding of the International Joint Conference on Artificial Intelligence*, 1, 162–164.
21. Syswerda, G. (1990). Schedule Optimization Using Genetic Algorithms, *Handbook of Genetic Algorithms*, Van Nostran Reinhold, New York.
22. Cheng, R. & Gen, M. (1994). Evolution program for resource constrained project scheduling problem, *Proceedings of IEEE International Conference of Evolutionary Computation*, 736-741.
23. Gen, M., Cheng, R. & Wang, D. (1997). Genetic algorithms for solving shortest path problems, *Proceedings of IEEE International Conference of Evolutionary Computation*, pp. 401-406.
24. Lin, L. & Gen, M. (2007). Bicriteria network design problem using interactive adaptive-weight GA and priority-based encoding method. *IEEE Transactions on Evolutionary Computation* in Reviewing.
25. Lindman, H. R. (1974). *Analysis of variance in complex experimental designs*. San Francisco: W. H. Freeman & Co.,.
26. Thomas, H., Charles, E. & Rivest, L. (1996) *Introduction to Algorithms*, MIT Press.
27. Gen, M., Cheng, R. & Oren, S. S. (1991). Network design techniques using adapted genetic algorithms. *Advances in Engineering Software*,32(9), 731–744.

28. Gen, M., Kumar, A. & Kim, R. (2005). Recent network design techniques using evolutionary algorithms. *International Journal Production Economics*, 98(2), 251–261.
29. Gen, M. (2006). *Study on Evolutionary Network Design by Multiobjective Hybrid Genetic Algorithm*. Ph.D dissertation, Kyoto University, Japan.
30. Abuali, F., Wainwright, R. & Schoenefeld, D. (1995). Determinant factorization: a new encoding scheme for spanning trees applied to the probabilistic minimum spanning tree problem. *Proceeding of 6th International Conference Genetic Algorithms*, 470–475.
31. Zhou, G. & Gen, M. (1998). An effective genetic algorithm approach to the quadratic minimum spanning tree problem. *Computers and Operations Research*, 25(3), 229–247.
32. Zhou, G. & Gen, M. (1998). Leaf-constrained spanning tree problem with genetic algorithms approach. *Beijing Mathematics*, 7(2), 50–66.
33. Fernandes, L. M. & Gouveia, L. (1998). Minimal spanning trees with a constraint on the number of leaves. *European Journal of Operational Research*, 104, 250–261.
34. Raidl, G. & Drexe, C. (2000). A predecessor coding in an evolutionary algorithm for the capacitated minimum spanning tree problem. *Proceeding of Genetic and Evolutionary Computation conference*, 309–316.
35. Ahuja, R. & Orlin, J. (2001). Multi-exchange neighborhood structures for the capacitated minimum spanning tree problem. *Mathematical Programming*, 91, 71–97.
36. Zhou, G. & Gen, M. (1997). Approach to degree-constrained minimum spanning tree problem using genetic algorithm. *Engineering Design and Automation*, 3(2), 157–165.
37. Zhou, G. & Gen, M. (1997). A note on genetic algorithm approach to the degree-constrained spanning tree problems. *Networks*,30, 105–109.
38. Raidl, G. (2000). A weighted coding in a genetic algorithm for the degree-constrained minimum spanning tree problem. *Proceedings Symposium on Applied*, 1, 440–445.
39. Raidl, G. (2000). An efficient evolutionary algorithm for the degree-constrained minimum spanning tree problem. *Proceedings Congress on Evolutionary Computation*, 1, 104–111.
40. Knowles, J. & Corne, D. (2000). A new evolutionary approach to the degree-constrained minimum spanning tree problem. *IEEE Transaction Evolutionary Computation*, 4(2), 125–134.
41. Chou, H., Premkumar, G. & Chu, C. (2001). Genetic algorithms for communications network design - an empirical study of the factors that influence performance. *IEEE Transaction Evolutionary Computation*, 5(3), 236–249.
42. Raidl, G. R. & Julstrom, B. A. (2003). Edge Sets: An Effective Evolutionary Coding of Spanning Trees. *IEEE Transaction on Evolutionary Computation*, 7(3), 225–239.
43. Raidl, G. R. & Julstrom, B. (2003). Greedy heuristics and an evolutionary algorithm for the bounded-diameter minimum spanning tree problem. *Proceeding SAC*, 747–752.
44. Julstrom, B. (2003). A permutation-coded evolutionary algorithm for the bounded-diameter minimum spanning tree problem. *Proceeding of Genetic and Evolutionary Computation Conference*, 2–7.
45. Zhou, G. & Gen, M. (1999). Genetic algorithm approach on multi-criteria minimum spanning tree problem. *European Journal of Operational Research*, 114(1), 141–152.
46. Lo, C. & Chang, W. (2000). A multiobjective hybrid genetic algorithm for the capacitated multipoint network design problem. *IEEE Transaction on Systems, Man and Cybernestic*, 30(3).
47. Neumann, F. (2004). Expected runtimes of a simple evolutionary algorithm for the multi-objective minimum spanning tree problem. *Proceeding 8th International Conference of Parallel Problem Solving from Nature*, 80–89.
48. Julstrom, B. & Raidl, G. (2001). Weight-biased edge-crossover in evolutionary algorithms for two graph problems. *Proceeding the 16th ACM Symposium on Applied Computing*, 321–326.
49. Schindler, B., Rothlauf, F. & Pesch, H. (2002). Evolution strategies, network random keys, and the one-max tree problem. *Proceeding Application of Evolutionary Computing on EvoWorkshops*, 143–152.
50. Rothlauf, F., Goldberg,D. & Heinz, A. (2002). Network random keys a tree network representation scheme for genetic and evolutionary algorithms. *Evolutionary Computation*, 10(1), 75–97.

51. Rothlauf, F., Gerstacker, J. & Heinzl, A. (2003). On the optimal communication spanning tree problem. *IlliGAL Technical Report*, University of Illinois.
52. Chankong, V. & Haimes, Y. Y. (1983). *Multiobjective Decision Making Theory and Methodology*. North-Holland, New York.
53. Davis, L., Orvosh, D., Cox, A. & Y. Qiu. (1993). A genetic algorithm for survivable network design. *Proceeding 5th International Conference Genetic Algorithms*, 408–415.
54. Piggott, P. l. & Suraweera, F. (1995). Encoding graphs for genetic algorithms: an investigation using the minimum spanning tree problem. *Progress in Evolutionary Computation.*, X. Yao, Ed. Springer, New York, 956, LNAI, 305–314.
55. Bean, J. C. (1994). Genetic algorithms and random keys for sequencing and optimization. *ORSA J. Computing*, 6(2), 154–160.
56. Cayley, A. (1889). A theorem on tree. *Quarterly Journal of Mathematics & Physical sciences*, 23, 376–378.
57. Prüfer, H. (1918). Neuer bewis eines Satzes über Permutationnen. *Archives of Mathematical Physica*, 27, 742–744.
58. Munakata, T. & Hashier, D. J. (1993). A genetic algorithm applied to the maximum flow problem *Proceeding 5th Internatioanl Conference Genetic Algorithms*, 488–493.
59. Akiyama, T., Kotani, Y. & Suzuki, T. (2000). The optimal transport safety planning with accident estimation process. *Proceeding of the 2nd international Conference on Traffic and Transportation Studies*, 99–106.
60. Ericsson, M., Resende, M.G.C. & Pardalos, P.M. (2001). A genetic algorithm for the weight setting problem in OSPF routing. *Journal of Combinatorial Optimization*, 6:299–333.
61. OR-Library. [Online]. Available: http://people.brunel.ac.uk/mastjjb/jeb/info.html
62. Beasley, J. E., & Christofides, N. (1989). An algorithm for the resource constrained shortest path problem. *Networks*, 19, 379-394.
63. Garey, M. R. & Johnson, D. S. *Computers and Intractability: a guide to the theory of NP-completeness*, W. H. Freeman, San Francisco, 1979.
64. Hansen, P. (1979). Bicriterion path problems. *Proceeding 3rd Conference Multiple Criteria Decision Making Theory and Application*, 109–127.
65. Skriver, A. J. V. & Andersen, K. A. (2000). A label correcting approach for solving bicriterion shortest-path problems. *Computers & Operations Research*, 27(6) 507–524.
66. Azaron, A. (2006). Bicriteria shortest path in networks of queues. *Applied Mathematics & Comput*, 182(1) 434–442.
67. Marathe, M. V., Ravi, R., Sundaram R., Ravi S. S., Rosenkrantz, D. J. & Hunt, H. B. (1998). Bicriteria network design problems.*Journal of Algorithms*, 28(1) 142–171.
68. Balint, V. (2003). The non-approximability of bicriteria network design problems. *Journal of Discrete Algorithms*, 1(3,4) 339–355.
69. Lee, H. & Pulat, P. S. (1991). Bicriteria network flow problems: continuous case. *European Journal of Operational Research*, 51(1) 119–126.
70. Yuan, D. (2003). A bicriteria optimization approach for robust OSPF routing. *Proceeding IEEE IP Operations & Management*, 91–98.
71. Yang, H., Maier, M., Reisslein, M. & Carlyle, W. M. (2003). A genetic algorithm-based methodology for optimizing multiservice convergence in a metro WDM network. J. Lightwave Technol., 21(5) 1114–1133.
72. Raghavan, S., Ball, M. O. & Trichur, V. (2002). Bicriteria product design optimization: an efficient solution procedure using AND/OR trees. *Naval Research Logistics,* 49, 574–599.
73. Gen, M., Cheng, R. & Oren, S.S. (2001). Network Design Techniques using Adapted Genetic Algorithms. *Advances in Engineering Software*, 32(9), 731–744.
74. Wu, W. & Ruan, Q. (2004). A gene-constrained genetic algorithm for solving shortest path problem. *Proceeding 7th International Conference Signal Processing*, 3, 2510–2513.
75. Li, Q., Liu, G., Zhang, W., Zhao, C., Yin Y. & Wang, Z. (2006). A specific genetic algorithm for optimum path planning in intelligent transportation system. *Proceeding 6th International Conference ITS Telecom*,140–143.

76. Kim, S. W., Youn, H. Y., Choi, S. J. & Sung, N. B. (2007). GAPS: The genetic algorithm-based path selection scheme for MPLS network. *Proceeding of IEEE International Conference on Information Reuse & Integration*, 570–575.
77. Hasan, B. S., Khamees, M. A. & Mahmoud, A. S. H. (2007). A heuristic genetic algorithm for the single source shortest path problem. *Proceeding IEEE/ACS International Conference on Computer Systems & Applications*, 187–194. d
78. Ji, Z., Chen, A. & Subprasom, K. (2004). Finding multi-objective paths in stochastic networks: a simulation-based genetic algorithm approach. *Proceedings of IEEE Congress on Evolutionary Computation*, 1, 174–180.
79. Chakraborty, B., Maeda, T. & Chakraborty, G. (2005). Multiobjective route selection for car navigation system using genetic algorithm. *Proceeding of IEEE Systems, Man & Cybernetics Society*, 190–195.
80. Garrozi, C. & Araujo, A. F. R. (2006). Multiobjective genetic algorithm for multicast routing. *Proceeding IEEE Congress on Evolutionary Computation*, 2513–2520.
81. Kleeman, M. P.,Lamont, G. B., Hopkinson, K. M. & Graham, S. R. (2007). Solving multicommodity capacitated network design problems using a multiobjective evolutionary algorithm.*Proceeding IEEE Computational Intelligence in Security & Defense Applications*, 33–41.
82. Zhou, G., Min, H. & Gen, M. (2003). A genetic algorithm approach to the bi-criteria allocation of customers to warehouses. *International Journal of Production Economics*, 86, 35–45.

Chapter 3
Logistics Network Models

With the development of economic globalization and extension of worldwide electronic marketing, global enterprise services supported by universal supply chain and world-wide logistics become imperative for the business world. How to manage logistics system efficiently thas hus become a key issue for almost all of the enterprises to reduce their various costs in today's keenly competitive environment of business, especially for many multinational companies. Today's pervasive internet and full-fledged computer aided decision supporting systems (DSS) certainly provide an exciting opportunity to improve the efficiency of the logistics systems. A great mass of research has been done in the last few decades. However, weltering in giving perfect mathematical representations and enamored with developing various type of over-intricate techniques in solution methods, most researchers have neglected some practical features of logistics.

In this chapter, the logistics network models are introduced, consolidating different aspects in practical logistics system. A complete logistics system covers the entire process of shipping raw materials and input requirements from suppliers to plants, the conversion of the inputs into products at certain plants, the transportation of the products to various warehouse of facilities, and the eventual delivery of these products to the final customers. To manage the logistics system efficiently, the dynamic and static states of material flows – transportation and storage – are key points that we need to take into consideration.

3.1 Introduction

A supply chain is a set of facilities, supplies, customers, products and methods of controlling inventory, purchasing, and distribution. The chain links suppliers and customers, beginning with the production of raw material by a supplier, and ending with the consumption of a product by the customer. In a supply chain, the flow of goods between a supplier and customer passes through several stages, and each stage may consist of many facilities [1]. In recent years, the *supply chain network* (SCN) design problem has been gaining importance due to increasing competitiveness introduced by market globalization [2]. Firms are obliged to maintain high customer service levels while at the same time they are forced to reduce cost and maintain profit margins. Traditionally, marketing, distribution, planning, manufacturing, and purchasing organizations along the supply chain operated independently. These organizations have their own objectives and these are often conflicting. But, there is a need for a mechanism through which these different functions can be integrated together. *Supply chain management* (SCM) is a strategy through which such integration can be achieved. Illustration of a supply chain network is shown in Fig. 3.1.

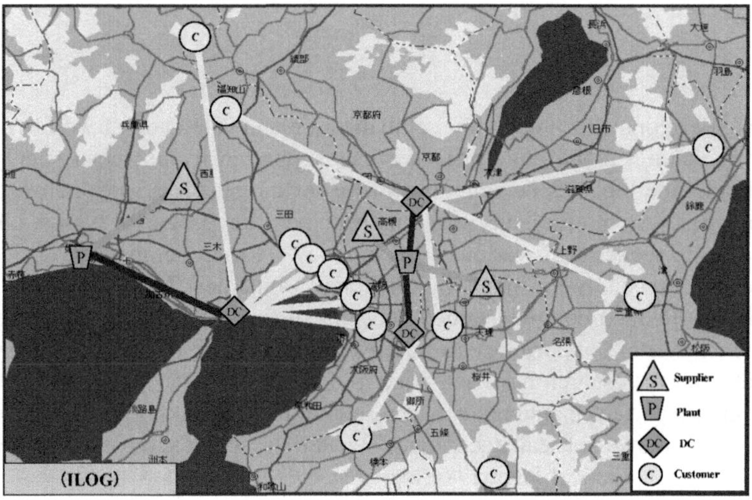

Fig. 3.1 Illustration of a supply chain network (ILOG)

The *logistics network design* is one of the most comprehensive strategic decision problems that need to be optimized for long-term efficient operation of whole supply chain. It determines the number, location, capacity and type of plants, warehouses and distribution centers (DCs) to be used. It also establishes distribution channels, and the amount of materials and items to consume, produce and ship from suppliers to customers. The *logistics network models* cover a wide range of formulations ranging from simple single product type to complex multi-product ones, and from linear deterministic models to complex non-linear stochastic ones. In the literature there are different studies dealing with the design problem of logistics networks and these studies have been surveyed by Vidal and Goetschalckx [3], Beamon [4], Erenguc *et al.* [5], and Pontrandolfo and Okogbaa [6]. Illustration of a simple network of three stages in supply chain network is shown in Fig. 3.2.

An important component in logistics network design and analysis is the establishment of appropriate performance measures. A performance measure, or a set of

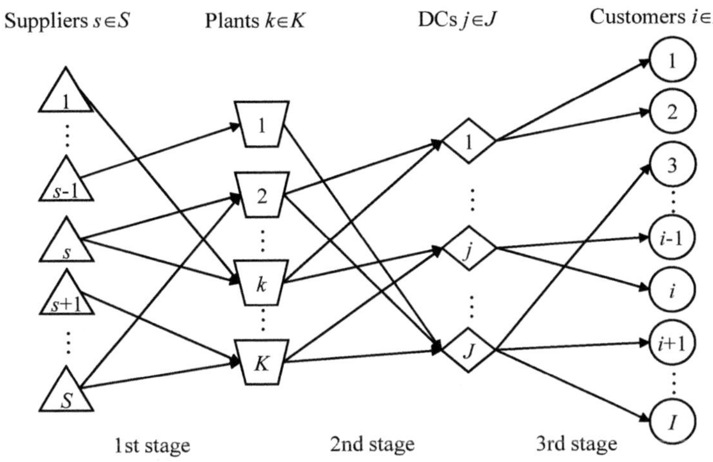

Fig. 3.2 A simple network of three stages in a supply chain network

performance measures, is used to determine efficiency and/or effectiveness of an existing system, to compare alternative systems, and to design proposed systems. These measures are categorized as qualitative and quantitative. Customer satisfaction, flexibility and effective risk management belong to qualitative performance measures. Quantitative performance measures are also categorized by (1) objectives that are based directly on cost or profit such as cost minimization, sales maximization, profit maximization, *etc.* and (2) objectives that are based on some measure of customer responsiveness such as fill rate maximization, customer response time minimization, lead time minimization, *etc.* [4].

In traditional logistics system, the focus of the integration of logistics system is usually on a single objective such as minimum cost or maximum profit. For example, Jayaraman and Prikul [7], Jayaraman and Ross [8], Yan *et al.* [9], Syam [10], Syarif *et al.* [11], Amiri [12], Gen and Syarif [13], Truong and Azadivar [14], and Gen *et al.* [15] had considered total cost of logistics as an objective function in their studies. However, there are no design tasks that are single objective problems. The design/planning/scheduling projects usually involve trade-offs among different incompatible goals. Recently, multi-objective optimization of logistics has been considered by different researchers in literature. Sabri and Beamon [1] developed an integrated multi-objective supply chain model for strategic and operational supply chain planning under uncertainties of product, delivery and demand. While cost, fill rates, and flexibility were considered as objectives, and constraint methods had been used as a solution methodology. Chan and Chung [16] proposed a multi-objective genetic optimization procedure for the order distribution problem in a demand driven logistics. They considered minimization of total cost of the system, total delivery days and the equity of the capacity utilization ratio for manufacturers as objectives. Chen and Lee [17] developed a multi-product, multi-stage, and multi-period scheduling model for a multi-stage logistics with uncertain demands and product prices.

As objectives, fair profit distribution among all participants, safe inventory levels and maximum customer service levels, and robustness of decision to uncertain demands have been considered, and a two-phased fuzzy decision-making method was proposed to solve the problem. Erol and Ferrell [18] proposed a model that assigning suppliers to warehouses and warehouses to customers. They used a multi-objective optimization modeling framework for minimizing cost and maximizing customer satisfaction. Guillen *et al.*, [19] formulated the logistics network model as a multi-objective stochastic mixed integer linear programming model, which was solved by e-constraint method, and branch and bound techniques. Objectives were SC profit over the time horizon and customer satisfaction level. Chen *et al.* [20] developed a hybrid approach based on a genetic algorithm and Analytic Hierarch Process (AHP) for production and distribution problems in multi-factory supply chain models. Operating cost, service level, and resources utilization had been considered as objectives in their study. Altiparmak *et al.* [21, 22] developed multiobjective genetic algorithms for a single-source, single and multi-product, multi-stage SCN design problems. The studies reviewed above have found a Pareto-optimal solution or a restrictive set of Pareto-optimal solutions based on their solution approaches for the problem.

During the last decade, there has been a growing interest in using genetic algorithms (GA) to solve a variety of single and multi-objective problems in production and operations management that are combinatorial and NP hard (Gen and Cheng [23], Dimopoulos and Zalzala [24], Aytug *et al.* [25]). Based on the model structure, the core models of logistics network design are summarized in Fig. 3.3.

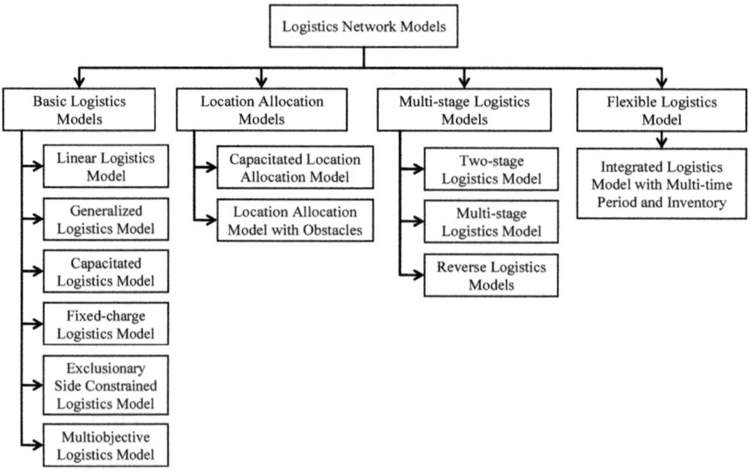

Fig. 3.3 The core models of logistics network design introduced in this chapter

3.2 Basic Logistics Models

It has been said by Peter Deruke, an American management specialist that logistics is the last frontier for cost reduction and the third profit source of enterprises [26]. The interest in developing effective logistics system models and efficient optimization methods has been stimulated by high costs of logistics and potentially capable of securing considerable savings. Aiming at the practicability of logistics network, we will concentrate on developing a flexible logistics network model and extending it to handle some features practically needed.

The term logistics means, in a broad sense, the process of managing and controlling flows of goods, energy, information and other resources like facilities, services, and people, from the source of production to the marketplace, and the relative techniques for it. It involves the integration of information, transportation, inventory, warehousing, material handling, and packaging. The responsibility of logistics is the geographical repositioning of raw materials, work in process, and inventories to where required at the lowest possible cost. Logistics plays a role in keen competitions among business entities of today as crucial as it does in wars, from which the word was generated. It is difficult to accomplish any marketing or manufacturing without logistical support. In spite of the various contents of the logistics system, in this chapter we will restrict it to the production-distribution-inventory system, and focus on the logistics network, trying to fill some gaps between research and application in this field.

Nowadays the design and optimization of logistics systems provides a most exciting opportunity for enterprises to reduce significantly various costs in the supply chain. Tremendous strides have been made in automating transaction processing and data capture related to logistics operations of business in the last three decades. Notwithstanding that these innovations have cut down costs in their own right by reducing manual effort, their greatest impact is yet to come – they pave the way for optimizing logistics decisions with the help of computer-aided technology. Although logistics optimization is neither facile nor low-cost, for most companies it can provide a chance to add to their profit margin by making optimized decisions.

3.2.1 Mathematical Formulation of the Logistics Models

Let us consider a directed network $G = (N, A)$, which consists of a finite set of *nodes* $N = \{1, 2, \dots, n\}$ and a set of directed *arcs* $A = \{(i, j), (k, l), \dots, (s, t)\}$, which joins pairs of nodes in N.

3.2.1.1 Linear Logistics Model

The linear logistics models (or transportation problem: TP) contain two main sets of constraints: one set of I constraints associated with the source nodes denoted as

S and one set of J constraints associated with the destination nodes denoted as D. There are IJ variables in the problem, each corresponding to an arc from a source node to a destination node. Therefore, the underlying graph is a direct graph and it is characterized as a bipartite graph, that is, the nodes of this network model can be partitioned into two sets S and D. It is graphically illustrated in Fig. 3.4.

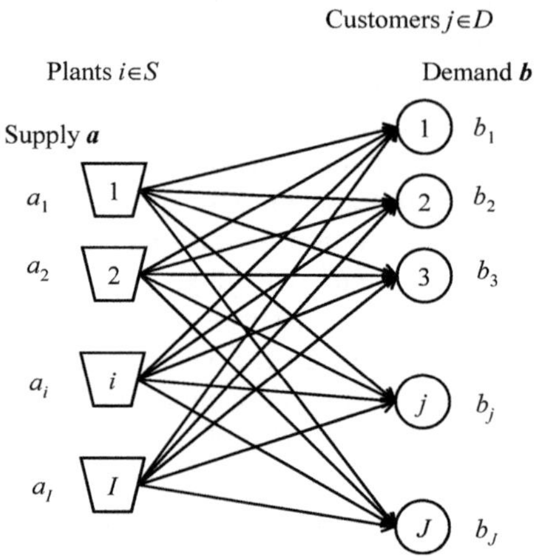

Fig. 3.4 Bipartite graph

Following assumptions describe the property of linear logistics model:

A1. Each supply node in S has a fixed quantity of supply.
A2. Each demand node in D has certain quantity of demand.
A3. For each arc (i, j) in A, we have $i \in S$ and $j \in D$. It means that each arc corresponds to one connection between the supply node and the demand node.

The objective of the TP model is to find the minimal cost pattern of shipment. It can be formulated as follows:

$$\min z(x) = \sum_{i=1}^{I} \sum_{j=1}^{J} c_{ij} x_{ij} \tag{3.1}$$

$$\text{s. t. } \sum_{j=1}^{J} x_{ij} \leq a_i, \quad \forall\, i \tag{3.2}$$

$$\sum_{i=1}^{I} x_{ij} \geq b_j, \quad \forall\, j \tag{3.3}$$

$$x_{ij} \in 0, \quad \forall\, i, j \tag{3.4}$$

where x_{ij} is the unknown quantity shipping from source i to destination j. c_{ij} is the cost of shipping 1 unit from source$i$ to destination j, a_i is number of units available at source i. b_j is number of units demanded at destination j. Note that such a formulation is consistent only if

$$\sum_{i=1}^{I} a_i = \sum_{j=1}^{J} b_j$$

That is, the rim conditions must balance. Although this seems to be highly restrictive, it may be circumvented easily by the employment of what are known as dummy sources or dummy sinks.

A solution for a linear transporation problem can be also represented with a transportation tableau, as shown in Fig. 3.5. The solution has the following common characteristics:

(1) It has $I+J-1$ basic variables (called, basis), and each variable corresponds to some cell in the transportation tableau
(2) The basis must be a transportation tree, that is, there must be at least one basic cell in each row and in each column of the transportation tableau
(3) The basis should not contain a cycle

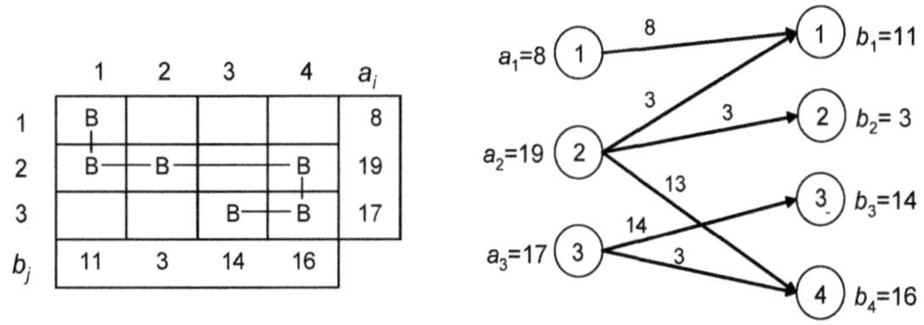

Fig. 3.5 Sample of transportation tableau and its transportation graph

3.2.1.2 Generalized Logistics Model

The generalized logistics model is an extention of logistics model with the consideration of changes of shipment quantities, which is usually formulated in the following form:

$$\min z(x) = \sum_{i=1}^{I} \sum_{j=1}^{J} c_{ij} x_{ij} \tag{3.5}$$

$$\text{s. t.} \sum_{j=1}^{J} x_{ij} \leq a_i, \quad \forall\, i \tag{3.6}$$

$$\sum_{i=1}^{I} p_{ij}x_{ij} \geq b_j, \quad \forall\, j \tag{3.7}$$

$$x_{ij} \geq 0, \quad \forall\, i, j \tag{3.8}$$

where p_{ij} is unit of quantity to be transported. Generalized logistics model is widely used to treat such a problem, in which the quantity of products may change during the transportation for a variety of reasons, such as loss or rot of foods.

3.2.1.3 Capacitated Logistics Model

The capacitated logistics model (or capacitated transportation problem: cTP) is an extention of the logistics model with the constraints of the transportation capacity on each route. The mathematical model is formulated as follows:

$$\min z(x) = \sum_{i=1}^{I}\sum_{j=1}^{J} c_{ij}x_{ij} \tag{3.9}$$

$$\text{s. t. } \sum_{j=1}^{J} x_{ij} \leq a_i, \quad \forall\, i \tag{3.10}$$

$$\sum_{i=1}^{I} p_{ij}x_{ij} \geq b_j, \quad \forall\, j \tag{3.11}$$

$$0 \leq x_{ij} \leq u_{ij}, \quad \forall\, i, j \tag{3.12}$$

where u_{ij} is the upper bound of transportation capacity.

3.2.1.4 Fixed-charge Logistics Model

The fixed-charge logistics model (or fixed-charge transportation problem: fcTP) is an extension of the logistics model. Many practical transportation and distribution problems, such as the minimum cost network flow (transshipment) problem with a fixed-charge for logistics, can be formulated as fixed-charge logistics models. For instance, in a logistics model, a fixed cost may be incurred for each shipment between a given plant and a given customer, and a facility of a plant may result in a fixed amount on investment. The fcTP takes these fixed-charges into account, so that the TP can be considered as an fcTP with equal fixed costs of zero for all routes. The fcTP is much more difficult to solve due to the presence of fixed costs, which cause discontinuities in the objective function.

In the fixed-charge logistics model, two types of costs are considered simultaneously when the best course of action is selected: (1) variable costs proportional to the activity level; and (2) fixed costs. The fcTP seeks the determination of a minimum cost transportation plan for a homogeneous commodity from a number of plants to a number of customers. It requires the specification of the level of supply at each plant, the amount of demand at each customer, and the transportation cost and fixed cost from each plant to each customer. The goal is to allocate the supply available at each plant so as to optimize a criterion while satisfying the demand at each customer. The usual objective function is to minimize the total variable cost and fixed costs from the allocation. It is one of the simplest combinatorial problems involving constraints. This fixed-charge logistics model with I plants and J customers can be formulated as follows:

$$\min z = \sum_{i=1}^{I}\sum_{j=1}^{J}(c_{ij}x_{ij} + f_{ij}\delta(x_{ij})) \tag{3.13}$$

$$\text{s. t. } \sum_{j=1}^{J} x_{ij} \leq a_i, \quad \forall\, i \tag{3.14}$$

$$\sum_{i=1}^{I} x_{ij} \geq b_j, \quad \forall\, j \tag{3.15}$$

$$x_{ij} \geq 0, \quad \forall\, i, j \tag{3.16}$$

where

$$\delta(x) = \begin{cases} 1, \text{ if } x > 0 \\ 0, \text{ if } x \leq 0 \end{cases} \text{ is the step function;}$$

i is index of plant $(i = 1, 2, \ldots, I)$, j is index of customer $(j = 1, 2, \ldots, J)$, a_i is number of units available at plant i, b_j is number of units demanded at customer j, c_{ij} is the cost of shipping one unit from plants i to customer j, d_{ij} is the fixed cost associated with route (i, j), x_{ij} is the unknown quantity to be transported on route (i, j), $f_{ij}(x)$ is the total transportation cost for shipping per unit from plant i to customer j in which f_{ij} (x)=$c_{ij}x_{ij}$ will be a cost function if it is linear.

3.2.1.5 Exclusionary Side Constrained Logistics Model

For specific real world applications, it is often that the logistics problem is extended to satisfy additional constraints. Sun introduced the problems called the exclusionary side constraint logistics model (or exclusionary side constraint transportation problem: escTP). In this model, the logistics model is extended to satisfy the additional constraint in which the simultaneous shipment from some pairs of source center is prohibited.

In our real life, this kind of problem represents many applications. For example, some chemicals cannot be stored in the same room, food and poison products cannot be stored together, although they may be transported by the same distribution system. With this situation, the objective is to find a reasonable and feasible assignment strategy so that the total cost can be minimized while satisfying the source and demand requirements without shipping from any pairs of prohibited source simultaneously. With this additional side constraint, the difficulty of the problem is enormously increased, yet the relevance for the real world applications also increases significantly. Moreover, since this side constraint is nonlinear, it is impossible to solve this problem using a traditional linear programming software package (*i.e.*, LINDO)

Let a homogeneous product be transported from each of I source centers to J destinations. The source centers are supply points, characterized by available capacities a_i, for $i \in S = \{1,\ldots,I\}$. The destinations are demand points, characterized by required level of demand b_j, for $j \in D = \{1,\ldots,J\}$. A cost function c_{ij} is associated with transportation of a unit product from source i to destination j. The exclusionary side constraint logistics model is formulated as follows:

$$\min z(x) = \sum_{i=1}^{I}\sum_{j=1}^{J} c_{ij}x_{ij} \tag{3.17}$$

$$\text{s. t.} \sum_{j=1}^{J} x_{ij} \le a_i, \quad \forall\, i \tag{3.18}$$

$$\sum_{i=1}^{I} x_{ij} \ge b_j, \quad \forall\, j \tag{3.19}$$

$$x_{ij}x_{kj} = 0, \quad \forall\ ((i,k) \in D_j, \quad \forall\, j) \tag{3.20}$$

$$0 \le x_{ij} \le u_{ij}, \quad \forall\, i,j \tag{3.21}$$

where, D_j = good from source i and k cannot be simultaneously shipped to destination j . In this model, the constraint at Eq. 3.18 is the capacity constraint. Using the constraint at Eq. 3.19, it is guaranteed that all demands are met. The constraint at Eq. 3.20 shows that two different sources i and k are not allowed to serve the destination j simultaneously. In this problem, one must determine the amount x_{ij} of the product to be transported from each source i to each destination j so that all constraint are satisfied and the total transportation cost can be minimized.

3.2.1.6 Multiobjective Logistics Model

Often the representation of more than one goal or objective in models of economic decisions is desired. In a general sense we can think of the maximization of the utility framework as encompassing all objectives. However, the operational use of models often requires the specification of instrumental or intermediate goals. Or perhaps, the relative importance of the goals, or the specification of a minimum (or maximum) level of attainment required (or allowed) that are to be minimized with respect to one or more of the objectives, must be designed.

In the general logistics model, the objective is to minimize total transportation costs. The basic assumption underlying this method is that management is concerned primarily with cost minimization. However, this assumption is not always valid. In the logistics model, there may be multiple objectives, such as the fulfilment of transportation schedule contracts, fulfilment of union contracts, provision for a stable employment level in various plants and transportation fleets, balancing of workload among a number of plants, minimization of transportation hazards, and, of course, minimization of cost.

The bicriteria linear logistics model (or bicriteria transportation problem: bTP) is a special case of multiobjective logistics model since the feasible region can be depicted with a two dimensional criteria space. The following two objectives are considered: minimizing the total shipping cost and minimizing the total delivery time. Assume that there are i origins and j destinations. The bTP can be formulated as follows:

$$\min z_1 = \sum_{i=1}^{I} \sum_{j=1}^{J} c_{ij}^1 x_{ij} \tag{3.22}$$

$$\min z_2 = \sum_{i=1}^{I} \sum_{j=1}^{J} c_{ij}^2 x_{ij} \tag{3.23}$$

$$\text{s. t. } \sum_{j=1}^{J} x_{ij} \le a_i, \quad \forall\, i \tag{3.24}$$

$$\sum_{i=1}^{I} x_{ij} \ge b_j, \quad \forall\, j \tag{3.25}$$

$$x_{ij} \ge 0, \quad \forall\, i, j \tag{3.26}$$

where, x_{ij} is the amount of units shipped from origin i to destination j, c_{ij}^1 is the cost of shipping one unit from origin i to destination j, c_{ij}^2 is the deterioration of shipping one unit from origin i to destination j, a_i is the number of units available at origin i, and b_j is the number of units demanded at destination j.

The fixed-charge logistics model with one more objective function, for instance minimization of total delivery time, can be represented by a bicriteria fixed-charge transportation model (b-fcTP). The formulation of the b-fcTP can be given as follows:

$$\min z_1 = \sum_{i=1}^{I} \sum_{j=1}^{J} (c_{ij} x_{ij} + f_{ij} \delta(x_{ij})) \tag{3.27}$$

$$\min z_2 = \sum_{i=1}^{I} \sum_{j=1}^{J} t_{ij}(x_{ij}) \tag{3.28}$$

$$\text{s. t. } \sum_{j=1}^{J} x_{ij} \le a_i, \quad \forall\, i \tag{3.29}$$

$$\sum_{i=1}^{I} x_{ij} \geq b_j, \quad \forall\, j \tag{3.30}$$

$$x_{ij} \geq 0, \quad \forall\, i, j \tag{3.31}$$

where

$$\delta(x) = \begin{cases} 1, \text{ if } x > \ 0 \\ 0, \text{ if } x \leq \ 0 \end{cases} \text{ is the step function;}$$

t_{ij} is per unit delivery time from plant i to consumer j, and $t_{ij}(x)$ can be viewed as a function of delivery time of the form $t_{ij}(x)$=$t_{ij}x_{ij}$.

3.2.2 Prüfer Number-based GA for the Logistics Models

Let P(t) and C(t) be parents and offspring in current generation t, respectively. The overall procedure of Prüfer number-based GA (puGA) for solving bTP is outlined as follows:

---

procedure: puGA for bTP models
input: problem data, GA parameters (*popSize*, *maxGen*, p_M, p_C)
output: Pareto optimal solutions E
begin
  $t \leftarrow 0$;
  initialize $P(t)$ by Prüfer number-based encoding routine;
  calculate objectives $zi(P), i = 1,\ldots,q$ by Prüfer number-based decoding routine;
  create Pareto $E(P)$;
  evaluate $eval(P)$ by adaptive-weight fitness assignment routine;
  while (not terminating condition) **do**
    create $C(t)$ from $P(t)$ by one cut-point crossover routine;
    create $C(t)$ from $P(t)$ by replacement mutation routine;
    calculate objectives $zi(C)$, $i = 1,\ldots,q$ by Prüfer number-based decoding routine;
    update Pareto $E(P,C)$;
    evaluate eval(P,C) by adaptive-weight fitness assignment routine;
    select $P(t+1)$ from $P(t)$ and $C(t)$ by RWS and $(\mu+\lambda)$-selection routine;
    $t \leftarrow t+1$;
  end
  output the best solutions $E(P,C)$
end

3.2.2.1 Genetic Representation

GA-based function optimizers have demonstrated their usefulness over a wide range of difficult problems. To solve a given problem, first a proper representation of the solution is needed. GAs usually encode the solutions for the optimization problems as a finite-length string over some finite alphabet. It is noted that the choice of representation can itself affect the performance of a GA-based function optimizer. Various encoding methods have been created for particular problems to provide effective implementation of GAs. Tree-based representation is known to be one way for representing network problems. Basically, there are three ways of encoding tree:

1. Edge-based encoding
2. Vertex-based encoding
3. Edge-and-vertex encoding [2]

Matrix-based Encoding: Michalewicz *et al.* [27] were the first researchers, who used GA for solving linear and non-linear transportation/distribution problems. In their approach, a matrix-based representation had been used. Suppose $\mid K \mid$ and $\mid J \mid$ are the number of sources and depots, respectively; the dimension of matrix will be $\mid K \mid . \mid J \mid$. Although the representation is very simple, there is a need to devise special crossover and mutation operators for obtaining feasible solutions.

A matrix is perhaps the most natural representation of a solution for a logistics problem. The allocation matrix of a logistics problem can be written as follows:

$$X_k = [x_{ij}] = \begin{bmatrix} x_{11} & x_{12} & \cdots & x_{1J} \\ x_{21} & x_{22} & \cdots & x_{2J} \\ \cdots & \cdots & \cdots & \cdots \\ x_{I1} & x_{I2} & \cdots & x_{IJ} \end{bmatrix}$$

where X_k denotes the k-th chromosome (solution) and the element of i_t, x_{ij}, is the corresponding decision variable.

Prüfer Number-based Encoding: The use of Prüfer number-based GA (puGA) for solving some network problems was introduced in [23, 28]. They employed a Prüfer number in order to represent a candidate solution to the problems and developed feasibility criteria for a Prüfer number to be decoded into a spanning tree. They noted that the use of a Prüfer number was very suitable for encoding a spanning tree, especially in some research fields, such as transportation problems, minimum spanning tree problems, and so on. Also, it is shown that is the approach needs only $\mid K \mid +$ $\mid J \mid$ -2 digit number to represent uniquely a distribution network with $\mid K \mid$ sources and $\mid J \mid$ depots, where each digit is an integer between 1 and $\mid K \mid + \mid K \mid$. Although Prüfer number encoding of the spanning tree has been successfully applied to transportation problems [29], it needs some repair mechanisms. Since a transportation tree is a special type of spanning tree, Prüfer number encoding may correspond to an infeasible solution, *i.e.* it doesn't represent a transportation tree. From this point, there is a need to design a criterion for checking the feasibility of chromosomes. If a chromosome does not pass this checking, it will be repaired. This is the first repair. The second repair may be also necessary after decoding process of chromosomes. When a Prüfer number is decoded, some links on the corresponding spanning tree can have 0 flows. In this situation, these links are removed from the tree and new ones are added to obtain a spanning tree. After this regulation, the current Prüfer number will not be representative of a feasible spanning tree. Therefore, there is a need to re-encode the given Prüfer number according to the new spanning tree.

Denote plants 1, 2,..., m as the components of the set $S=\{1,2,\ldots,m\}$ and warehouses 1, 2,..., n as the components of the set $D=\{m+1,...,m+n\}$. Obviously, the logistics model has m+n nodes and $m \times n$ edges in its transportation tree.

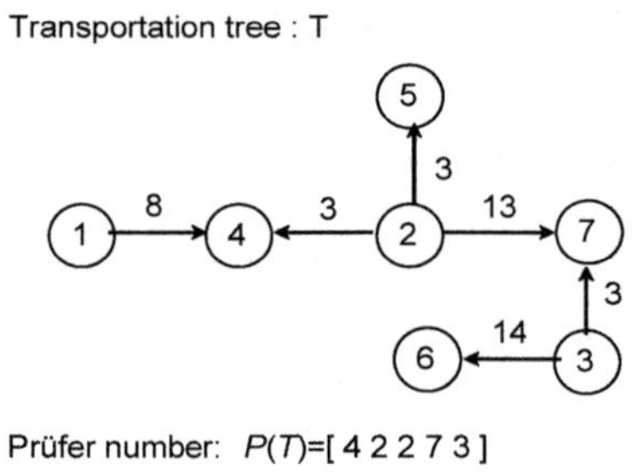

Fig. 3.6 Sample of spanning tree and its Prüfer number.

The transportation graph can be represented as a spanning tree as shown in Fig. 3.6. For a complete graph with p nodes, there are $p^{(p-2)}$ distinctly labeled trees. This means that we can use permutations of only $p-2$ digits in order to represent uniquely a tree with p nodes, where each digit is an integer between 1 and p inclusive. For any tree there are always at least two leaf nodes.

The Prüfer number is one of the nodes encoding a tree as shown in Fig. 3.7. Therefore, it can be used to encode a transportation tree. The constructive procedure for a Prüfer number according to a tree is as follows.

The illustration of converting the transportation tree to a corresponding Prüfer number is shown in Fig. 3.8 The decoding procedure from a Prüfer number to a transportation tree is illustrated in Fig. 3.9.

Each node in the logistics model has its quantity of supply or demand, which are characterized as constraints. Therefore, to construct a transportation tree, the constraint of nodes must be considered. From a Prüfer number, a unique transportation tree also can be generated by the following procedure (in Fig. 3.10).

Prüfer Number-based Initialization: The initialization of a chromosome (a Prüfer number) is performed from randomly generated m+n–2 digits in range $[1,m+n]$. However, it is possible to generate an infeasible chromosome, which will not yield to a transportation tree. Prüfer number encoding can not only uniquely represent all possible spanning trees, but also explicitly contains the information of node degree, that is, any node with degree d will appear exactly d–1 times in the encoding. This means that when a node appears d times in a Prüfer number, the node has d+1

procedure: Encoding a Tree to a Prüfer Number
input: tree data set
output: a Prüfer number *P*
step 1: Let *i* be the lowest numbered leaf node in tree *T*.
Let *j* be the node which is the predecessor of *i*.
Then *j* becomes the rightmost digit of Prüfer number *P*(*T*),
P(*T*) is built up by appending digits to the right;
step 2: Remove *i* and the substep (*i, j*) from further consideration.
Thus, *i* is no longer considered at all and if *i* is the only
successor of *j*, then *j* becomes a leaf node.
step 3: If only two nodes remain to be considered, then Prüfer number
P(*T*) has been formed, stop; Otherwise return to Step 1.

Fig. 3.7 Pseudocode of encoding a tree to a Prüfer number.

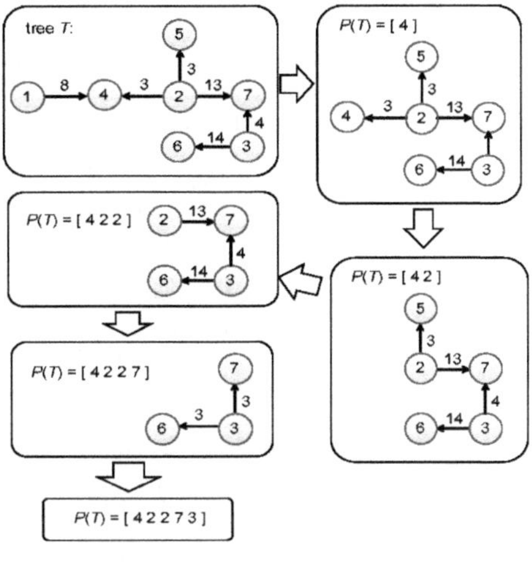

Fig. 3.8 Encoding procedure

connections with other nodes. This can be used to establish a criterion for checking the feasibility of an encoding.

The criterion for handling the feasibility of a chromosome is proposed as follows. Denote $\overline{P}(T)$ as the set of nodes that are not included in the Prüfer number $P(T)$. Let L_S be the sum of connections of nodes which are included in set $|S \cap P(T)|$, and L_D be the sum of connections of nodes which are included in set $|D \cap P(T)|$. Also let $\overline{L}_S$ be the times of appearances of those nodes in $P(T)$ and included in set S, and let $\overline{L}_D$ be the times of appearances of those nodes in $P(T)$ and included in set D, respectively.

Feasibility of a chromosome: If $L_S+\overline{L}_S=L_D+\overline{L}_D$, then $P(T)$ is feasible. Otherwise $P(T)$ is infeasible.

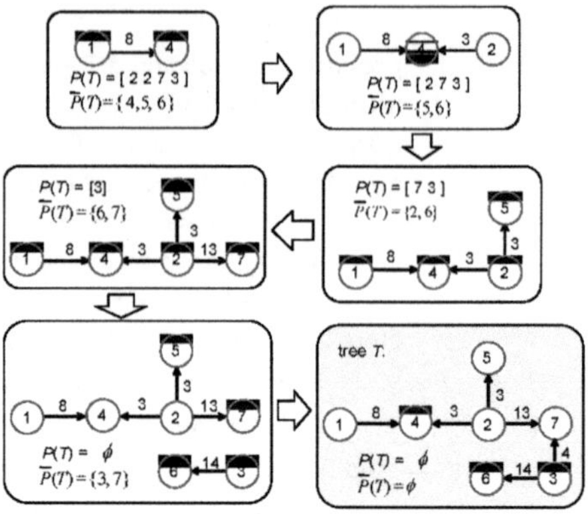

Fig. 3.9 Decoding procedure

---

procedure: Decoding a Prüfer Number to a Transportation Tree
input: a Prüfer number *P*
output: tree data set

step 1: Let $P(T)$ be the original Prüfer number and let $\overline{P}(T)$ be the set of all nodes that are not part of $P(T)$ and are designated as eligible for consideration.

step 2: Repeat the substep 2.1 - 2.5 until no digit left in $P(T)$.

2.1 Let i be the lowest numbered eligible node in $\overline{P}(T)$ and j be the leftmost digit of $P(T)$.

2.2 If i and j are not in the same set S or D, add the edge (i, j) to tree T. Otherwise, select the next digit k from $P(T)$ that not included in the same set with i, exchange j with k, add the edge (i, k) to the tree T.

2.3 Remove j (or k) from $P(T)$ and i from $\overline{P}(T)$. If j (or k) does not occur anywhere in the remaining part of $P(T)$, put it into. Designate i as no longer eligible.

2.4 Assign the available amount of units to $x_{ij} = \min\{a_i, b_j\}$ (or $x_{ik} = \min\{a_i, b_k\}$) to edge (i, j) [or (i,k)], where $i \in S$ and $j, k \in D$.

2.5 Update availability $a_i = a_i - x_{ij}$ and $b_j = b_j - x_{ij}$ (or $b_k = b_k - x_{ik}$).

step 3: If no digits remain in $P(T)$ then there are exactly two nodes, i and j, still eligible in $\overline{P}(T)$ for consideration. Add edge (i, j) to tree T and form a tree with $m+n-1$ edges.

step 4: If no available amount of units to assign, then stop. Otherwise, there are remaining supply r and demand s, add edge (r, s) to tree and assign the available amount of units $x_{rs}=a_r=b_s$ to edge. If there exists a cycle, remove the edge that is assigned zero flow. A new spanning tree is formed with $m+n-1$ edges.

Fig. 3.10 Pseudocode of decoding Prüfer number to a tree

For instance, the Prüfer number $P(T)$ is [4 2 2 7 3] and $\overline{P}(T)$ is $\{1,5,6\}$ in Fig. 3.5. Now, $\overline{LS}$ is the sum of connections of nodes 2 and 3 included in set S from $P(T)$. There are 2+1 for node 2 and 1+1 for node 3, so $\overline{LS}$=5. Similarly L_D is the sum of connections for nodes 4 and 7 included in set D, and L_D=4. $\overline{LS}$ is the times of appearances of node 1 in $\overline{P}(T)$, so $\overline{LS}$=1. $\overline{LD}$ is the times of appearances of nodes 5 and 6, so $\overline{LD}$=2. So, L_S+$\overline{LS}$=6 and LD+$\overline{LD}$=6, and therefore this Prüfer number satisfies the criterion of feasibility.

3.2.2.2 Genetic Operators

Crossover: The one-cut point crossover operation is used as illustrated in Fig. 3.11. To avoid unnecessary decoding from which an infeasible chromosome (a Prüfer number) may be generated after using the crossover operator, we add the feasibility checking procedure into the crossover operation with the criterion given above. A Prüfer number generated by the crossover operation is always feasible and corresponding to a transportation tree.

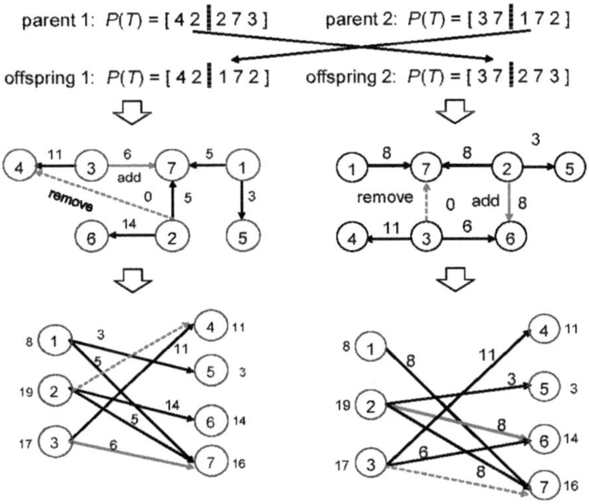

Fig. 3.11 One-cut point crossover and offspring.

Mutation: The inversion mutation and displacement mutation can be used. With the inversion mutation, two positions within a chromosome are selected at random, and then the substring between these two positions is inverted. With the displacement mutation, a substring is selected at random, and is inserted at a random position.

Selection: A mixed strategy of combining roulette wheel selection (RWS) with the $(\mu + \lambda)$-selection is used as the selection procedure. The strategy can enforce the best chromosomes into the next generation and avoid the premature convergence of the evolutionary process. The strategy first selects the μ best chromosomes from μ parents and λ offspring. If there are not μ different chromosomes available, the vacant pool of the population for next generation is filled up with the RWS.

3.2.3 Numerical Experiments

3.2.3.1 Linear Logistics Problems

First, to confirm the effectiveness of the Prüfer number-based GA for the linear logistics problems, five numerical examples were tested. These examples were randomly generated where the supplies and demands are generated randomly and uniformly distributed over [10, 100], and the transportation costs over [5, 25].

The simulations were carried out on each example with their best parameter setting. Table 3.1 shows average simulation results from 20 times running by the Prüfer number-based GA (puGA) and matrix-based GA (mGA). From Table 3.1 we know that the optimal solution was found by the puGA with less time than mGA in each running. This means that puGA is more appropriate than mGA for solving the linear logistics problems.

Table 3.1 Average results by the mGA and puGA approach for TP

Problem size: $m \times n$	Parameter		Matrix-based GA		Prüfer number-based GA	
	popSize	maxGen	Percent	ACT (s)	Percent	ACT (s)
3×4	10	200	100 (%)	1.02	100 (%)	1.07
4×5	20	500	100 (%)	12.75	100 (%)	8.77
5×6	20	500	100 (%)	18.83	100 (%)	9.99
6×7	100	1000	100 (%)	1038.25	100 (%)	413.63
8×9	200	2000	100 (%)	4472.19	100 (%)	848.11

Percent: the percentage of running times that obtained optimal solution
ACT: the average computing time

3.2.3.2 Bicriteria Logistics Problems

The performance of the Prüfer number-based GA for a bicriteria logistics problem is examined with four numerical examples. The first and second examples are the bicriteria logistics problem, in which one has three origins and four destinations, and another one has seven origins and eight destinations where the supplies, demands and coefficients of objective functions shown in Table 3.2. For the first example, the parameters for the algorithm are set as follows: population size *popSize*=30; crossover probability p_C=0.2; mutation probability p_M=0.4; maximum generation *maxGen*=500; and run by 10 times. The Pareto optimal solutions can be found at all times. The obtained solutions (143, 265), (156, 200), (176, 175), (186, 171), (208,167) are known as Pareto optimal solutions which are formed in Pareto frontier with the exact extreme points (143, 265) and (208, 167). The average computing time, on the puGA and mGA are 5.45 and 10.42, respectively.

Table 3.2 Supplies, demands and coefficients of objectives for bTP

From O_i	c_{ij}^1								c_{ij}^2								
To D_j	8	9	10	11	12	13	14	15	8	9	10	11	12	13	14	15	a_i
1	1	2	7	7	8	10	9	2	4	4	3	4	5	8	9	10	10
2	1	9	3	4	3	5	7	1	6	2	5	1	7	4	12	4	8
3	8	9	4	6	4	1	6	9	2	9	1	8	9	1	4	0	12
4	2	4	5	5	3	2	3	2	3	5	5	3	2	8	3	3	16
5	5	4	5	1	9	9	1	6	1	4	12	2	1	5	4	9	21
6	8	3	3	2	2	3	6	7	2	23	4	4	6	2	4	6	15
7	1	2	6	4	5	9	3	5	1	2	1	9	0	13	2	3	7
b_j	9	7	15	10	13	16	7	12	9	7	15	10	13	16	7	12	89

The other two examples are randomly generated with three objectives. The supplies and demands are also generated randomly and uniformly distributed over $[10, 100]$ and the coefficients of each objective function are over $[5, 25]$. The problem size and the simulation results with the best parameters (p_C=0.2 and p_M=0.4 for puGA) are shown in Table 3.3 The average computing time was computed on 10 running times.

Table 3.3 Comparison with puGA and mGA

Problem size			Memory for a solution		ACT (s)		Parameter	
m	n	Q	mGA	puGA	mGA	puGA	*popSize*	*maxGen*
3	4	2	12	5	10.42	5.45	30	500
7	8	2	56	13	1632.75	863.01	200	2000
8	9	3	72	15	6597.41	2375.53	200	2000
10	15	3	150	23	19411.08	11279.30	300	2000

m: number of origins n: number of destinations Q: number of objectives
mGA: matrix-based GA puGA: Prüfer number-based GA

Figure 3.12 shows that the comparison between the mGA and puGA with parameters p_C=0.4, p_M=0.2 for the mGA and p_C=0.2, p_M=0.4 for the puGA on example 2. From Fig. 3.12, we can see that the results obtained by the puGA are much better than those by the mGA in the sense of Pareto optimality because most of them are not dominated by those obtained in the mGA. Therefore, we can conclude that they are closer to the real Pareto frontier than those by the mGA.

Besides, in Table 3.3 the average computing time on the puGA is shorter than that on the mGA for each example. Obviously, the puGA is more efficien than the mGA in the evolutionary process. It is because the Prüfer number-based encoding only requires $m+n-2$ memory spaces for each chromosome but the matrix-based encoding requires $m \times n$ memory spaces for representing a chromosome. As a result, it takes more time on genetic operations for mGA compared with the puGA. Therefore, for a large-scale problem, the puGA will be surely more time-saving than the mGA.

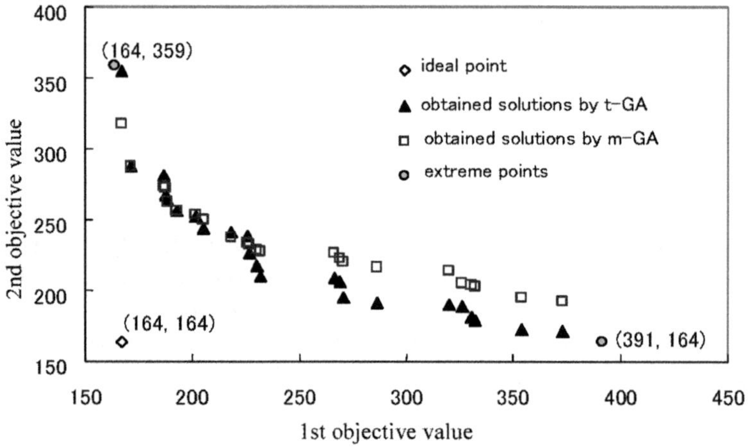

Fig. 3.12 Comparison between puGA and mGA on the obtained solutions for second tested problem

3.3 Location Allocation Models

The efficient and effective movement of goods from raw material sites to processing distribution centers (DCs), component fabrication plants, finished goods assembly plants, DCs, retailers and customers is critical in today's competitive environment [30]. Within individual industries, the percentage of the cost of a finished delivered item to the final consumer can easily exceed this value. *Supply chain management* (SCM) entails not only the movement of goods but also decisions about (1) where to produce, what to produce, and how much to produce at each site, (2) what quantity of goods to hold in inventory at each stage of the process, (3) how to share information among parties in the process and finally, (4) where to locate plants and distribution centers.

Council of Logistics Management (CLM) has defined logistics management as that part of Supply Chain Management that plans, implements, and controls the efficient, effective forward and reverses flow and storage of goods, services and related information between the point of origin and the point of consumption in order to meet customers' requirements [31]. Logistics management problems can be classified as:

- Location problems
- Allocation problems
- Location allocation problems

Location problems involve determining the location of one or more new DCs in one or more of several potential sites. Obviously, the number of sites must at least equal the number of new DCs being located. The cost of locating each new DC at each of the potential sites is assumed to be known. It is the fixed cost of locating a new DC at a particular site plus the operating and transportation cost of serving customers from this DC-site combination. *Allocation problems* assume that the number and location of DCs are known *a priori* and attempt to determine how each customer is to be served. In other words, given the demand for goods at each customer center, the production or supply capacities at each DC, and the cost of serving each customer from each DC, the allocation problem determines how much each DC is to supply to each customer center. *Location allocation problems* involve determining not only how much each customer is to receive from each DC but also the number of DCs along with their locations and capacities.

Location decisions may be the most critical and most difficult of the decisions needed to realize an efficient supply chain. Transportation and inventory decisions can often be changed at relatively short notice in response to changes in the availability of raw materials, labor costs, component prices, transportation costs, inventory holding costs, exchange rates and tax codes. Information sharing decisions are also relatively flexible and can be altered in response to changes in corporate strategies and alliances. Thus, transportation, inventory, and information sharing decisions can be readily re-optimized in response to changes in the underlying conditions of the supply chain. Decisions about production quantities and locations are, perhaps, less flexible, as many of the costs of production may be fixed in the short term. Labor costs, for example, are often dictated by relatively long-term contracts. Also, plant capacities must often be taken as fixed in the short-term. Nevertheless, production quantities can often be altered in the intermediate term in response to changes in material costs and market demands. DC location decisions, on the other hand, are often fixed and difficult to change even in the intermediate term. The location of a multibillion-dollar automobile assembly plant cannot be changed as a result of changes in customer demand, transportation costs, or component prices. Modern DCs with millions of dollars of material handling equipment are also difficult, if not impossible, to relocate except in the long term. Inefficient locations for production and assembly plants as well as DCs will result in excess costs being incurred throughout the lifetime of the DCs, no matter how well the production plans, transportation options, inventory management, and information sharing decisions are optimized in response to changing conditions.

However, the long-term conditions under which production plants and DCs will operate are subject to considerable uncertainty at the time these decisions must be made. Transportation costs, inventory carrying costs (which are affected by interest rates and insurance costs), and production costs, for example, are all difficult to predict. Thus, it is critical that planners recognize the inherent uncertainty associated with future conditions when making DC location decisions.

3.3.1 Mathematical Formulation of the Logistics Models

The DC location problem concerns the optimal number and location of DCs needed to provide some service to a set of customers. The optimal solution must balance two types of costs – the cost of establishing a DC at a particular location and the total cost of providing service to each of the customers from one of the opened DCs. In its simplest form, if each opened DC can serve only a limited number of customers the problem is called capacitated.

3.3.1.1 Capacitated Location Allocation Model

Cooper was the first author formally to recognize and state the *multi-Weber problem.* He proposed a heuristics called *alternative location allocation*, which is the best heuristic available. With the development of nonlinear programming techniques, relaxing integer allocation constraints and treating the location variables and allocation variables simultaneously, some new methods have been developed [32].

In *Cooper's location allocation model*, it is assumed that a DC has an infinite service capacity. However, this assumption is not the case in practice. The capacity constraint is an important factor in DC location analysis.

The most general version of location allocation model is the *capacitated location allocation problem* (cLAP) as follow.

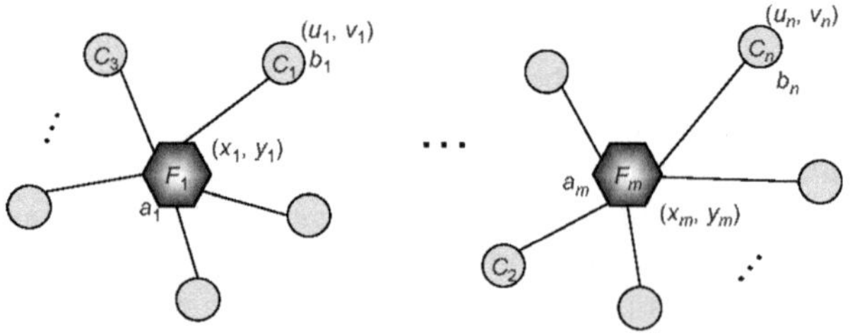

Fig. 3.13 Capacitated location allocation model

In this section, we are interested in finding the location of m DCs in continuous space in order to serve customers at n fixed points as well as the allocation of each customer to the DCs so that total distance sums are minimized. We assume there is restriction on the capacity of the DCs. Each customer has a demand b_j $(j = 1,2,\dots,n)$, and each DC has a service capacity a_i $(i = 1,2,\dots,m)$. The cLAP can be illustrated in Fig. 3.13.

The capacitated location allocation problem can be modeled as a *mixed integer programming model* as follows:

Notation

Indices

i: index of DC $(i = 1,2,\dots,m)$
j: index of customer $(j = 1,2,\dots,n)$

Parameters

m: total number of DCs
n: total number of customers
$C_j = (u_j, v_j)$: j-th customer $j = 1,2,\dots,m$
a_i: capacity of i-th DC
b_j: demand of j-th customer
$t(F_i, C_j)$: the Euclidean distance from the location of DC i, (x_i, y_i) to the location of a customer at fixed point j, (u_j, v_j)
$t(F_i, C_j) = \sqrt{(x_i - u_j)^2 + (y_i - v_j)^2}$

Decision Variables

z_{ij}: 0–1 decision allocation variables
$z_{ij} = 1$, representing the j-th customer is served by i-th DC. Otherwise 0
$F_i = (x_i, y_i)$: unknown location of the i-th DC

$$\min f(F,z) = \sum_{i=1}^{m} \sum_{j=1}^{n} t(F_i, C_j) \cdot z_{ij} \tag{3.32}$$

$$\text{s. t. } g_i(z) = \sum_{j=1}^{n} b_j z_{ij} \le a_i \quad i = 1,2,\dots,m \tag{3.33}$$

$$g_{m+j}(Z) = \sum_{i=1}^{m} z_{ij} = 1 \quad j = 1,2,\dots,n \tag{3.34}$$

$$z_{ij} = 0 \text{ or } 1, \quad i = 1,2,\dots,m, \quad j = 1,2,\dots,n \tag{3.35}$$

$$(x_i, y_i) \in R \times R, \quad \forall\, (i, j) \tag{3.36}$$

where $t(F_i, C_j)$ is the Euclidean distance from the location (coordinates) of DC i,(x_i, y_i) (the decision variable), to the location of a customer at fixed point j,(u_j, v_j); and z_{ij} is the 0–1 allocation variable,z_{ij}=1 representing that customer j is served by DC i and z_{ij}=0 otherwise$(i = 1,2,\ldots,m, j = 1,2,\ldots n)$. The constraint at Eq. 3.33 reflects that the service capacity of each DC should not be exceeded. The constraint at Eq. 3.34 reflects that every customer should be served by only one DC.

The cLAP is a well-known combinatorial optimization problem with applications in production and distribution system. It is more complex because the allocation sub-problem is a general assignment problem known as the NP-hard combinatorial optimization problem. Although the approaches for the location allocation problem can be extended to cLAP, there is no method capable of finding a global or near global optimal solution, especially for practical scale problems. It is important and necessary to develop an efficient method to find the global or near global optimal solution for cLAP.

3.3.1.2 Location Allocation Model with Obstacles

In practice, obstacle or forbidden region constraint usually needs to be considered. For example, lakes, rivers, parks, buildings, cemeteries and so on, provide obstacles to DC location design. There are two kinds of obstacles: (1) the prohibition of locations; (2) the prohibition of connecting paths. In the formes case, a new DC should not be located within the region of obstacles. In the latter case, the connecting path between customer and DC should avoid passing through the region of obstacles. Network location model can solve obstacle problems if a detailed network structure is given in advance; however, it is difficult to give such a kind of network. *Obstacle location allocation problem* (oLAP) is more realistic than traditional location allocation problems, because the obstacle must be taken into consideration in many practical cases.

There are n customers with known locations, and m DCs must be built to supply some kind of service to all customers. There are also q obstacles representing some forbidden areas. The oLAP is to choose the best locations for DCs so that the sum of distances among customers and their serving DCs is minimal. It is also assumed that all obstacles can be represented as a convex polygon. We illustrate this case problem in Fig. 3.14.

The obstacle location allocation problem can be formulated as a nonlinear integer programming model as follows:

$$\min f(D, z) = \sum_{i=1}^{m} \sum_{j=1}^{n} t(F_i, C_j) \cdot z_{ij} \tag{3.37}$$

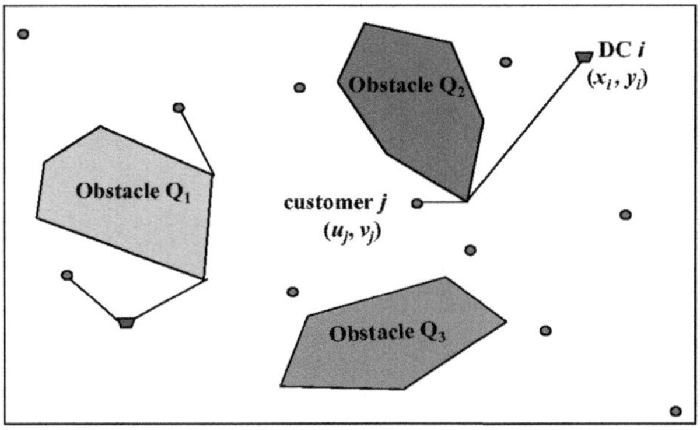

Fig. 3.14 Illustration of the obstacle location allocation problem

$$\text{s. t. } \sum_{j=1}^{n} d_j z_{ij} \leq q_i, \quad i = 1, 2, \ldots, m \tag{3.38}$$

$$\sum_{i=1}^{m} z_{ij} = 1 \quad j = 1, 2, \ldots, n \tag{3.39}$$

$$D_i = (x_i, y_i) \notin \{Q_k \mid k = 1, 2, \ldots, q\}, \quad i = 1, 2, \ldots, m \tag{3.40}$$

$$(x_i, y_i) \in R_T, \quad i = 1, 2, \ldots, m \tag{3.41}$$

$$z_{ij} = 1 \text{ or } 0, \quad i = 1, 2, \ldots, m, \ j = 1, 2, \ldots, n \tag{3.42}$$

where, Q_k is an obstacle $(k = 1, 2, \ldots, q)$.

The constraint at Eq. 3.38 represents that service capacity of each facility should not be exceeded; the constraint at Eq. 3.39 represenst that each customer should be served by only one facility; the constraint at Eq. 3.40 represents that locations of DCs should not be within any obstacle. When obstacles are ignored, $t(F_i, C_j)$ becomes Euclidean distance between DC_i and customer C_j . But if obstacles are considered, $t(F_i, C_j)$ is the function of DC_i and C_j as well as the vertices of all obstacles. It cannot be expressed in an analytical form.

3.3.2 Location-based GA for the Location Allocation Models

Let $P(t)$ and $C(t)$ be parents and offspring in current generation t, respectively. The overall procedure of location-based GA (lGA) for solving location allocation models is outlined as follows:

procedure: lGA for LAPs models
input: LAP data, GA parameters $(popSize, maxGen, p_M, p_C)$
output: the best solution
begin
  $t \leftarrow 0$;
  iinitialize $P(t)$ by position-based encoding routine;
  check the feasibility of $P(t)$ for oLAP;
  evaluate $P(t)$ by nearest neighbor algorithm routine;
  while (**not** terminating condition) **do**
    create $C(t)$ from $P(t)$ by linear convex crossover routine;
    create $C(t)$ from $P(t)$ by subtle mutation routine;
    create $C(t)$ from $P(t)$ by violent mutation routine;
    check the feasibility of $C(t)$ for oLAP;
    evaluate $C(t)$ by nearest neighbor algorithm routine;
    select $P(t+1)$ from $P(t)$ and $C(t)$ by ES–$(\mu+\lambda)$ selection routine;
    $t \leftarrow t+1$;
  end
  output the best solution
end

3.3.2.1 Genetic Representation

In continuous location problems, a binary representation may result in locating two DCs which are very close to each other. As in [28] we also use a real number representation where a chromosome consists of m (x, y) pairs representing the sites of DCs to be located, and p is the number of DCs. For instance this is represented as follows:

$$\text{chromosome } v = [(x_1, y_1), (x_2, y_2), \ldots, (x_i, y_i), \ldots, (x_m, y_m)]$$

where the coordinate (x_i, y_i) denotes the location of the i-th DC, $i = 1, \ldots, m$. Illustration of a simple chromosome v is showed as follows:

$$\begin{aligned}\text{chromosome } v &= \left[(x_1^1, y_1^1), (x_2^2, y_2^2), (x_3^3, y_3^3)\right] \\ &= [(1.3, 3.8), (2.5, 0.4), (3.1, 1.7)]\end{aligned}$$

The initial population is usually constructed by choosing the real numbers randomly between the lower and upper bound of the abscissa and the ordinate of the customer locations. The initial population can be constructed as follows:

$$v_k = \left[(x_1^k, y_1^k), (x_2^k, y_2^k), \ldots, (x_i^k, y_i^k), \ldots, (x_m^K, y_m^K)\right], \forall\, k = 1, 2, \ldots, popSize$$

where (x_{ik}, y_{ik}) are the (x, y) coordinates of the i-th selected DC,$i = 1, \ldots, m$, of the k-th chromosome,$x_i^k \in [x_{min}, x_{max}]$ and $y_i^k \in [y_{min}, y_{max}]$. The position-based encoding procedure is showed in Fig. 3.15.

```
procedure: position-based encoding
input: C_j=(u_j, v_j), total number of DCs m, total number of customers n
output: chromosome v_k, k=1,2,...,popSize
begin
    x_min = min_j {u_j | j = 1,...,n};  x_max = max_j {u_j | j = 1,...,n};
    y_min = min_j {v_j | j = 1,...,n};  y_max = max_j {v_j | j = 1,...,n};
    for k=1 to popSize
        for i=1 to m
            v_k(2i -1)=random [x_min, x_max];
            v_k(2i)=random [y_min, y_max];
        end
    end
    output chromosome v_k, k=1,2,...,popSize;
end
```

Fig. 3.15 Pseudocode of location-based encoding

3.3.2.2 Feasibility Checking for oLAP

Unlike cLAPs, the constraint functions on location variables in this problem are not a mathematical formula but a geometrical one. So a computationally efficient algorithm is important and necessary. It can be seen that the basic algorithm to check the feasibility of a solution is equivalent to checking if a point on the plane falls within a polygon or not. In order to develop such a basic algorithm, some results of the planar and coordinate geometry are necessary.

Proposition 1: Given three points $P1(x1, y1)$, $P2(x2, y2)$, and $P3(x3, y3)$ on the coordinate plane, then the square of the triangle $\triangle P1P2P3$ can be calculated as:

$$S_{\triangle P_1P_2P_3} = \frac{1}{2}\left|\det = \begin{bmatrix} x_1 & y_1 & 1 \\ x_2 & y_2 & 1 \\ x_3 & y_3 & 1 \end{bmatrix}\right|$$

For example: there are three points $P1(2,2)$, $P2(1.5,1)$, and $P3(3,0.5)$ on the coordinate plane. According to Proposition 1, the square of this triangle can be calculated as

$$\begin{aligned}
S_{\triangle p_1 p_2 p_3} &= \frac{1}{2}\left| \det = \begin{bmatrix} x_1 & y_1 & 1 \\ x_2 & y_2 & 1 \\ x_3 & y_3 & 1 \end{bmatrix} \right| \\
&= \frac{1}{2}\left| (-1)^{(1+3)} \det = \begin{bmatrix} x_2 & y_2 \\ x_3 & y_3 \end{bmatrix} \right. \\
&\quad + (-1)^{(2+3)} \det = \begin{bmatrix} x_1 & y_1 \\ x_3 & y_3 \end{bmatrix} \\
&\quad \left. + (-1)^{(3+3)} \det = \begin{bmatrix} x_1 & y_1 \\ x_2 & y_2 \end{bmatrix} \right| \\
&= \frac{1}{2} \mid x_2y_3 - x_3y_2 - x_1y_3 + x_3y_1 + x_1y_2 + x_2y_1 \mid \\
&= \frac{1}{2} \mid x_1(y_2 - y_3) + x_2(y_3 - y_1) + x_3(y_1 - y_2) \mid \\
&= \frac{1}{2} \mid 1.75 \mid = 0.875
\end{aligned}$$

Proposition 2: Let P_{k1}, P_{k2}, ... P_{kp_k} be vertices of a p_k-polygon and D_i be a point on the same place, be the square of the polygon P_{k1}, P_{k2}, be the square of triangle $D_iP_{kr}P_{k(r+1)}$ $(r = 1, 2, \ldots, p_k)$. Consisting of point D_i and the edge of the polygon (where P_{k1}), then

(1) Point D_i is at the outside the polygon P_{k1}, P_{k2} if and only if

$$S_{Qk} < \sum_{r=1}^{P_k} S_{\triangle D_i P_{kr} P_{k(r+1)}}$$

(2) Point D_i is strictly at the inside of the polygon P_{k1}, P_{k2} if and only if

$$S_{Qk} = \sum_{r=1}^{P_k} S_{\triangle D_i P_{kr} P_{k(r+1)}}$$

$$S_{\triangle D_i P_{kr} P_{k(r+1)}} > 0, \quad r = 1, 2, \ldots, p_k$$

(3) Point D_i is on an edge of the polygon P_{k1}, P_{k2} if and only if

$$S_{Qk} = \sum_{r=1}^{P_k} S_{\triangle D_i P_{kr} P_{k(r+1)}}$$

$$S_{\triangle D_i P_{kr} P_{k(r+1)}} = 0, \qquad \exists r_0 \in 1, 2, \ldots, p_k$$

For example, in order to explain these propositions, assume that there is k-th polygon that including four vertices $P_{k1}, P_{k2}, P_{k3}, P_{k4}$ $(p_k = 4)$ to represent an obstacle in the area [5, 4]; three cases of DCs location are shown in Fig. 3.16.

Table 3.4 Locations of DCs and vertices of obstacle

Locations of DCs and vertices of obstacle

DC_i			obstacle Q_k		
i	x_i	y_i	r	a_{kr}	b_{kr}
1	2	3	1	2	2
			2	3	3
2	4	2.2	3	4.5	2
3	3.5	1.75	4	2.5	1.5

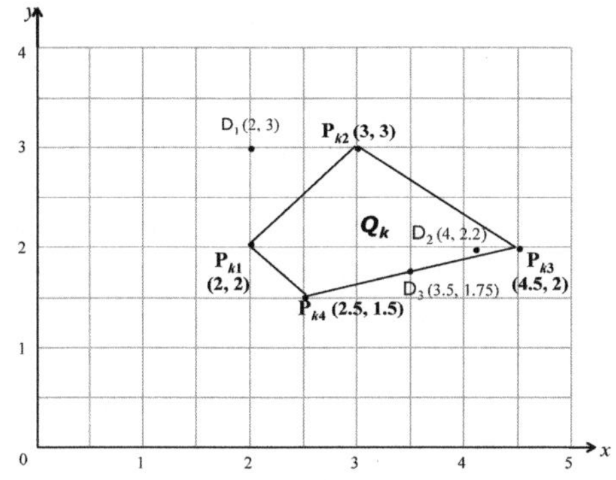

Fig. 3.16 Different situations of DCs

Repairing Infeasibility: A simple way is used for repairing the infeasible chromosomes. According the characteristics of oLAP, the computational complexity of this basic algorithm is

$$Q_k(P_{k1}, P_{k2}, \ldots, P_{kr}, \ldots, P_{kp_k})$$

where p_k is the number of vertices of the k-th polygon.

If location of a distribution center in a solution where in each solution there are m locations for the distribution centers is infeasible, *i.e.,* the point falls inside of some obstacles, it is replaced by one vertex of the obstacle to move to the nearest vertices of the obstacle as shown in Fig. 3.17.

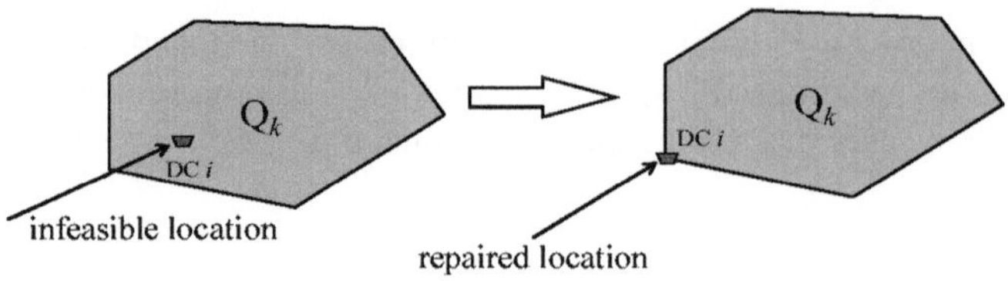

Fig. 3.17 Illustration of repairing infeasibility

3.3.2.3 Shortest Path Avoiding Obstacle

Because there are some obstacles in the locating area, we need to consider the condition that connecting paths between distribution centers and customers should not be allowed to pass through any of obstacle.

Given two points on the plane representing the location of a DC and the location of a customer, the shortest connecting path is the direct line segment between them if obstacles are ignored. However, when obstacles are considered the shortest connecting path is no longer line segment since no obstacle is allowed to be broken through, *i.e.*, the connecting path should avoid obstacles. How to find the shortest connecting path becomes a complicated problem especially when number of obstacles large and their positional relations are complex. Many researchers have concentrated on the path avoiding obstacle problem, especially in robot navigation. Here the visible graph idea is adopted; however, some differences are considered in order to develop the special algorithm to oLAP. According to experience, only a small number of obstacles have relations to the connecting path linking the specified two points on the plane. Thus it is not necessary to construct the whole visible graph including all obstacles, which would lead to a large-scale graph. An algorithm that only considers the relevant obstacles to construct a small-size visible graph is given, and then the Dijkstra's shortest path algorithm is applied to the visible graph to obtain the shortest path avoiding obstacles.

Notice that the connecting paths between a DC and a customer which cross one or more of the obstacles should be a piece-wise line segment and the middle points correspond to some vertical points of the obstacles, as shown in Fig. 3.18.

The visible network nodes consist of the location points corresponding to the distribution center and the customer and all vertices of the obstacles, and edges are all line segment between all pairs of points which do not break through any obstacle. An example is shown in Figs. 3.19 and 3.20:

After the visible network has been constructed, use Dijkstra's shortest algorithm to find the shortest path. It is very easy to find the shortest path between the two points while at the same time avoiding obstacles.

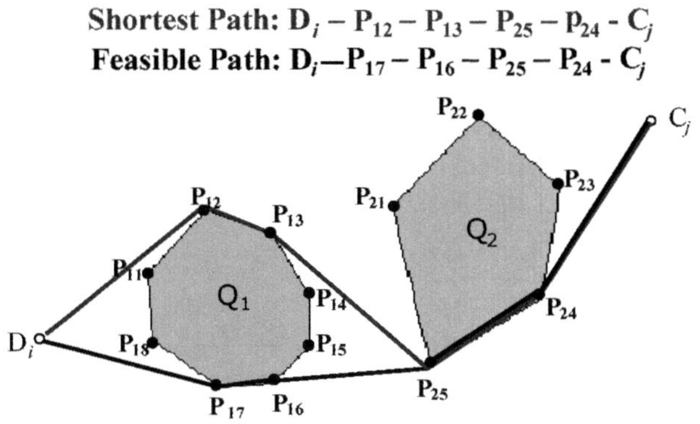

Fig. 3.18 Feasible connecting paths

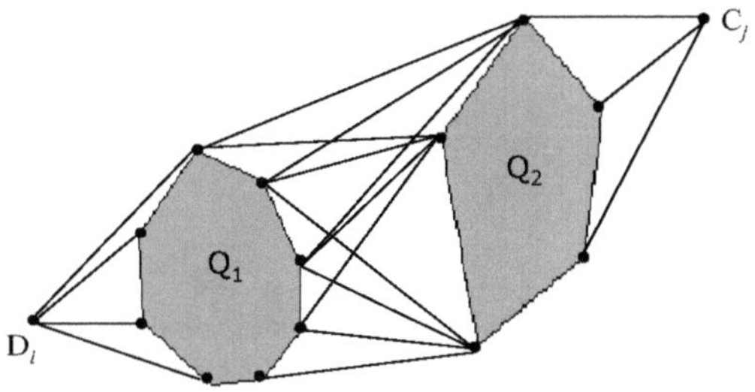

Fig. 3.19 Illustration of the visible network

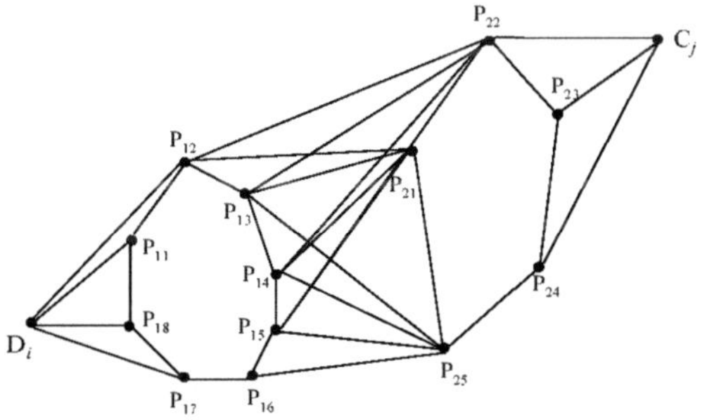

Fig. 3.20 Illustration of the network

3.3.2.4 Construction Heuristic (Nearest Neighbor Algorithm) for Allocation Customers

Nearest neighbor algorithm (NNA) is used to allocate customers based on customer demand and minimum distance between customers so that NNA preserves each DC and does not exceed its capacity by customer demand. Allocating procedure is shown in Fig. 3.21.

```
procedure: Allocating customers
input: capacity of DCs a_i, t(Fi,Cj), d_j*j,
       set of customers V, number of DCs m
       customer demand b_j
output: allocated customer to S_i
step 0: Set S_i ← ø, e_i ← 0, i=1,2,...,m
        i ← 1;
step 1: Find the closest customer j* from located DC, F_i; j* ← argmin{ t(F_i,C_j)| i=1, j∈V };
        Assign customer j* to S_i: S_i ← S_i∪{ j* }
        e_i ← e_i + b_j*;
        Delete customer j* in set of customers V: V ← V\{ j* };
step 2: i ← i +1;
        if i ≤ m then goto step 1;
step 3: i ← 1;
step 4: if no unselected customer exists then stop;
step 5: :Find the closest customer k* from previous assigned customer j*:
        k* ← argmin{ d_j*j | j∈V };
        Find the closest DC, i* from customer k*
        if e_i+ b_k* ≤ a_i
        Assign customer j* to S_i: S_i ← S_i∪{ j* }
        e_i ← e_i + b_k*;
        Delete customer j* in set of customers V: V ← V\{ j* };
step 6: i ← i +1;
        if i ≤ m then goto step 4;
        otherwise goto step 3
```

Fig. 3.21 Pseudocode of allocating customers

After the allocation procedure, a feasible allocation can be obtained and it is the solution of the allocation subproblem. There is an example of how to cluster, allocation solution in Fig. 3.22.

3.3.2.5 Genetic Operators

Linear Convex Crossover: Crossover depends on how to choose the parents and how parents produce their children. GA usually selects parents by spinning the

Table 3.5 Data set (customer locations (u_j, v_j) /demands b_j and DCs capacity a_i)

j	(u_j,v_j)	b_j	j	(u_j,v_j)	b_j	F_i	(x_i,y_i)	Capacity a_i
1	(4.0, 4.8)	5	6	(3.9, 5.8)	2	1	(2.8, 5.8)	20
2	(5.7, 5.0)	9	7	(3.3, 7.2)	2	2	(4.0, 2.4)	15
3	(1.2, 5.6)	8	8	(2.0, 7.5)	5	3	(4.6, 3.7)	25
4	(1.8, 3.6)	9	9	(2.2, 2.7)	1			
5	(3.0, 2.5)	3	10	(6.8, 5.5)	2			

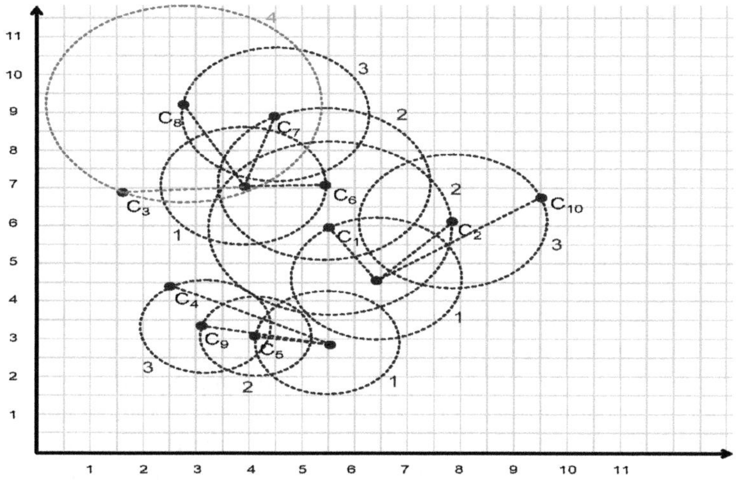

Fig. 3.22 Example of allocation solution

Table 3.6 Trace table for clustering customers

i	V	j^*	k^*	$t(F_i,C_j)$	d_{j^*j}	S_i	e_i	a_j
1	{1,2,3,4,5,6,7,8,9,10}	6	-	1.1	-	S_1={6}	2	20
2	{1,2,3,4,5,7,8,9,10}	5	-	1	-	S_2={5}	3	15
3	{1,2,3,4,7,8,9,10}	1	-	1.2	-	S_3={1}	5	25
1	{2,3,4,7,8,9,10}	6	7	-	1.5	S_1={6,7}	2+2=4	20
2	{2,3,4,8,9,10}	5	9	-	1.4	S_2={5,9}	3+1=4	15
3	{2,3,4,8,10}	1	2	-	1.7	S_3={1,2}	5+9=14	25
1	{3,4,8,10}	7	8	-	1.3	S_1={6,7,8}	2+2+5=9	20
2	{3,4,10}	9	4	-	0.4	S_2={5,9,4}	3+1+9=13	15
3	{3,10}	2	10	-	0.8	S_3={1,2,10}	5+9+2=16	25
1	{3}	8	3	-	2.1	S_1={6,7,8,3}	2+2+5+8=17	20
2	ø	-	-	-	-	-	-	-

weight roulette wheel rule and ES such as $(\mu+\lambda)$ which aims at numerical optimization by giving every member in the population pool equal probability to produce a child and let evolution be done by selection procedure. We adopt the idea of ES-$(\mu+\lambda)$ to select parents to produce child. Suppose the two chromosomes are:

$$\begin{aligned} v_1 &= \left[(x_1^1, y_1^1), (x_2^1, y_2^1), \ldots, (x_i^1, y_i^1), \ldots, (x_m^1, y_m^1)\right] \\ v_2 &= \left[(x_1^2, y_1^2), (x_2^2, y_2^2), \ldots, (x_i^2, y_i^2), \ldots, (x_m^2, y_m^2)\right] \end{aligned}$$

One child is generated by v_1 and v_2 as follows:

$$\bar{v} = \left[(\bar{x_1}, \bar{y_1}), (\bar{x_2}, \bar{y_2}), \ldots, (\bar{x_i}, \bar{y_{i})}, \ldots, (\bar{x_m}, \bar{y_m})\right]$$

and each gene in the chromosomes of children are decided by following equations:

$$\begin{aligned} \bar{x_i^1} &= \alpha_i \cdot x_i^1 + (1-\alpha_i) \cdot x_i^2 \\ \bar{y_i^1} &= \alpha_i \cdot y_i^1 + (1-\alpha_i) \cdot y_i^2 \\ \bar{x_i^2} &= (1-\alpha_i) \cdot x_i^1 + \alpha_i \cdot x_i^2 \\ \bar{y_i^2} &= (1-\alpha_i) \cdot y_i^1 + \alpha_i \cdot y_i^2 \end{aligned}$$

where i is a random number in (0, 1) $(i = 1, 2, \ldots, m)$. We use this crossover operator to produce p_C as shown in Fig. 3.23.

Subtle and Violent Mutation: Mutation is very important to introduce a new gene and prevent the premature of genetic process. For any chromosome in a population an associated real value, $0 \le \rho \le 1$, is generated randomly. If ρ is less than the predefined mutation threshold, p_M, a mutation operator is applied to this chromosome. Considering the characteristics of the original problem, we suggest two kinds of mutation operators. One is subtle mutation which only gives a small random disturbance to a chromosome to form a new child chromosome. Another is violent mutation which gives a new child chromosome totally randomly the same as the initialization. We also use the above two kinds of mutation operators alternatively in the evolutionary process.

Suppose the chromosome to be mutated is as follows:

$$v_k = \left[(x_1^k, y_1^k), (x_2^k, y_2^k), \ldots, (x_i^k, y_i^k), \ldots, (x_m^k, y_m^k)\right]$$

Then the child produced by the subtle mutation is as follows:

$$v' = \left[(x_1', y_1'), (x_2', y_2'), \ldots, (x_i', y_i'), \ldots, (x_m', y_m')\right]$$

$$\begin{aligned} x_i' &= x_i^k + \text{random value in } [-\varepsilon, \varepsilon] \\ y_i' &= y_i^k + \text{random value in } [-\varepsilon, \varepsilon] \end{aligned}$$

where ε is a small positive real number. The child produced by violent mutation is as follows:

```
procedure: Linear convex crossover
input: parents, crossover probability p_C
output: offspring
begin
    k ← 1
    do while k ≤ popSize p_C
        repeat
            i ←random[1, popSize]
            j ←random[1, popSize]
        until i≠j
        α=random (0 ,1)
        for l=1 to m
            x̄_l^k = α · x_l^i + (1 − α) · x_l^j
            ȳ_l^k = α · y_l^i + (1 − α) · y_l^j
            x̄_l^{k+1} = (1 − α) · x_l^i + α · x_l^j
            ȳ_l^{k+1} = (1 − α) · y_l^i + α · y_l^j
        end
        k ← k +2
    end
end
```

Fig. 3.23 Pseudocode of linear convex crossover

$$x_i' = \text{random value in } [x_{min}, x_{max}]$$
$$y_i' = \text{random value in } [y_{min}, y_{max}]$$

$$v' = [(x_1', y_1'), (x_2', y_2'), \ldots, (x_i', y_i'), \ldots, (x_m', y_m')]$$

$ES-(\mu+\lambda)$ selection: It is adopted to select the better individuals among parents and their children to form the next generation. However, the strategy usually leads to degeneration of the genetic process. In order to avoid this degeneration, a new selection strategy called relative prohibition is suggested.

Give two positive parameters α and γ , the neighborhood for a chromosome v_k is defined as follows:

$$\Omega(v_k, \alpha, \gamma) \cong \{s \mid \| s - s_k \| \leq \gamma, D(s_k) - D(s) < \alpha, s \in R^{2m}\}$$

In the selection process, once s_k is selected into the next generation, any chromosome falling within its neighborhood is prohibited from selection. The value of γ defines the neighborhood of s_k in terms of location, which is used to avoid selecting individuals with very small difference in location. The value of α defines the neighborhood of s_k in terms of fitness, which is used to avoid selecting individuals with very small difference in fitness. The neighborhood defined by the above equation is called a prohibited neighborhood.

3.3.3 Numerical Experiments

3.3.3.1 Capacitated Location Allocation Problems

In the first experiment, the test problem consists of 3 DCs and 16 customers (Gen and Cheng [28]). The demand of each customer and the coordinates of each customer and demand of customer are shown in Table 3.7. We assume the capacity of each DC as in Table 3.8. The GA parameters for this problem were set as: *popSize* =100; *maxGen* = 1000; p_C= 0.5; p_M= 0.5; and $[x_{min}, x_{max}] \times [y_{min}, y_{max}] \rightarrow [0, 12000] \times [0, 10000]$

Table 3.7 Coordinates and demands of customers

j	(u_j,v_j)	b_j	j	(u_j,v_j)	b_j
1	(0,0)	100	9	(4000,10000)	100
2	(0,500)	100	10	(5000,1000)	100
3	(1000,4000)	100	11	(7000,6000)	100
4	(1000,9000)	100	12	(8000,1000)	100
5	(2000,2000)	100	13	(8000,10000)	100
6	(2000,6000)	100	14	(10000,7000)	100
7	(4000,4000)	100	15	(11000,2000)	100
8	(4000,8000)	100	16	(12000,10000)	100

Table 3.8 Capacity of DCs

i	a_1	a_2	a_3
1	1800		
2	1000	1000	
3	800	600	600

We took the best result from 100 runs and then compared with ALA [32, 33], EP [33] and location-based GA in Fig. 3.24. The best result of this experiment by using location-based GA is summarized in Table 3.9 and Fig. 3.25.

Table 3.9 The best solution by location-based GA

	i a_i	Location	Allocation	Distance
	a_1=18000	(1937.50,5312.24)	1,2,3,4,5,6,7,8,9,10,11,12,13,14,15,16	76525.85
	a_1=1000	(2300.00,4900.00)	1,2,3,4,5,6,7,8,9,10	55298.45
	a_2=1000	(9333.33,6000.00)	11,12,13,16,45,16,	
	a_1=800	(2731.00,8466.00)	4,6,8,9	43324.09
	a_2=600	(9281.00,6538.00)	11,12,13,14,15,16	
	a_3=600	(1739.00,1961.00)	1,2,3,5,7,10	

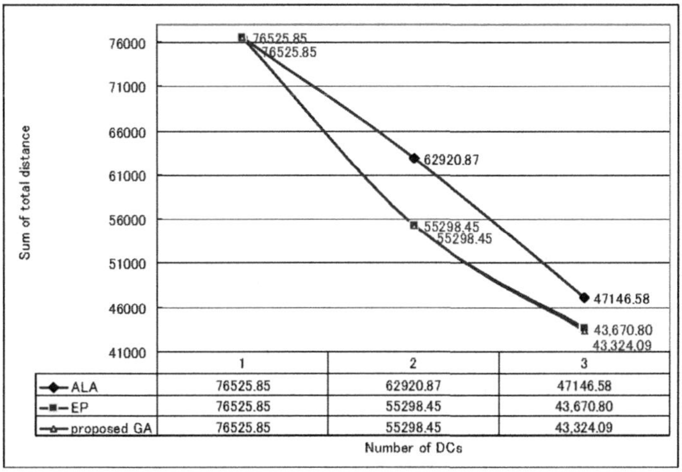

Fig. 3.24 Comparison result

3.3.3.2 Location Allocation Problem with Obstacles

In the second experiment, we develop two test problems randomly generated. In test 1 there are 14 customers whose coordinates of locations and demands are shown in Table 3.10. There exist four obstacles: two small towns, one large lake, and one forbidden region as shown in Table 3.11. The capacities of each DC are shown in Table 3.12. The GA parameters are: *popSize* =20, p_C=0.5,p_M=0.5, *maxGen*=1000. The comparative results are given Table 3.13 and Fig. 3.26.

In test 2, there are 24 customers whose coordinates of locations and demands are shown in Table 3.14. The capacities of each DC are shown in Table 3.15. There exist four obstacles: two small towns, one large lake, and one forbidden region as shown in Table 3.16. The GA parameters are: *popSize* =20, p_C=0.5, p_M=0.5, *maxGen*=1000. The comparative results are given Table 3.17 and Fig. 3.27.

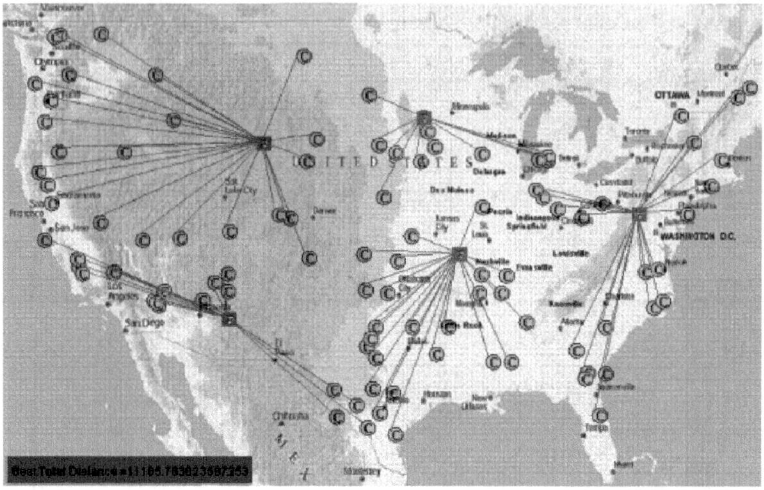

Fig. 3.25 Illustration of the result

Table 3.10 Coordinates and demand of customers

j	(u_j, v_j)	d_j	j	(u_j, v_j)	d_j
1	(4.0, 2.0)	1.0	8	(28.0, 22.0)	1.0
2	(8.0, 6.0)	1.0	9	(32.0, 32.0)	1.0
3	(14.0, 4.0)	1.0	10	(26.0, 40.0)	1.0
4	(11.0, 14.0)	1.0	11	(16.0, 38.0)	1.0
5	(18.0, 16.0)	1.0	12	(12.0, 42.0)	1.0
6	(30.0, 18.0)	1.0	13	(8.0, 38.0)	1.0
7	(36.0, 10.0)	1.0	14	(2.0, 14.0)	1.0

Table 3.11 Locations of vertices of obstacles

k	p_k	Vertex coordinates of obstacles (a_{kr}, b_{kr}) r=1,2,…,p_k
1	5	(8.0, 3.0) (12.0, 3.0) (14.0, 6.0) (10.0, 8.0) (7.0, 5.0)
2	3	(14.0, 8.0) (16.0, 5.0) (15.0, 14.0)
3	4	(6.0, 10.0) (9.0, 9.0) (11.0, 16.0) (7.0, 18.0)
4	5	(13.0, 17.0) (22.0, 20.0) (20.0, 34.0) (18.0, 36.0) (11.0, 25.0)

Table 3.12 Capacity of DCs

DC i	Capacity q_i
1	4.0
2	6.0
3	4.0

Table 3.13 Comparision of heuristic and location-based GA

Method	Total length Sum	DC locations	Customer allocations
Heuristic [5]	107.0858	1.(15.5, 39.5) 2.(7.0, 6.59) 3.(25.8333, 18.6667)	10,11,12,13 1,2,3,14 4,5,6,7,8,9
hGA	97.422	1.(15.2, 38.9) 2.(9.1, 8.5) 3.(30.1, 18.7)	10,11,12,13 1,2,3,4,5,14 6,7,8,9

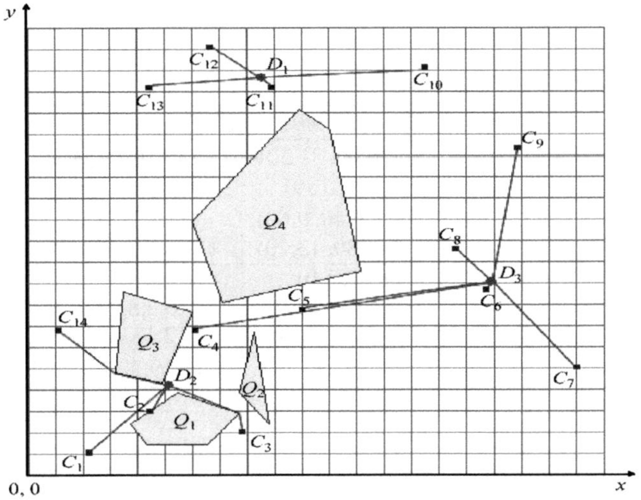

Fig. 3.26 Location results of this problem by location-based GA

Table 3.14 Coordinates and demand of customers

j	(u_j, v_j)	d_j	j	(u_j, v_j)	d_j
1	(1.2,2.4)	1.0	13	(28.8,33.6)	1.0
2	(3.6,6.0)	1.0	14	(24.0,3.6)	1.0
3	(9.6,2.4)	1.0	15	(22.8,30.0)	1.0
4	(12.0,9.6)	1.0	16	(18.0,34.0)	1.0
5	(18.0,6.0)	1.0	17	(14.4,25.2)	1.0
6	(24.1,3.6)	1.0	18	(12.0,18.0)	1.0
7	(33.6,2.4)	1.0	19	(7.2,26.4)	1.0
8	(34.8,6.0)	1.0	20	(6.0,19.2)	1.0
9	(25.2,14.4)	1.0	21	(2.4,13.2)	1.0
10	(30.0,19.2)	1.0	22	(2.4,22.8)	1.0
11	(28.8,25.2)	1.0	23	(10.8,34.8)	1.0
12	(33.6,31.2)	1.0	24	(1.2,33.6)	1.0

Table 3.15 Capacity of DCs

DC i	Capacity q_i
1	4.0
2	6.0
3	6.0
4	8.0

Table 3.16 Locations of vertices of obstacles

k	p_k	Vertex coordinates of obstacles (a_{kr}, b_{kr}) r=1,2,…,p_k
1	3	(6.0, 7.0) (12.0, 8.4) (4.8, 13.2)
2	4	(10.8, 26.4) (14.4, 31.2) (8.4, 33.6) (6.0, 31.2)
3	4	(25.2,18.0) (20.4,27.6) (16.8, 18.0) (18.0, 9.6)
4	6	(27.6,3.6) (30.0, 3.6) (32.4, 9.6) (31.2, 15.6) (28.8, 15.6) (26.4, 9.6)

Table 3.17 Comparision of heuristic and location-based GA

Method	Total length Sum	DC locations	Customer allocations
Heuristic [4]	184.21	1.(6.00, 31.20)	17,19,23,24
		2.(7.0, 6.59)	11,12,13,14,15,16
		3.(26.40, 9.60)	5,6,7,8,9,10
		4.(4.80, 13.20)	1,2,3,4,18,20,21,22
hGA	164.07	1.(5.2, 5.0)	1,2,3,4
		2.(25.2, 29.6)	5,6,7,8,9,10
		3.(26.6, 11.7)	11,12,13,14,15,16
		4.(7.6, 22.3)	17,18,19,20,21,22,23,24

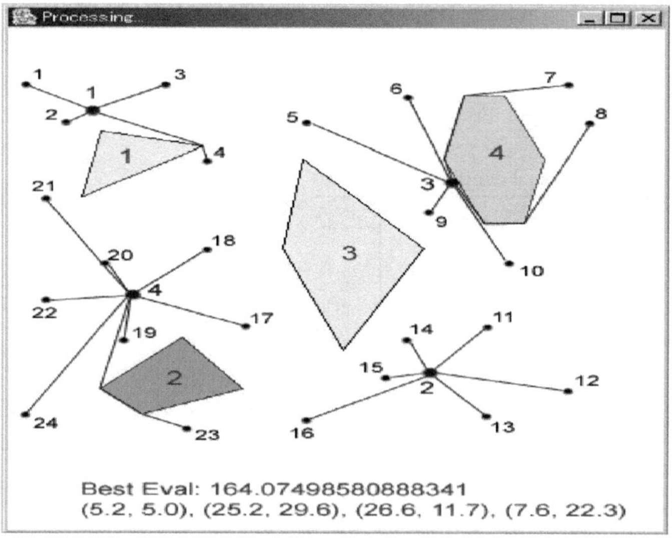

Fig. 3.27 Location results of this problem with location-based GA

3.4 Multi-stage Logistics Models

Supply Chain Management (SCM) describes the discipline of optimizing the delivery of goods, services and information from supplier to customer. Typical supply chain management goals include transportation network design, plant/DC location, production schedule streamlining, and efforts to improve order response time. Logistics network design is one of the most important fields of SCM. It offers great potential to reduce costs and to improve service quality. The transportation problem (TP) is a well-known basic network problem which was originally proposed by Hitchcock [34]. The objective is to find the way of transporting homogeneous product from several sources to several destinations so that the total cost can be minimized. For some real-world applications, the transportation problem is after extended to satisfy several other additional constraints or performed in several stages. Logistics is often defined as the art of bringing the right amount of the right product to the right place at the right time and usually refers to supply chain problems (Tilanus, [35]). The efficiency of the logistic system is influenced by many factors; one of them is to decide the number of DCs, and find the good location to be opened, in such a way that the customer demand can be satisfied at minimum DCs' opening cost and minimum shipping cost.

In literature, multi-stage logistics problem and its different variations have been studied. Geoffrion and Graves [36] were the first researchers studied on two-stage distribution problem. They employed Benders' partitioning procedure to solve the problem. Pirkul and Jayaraman [37] presented a new mathematical formulation called PLANWAR to locate a number of production plants and warehouses and to design distribution network so that the total operating cost can be minimized. They developed an approach based on Lagrangian relaxation to solve the problem. Heragu [38] introduced the problem called the two-stage transportation problem (tsTP) and gave the mathematical model for this problem. The model includes both the inbound and outbound transportation cost and aims to minimize the overall cost. Heragu stated that this model did not require the plant capacity to be known or fixed. Hindi *et al.* [39] addressed a two-stage distribution-planning problem. They considered two additional requirements on their problem. First, each customer must be served from a single DC. Second, it must be possible to ascertain the plant origin of each product quantity delivered. The authors gave a mathematical model for the problem and developed a branch and bound algorithm to solve the problem. Tragantalerngsak *et al.* [40] considered a two-echelon facility location problem in which the facilities in the first echelon were uncapacitated while the facilities in the second echelon were capacitated. The goal was to determine the number and locations of facilities in both echelons in order to satisfy customer demand of the product. They developed a Lagrangian relaxation based-branch and bound algorithm to solve the problem. Amiri [41] studied a distribution network design problem in a supply chain system involving the location of plants and warehouses, determining the best strategy for distributing the product from the plants to warehouses and from the warehouses to customers. Different from other studies in the literature, the author considered multiple levels of capacities for the plants and warehouses, and developed a mathematical model for the problem. The author had proposed a heuristic solution approach based on Lagrangian relaxation to solve the problem.

Recently, GAs have been successfully applied to logistics network models. Michalewicz *et al.* [27] and Viagnaux and Michalewicz [42] firstly discussed the use of GA for solving linear and nonlinear transportation problems. In their study, while matrix representation was used to construct a chromosome, the matrix-based crossover and mutation had been developed. Another genetic algorithm (GA) approach for solving solid TP was given by Li *et al.* [43]. They used the three-dimensional matrix to represent the candidate solution to the problem. Syarif and Gen [44] considered production/distribution problem modeled by tsTP and propose a hybrid genetic algorithm. Gen *et al.* [15] developed a priority-based Genetic Algorithm (priGA) with new decoding and encoding procedures considering the characteristic of tsTP. Altiparmak *et al.* [21] extended priGA to solve a single-product, multi-stage logistics design problem. The objectives are minimization of the total cost of supply chain, maximization of customer services that can be rendered to customers in terms of acceptable delivery time (coverage), and maximization of capacity utilization balance for DCs (*i.e.* equity on utilization ratios). Futhermore, Altiparmak *et al.* [22] also apply the priGA to solve a single-source, multi-product, multi-stage logistics design problem. As an extended multi-stage logistics network model, Lee *et al.* [45] apply the priGA to solve a multi-stage reverse logistics network problem (mrLNP), minimizing the total of costs to reverse logistics shipping cost and fixed cost of opening the disassembly centers and processing centers.

3.4.1 Mathematical Formulation of the Multi-stage Logistics

3.4.1.1 Two-stage Logistics Model

The two-stage logistics model (or two-stage transportation problem: tsTP) is considered to determine the distribution network to satisfy the customer demand at minimum cost subject to the plant and DCs capacity and also the minimum number of DCs to be opened. The tsTP can be defined by the following assumptions:

A1. The customer locations and their demand were known in advance.
A2. The numbers of potential DC locations as well as their maximum capacities were also known.
A3. Each facility has a limited capacity and there is a limit about the number of facility to be opened.

Notation

Indices

i: index of plants $(i = 1,2,\dots,I)$
j: index of DC $(j = 1,2,\dots,J)$
k: index of customer $(k = 1,2,\dots,K)$

Parameters

I: total number of plants
J: total number of DCs
K: total number of customers
a_i: capacity of plant i
b_j: capacity of distribution center j
d_k: demand of customer k
t_{ij}: unit cost of transportation from plant i to distribution center j
c_{jk}: unit cost of transportation from distribution center j to customer k
g_j: fixed cost for operating distribution center j
W: an upper limit on total number of DCs that can be opened

Decision Variables

x_{ij}: the amount of shipment from plant i to distribution center j
y_{jk}: the amount of shipment from distribution center j to customer k
z_j: 0–1 variable that takes on the value 1 if DC j is opened

The tsTP includes the transportation problem and facility location problem. The problem is referred to as the capacitated p-median facility location problem and it is an NP-hard problem (Garey and Johnson [46]). The illustration of the two-stage logistics problem and network structure are showed in Figs. 3.28 and 3.29. The mathematical model of the problem is

$$\min Z = \sum_{i=1}^{I}\sum_{j=1}^{J} t_{ij}x_{ij} + \sum_{j=1}^{J}\sum_{k=1}^{K} c_{jk}y_{jk} + \sum_{j=1}^{J} g_j z_j \tag{3.43}$$

$$\text{s. t.} \sum_{j=1}^{J} x_{ij} \le a_i, \quad \forall\, i \tag{3.44}$$

$$\sum_{k=1}^{K} y_{jk} \le b_j z_j, \quad \forall\, j \tag{3.45}$$

$$\sum_{j=1}^{J} z_j \le W \tag{3.46}$$

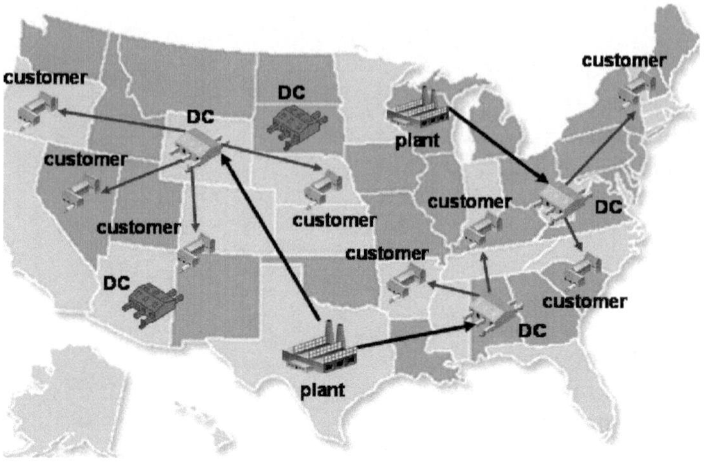

Fig. 3.28 Illustration of two-stage logistics problem

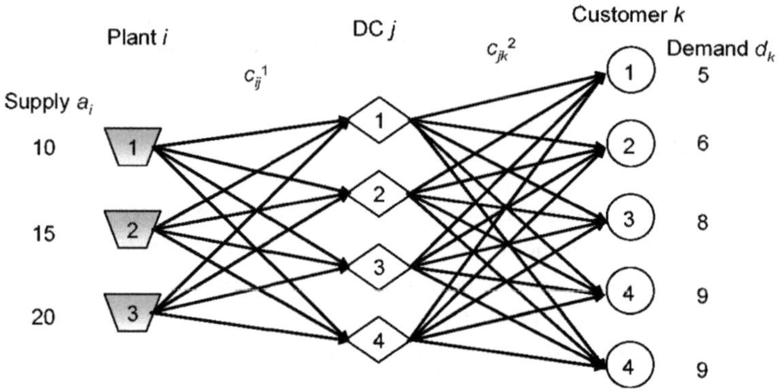

Fig. 3.29 Illustration of two-stage logistics model

$$\sum_{j=1}^{J} y_{jk} \geq d_k, \quad \forall\, k \tag{3.47}$$

$$\sum_{i=1}^{I}\sum_{j=1}^{J} x_{ij} = \sum_{j=1}^{J}\sum_{k=1}^{K} y_{jk} \tag{3.48}$$

$$x_{ij}, y_{jk} \geq 0, \quad \forall\, i, j, k \tag{3.49}$$

$$z_j = 0, 1 \quad \forall\, j \tag{3.50}$$

where the constraints at Eqs. 3.44 and 3.45 ensure that the plant-capacity constraints and the distribution center-capacity constraints, respectively, the constraint at Eq.

3.46 satisfies the opened DCs not exceeding their upper limit. This constraint is very important when a manager has limited available capital. Constraint at Eq. 3.47 ensures that all the demands of customers are satisfied by opened DCs; constraints at Eqs. 3.49 and 3.50 enforce the non-negativity restriction on the decision variables and the binary nature of the decision variables used in this model. Without loss of generality, we assume that this model satisfies the balanced condition, since the unbalanced problem can be changed to a balanced one by introducing dummy suppliers or dummy customers.

3.4.1.2 Three-stage Logistics Model

There are now expanded multi-stage logistics models for applications to real-world cases so the SCN design problem is referred to as three-stage, *i.e.*, suppliers, plants, DCs and customers. The supply chain network design problem has been gaining importance due to increasing competitiveness introduced by market globalization. Firms are obliged to maintain high customer service levels while at the same time they are forced to reduce cost and maintain profit margins. A supply chain is a network of facilities and distribution options that performs the functions of procurement of materials, transformation of these materials into intermediate and finished products, and distribution of these finished products to customers.

Notation

Index

i: index of suppliers $(i = 1,2,\ldots,I)$
j: index of plants $(j = 1,2,\ldots,J)$
k: index of DC $(k = 1,2,\ldots,K)$
l: index of customer $(l = 1,2,\ldots,L)$

Parameters

a_i: capacity of supplier i
b_j: capacity of plant j
c_k: capacity of DC k
d_l: demand of customer l
s_{ij}: unit cost of production in plant j using material from supplier i
t_{jk}: unit cost of transportation from plant j to DC k
u_{kl}: unit cost of transportation from DC k to customer l
f_j: fixed cost for operating plant j

g_k: fixed cost for operating DC k
W: an upper limit on total number of DCs that can be opened
P: an upper limit on total number of plants that can be opened

Decision Variables

x_{ij}: quantity produced at plant j using raw material from supplier j
y_{jk}: amount shipped from plant j to DC k
z_{kl}: amount shipped from DC k to customer l
w_j: 0–1 variable that takes on the value 1 if production takes place at plant j
z_k: 0–1 variable that takes on the value 1 if DC k is opened

The design task involves the decision of choosing to open a facility (plant or DCs) or not, and the distribution network design to satisfy the customer demand at minimum cost. The three-stage logistics network model is formulated as follows:

$$\min \sum_i \sum_j s_{ij} x_{ij} + \sum_j \sum_k t_{jk} y_{jk} + \sum_k \sum_l u_{kl} z_{kl} + \sum_j f_j w_j + \sum_k g_k z_k \tag{3.51}$$

$$\text{s. t. } \sum_j x_{ij} \le a_i, \quad \forall\, i \tag{3.52}$$

$$\sum_k y_{jk} \le b_j w_j, \quad \forall\, j \tag{3.53}$$

$$\sum_j w_j \le P \tag{3.54}$$

$$\sum_l z_{kl} \le c_k z_k, \quad \forall\, k \tag{3.55}$$

$$\sum_k z_k \le W \tag{3.56}$$

$$\sum_k z_k \ge d_l, \quad \forall\, l \tag{3.57}$$

$$w_j, z_k = 0, 1 \quad \forall\, j, k \tag{3.58}$$

$$x_{ij}, y_{jk}, z_{ki} \ge 0 \quad \forall\, i,\ j,\ k,\ l \tag{3.59}$$

3.4.1.3 Reverse Logistics Model

Recently, due to the great mumber of resources and environment problems, the number of trade regulations which are environmentally related have also increased through international organization and widespread agreement on environment preservation. So the increase of interest about the collection of disused products for recovering of resources is understandable.

Beyond the current interest in supply chain management, recent attention has been given to extending the traditional forward supply chain to incorporate a reverse logistic element owing to liberalized return policies, environmental concern, and a growing emphasis on customer service and parts reuse. Implementation of reverse logistics especially in product returns would allow not only for savings in inventory carrying cost, transportation cost, and waste disposal cost due to returned products, but also for the improvement of customer loyalty and futures sales.

Reverse logistics is defined by the European Working Group on Reverse Logistics (REVLOG) as the proposal of planning, implementing and controlling flows of raw materials, in process inventory and finished goods, from the point of use back to the point of recovery or point of proper disposal [47]. In a broader sense, reverse logistics refers to the distribution activities involved in product returns, source reduction, conservation, recycling, substitution, reuse, disposal, refurbishment, repair and remanufacturing [48].

Concerning reverse logistics, a lot of researches have been made on various fields and subjects such as reuse, recycling, remanufacturing logistics etc. In reuse logistics models, Kroon *et al.* [49] reported a case study concerning the design of a logistics system for reusable transportation packages. The authors proposed an mIP (mixed integer programming) model, closely related to a classical un-capacitated warehouse location model.

In recycling models, Pati *et al.* [50] developed an approach based on a mixed integer goal programming model to solve the problem. The model studies the interrelationship between multiple objectives of a recycled paper distribution network. The authors considered the minimization of reverse logistics cost as the objective and proposed the reverse logistics network of the remanufacturing process.

In remanufacturing models, Kim *et al.* [51] discussed a notion of remanufacturing systems in reverse logistics environment. They proposed a general framework of regarding supply planning and developed a mathematical model to optimize the supply planning function. The model determines the quantity of product parts processed in the remanufacturing facilities and subcontractors, and the amount of parts purchased from the external suppliers while maximizing the total remanufacturing cost saving. Further, Lee *et al.* [52] have also formulated a reverse logistics network problem (rLNP) minimizing total reverse logistics various shipping costs.

As shown in Fig. 3.30, the reverse logistics problem can be formulated as a multi-stage logistics network model. First, recovered products from customers are transported to disassembly center, in the first stage. Then disassembled parts are transported to processing center, in the second stage. Last, in the third stage, processed parts are transported to manufacturer. In the third stage, if the quantity of provided parts from processing center is not enough for requirement of manufacturer, then manufacturer must buy parts from supplier. In case of opposition, then exceeded capacities are distributed in order of recycling and disposal. The illustrative of the multi-stage reverse logistics network model is showed in Fig. 3.31. The reverse logistics model can be defined by the following assumptions:

A1. If the quantity of provided parts from processing center is not enough for the requirement of manufacturer, then manufacturer must buy parts from the supplier.

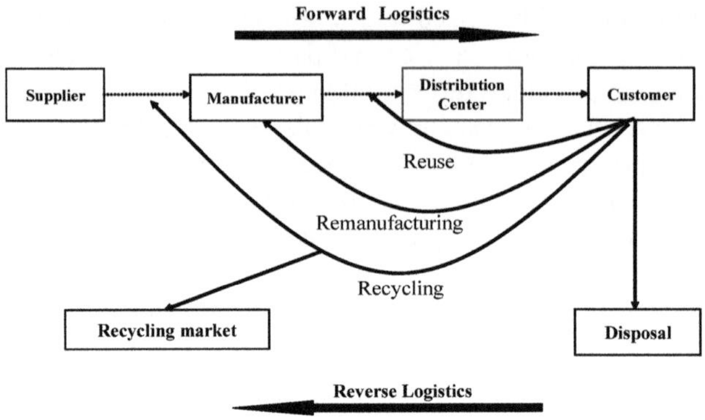

Fig. 3.30 Logistics flow of forward and reverse

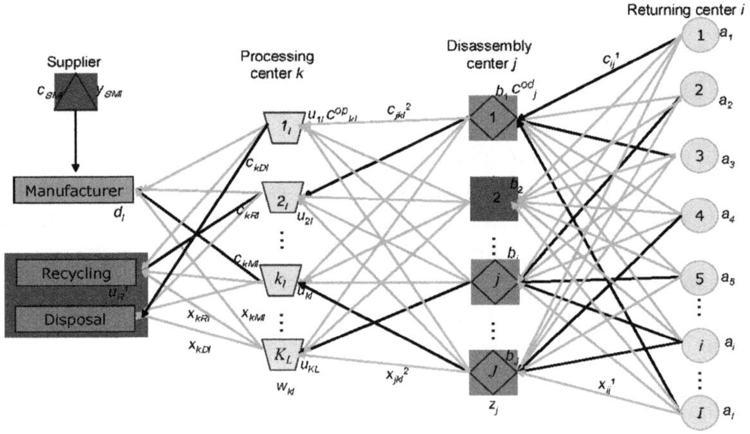

Fig. 3.31 Multi-stage reverse logistics network model

A2. If the quantity of provided parts from processing center exceeds the requirement of manufacturer, then exceeded capacities are distributed in order of recycling and disposal.

A3. Not considering the inventories.

A4. The parts demands of manufacturer are known in advance.

A5. Returning centers, disassembly centers and processing centers maximum capacities are also known.

A6. Disassembled parts are processed in order to part A and B in the processing centers because it is complex to consider part A and B at the same time.

Notation

Index

i: index of returning center $(i = 1, 2, \ldots, I)$
j: index of disassembly center $(j = 1, 2, \ldots, J)$
k: index of processing center $(k = 1, 2, \ldots, K)$
l: index of part $(l = 1, 2, \ldots, L)$

Parameters

I: number of returning centers
J: number of disassembly centers
K: number of processing centers
L: number of parts
a_i: capacity of returning center i
b_j: capacity of disassembly center j
u_{kl}: capacity of k for parts l
u^1_{kR}: capacity of R
d_l: demand of parts l in M
n_l: the number of l from disassembling one unit of product
c^1_{ij}: unit cost of transportation from i to j
c^2_{jkl}: unit cost of transportation from j to k
c_{kMl}: unit cost of transportation from k to M
c_{kRl}: unit cost of transportation from k to R
c_{kDl}: unit cost of transportation from k to D
c_{SMl}: unit cost of transportation from S to M
c^{od}_j: the fixed opening cost for j
c^{op}_{kl}: the fixed opening cost for k

Decision Variables

x^1_{ij}: amount shipped from i to j
y^2_{jkl}: amount shipped from j to k
x_{kMl}: amount shipped from k to M
x_{kRl}: amount shipped from k to R
x_{kDl}: amount shipped from k to D
y_{kl}: amount shipped from S to M
z_j: 1, if disassembly center j is open; 0, otherwise
w_{kl}: 1, if processing center j is open; 0, otherwise

The mathematical model of the problem is

$$\min z = \sum_{j=1}^{J} c_j^{od} z_j + \sum_{k=1}^{K}\sum_{l}^{L} c_{kl}^{op} w_{kl} + \sum_{i=1}^{I}\sum_{j=1}^{L} c_{ij}^{1} x_{ij}^{1} + \sum_{j=1}^{J}\sum_{k=1}^{K}\sum_{l=1}^{L} c_{jkl}^{2} x_{jkl}^{2}$$
$$+ \sum_{k=1}^{K}\sum_{l=1}^{L} c_{kMl} x_{kMl} - \sum_{k=1}^{K}\sum_{l=1}^{L} c_{kRl} x_{kRl} + \sum_{k=1}^{K}\sum_{l=1}^{L} c_{kDl} x_{kDl}$$
$$+ \sum_{l=1}^{L} c_{SMl} x_{SMl} \tag{3.60}$$

$$\text{s. t. } \sum_{j=1}^{J} x_{ij}^{1} \le a_i, \quad \forall\, i \tag{3.61}$$

$$\sum_{k=1}^{K} x_{jkl}^{2} \ge b_j Z_j, \quad \forall\, j,l \tag{3.62}$$

$$x_{kMl} + x_{kRl} + x_{kDl} \ge w_{kl}, \quad \forall\, k,l \tag{3.63}$$

$$\sum_{j=1}^{J} z_j \le J \tag{3.64}$$

$$\sum_{k=1}^{K} w_{kl} \le K, \quad \forall\, l \tag{3.65}$$

$$y_{SMl} + \sum_{k=1}^{k} x_{kMl} \ge d_l, \quad \forall\, l \tag{3.66}$$

$$x_{ij}^{1},\ x_{jkl}^{2},\ x_{kMl},\ x_{kRl},\ x_{kDl},\ y_{SMl} \ge 0, \quad \forall\, i,j,k \tag{3.67}$$

$$z_j = \{0,1\} \quad \forall\, j \tag{3.68}$$

$$w_{kl} = \{0,1\} \quad \forall\, k,l \tag{3.69}$$

(3.70)

where, *R*, *D*, *M* and *S* represent recycling, disposal, manufacturer and supplier. The objective function at Eq. 3.60 is to minimize total reverse logistics, *i.e.*, shipping cost and fixed cost of opening the disassembly centers and processing centers until used products which came from collection become reusable parts. Constraints at Eqs. 3.61–3.63 are the returning center-capacity constraints, disassembly center-capacity constraints and processing center-capacity constraints. Constraints at Eqs. 3.64 and 3.65 limit the number of disassembly centers and processing centers that can be opened. The constraint at Eq. 3.66 gives the demand of parts, the constraint at Eq. 3.67 enforces the non-negativity restriction on the decision variables, and Eqs. 3.68 and 3.69 impose the integrality restriction on the decision variables used in this model.

3.4.2 Priority-based GA for the Multi-stage Logistics

Let $P(t)$ and $C(t)$ be parents and offspring in current generation t, respectively. The overall procedure of priority-based GA (priGA) for solving multi-stage logistics models (msLP) is outlined as follows.

---

procedure: priGA for msLP
input: msLP data, GA parameters ($popSize$, $maxGen$, p_M, p_C)
output: he best solution E
begin
  $t \leftarrow 0$;
  initialize $P(t)$ by priority-based multi-stage encoding routine;
  evalute $P(t)$ by priority-based multi-stage decoding routine;
  while (**not** terminating condition) **do**
    create $C(t)$ from $P(t)$ by weight mapping crossover routine;
    create $C(t)$ from $P(t)$ by swap mutation routine;
    create $C(t)$ from $P(t)$ by immigration routine;
    evalute $C(t)$ by priority-based multi-stage decoding routine;
    select $P(t+1)$ from $P(t)$ and $C(t)$ by roulette wheel selection routine;
    $t \leftarrow t+1$;
  end
  output the best solution
end

3.4.2.1 Transportation Tree Representation

For a transportation problem, a chromosome consists of priorities for sources and depots to a obtain transportation tree. Its length is equal to the total number of sources ($|K|$) and depots ($|J|$), *i.e.*, $|K|+|J|$. For a given chromosome, the transportation tree is generated by a sequential arc appending procedure between sources and depots. At each step, only one arc is added to the tree by selecting a source (depot) with the highest priority and connecting it to a depot (source) with the minimum cost. Figure 3.32 represents a transportation tree with three sources and four depots, its cost matrix and the corresponding priority based encoding.

The decoding algorithm of the priority-based encoding and its trace table are given in Fig. 3.33 and Table 3.18, respectively. As is seen in the trace table, at the first step of the decoding procedure, an arc between Depot 1 and Source 1 is added to the transportation tree since Depot 1 has the highest priority in the chromosome and the lowest cost between Source 1 and Depot 1. After determining the amount of shipment that is g_{11} = min550, 300 = 300, capacity of source and demand of depot are updated as a_1 = 550–300 = 250, b_1 = 300–300 = 0, respectively. Since $b_1 = 0$, the priority of Depot 1 is set to 0, and Depot 4 with the next highest priority is selected.

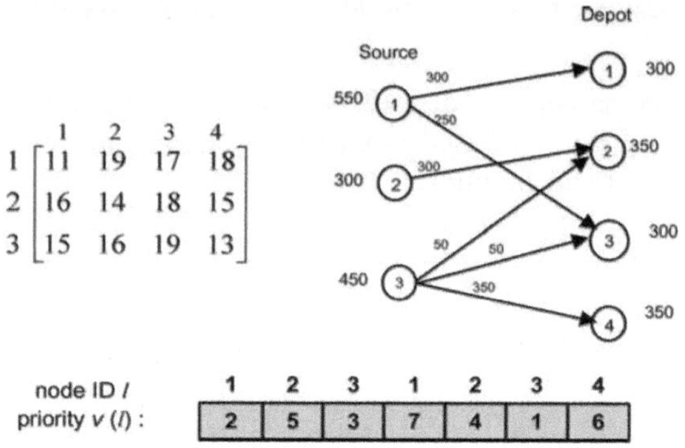

Fig. 3.32 A sample of transportation tree and its encoding

After adding arc between Depot 4 and Source 3, the amount of shipment between them is determined and their capacity and demand are updated as explained above, and this process repeats until demands of all depots are met.

---

procedure: Decoding of the chromosome for transportation tree
input: K: set of sources, J: set of depots,
    b_j: demand on depot j, $\forall j \in J$,
    a_k: capacity of source k, $\forall k \in K$,
    c_{kj}: transportation cost of one unit of product from source k to depot j, $\forall k \in K, \forall j \in J$,
    $v\,(k+j)$: chromosome, $\forall k \in K$, $\forall j \in J$,
output : g_{kj}: the amount of product shipped from source k to depot j
step 1. $g_{kj} \leftarrow 0$, $\forall k \in K$, $\forall j \in J$,
step 2. $l \leftarrow \arg\max\{v(t),\ t \in |K|+|J|\}$; select a node
step 3. if $l \in K$, then $k^* \leftarrow l$; select a source
    $j^* \leftarrow \arg\min\{c_{kj} \mid v(j) \neq 0, j \in J\}$; select a depot with the lowest cost
  else $j^* \leftarrow l$; select a depot
    $k^* \leftarrow \arg\min\{c_{kj} \mid v(j) \neq 0, k \in K\}$; select a source with the lowest cost
step 4. $g_{k^*j^*} \leftarrow \min\{a_{k^*}, b_{j^*}\}$; assign available amount of units
  Update availabilities on source (k^*) and depot (j^*)
    $a_{k^*} =; a_{k^*} - g_{k^*j^*} \quad b_{j^*} = b_{j^*} - g_{k^*j^*}$
step 3. if $a_{k^*} = 0$ then $v(k^*) = 0$
  if $b_{j^*} = 0$ then $v(j^*) = 0$
step 5. if $v(|K|+j) = 0$, $\forall j \in J$, then calculate transportation cost and return,
else goto Step 1.

Fig. 3.33 Decoding algorithm for the priority-based encoding

Table 3.18 Trace table of decoding procedure

Iteration	$v(k+j)$	a	b	k	j	g_{kj}
0	[2 5 3 | 7 4 1 6]	(550, 300, 450)	(300, 350, 300, 350)	1	1	300
1	[2 5 3 | 0 4 1 6]	(250, 300, 450)	(0, 350, 300, 350)	3	4	350
2	[2 5 3 | 0 4 1 0]	(250, 300, 100)	(0, 350, 300, 0)	2	2	300
3	[2 0 3 | 0 4 1 0]	(250, 0, 100)	(0, 50, 300, 0)	3	2	50
4	[2 0 3 | 0 0 1 0]	(250, 0, 50)	(0, 0, 300, 0)	3	3	50
5	[2 0 0 | 0 0 1 0]	(250, 0, 0)	(0, 0, 250, 0)	1	3	250
6	[0 0 0 | 0 0 0 0]	(0, 0, 0)	(0, 0, 0, 0)			

To obtain priority-based encoding for any transportation tree, the encoding algorithm is proposed as shown in Fig. 3.34. Priority assignment to nodes starts from the highest value, *i.e.*, $|K| + |J|$, and it is reduced by 1 until assigning priority to all nodes. In each iteration, first we select a node i (depot or customer) with highest priority, second we select another node j that connects with the node i and the shipment cost from i to j (or j to i) is lowest. Table 3.19 gives the trace table for the transportation tree in Fig. 3.32 to illustrate how its priority-based encoding is obtained. As is seen in Fig. 3.32, there are three leaf nodes. These are Source 2, Depot 1 and Depot 3. The corresponding costs to (from) depot (source) are 14, 11, and 13, respectively. Since the lowest cost is between Depot 1 and Source 1, the highest priority, *i.e.*, 7, is assigned to Depot 1. After removing the corresponding arc from the transportation tree, a new leaf node is selected and the process repeats as defined above.

procedure: Encoding of transportation tree,
input: K: set of sources, J : set of depots,
  b_j: demand on depot j, $\forall j \in J$,
  a_k: capacity of source k, $\forall k \in K$,
  c_{kj}: transportation cost of one unit of product from source k to depot j, $\forall k \in K, \forall j \in J$,
  g_{kj}: amount of shipment from source k to depot j
output: $v[t]$: chromosome
step 1: priority $p \leftarrow (|J|+|K|)$, $v[t] \leftarrow 0$, $\forall t \in |J|+|K|$;
step 2: $(k', j') \leftarrow$ **argmin**$\{c_{kj} \mid y_{kj} \neq 0 \;\&\; (b_j = y_{kj} \| a_k = y_{kj})\}$;
step 3: $b_{j'} = b_{j'} - y_{k'j'}$, $a_{k'} = a_{k'} - y_{k'j'}$;
step 4: **if** $b_{j'} = 0$ **then** $v[j'] \leftarrow p$, $p \leftarrow p-1$;
  if $a_{k'} = 0$ **then** $v[|J|+k'] \leftarrow p$, $p \leftarrow p-1$;
step 4: **if** $(b_{j'} = 0\ \forall j' \in J)$ & $(a_{k'} = 0, \forall k' \in K)$ **then goto** step 5;
  else return step 1;
step 5: **for** l=1 **to** p **do**
    $v[t] \leftarrow l$, t=**random**$[1, (|J|+|K|)]$ & $v[t]$=0;
step 6: **output** encoding $v[t]$, $\forall t \in |J|+|K|$

Fig. 3.34 Encoding algorithm for transportation tree

Table 3.19 Trace table of encoding procedure

Iteration	*p*	*a*	*b*	(*k*, *j*)	*v*(*k*+*j*)
0	7	(550, 300, 450)	(300, 350, 300, 350)	(1, 1)	[0 0 0 | 7 0 0 0]
1	6	(250, 300, 450)	(0, 350, 300, 350)	(3, 4)	[0 0 0 | 7 0 0 6]
2	5	(250, 300, 100)	(0, 350, 300, 0)	(2, 5)	[0 5 0 | 7 0 0 6]
3	4	(250, 0, 100)	(0, 50, 300, 0)	(2, 2)	[0 5 0 | 7 4 0 6]
4	3	(250, 0, 50)	(0, 0, 300, 0)	(3, 2)	[0 5 3 | 7 4 0 6]
5	2	(250, 0, 0)	(0, 0, 250, 0)	(1, 3)	[2 5 3 | 7 4 0 6]
6		(0, 0, 0)	(0, 0, 0, 0)		[2 5 3 | 7 4 1 6]

3.4.2.2 Multi-stage Encoding and Decoding

For a two-stage logistics model, the two priority-based encodings are used to represent the transportation trees on stages. This means that each chromosome in the population consists of two parts. While the first part (*i.e.*, the first priority-based encoding) represents transportation tree between plants and DCs, the second part (*i.e.*, the second priority-based encoding) represents transportation tree between DCs and customers. Fig. 3.35 gives overall decoding procedure for priority-based encoding on tsTP. As shown in Fig. 3.36, we consider a problem that has three feasible plants, four feasible DCs and five customers. When the upper limit of opened DCs is taken as three, the first part of chromosome consists of seven digits, the second part consists of nine digits. The total capacity of DCs which will be opened has to satisfy the total demand of customers. Considering this property, we first decode the second part of chromosome. In this phase, transportation tree on the second stage and decision related with which DCs will be opened are obtained simultaneously. After that the transportation tree on the first stage is obtained by considering DCs which were opened in the second stage of the decoding procedure. In the second stage, it is possible to obtain a transportation tree that doesn't satisfy the upper limit of opened DCs, since connection between source and depot is realized considering minimum cost. This means that chromosome represents an infeasible solution.

3.4.2.3 Genetic Operators

Crossover: In this section, we adopt the weight mapping crossover (WMX) proposed in Chapter 2 (page 67). WMX can be viewed as an extension of one-cut point crossover for permutation representation. As one-cut point crossover, two chromosomes (parents) would be chose a random-cut point and generate the offspring by using segment of own parent to the left of the cut point, then remapping the right segment based on the weight of other parent of the right segment.

Mutation: Similar to crossover, mutation is done to prevent the premature convergence and it explores new solution space. However, unlike crossover, mutation is

---

procedure : Decoding for two-stage logistics model

step 0 : $z_j \leftarrow 0, \ \forall j \in J$

step 1 : find $[y_{kj}]$ by **decoding procedure**;

step 2 : **for** $j = 1$ **to** $|J|$

  if $b_j' = 0$ **then** $v_2(j) \leftarrow 0$ **else** $z_j \leftarrow 1$

step 3 : **if** $\sum_{j=1}^{J} z_j \le W$, **then** find $[x_{ij}]$ by **decoding procedure**;

  else discard the chromosome

step 6 : **calculate** and **output** the total cost:

$$Z = \sum_{i=1}^{I}\sum_{j=1}^{J} t_{ij}x_{ij} + \sum_{j=1}^{J}\sum_{k=1}^{K} c_{jk}y_{jk} + \sum_{j=1}^{J} g_j z_j$$

Fig. 3.35 Decoding procedure for two-stage encoding

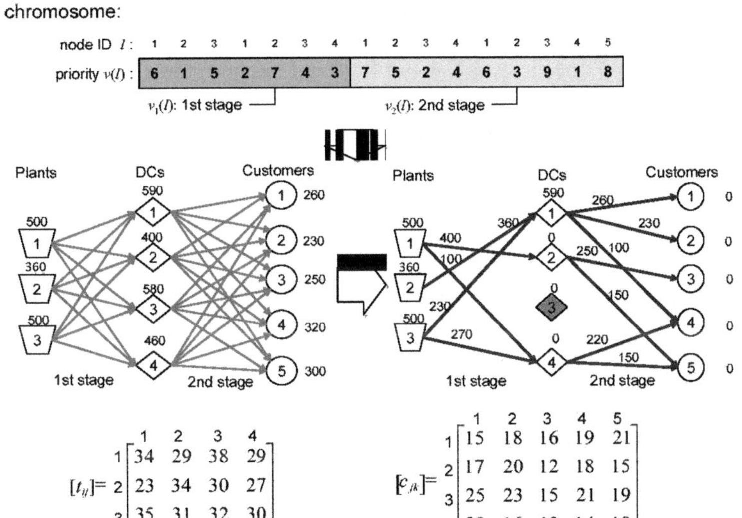

Fig. 3.36 An illustration of chromosome, transportation trees and transportation costs for each stage on tsTP

usually done by modifying a gene within a chromosome. We also investigated the effects of two different mutation operators on the performance of GA. Insert and swap mutations were used for this purpose. While a digit is randomly selected and it is inserted in a randomly selected new position in the insert mutation, two digits are randomly selected and their positions are exchanged in the swap operator.
Selection: We adopt roulette wheel selection (RWS). This is to determine selection probability or survival probability for each chromosome proportional to the fitness value. A model of the roulette wheel can be made displaying these probabilities. The detailed introduction is shown in Chapter 1 (page 12).

3.4.3 Numerical Experiments

To investigate the effectiveness of the developed GA with the new priority-based encoding method (priGA), we compare it with Prüfer number-based GA (puGA) proposed by Syarif and Gen [44]. Seven different test problems were considered. Since the problem is NP-hard, optimum results were only obtained for the first four problems. As seen in Table 3.20, while the number of plants changes between 3 and 40 in these problems, the number of DCs and the number of customers change between 4 and 70 and 5 and 100, respectively. The data in test problems were generated randomly. The cost coefficients for the first and second stage of the transportation problem were generated from a uniform distribution between 1 and 20. The demand requirements of customers were drawn from a uniform distribution between 5 and 60. The capacities of plants and DCs were also randomly generated considering constraints in the model. The interested reader can reach the data set for test problems from authors.

Table 3.20 The size of test problems

Problem ID	Number of Plants (I)	Number of DCs (J)	Number of Customers (K)	Optimum
1	3	4	5	1089
2	4	5	10	2283
3	4	5	15	2527
4	8	10	20	2886
5	10	20	40	-
6	15	25	50	-
7	40	70	100	-

In the first experiment, a preliminary study was realized to investigate the effects of different combinations of crossover and mutation operators on the performance of the priGA considering the first four problems with optimum solution. priGA was run 10 times using different initial seeds and crossover and mutation rates were taken as 0.5 in all experiments. Table 3.21 gives best, average, worst costs and average computation time (ACT) of 10 runs for each combination of crossover and mutation operators. Although all combinations of crossover and mutation operators find optimum solution for the first three small-scale problems, the percent deviations from optimum solution and the average performances increase based on problem size. The results for the combinations of crossover and mutation operators are slightly different from each other; however the WMX-swap exhibits better performance than others according to average performances. From ACT columns, we can see that while ACT increases based on problem size, ACT on combinations of crossover and mutation operators at same problem size are very close to each other. Since time differences are very small in each combination of crossover and mutation operators and they are also at a tolerable level, we selected WMX and swap as crossover and mutation operators in our GA approach considering average performance of the combination.

Table 3.22 gives computational results for the puGA and priGA based on Prüfer number encoding and priority-based encoding methods, respectively, for seven test problems. In puGA, one-cutpoint crossover and insertion mutation operators were used as genetic operators and rates were taken as 0.5. Each test problem was run 10 times using GA approaches. To make comparison between puGA and priGA according to solution quality and computational burden, we consider again best, average and worst costs and also ACT. In addition, each test problem is divided into three numerical experiments to investigate the effects of population size and number of generations on the performance of puGA and priGA. When we compare columns of the best cost of the puGA and priGA, it is possible to see that priGA developed in this study reaches optimum solutions for the first four test problems, while puGA finds optimum solution for only the first problem. In addition, average percent deviation from optimum solution on puGA changes between 2.3 and 3 except for the first problem. For large size problems, *i.e.*, the last three problems, the best costs of priGA are always smaller than found with puGA. We can generalize this situation on average and worst costs of priGA. As a result, priGA exhibits better performance than puGA according to solution quality. When we investigate the effects of population size and number of generations, we can reach an expected result that the solution quality of puGA and priGA improves with an increasing population size and the number of generations.

However, there is always a trade-off between solution quality and solution time. As is seen from Table 3.22, ACTs of puGA and priGA increase based on population size and number of generations. We can also observe that ACTs of puGA are smaller than the ACTs of priGA. To see whether the differences between puGA and priGA according to solution quality and computation burden are statistically significant or not, a statistical analysis using sign test, which is a nonparametric version of paired-

Table 3.21 The computational results for different combinations of crossover and mutation operators

Parameters			WMX+Swap				WMX+Insertion			
ID	*popSize*	*maxGen*	Best	Average	Worst	ACT*	Best	Average	Worst	ACT*
1	10	300	1089	1089.0	1089	0.12	1089	1089.0	1089	0.12
2	20	1000	2283	2283.2	2285	0.78	2283	2283.8	2285	0.78
3	30	1500	2527	2527.0	2527	2.04	2527	2527.0	2527	2.05
4	75	2000	2886	2891.2	2899	12.99	2890	2893.7	2899	13.18
			PMX+Swap				PMX+Insertion			
			Best	Average	Worst	ACT*	Best	Average	Worst	ACT*
1	10	300	1089	1089.0	1089	0.12	1089	1089.0	1089	0.12
2	20	1000	2283	2284.0	2285	0.77	2283	2283.6	2285	0.78
3	30	1500	2527	2527.0	2527	2.01	2527	2527.0	2527	2.02
4	75	2000	2890	2894.0	2899	12.96	2890	2893.5	2899	13.15
			OX+Swap				OX+Insertion			
			Best	Average	Worst	ACT*	Best	Average	Worst	ACT*
1	10	300	1089	1089.0	1089	0.11	1089	1089.0	1089	0.11
2	20	1000	2283	2284.2	2285	0.63	2283	2283.8	2285	0.64
3	30	1500	2527	2527.0	2527	1.60	2527	2527.0	2527	1.62
4	75	2000	2886	2895.8	2905	9.94	2890	2895.9	2900	10.14
			PX+Swap				PX+Insertion			
			Best	Average	Worst	ACT*	Best	Average	Worst	ACT*
1	10	300	1089	1089.0	1089	0.12	1089	1089.0	1089	0.11
2	20	1000	2283	2283.6	2285	0.66	2283	2284.6	2285	0.66
3	30	1500	2527	2527.0	2527	1.69	2527	2527.0	2527	1.72
4	75	2000	2892	2898.0	2904	10.56	2895	2896.9	2901	10.64

t test, was carried out considering 21 cases given in Table 3.22. In the test, two different hypotheses were constructed. These are:

$$H_0^{(1)} : \text{Mean difference between average costs of puGA and priGA equal to zero}$$

$$H_0^{(2)} : \text{Mean difference between ACTs of puGA and priGA equal to zero}$$

At the significant level of $\alpha = 0.05$, sign tests should that, while priGA outperformed puGA with a p-value of 0.000 and mean difference of 2353.02, its computation was bigger than that of puGA with a p-value of 0.000 and mean difference of 42.43. The statistical difference between priGA and puGA according to computational burden comes from the crossover operators used in priGA. Priority-based encoding is in the class of permutation encodings and it needs a special crossover operator such as WMX, OX or PMX to obtain feasible solutions. As expected, these crossover operators consume more time than the one-cutpoint crossover operator which was used in puGA.

Table 3.22 Computational results for puGA and pbGA

Parameters			puGA				priGA			
ID	*popSize*	*maxGen*	Best	Average	Worst	ACT	Best	Average	Worst	ACT
	10	300	1089	1175.4	1339	0.07	1089	1089.0	1089	0.12
1	15	500	1089	1091.8	1099	0.16	1089	1089.0	1089	0.23
	20	1000	1089	1089.0	1089	0.35	1089	1089.0	1089	0.57
	20	1000	2341	2402.5	2455	0.48	2283	2283.2	2285	0.78
2	30	1500	2291	2375.2	2426	1.06	2283	2283.0	2283	1.76
	50	2000	2303	2335.8	2373	2.42	2283	2283.0	2283	4.10
	30	1500	2781	2874.4	2942	1.25	2527	2527.0	2527	2.04
3	50	2500	2719	2787.1	2874	3.43	2527	2527.0	2527	5.91
	100	4000	2623	2742.2	2796	11.85	2527	2527.0	2527	21.32
	75	2000	3680	3873.8	4030	7.78	2886	2891.2	2899	12.99
4	100	3000	3643	3780.4	3954	15.93	2886	2892.6	2899	26.85
	150	5000	3582	3712.5	3841	41.41	2886	2890.0	2893	71.76
	75	2000	5738	5949.1	6115	18.29	2971	2985.3	3000	29.07
5	100	3000	5676	5786.1	5889	36.88	2967	2980.6	2994	59.13
	150	5000	5461	5669.4	5835	94.33	2952	2973.2	2989	153.02
	100	2000	7393	7705.6	8067	36.27	2975	2999.0	3025	56.32
6	150	3000	7415	7563.8	7756	76.23	2963	2994.3	3005	130.29
	200	5000	7068	7428.5	7578	188.37	2962	2984.9	3000	295.28
	100	2000	10474	11083.1	11306	177.03	3192	3204.2	3224	241.74
7	150	3000	10715	10954.7	11146	395.52	3148	3184.3	3207	548.30
	200	5000	10716	10889.4	11023	875.03	3136	3179.6	3202	1213.65

3.5 Flexible Logistics Model

With the development of economic globalization and extension of global electronic marketing, global enterprise services supported by universal supply chains and worldwide logistics become imperative for the business world. How to manage logistics system efficiently has thus become a key issue for many companies to control their costs. That is also why an elaborately designed logistics network under the inference today's fully-fledged information technology is catching more and more attention of business entities, especially of many multinational companies. However, it seems to be quite difficult to do successfully for these companies due to their huge and extremely complicated logistics networks, though they usually have a strong desire to cut down their logistics cost. In this section, we introduce a new flexible multistage logistics network (fMLN) model, which can overcome this kind of difficulty in practice and theory, and its efficiency is proved mathematically.

The logistics network design problem is defined as follows: given a set of facilities including potential suppliers, potential manufacturing facilities, and distribution centers with multiple possible configurations, and a set of customers with deterministic demands, determine the configuration of the production-distribution system between various subsidiaries of the corporation such that seasonal customer demands and service requirements are met and the profit of the corporation is maximized or the total cost is minimized [53]. It is an important and strategic operations management problem in supply chain management (SCM), and solving it systemically is to provide an optimal platform for efficient and effective SCM. In this field, usable researche is conducted [54, 15, 55, 21]. However, to date the structures of the logistics network studied in all the literatures are in the framework of the traditional multistage logistics network (tMLN) model. The network structure described by tMLN model is like that shown in Fig. 3.37 There are usually several facilities organized as four echelons (*i.e.*, plants, distribution centers (DCs), retailers and customers [56] in some certain order, and the product delivery routes should be decided during three stages between every two adjoining echelons sequentially.

Although the tMLN model and its application had made a big success in both theory and business practices, as time goes on, some faults of the traditional structure of logistics network came to light, making it impossible to fit the fast changing competition environments or meet the diversified customer demands very well. For multinational companies, running their business in a large geographical area, in one or several continents, for instance, these disadvantages are more distinct. For easy understanding, let's suppose a multinational company (called ABC Co. Ltd., hereafter) is selling personal computers throughout Europe. If ABC establishes its logistics network serving its customers all over Europe following the traditional three-stage logistics network model, its distribution paths will be very long. It leads to

(1) High transportation costs and other related costs, including loading/ unloading cost, fixed operating cost of facilities, and labor cost;
(2) Long response time, which slows down the response speed to changes in demands and decreases customer satisfaction;
(3) Bullwhip effect [57], since lots of facilities in the multistage play a role as mid-process;
(4) The parts demands of manufacturer are known in advance.

Today's energetic business competition presses enterprises to build up a logistics network productively and flexibly. At this point, Dell's success in SCM may give us some clear ideas on designing cost-effective logistics networks [58]. Nevertheless, the company's entire direct-to-consumer business model, not just its logistics approach, gives a remarkable edge over its competitors. Here, however, we pay special attention to its distribution system. The delivery modes employed by Dell and other companies are illustrated in Fig. 3.38, based on which they realize their strategy in distribution–skipping the mid-process; being close to customers reduces transportation cost and delivery time and increasing customer satisfaction. This is how Dell

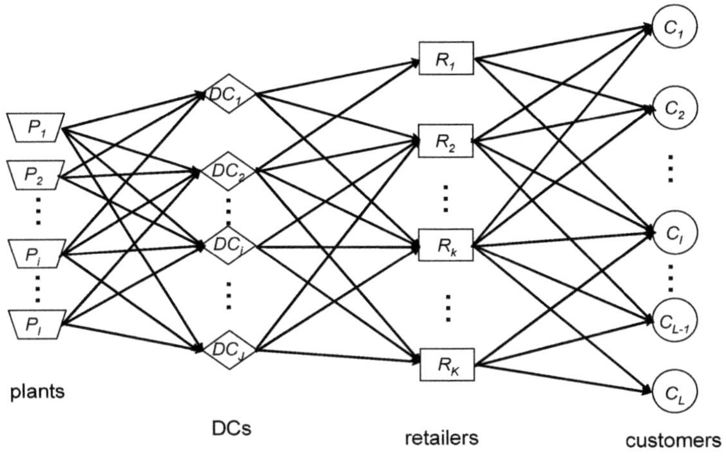

Fig. 3.37 The structure of traditional multistage logistics network (tMLN)

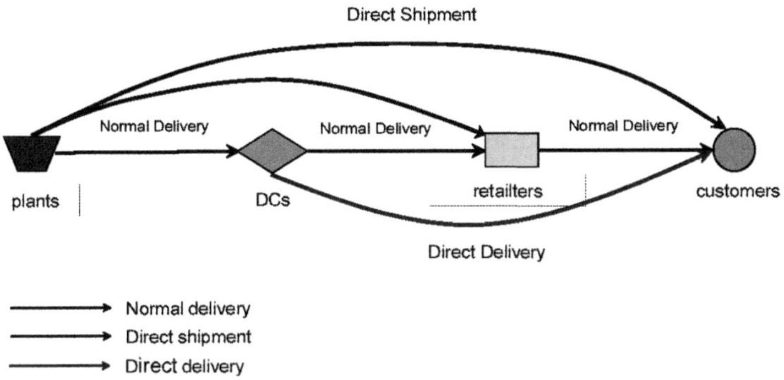

Fig. 3.38 Three kinds of delivering modes in flexible multistage logistics network model

beats its competitors. In more detail, they are: normal delivery, to deliver products from one stage to another adjoining one. direct shipment, to transport products from plants to retailers or customers directly; direct delivery, to deliver products from DCs to customers not *via* retailers. In a network, the direct shipment and direct delivery play the role of arcs connecting two nonadjacent echelons.

By introducing this kind of direct shipment and direct delivery, we extend the tMLN model to a new logistics network (we name it nonadjacent multistage logistics network or flexible multistage logistics network; fMLN for abbreviation hereafter) model, shown in Fig. 3.39. Its application provides a new potential way to shorten the length between the manufactures and final customers, and to serve the customers flexibly. With its help, Dell avoided the problems that other computer makers encounter such as the impracticable long supply chains, delivery delays and risks of missing sudden changes in demand or obsolescence of evolving its products.

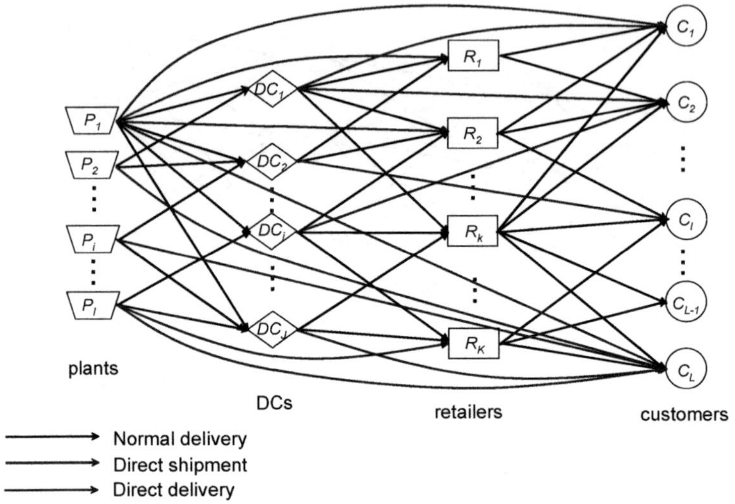

Fig. 3.39 The structure of flexible multistage logistics network (fMLN) models

The trade-off between direct distribution and indirect distribution is also a critical issue in decision making in distribution channel in sales and marketing [59, 60]. However, this kind of flexible logistics network model, with great practical value, has been studied from the viewpoint of logistics network design using the latest soft computing techniques. Basically, the multistage transportation problem involving the open/close decision is NP-hard. The existence of the new delivery modes, which distinguish the fMLN from its traditional counterpart, makes the solution space to the problem much larger and more complex. The conventional methods are unable to solve with acceptable computation cost.

3.5.1 Mathematical Formulation of the Flexible Logistics Model

The flexible logistics model can be defined by the following assumptions:

- A1. In this study, we consider the single product case of a logistics network optimization problem.
- A2. We consider a single time period, such as one week or one month.
- A3. In the logistics network, there are maximum four echelons: plants, DCs, retailers and customers.
- A4. There are three delivery modes: normal delivery, direct shipment and direct delivery, as mentioned above.
- A5. Each customer is served by only one facility.
- A6. Customer demands are known in advance.

A7. Customers will get products at the same price, no matter where she/he gets them. It means that the customers have no special preferences.

The following notations are used to formulate the mathematical model

Notation

Indices

i: index of plant $(i = 1,2,\ldots,I)$
j: index of DC $(j = 1,2,\ldots,J)$
k: index of retailer $(k = 1,2,\ldots,K)$
l: index ofcustomer $(l = 1,2,\ldots,L)$

Parameters

I: number of plants
J: number of DCs
K: number of retailers
L: number of customers
P_i: plant i
DC_j: DC j
R_k: retailer k
C: customer l
b: output of plant i
d_l: demand of customer l
c_{1ij}: unit shipping cost of product from P_i to DC_j
c_{2jk}: unit shipping cost of product from DC_j to R_k
c_{3kl}: unit shipping cost of product from R_k to C_l
c_{4il}: unit shipping cost of product from P_i to C_l
c_{5jl}: unit shipping cost of product from DC_j to C_l
c_{6ik}: unit shipping cost of product from P_i to R_k
u_j^D: upper bound of the capacity of DC_j
u_k^R: upper bound of the capacity of R_k
f_j^F: fixed part of the open cost of DC_j
c_j^{1V}: variant part of the open cost (lease cost) of DC_j
q_j^1: throughput of DC_j

$$q_j^1 = \sum_{i=1}^{I} x_{1ij}, \quad \forall\, j$$

f_j: open cost of DC_j

$$f_j = f_j^F + c_j^W q_j^1, \quad \forall\, j$$

g_k^F: fixed part of the open cost of R_k
c_k^{2V}: variant part of the open cost (lease cost) of DC_j

q_k^2: throughput of R_k
$q_k^2 = \sum_{l=1}^{L} x_{3kl}, \quad \forall\, k$
g_k: open cost of R_k
$q_k = g_k^F + c_k^{2V} q_k^2, \quad \forall\, k$

Decision Variable

x_{1ij}: transportation amoun from P_i to DC_j
x_{2jk}: transportation amoun from DC_j to R_k
x_{3kl}: transportation amoun from R_k to C_l
x_{4il}: transportation amoun from P_i to C_l
x_{5jl}: transportation amoun from DC_j to C_l
x_{6ik}: transportation amoun from P_i to R_k
y_j^1: 1, if DC_j is opened; 0, otherwise
y_k^2: 1, if R_k is opened, 0, otherwise

The flexible logistics problem can be formulated as a location-allocation problem as mixed integer programming model (model 1), and the objective function is to minimize the total logistics cost. The open cost of DCs and retailers comprise two parts: fixed cost and variant cost. When one facility is set, the fixed cost is incurred. The variant cost is dependent on the throughput of the facility, which indicates how large the scale (the capacity) of the facility needs to be established.

$$\begin{aligned}\max z = & \sum_{i=1}^{I}\sum_{j=1}^{J} c_{1ij}x_{1ij} + \sum_{j=1}^{J}\sum_{k=1}^{K} c_{2jk}x_{2jk} + \sum_{k=1}^{K}\sum_{l=1}^{L} c_{3kl}^{T}x_{3kl} + \sum_{i=1}^{I}\sum_{l=1}^{L} c_{4il}x_{4il} \\ & + \sum_{j=1}^{J}\sum_{l=1}^{L} c_{5jl}x_{5jl} + \sum_{i=1}^{I}\sum_{k=1}^{K} c_{6ik}x_{6ik} \\ & + \sum_{j=1}^{J} f_j y_j^1 + \sum_{k=1}^{K} g_k y_k^2 \end{aligned} \tag{3.71}$$

$$\text{s. t. } \sum_{j=1}^{J} x_{1ij} + \sum_{l=1}^{L} x_{4il} + \sum_{k=1}^{K} X_{6ik} \le b_i, \quad \forall\, i \tag{3.72}$$

$$\sum_{i=1}^{I} x_{1ij} = \sum_{k=1}^{K} x_{2jk} + \sum_{l=1}^{L} x_{5jl}, \quad \forall\, j \tag{3.73}$$

$$\sum_{j=1}^{J} x_{2jk} + \sum_{i=1}^{I} x_{6ik} = \sum_{l=1}^{L} x_{3kl}, \quad \forall\, k \tag{3.74}$$

$$\sum_{l=1}^{L} x_{4il} + \sum_{j=1}^{J} x_{5jl} + \sum_{k=1}^{K} X_{3kl} \ge d_l, \quad \forall\, l \tag{3.75}$$

$$\sum_{i=1}^{I} x_{1ij} \leq u_j^D y_j^1, \quad \forall\, j \tag{3.76}$$

$$\sum_{l=1}^{L} x_{3kl} \leq u_k^R y_k^2, \quad \forall\, k \tag{3.77}$$

$$x_{1ij},\ x_{2jk},\ x_{3kl},\ x_{4il},\ x_{5jl},\ x_{6ik} \geq 0 \quad \forall\, i,\ j,\ k,\ l \tag{3.78}$$

$$y_j^1, y_k^2 \in 0,1, \quad \forall\, j,k \tag{3.79}$$

where the objective function Eq. 3.71 means to minimize the total logistics cost, including shipping cost and open cost of facilities. The fourth, fifth and sixth terms are newly introduced to describe the costs of the new delivery modes. The constraint (3.72) means the production limit of plants. The constraints at Eqs. 3.73 and 3.74 are the flow conservation principle. The constraint at Eq. 3.75 ensures that the customers' demand should be satisfied. The constraints Eqs. 3.76 and 3.77 make sure the upper bound of the capacity of DCs and retailers cannot be surpassed.

For tMLN model, the mathematical model (model 2) can be described as follows:

$$\min z = \sum_{i=1}^{I}\sum_{j=1}^{J} c_{1ij}x_{1ij} + \sum_{j=1}^{J}\sum_{k=1}^{K} c_{2jk}x_{2jk} + \sum_{k=1}^{K}\sum_{l=1}^{L} c_{3kl}^{T}x_{3kl} + \sum_{j=1}^{J} f_j y_j^1 + \sum_{k=1}^{K} g_k y_k^2 \tag{3.80}$$

$$\text{s. t.} \ \sum_{j=1}^{J} x_{1ij} \leq b_i, \quad \forall\, i \tag{3.81}$$

$$\sum_{i=1}^{I} x_{1ij} = \sum_{k=1}^{K} x_{2jk}, \quad \forall\, j \tag{3.82}$$

$$\sum_{j=1}^{J} x_{2jk} = \sum_{l=1}^{L} x_{3kl}, \quad \forall\, k \tag{3.83}$$

$$\sum_{i=1}^{I} x_{1ij} \leq u_j^D y_j^1, \quad \forall\, j \tag{3.84}$$

$$\sum_{l=1}^{L} x_{3kl} \leq u_k^R y_k^2, \quad \forall\, k \tag{3.85}$$

$$x_{1ij},\ x_{2jk},\ x_{3kl} \geq 0 \quad \forall\, i,j,k,l \tag{3.86}$$

$$y_j^1, y_k^2 \in 0,1, \quad \forall\, j,k \tag{3.87}$$

This model is quite similar to that of fMLN. Each formula can find its counterpart in the previous model except for the effects of direct shipment and direct delivery.

Yesikokcen *et al.* [61] proved that to allocate all the customers to available plants can provide a lower bound on the optimal objective function value of model 2. However, his research is conducted in continuous Euclidian space, and the transportation cost is directly proportional to the Euclidian distances between points. In fact data of real-world problems do not satisfy these requirements. (For instance, the direct transportation between two points may be very difficult or expensive due to the traffic situation limits; in international trade when the shipment strides the border among countries, duty can be requested.) All of these make the transportation cost out of direct proportion to the Euclidian distance.

In order to understand the exact effects of direct shipment and direct delivery, let's conduct a brief survey on the bounds of the objective functions. First of all, let's consider the transportation costs in both formulas. According to constraint 3.75, to optimize the transportation flow to each customer we usually have the following balancing condition:

$$\sum_{l=1}^{L} x_{4il} + \sum_{j=1}^{J} x_{5jl} + \sum_{k=1}^{K} X_{3kl} = d_l, \ \forall\, l \tag{3.88}$$

Considering all the available delivery paths, the transportation cost of the product for customer l is

$$c_l^T = \min\left\{c_{4il}d_l, (c_{1ij}+c_{5jl})d_l, (c_{6ik}+c_{3ld})d_l, c_{1ij}+(c_{2jk}+c_{3kl})d_l\right\}, \quad \forall\, l \tag{3.89}$$

Here we suppose the delivery paths go through plant I, DC j and retailer k. While for the tMLN model, from Eqs. 3.12 – 3.14 we have the following balancing condition for tMLN model:

$$\sum_{l=1}^{I} x_{1ij} = \sum_{j=1}^{J} x_{2jk} = \sum_{k=1}^{K} x_{3kl} = d_l, \quad \forall\, l \tag{3.90}$$

The transportation cost of the product for customer l is

$$c_l^T{}' = (c_{1ij}+c_{2jk}+c_{3kl})d_l, \quad \forall\, l \tag{3.91}$$

(*Here we use the same symbols added with to represent the quantity in the tMLN model as those in fMLN model.) Comparing 3.91 with 3.83 we can easily get

$$c_l^T \le c_l^T{}', \quad \forall\, l \tag{3.92}$$

Next, let's study on the open cost of facilities. From Eq. 3.73, we can get

$$q_j^1 = \sum_{i=1}^{I} x_{1ij} = \sum_{k=1}^{K} x_{2jk} + \sum_{l=1}^{L} x_{5jl} \leq \sum_{l=1}^{L} x_{5jl}, \quad \forall j$$

Substitute it into Eq. 3.88 we can learn

$$\sum_{j=1}^{J} q_j^1 \leq \sum_{l=1}^{L} d_l \tag{3.93}$$

In the same way, we can get

$$\sum_{k=1}^{K} q_k^2 \leq \sum_{l=1}^{L} d_l \tag{3.94}$$

While from Eq. 3.90 in the tMLN model and definitions of throughputs of facilities, we thus have

$$\sum_{j=1}^{J} q_j^1{}' = \sum_{j=1}^{J} q_j^2{}' = \sum_{l=1}^{L} d_l$$

From the definitions of the open costs, by selecting the appropriate delivery paths, it's reasonable to say that

$$\sum_{j=1}^{J} f_j y_j^1 + \sum_{k=1}^{K} g_k y_k^2 \leq \sum_{j=1}^{J} f_j{}' y_j^1{}' + \sum_{k=1}^{K} g_k{}' y_k^2{}' \tag{3.95}$$

It means that the total facility open costs of fMLN model are no more than that of tMLN. In summary, as explained before by taking the two parts of the total logistics cost into consideration we can draw conclusion that the optimal value of objective function of fMLN model is no more than that of the tMLN model.

From the mathematical analysis, we conclude that the utilization of nonadjacent model can provides a more cost-effective platform for logistics network design and the optimization problem. In another words, selecting more appropriate delivery paths provides chances to cut down the transportation cost and to shorten the length of the distribution channel. On the other hand, the fMLN model can improve the efficiency of the logistics network by guaranteeing more customers an supported using limited resources (the capacitated facilities).

3.5.2 Direct Path-based GA for the Flexible Logistics Model

Let $P(t)$ and $C(t)$ be parents and offspring in current generation t, respectively. The overall procedure of direct path-based GA (dpGA) for solving flexible logistics model (or fMLN) is outlined as follows.

---

procedure: dpGA for fMLP model
input: FMLP data, GA parameters ($popSize$, $maxGen$, p_M, p_C)
output: the best solution E
begin
  $t \leftarrow 0$;
  initialize $P(t)$ by random path-based direct encoding routine;
  evalute $P(t)$ by local optimization based decoding routine;
  while (**not** terminating condition) **do**
    create $C(t)$ from $P(t)$ by two-cut point crossover routine;
    create $C(t)$ from $P(t)$ by combinatorial crossover routine;
    create $C(t)$ from $P(t)$ by insertion mutation routine;
    create $C(t)$ from $P(t)$ by shift mutation routine;
    evalute $C(t)$ by local optimization based decoding routine routine;
    select $P(t+1)$ from $P(t)$ and $C(t)$ by roulette wheel selection routine;
    $t \leftarrow t+1$;
  end
  output the best solution
end

3.5.2.1 Genetic Representation

Although using fMLN model guarantees a flexible and efficient logistics network, the existence of the new delivery modes within makes the problem much more complex. Concretely the complexity of tMLN model is $O(I^J * J^K * K^L)$, while that is $O(I^J * (I+J)^K * (I+J+K)L)^L$ for the non-adjoining MLN model.

For solving fMLN, the random path-based direct encoding method is adopted, the ability of which to handle candidate solutions has been developed by [23]. As shown in Fig. 3.40, in the genes with length 3*L (L is the total number of customers), every three loci constitute one unit, each of which represents a delivery route to a customer, from the plant *via* DC and retailer. The allele on first locus of each unit means the ID of a plant, the start node of the delivery path. The second is the ID of DCs, and the last one indicates the ID of retailers on the path.

One randomly assigns the ID of plant, DC or retailer to the relevant locus as described in Fig. 3.41. Using this encoding method, an infeasible solution may be generated, which violates the facility capacity constraints, so a repairing procedure is needed. In the delivery route to a customer, at least one plant is necessary; if the total demand to the plant exceeds its supply capacity, the customer is assigned

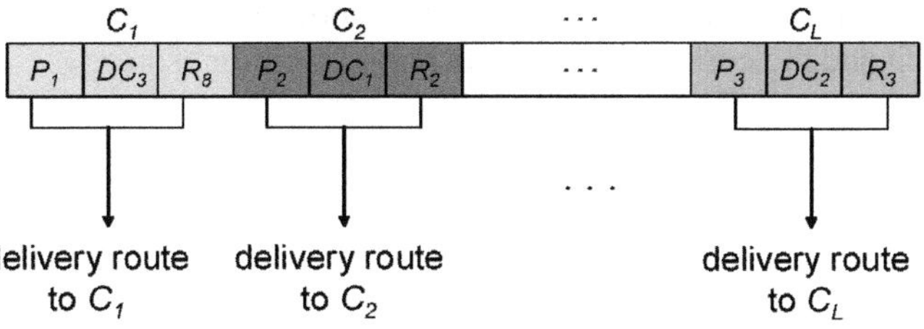

Fig. 3.40 Representation of random path-based direct encoding

to another plant with sufficient products supply, the transportation price between which and the customer is the lowest.

```
procedure: random path-based direct encoding
input: number of plants (I), DCs (J), retailers (K) and customers (L)
output: the chromosome v_k[·]
step 0: for i=1 to L
            v_k[3*j] ← random (1, I);
            v_k[3*j+1] ← random (0, J);
            v_k[3*j+2] ← random (0, K);
step 1: output the chromosome v_k[·];
```

Fig. 3.41 Pseudocode of random path-based direct encoding

3.5.2.2 Random Path-based Direct Decoding

Using the random path-based direct encoding method we can conveniently obtain the delivery route. By doing this, computation time can be greatly cut down. Additionally, in order to improve the quantity of the result, many kinds of local optimization methods have been developed including the hill climbing method, *etc.* ([62, 63]). For solving fMLP, the local optimization technique is adopted based on the concept of neighborhood.

Neighborhood: In each gene unit consisting of three loci, four delivery paths can be formed including normal delivery, direct shipment and direct delivery. All of them form a neighborhood. For instance, we can obtain the neighborhood given in Table 3.23from the sample of gene unit shown in Fig. 3.42, which represents the delivery route to customer 1. In this way, the neighborhood can be decided according to the customer ID and the corresponding gene unit in a chromosome.

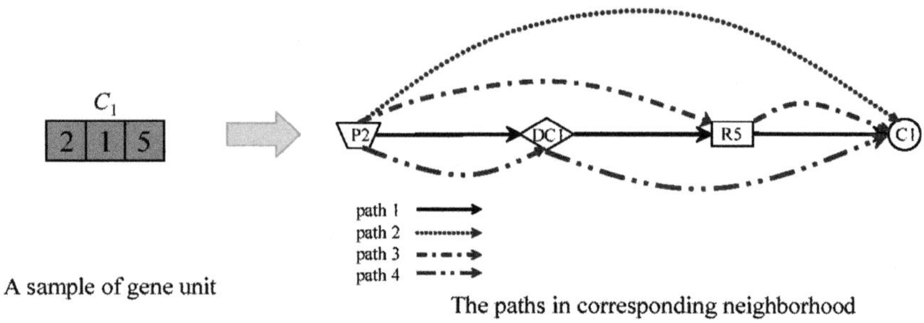

Fig. 3.42 A sample of gene unit and its corresponding neighbor

Table 3.23 The delivery paths in corresponding neighborhood

#	Delivery Route
1	{P2-DC1-R5-C1}
2	{P2-C1}
3	{P2-R5-C1}
4	{P2-DC1-C1}

Local Optimization: We assign the most economic route in a neighborhood to the customer by calculating and comparing the total transportation through each of these routes. By confining the local optimization process as shown in Fig. 3.43, in a relatively narrow area, we can control well the trade-off between quantity of the solutions and consumption of computation time.

Decoding: As shown in Fig. 3.44, using the local optimization-based decoding, we can also deal with the trade-off between local search and global search well. The process is described in following procedure.

3.5.2.3 Genetic Operators

Two-cut Point Crossover: We randomly select two cutting points and then exchange the substrings between the two parents.
Combinatorial Crossover: Before explaining the Combinatory Crossover, we first introduce the definition of set of plants, DCs and retailers in a chromosome, which

procedure: local optimization
input: problem data and gene unit u_k
output: a best delivery path to a customer
step 0: form all the available paths from u_k;
step 1: calculate the total transportation cost through each path;
step 2: select the most economic one from all the paths
as the delivery route for a customer;
step 3: **output** the delivery path;

Fig. 3.43 Pseudocode of local optimization

procedure: Local optimization based decoding
input: problem data and chromosome $v_k[\cdot]$
output: a candidate solution
step 0: **for** i=1 to L
  determine the delivery path to customer i using **local optimization procedure**;
step 1: **output** a solution;

Fig. 3.44 Pseudocode of local optimization-based decoding

are assemblies of the plants, DCs and retailers in each unit in the chromosome separately and are illustrated in Figs. 3.45 and 3.46. Then we randomly select some sub-sets from one or more of these sets, and exchange them between the two parents.

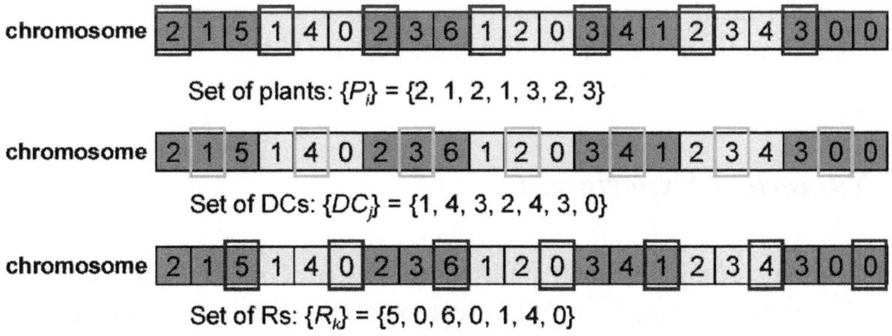

Fig. 3.45 Example of set of plants, DCs and retailers

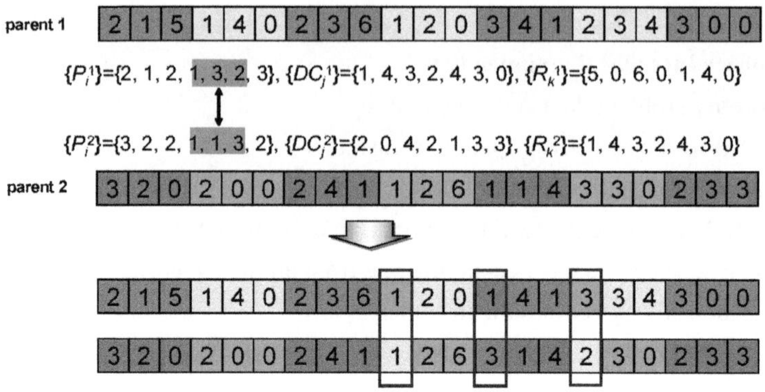

Fig. 3.46 Example of combinatory crossover

Insertion Mutation: We randomly select a string which consists of some gene units, and then insert them into another selected locus (in Fig. 3.47).

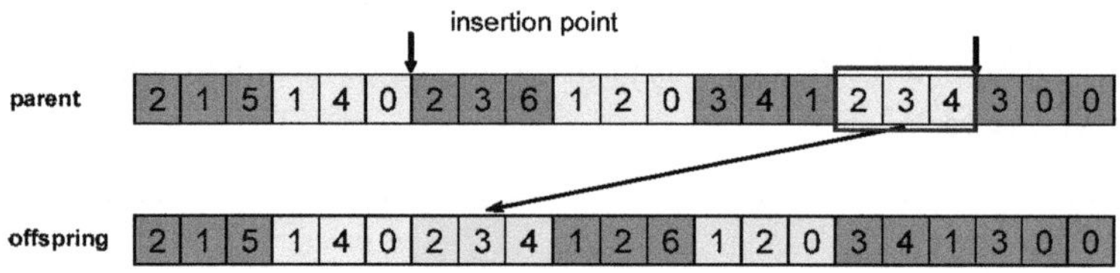

Fig. 3.47 Illustration of insertion mutation

Shift Mutation: We select some units at random, and then randomly assign a new value to each locus in it.
Selection: We adopt roulette wheel selection (RWS). It is to determine selection probability or survival probability for each chromosome proportional to the fitness value. A model of the roulette wheel can be made displaying these probabilities. The detailed introduction is shown in Chap. 1 (page 12).

3.5.3 Numerical Experiments

The experiments compare direct path-based GA (dpGA) and priGA proposed by Gen *et al.* [15] for solving four fMLN test problems. The size of the test problems in terms of number of plant, DC, retailer and customer vary from a relatively small one (2, 3, 5, 20) to an excitingly large one (5, 20, 30, 150) for real-world cases, respectively. The size of these problems and their computing complexity are given in Table 3.24. The graphical illustration of problem 1 is shown in Fig. 3.48. The problems are generated in the following steps. (1) Randomly generate coordinates of sites of plants, DCs retailers and customers (for problems 2, 3 and 4). (2) Calculate the Euclidian distances between relative sites. (3) Because of the parameter sensitivity of the problem, set the cost coefficients reasonably; this is very critical to the examples. The relative coefficient combines with the following formula:

$$c_{xy} = a_t d_{xy} + b_t$$

where c_{xy} is the unit cost coefficient between site x and y, t is the type considering the combination of delivery modes and starting site a_t represents affected unit transportation cost by distance, mainly dependent on the delivery mode, b_t is the part of unit transportation cost independent of distance, generated from a uniform distribution function between 0 and b_t^{max}. The demands of customers are generated randomly between 10 and 100 following the uniform distribution.

Table 3.24 The size of test problems

Problem #	Number of plants I	Number of DCs J	Number of retailers K	Number of customers L	Traditional MLN model		Non-adjacent MLN model	
					Number of decision variables	Comp. complexity	Number of decision variables	Comp. complexity
1	2	5	8	20	223	$1.44*10^{25}$	379	$6.13*10^{31}$
2	3	10	15	60	1105	$2.17*10^{90}$	1930	$2.04*10^{108}$
3	5	20	30	150	5200	$3.79*10^{274}$	9150	$9.38*10^{316}$
4	5	20	30	150	5200	$3.79*10^{274}$	9150	$9.38*10^{316}$

The computational results of the test problems include the average of best solutions (ABS), and standard deviation (SD) of 50 runs. Figures 3.49 and 3.50 show the curves of ABS and the curves of SD considering all test problems.

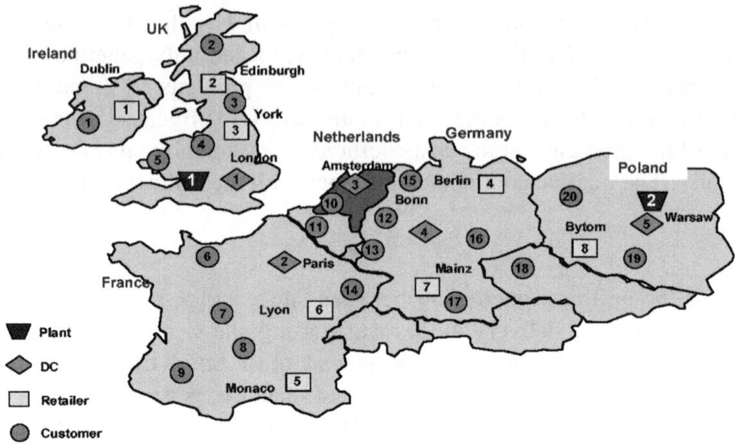

Fig. 3.48 Graphical illustration of problem 1

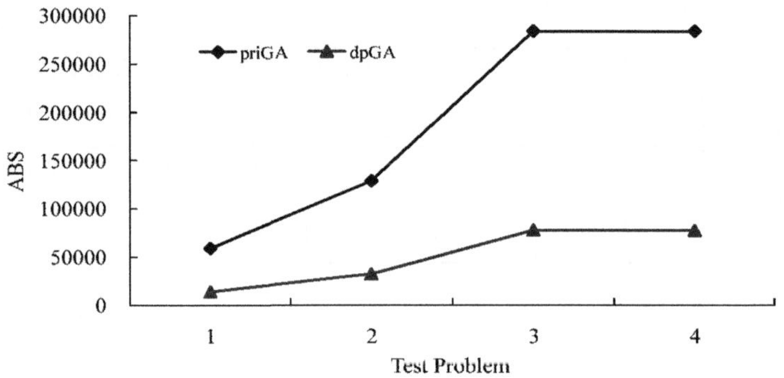

Fig. 3.49 The ABS of different approaches in all test problems

3.6 Integrated Logistics Model with Multi-time Period and Inventory

Today's business competition energetic of presses enterprises to build up a logistics network productively and flexibly. At this point, Dell's success in SCM may give us some clear ideas on designing cost-effective logistics networks [58]. Nevertheless, the company's entire direct-to-consumer business model, not just its logistics approach, gives a remarkable edge over its competitors. Here, however, we pay special attention to its distribution system. The delivery modes employed by Dell and other companies are illustrated in flexible logistics model discussed in the above section, based on which they realize their strategy in distribution-skipping the mid-process, being close to customers reduces transportation cost and delivery time and increas-

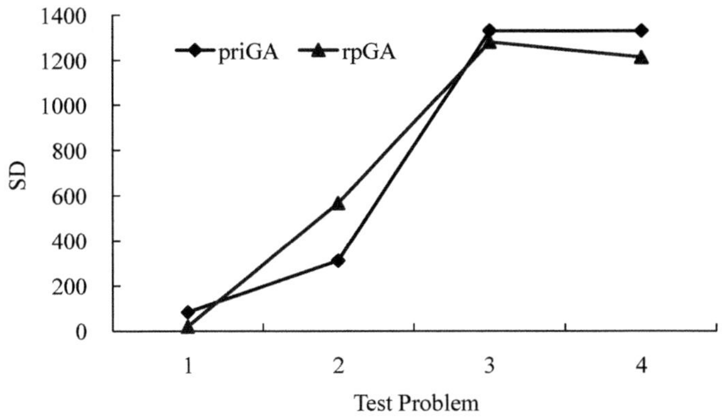

Fig. 3.50 The SD of different approaches in all test problems

ing customer satisfaction. This is how Dell beats its competitors. In more detail, they are: *normal delivery*, to deliver products from one stage to another adjoining one, direct shipment, to transport products from plants to retailers or customers directly, direct delivery, to deliver products from DCs to customers not *via* retailers. In a network, the direct shipment and direct delivery play role of arcs connecting two nonadjacent echelons.

Generally, logistics network design focuses on optimizing the configuration of logistics network from manufacturers (plants) *via* some other facilities to consumers (customers) to minimize the total transportation cost (loading/unloading cost included) and open cost of facilities opened (part of administration cost). While, the *inventory control* (IC) aims at cutting down the inventory cost (storage cost) by adopting the optimal control policy to satisfy some particular pattern of demand. Traditionally in research on IC, the network structures are usually given and fixed. However, in the research on logistics network design it is implicitly assumed that the nodes (facilities) in the network have no storage function, the products just 'flow' through the nodes toward the nodes in a lower level. However, in real-world problems, the two aspects of logistics are tightly connected to each other in the SCM decision making process for enterprises.

In this section, we introduce and a extend flexible logistics network model by considering the time- period and inventory control, called integrated multistage logistics network model (iMLN). Its application provides a new potential way to shorten the distance between the manufactures and final customers, and to serve the customers flexibly. With its help, Dell avoided the problems that other computer makers encounter such as the impracticable long supply chains, delivery delays and risks of missing sudden changes in demand or obsolescence of evolving its products. The trade-off between direct distribution and indirect distribution is also a critical issue in decision making in distribution channel in sales and marketing [59]. In addition, multistage logistics problem involving the open/close decision is NP-hard [60]. This integrated multistage logistics network problem is an extended version of the multistage logistics model. Its solution space is larger than the tMLN model and its structure is more complex.

3.6.1 Mathematical Formulation of the Integrated Logistics Model

The integrated logistics model can be defined by the following assumptions:

A1. We consider logistics network in which the single product is delivered.
A2. We consider the inventory control problem over finite planning horizons.
A3. In the logistics network, there are a maximum four echelons: plants, DCs, retailers and customers.
A4. There are three delivery modes: normal delivery, direct shipment and direct delivery, as mentioned above.
A5. Each customer will be served by only one facility.
A6. Customer demands are deterministic and time-varying.
A7. The replenishment rate is infinite.
A8. The leading time is fixed and known.
A9. Lot-splitting is not allowed.
A10. Shortages are not allowed.
A11. Customers will get products at the same price, no matter where he/she gets them. It means that the customers have no special preferences.

The following notation is used to formulate the mathematical model:

Notation

Indices

i: index of plant $(i = 1,2,\ldots,I)$
j: index of DC $(j = 1,2,\ldots,J)$
k: index of retailer $(k = 1,2,\ldots,K)$
l: index ofcustomer $(l = 1,2,\ldots,L)$
t,t: index of time period $(t,t = 1,2,\ldots,T)$

Parameters

I: number of plants
J: number of DCs

K: number of retailers
L: number of customers
T: planning horizons
P_i: plant i
DC_j: DC j
R_k: retailer k
C_l: customer l
$b_i(t)$: amount of product produced in plant i
$d_l(t)$: demand of product of customer l
u_j^D: upper bound of the capacity of DC_j
u_k^R: upper bound of the capacity of R_k
c_{1ij}: unit shipping cost of product from P_i to DC_j
c_{2jk}: unit shipping cost of product from DC_j to R_k
c_{3kl}: unit shipping cost of product from R_k to C_l
c_{4il}: unit shipping cost of product from P_i to C_l
c_{5jl}: unit shipping cost of product from DC_j to C_l
c_{6ik}: unit shipping cost of product from P_i to R_k
c_j^{1H}: unit holding cost of inventory per period at DC_j
c_k^{2H}: unit holding cost of inventory per period at R_k
A_j^1: fixed charge when any products transported to DC_j
A_k^2: fixed charge when any products transported to R_k
f_j^F: fixed part of open cost of DC_j
c_j^{1V}: variant part of open cost (lease cost) of DC_j
f_j: open cost of DC_j

$$f_j = f_j^F + c_j^{1V} \max_t \{y_j, (t)\}, \forall\, j$$

g_k^F: fixed part of the open cost of R_k
c_k^{2V}: variant part of the open cost (lease cost) of R_j
g_k: open cost of R_k

$$g_k = g_k^F + c_k^{2V} \max_t \{y_k, (t)\}, \forall\, k$$

t_{1ij}^L: leading time for transporting product from P_i to DC_j
t_{2jk}^L: leading time for transporting product from DC_j to R_k
t_{3kl}^L: leading time for transporting product from R_k to C_l
t_{4il}^L: leading time for transporting product from P_i to C_l
t_{5jl}^L: leading time for transporting product from DC_j to C_l
t_{6jk}^L: leading time for transporting product from P_i to R_k
$\delta(x)$: $1, x \neq 0; 0,$ otherwise

Decision Variables

$x_{1ij}(t)$: transportation amoun from P_i to DC_j at time t
$x_{2jk}(t)$: transportation amoun from DC_j to R_k at time t
$x_{3kl}(t)$: transportation amoun from R_k to C_l at time t

$x_{4il}(t)$: transportation amoun from P_i to C_l at time t
$x_{5jl}(t)$: transportation amoun from DC_j to C_l at time t
$x_{6jl}(t)$: transportation amoun from P_i to DC_j at time t
$y_j^1(t)$: inventory in DC_j at time t
$y_j^2(t)$: inventory in R_k at time t
z_j: 1, if DC_j is opened; 0, otherwise
w_k: 1, if R_k is opened; 0, otherwise

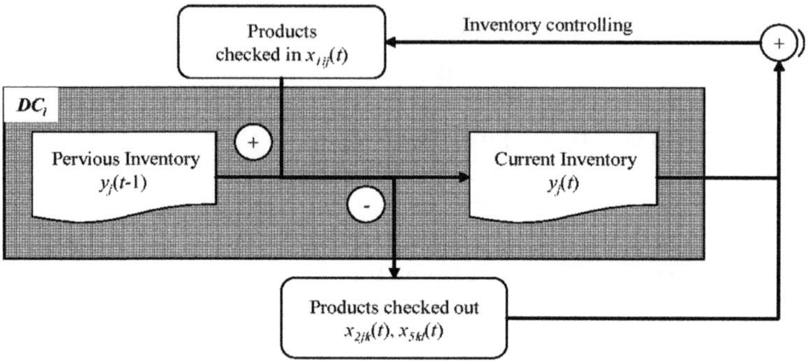

Fig. 3.51 Inventory in DC I at time t

We combine the location-allocation model with lot sizing to treat this iMLN problem and formulate it as a mixed integer programming model. Inventory in DC i at time t is described in Fig. 3.51. The objective function is to minimize the total logistics cost, including transportation costs, holding cost, ordering cost and open cost of facilities. The open costs of DCs and retailers comprise two parts: fixed cost and variant cost. When one facility is set, the fixed cost is incurred. The variant cost is dependent on the throughput of the facility, which indicates how large the scale (the capacity) of the facility needs to be established:

$$
\begin{aligned}
\min z = \sum_{t=0}^{T} \Bigg[& \sum_{i=1}^{I}\sum_{j=1}^{J} c_{1ij}x_{1ij}(t) + \sum_{j=1}^{J}\sum_{k=1}^{K} c_{2jk}x_{2jk}(t) + \sum_{k=1}^{K}\sum_{l=1}^{L} c_{3kl}x_{3kl}(t) \\
& + \sum_{i=1}^{I}\sum_{l=1}^{L} c_{4il}x_{4il}(t) + \sum_{j=1}^{J}\sum_{l=1}^{L} c_{5jl}x_{5jl}(t) + \sum_{i=1}^{I}\sum_{k=1}^{K} c_{6ik}x_{6ik}(t) \\
& + \sum_{i=1}^{I}\sum_{j=1}^{J} A_j^1\theta(x_{1ij}(t)) + \sum_{i=1}^{I}\sum_{k=1}^{K} A_j^2\theta(x_{6ik}(t)) + \sum_{j=1}^{J}\sum_{k=1}^{K} A_k^2\theta(x_{2jk}(t)) \\
& + \sum_{j=1}^{J} c_j^1 y_j^1(t) + \sum_{k=1}^{K} c_k^2 y_k^2(t) \Bigg] + \sum_{j=1}^{J} f_j z_j + \sum_{k=1}^{K} g_k w_k
\end{aligned} \tag{3.96}
$$

$$\text{s. t.} \sum_{\gamma=1}^{t} \left(\sum_{j=1}^{J} x_{1ij}(\tau) + \sum_{l=1}^{L} x_{4il}(\tau) + \sum_{k=1}^{K} x_{6ik}(\tau) \right) \leq \sum_{\tau=1}^{t} b_i(\tau), \quad \forall\, i,t \tag{3.97}$$

$$y_j^1(t-1) + \sum_{i=1}^{I} x_{1ij}\left(t - t_{1ij}^L\right) - \sum_{k=1}^{K} x_{2jk}\left(t - t_{2jk}^L\right) - \sum_{l=1}^{L} x_{5jl}\left(t - t_{5il}^L\right) = y_j^1(t), \quad \forall\, j,t \tag{3.98}$$

$$y_j^2(t-1) + \sum_{i=1}^{I} x_{6ik}\left(t - t_{6ik}^L\right) + \sum_{j=1}^{J} x_{2jk}\left(t - t_{2jk}^L\right) - \sum_{l=1}^{L} x_{3kl}\left(t - t_{2kl}^L\right) = y_k^2(t), \quad \forall\, k,t \tag{3.99}$$

$$\sum_{l=1}^{L} x_{4il}\left(t - t_{4il}^L\right) + \sum_{j=1}^{J} x_{5jl}\left(t - t_{5jl}^L\right) + \sum_{k=1}^{K} x_{3kl}\left(t - t_{3kl}^L\right) \geq d_l^t, \quad \forall\, l,t \tag{3.100}$$

$$y_j^1(t) \leq u_j^D z_j, \quad \forall\, j,t \tag{3.101}$$

$$y_k^2(t) \leq u_k^R w_k, \quad \forall\, k,t \tag{3.102}$$

$$x_{1ij}(t), x_{2jk}(t), x_{3kl}(t), x_{4il}(t), x_{5jl}(t), x_{6ik}(t) \geq 0, \quad \forall\, i,j,k,l,t \tag{3.103}$$

$$y_j(t), y_k^2(t) \geq 0, \quad \forall\, j,k,t \tag{3.104}$$

$$z_j, w_k \in \{0,1\}, \quad \forall\, j,k \tag{3.105}$$

where the objective function Eq. 3.96 is to minimize the total logistics cost, including the amount dependent transportation cost (the first 6 terms) and set up cost of orders (the 7th, 8th and 9th terms), the inventory holding cost (the 10th and 11th) and the open cost of facilities (the last 2 terms). The constraint at Eq. 3.97 makes sure the production limit of plants. The constraint at Eq. 3.98 specifies the inventory of DCs. The constraint at Eq. 3.99 specifies the inventory of retailers. The constraint at Eq. 3.100 ensures that the customers' demands should be satisfied. The constraints at Eqs. 3.101 and 3.102 make sure the DCs and retailers cannot be overloaded. The constraints at Eqs. 3.97, 3.98 and 3.99 together with Eq. 3.100 can ensure no backlogs taking place.

3.6.2 Extended Priority-based GA for the Integrated Logistics Model

Let $P(t)$ and $C(t)$ be parents and offspring in current generation t, respectively. The overall procedure of extended priority-based GA (epGA) for solving integrated multistage logistics model (or iMLN) is outlined as follows.

---

procedure: epGA for iMLP model
input: iMLP data, GA parameters ($popSize$, $maxGen$, p_M, p_C)
output: the best solution E
begin
  $t \leftarrow 0$;
  initialize $P(t)$ by extended priority-based encoding routine;
  evalute $P(t)$ by multi-period integrated decoding routine;
  while (**not** terminating condition) **do**
    create $C(t)$ from $P(t)$ by PMX routine;
    create $C(t)$ from $P(t)$ by two-cut point crossover routine;
    create $C(t)$ from $P(t)$ by inversion mutation routine;
    create $C(t)$ from $P(t)$ by insertion mutation routine;
    climb $C(t)$ by local search routine;
    create $C(t)$ from $P(t)$ by immigration routine;
    evalute $C(t)$ by multi-period integrated decoding routine;
    if $t > u$ **then**
      auto-tuning p_M, p_C and p_I by FLC;
    select $P(t+1)$ from $P(t)$ and $C(t)$ by roulette wheel selection routine;
    $t \leftarrow t+1$;
  end
  output the best solution
end

3.6.2.1 Genetic Representation

A chromosome must have the necessary gene information for solving the problem. Chromosome representation will greatly affect the performance of GAs. Hereby, the first step of applying GA to a specific problem is to decide how to design a chromosome.

The priority-based encoding with random number-based encoding is combined by dividing a chromosome into two segments, as shown in Fig. 3.52. The first segment is encoded by using a priority-based encoding method which can escape the repairing mechanisms in the searching process of GA, developed by [15]. This method is an indirect approach: to encode some guiding information for constructing a delivery path, but not a path itself, in a chromosome. In this method a gene contains two kinds of information: the locus, the position of a gene located within the structure of a chromosome, and the allele, the value taken by the gene. The locus is also used to represent the ID of the node in a graph and the allele is used to represent the priority of the node for constructing a path in the network. The second segment of a chromosome consists of two parts: the first part with K loci containing the guide information about how to assign retailers in the network, and the other with length

L including information of customers. Each locus is assigned an integer in the range from 0 to 2. The procedure of initialization by extended priority-based encoding is shown in Fig. 3.53.

DC_1	…	DC_J	R_1	…	R_K	C_1	…	C_L
10	…	4	11	…	1	3	…	2

(a) 1st segment: Priority-based encoding for traditional multistage logistics network

R_1	…	R_K	C_1	…	C_L
1	…	0	2	…	0

(b) 2nd segment: Extended encoding for direct shipment and direct delivery

Fig. 3.52 Representation of extended priority-based encoding method

```
procedure: extended priority-based encoding
input: DCs (J) retailers (K) and customers (L)
output: the chromosome v_k[]
step 0: for i=1 to N                        // N = J+K+L; 1st segment of chromosome
          v_k[i] ← i;
step 1: for i=1 to ⌈N_1 /2⌉
              repeat
                  j ← random (1,N);
                  l ← random (1,N);
              until l≠j
              swap (v_k[j], v_k[l]);
step 2: for i= N+1 to N+K+L                 // 2nd segment of chromosome
              v_k[i] ← random (0,2);
step 3: output the chromosome v_k[];
```

Fig. 3.53 The procedure of initialization by extended priority-based encoding

3.6.2.2 Extended Priority-based Decoding

Decoding is the mapping from chromosomes to candidate solutions to the problem. A distinct difference between the traditional priority-based decoding method and the method employed in this chapter is that we use the information in the second segment to give the process for deciding the paths of direct shipment and direct delivery. The detailed procedure is shown in Fig. 3.54.

```
procedure: extended priority-based decoding
input: chromosome v_k and problem data
output: solution candidate to the problem
step 1: Allocate customers
    repeat
        Select the customer l with the highest priority
            in 1st segment of chromosome v_k[(J+K+1):( J+K+L)];
        Assign p←v_k[J+K+L+K+l];   // the allele of 2nd segment of chromosome
        if (p = 0) then Decide a direct shipment form plant i to customer l
            Which the plant i has sufficient capacity and the cheapest transportation cost;
            Update demand to plant i;
        if (p = 1) then Decide a direct delivery form DC j to customer l
            Which DC j has sufficient capacity and the cheapest transportation cost;
            Update demand to DC j;
        if (p = 2) then Decide a transportation form retailer k to customer l
            Which retailer k has sufficient capacity and the cheapest transportation cost;
            Update demand to the retailer;
        Assign v_k[J+K+L+K+l]←0;
    until all the customers have been assigned;
step 2: Calculate the throughput of each retailer.
step 3: Allocate retailers    // similar process with step 1
step 4: Calculate the throughput of each DC;
step 5: Allocate DCs;
    repeat
        Select the DC j with the highest priority in 1st segment of chromosome v_k[1: J];
        if (throughput of DC i > 0) then Decide a transportation form plant i to DC j
            Which plant i has sufficient capacity and the cheapest transportation cost;
            Update demand to plant i;
        else close the DC;
            v_k(j) ←0;
    until all the DCs are assigned;
step 6: output solution candidate to the problem;
```

Fig. 3.54 The procedure of extended priority-based decoding

3.6.2.3 Wagner and Whitin Algorithm

As the extension of economic order quantity (EOQ), the Wagner and Whitin (WW) algorithm was developed based on dynamic programming to treat the determinate time-varying demand problem. We present the case of some DC, DC i, as an example and use the notation described previously to summarize it briefly. The problem can be stated as follows: given a fixed ordering cost A_i, unit holding cost c_i^h and demand $d(t)$ for T periods, determine a replenishment schedule that minimizes the total ordering and holding costs in the T periods, assuming zero starting inventory and no backlogging.

Let $C(t)$ denote the cost of an optimal replenishment schedule for the first t periods and $Z(t;t)$ the minimal cost for the first t periods with the last replenishment in period t. For optimal ordering time t_t, $1 < t < T$, gives

$$\tau_{t-1} \leq \tau_t$$

In each step of the procedure $C(t)$ is determined by
and $Z(t;t)$ is given by

$$Z(\tau;t) = Z(\tau;t-1) + (t-\tau)c_j^1 d(t), \tau_{t-1} < \tau < t$$
$$Z(\tau;t) = A + Z*(\tau;y-1), \quad \tau = t$$

with $C(1) = Z*(1;1) = A$ and t_1 =1. After T steps the minimal cost is $C(T)$ and the associated decisions can be found by backtracking. The last replenishment is placed in period t_T and no inventory is left at the end of period T.

3.6.2.4 Multi-period Integrated Decoding

Multi-period integrated decoding integrates the extended priority-based decoding method and the WW algorithm to decide the allocation of facilities in multiple periods and the relative replenishment schedule (reorder time and quantity included). The WW algorithm is developed for uncapacitated lot sizing decision, but the facilities are capacitated, so here we use nonstationary network configuration to deal with the problem.

---

procedure 3: Multi-period Integrated Decoding
input: problem data and chromosome v_k
output: network configuration, reorder schedule of facilities
step 0: Allocate customers by repeating hybrid extended priority-based decoding T times;
step 1: Decide reorder schedule of retailers using WW algorithm;
step 2: Allocate retailers by repeating hybrid extended priority-based decoding T times;
step 3: Decide reorder schedule of DCs using WW algorithm;
step 4: Allocate DCs by repeating hybrid extended priority-based decoding T times;
step 5: output network configuration, reorder schedule of facilities;

Fig. 3.55 The procedure of multi-period integrated decoding

For easy understanding, an example is given, based on what the decoding procedure explained. The problem size is two plants, three DCs, three retailers and five customers. According to the decoding procedure, we can get the logistics network configurations in Fig. 3.57. Then replenishment schedule is shown in Table 3.25 from the chromosome given in Fig. 3.56.

node ID	DC_1	DC_2	DC_3	R_1	R_2	R_3	C_1	C_2	C_3	C_4	C_5	R_1	R_2	R_3	C_1	C_2	C_3	C_4	C_5
priority v_k	3	5	8	12	2	4	11	7	1	6	9	2	0	1	1	0	1	2	0

Fig. 3.56 Example of chromosome of epGA

Table 3.25 Replenishment schedules of the facilities

Facility \ Period	1	2	3
Retailer 2	25	0	15
DC 1	35	0	20
DC 2	0	10	0

3.6.3 Local Search Technique

GA can do a global search in entire space, but there is no way for exploring the search space within the convergence area generated by the GA loop. Therefore, it is sometimes impossible or insufficient for GA to locate an optimal solution in the optimization problems with complex search spaces and constraints. To overcome this weakness, various methods for hybridizing GA using local search techniques have been suggested [23, 64, 65]. One of the common forms of hybridized GA is to incorporate a local search technique to a GA loop. With this hybrid approach, a local search technique is applied to each newly generated offspring to move it to a local optimal solution before injecting it into the new population [28].

The iterative hill climbing method [62] is one of the most popular local search methods. But in the original version of the iterative hill climbing method, the best chromosome is selected from the randomly generated chromosomes in the neighborhood of original chromosome v_c; if its fitness is larger than that of the original one, it takes the place of the original one, else the hill climbing stops. It is not easy to get a better chromosome in a single time of iteration; we thus extend this method by permitting its logjam several times. In our approach the patient iterative hill climbing (piHC) method is incorporated in a GA loop. Therefore, GA carries out a global search and the piHC method carries out local search around the convergence area by

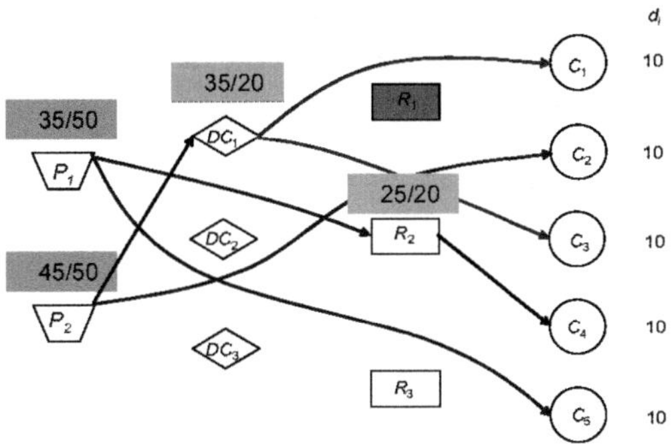

(a) The logistics network configuration at t=1

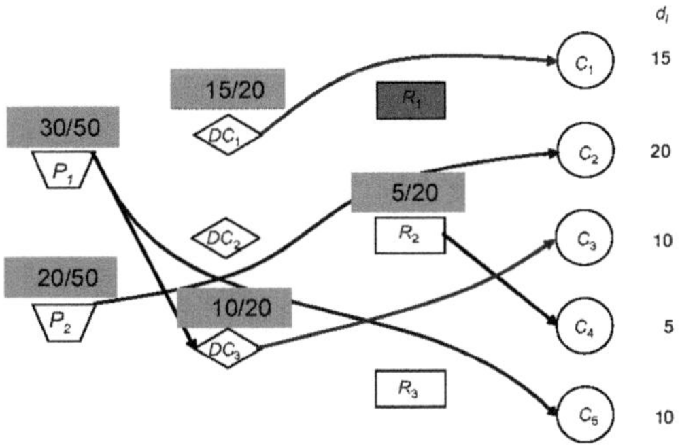

(b) The logistics network configuration at

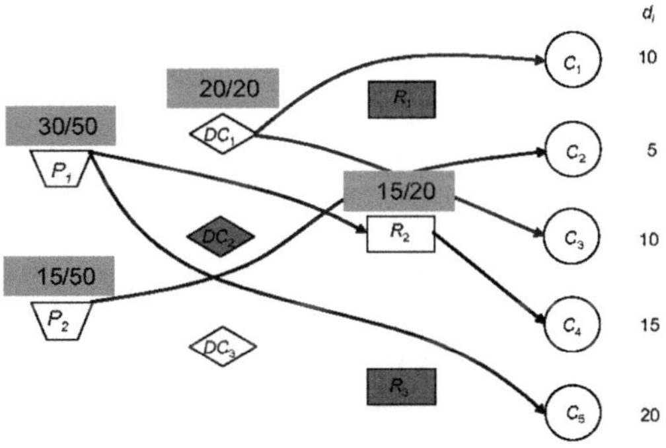

(c) The logistics network configuration at t=3

Fig. 3.57 An example of decoding procedure

GA loop. The latter can guarantee the desired properties of local search technique for hybridization with GA [23]. The detailed procedure of piHC is shown in Fig. 3.58.

```
procedure: Patient Iterative Hill Climbing (piHC)
input: problem data, population P(t), EA parameters
output: improved population P'(t)
begin
  for i=0 to popSize
    r←random(0,1);
    if r<p_c                         //p_c: probability of crossover
      t←0;
      while t≤Max do
           generate 20 new chromosomes from v_c(i) by shift mutation;
           evaluate fitness of the chromosomes newly generated;
           select the best one v_n(i) form the chromosomes;
           if f(v_n)>f(v_c) then
              v_c←v_n;
           else
              t++;
  output improved population P'(t);
end
```

Fig. 3.58 The procedure of patient iterative hill climbing

3.6.3.1 Genetic Operators

Partial-Mapped Crossover (PMX): It works by four steps: step 1, select the substrings at random from parents; step 2, exchange substrings between parents; step 3, determine mapping relationship between the alleles in the substrings; step 4, legalize offspring by changing the alleles out of the substring according to the mapping relationship.

Two-cut Point Crossover: We randomly select two cutting points within the second segments of two parents and exchange the substrings in the middle of the two cutting points between parents.

Inversion Mutation: We randomly select a substring in the first segment, and then reverse the locus in the substring.

Insertion Mutation: We select a string in the second segment at random, and then insert them into another randomly selected locus.

Immigration: The algorithm is modified to (1) include immigration routine, ineach generation, (2) generate and (3) evaluate popSize $\cdot$ p_I random members, and (4)

replace the popSize $\cdot$ p_I worst members of the population with the popSize $\cdot$ p_I I random members (pI, called the immigration probability).
Selection: We adopt the roulette wheel selection (RWS). It is to determine selection probability or survival probability for each chromosome proportional to the fitness value. A model of the roulette wheel can be made displaying these probabilities.
Auto-tuning Scheme: The probabilities of immigration and crossover determined the weight of exploitationfor the search space. Its main scheme is to use two FLCs: auto-tuning for exploration and exploitation $T\,[p_M \wedge (p_C \wedge p_I)]$ and auto-tuning for genetic exploitation and random exploitation $(T[p_C \wedge p_I])$; there are implemented independently to adaptively regulate the genetic parameters during the genetic search process. The detailed scheme is showed in Chap. 1.

3.6.4 Numerical Experiments

In this section, three test problems are used to compare the performance of epGA approaches. The size in terms of the number of plant, DC, retailer, customer and total decision variables is described in Table 3.26. We also give the graphical illustration of problems 1 and 2 in Fig. 3.59, and the total product demand and supply in the three problems are given in Fig. 3.60. Interested readers can access the data set for the test problems from the authors.

To investigate the effectiveness of the proposed FLC, a heuristic rule to control the mutation rate suggested by [66] is also applied to the test problems as a comparison. In their method the mutation rate p_M is dynamically regulated by using a linear function of the number of generation. (We call it linear parameter control policy LPP.)

Table 3.26 The size of the test problems

Problem #	Number of plants I	Number of DCs J	Number of retailers K	Number of customers L	Total planning horizons T	Number of decision variables
1	2	5	8	20	20	7,593
2	2	5	8	20	52	18,721
3	3	10	15	60	52	100,385

The parameter value through experiments for all the approaches is shown as follows: *popSize*=100; *maxGen*=500. In epEA and epEA+LS the probability of crossover and probability of mutation need to be tuned by hand; we set p_C=0.6 and p_M=0.6. (When employing two crossovers, we use the same crossover rate for them; the mutations are the same). While in epEA+LS+FLC, initial parameter values is p_C =0.9 and p_M=0.9, and p_C =0.6 and p_M =0.9 in epEA+LS+LPP.

The computational results are shown in Tables 3.27–3.29; in these tables we give best value of objective function (best value), average of best solutions (AVG), stan-

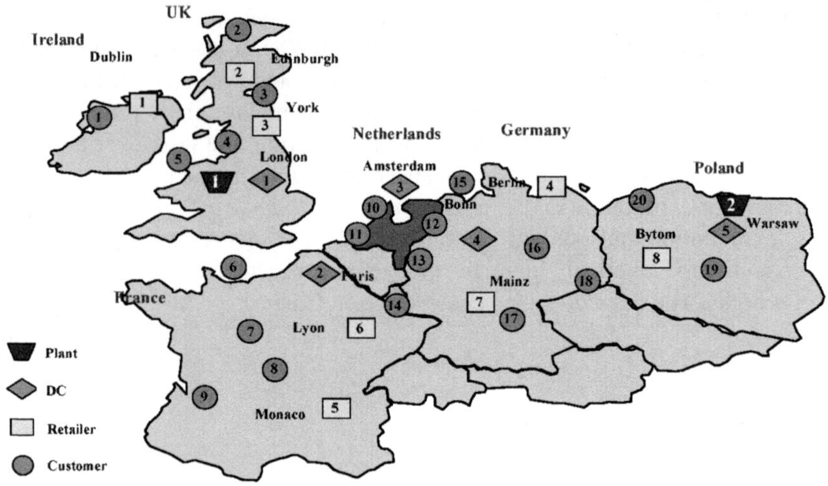

Fig. 3.59 Graphical illustrations of the test problems 1 & 2.

dard deviation of the 50 best solution (SD), and average computation time (ACT) of 50 best solutions obtained by the 50 runs of each program. In order to judge whether the proposed LS technique and FLC are effective to improve the performance of epEA, a statistical analysis using *t* tests was carried out to exam whether the differences between AVGs of these approaches are statistically significant at the significant level $\alpha = 0.05$. The *t*-values yielded by the tests are also given in the tables. According to these results of the *t* test we can conclude that epEA+LS is more effective than epEA; what's more the performance of epEA+LS+FLC is better than that of epEA+LS and epEA+LS+LPP. Additionally, Fig. 3.60 shows the convergence behavior of the four methods for test problem 2. The epEA+LS+LPP gives a smaller best value than epEA+LS, but its convergence behavior is not always superior to that of epEA+LS. The effectiveness of the proposed LS and FLC are thereby proved.

3.7 Summary

In this chapter, we examined the logistics network models using GAs. We focused on the basic logistics models and the practicability of logistics system, *i.e.* to extend the existing models for addressing real-world problem more appropriately and conveniently, and to develop efficient methods for the problems that can be easily used in computer-aided decision support systems (DSS). We introduced basic lo-

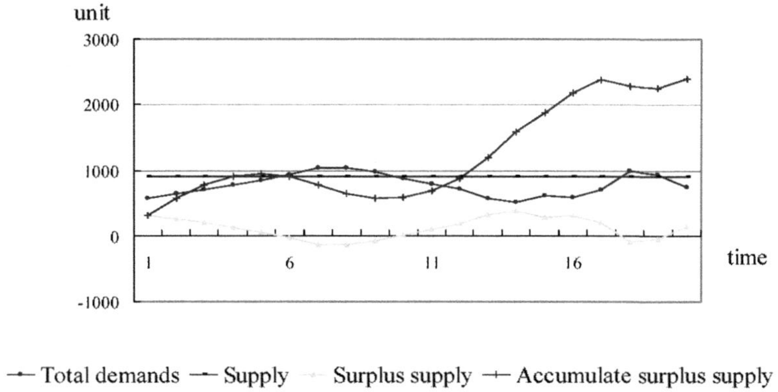

(a) Customer demand and product supply in problem 1

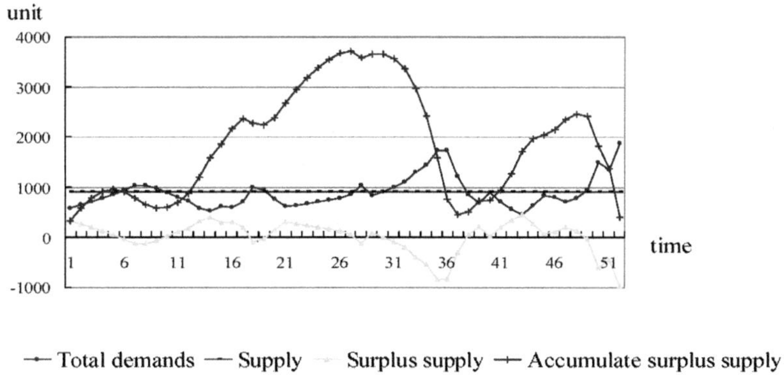

(b) Customer demand and product supply in problem 2

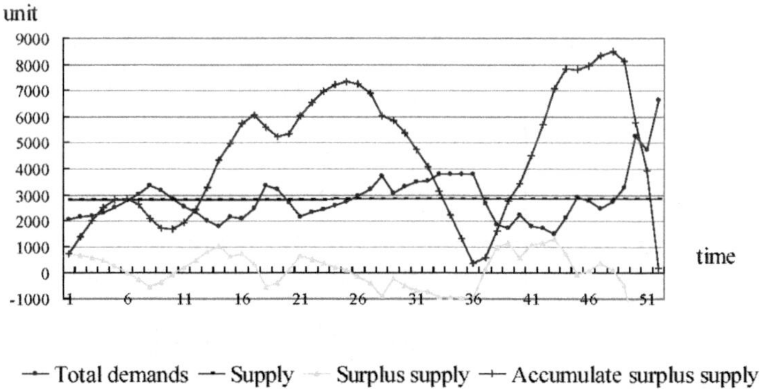

(c) Customer demand and product supply in problem 3

Fig. 3.60 Customer demand and product supply for the test problems

Table 3.27 Computational results of the test problems with different approaches for problem 1

	epEA	epEA+LS	epEA+LS+LPP	epEA+LS+FLC
Best value :	52501.21	42725.70	52501.21	28256.59
AVG:	81159.04	66875.03	64875.03	60608.75
SD:	25182.13	11382.92	11382.92	8070.29
ACT (s):	34.32	198.02	189.02	177.23
AVG difference (comparing with epEA)	-	14284.01	-	20550.29
t value (compared with epEA)	-	3.65	-	5.49
AVG comparing with epGA+LS+FLC	-	6266.28	4266.28	-
t value (comparing with epGA+LS+FLC)	-	3.17	2.16	-

Table 3.28 Computational results of the test problems with different approaches for problem 2

	epEA	epEA+LS	epEA+LS+LPP	epEA+LS+FLC
Best value :	129540.15	79745.95	71251.11	71251.11
AVG:	252018.43	222351.25	220157.14	197157.00
SD:	62820.00	57621.04	52115.96	53461.36
ACT (s):	119.53	530.73	525.22	488.93
AVG difference (comparing with epEA)	-	29667.17	-	54861.43
t value (compared with epEA)	-	2.46	-	4.70
AVG comparing with epGA+LS+FLC	-	25194.25	23000.14	-
t value (comparing with epGA+LS+FLC)	-	2.26	2.17	-

Table 3.29 Computational results of the test problems with different approaches for problem 2

	epEA	epEA+LS	epEA+LS+LPP	epEA+LS+FLC
Best value :	696173.02	626941.51	618909.17	445580.48
AVG:	2016374.99	1682993.65	1658632.77	1398932.71
SD:	897517.89	767034.86	556739.53	538937.53
ACT (s):	10.85	61.26	58.26	67.09
AVG difference (comparing with epEA)	333381.34	333381.34	-	617442.27
t value (compared with epEA)	2.19	2.19	-	4.13
AVG comparing with epGA+LS+FLC	-	284060.93	259700.06	-
t value (comparing with epGA+LS+FLC)	-	2.47	2.33	-

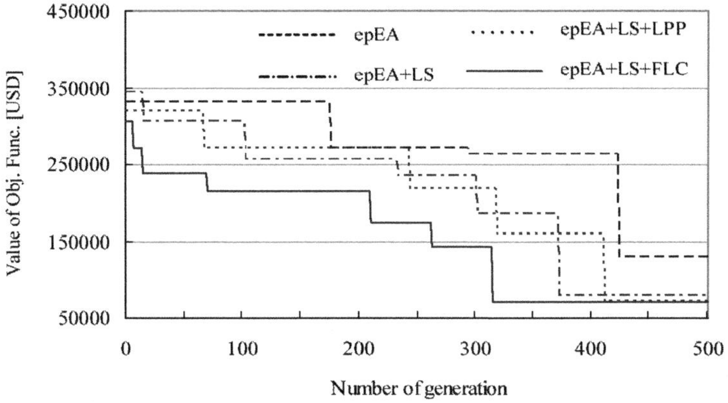

Fig. 3.61 Convergence behavior of different approaches in problem 2

gistics models, location allocation models, multistage logistics models, a flexible multistage logistics network model, and an integrated logistics model.

To solve these logistics models, we introduced different GA approaches, such as matrix-based encoding method, Prüfer number-based encoding method, priority-based encoding method, extended priority-based GA, and random path-based GA. The effectiveness and efficiency of GA approaches were investigated with various scales of logistics problems by comparing with recent related researches.

References

1. Sabri, E. H. & Beamon, B. M. (2000). A multi-objective approach to simultaneous strategic & operational planning in supply chain design, *Omega*, 28, 581–598.
2. Thomas, D. J. & Griffin, P. M. (1996). Coordinated supply chain management, *European Journal of Operational Research*, 94, 1–115.
3. Vidal, C. J. & Goetschalckx, M. (1997). Strategic production-distribution models: A critical review with emphasis on global supply chain models, *European Journal of Operational Research*, 98, 1–18.
4. Beamon, B. M. (1998). Supply chain design and analysis: models and methods, *International Journal of Production Economics*, 55, 281–294.
5. Erenguc, S. S., Simpson, N. C. & Vakharia, A. J. (1999). Integrated production/distribution planning in supply chains: an invited review, *European Journal of Operational Research*, 115, 219–236.
6. Pontrandolfo, P. & Okogbaa, O. G. (1999). Global manufacturing: a review and a framework for planning in a global corporation, *International Journal of Production Economics*, 37(1), 1–19.
7. Jayaraman, V. & Pirkul, H. (2001). Planning and coordination of production and distribution facilities for multiple commodities, *European Journal of Operational Research*, 133, 394–408.

8. Jayaraman, V. & Ross, A. (2003). A simulated annealing methodology to distribution network design & management, *European Journal of Operational Research*, 144, 629–645.
9. Yan, H., Yu Z. & Cheng, T. C. E. (2003). A strategic model for supply chain design with logical constraints: formulation and solution, *Computers & Operations Research*, 30(14), 2135–2155.
10. Syam, S. S. (2002). A model and methodologies for the location problem with logistical components, *Computers and Operations Research*, 29, 1173–1193.
11. Syarif, A., Yun, Y. & Gen, M. (2002). Study on multi-stage logistics chain network: a spanning tree-based genetic algorithm approach, *Computers & Industrial Engineering*, 43, 299–314.
12. Amiri, A. (2004). Designing a distribution network in a supply chain system: formulation and efficient solution procedure, *European Journal of Operational Research*, in press.
13. Gen, M. & Syarif, A. (2005). Hybrid genetic algorithm for multi-time period production / distribution planning, *Computers & Industrial Engineering*, 48(4), 799–809.
14. Truong, T. H. & Azadivar, F. (2005). Optimal design methodologies for configuration of supply chains, *International Journal of Production Researches*, 43(11), 2217–2236.
15. Gen, M., Altiparamk, F. & Lin, L. (2006). A genetic algorithm for two-stage transportation problem using priority-based encoding, *OR Spectrum*, 28(3), 337–354.
16. Chan, F. T. S. & Chung, S. H. (2004). A multi-criterion genetic algorithm for order distribution in a demand driven supply chain, *International Journal of Computer Integrated Manufacturing*, 17(4), 339–351.
17. Chen, C. & Lee, W. (2004). Multi-objective optimization of multi-echelon supply chain networks with uncertain product demands & prices, *Computers and Chemical Engineering*, 28, 1131–1144.
18. Erol, I. & Ferrell Jr. W. G. (2004). A methodology to support decision making across the supply chain of an industrial distributor, *International Journal of Production Economics*, 89, 119–129.
19. Guillen, G., Mele, F. D., Bagajewicz, M. J., Espuna, A. & Puigjaner, L. (2005). Multiobjective supply chain design under uncertainty, *Chemical Engineering Science*, 60, 1535–1553.
20. Chan, F. T. S., Chung, S. H. & Wadhwa, S. (2004). A hybrid genetic algorithm for production and distribution, *Omega*, 33, 345–355.
21. Altiparmak, F., Gen, M., Lin, L. & Paksoy, T. (2006). A genetic algorithm approach for multi-objective optimization of supply chain networks, *Computers & Industrial Engineering*, 51(1), 197–216.
22. Altiparmak, F., Gen, M., Lin, L. & Karaoglan, I. (2007). A steady-state genetic algorithm for multi-product supply chain network design, *Computers and Industrial Engineering*, & [online].Available :
23. Gen, M. & Cheng, R. (2000). *Genetic algorithms & engineering optimization*, Wiley: New York.
24. Dimopoulos, C. & Zalzala, A. M. S. (2000). Recent developments in evolutionary computation for manufacturing optimization: problems, solutions and comparisons, *IEEE Transactions on Evolutionary Computation*, 4(2), 93–113.
25. Aytug, H., Khouja, M. & Vergara, F. E. (2003). Use of genetic algorithms to solve production and operations management: a review, *International Journal of Production Researches*, 41(17), 3955–4009.
26. Guo, J. (2001). Third-party Logistics - Key to rail freight development in China, Japan Railway & Transport Review, 29, 32–37. [Online]. available : http://www.jrtr.net/jrtr29/f32_jia.html.
27. Michalewicz, Z., Vignaux, G. A. & Hobbs, M. (1991). A non-standard genetic algorithm for the nonlinear transportation problem, *ORSA Journal on Computing*, 3(4), 307–316.
28. Gen, M. & Cheng, R. W. (1997). *Genetic Algorithms and Engineering Design*, New York: John Wiley & Sons.
29. Gen, M., Li, Y. Z. & Ida, K. (1999). Solving multiobjective transportation problems by spanning tree-based genetic algorithm, *IEICE Transaction on Fundamentals*, E82-A(12), 2802–2810.

30. Simchi-Levi, D., Kaminsky, P. & Simchi-Levi, E. (2003). *Designing and Managing the Supply Chain: Concepts Strategies & Case Studies*, Second Edition, McGraw-Hill, Irwin, Boston, MA.
31. Council of Supply Chain Management Professionals, [Online]. Available : http://www.cscmp.org/Website/AboutCSCMP/Definitions/Definitions.asp
32. Cooper, L. (1963). Location-allocation problems, *Operations Research*, 11, 331–343.
33. Gong, D., Yamazaki, G. & Gen, M. (1996). Evolutionary program for optimal design of material distribution system, *Proceeding of IEEE International Conference on Evolutionary Computation*, 131–134.
34. Hitchcock, F. L. (1941). The distribution of a product from several sources to numerous localities, *Journal of Mathematical Physics*, 20, 24–230.
35. Tilanus, B. (1997). Introduction to information system in logistics and transportation. In B. Tilanus, *Information systems in logistics and transportation*, Amsterdam: Elsevier, 7–16.
36. Geoffrion, A. M. & Graves, G. W. (1974). Multicommodity distribution system design by benders decomposition, *management science*, 20, 822–844.
37. Pirkul, H. & Jayaraman, V. (1998). A multi-commodity, multi-plant capacitated facility location problem: formulation and efficient heuristic solution, *Computer Operations Research*, 25(10), 869–878.
38. Heragu, S. (1997). *Facilities Design*, PSW Publishing Company.
39. Hindi, K. S., Basta, T. & Pienkosz, K. (1998). Efficient solution of a multi-commodity, two-stage distribution problem with constraints on assignment of customers to distribution centers, *International Transactions on Operational Research*, 5(6), 519–527.
40. Tragantalerngsak, S., Holt, J. & Ronnqvist, M. (2000). An exact method for two-echelon, single-source, capacitated facility location problem, *European Journal of Operational Research*, 123, 473–489.
41. Amiri, A. (2005). Designing a distribution network in a supply chain system: formulation and efficient solution procedure, *in press*, *European Journal of Operational Research*.
42. Vignaux, G. A. & Michalewicz, Z. (1991). A genetic algorithm for the linear transportation problem, *IEEE Transactions on Systems, Man, and Cybernetics*, 21(2), 445–452.
43. Li, Y. Z., Gen, M. & Ida, K. (1998). Improved genetic algorithm for solving multi-objective solid transportation problem with fuzzy number, *Japanese Journal of Fuzzy Theory and Systems*, 4(3), 220–229.
44. Syarif, A. & Gen, M. (2003). Double spanning tree-based genetic algorithm for two stage transportation problem, *International Journal of Knowledge-Based Intelligent Engineering System*, 7(4).
45. Lee, J. E., Gen, M. & Rhee, K. G. (2008). A multi-stage reverse logistics network problem by using hybrid priority-based genetic algorithm, *IEEJ Transactions on Electronics, Information & Systems*, in Press.
46. Garey, M. R. & Johnson, D. S. (1979). *Computers and intractability: a guide to the theory of NP-completeness*, Freeman.
47. [Online]. available : http://www.fbk.eur.nl/OZ/REVLOG/PROJECTS/TEMMINOLOGY/def-reverselogistics.html.
48. Stock, J. K. (1992). *Reverse logistics* : *White paper*, Council of Logistics Management, Oak Brook, IL.
49. Kroon, L. & Vrijens, G. (1995). Returnable containers : An example of reverse logistics, *International Journal of Physical Distribution and Logistics Management*, 25(2), 56–68.
50. Pati, R. K., Vrat, P. & Kumar, P. A. (2008). Goal programming model for paper recycling system, *International Journal of Management Science*, Omega, 36(3), 405–417.
51. Kim, K. B., Song, I. S. & Jeong, B. J. (2006). Supply planning model for remanufacturing system in reverse logistics environment, *Computer & Industrial Engineering*, 51(2), 279–287.
52. Lee, J. E., Rhee, K. G. & Gen, M. (2007). Designing a reverse logistics network by priority-based genetic algorithm, *Proceeding of International Conference on Intelligent Manufacturing Logistics System*, 158–163.

53. Goetschalckx, M., Vidal, C. J. & Dogan, K. (2002). Modeling & design of global logistics systems: A review of integrated strategic & tactical models & design algorithms, *European Journal of Operational Research*, 143, 1–18.
54. Gen, M. & Syarif, A. (2005). Hybrid genetic algorithm for multi-time period production/distribution planning, *Computers & Industrial Engineering*, 48(4), 799–809.
55. Nakatsu, R. T. (2005). Designing business logistics networks using model-based reasoning & heuristic-based searching, *Expert Systems with Applications*, 29(4), 735–745.
56. Harland, C. (1997). Supply chain operational performance roles, *Integrated Manufacturing Systems*, 8, 70–78.
57. Lee, H. L., Padmanabhan, V. & Whang, S. (1997). The bullwhip effect in supply chains, *Sloan Management Review*, 38(3), 93–102.
58. Logistics & Technology. [Online].Available:http://www.trafficworld.com/news/log/112904a.asp.
59. Johnson, J. L. & Umesh, U. N. (2002). The interplay of task allocation patterns and governance mechanisms in industrial distribution channels, *Industrial Marketing Management*, 31(8), 665–678.
60. Tsay, A. A. (2002). Risk sensitivity in distribution channel partnerships: implications for manufacturer return policies, *Journal of Retailing*, 78(2), 147–160.
61. Yesikökçen, G. N. & Wesolowsky, G. O. (1998). A branch-and-bound algorithm for the supply connected location-allocation problem on network, *Location Science*, 6, 395–415.
62. Michalewicz, Z. (1996). *Genetic Algorithms + Data Structures = Evolution Program*, 3rd ed., New York: Spring-Verlag.
63. Yun, Y. S. & Moon, C. U. (2003). Comparison of adaptive genetic algorithms for engineering optimization problems, *International Journal of Industrial Engineering*, 10(4), 584–590.
64. Lee, C. Y., Yun, Y. S. & Gen, M. (2002). Reliability optimization design for complex systems by hybrid GA with fuzzy logic control and local search. *IEICE Transaction on Fundamentals of Electronics Communications & Computer Sciences*, E85-A(4), 880–891.
65. Lee, C. Y., Gen, M. & Kuo, W. (2001). Reliability optimization design using a hybridized genetic algorithm with a neural-network technique. *IEICE Transacion on Fundamentals of Electronics Communications & Computer Sciences*, E84-A(2), 627–635.
66. Eiben, A. E., Hinterding, R. & Michalewicz, Z. (1999). Parameter control in evolutionary algorithms, *IEEE Transactions on Energy Conversion*, 3(2), 124 –141.

Chapter 4
Communication Network Models

Modeling and design of large communication and computer networks have always been an important area to both researchers and practitioners. The interest in developing efficient design models and optimization methods has been stimulated by high deployment and maintenance costs of networks, which make good network design potentially capable of securing considerable savings. For many decades, the area of network design had been relatively stable and followed the development of telephone networks and their smooth transition from analog to digital systems. In the past decade, however, networks have undergone a substantial change caused by the emergence and rapid development of new technologies and services, an enormous growth of traffic, demand for service availability and continuity, and attempts to integrate new networking and system techniques and different types of services in one network. As a consequence, today's network designers face new problems associated with diverse technologies, complicated network architectures, and advanced resource and service protection mechanisms.

4.1 Introduction

Communication networks were born more than 100 years ago. There has been an explosive growth in computer networks since the 1980s. Communication network are pervasive. They arise in numerous application settings and in many forms. Communication networks are perhaps the most visible in our everyday lives, from automated teller machines, to airline reservation systems, to electronic mail services, to electronic bulletin boards, to internet services. These computer networks consist of many personal computers, workstations, mainframes, and some network devices (*e.g.*, network cables, bridges, routers, gateways, HUBs, and so on). The use of these networks systems has also been rapidly increasing in order to share expensive hardware/software resources and provide access to main systems from distant locations. The dissemination of these networks systems contents and service quantity from users means that expansion, and reconstruction of the network systems are

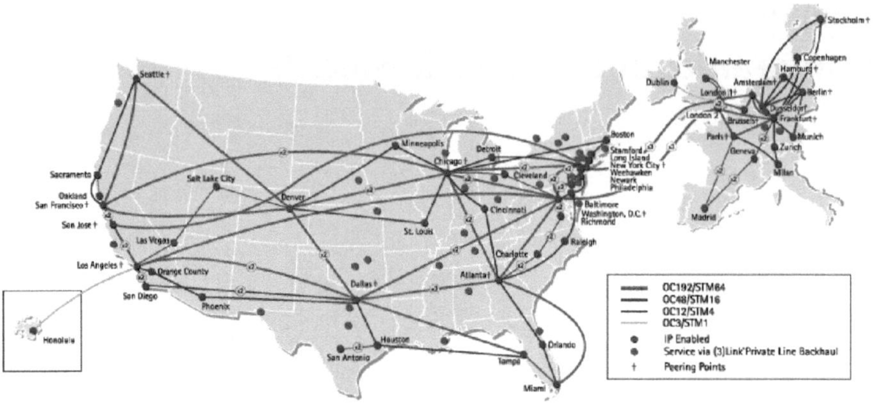

Fig. 4.1 An example of digital fiber-optic network [1]

required. When establishing and reconstructing network systems, the problems of how to design effectively a network, is which certain constraints are met and objectives (connection cost, message delay, traffic, reliability, capacitated, and so on) are optimized, are very important in many real world applications.

The topology design problems for these communication networks systems, *e.g.*, *local area networks* (LANs), *metropolitan area networks* (MANs), *wide area networks* (WANs), *INTERNET*, and so on, have been received much attention from many related researchers, such as network designers, network analysts, network administrators; accordingly the scale of communication network is larger and the requested contents (or services) from users are suddenly increasing.

Figure 4.1 shows an example of communication networks [1]. The backbone network, which connects local access networks to each other, is characterized by distributed traffic requirements and is generally implemented using packet switching techniques. There are two views of the communication network design.

Network View

- Different network design rules are applicable for different types of networks with specific routing, demand flow, and link capacity representation; thus, it is important to understand the subtle differences in the modeling so that proper frameworks can be used.
- Network failure is an important component of network design, and should be taken in to account, where applicable.
- Networks are multi-layered (this is a fact).
- Networks tend to be large.

Approach/Algorithm/Theory View

- We present different ways to formulate problems so that you can choose which ones to use and where to use approximations in the modeling.
- Even for the same type of network, different network providers may use different rules and constraints in network design. The modeling formulation will help you to see how different rules might be incorporated as constraints.
- Many practical size problems can be solved with commercial optimization tools. In many cases, this may be enough to solve the problems you might face.
- As soon as a network gets to be bigger than a few nodes, running a network design problem with a canned approach may sometimes take a considerable amount of time; hence, some idea about developing efficient algorithms is necessary. These are important points to note. However, depending on your interest, you can concentrate primarily on the material, on modeling and/or existing algorithms.
- In many cases, the optimal solution to the last bit rule is not necessary; in practice, an approximate or heuristic approach is sufficient.
- Some fundamental principles are important. A simple case in point is linear programming duality. Who would have thought that for IP networks running *open shortest path first* (OSPF) protocol, the determination of the link weight is related to solving the dual linear programming problem of a multi-commodity based network flow allocation problems.

Simply put, we ask ourselves how to design cost-effective core/backbone networks taking into consideration properties of the network, or in short, how to do network design. Interestingly, many of the problems posed here can be formulated into mathematical models and efficient optimization procedures (or algorithms) can be developed to solve these problems.

An important aspect that encompasses all networks is network management [2]. Different networks may have different network management cycles. Certainly, network management was born the day telephone connection was manned by people manually connecting calls. While the term network management sometimes has a different meaning to different people, we use this term here to mean the entire process from planning a network to deploying a network management systems and protocols may be necessary. In essence, a similar functionality is needed for communication and computer networks which are broadly referred to as network management.

Going beyond that (*i.e.*, from minutes time frame to days/weeks time frame to month/years' time frame), networks typically face what can be best classified as routing information update, capacity management, or reliability design. At this point, the routing information update, capacity management and reliability design are already installed. Based on short-term traffic trend/forecasting, some arrangement or rearrangement may allow better usage of the network (*i.e.*, network utilization. Thus, this may involve routing information update; on metaphase-term traffic trend/forecasting, some arrangement or re-arrangement may involve capacity management (from one service to another, and, within a service, from one demand to

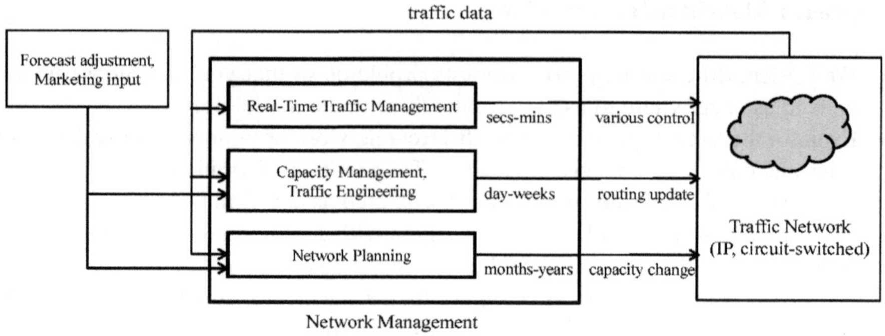

Fig. 4.2 Network management cycle and feedback process for traffic/service networks [2]

another) and on short-term traffic trend/forecasting, reliability design is involved perhaps. Thus, in some instances, routing information update, capacity management and reliability design may be necessary and possible. Finally, going from weeks to months (and sometimes to years), the overall network planning and engineering issues need to be addressed.

As we know, there are networks underneath the traffic network which are transport networks. These do not have the same type of real-time traffic management issues faced by service networks. Restoring a path or a route in case of an actual failure (especially in the absence of a pre-planned stand-by route) would be the type of real-time management problems in transport networks. Pioro and Medhi [2] provide the network management cycle of transport networks in Fig. 4.2. A cost-

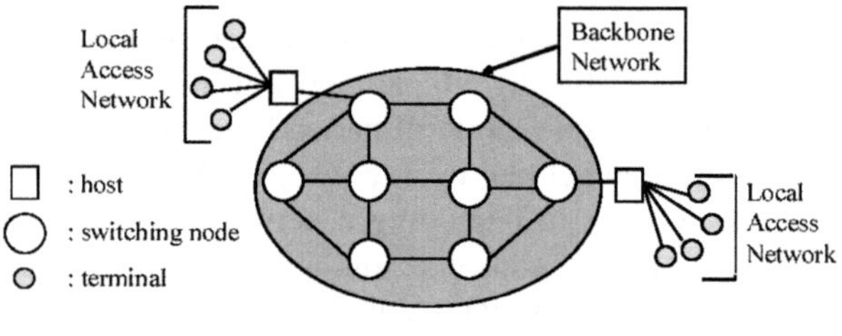

Fig. 4.3 Hierarchical structure of communication networks

effective structure for a large communication network is a *multilevel hierarchical structure* consisting of a backbone network (high level) and local access networks (low level) [3]. Figure 4.3 shows the hierarchical structure of communication networks. The backbone network, which connects local access networks to each other, is characterized by distributed traffic requirements and is generally implemented using packet switching techniques. In packet switching techniques, messages are broken into blocks of a certain size called packets; the packets, when they contain

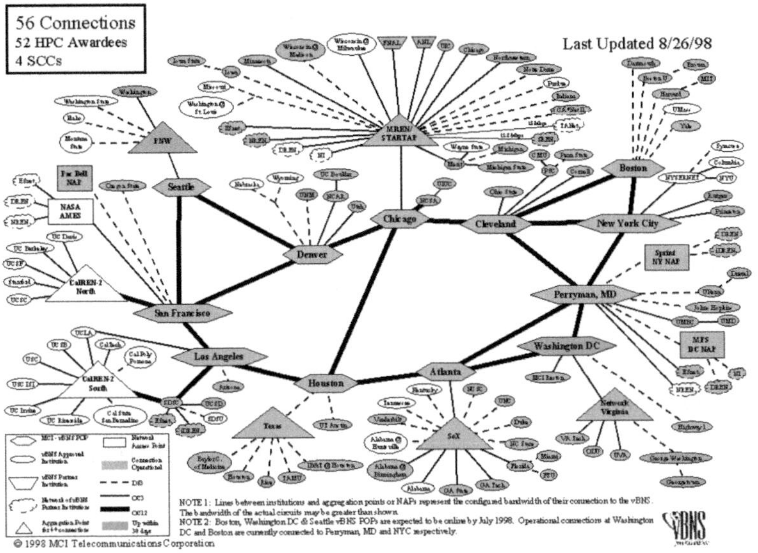

Fig. 4.4 vBNS logical network map

the destination address, can follow different routes toward their destination. The backbone network itself may be multilevel, incorporating, for example, terrestrial and satellite channels. Local access networks have in general centralized traffic patterns (most traffic is to and from the gateway backbone node) and are implemented with centralized techniques such as multiplexing, concentrating, and polling to share data coming from several terminals having lower bit rates on a single high capacity link. In special cases, the network may consist primarily of either centralized or distributed portions exclusively. Figure 4.4 show vBNS (very high speed backbone network system) backbone network and its logical network map funded in part by NSF. vBNS is a research platform for advancement and development of high-speed scientific and engineering applications, data routing and data switching capabilities.

The topological design problem for a large hierarchical network can be formulated as follows:

Given

- Terminal and host locations (terminal-to-host and host-to-host)
- Traffic requirements (terminal-to-host and host-to-host)
- Candidate sites for backbone nodes
- Cost elements (line tariff structures, nodal processor costs, hardware costs, *etc.*)

Minimize

- Total communications cost= (backbone line costs) + (backbone node costs) + (local access line costs) + (local access hardware costs) such that average delay and reliability requirements, which are index of quality of service, are met. Average packet delay in a network can be defined as the mean time taken by a packet to travel from a source to a destination node. Reliability is concerned with the ability of a network to be available for the desired service to the end-users. Reliability of a network can be measured using deterministic or probabilistic connectivity measures [4].

The global design problem consists of two sub-problems: the design of the backbone and the design of the local distribution networks. The two sub-problems interact with each other through the following parameters:

1. Backbone node number and locations
2. Terminal and host association to backbone nodes
3. Delay requirements for backbone and local networks
4. Reliability requirement for backbone and local networks

Once the above variables are specified, the subproblems can be solved independently [3]. Topological design of backbone and local access networks can be viewed as a search for topologies that minimize communication costs by taking into account delay and reliability constraints. This is an NP hard problem which is usually solved by means of heuristic approaches. For example, branch exchange, cut saturation algorithms, concave branch elimination are well known greedy heuristic approaches for backbone network design [3], [5]–[7]. Easu and Williams algorithm [8], Sharma's algorithm [9] and the unified algorithm [10] are known greedy heuristics for centralized network design. Recently, genetic algorithms have been successfully applied to network design problems. The studies for centralized network design are given as: Abuali *et al.* [11], Elbaum and Sidi [12], Kim [13], Kim and Gen [14, 15], Lo and Chang [16], Palmer and Kershenbaum [17], and Zhou and Gen [18]–[20]. The studies for backbone design are given as follows: Davis and Coombs [21], Kumar *et al.* [22, 23], Ko *et al.* [24], Dengiz *et al.* [25, 26], Deeter and Smith [27], Pierre and Legault [28], Konak and Smith [29], Cheng [30], Liu and Iwamura [31], and Altiparmak *et al.* [32, 33].

In this chapter, we introduce GA applications on centralized network design, which is basic for local access networks, backbone network design and reliable design of networks summarized in Fig. 4.5.

4.2 Centralized Network Models

The problem of effectively transmitting data in a network involves the design of communication sub-networks, *i.e.,* local access networks (LAN). Local access networks are generally designed as centralized networks. A centralized network is a network where all communication is to and from a central site (backbone node). In such networks, terminals are connected directly to the central site. Sometimes multipoint lines are used, where groups of terminals share a tree to the center and each multipoint line is linked to the central site by one link only. This means that optimal

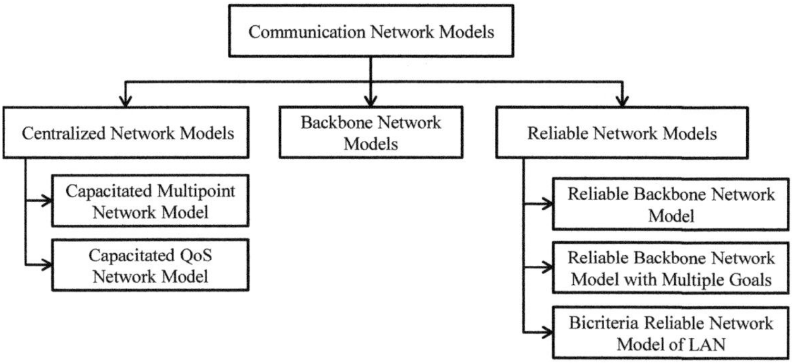

Fig. 4.5 The core models of logistics network design introduced in this chapter

topology for this problem corresponds to a tree in a graph $G = (V, E)$ with all but one of the nodes in V corresponding to the terminals. The remaining node refers to the central site, and edges in E correspond to the feasible communication wiring. Each subtree rooted in the central site corresponds to a multipoint line. Usually, the central site can handle, at most, a given fixed amount of information in communication. This, in turn, corresponds to restricting the maximum amount of information flowing in any link adjacent to the central site to that fixed amount. In the combinatorial optimization literature, this problem is known as the constrained minimum spanning tree problem. These constrained MST can be summarized as follows: *capacitated MST* (cMST) model, *degree-constrained MST* (dMST) model, *stochastic MST* (sMST) model, *quadratic MST* (qMST) model, *probabilistic MST* (pMST) model, *leaf-constrained MST* (lcMST) model and *multicriteria MST* (mMST) model.

The constrained MST models have been shown to be NP-hard by Papdimitriou [34]. Chandy and Lo [35], Kershenbaum [10] and Elias and Ferguson [36] had proposed heuristics approaches for this problem. Gavish [37] also studied a new formulation and its several relaxation procedures for the capacitated minimum directed tree problem. Recently, GA has been successfully applied to solve this problem and its variants;; see Abuali *et al.* [11], Lo and Chang [16], Palmer and Kershenbaum [17], and Zhou and Gen [18]–[20]. The detailed description of GAs for MST problems is introduced in Section 2.3.

4.2.1 Capacitated Multipoint Network Models

The minimum spanning tree (MST) problem is one of the best-known network optimization problems which attempt to find a minimum cost tree network that connects all the nodes in the communication network. The links or edges have associated costs that could be based on their distance, capacity, quality of line, *etc*. Illustration of a network with spanning tree structure is shown in Fig. 4.6.

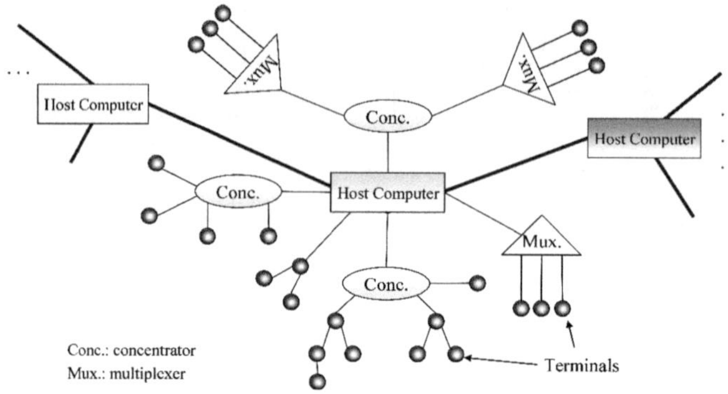

Fig. 4.6 Illustration of a network with spanning tree structure

In the real world, the MST is often required to satisfy some additional constraints for designing communication networks such as the capacity constraints on any edge or node, degree constraints on nodes, and type of services available on the edge or node. These additional constraints often make the problem NP-hard. In addition, there are usually such cases that one has to consider simultaneously multicriteria in determining an MST, because there are multiple attributes defined on each edge; this has become subject to considerable attention. Almost every important real-world decision making problem involves multiple and conflicting objectives. *Capacitated multipoint network* (CMN) model is a typical bicriteria communication network model, which is meant to find a set of links with the two conflicting objectives of minimizing communication cost and minimizing the transfer delay and the constraint of network capacity. This problem can be formulated as the *multiobjective capacitated minimum spanning tree* (mcMST) problem, and is NP-hard.

4.2.1.1 Mathematical Formulation of the CMN

The communication network is modeled using an edge-weighted undirected graph $G = (V,E,C,D,W)$ with $n = |V|$ nodes and $m = |E|$ edges; $c_{ij} \in C = [c_{ij}], d_{ij} \in D = [d_{ij}]$ and $w_{ij} \in W = [w_{ij}]$ represent the cost, delay and weight of each edge $(i, j) \in E$, where the variable is restricted to be a nonnegative real number, respectively. And u_i is given weight capacity of each node i. Figure 4.7 and Table 4.1 present a simple network with a data set of 12 nodes and 40 arcs.

Indices

$i, j = 1,2, \cdots, n$, index of node

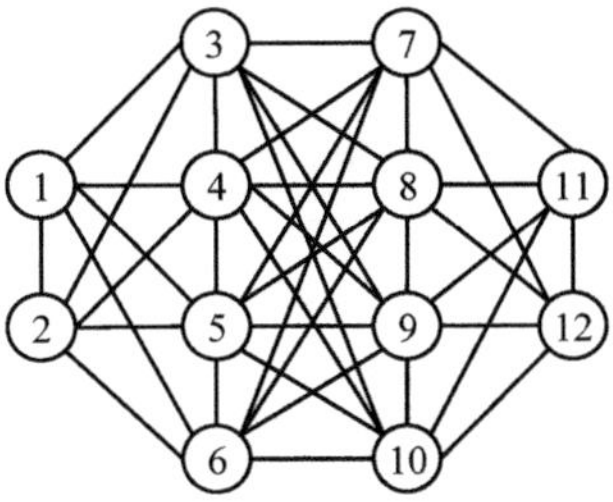

Fig. 4.7 A simple network with 12 nodes and 40 edges [38]

Table 4.1 The network data set of 12 nodes and 40 edges

k	Edge (i, j)	Weight w_{ij}	k	Edge (i, j)	Weight w_{ij}	k	Edge (i, j)	Weight w_{ij}	k	Edge (i, j)	Weight w_{ij}
1	(1, 2)	35	11	(3, 7)	51	21	(5, 7)	26	31	(7, 12)	41
2	(1, 3)	23	12	(3, 8)	23	22	(5, 8)	35	32	(8, 9)	62
3	(1, 4)	26	13	(3, 9)	64	23	(5, 9)	63	33	(8, 11)	26
4	(1, 5)	29	14	(3, 10)	28	24	(5, 10)	23	34	(8, 12)	30
5	(1, 6)	52	15	(4, 5)	54	25	(6, 7)	27	35	(9, 10)	47
6	(2, 3)	34	16	(4, 7)	24	26	(6, 8)	29	36	(9, 11)	68
7	(2, 4)	23	17	(4, 8)	47	27	(6, 9)	65	37	(9, 12)	33
8	(2, 5)	68	18	(4, 9)	53	28	(6, 10)	24	38	(10, 11)	42
9	(2, 6)	42	19	(4, 10)	24	29	(7, 8)	38	39	(10, 12)	26
10	(3, 4)	23	20	(5, 6)	56	30	(7, 11)	52	40	(11, 12)	51

Parameters

$n = |V|$: number of nodes
$m = |E|$: number of edges
$w_{ij} \in W$: weight of each edge $(i, j) \in E$
$c_{ij} \in C$: cost of each edge $(i, j) \in E$
$d_{ij} \in D$: delay of each edge $(i, j) \in E$
u_i: weight capacity of each node i
d_i: degree constraint on node i

Decision Variables

x_{ij}: the 0,1 decision variable; 1, if edge $(i, j) \in E$ is selected, and 0, otherwise.
y_{ij}: degree value on edge (i, j)

The decision variables in the *capacitated multipoint network* (CMN) model are the minimum cost z_1 with minimum delay z_2. Mathematically, the problem is reformulated as a constrained MST problem is as follows:

$$\min z_1(\boldsymbol{x}) = \sum_{(i,j)\in E} c_{ij}x_{ij} \tag{4.1}$$

$$\min z_2(\boldsymbol{x}) = \sum_{(i,j)\in E} d_{ij}x_{ij} \tag{4.2}$$

$$\text{s. t.} \sum_{(i,j)\in E} x_{ij} = n-1 \tag{4.3}$$

$$\sum_{(i,j)\in A_S} x_{ij} \le |S|-1 \quad \text{for any set } S \text{ of nodes} \tag{4.4}$$

$$\sum_{j=1}^{n} w_{ij}x_{ij} \le u_i, \quad \forall\, i \tag{4.5}$$

$$x_{ij} \in \{0,1\}, \quad \forall\,(i,j)\in E \tag{4.6}$$

In this formulation, the 0–1 variable x_{ij} indicates whether we select edge (i,j) as a part of the chosen spanning tree (note that the second set of constraints with $|S|=2$ implies that each $x_{ij} \le 1$). The constraint at Eq. 4.3 is a cardinality constraint implying that we choose exactly n–1 edges, and the packing constraint at Eq. 4.4 implies that the set of chosen edges contain no cycles (if the chosen solution contained a cycle, and S were the set of nodes on a chosen cycle, the solution would violate this constraint). The constraint at Eq. 4.5 guarantees that the total link weight of each node i does not exceed the upper limit W_i.

Note that this is a somewhat simplified model because in a realistic communication networks, delays consist of delays due to propagation, link bandwidth, and queuing at intermediate nodes. This simplified model has been (and still is) widely used in the scientific literature on computer communications and networks because many efficient algorithms for the more complicated model are based on efficient algorithms for this simplified model [39].

4.2.1.2 PrimPred-based GA for the CMN

Let $P(t)$ and $C(t)$ be parents and offspring in current generation t, respectively. The overall procedure of *PrimPred-based GA* (primGA) for solving CMN is outlined as follows.

PrimPred-based Encoding and Decoding

When a GA searches a space of spanning trees, its initial population consists of chromosomes that represent random trees. It is not as simple as it might seem to choose spanning trees of a graph so that all are equally likely. Applying Prim's algorithm to the EAs chooses each new edge at random rather than according to its cost [38]. The encoding procedure and decoding procedure are shown in Subsect. 2.3.2.1 (on page 91).

procedure: primGA for CMN model
input: CMN data, GA parameters $(popSize, maxGen, p_M, p_C)$
output: Pareto optimal solutions E
begin
  $t \leftarrow 0$;
  initialize $P(t)$ by PrimPred-based encoding routine;
  calculate objectives $z_1(P)$and$z_2(P)$ by PrimPred-based decoding routine;
  create Pareto $E(P)$;
  evaluate $eval(P)$ by interactive adaptive-weight fitness assignment routine;
  while (**not** terminating condition) **do**
    create $C(t)$ from $P(t)$ by Prim-based crossover routine;
    create $C(t)$ from $P(t)$ by LowestCost mutation routine;
    create $C(t)$ from $P(t)$ by immigration routine;
    calculate objectives $z_1(C)$and$z_2(C)$ by PrimPred-based decoding routine;
    update Pareto $E(P,C)$;
    evaluate $eval(P,C)$ by interactive adaptive-weight fitness assignment routine;
    select $P(t+1)$ from $P(t)$ and $C(t)$ by roulette wheel selection routine;
    $t \leftarrow t+1$;
  end
  output Pareto optimal solutions $E(P,C)$
end

Prim-based Crossover

To provide good heritability, a crossover operator must build an offspring spanning tree that consists mostly or entirely of edges found in the parents. In this chapter, we also applying Prim's algorithm described in the previous section to the graph $G' = (V, T_1 \cup T_2)$, where T_1 and T_2 are the edge sets of the parental trees. The pseudocode of crossover process is given in Subsection 2.3.2.1.

LowestCost Mutation

For considering represent a feasible solutions after mutated, improving the heritability of offspring, we proposed a new mutation operator, first removing a randomly chosen edge, determining the separated components, then reconnecting them by the lowest cost edge that recon. The pseudocode of mutation process is given in Subsect. 2.3.2.1.

Interactive Adaptive-weight Fitness Assignment

For solving CMN, first two extreme points are defined as follows: the maximum extreme point $z^+ \leftarrow [z_1^{\max,} z_2^{\max}]$ and the minimum extreme point $z^- \leftarrow [z_1^{\min}, z_2^{\min}]$ in criteria space, where $z_1^{\max}, z_2^{\max}, z_1^{\min}$ and $z_2^{\min}$ are the maximum value and minimum value for objective 1 (minimizing cost) and the reciprocal of objective 2 (minimizing delay) in the current population. Calculate the adaptive weight $w_1 = z_1^{\min} z_1^{\max}/(z_1^{\max} - z_1^{\min})$ for objective 1 and the adaptive weight $w_2 = z_2^{\min} z_2^{\max}/(z_2^{\max} - z_2^{\min})$ for objective 2. Afterwards, calculate the penalty term $p(v_k)$=0, if v_k is a nondominated solution in the nondominated set P. Otherwise $p(v_{k'})$=1 for a dominated solution $v_{k'}$. Last, calculate the fitness value of each chromosome by combining the method as follows; the pseudocode of interactive adaptive-weight fitness assignment is showed in Subsection 2.3.2.1:

$$eval(v_k) = w_1(1/z_1^k - 1/z_1^{\max}) + w_2(1/z_2^k - 1/z_2^{\max}) + p(v_k), \ \forall\, k \in popSize \quad (4.7)$$

4.2.1.3 Numerical Experiments

In this experimental study, the different multiobjective GAs are compared for solving CMP by different fitness assignment approaches; there are *strength Pareto Evolutionary Algorithm* (spEA), *non-dominated sorting Genetic Algorithm II* (nsGA II), *random-weight Genetic Algorithm* (rwGA) and *interactive adaptive-weight Genetic Algorithm* (i-awGA). The data in the test problem was generated randomly. All the simulations were performed with Java on a Pentium 4 processor (2.6-GHz clock). In each GA approach, PrimPred-based encoding was used, and Prim-based crossover and LowestCost mutation were used as genetic operators. GA parameter settings were taken as follows:

Population size: $popSize$ =20;
Crossover probability: $p_C = 0.70$;
Mutation probability: p_M =0.50;
Stopping criteria: evaluation of 5000 solutions

Performance Measures

In these experiments, the following performance measures are considered to evaluate the efficiency of the different fitness assignment approaches. The detailed explanations have introduced in Sect. 1.4.4.

Number of obtained solutions $|S_j|$: Evaluate each solution set depending on the number of obtained solutions.

Ratio of nondominated solutions $R_{NDS}(S_j)$: This measure simply counts the number of solutions which are members of the Pareto-optimal set S^*.

Average distance $D1_R(S_j)$: Instead of finding whether a solution of S_j belongs to the set S^* or not, this measure finds an average distance of the solutions of S_j from S^*.

Reference Solution Set

For finding the reference solution set S^*, we used the following GA parameter settings in all of four fitness assignment approaches (spEA, nsGA II, rwGA and i-awGA): population size, *popSize*=50; crossover probability, p_C=0.90; mutation probability, p_M=0.90; Stopping criteria: evaluation of 100,000 generations.

The i-awGA with spEA, nsGA II and rwGA are compared through computational experiments on the 40-node/1560-arc test problem under the same stopping condition (*i.e.*, evaluation of 5000 solutions). Each algorithm was applied to each test problem 10 times and gives the average results of the 3 performance measures (*i.e.*, the number of obtained solutions $|S_j|$, the ratio of nondominated solutions $R_{NDS}(S_j)$, and the average distance D_{1R} measure). In Table 4.2 and Fig. 4.8, better results of all performance measures were obtained from the i-awGA than other fitness assignment approach.

Table 4.2 Performance evaluation of fitness assignment approaches for the 40-node/1560-arc test problem

# of	$\lvert S_j \rvert$				$R_{NDS}(S_j)$				$D1_R(S_j)$			
eval. solut.	spEA	nsGA II	rwGA	i-awGA	spEA	nsGA II	rwGA	i-awGA	spEA	nsGA II	rwGA	i-awGA
50	31.45	30.40	32.60	36.20	0.34	0.31	0.36	0.39	178.85	200.47	182.03	162.57
500	42.40	45.60	43.20	47.60	0.42	0.45	0.40	0.52	162.97	151.62	160.88	157.93
2000	46.60	52.20	45.30	55.50	0.54	0.61	0.58	0.66	118.49	114.60	139.40	92.41
5000	51.20	54.40	50.30	60.70	0.64	0.70	0.62	0.73	82.70	87.65	117.48	77.98

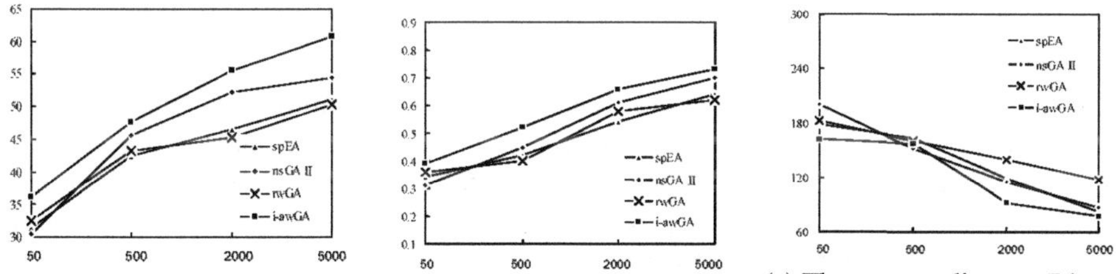

(a) The number of obtained solutions $|S_j|$

(b) The ratio of nondominated solutions $R_{NDS}(S_j)$

(c) The average distance $D1_R$

Fig. 4.8 Performance evaluation of fitness assignment approaches using the performance measures for the 40-node/1560-arc test problem.

4.2.2 *Capacitated QoS Network Model*

As the network speed becomes faster and requirements of various services are increased, many researches are currently developing technologies aimed at evolving and enhancing the capabilities of existing networks. A *next generation network* (NGN) is a multi-service functional network, capable of differentiating between users and applications through policies of *quality of service* (QoS), access and security guidelines, providing *virtual private networks* (VPNs). The NGN is characterized by the following main aspects [71]: (1) support for a wide range of services, applications and mechanisms based on service building blocks (including real time/ streaming/ non-real time services and multi-media); (2) broadband capabilities with end-to-end QoS and transparency; (3) unrestricted access by users to different service providers; (4) a variety of identification schemes which can be resolved to IP addresses for the purposes of routing in IP networks, *etc*.

However, current Internet routing protocols such as *open shortest path first* (OSPF), *routing information protocol* (RIP), and *border gateway protocol* (BGP) are called "best-effort" routing protocols, which means they will try their best to forward user traffic, but can provide no guarantees regarding loss rate, bandwidth, delay, delay jitter, *etc*. It's intolerable for NGN services, for example videoconferencing and video on-demand, which require high bandwidth, low delay, and low delay jitter, and provid the different types of network services at the same time is very difficult. Thus, the study of QoS is very important nowadays [72]. To provide QoS in NGN, many techniques have been proposed and studied, including integrated services [69], differential services [68], *MultiProtocol label switching* (MPLS) [70], traffic engineering and QoS-based routing [72]. And most problems can be formulated as the optimization models, such as the network reliability optimization model, shortest path routing model and constrained minimum spanning tree model *etc*.

Lin and Gen [67] formulate the problem as an *extended capacitated minimum spanning tree* (cMST) problem, the objective of which is minimizing the communication cost (defined as a kind of performance measures for NGN's QoS) while considering the following constraint: (1) consider the capabilities of the network; (2) define different priority for different types of services; (3) dynamic environment. As we know, this cMST is an NP-hard problem. In addition, the complex structures and complex constraints of this problem are to be handled simultaneously, which makes the problem intractable to traditional approaches.

4.2.2.1 Mathematical Formulation of the Capacitated QoS Network

The communication network is modeled using an edge-weighted undirected graph $G = (V, E, Q, U)$ with n nodes and m edges. Figure 4.9 presents a simple network with 12 nodes and 40 arcs.

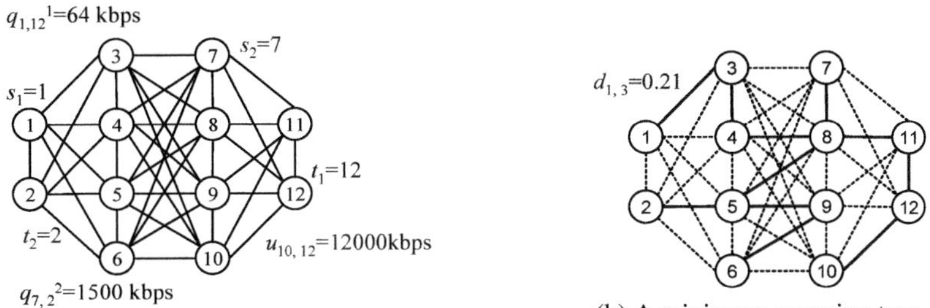

(a) Simple example of network

(b) A minimum spanning tree

Fig. 4.9 A simple network with 12 nodes and 40 edges

Indices

$i, j, k = 1, 2, \cdots, n$, index of node

Parameters

$n = |V|$: number of nodes
$m = |E|$: number of edges
$q^l_{st} \in Q$: requirement of type l form source node s to sink node t
$u_{ij} \in U$: capacity of edge (i, j)
$w_l \in W$: weight (priority) of type l communication service
$d_{ij} \in D$: delay of edge (i, j) (or defined as a kind of performance measures for NGN's QoS),where $d_{ij} = \sum_l w_l \Gamma(q^l_{ij} - u_{ij})$
$\Gamma(q^l_{ij} - u_{ij})$: a function for delay defination of service type l

Decision Variables

y_{ij}: the amount of requirement through arc (i, j)
x_{ij}: 0–1 decision variable

The objective is minimizing the total average delay. Mathematically, the problem is reformulated as a *capacitated minimal spanning tree* (cMST) model is as follows:

$$\min f = \sum_{(i,j)\in E} \left(\sum_{l=1}^{L} w_l \cdot \Gamma\left(\min\left\{0, |y_{ij} - u_{ij}|\right\}\right) \right) \tag{4.8}$$

$$\text{s. t.} \quad \sum_{(i,j)\in E} x_{ij} = n - 1 \tag{4.9}$$

$$\sum_{(i,j)\in A_S} x_{ij} \le |S| - 1 \quad \text{for any set } S \text{ of nodes} \tag{4.10}$$

$$\sum_{(i,j)\in E} y_{ij} - \sum_{(k,i)\in E} y_{ki} = \begin{cases} q^l_{st} & i = s \\ 0 & i \in V - s,t, \quad \forall\, (s,t) \in q^l_{st},\ \forall\, l \\ -q^l_{st} & i = t \end{cases} \tag{4.11}$$

$$y_{ij} \geq 0 \quad \forall\, i,\ j \tag{4.12}$$

$$x_{ij} \in \{0,1\}, \quad \forall\, (i,j) \in E \tag{4.13}$$

In this formulation, the 0–1 variable x_{ij} indicates whether we select edge (i,j) as part of the chosen spanning tree (note that if $y_{ij} > 0$ then x_{ij} =1, else x_{ij} =0). The constraint at Eq. 4.9 is a cardinality constraint implying that we choose exactly n–1 edges, and the packing constraint at Eq. 4.10 implies that the set of chosen edges contain no cycles (if the chosen solution contained a cycle, and S were the set of nodes on a chosen cycle, the solution would violate this constraint); the constraint at Eq. 4.11 implies a flow conservation law dependent communication requirement to be observed at each of the nodes other than s or t.

4.2.2.2 PrimPred-based GA for the Capacitated QoS Network

Let $P(t)$ and $C(t)$ be parents and offspring in current generation t, respectively. The overall procedure of *PrimPred-based GA* (primGA) for solving capacitated QoS network is outlined as follows.

---

procedure: PrimPred-based GA for capacitated QoS models
input: network data ($V, E, C, D, \boldsymbol{u}$),
    GA parameters (*popSize*, *maxGen*, p_M, p_C)
output: the best solution
begin
  $t \leftarrow 0$;
  initialize $P(t)$ by PrimPred-based encoding routine;
  evaluate $P(t)$ by PrimPred-based decoding routine;
  while (**not** terminating condition) **do**
    create $C(t)$ from $P(t)$ by Prim-based crossover routine;
    create $C(t)$ from $P(t)$ by LowestCost mutation routine;
    create $C(t)$ by immigration routine
    evaluate $C(t)$ by PrimPred-based decoding routine;
    if $t > u$ **then**
      auto-tuning p_M, p_C and p_I by FLC;
    select $P(t+1)$ from $P(t)$ and $C(t)$ by roulette wheel selection routine;
    $t \leftarrow t+1$;
  end
  output the best solution
end

PrimPred-based Encoding and Decoding

When a GA searches a space of spanning trees, its initial population consists of chromosomes that represent random trees. It is not as simple as it might seem to choose spanning trees of a graph so that all are equal likely. For applying Prim's algorithm to the EAs chooses each new edge at random rather than according to its cost [38]. The encoding procedure and decoding procedure are shown in Subsect. 2.3.2.1 (on page 91).

Prim-based Crossover

To provide good heritability, a crossover operator must build an offspring spanning tree that consists mostly or entirely of edges found in the parents. In this paper, we also applying Prim's algorithm described in the previous section to the graph $G' = (V, T_1 \cup T_2)$, where T_1 and T_2 are the edge sets of the parental trees. The pseudocode of crossover process is given in Subsect. 2.3.2.1.

LowestCost Mutation

For considering representing a feasible solutions after mutated, improving the heritability of offspring, we proposed a new mutation operator, first removing a randomly chosen edge, determining the separated components, then reconnecting them by the lowest cost edge that recon. The pseudocode of mutation process is given in Subsect. 2.3.2.1.

Immigration

The immigration is modified to (1) include immigration routine, in each generation, (2) generate and (3) evaluate popSize p_{I} random chromosomes, and (4) replace the $popSize\, p_{\text{I}}$ worst chromosomes of the current population (p_{I} , called the immigration probability). The detailed introduction is showed in Subsect. 1.3.2.4.

Auto-tuning Scheme

The probabilities of immigration and crossover determined the weight of exploitation for the search space. Its main scheme is to use two FLCs: auto-tuning for exploration and exploitation $T[p_{\text{M}} \wedge (p_{\text{C}} \vee p_{\text{I}})]$ and auto-tuning for genetic exploitation and random exploitation ($T[p_{\text{C}} \wedge p_{\text{I}}]$) are implemented independently to regulate adaptively the genetic parameters during the genetic search process. The detailed scheme is shown in Subsect. 1.3.2.4.

4.2.2.3 Numerical Experiments

In the experiment, PrimPred-based GA is compared with various GAs (Zhou and Gen [19]; Raidl and Julstrom [38]) and traditional Prim's algorithm (without considering the capacity constraint). The three complete network structures (# of nodes n=20) with three kinds of service (Type-1: cable television; Type-2: IP phone; Type-3: data) are combined. The weight (priority) of these three types is 0.60, 0.30 and 0.10 respectively. The capacity of each edge (i, j) are represented as random variables depending on the uniform distribution run if(m, 100, 1000). The 30 time-period requirements of different service types from node s to node t are represented as random variables depending on the distribution functions (Type-1: exponential distribution rexp($|Q|$, 0.03); Type-2: lognormal distribution: 0.1rlnorm($|Q|$, 0.1, 0.1); Type-3: normal distribution: rnorm($|Q|$, 0.01, 0.001)), where $|Q|$=100. The genetic parameter is initialized as follows: population size 20, crossover probability 0.50, mutation probability 0.50. There are two different termination conditions of per time-period employed. One of them is computation time equal to 0.80 s (depending on the updating time of communication requirements). Another one is the objective function equal to 0 (all requirements can be sufficed). The comparisons results are summarized in Fig. 4.10.

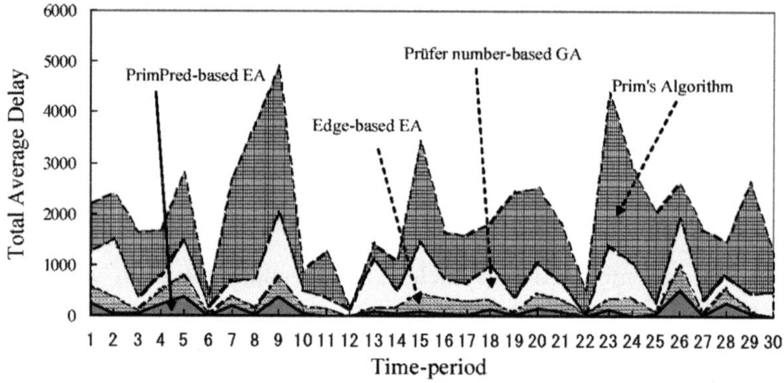

Fig. 4.10 Comparisons results of three different EA approaches and Prim's algorithm

4.3 Backbone Network Model

As explained at the beginning of this chapter, for cost-effectiveness, communication networks are designed in a *multilevel hierarchical structure* consisting of *backbone networks* and *local access networks* (LAN). A simple model for a backbone network is an undirected graph $G = (V,E)$ with node set V and edge set E. In this model, the nodes represent the connection points where LANs are hooked up to the backbone network *via* gateways. In addition to being connection points, the nodes are the processing units that carry out traffic management on the network by forwarding data packets to the nodes along their destinations (*i.e.*, known as routers). The edges represent the high capacity multiplexed lines on which data packets are transmitted bi-directionally between the node pairs. Each network link is characterized by a set of attributes which principally are the flow and the capacity. For a given link between a node pair (i,j), the flow f_{ij} is defined as the effective quantity of information transported by this link, while its capacity C_{ij} is a measure of the maximal quantity of information that it can transmit. Flow and capacity are both expressed in bits/s (bps). The traffic γ_{ij} between a node pair (i,j) represents the average number of packets/s sent from source i to destination j. The flow of each link that composes the topological configuration depends on the traffic matrix. Indeed, this matrix varies according to the time, the day and applications used [6]. Although some designs are based on the rush hour traffic between two nodes as in the study of Dutta and Mitra [47], generally the average traffic is considered in designs [3, 5, 6].

The topological design of backbone networks can be formulated as follow: given the geographical location of nodes, the traffic matrix $M = (\gamma_{ij})$, the capacity options and their cost, the maximum average delay $T_{\max}$, minimize the total cost over topological configuration together with flow and capacity assignment, subject to delay and reliability constraints. In addition to this formulation, some formulations associate a cost with the waiting time of a packet in the network, or seek to optimize the cost and the average delay by making abstraction of the reliability constraint [48]. Others seek to minimize the cost under delay and flow constraints [63]. The delay constraint is expressed as follows: $T < T_{\max}$. The flow and capacity assignments depend on the traffic and the average length of packets. The optimum capacity assignment can be obtained by minimizing the average delay [61]. The average packet delay (T) of a network is given as

$$T = \frac{1}{\gamma} \sum_{(i,j)\in E} \frac{f_{ij}}{C_{ij} - f_{ij}} \tag{4.14}$$

The average traffic flow on a link is a function of routing algorithm and the traffic load distribution on the network [5]. The reliability is concerned with the ability of a network to be available for the desired service to the end-users. In the literature, reliability measures to define connectivity level of a network are classified as probabilistic and deterministic. Probabilistic measures give the probability of a network being connected [4]. Deterministic measures are related with the vulnerability of a network such as k-connectivity (k-node disjoint paths between each pair of node), articulation, cohesion, *etc.* [57]. Generally, deterministic measures are considered as network reliability constraints, because they are computationally more tractable than the probabilistic measures and ensure minimum reliability.

Since topological design of communication networks is an NP-hard problem, heuristics techniques has been applied such as *branch exchange method* (BXC), *concave branch elimination* (CBE), *cut-saturation algorithm* (CSA) [3]–[7]. Newport and Varshney [57] modified CSA considering network survivability into design process. Kershenbaum *et al.* [54] proposed another heuristic called MENTOR. Gavish [48] formulated an overall deign problem, which includes both local access network and backbone design together, as a nonlinear optimization problem. The solution methodology depends on Lagrangean relaxation and subgradient optimization procedure. Dutta and Mitra [47] and Sykes and White [62] have proposed heuristic approaches introducing the artificial intelligence concept of knowledge-based system for the problem. Recently, metaheuristic techniques have been applied to design of backbone networks. While Coan *et al.* [46], Pierre *et al.* [58] and Rose [60] have used simulated annealing as a solution approach, Beltran and Skorin-Kapov [43], Glover *et al.* [51], Koh and Lee [55] and Pierre and Elgibaoui [59] propose tabu search algorithm for the problem. Genetic algorithms have also been applied successfully to the design of backbone networks. Davis and Coombs [21], Ko *et al.* [24], Konak and Smith [29] and Pierre and Legault [28] have used GA as a solution approach.

4.3.1 Pierre and Legault's Approach

Pierre and Legault [28] developed a *binary-based GA* for designing backbone networks considering a deterministic reliability measure. Let $P(t)$ and $C(t)$ be parents and offspring in current generation t, respectively. The overall procedure of binary-based GA (bGA) for solving backbone network model is outlined as follows.

---

procedure: bGA for backbone network model
input: problem data, GA parameters ($popSize$, $maxGen$, p_M, p_C)
output: the best solution
begin
  $t \leftarrow 0$;
  initialize $P(t)$ by binary-based encoding routine;
  evaluate $P(t)$ by binary -based decoding routine;
  while (**not** terminating condition) **do**
    create $C(t)$ from $P(t)$ by one cut-point crossover routine;
    create $C(t)$ from $P(t)$ by bit-flip mutation routine;
    create $C(t)$ from $P(t)$ by inversion mutation routine;
    evaluate $C(t)$ by binary-based decoding routine;
    select $P(t+1)$ from $P(t)$ and $C(t)$ by roulette wheel selection routine;
    $t \leftarrow t+1$;
  end
  output the best solution
end

4.3.1.1 Genetic Representation

Binary encoding had been used to represent a network. In a chromosome, 1s and 0s symbolize the existence or nonexistence of a link. The length of chromosomes is directly linked to the number of nodes in the network. Thus, a network composed of n nodes will be represented by chromosomes of length $n(n-1)/2$, that is, the maximum number of links that compose a network of n nodes. Figure 4.11a shows a three-node-connected topology, with its associated chromosome, for a network. Consider the example with eight nodes shown in Fig. 4.12. Note that there are $(8\times 7)/2 = 28$ possible links for this example but only 13 are included; the other 15 are not in the topology. The chromosome representation of the topology (x) is given in Fig. 4.11b. It is possible to see that the index of the topology is represented by the following equation:

$$\text{index of } k = \frac{n(n-1)}{2} - \frac{(n-i)(n-i+1)}{2} + (j-i) \tag{4.15}$$

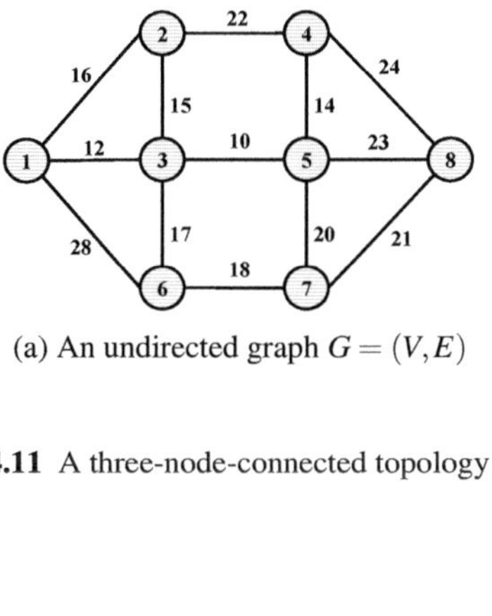

(a) An undirected graph $G = (V, E)$

$$\begin{array}{c|cccccccc} & 1 & 2 & 3 & 4 & 5 & 6 & 7 & 8 \\ \hline 1 & - & 1 & 1 & 0 & 0 & 1 & 0 & 0 \\ 2 & 1 & - & 1 & 1 & 0 & 0 & 0 & 0 \\ 3 & 1 & 1 & - & 0 & 1 & 1 & 0 & 0 \\ 4 & 0 & 1 & 0 & - & 1 & 0 & 0 & 1 \\ 5 & 0 & 0 & 1 & 1 & - & 0 & 1 & 1 \\ 6 & 1 & 0 & 1 & 0 & 0 & - & 1 & 0 \\ 7 & 0 & 0 & 0 & 0 & 1 & 1 & - & 1 \\ 8 & 0 & 0 & 0 & 1 & 1 & 0 & 1 & - \end{array}$$

(b) Matrix representation

Fig. 4.11 A three-node-connected topology

locus:	1	2	3	4	5	6	7	8	9	10	11	12	13	14	15	16	17	18	19	20	21	22	23	24	25	26	27	28
Link ID:	1	1	0	0	1	0	0	1	1	0	0	0	0	0	1	1	0	0	1	0	0	1	0	1	1	1	0	1

Fig. 4.12 Chromosome representation of Fig. 4.11

Initialization

Since network reliability constraint is k-node-connectivity, the initial population is generated in two phases. In a first phase, an initial topology of degree k is produced and then other chromosomes are created by modifying the initial topology, *i.e.*, chromosomes are created with adding new links between 1 and $\sqrt{n(n-1)/2}$. The creation of the initial topology is done deterministically. The link selection criterion is based on Euclidean distances between nodes. First, the $\lceil kn/2 \rceil$ links, whose distance between nodes is the shortest, are added to chromosome and chromosome is checked whether it is k-connected, *i.e.*, whether all nodes have at least k incident links, or not. If not, other links are added to the chromosome until the resulting topology is k-connected. The procedure of initial population is given in Fig. 4.13.

Procedure: binary-based encoding
Input: a k-connected network
Output: binary-based chromosomes
Step 1. Sequence the links in increasing order of their distance
Step 2. Create the initial topology
    (2.1) Add the first shortest $\lceil kn/2 \rceil$ links to topology.
    (2.2) Check its connectivity. If it is not k-connected, add one or more shortest links from remaining link set until satisfy k-connectivity.
Step 3. Repeat following steps until reach to population size
    (3.1) Determine the number of links (l) between 1 and $\sqrt{n(n-1)/2}$
    (3.2) Obtain new topology with adding l shortest links from remaining link set to initial topology.

Fig. 4.13 The pseudocode of binary-based encoding

Evaluation

The problem is to design a backbone network for minimizing total communication cost under maximum allowable average delay and k-node-connectivity constraints. Fitness function is the objective function of the problem. Once a candidate network with k-node-connectivity is obtained, first capacity assignment of its links is realized considering capacity options, and then average delay using equation (1) and cost are calculated. Link cost is function of its capacity and distance. Therefore, the cost of a link comprises two components: a permanent cost related to the capacity of link, and a variable cost related to the physical length of this link. Let $d_{ij}(l_{ij}, C_{ij})$ be the leasing cost of a link between nodes i and j, of capacity C_{ij} and of length l_{ij}; the total link cost D of a candidate network is calculated as follows:

$$D = \sum_{(i,j)\in E} d_{ij}(l_{ij}, C_{ij}) \tag{4.16}$$

In the study, six different capacity options had been considered. Table 4.3 gives the leasing costs of different types of link. The capacity of each link between node pairs for a topology had been assigned considering related link flow and capacity options.

Table 4.4 shows the link flows and capacities associated with the topology of Fig. 4.11.

Table 4.3 Capacity options and their costs

Capacity (Kbps)	Variable cost ((\$/month)/km)	Fixed cost (\$/month)
9.6	3.0	10.0
19.2	5.0	12.0
56.0	10.0	15.0
100.0	15.0	20.0
200.0	25.0	25.0
560.0	90.0	60.0

Table 4.4 Link flows and capacities associated with the network of Fig. 4.11

Link (i, j)	Flow f_{ij} (Kbps)	Capacity c_{ij} (Kbps)
(1, 2)	20.00	56.00
(1,3)	40.00	56.00
(1, 6)	10.00	19.20
(2, 3)	40.00	56.00
(2, 4)	30.00	56.00
(3, 5)	100.00	200.00
(3, 6)	50.00	56.00
(4, 5)	40.00	56.00
(4, 8)	20.00	56.00
(5, 7)	40.00	56.00
(5, 8)	30.00	56.00
(6, 7)	20.00	56.00
(7, 8)	20.00	56.00

As an example, it is possible to calculate average delay and cost of the topology given in Fig. 4.11. The values on the links in Fig. 4.11 represent link lengths in kilometers. Table 4.4 provides the values required for calculating $d_{ij}(l_{ij}, C_{ij})$. When the average packet size and the traffic requirements are considered as 1000 bits and $\gamma_{ij} = 5$ packets/s, the average delay and the cost of the topology are $T = 89 : 1\text{ms}$ and $D = \$2612$/month, respectively.

4.3.1.2 Genetic Operators

One-point crossover, bit-flip mutation and inversion are used as genetic operators. In crossover operators, chromosomes selected with roulette wheel selection mechanism are mated with the best chromosome of the current generation to obtain offspring. As a second phase of crossover operator, inversion is applied to offspring, *i.e.*, the gene at position 1 is transferred to position l (length of the chromosome), while the gene at position l replaces gene 1; the gene at position 2 is transferred to position $l-1$, and the gene in position $l-1$ is transferred to position 2, and so on.

4.3.2 Numerical Experiments

To measure the efficiency of the GA and the quality of its solutions, different networks sizes between 6 and 30 nodes were considered. Population size, probability of crossover, probability of mutation, inversion rate and number of generations were set to 80, 0.95, 0.12, 0.23 and 40, respectively. While average packet size was taken as 1000 bits, uniform traffic was set to 5 packets/s.

First, Pierre and Legault had examined the GA performance according to initial topology considering three-node-connectivity. Table 4.5 gives % of improvement of the GA. They showed that while GA improved the solutions only very slightly for the networks with less than 15 nodes, after that it realized at least 15% improvement on the initial topology. Also they explained deterioration of the solution of problem 2 as particularities of the initial population generated. They also compared the result of GA with optimum solution for the network with six nodes and showed that there was 4.3% of deviation from optimum solution.

Table 4.5 Cost *vs* size of the network

No.	n	C_L ($/month) Initial topology	C_L ($/month) GA	Standard deviation	% of improvement
1	6	21632	20733	858.255	4%
2	10	35330	35462	671.501	-
3	12	47810	43642	4397.09	9%
4	15	89856	72873	2954.07	19%
5	25	320134	272539	16095.81	15%
6	30	408684	346529	1046.45	15%

The performance of GA had also been compared with simulated annealing (SA) and cut saturation (CS) algorithms. In the analysis, number of generations, probabilities of crossover, mutation and inversion had been taken as 25, 0.80, 0.10 and 0.20, respectively. Results are given in Tables 4.6 and 4.7. In Table 4.6, the last two columns represent the improvement in percentage of costs and the average delay with GA according to results of SA. It is seen that the GA offers a better solution than SA in terms of network cost (D) and average delay (T). When GA was compared with CS in Table 4.7, maximum delay had been set to 120 ms. Comparative results have showed that the GA appears to be more effective when network size increases. In addition, Pierre and Legault noted that CS generally took less running time than GA for converging to a good solution.

Table 4.6 Comparative results SA *vs* GA

No.	N	K	SA		GA			% C_L	%T
			C_L (\$/month)	T (ms)	C_L (\$/month)	T (ms)	T_{max} (ms)		
1	15	3	74100	123.02	70198	86.88	125	6%	42%
2	18	3	138805	114.04	101846	91.55	115	36%	25%
3	18	4	119238	86.97	110030	84.23	90	8%	3%
4	18	5	111725	103.42	112287	89.21	105	-1%	16%
5	20	3	189570	105.90	128524	84.58	106	47%	25%
6	20	4	140863	79.28	133427	75.90	80	6%	4%
7	20	5	156872	99.83	134441	94.73	100	17%	5%
8	25	3	284797	100.60	229198	67.91	101	24%	48%
9	25	4	323333	105.62	201174	95.41	106	61%	11%
10	25	5	284797	100.60	206775	83.14	101	38%	21%

Table 4.7 Comparative results CS *vs* GA

No.	N	CS		GA		% C_L
		C_L (\$/month)	T (ms)	C_L (\$/month)	T (ms)	
1	10	96908	76.9	102723	84.3	-6.0%
2	12	112427	77.2	114804	86.7	-2.1%
3	14	124215	82.6	123198	92.4	0.8%
4	15	130788	82.8	123987	100.3	5.2%
5	16	130884	76.4	122226	102.4	6.6%
6	18	138228	92.4	128730	98.7	6.9%
7	20	142643	86.8	129436	109.7	9.3%
8	22	147208	92.3	132842	108.5	9.8%
9	24	148242	95.4	135708	117.4	8.5%
10	25	155430	97.2	139315	115.8	10.4%

4.3.3 Konak and Smith's Approach

Konak and Smith [29] proposed an adjacency *matrix-based GA* (amGA) for this problem. Although GA performs a search by exploiting information sampled from different regions of the solution space, crossover and mutation are generally blind operators, *i.e.*, they do not use any problem specific information. This property causes the GA slow convergence before finding an accurate solution and failure to exploit local information [49]. In this subsection, CS, a well known heuristic method for design problem, was combined with mutation operator to improve candidate networks in search process.

Let $P(t)$ and $C(t)$ be parents and offspring in current generation t, respectively. The overall procedure of amGA for solving backbone network model is outlined as follows.

procedure: amGA for backbone network model
input: problem data, GA parameters ($popSize$, $maxGen$, p_M, p_C)
output: the best solution
begin
  $t \leftarrow 0$;
  initialize $P(t)$ by adjacency matrix encoding routine;
  evaluate $P(t)$ by adjacency matrix decoding routine;
  while (**not** terminating condition) **do**
    create $C(t)$ from $P(t)$ by uniform crossover routine;
    create $C(t)$ from $P(t)$ by cut saturation mutation routine;
    evaluate $C(t)$ by adjacency matrix decoding routine;
    select $P(t+1)$ from $P(t)$ and $C(t)$ by elite-based selection routine;
    $t \leftarrow t+1$;
  end
  output the best solution
end

4.3.3.1 Genetic Representation

Adjacency matrix had been used to represent a network design. In this representation, a network is stored $n \times n$ matrix $A = \{a_{ij}\}$, where $n = |V|$ and $a_{ij} = 1$ if link $\{i, j\} \in E'$ (E' is set of links in a topology), $a_{ij} = 0$ otherwise. Since there are link alternatives having different capacities to assign between two nodes in the network, Konak and Smith made a slight modification on adjacency matrix to handle link alternatives. According to this modification, if there is a link between i, j, a_{ij} equal the integer values corresponding to link alternatives; otherwise $a_{ij} = 0$. For example, $a_{12} = 2$ means that link 1, 2 exists and it is a type 2 link. Since links are bidirectional only the upper triangular of A is needed for representation. Initial population contains randomly generated solutions that satisfy k-node-connectivity. Let the number of link alternatives be 3. While Fig. 4.14a shows a three-connected network, its chromosome representation is given in Fig. 4.14b. The number on the link in Fig. 4.14 represents link type.

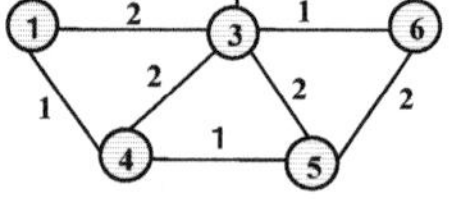

(a) An undirected graph $G = (V, E)$

$$
\begin{array}{c} \\ 1 \\ 2 \\ 3 \\ 4 \\ 5 \\ 6 \end{array}
\begin{array}{c}
\begin{array}{cccccc} 1 & 2 & 3 & 4 & 5 & 6 \end{array} \\
\begin{bmatrix}
- & 1 & 2 & 1 & 0 & 0 \\
 & - & 3 & 0 & 0 & 1 \\
 & & - & 2 & 2 & 0 \\
 & & & - & 1 & 0 \\
 & & & & - & 2 \\
 & & & & & -
\end{bmatrix}
\end{array}
$$

(b) adjacency matrix-based encoding

Fig. 4.14 Sample network topology and chromosome representation

Evaluation

The problem considered in this study is to design a communication backbone network for minimizing cost subject to maximum allowable average delay and k-node-connectivity. The evaluation of candidate networks depends on the routing and the capacity assignment algorithm. In the subsection, the shortest path policy for static routing (where routing policy is determined once and does not change with changing traffic patterns) had been used. In this policy, the packets between two nodes are sent *via* the shortest path between them. In amGA, the shortest paths are determined by Dijkstra's algorithm and k-node-connectivity is tested by using the shortest path augmenting algorithm. In the subsection, objective function is considered as fitness function. Based on the these explanations, the general procedure for the evaluation of the networks is as follows:

1. Determine the routing schema;
2. Calculate the average flows on the links;
3. Determine the capacity of the links;
4. Calculate average delay of the network and test k-node-connectivity;
5. Calculate cost of the network.

4.3.3.2 Genetic Operators

A uniform crossover operator with k-node-connectivity repair algorithm was used to obtain offspring. For crossover, two parents are selected from a population with q-tournament selection with $q = 2$. The selected parents produce only one child. If both parents possess a link and types of link are the same in both parents, the child will have that link. If the type of link is different in both parents, the type of link in the child is selected from parents with 0.5 probability. If both parents do not possess the link, the child will not. If one parent has the link and the other has not, the child will have the link with 0.5 probability. This crossover operator ensures that the child inherits the common topological properties of the parents. After crossover, if the child is not k-node-connected, it is brought to k-node-connectivity by directly connecting the nodes violating k-node-connectivity. The procedure of crossover operator is given below where P_1 and P_2 represent the parent upper triangular matrix and C represents the child upper triangular matrix. Figure 4.16 represents the uniform crossover operator.

The mutation operator is a modified version of CS and serves as a local search operator to improve solutions in the population. But it is a computationally expensive procedure in amGA. Therefore, it is only applied to good solutions, which are most likely to improve upon the best feasible solution. Since the parents in a population are the best solutions of the last population, the mutation operator is applied to them. After mutation, the original network is replaced with the mutated one.

procedure: uniform crossover
input: two parents P_1 and P_2
output: offspring C
step 1. Select two parents (P_1 and P_2) with q-tournament selection procedure
step 2. Set $i \leftarrow 1$ and $j \leftarrow i+1$
step 3. If $P_1(i,j) = P_2(i,j)$ then goto step 4, else goto step 5.
step 4. $C(i,j) \leftarrow P_1(i,j)$ goto step 7
step 5. Generate a random number rand(.)$\in [0,1]$
step 6. If rand(.) ≤ 0.50 then $C(i,j) \leftarrow P_1(i,j)$ else $C(i,j) \leftarrow P_2(i,j)$
step 7. $j \leftarrow j+1$, if $j \leq n$ goto step 3
step 8. $i \leftarrow i+1$, if $i \leq n$ then $j \leftarrow i+1$ and goto step 3 else stop

Fig. 4.15 The pseudocode of uniform crossover

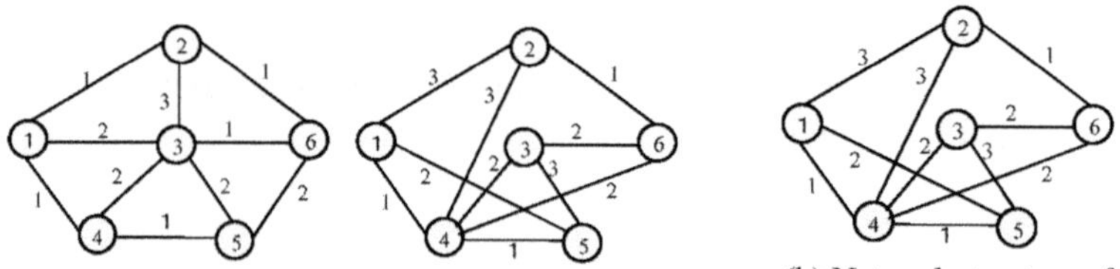

(a) Network structure of parents P_1 and P_2

(b) Network structure of offspring C

	1	2	3	4	5	6
1	-	1	2	1	0	0
2		-	3	0	0	1
3			-	2	2	1
4				-	1	0
5					-	2
6						-

	1	2	3	4	5	6
1	-	3	0	1	2	0
2		-	0	3	0	1
3			-	2	3	2
4				-	1	2
5					-	0
6						-

(c) Adjacency matrix of parents P_1 and P_2

	1	2	3	4	5	6
1	-	3	2	1	0	0
2		-	0	3	0	1
3			-	2	2	2
4				-	1	2
5					-	0
6						-

(d) Adjacency matrix of offspring C

Fig. 4.16 Illustration of uniform crossover operator

4.3.3.3 Selection Mechanism

After the mutation operator, it is possible to obtain infeasible networks which do not meet k-node connectivity and maximum delay requirements. Discarding these solutions is not desirable, because an infeasible solution might be closer to the optimum than many feasible solutions. In the selection routine, first, the offspring and mutated solutions are sorted according to the following two rules:

- Between two feasible or two infeasible solutions with respect to the delay constraint only, the one with the lower cost is superior to other one.

- Between one feasible and one infeasible solution, if $T_{infeasible} < (1+\alpha)T_{max}$, where α is a small number less than 1, the infeasible solution is treated as if it was feasible and the first rule is applied. Otherwise, the feasible solution is superior to the infeasible one, regardless of cost.

After μ solutions generated randomly for the initial population, μ offspring are created by crossover in each generation of amGA. The offspring are added to the population, which increases population size to 2μ. The original members of the population (μ parents) are mutated and replaced with mutated ones. Then the final population, which includes μ offspring and μ mutated solutions, is sorted according to the rules given before. Finally, the μ solutions of the sorted population survive for the next generation.

4.3.4 Numerical Experiments

To investigate the performance of amGA, the problem set given in Pierre and Elgibaoui [59] was used. In this reference, tabu search (TS) had been employed to solve this problem. Table 4.8 gives the comparative results. Konak and Smith compared the average results of the amGA with the best results of TS by using a paired t-test. The p value indicates the significance of the hypothesis that amGA generates lower cost than the results of TS. Based on the statistical analysis, they showed that amGA managed to improve on the results of the TS with reasonable CPU times in almost all cases. In addition, paired t-test was applied to compare the best results of amGA and TS and it was demonstrated that the best results of amGA excluding problem 9 were statistically better than the results of TS at a significance level of 0.043.

4.4 Reliable Network Models

An important problem appearing in communication networks is to design an optimal topology for balancing system reliability and cost. The most general network reliability problems in the literature have two problem types:

1. k-terminal network reliability problem
2. All-terminal network reliability problem

With a k-terminal network reliability problem, given a set of target nodes $k \in V$ with $k = |K|$, the k-terminal reliability is Pr(for a given node $s \in K$, there exists at least 1 working path from s to all other nodes in k). This reliability is the sum of the probabilities of disjoint success paths. Because the complexity of identifying all disjoint success paths is exponential, it is a well-known NP-hard problem. An all-terminal network reliability problem has the probability that all nodes (computers or terminals) on the network can communicate with all others. Exact calculation

Table 4.8 The results of the computational experiments (T=ms, Cost=\$/month, CPU=min)

Problem		amGA parameters	Reference Result		amGA Average			amGA Best		amGA vs. TS	
#	$\{\|v\|, K, T_{max}\}$	$\{\mu, NG, \alpha\}$	Cost	T	Cost	T	CPU	Cost	T	%ΔCost	p
1	{10, 2, 100}	{10, 100, 0.1}	25272	69	24923	94	0.78	24171	89	1.3	0.076
2	{10, 2, 100}	{10, 200, 0.1}	25272	69	24901	95	1.5	24507	99	1.4	0.069
3	{10, 2, 100}	{15, 200, 0.1}	25272	69	24646	98	2.3	24137	99	2.4	0.004*
4	{15, 2, 150}	{15, 100, 0.1}	65161	142	55985	92	2.9	55008	11	14.0	0.000*
5	{15, 2, 100}	{15, 100, 0.1}	N/A		55510	91	3.6	54838	98		
6	{15, 2, 100}	{15, 100, 0.0}	N/A		56369	79	3.6	55112	83		
7	{15, 3, 200}	{10, 100, 0.1}	60907	79	57946	92	2.8	57239	79	4.8	0.000*
8	{15, 4, 200}	{10, 100, 0.1}	60013	93	59293	101	3.7	57334	122	1.2	0.162
9	{15, 4, 200}	{10, 100, 0.1}	41626	90	59293	101	3.7	57334	122	-29.7	
10	{20, 2, 150}	{15, 100, 0.1}	144131	104	102272	79	8.4	98017	74	29.0	0.000
11	{20, 3, 250}	{15, 100, 0.1}	105495	112	101439	85	12.3	99759	102	3.8	0.001*
12	{20, 4, 250}	{15, 100, 0.1}	103524	88	101166	90	16.1	100054	101	2.2	0.002*
13	{20, 5, 250}	{15, 100, 0.1}	106122	83.1	104665	83	10.9	102617	96	1.3	0.074
14	{20, 3, 50}	{15, 100, 0.1}	N/A		107846	49	7.8	106413	49		
15	{20, 3, 50}	{15, 100, 0.0}	N/A		108454	49	9.9	105337	49		
16	{23, 2, 80}	{15, 100, 0.1}	210070	58	139925	64	12.8	137270	70	34.6	0.000*
17	{23, 3, 190}	{15, 100, 0.1}	148340	91	132774	75	20.4	131096	86	10.5	0.000*
18	{23, 4, 190}	{15, 100, 0.1}	163129	90	133637	81	23.0	132002	67	18.0	0.000*
19	{23, 5, 190}	{15, 100, 0.1}	139154	86	134741	89	26.0	133411	80	2.5	0.000*

*Significant results at a level of 0.05

of all-terminal reliability is an NP-hard problem, precluding its use during optimal network topology design, where this calculation must be made thousands or millions of times.

This problem and related versions have been studied in the literature with two approaches:

1. Enumerative-based method
2. Heuristic-based method

Jan *et al.* [53] developed an algorithm using decomposition based on branch and bound to minimize link costs with a minimum network reliability constraint; this is computationally tractable for fully connected networks up to 12 nodes. Using a greedy heuristic, Aggarwal *et al.* [74] maximized reliability given a cost constraint for networks with differing link reliabilities and all-terminal metric reliability. Ventetsanopoulos and Singh [64] used a two-step heuristic procedure for the problem of minimizing a network's cost subject to a reliability constraint. The algorithm first used a heuristic to develop an initial feasible network configuration, and then a branch and bound approach was used to improve this configuration.

A deterministic version of simulated annealing was used by Atiqullah and Rao [40] with exact calculation of network reliability to find the optimal design of very small networks (five nodes or less). Pierre *et al.* [58] also used simulated annealing to find optimal designs for packet switch networks where delay and capacity were considered, but reliability was not. Tabu search was used by Glover *et al.* [51] to choose network design when considering cost and capacity, but not reliability. Another tabu search approach by Beltran and Skorin-Kapov [43] was used to design reliable networks by searching for the least cost spanning two-tree, where the two-tree objective was a coarse surrogate for reliability. Koh and Lee [55] also used tabu search to find telecommunication network designs that required some nodes (special offices) having more than one link while others (regular offices) required only one link, while also using this link constraint as a surrogate for network reliability.

Since genetic algorithms (GAs) (and other evolutionary algorithms) promise solutions to such complicated problems, they have been used successfully in various practical applications [66]. Hosam used Chat [75] proposed a novel computationally practical method for estimating all-terminal network reliability. It shows how a neural network can be used to estimate all-terminal network reliability by using the network topology, the link reliabilities and an upperbound on all-terminal network reliability as inputs. And Altiparmak and Bulgak [76] proposed artificial neural network (ANN) as an alternative way to estimate network reliability.

Kumar *et al.* [23] developed a GA considering diameter, average distance, and computer network reliability and applied it to four test problems of up to nine nodes. They calculated all-terminal network reliability exactly and used a maximum network diameter (minimal number of links between any two nodes) as a constraint. The same authors used this GA to expand existing computer networks [77]. Davis *et al.* [78] approached a related problem considering link capacities and rerouting upon link failure using a customized GA. Abuali *et al.* [79] assigned terminal nodes to concentrator sites to minimize costs while considering capacities using a GA, but no reliability was considered. The same authors in [80] solved the probabilistic minimum spanning tree problem where inclusion of the node in the network is stochastic and the objective is to minimize connection (link) costs, again without regard to reliability. Walters and Smith [81] used a GA to address optimal design of a pipe network that connects all nodes to a root node using a nonlinear cost function. Reliability and capacity were not considered, making this a somewhat simplistic approach. Deeter and Smith [82] presented a GA approach for a small (five nodes) minimum cost network design problem with alternative link reliabilities and an all-terminal network reliability constraint. Dengiz *et al.* [25] addressed the all-terminal network design problem on a test suite of 20 problems using a fairly standard GA implementation.

4.4.1 Reliable Backbone Network Model

A communication network is modeled by a probabilistic graph $G = (V, E, p, q)$, in which V and E are the set of nodes and links that correspond to communication center and communication links, p and q are the set of the reliabilities and unreliabilities of links, respectively. $V = \{1, 2, \cdots, n\}$ and a set of directed arcs $E = \{(i, j), (k, l), \cdots, (s, t)\}$ Link (i, j) is said to be incident with nodes i and j, and the number of links is e. The design problem is to select a topology to either maximize reliability given a cost constraint or to minimize cost. This problem is an NP-hard problem. In this subsection, the mathematical formulation and assumptions are considered (Dengiz *et al.* [25]). The following typically define the problem assumptions:

A1. The location of each network node is given
A2. Nodes are perfectly reliable
A3. Link costs and reliabilities are fixed and known
A4. Each link is bi-directional
A5. There are no redundant links in the network
A6. Links are either operational or failed
A7. The failures of links are independent
A8. No repair is considered

Notation

G: a probability graph
V: set of nodes
E: set of links
V': set of operational nodes
E': set of operational links
p,q: [reliability, unreliability] for all links, $p+q \equiv 1$
R_0: network reliability requirement
(i,j): link between nodes i and j, $i,j = 1,2,...,N$
c_{ij}: the cost of link (i,j)
x_{ij}: decision variable, $x_{ij} \in 0,1$
Ω: all operations l states, E'

When minimizing cost subject to a minimum reliability constraint, the optimization problem is formulated as follow:

$$\min\ c(\boldsymbol{x}) = \sum_{i=1}^{n-1} \sum_{j=i+1}^{n} c_{ij} x_{ij} \tag{4.17}$$

$$\text{s. t. } R(\boldsymbol{x}) \geq R_0 \tag{4.18}$$

$$x_{ij} \geq 0, \qquad \forall\ i,j \tag{4.19}$$

where $x_{ij} \in \{0,1\}$ is the decision variables, c_{ij} is the cost of (i,j) link, $\boldsymbol{x}$ is the link topology of $\{x_{11}, x_{12}, \cdots, x_{ij}, \cdots, x_{n-1,n}\}$, $R(\boldsymbol{x})$ is the network reliability and R_0 is the minimum reliability requirement.

Under the above assumptions, at any instant of time, only some links of G might be operational. A state of G is sub-graph $G' = (V', E')$, where V' and E' is the set of operational (not failed) nodes and links such that $V' \subseteq V$ and $E' \subseteq E$, respectively. Note that $V' = V$ for all-terminal reliability. The network reliability of states $E' \subseteq E$ is

$$R(\boldsymbol{x}) = \left(\sum_{\Omega} \left[\prod_{l \in E'} p \right] \left[\prod_{l \in (E/E')} q \right] \right) \tag{4.20}$$

where Ω = all operational l states, E'. The exact calculation of network reliability is an NP-hard problem because Ω grows exponentially with network size. Therefore, a variety of simulation methods and bounds – Ball and Provan [42]; Colbourn [83]; Jan [52] – have been proposed for efficiently estimating all-terminal, k-terminal and two-terminal reliability. When designing larger size networks, generally simulation and bounds (lower and upper) as part of the search process are used to estimate reliability of networks.

4.4.1.1 KruskalRST-based GA for the Reliable Backbone Network Model

Lin and Gen [73] proposed a Kruskalrst-based GA for solving an all-terminal reliable backbone network problem. Let $P(t)$ and $C(t)$ be parents and offspring in current generation t, respectively. The overall procedure of KruskalRST-based GA (kruskalGA) for solving reliable backbone network model is outlined as follows.

---

procedure: kruskalGA for reliable backbone network model
input: network data (V, E, p, q), cost set $\boldsymbol{c}$, requirement R_0,
    GA parameters ($popSize$, $maxGen$, p_M, p_C,)
output: the best solution
begin
  $t \leftarrow 0$;
  initialize $P(t)$ by KruskalRST-based encoding routine;;
  evaluate $P(t)$ by KruskalRST-based decoding routine;
  while (**not** terminating condition) **do**
    create $C(t)$ from $P(t)$ by KruskalRST-based crossover routine;
    create $C(t)$ from $P(t)$ by LowestCost mutation routine;
    create $C(t)$ from $P(t)$ by immigration routine;
    feasibility check and repairing by repairing routine;
    evaluate $C(t)$ by KruskalRST-based decoding routine;
    if $t > u$ **then**
      auto-tuning p_M, p_C and p_I by FLC;
    select $P(t+1)$ from $P(t)$ and $C(t)$ by roulette wheel selection routine;
    $t \leftarrow t+1$;
  end
  output the best solution
end

KruskalRST-based Encoding and Decoding

How to encode a solution of the network design problems into a chromosome is a key issue for genetic algorithms. Special difficulty arises from (1) an operational sub-graph contains a variable number of links, and (2) all-terminal reliability is the probability that there exists at least one path from each node to every node in the network, *i.e.*, the network has at least one spanning tree.

Generally edge-based encoding method (*i.e.*, binary encoding or adjacency matrix) has been used to represent a candidate network as a chromosome in different studies [22, 25, 31]. After applying genetic operators, it is possible to obtain an unconnected network (illegal chromosome) with this encoding method. In the literature, this problem is handled in different ways such as allowing infeasible chromosomes to be present in the population pool, correcting this with a repair strategy, discarding infeasible chromosomes or developing special crossover and mutation operators.

Altiparmak *et al.* [85] proposed a new representation based on Prüfer numbers for network reliability problems. A chromosome consists of two parts. While the first part includes Prüfer numbers to represent spanning tree of a candidate network, the second part consists of pairs of nodes. Each pair of nodes in the second part represents a link except to links of the spanning tree in the network.

Recently, researchers investigated and developed edge-based encoding methods in genetic algorithms for designing several MST-related problems. Considering the disadvantage of edge-based encoding, the heuristic method is adopted in chromosome generating procedures. These methods can insure feasibility; all chromosomes can be represented by feasible solutions [86]–[88]. In this study, as a kind of edge-based encoding methods, Raidl and Julstrom [38] proposed KruskalRST is developed for network reliability problems.

Kruskal's algorithm also applies a greedy strategy to build a minimum spanning tree of a graph. It examines edges in order of increasing cost and includes in the tree those that connect previously unconnected components. Examining the edges in random order yields a procedure Raidl and Julstrom call KruskalRST. As discussed above, all-terminal reliability is the probability that there exists at least one spanning tree. In this encoding method, a chromosome consists of two parts. The first part with KruskalRST represents spanning tree for a candidate network, the second part consists of edges that do not exist in the spanning tree. Thus, the length of chromosome is changed by the number of edges. In this encoding method, there is no need to check network connectivity after applying genetic operators. Because of this property, it is possible to save computation time for solving network design problem. Fig. 4.17 shows a simple network whose base graph consists of 6 nodes and 15 possible edges, but with 7 edges present.

The edge-based representation of this network is shown as follows and the pseudocode of edge-based encoding process is given in Fig. 4.18

$$[(1,2),(2,5),(3,5),(4,5),(4,6)|(1,3),(3,6)]$$

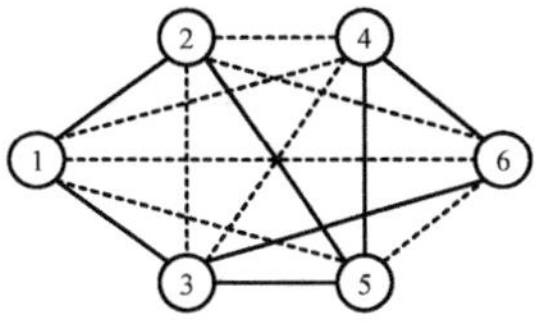

(a) A simple network

$$[c_{ij}] = \begin{array}{c} \\ 1 \\ 2 \\ 3 \\ 4 \\ 5 \\ 6 \end{array} \begin{array}{c} \begin{array}{cccccc} 1 & 2 & 3 & 4 & 5 & 6 \end{array} \\ \left[\begin{array}{cccccc} - & 29 & 54 & 58 & 25 & 40 \\ & - & 34 & 62 & 45 & 89 \\ & & - & 36 & 52 & 72 \\ & & & - & 32 & 51 \\ & & & & - & 54 \\ & & & & & - \end{array}\right] \end{array}$$

(b) The cost of link (i, j)

Fig. 4.17 Typical network

---

procedure: KruskalRST-based Encoding
input:Network Data (V, E, p, q),
  number of nodes n, number of links e
output: chromosome $v = E'$
begin
  $E' \leftarrow \phi$; // First routine: build a spanning tree
  $A \leftarrow E$;
  while $(|E'| < n)$
    choose edge $(i, j) \in A$ by KruskalRST;
    $A \leftarrow A - \{(i, j)\}$;
    if i and j are not yet connected in E' **then**
      $E' \leftarrow E' \cup \{(i, j)\}$
  end
  $n_{\text{add}} \leftarrow$ **random**$[n, e-1]$; // Second routine: add edges that does exist in E'
  for $l = 1$ **to** n_{add}
    choose edge $(i, j) \in A$ at random;
    $E' \leftarrow E' \cup \{(i, j)\}$
    $A \leftarrow A - \{(i, j)\}$;
  end
  output chromosome $v \leftarrow E'$;
end

Fig. 4.18 Pseudocode of KruskalRST-based encoding method

Evaluation and Repair Algorithm

In this algorithm, population consists of feasible networks, *i.e.*, meeting reliability requirement. If a network in a population doesn't meet the reliability requirement it is repaired by the repair algorithm. Since a minimization problem is considered, the fitness function has been taken as the inverse of the objective function. For each candidate network, Jan's [53] upper bound is used to compute the upper bound of reliability, $R_U(\boldsymbol{x})$. If the upper bound of reliability of the candidate network is lower than the reliability requirement, then the candidate network is repaired. Networks which have $R_U(\boldsymbol{x}) > R_0$ and the lowest cost so far are sent to the simulation subroutine for precise estimation of network reliability using a Monte Carlo technique by Yeh *et al.* [66].

After determining the sets of nodes with the first and second smallest nodes degrees in the candidate network, repair is made based on the number of nodes in the first set. The purpose is to realize maximum increase of reliability while reducing additional link cost. The steps of algorithm are given in Fig. 4.19, an illustrative example can be seen in Fig. 4.20 and link costs as in Fig. 4.17.

```
procedure : Repair Algorithm
input : A network states (V, E', p, q), edge cost set c
output : Repaired chromosome v' = E''
begin
    node degree d_k ← degree(V, E'), ∀k ∈ V;
    set D_1 ← argmin{d_k | k = 1,...,n};
    set D_2 ← argmin{d_k | d_k ∉ D_1 & k = 1,...,n};
    if min{d_k} = 1 then
        set C ← {(i, j) | i ∈ D_1 & ∃j ∈ D_2 & (i, j) ∉ E'}
        (i', j') ← argmin{c_ij | (i, j) ∈ C};
    else
        (i', j') ← argmin{c_ij | (i, j) ∈ D_1 ∪ D_2 & (i, j) ∉ E'};
    end
    E'' ← E' ∪ {(i', j')};
    output chromosome v ← E'';
end
```

Fig. 4.19 Pseudocode of repair algorithm

KruskalRST-based Crossover

Crossover is the main genetic operator. It operates on two parents (chromosomes) at a time and generates offspring by combining both chromosomes' features. In network design problems, crossover plays the role of exchanging each partial route of two chosen parents in such a manner that the offspring produced by the crossover represents.

(a) Original network state

(b) Repaired network where the edge between nodes 1 and 2 is added

Fig. 4.20 Illustrative example of network repair

To provide good heritability, a crossover must build an offspring that consists mostly or entirely of edges found in the parents. In this chapter, we also apply the KruskalRST algorithm described in the previous section. First, we generate a new graph $G' = (V, E_1' \cup E_2')$, where E_1' and E_2' are the link sets of the parental network states, then build more network states by KruskalRST-based encoding. Figure 4.21 gave the pseudocode of the crossover process. An illustrative example can be seen in Fig. 4.22.

---

procedure : KruskalRST-based Crossover
input:parent $v_1 = E_1', v_2 = E_2'$
output: offspring $v' = E''$
begin
    edge set $E \leftarrow E_1' \cup E_2'$;
    subgraph $G \leftarrow (V, E)$;
    offspring $v'(E'') \leftarrow$ **KruskalRST - based Encode**(G);
        // routine:KruskalRST-based Encoding
    output offspring $v' \leftarrow E''$;
end

Fig. 4.21 Pseudocode of crossover

Parent v_1=[(1, 2), (1, 3), (2, 5), (3, 6), (4, 5) | (3, 5), (4, 6)]
Parent v_2=[(1, 2), (1, 3), (2, 4), (3, 5), (4, 6) | (3, 4), (5, 6)]
↓
Subgraph $G \leftarrow$ {(1, 2), (1, 3), (2, 4), (2, 5), (3, 4), (3, 5), (3, 6), (4, 5), (4, 6), (5, 6)}
↓ KruskalRST-based Encod (G)
Offspring v' ←[(1, 2), (1, 3), (2, 4), (3, 5), (5, 6)|(3, 4), (3, 6)]

Fig. 4.22 Illustrative example of crossover

Heuristic Mutation

The idea of combining GAs and local search heuristics for solving optimization problems has been investigated extensively during the past decade, and various methods of hybridization have been proposed. Here we propose an HDLC (highest degree lowest cost) mutation that it is a kind of heuristic strategy which exploits the best solution for possible improvement ignoring the exploration of the search space.

For solving network reliability problems, HDLC mutation takes the form of a randomized greedy local search operator. HDLC mutation is applied differently according to node degrees of the network. This mutation process is shown in Fig. 4.23 and an illustrative example can be seen in Fig. 4.24 and link costs as in Fig. 4.17.

---

procedure : HDLC Mutation
input : Chromosome $v = E'$, edge cost set $\boldsymbol{c}$
output : offspring $v' = E''$
begin
  $E'' \leftarrow E'$;
  node degree $d_k \leftarrow$ **degree**(V, E''), $\forall k \in V$;
  set $D \leftarrow$ **argmax**$\{d_k \mid k = 1,\ldots,n\}$;
  $(k,l) \leftarrow$ **argmax**$\{c_{kl} \mid k \in D, d_l \geq 2 \,\&\, (k,l) \in E''\}$;
  $E'' \leftarrow E'' - \{(k,l)\}$;
  $(i,j) \leftarrow$ **argmin**$\{c_{ij} \mid (i,j) \notin E''\}$;
  $E'' \leftarrow E'' \cup \{(i,j)\}$;
  output chromosome $v \leftarrow E''$;
end

Fig. 4.23 Pseudocode of HDLC mutation

Parent v=[(1, 3), (2, 4), (2, 5), (3, 5), (3, 6) | (4, 5), (4, 6)]
↓
Offspring v' ←[<u>(1, 2)</u>, (1, 3), (2, 5), (3, 5), (3, 6)|(4, 5), (4, 6)]

Fig. 4.24 Illustrative example of HDLC mutation

Immigration

The immigration is modified to (1) include immigration routine, in each generation, (2) generate and (3) evaluate popSize p_I random chromosomes, and (4) replace the $popSize\, p_I$ worst chromosomes of the current population (p_I , called the immigration probability). The detailed introduction is shown in Sect. 1.3.2.4.

Auto-tuning Scheme

The probabilities of immigration and crossover determined the weight of exploitation for the search space. Its main scheme is to use two FLCs: auto-tuning for exploration and exploitation $T[p_M \wedge (p_C \vee p_I)]$ and auto-tuning for genetic exploitation and random exploitation ($T[p_C \wedge p_I]$) are implemented independently to regulate adaptively the genetic parameters during the genetic search process. The detailed scheme is shown in Sect. 1.3.2.4.

4.4.1.2 Numerical Experiments

In the experiments, KruskalRST-based encoding is compared with Dengiz *et al.*'s algorithm [25] and Altiparmak *et al.*'s algorithm [85] on nine test problems. In KruskalRST-based GA, KruskalRST-based crossover, HDLC (Highest Degree Lowest Cost) mutation and immigration are used as genetic operators, with added self-controlled strategy by Fuzzy Logic Control (FLC). Each test problem was run 30 times (it allows the central limit theorem to kick in) for each test. In addition, the effects of auto-tuning strategy is also investigated by comparing with the fixed GA parameter setting. All the simulations were performed with Java on a Pentium 4 processor (3.40-GHz clock) and 3.00 GB RAM. The initializations of GA parameter setting are:

population size, $popSize = 20$;
maximum generation, $maxGen = 1000$;
crossover probability, $p_C = 0.4$;
mutation probability, $p_M = 0.4$;
immigration probability, $p_I = 0.2$;
self-controlled parameter, $u = 3, \alpha = 2$.

Figure 4.25 shows the best solutions of 30 runs for each approach with the largest test problem, and Fig. 4.26 shows the auto-turning of genetic operators by proposed approach. The *flc-aGA* was proposed by Yun and Gen [89]; The *simple GA* combines KruskalRST-based encoding, KruskalRST-based crossover and HDLC mutation. The *simple GA without local search* combines KruskalRST-based encoding, KruskalRST-based crossover and swap mutation. The convergence of KruskalRST-based GA is faster than the others. We can show that the search schemes of the KruskalRST-based GA for locating the global optimum is more efficient than the others. In addition, the convergence of simple GA is very slow, and stopped improvement after 400 generations, and simple GA without local search could barely converge; this means that the global optimum is not found by fixed parameters setting. Compared with two different GAs with parameter tuning schemes, the convergence of KruskalRST-based GA with auto-tuning scheme (atGA) is more efficient than flc-aGA. The parameter tuning process of genetic operators by atGA is shown in Fig. 4.25. Another experiment shows the comparison results using the kbsGA (Dengiz *et al.* [25]), nbGA (Altiparmak *et al.* [85]) and atGA. As shown in Table 4.9, the atGA give the best performance based on the mean cost of 30 times runs except to #6 and #8 test problems, though its result is slightly best than kbsGA and nbGA. It is proves that the atGA has a higher probability for locating the global optimum than the others.

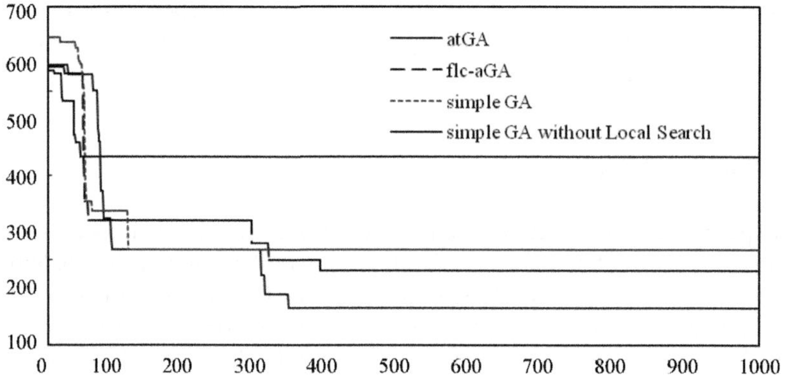

Fig. 4.25 Typical convergence plot of atGA, flc-aGA and simple GAs

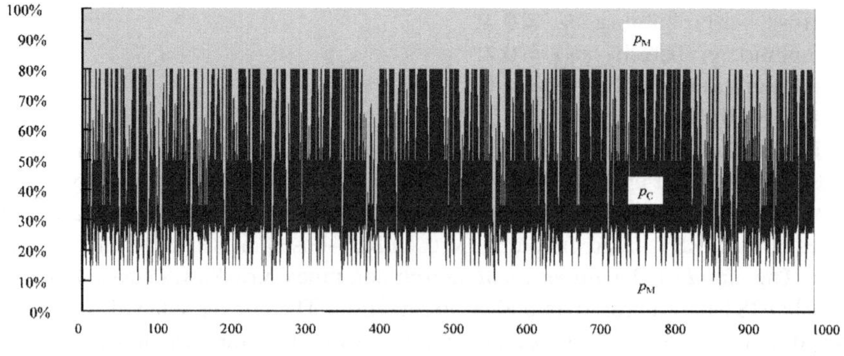

Fig. 4.26 Auto-turning of genetic operators by proposed approach

Table 4.9 Comparison of results (kbsGA: Dengiz *et al.* [25], nbGA: Altiparmak *et al.* [85])

Networks	p	R_0	Opti. Cost	kbsGA		nbGA		scGA	
				Best	Mean	Best	Mean	Best	Mean
G=(8,28)	0.90	0.90	208	208	210.1	208	208.0	208	208.0
	0.90	0.95	247	247	249.5	247	247.0	247	247.0
	0.95	0.95	179	179	180.3	179	180.1	179	179.6
G=(9,36)	0.90	0.90	239	239	245.1	239	241.6	239	240.2
	0.90	0.95	286	286	298.2	286	288.1	286	287.5
	0.95	0.95	209	209	227.2	209	213.3	209	214.7
G=(10,45)	0.90	0.90	154	156	169.8	154	161.6	154	157.7
	0.90	0.95	197	205	206.6	197	197.0	197	197.5
	0.90	0.95	136	136	150.4	136	149.1	136	138.1

4.4.2 Reliable Backbone Network Model with Multiple Goals

During the design of communication networks, it is possible to consider the reliability between a subset of nodes, which are designated as critical (*e.g.,* the cost computer in the network or its backbone), in addition to all-terminal reliability in the network. Liu and Iwamura had considered this situation and they modeled the topological design problem of communication networks as a dependent chance goal programming model. *Chance constrained programming* was pioneered by Charnes and Cooper [45] as a means of handling uncertainty by specifying a confidence level at which it is desired that the stochastic constraint hold. Sometimes a complex stochastic decision system undertakes multiple tasks, called events, and decision-maker wishes to maximize the chance functions which are defined as the probabilities of satisfying these events. In order to model this type of problem, Liu [56] initiated a theoretical framework of dependent-chance programming. Liu and Iwamura [31] defined a dependent-chance goal programming model according to priority structure and target levels set by the decision maker for reliable design of communication networks:

$$\min \sum_{i=1}^{m} \sum_{j=1}^{l} p_j \left(u_{ij} d_i^+ + v_{ij} d_i^-\right) \tag{4.21}$$

$$\text{s. t. } R(K_i, x) + d_i^- - d_i^+ = R_i, \quad \forall\, i \tag{4.22}$$

$$\sum_{i=1}^{m-1} \sum_{j=i+1}^{m} c_{ij} x_{ij} \leq c_0 \tag{4.23}$$

$$d_i^-, d_i^+ \geq 0, \quad \forall\, i \tag{4.24}$$

where p_j = the pre-emptive priority factor which expresses the relative importance of various goals, $p_j \gg p_{j+1}$, for all j, where u_{ij} = weighting factor corresponding to positive deviation for goal i with priority j assigned, v_{ij} = weighting factor corresponding to negative deviation for goal i with priority j assigned, d^+ = positive deviation from the target of goal i, d^- = negative deviation from the target of goal i, R_i = the target reliability level of the set K_i, l = number of priorities, m = number of goal constraints, and $R(K_i,x)$ is k-terminal reliability.

4.4.2.1 Liu and Iwamura's Approach

Liu and Iwamura had used binary encoding to represent a candidate network design (*i.e.*, link topology $\boldsymbol{x}$). Let $P(t)$ and $C(t)$ be parents and offspring in current generation t, respectively. The overall procedure of Liu and Iwamura's binary-based GA (bGA) for solving reliable backbone network model with multiple goals is outlined as follows.

---

procedure: bGA for reliable backbone network model with multiple goals
input: network data, GA parameters (*popSize*, *maxGen*, p_M, p_C)
output: the best solution
begin
  $t \leftarrow 0$;
  initialize $P(t)$ by binary-based encoding routine
  evaluate $P(t)$ by binary-based decoding routine;
  while (**not** terminating condition) **do**
    create $C(t)$ from $P(t)$ by two cut-point crossover routine;
    create $C(t)$ from $P(t)$ by replacement mutation routine;
    evaluate $C(t)$ by binary-based decoding routine;
    select $P(t+1)$ from $P(t)$ and $C(t)$ by roulette wheel selection routine;
    $t \leftarrow t+1$;
  end
  output the best solution
end

Binary-based Encoding

In binary encoding, each chromosome consists of $n(n-1)/2$ dimensional vector $Y = (y_1, y_2, \cdots, y_{n(n-1)12})$ where y_i is taken as 0 or 1 for $1 < i < n(n-1)/2$. Then the relationship between a link topology $\boldsymbol{x}$ and a chromosome Y is given as follows:

$$x_i j = y_{2n-i)(i-1)/2+j-i}, \qquad 1 \le i \le n-1, \quad i+1 \le j \le n \tag{4.25}$$

Initial population is randomly generated considering feasibility, *i.e.*, chromosomes represent connected network and satisfy cost constraint. If any chromosome does not satisfy at least one of the requirements, then generation procedure is repeated until they are satisfied.

Evaluation and Selection

Monte Carlo simulation is used to estimate k-terminal reliability with respect to prescribed target set for each candidate network in the population. Procedure of Monte Carlo simulation is given in Fig. 4.27, where N is the number of runs:

---

procedure: Monte Carlo Simulation
step 1. Set counter $N' = 0$.
step 2. Randomly generate an operational link set E' from
the link topology x considering link reliabilities.
step 3. If $G = (V, E')$ is k-connected, then $N' \leftarrow N' + 1$.
step 4. Repeat the second and third steps N times.
step 5. $R(k, x) = N'/N$.

Fig. 4.27 Pseudocode of Monte Carlo simulation

In any generation, after calculating the objective function values of the chromosome, chromosomes in the population are rearranged from good to bad according to their objective function value. Since the problem is formulated goal programming, Liu and Iwamura used the following order relationship for the chromosomes: for any two chromosome, if the higher-priority objectives are equal, then, at the current priority level, the one with a minimal objective value is better, and if two different chromosomes have the same objective values at every level, then there is nodifference between them. Based on this explanation, rank-based fitness function has been defined as fitness $eval(Y_k) = \alpha(1-\alpha)i-1,\ k = 1,2,\cdots,\ popSize$. where $\alpha \in (0,1)$, i=1 means the best individual, $i = popSize$ the worst individual.

In each generation, after calculating the fitness values of the chromosomes, roulette wheel selection mechanism is used to select parents for next generation.

Two Cut-point Crossover

In this operator, first two crossover positions n_1 and n_2 between 1 and $n(n-1)/2$ such that $n_1 < n_2$ are randomly determined, and the genes between these two points of the chromosomes (*i.e.*, parents) are exchanged and two children are obtained. After the crossover operator, first a check is made whether children are feasible (*i.e.*, each child represents a topology and satisfies cost constraint) or not and then the parents are replaced with the feasible children.

Replacement Mutation

First two mutation positions n_1 and n_2 between 1 and $n(n-1)/2$ such that $n_1 < n_2$ are randomly determined, a new chromosome is obtained with regeneration of the genes between the two points from 0, 1. If offspring is feasible then it is replaced with the parents, if not the process defined above is repeated until a feasible offspring is obtained.

4.4.2.2 Numerical Experiments

To represent the effectiveness of the GA approach, Liu and Iwamura had given two examples with 10 nodes and 20 nodes in their studies. In these examples, population size, probabilities of crossover and mutation were taken as 30, 0.3 and 0.2, respectively, and Monte Carlo simulation had been run 2000 times to estimate reliability of each candidate network, the value of α in the rank-based fitness function was set to 0.05. Here we introduce the results for 10 nodes fully-connected network (*i.e.*, G = (10,45)). For this network, it is assumed that the reliabilities of all links are 0.90, nodes are perfectly reliable. Table 4.10 gives the cost matrix. Target levels and priority structure are given below:

Table 4.10 Cost matrix for 10-node network

	1	2	3	4	5	6	7	8	9	10
1										
2	30	-								
3	43	26	-							
4	45	76	38	-						
5	50	45	17	35	-					
6	62	25	30	28	15	-				
7	25	46	30	16	25	38	-			
8	15	45	13	20	37	40	36	-		
9	51	15	45	10	34	10	46	42	-	
10	45	25	45	15	37	40	16	24	45	-

Priority 1. For the subset of nodes K_1 = 1, 3, 6, 7, the K_1-terminal reliability $R(K_1,x)$ should achieve 99%:

$$R(K_1,x) + d_1^- - d_1^+ = 99\%$$

where d_1^- will be minimized.

Priority 2. For the subset of nodes K_2 = 2, 4, 5, 9, the K_2-terminal reliability $R(K_2,x)$ should achieve 95%:

$$R(K_2,x)+d_2^- - d_2^+ = 95\%$$

where d_2^- will be minimized.

Priority 3. For the subset of nodes K_3 = (1,2, 3, 4, 5, 6, 7, 8, 9, 10), the K_3-terminal reliability $R(K_3,x)$ (here the overall reliability) should achieve 90%:

$$R(K_3,x)+d_3^- - d_3^+ = 90\%$$

where d_3^- will be minimized.

Total available capital is 250. This is given as

$$\sum_{i=1}^{9}\sum_{j=i+1}^{10} c_{ij}x_{ij} \le 250$$

Under these definitions, topological optimization model for communication network reliability is

$$\text{lexmin } \{d_1^-, d_2^-, d_3^-\} \quad (4.26)$$
$$\text{s. t. } R(K_1,x)+d_1^- - d_1^+ = 99\% \quad (4.27)$$
$$R(K_2,x)+d_2^- - d_2^+ = 95\% \quad (4.28)$$
$$R(K_3,x)+d_3^- - d_3^+ = 90\% \quad (4.29)$$
$$\sum_{i=1}^{9}\sum_{j=i+1}^{10} c_{ij}x_{ij} \le 250 \quad (4.30)$$
$$x_{ij} = 0 \text{ or } 1, \quad \forall\, i,j \quad (4.31)$$
$$d_i^-, d_i^+ \ge 0, \quad i = 1,2,3 \quad (4.32)$$

After 100 generations, the optimal link topology, which satisfies the three goals, is given in Fig. 4.28. The terminal reliability levels are $R(K_1,x^*) = 0.991$, $R(k2,x*) = 0.956$, $R(K_3,x^*) = 0.938$, respectively and total cost is 242.

If the total available capital is 210, then the optimal link topology obtained by GA with 600 generations is given in Fig. 4.29. While this topology satisfies the first goal, the deviations of the second and third goals are 0.08 and 0.15, respectively. The terminal reliability levels for the topology are $R(K_1,x^*)$ = 0.99, $R(K_2,x^*)$ = 0.87, $R(K_3,x^*)$ = 0.75 and the total cost is 207.

$$x^* = \begin{pmatrix} - & 0 & 0 & 0 & 0 & 0 & 1 & 1 & 0 & 0 \\ & - & 1 & 0 & 0 & 1 & 0 & 0 & 0 & 1 \\ & & - & 0 & 1 & 0 & 0 & 1 & 0 & 0 \\ & & & - & 0 & 0 & 0 & 1 & 1 & 0 \\ & & & & - & 1 & 1 & 0 & 0 & 0 \\ & & & & & - & 0 & 0 & 1 & 0 \\ & & & & & & - & 0 & 0 & 1 \\ & & & & & & & - & 0 & 0 \\ & & & & & & & & - & 0 \\ & & & & & & & & & - \end{pmatrix}$$

Fig. 4.28 Optimal link topology when total available capital is 250

$$x^* = \begin{pmatrix} - & 0 & 0 & 0 & 0 & 0 & 0 & 1 & 0 & 0 \\ & - & 1 & 0 & 0 & 1 & 0 & 0 & 0 & 1 \\ & & - & 0 & 1 & 0 & 0 & 1 & 0 & 0 \\ & & & - & 0 & 0 & 1 & 1 & 1 & 1 \\ & & & & - & 1 & 0 & 0 & 0 & 0 \\ & & & & & - & 0 & 0 & 1 & 0 \\ & & & & & & - & 0 & 0 & 0 \\ & & & & & & & - & 0 & 0 \\ & & & & & & & & - & 0 \\ & & & & & & & & & - \end{pmatrix}$$

Fig. 4.29 Optimal link topology when total available capital is 210

4.4.3 Bicriteria Reliable Network Model of LAN

In communication networks, local area networks (LANs) are commonly used as the communication infrastructure that meets the demand of the user in the local environment. These networks typically consist of several LAN segment connected together *via* bridges. The use of these transparent bridges requires loop-free paths between LAN segments [12]. Therefore, only spanning tree topologies can be used as active LANs configurations. Figure 4.30 shows a typical LANs structure. The performance criteria (*i.e.*, cost, delay, reliability, *etc.*) of these systems are largely affected by network topology. The network topology design problem includes two primary issues: clustering and routing. The clustering problem consists of two issues, *i.e.*, how many segments (clusters) should the LAN be divided into and how to allocate the users (stations) to the LAN segment (cluster). The routing problem is defined as the determination of segments interconnection spanning tree topology. The selection of optimal LAN configuration is an NP-hard problem. Therefore, heuristic methods can be employed to solve the problem. Elbaum and Sidi [12] used GA based on Huffman tree for topological designing of LANs. Gen *et al.* [50] proposed a GA for solving bicriteria network design problems considering connection cost and average message delay. Kim [13] and Kim and Gen [14] proposed a spanning tree-based GA for solving bicriteria LAN design problem to minimize connection cost and average delay considering network reliability. Before giving the details of this study, we introduce the mathematical model.

Consider a LAN that connects m users (stations). For example, Fig. 4.30 shows LANs with 5 service centers and 18 users. The communication traffic demands between users are given by an $m \times m$ matrix U which is called the user traffic matrix. An element u_{ij} of matrix U represents the traffic from users i to users j. In this study, it is assumed that traffic characteristics are known and summarized in the user traffic matrix U. To represent the connection between service centers, an $n \times n$ service center topology matrix X_1 is defined. An element x_{1ij} is represented as

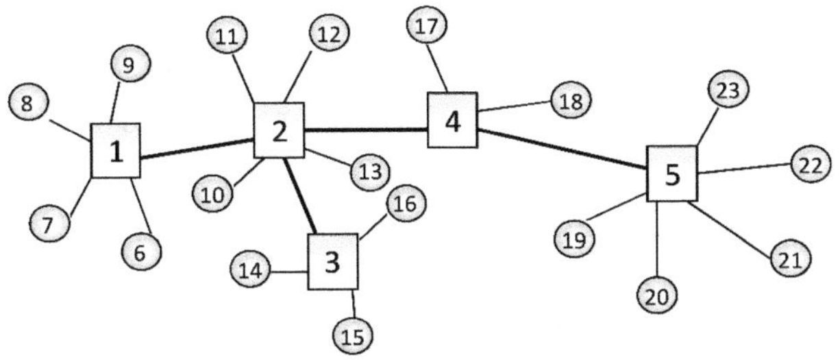

Centers: {1,2,3,4,5}, i=1,2, ..., n+m-2

Users: {6,7,8,9,10,11,12,13,14,15,16, 17,18,19,20,21,22,23}

Fig. 4.30 A sample LANs structure

$$x_{1ij} = \begin{cases} 1 \text{ if the centers } i \text{ and } j \text{ are connected} \\ 0 \text{ otherwise} \end{cases} \tag{4.33}$$

If it is assumed that the LAN is partitioned into n segments (service centers or clusters), the users are distributed over those n service center. The $n \times m$ clustering matrix X_2 specifies which users belong to which center. Thus

$$x_{2ij} = \begin{cases} 1 \text{ if user } j \text{ belongs to center } j \\ 0 \text{ otherwise} \end{cases} \tag{4.34}$$

A user belongs only to one center: thus $\forall\ j = 1,2,\cdots,m, \sum_{i=1}^{n} x_{2ij} = 1$. Then the spanning tree matrix $X([X_1,X_2])$ has a dimension with $n \times (n+m)$. T called the service center traffic matrix. An element t_{ij} of the matrix represents the traffic forwarded from users in center i to users in center $j(\cdot)$.

A center is a LAN segment with a known capacity. A LAN segment can be token-ring, Ethernet, or other kinds of similar architectures. It is clear that the behavior and the performances of each kind of LAN segment are different from each other. Nevertheless, the average delay in all those architectures responds qualitatively in the same manner to the load growth. The average delay increases as the load over the segment builds up. When the load approaches the segment capacity, the delay approaches infinity.

Notation

B_{ij}: the delay per bit due to the link between center i and j
C_i: the traffic capacity of center i
g_i: the maximum number of users capable of connecting to center i

An M/M/1 model (Bertsekas and Gallager [44]) had been used to describe a single cluster (segment) behavior. The bicriteria network topology design problem is

$$\min \frac{1}{\Gamma}\left[\sum_{i=1}^{n}\frac{F_i(X)}{C_i - F_i(X)} + \sum_{i=1}^{n}\sum_{j=1}^{n}\beta_{ij}f_{ij}(X)\right] \tag{4.35}$$

$$\min \sum_{i=1}^{n-1}\sum_{j=i+1}^{n} w_{1ij}x_{1ij} + \sum_{i=1}^{n}\sum_{j=1}^{m} w_{2ij}x_{2ij} \tag{4.36}$$

$$\text{s. t. } R(X) > R_{\min} \tag{4.37}$$

$$\sum_{j=1}^{m} x_{2ij} \le g_i, \quad i = 1,2,\cdots,n \tag{4.38}$$

$$\sum_{i=1}^{n} x_{2ij} = 1, \quad j = 1,2,\cdots,m \tag{4.39}$$

$$F_i(X) < C_i, \quad i = 1,2,\cdots,n \tag{4.40}$$

where $R(x)$ is the network reliability, w_{1ij} is the cost of the link between centers i and j, w_{2ij} is the cost of the link between center i and user j, and g_i is the maximum number of users capable of connecting to center i. The total offered traffic Γ is

$$\Gamma = \sum_{i=1}^{n}\sum_{j=1}^{n} t_{ij} \tag{4.41}$$

The total traffic at center k, $F_k(X)$, is

$$F_k(X) = \sum_{i=1}^{n}\sum_{j=1}^{n} t_{ij}a_{ij}^{k}(X), \quad k = 1,2,\cdots,n \tag{4.42}$$

where

$$a_{ij}^{k}(X) = \begin{cases} 1 & \text{if traffic from center } i \text{ to center } j \text{ through center } k \text{ exists} \\ 0 & \text{otherwise} \end{cases} \tag{4.43}$$

The total traffic through link (k,l), $f_{kl}(X)$, is

$$f_{kl}(X) = \sum_{i=1}^{n}\sum_{j=1}^{n} t_{ij}b_{ij}^{(k,l)}(X), \quad k = 1,2,\cdots,n, \quad l = 1,2,\cdots n \tag{4.44}$$

where

$$b_{ij}^{(k,l)}(X) = \begin{cases} 1 \text{ if traffic from center } i \text{ to center } j \text{ passes through an} \\ \quad \text{existing center link connecting centers } k \text{ and } l \text{ exists} \\ 0 \text{ otherwise} \end{cases} \quad (4.45)$$

4.4.3.1 Kim and Gen's Approach

Let $P(t)$ and $C(t)$ be parents and offspring in current generation t, respectively. The overall procedure of Kim and Gen's Prüfer number-based GA (pnGA) for solving bicriteria reliable network model is outlined as follows:

---

procedure: pnGA for bicriteria reliable network model
input: network data, GA parameters (*popSize*, *maxGen*, p_M, p_C)
output: the best solution
begin
  $t \leftarrow 0$;
  initialize $P(t)$ by Prüfer number-based encoding routine
  feasibility check and repairing by repairing routine;
  evaluate $P(t)$ by Prüfer number-based decoding routine;
  while (**not** terminating condition) **do**
    create $C(t)$ from $P(t)$ by multi-cut point crossover routine;
    create $C(t)$ from $P(t)$ by swap mutation routine;
    feasibility check and repairing by repairing routine;
    evaluate $C(t)$ by Prüfer number-based decoding routine;
    select $P(t+1)$ from $P(t)$ and $C(t)$ by roulette wheel selection routine;
    $t \leftarrow t+1$;
  end
  output the best solution
end

Prüfer Number-based Encoding

Two types of data structure are suitable to represent LAN configurations: (1) both service centers and user are represented by Prüfer numbers, and (2) service centers are represented by Prüfer numbers and user are depicted by a clustering string that describes the distribution of users into service centers. Kim and Gen employed an encoding method used with Prüfer numbers and clustering string. Figure 4.31 shows an example of encoding method for LAN structure given in Fig. 4.30. The chromosomes are generated randomly in the initialization process. The service centers are composed of n–2 digits (Prüfer numbers) randomly generated in the range $[1,n]$, and users are made up of m digits (clustering string) randomly generated in the range $[1, n]$, which indicate how to allocate users to service centers so that each user belongs to a specific center.

	1	2	3
Centers:	2	2	4

	6	7	8	9	10	11	12	13	14	15	16	17	18	19	20	21	22	23
Users:	1	1	1	1	2	2	2	2	3	3	3	4	4	5	5	5	5	5

Fig. 4.31 Prüfer number-based encoding for Fig. 4.30

Evaluation

Since the problem is a bicriteria problem, Pareto optimal solutions have to be handled. Therefore, the weighted-sum method is used to construct the fitness function, which combines the bicriteria objective functions into one overall fitness function.

Step 1. Convert chromosomes represented by the Prüfer number and clustering string to spanning tree matrix X.

Step 2. Calculate the objective function values $[f_i(X), i = 1, 2]$.

Step 3. Choose solution points that contain the minimum $z_i^{\min}$ (or maximum $z_i^{\max}$) corresponding to each objective function value, and compare with the stored solution points in the preceding generation and select the best points to save again:

$$z_i^{\min(t)} = \min\left\{ (z_i^{\min(t-1)}, f_i^{(t)}(X_k)) | k = 1, 2, \cdots, chrSize \right\}, \quad i = 1, 2$$

$$z_i^{\max(t)} = \max\left\{ (z_i^{\max(t-1)}, f_i^{(t)}(X_k)) | k = 1, 2, \cdots, chrSize \right\}, \quad i = 1, 2$$

where $z_i^{\min(t)}$ and $z_i^{\max(t)}$ are the minimum and maximum value of the i-th objective function in generation t, respectively; $f_i^{(t)}(X_k)$ is the i-th objective function value of the k-th chromosome X_k in the generation t; and *chrSize* is equal to population size plus the offspring generated after genetic operations.

Step 4. Calculate the weight coefficient as follows:

$$\alpha_i = \frac{z_i^{\max(t)} - z_i^{\min(t)}}{z_i^{\max(t)}}, \qquad i = 1, 2$$

$$\beta_i = \frac{\alpha_i}{\alpha_1 + \alpha_2}, \qquad i = 1, 2$$

Step 5. Calculate the fitness function value for each chromosome as follows:

$$eval(X_k) = \sum_{i=1}^{2} \beta_i d_i(X_k), \qquad k = 1,2,\cdots,chrSize$$

where $d_i(X_k)$ is the normalized value of the i-th objective function and is represented as follows:

$$d_i(X_k) = \begin{cases} \frac{f_i^t(X_k) - z_i^{\min(t)} + \gamma}{z_i^{\max(t)} - z_i^{\min(t)} + \gamma} & \text{for maximization} \\ \frac{z_i^{\max(t)} - f_i^t(X_k) + \gamma}{z_i^{\max(t)} - z_i^{\min(t)} + \gamma} & \text{for minimization} \end{cases} \tag{4.46}$$

where γ is a small positive real number that is usually restricted within the open interval (0,1), used to prevent the equation from zero division, and makes it possible to adjust the selection behavior from fitness-proportional selection to pure random selection.

Tree-Based Reliability Calculation

In this study, the probability of all operative nodes (centers and users) being connected has been considered as the reliability measure. When reliability of nodes and links is known and spanning tree is considered as a rooted tree, a state vector is associated with the root of each subtree. The state vector contains all information about that node relevant to calculation. In this situation, it is possible to define a set of recursion relations that yield the state vector of a rooted tree given the state of its subtrees. For subtrees involving single nodes, the state is obvious. Then the rooted subtrees become larger and larger rooted subtrees using recursion relations until the state of the entire network is obtained (Kershenbaum [6]).

Deriving the recurrence relations is somewhat mechanical. It comes simply from considering the situation depicted in Fig. 4.32. In this figure, there are two subtrees, one with root i and the other having as its root j. It is assumed that the state of node i and node j are known and it is wanted to compute the state of j relative to the tree obtained by joining node i into node j with link (i,j), where node j is the father node of node i.

Assume that the probability of node failure p_i^f and the probability of the node being operative $p_i^o(= 1 - p_i^f)$ are associated with node i. Similarly, link (i,j) has probabilities l_i^f and l_i^o of the link (i,j) failing and being operative, respectively. Also, the state vectors for each subtree are defined: e_i is the probability that all nodes in the subtree are failed; o_i is the probability that the set of operative nodes, including the root of the subtree, are connected; and r_i is the probability that the root of the subtree is failed and the set of operative nodes in the subtree is connected.

For a tree with root node 1 and n nodes, the reliability of the tree, $R(X)$ *i.e.*, the probability of all operating nodes communicating is calculated as shown in Fig. 4.33:

Fig. 4.32 Recurrence relations

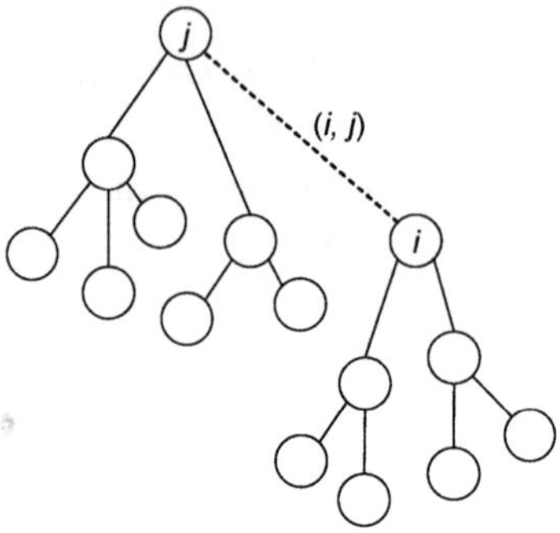

Procedure: Reliability Calculation
Step 1: Set $r_i = 0$, $o_i = p_i^o$, $e_i = p_i^f$, $i = 1,2,..., n$. Set $i = n$.
Step 2: If node j is the father node of node i, using the following recurrence relations, recalculate r_i, o_i, e_i:

$$r_j' = r_j\, e_i + r_i\, e_j + o_i\, e_j$$
$$o_j' = o_i\, o_j\, l_i + o_j\, e_i$$
$$e_j' = e_i\, e_j$$

Step 3: Set $i = i - 1$. If $i = 1$, go to step 4; otherwise go to step 2.
Step 4: Return $r_1 + o_1 + e_1$

Fig. 4.33 Pseudocode of reliability calculation

Selection

Roulette wheel and elitist selection mechanisms are combined in order to enforce the GA to search solution space freely. The roulette wheel selection, which is a fitness-proportional method, is used to reproduce randomly new generation and the elitist method is employed to preserve the best chromosome for the next generation and overcome the stochastic errors of sampling.

Genetic Operators

Multi-cut point crossover and swap mutation are used. This type of crossover is accomplished by selecting two parent solutions and randomly taking a component from one parent to form the corresponding component of the offspring. Figure 4.34 show the application of the crossover operator.

In swap mutation, randomly two positions are selected and their contents are swapped. Application of the swap operator is seen in Fig. 4.35.

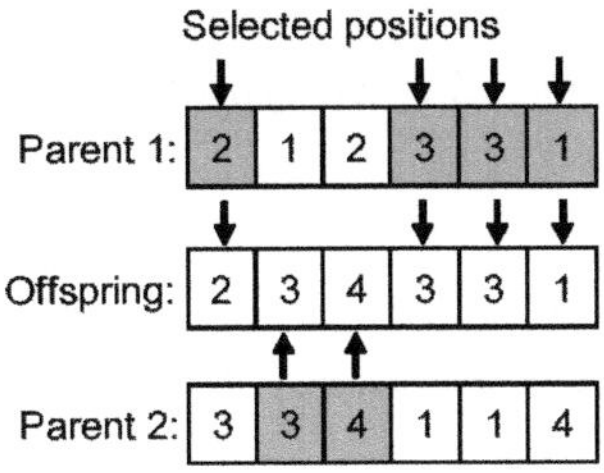

Fig. 4.34 Multi-cut point crossover operator

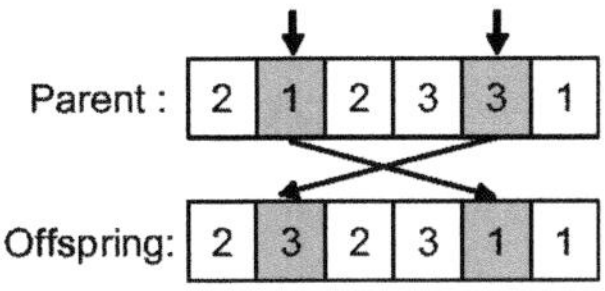

Fig. 4.35 Swap mutation operator

Repairing Algorithm

Because of the existence of a constraint on the maximum number that is capable of connecting to each center, the chromosomes generated randomly in the initial population and the offspring produced by crossover may be illegal in the sense of violating the maximum number of connections for each center. A repair strategy is used to modify the connection number for center in an illegal chromosome.

Let $\bar{G}$ be a set of centers whose maximum number of connections has not been checked and modified in a chromosome. If center i violates the constraint with the maximum number g_i of connections for center i, this means that the number of this center in the chromosome is more than $g_i - 1$. Then decrease the number of center by randomly replacing it with another center from $\bar{G}$.

4.4.3.2 Numerical Experiments

Kim and Gen used the following two instances to demonstrate the performance of their genetic algorithm. For each example, it is assumed that operative probability of centers is 0.95, the operative probability of users is 0.90, the operative probability of links between centers is 0.90, and the operative probability of links between center and users is 0.85. In addition, the cost matrix of the links between centers i and j (w_{1ij}) is randomly generated from [100, 250], and the cost matrix of the link between center i and user j (w_{2ij}) is randomly generated between [1,100]. The users' traffic matrix U was taken from Elbaum and Sidi [12]. Crossover and mutation rates are set to 0.3 and 0.7, respectively. The number of generations is considered as 500. To obtain Pareto optimal solutions, GA runs 20 times for each problem.

Example 4.1

The first example has 4 service center (n = 4), 8 users (m = 8), g_i = 3 the traffic capacity of center i (C_i) is 50. Population size is set to 100. From Fig. 4.36 and

Table 4.11, it is possible to see that Kim and Gen's encoding method obtained better results than those of the encoding method by proposed by Dengiz *et al.* [25].

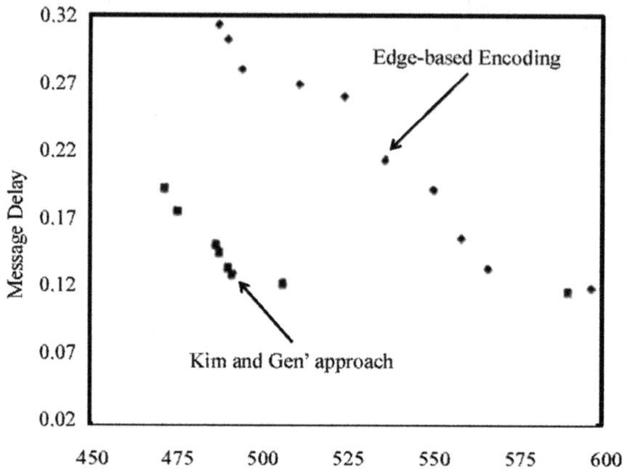

Fig. 4.36 Pareto solutions for Example 4.1

Table 4.11 Comparison of proposed encoding and edge-based encoding

Example no.	Kim and Gen's approach		Edge-based encoding	
	ACT (sec.)	Memory size n+m-2	ACT (sec.)	Memory size (n+m)(n+m-1)/2
1	5.07	10	7.12	66
2	32.29	34	-	630

ACT: Average Computation Time; -: Not executable

Example 4.2

The second example has 6 service centers ($n = 6$), 30 users ($m = 30$), $g_i = 10$ and the traffic capacity of center i (C_i) is 300. To determine the best compromise solution among the Pareto solutions, Kim and Gen applied the TOPSIS method. TOPSIS stands for the technique for order preference by similarity to ideal solution, which is based upon the concept that the chosen alternative should have the shortest distance from positive ideal solution and the farthest from the negative ideal solution. Figure 4.37 gives the best compromise solution obtained by TOPSIS. The best compromise solutions is connection cost 1377, message delay 0.020446 and network reliability 0.920843, and the chromosome is shown in Fig. 4.38.

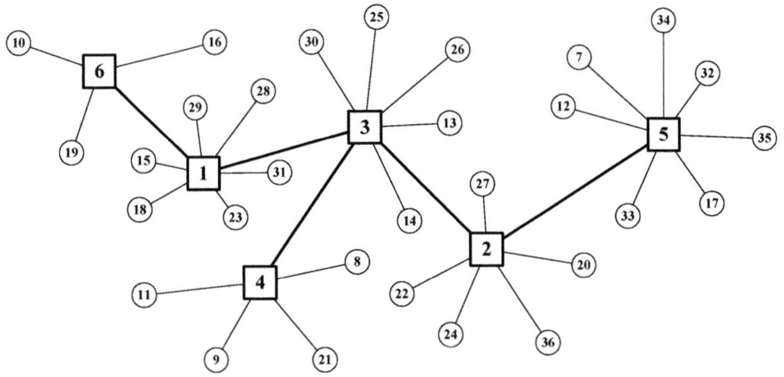

Fig. 4.37 The best compromise solution for Pareto solutions

Centers:

1	2	3	4
3	2	3	1

Users:

7	8	9	0	11	12	13	14	15	16	17	18	19	20	21
5	4	4	6	4	5	3	3	1	6	5	1	6	2	4
22	23	24	25	26	27	28	29	30	31	32	33	34	35	36
2	1	2	3	3	2	1	1	3	1	5	5	5	5	2

Fig. 4.38 A best chromosome for Example 4.2

4.4.4 Bi-level Network Design Model

The *bi-level programming problem* is a mathematical programming problem which is composed of upper-level problem and lower-level problem and can be translated into Nash game, when non-cooperatively optimizing upper-level problem and lower-level problem contrast with each other. The bicriteria reliable network model of LAN discussed at above subsection, also can be formulated as a bi-level programming model. An example of *bi-level network design problem* with 5 backbone routers and 18 leaf routers is showed in Fig. 4.39. This bi-level programming model is made up of the upper-level problem that the decision maker of backbone network minimizes connection cost and the lower-level problem that the decision maker of distribution network minimizes average message delay. Supposing that each decision maker are players in the game and players maximize their own benefit (optimize the objective function) with non-cooperation; then bi-level network design problems can be interpreted as Nash game.

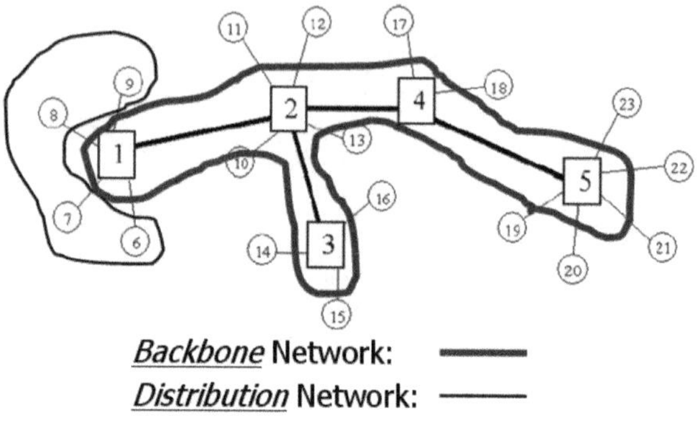

Fig. 4.39 A sample bi-level network structure

The communication traffic demands between routers are given by an $m \times m$ matrix U which is called the *routers traffic matrix.* An element u_{ij} of matrix U represents the traffic from router i to router j. Note that traffic between a pair of routers can vary from heavy traffic to no traffic. Moreover, this traffic can be burst or constant. The traffic peak rate is probably an overestimation of the traffic requirement, since most of the time the actual traffic is far below the peak. On the other hand, taking the average traffic rate as the requirement may yield poor results when heavy traffic has to be forwarded. For our purposes, we shall assume that traffic characteristics are known and summarized in the router traffic matrix U.

The $n \times n$ backbone router topology matrix X_1 represents the connected appearance of backbone routers. An element x_{1ij} is represented by

$$x_{1ij} = \begin{cases} 1 \text{ if the centers } i \text{ and } j \text{ are connected} \\ 0 \text{ otherwise} \end{cases} \tag{4.47}$$

Let the network be partitioned into n segments (backbone routers or clusters). The routers are distributed over those n backbone routers. The $n \times m$ clustering matrix X_2 specifies which router belongs to which backbone router. Thus,

$$x_{2ij} = \begin{cases} 1 \text{ if user } j \text{ belongs to center } j \\ 0 \text{ otherwise} \end{cases} \tag{4.48}$$

A router can belong to only one backbone router. We define $n(n+m)$ matrix X called the spanning tree matrix ($[X_1X_2]$) and define an $n \times n$ matrix T called the *backbone router traffic matrix.* An element t_{ij} of this matrix represents the traffic forwarded from routers in backbone router i to routers in backbone router j. Obviously, $T = X_2^T U X_2$.

The backbone routers are interconnected *via* fiber optic cables. The traffic from source backbone router i to backbone router j is forwarded through the cables. The path between backbone routers i and j may include a single cable which directly connects between backbone routers or multiple cables in case there is no direct connection between the backbone routers. In the latter case, the traffic travels through cables and backbone routers on the connecting path.

A backbone router is a segment with a known capacity. A segment can be token-ring, Ethernet, or other kinds of similar architectures. It is clear that the behavior and the performance of each kind of segment are different from each other. Nevertheless, the average message delay in all those architectures responds qualitatively in the same manner to the load growth. The average message delay increases as the load over the segment builds up. When the load approaches the segment capacity, the delay approaches infinity. We can formulate the bi-level network design problems as a same M/M/1 model with bicriteria reliable network model of LAN discussed at above subsection.

4.4.4.1 Kim and Lee's approach

Kim and Lee proposed a Nash GA for solving this bi-level network design model [90]. Let $P(t)$ and $C(t)$ be parents and offspring in current generation t, respectively. The overall procedure of Kim and Lee's Prüfer number-based Nash GA (pn-nGA) is outlined as follows:

Nash GA

The idea of Nash GA is to bring together genetic algorithms and Nash game in order to cause the GA to build the Nash equilibrium. Figure 4.40 show how such merging can be achieved with two players trying to optimize two different objectives.

Let $X = X_1 X_2$ be the string representing the potential solution for a dual objective optimization problem. Then X_1 denotes the subset of variables handled by Player 1 (P_1) and optimized along criterion 1. Similarly X_2 denotes the subset of variables handled by Player 2 (P_2) and optimized along criterion 2. According to Nash theory, P_1 optimizes X with respect to the first criterion by modifying X_1 while X_2 is fixed by P_2. Symmetrically, P_2 optimizes X with respect to the first criterion by modifying X_1 while X_1 is fixed by P_1. There are two different populations, *i.e.*, P_1's optimization task is performed by Population 1 (Pop1) whereas P_2's optimization task is performed by Population 2 (Pop2). Let $X_{1,k-1}$ be the best value found by P_1 at generation $k-1$ and $X_{2,k-1}$ be the best value found by P_2 at generation $k-1$. At generation k, P_1 optimizes X_{1k} while using $X_{2,k-1}$ in order to evaluate X. P_2 also optimizes X_{2k} while using $X_{1,k-1}$ in order to evaluate X. After the optimization process,

procedure: pnGA for bi-level network design model
input: network data, GA parameters (*popSize*, *maxGen*, p_M, p_C)
output: the best solution
begin
  $t \leftarrow 0$;
  initialize $P_1(t)$ by Prüfer number-based encoding routine
  initialize $P_2(t)$ by clustering string-based encoding routine
  feasibility check and repairing by repairing routine;
  evaluate $P_1(t)$ by Prüfer number-based decoding routine;
  evaluate $P_2(t)$ by clustering string-based-based decoding routine;
  while (**not** terminating condition) **do**
    create $C_1(t)\&C_2(t)$ from $P_1(t)\&P_2(t)$ by multi-cut point crossover routine;
    create $C_1(t)\&C_2(t)$ from $P_1(t)\&P_2(t)$ by replacement mutation routine;
    feasibility check and repairing by repairing routine;
    evaluate $C_1(t)$ by Prüfer number-based decoding routine;
    evaluate $C_2(t)$ by clustering string-based-based decoding routine;
    select $P_1(t+1)$ from $P_1(t)$ and $C_1(t)$ by tournament selection routine;
    select $P_2(t+1)$ from $P_2(t)$ and $C_2(t)$ by tournament selection routine;
    $t \leftarrow t+1$;
  end
  output the best solution
end

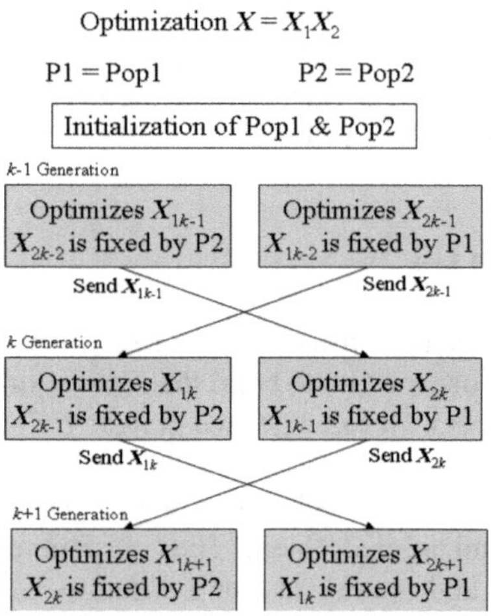

Fig. 4.40 The process of Nash GA

P_1 sends the best value X_{1k} to P_2 who will use it at generation $k+1$. Similarly, P_2 sends the best value X_{2k} to P_1 who will use it at generation $k+1$. Nash equilibrium is reached when neither P_1 nor P_2 can further improve their criteria [91, 92].

Genetic Representation

Two types of data structure are suitable to represent bi-level network design problem configurations: (1) the backbone routers are represented by Prüfer number, and (2) leaf routers are described by clustering string which describes distribution of leaf routers into the backbone routers. Kim and Lee employed an encoding method used with Prüfer numbers and clustering string. Figure 4.41 shows an example of encoding method for LAN structure given in Fig. 4.39. The chromosomes are generated randomly in the initialization process.

	1	2	3
Backbone Net:	2	2	4

	6	7	8	9	10	11	12	13	14	15	16	17	18	19	20	21	22	23
Distribution Net:	1	1	1	1	2	2	2	2	3	3	3	4	4	5	5	5	5	5

Fig. 4.41 A genetic representation for Fig. 4.39

Evaluation

Kim and Lee calculate the evaluation value of a chromosome as their own objective function for each decision maker:

$$f_1(v'_k) = f_1(v^i_{1k}, v^*_{2,k-1}) \tag{4.49}$$

$$f_2(v'_k) = f_2(v^*_{1k}, v^j_{2,k-1}) \tag{4.50}$$

where v^i_{1k} and v^j_{2k} are i-th and j-th chromosome of P_1 and P_2, respectively. $v^*_{1,k-1}$ and $v^*_{2,k-1}$ are best chromosome selected by adversary at $k-1$ generation, respectively. $f_1(\cdot)$ and $f_2(\cdot)$ means the objective function of each decision makers.

Genetic Operators

Crossover is implemented with uniform crossover operator (or multi-point crossover), which has been shown to be superior to traditional crossover strategies for combinatorial problem. Uniform crossover first generates a random crossover mask and then exchanges relative genes between parents according to the mask. A crossover mask is simply a binary string with the same size of chromosome. The parity of each bit in the mask determines, for each corresponding bit in an offspring, which parent it will receive that bit from (see Fig. 4.34).

A simple replacement mutation is used, which simply selects one position at random and random perturbation within the permissive range from 1 to n (the number of backbone routers) as shown in Fig. 4.42.

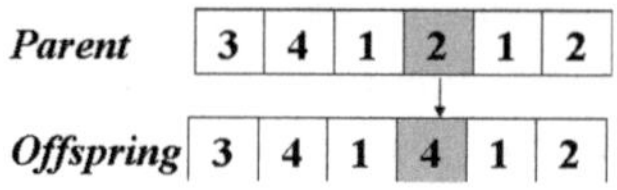

Fig. 4.42 Replacement mutation operator

Selection

Tournament selection is one of many methods of selection in GA which runs a tournament among a few individuals and selects the winner (the one with the best fitness). Selection pressure can be easily adjusted by changing the tournament size. If the tournament size is higher, weak individuals have a smaller chance of being selected. Tournament selection has several benefits: it is efficient to code, works on parallel architectures and allows the selection pressure to be easily adjusted. Also, the elitist method is employed to preserve the best chromosome for the next generation and overcome the stochastic errors of sampling. Using this selection process, Kim and Lee keep the best chromosome from the current generation to the next generation by the following process:

1. Select a group of N ($N \geq 2$) individuals, which N is tournament size.
2. Select the individual with the highest fitness from the group.

4.4.4.2 Numerical Experiments

The test problems are experimented on to confirm that the Kim and Lee's Nash GA can be used effectively for bi-level network design problems. The experimented device is PC (Windows XP SP2) with Pentium4 2.66GHz CPU and 512MB RAM. Data for problems are referred by reference (Elbaum and Sidi [12]) due to unavailability of real network systems information.

Example 4.1 has 4 backbone routers ($n = 4$), 8 routers ($m = 8$), the link weight between backbone router i and backbone router j (w_{1ij}) randomly generated in [100, 250], and the link weight between backbone router i and router j (w_{2ij}) randomly selected in [1, 100]. The traffic capacity of backbone router i (C_i) is set up to 50, the operative probability of backbone routers is 0.95, the operative probability of terminals is 0.9, the operative probability of link between backbone routers is 0.9, the operative probability of link between backbone router and router is 0.85, and Rmin is set up to 0.9. The router traffic matrix U for this experiment is established as follows:

The parameters for Nash GA are experimented on in the following method. Let

$$\begin{bmatrix} 0 & 3 & 3 & 3 & 0 & 0 & 0 & 1 \\ 3 & 0 & 3 & 3 & 0 & 0 & 1 & 0 \\ 3 & 3 & 0 & 3 & 0 & 1 & 0 & 0 \\ 3 & 3 & 3 & 0 & 1 & 0 & 0 & 0 \\ 0 & 0 & 0 & 1 & 0 & 3 & 3 & 3 \\ 0 & 0 & 1 & 0 & 3 & 0 & 3 & 3 \\ 0 & 1 & 0 & 0 & 3 & 3 & 0 & 3 \\ 1 & 0 & 0 & 0 & 3 & 3 & 3 & 0 \end{bmatrix}$$

the population size of backbone network be 20, the population size of distribution networks 40, the probability of crossover 0.8 and the probability of mutation 0.7, and the maximum generation number 500.

Example 4.2 has 6 backbone routers ($n = 6$), 30 routers ($m = 30$), the link weight between backbone router i and backbone router j (w_{1ij}) randomly generated in [100, 250], and the link weight between backbone router i and router j (w_{2ij}) randomly selected in [1, 100]. The traffic capacity of backbone router i is set up to 300, the operative probability of backbone routers is 0.95, the operative probability of routers is 0.9, the operative probability of link between backbone routers is 0.9, the operative probability of link between backbone router and router is 0.85, and Rmin is set up to 0.9. The users traffic matrix U for Example 4.2 is established,

$$\begin{bmatrix} U_3 & U_1 & O & U_1 & O & U_1 \\ U_1 & U_3 & U_1 & O & U_1 & O \\ O & U_1 & U_3 & U_1 & O & U_1 \\ U_1 & O & U_1 & U_3 & U_1 & O \\ O & U_1 & O & U_1 & U_3 & U_1 \\ U_1 & O & U_1 & O & U_1 & U_3 \end{bmatrix}$$

where O is zero matrix, U_1 and U_3 is setting, by the following matrix:

The parameters for Nash GA are also set as follows: the probability of crossover is 0.8, the probability of mutation is 0.7, the population size of P_1 is 30, the population size of P_2 is 150, and the maximum generation number is 2000.

Example 4.3 has 10 backbone routers (n=10), 50 routers (m=50), the link weight between backbone router i and backbone router j (w_{1ij}) randomly generated in [100, 250], and the link weight between backbone router i and router j (w_{2ij}) randomly selected in [1, 100]. The traffic capacity of backbone router i is set up to 300, the operative probability of backbone routers is 0.95, the operative probability of routers is

$$
\boldsymbol{U}_1 = \begin{bmatrix} 0 & 0 & 0 & 0 & 1 \\ 0 & 0 & 0 & 1 & 0 \\ 0 & 0 & 1 & 0 & 0 \\ 0 & 1 & 0 & 0 & 0 \\ 1 & 0 & 0 & 0 & 0 \end{bmatrix} \qquad \boldsymbol{U}_3 = \begin{bmatrix} 0 & 3 & 3 & 3 & 3 \\ 3 & 0 & 3 & 3 & 3 \\ 3 & 3 & 0 & 3 & 3 \\ 3 & 3 & 3 & 0 & 3 \\ 3 & 3 & 3 & 3 & 0 \end{bmatrix}
$$

0.9, the operative probability of link between backbone routers is 0.9, the operative probability of link between backbone router and router is 0.85, and Rmin is set up to 0.9. The users traffic matrix U for Example 4.3 is established as follows:

$$
\begin{bmatrix}
\boldsymbol{U}_3 & \boldsymbol{U}_1 & \boldsymbol{O} & \boldsymbol{U}_1 & \boldsymbol{O} & \boldsymbol{U}_1 & \boldsymbol{O} & \boldsymbol{U}_1 & \boldsymbol{O} & \boldsymbol{U}_1 \\
\boldsymbol{U}_1 & \boldsymbol{U}_3 & \boldsymbol{U}_1 & \boldsymbol{O} & \boldsymbol{U}_1 & \boldsymbol{O} & \boldsymbol{U}_1 & \boldsymbol{O} & \boldsymbol{U}_1 & \boldsymbol{O} \\
\boldsymbol{O} & \boldsymbol{U}_1 & \boldsymbol{U}_3 & \boldsymbol{U}_1 & \boldsymbol{O} & \boldsymbol{U}_1 & \boldsymbol{O} & \boldsymbol{U}_1 & \boldsymbol{O} & \boldsymbol{U}_1 \\
\boldsymbol{U}_1 & \boldsymbol{O} & \boldsymbol{U}_1 & \boldsymbol{U}_3 & \boldsymbol{U}_1 & \boldsymbol{O} & \boldsymbol{U}_1 & \boldsymbol{O} & \boldsymbol{U}_1 & \boldsymbol{O} \\
\boldsymbol{O} & \boldsymbol{U}_1 & \boldsymbol{O} & \boldsymbol{U}_1 & \boldsymbol{U}_3 & \boldsymbol{U}_1 & \boldsymbol{O} & \boldsymbol{U}_1 & \boldsymbol{O} & \boldsymbol{U}_1 \\
\boldsymbol{U}_1 & \boldsymbol{O} & \boldsymbol{U}_1 & \boldsymbol{O} & \boldsymbol{U}_1 & \boldsymbol{U}_3 & \boldsymbol{U}_1 & \boldsymbol{O} & \boldsymbol{U}_1 & \boldsymbol{O} \\
\boldsymbol{O} & \boldsymbol{U}_1 & \boldsymbol{O} & \boldsymbol{U}_1 & \boldsymbol{O} & \boldsymbol{U}_1 & \boldsymbol{U}_3 & \boldsymbol{U}_1 & \boldsymbol{O} & \boldsymbol{U}_1 \\
\boldsymbol{U}_1 & \boldsymbol{O} & \boldsymbol{U}_1 & \boldsymbol{O} & \boldsymbol{U}_1 & \boldsymbol{O} & \boldsymbol{U}_1 & \boldsymbol{U}_3 & \boldsymbol{U}_1 & \boldsymbol{O} \\
\boldsymbol{O} & \boldsymbol{U}_1 & \boldsymbol{O} & \boldsymbol{U}_1 & \boldsymbol{O} & \boldsymbol{U}_1 & \boldsymbol{O} & \boldsymbol{U}_1 & \boldsymbol{U}_3 & \boldsymbol{O} \\
\boldsymbol{U}_1 & \boldsymbol{O} & \boldsymbol{U}_1 & \boldsymbol{O} & \boldsymbol{U}_1 & \boldsymbol{O} & \boldsymbol{U}_1 & \boldsymbol{O} & \boldsymbol{O} & \boldsymbol{U}_3
\end{bmatrix}
$$

The parameters for Nash GA are also set as follows: the probability of crossover is 0.8, the probability of mutation is 0.7, the population size of P_1 is 50, the population size of P_2 is 200, and the maximum generation number is 2000.

As shown in Table 4.12, Kim and Lee can find that Nash equilibrium converges on the near-optimal of objective function for each decision maker (connection cost for P_1 and average message delay for P_2), because two kinds of decision makers optimize their own objective function with non-cooperation.

4.5 Summary

Modeling and design of large communication and computer networks have always been an important area to both researchers and practitioners. The interest in developing efficient design models and optimization methods has been stimulated by high deployment and maintenance costs of networks, which make good network design potentially capable of securing considerable savings. In this chapter, we introduce

Table 4.12 The Results of Numerical Experiments

Example		The Results from P1			The Results from P2			Time
		Cost	Delay	Reliability	Cost	Delay	Reliability	
1	Avg.	551.00	0.06182	0.91441	553.15	0.06104	0.91314	168.4
	Var.	26.38	0.00239	0.00945	25.56	0	0.00897	1.8
	Best	538.00	0.06104	0.91199	538.00	0.06104	0.91199	165.0
	Worst	625.00	0.06104	0.90227	625.00	0.06104	0.90227	171.0
2	Avg.	1307.80	0.03754	0.99958	1320.15	0.03448	0.99950	9638.1
	Var.	33.01	0.01078	0.00059	28.47	0.01064	0.00100	133.3
	Best	1306.00	0.04569	0.99989	1306.00	0.04569	0.99989	9763.0
	Worst	1379.00	0.02446	0.99951	1394.00	0.02218	0.99989	9440.0
3	Avg.	1755.05	0.02937	0.999996	1783.30	0.02211	0.999997	96364.5
	Var.	34.10	0.00800	2.922E-06	38.12	0.00417	2.291E-06	5222.1
	Best	1712.00	0.04351	0.999993	1715	0.02344	0.999999	105912
	Worst	1836	0.04254	0.999997	1888	0.02436	0.999998	94322

GA applications on centralized network design, which is basic for local access networks, backbone network design and reliable design of networks.

To solve these communication network models, we introduced different GA approaches, such as a PrimPred-based GA for solving capacitated multipoint network models and capacitated QoS network models, a binary-based GA for solving backbone network models and reliable backbone network models with multiple goals, an adjacency matrix-based GA for backbone network models, a KruskalRST-based GA for solving reliable backbone network model, and a Prüfer number-based GA for solving bicriteria reliable network models *etc*. The effectiveness and efficiency of GA approaches were investigated with various scales of communication network problems by comparing with recent related researches.

References

1. Coloquest [online]. http://www.coloquest.com/gigenet-network/
2. Pioro, M. & Medhi, D. (2004). *Routing, Flow, Capacity Design in Communication & Computer Networks*, Morgan Kaufmann.
3. Boorstyn, R. R. & Frank, H. (1977). Large-scale network topological optimization, *IEEE Transasctions on Communication*, COM-25(1), 29–47.
4. Colbourn, C. J. (1987). *The Combinatorics of Network Reliability*, Oxford University Press.
5. Gerla, M. & Kleinrock, L. (1977). On the topological design of distributed computer networks, *IEEE Transactions on Communications*, COM-25(1), 48–60.
6. Kershenbaum, A. (1993). Telecommunications Network Design Algorithms. *New York:McGraw-Hill*, 1993.
7. Tanenbaum, A. S. (1981) *Computer Networks*, Prentice Hall, New Jersey.
8. Easu, L. R. & Williams, K. C. (1966). On teleprocessing system design: a method for approximating the optimal network, *IBM System Journal*, 5, 142–147.

9. Sharma, R. & El-Bardai, M. (1970). Suboptimal communications networks synthesis, *Proceedings of the IEEE International Conference on Communications*, 19.11–19.16.
10. Kershenbaum, A. & Chou, W. (1974). A unified algorithm for designing multidrop teleprocessing networks, *IEEE Transactions on Communications*, 22, 1762–1772.
11. Abuali, F. N., Wainwright, R. L. & Schoenefeld, D. A. (1995). Determinant factorization: a new encoding scheme for spanning tree applied to the probability minimum spanning tree problem, *in Proceedings of 6th International Conference on Genetic Algorithms*, 470–477.
12. Elbaum, R. & Sidi, M. (1996). Topological design of local-area networks using genetic algorithms, *IEEE/ACM Transactions on Networking*, 4(5), 766–778.
13. Kim, J. R. (2000). *Study on advanced genetic algorithms for reliable network design*, Ph.D. Disertation, Ashikaga Institute of Technology.
14. Kim, J. R. & Gen, M. (1999). Genetic algorithm for solving bicriteria network topology design problem, *Proceedings of the 1999 Congress on Evolutionary Computation*, 2272–2279.
15. Kim, J. R. & Gen, M. (2000). A genetic algorithm for bicriteria communication network topology design, *Engineering Valuation & Cost Analysis*, 3, 351–363, 2000.
16. Lo, C. C. & Chang, W. H. (2000). A multiobjective hybrid genetic algorithm for the capacitated multipoint network design problem, *IEEE Transacion on System, Man & Cybernetics-Part B*, 30(3), 461–470.
17. Palmer, C. C. & Kershenbaum, A. (1995). An approach to a problem in network design using genetic algorithms, *Networks*, 26, 151–163.
18. Zhou, G. & Gen, M. (1997). Approach to degree-constrained minimum spanning tree problem using genetic algorithm, *Engineering Design & Automation*, 3(2), 157–165.
19. Zhou, G. & Gen, M. (1997). A note on genetic algorithm approach to the degree-constrained minimum spanning tree problem, Networks, 30, 105–109.
20. Zhou, G. & Gen, M. (2003). A genetic algorithm approach on tree-like telecommunication network design problem, *Journal of Operational Research Society*, 54(3), 248–254.
21. Davis, L. & Coombs, S. (1989). *Optimizing network link sizes with genetic algorithms, in Modeling and Simulation Methodology, Knowledge System's Paradigms*. Amsterdam, The Netherlands: Elsevier.
22. Kumar, A., Pathak, R. M. & Gupta, Y. P. (1995). Genetic algorithm - based reliability optimization for computer network expansion, *IEEE Transactions on Reliability*, 44(1), 63–72.
23. Kumar, A., Pathak, R. M., Gupta, Y. P. & Parsei, H. R. (1995). A genetic algorithm for distributed system topology design, *Computers and Industrial Engineering*, 28(3), 659–670.
24. Ko, K. T., Tang, K. S., Chan, C. Y., Man, K. F. & Kwong, S. (1997). Using genetic algorithm to design mesh networks, *IEEE Computer*.
25. Dengiz, B., Altiparmak, F. & Smith, A. E. (1997). Efficient optimization of all-terminal reliable networks, *IEEE Transactions on Reliability*, 41(1), 18–26.
26. Dengiz, B., Altiparmak, F. & Smith, A. E. (1997). Local search genetic algorithm for optimal design of reliable networks, *IEEE Transactions on Evolutionary Computation*, 1(3).
27. Deeter, D. L. & Smith, A. E. (1998). Economic design of reliable networks, *IIE Transactions*, 30, 1161-1174.
28. Pierre, S. & Legault, G. (1998). A genetic algorithm for designing distributed computer network topologies, *IEEE Transactions on Systems, Man & Cybernetics-Part B: Cybernetics*, 28(2), 249–257,
29. Konak, A. & Smith, A. E. (1999). A hybrid genetic algorithm approach for backbone design of communication networks, *Proceeding of the 1999 Congress on Evolutionary Computation*, 1817–1823.
30. Cheng, S. T. (1998), Topological optimization of a reliable communication network, *IEEE Transactions on Reliability*, 47(3), 225–233.
31. Liu, B. & Iwamura, K. (2000), Topological optimization models for communication network with multiple reliability goals, *Computers & Mathematics with Applications*, 39, 59–69.
32. Altiparmak, F., Dengiz, B. & Smith, A. E. (2003). optimal design of reliable computer networks: a comparison of metaheuristics, *Journal of Heuristics*, 9(6), 471–487.

33. Altiparmak, F., Gen, M., Dengiz, B. & Smith, A. E. (2003). Topological optimization of communication networks with reliibility constraint by an evolutionary approach, *Proceedings of International Workshop on Reliability and Its Applications*, 183–188.
34. Papadimitriou, C. H. (1978). The complexity of the capacitated tree problem, *Networks*, 8, 217-230.
35. Chandy, K. M. & Lo, T. (1973). The capacitated minimum spanning tree, *Networks*, 3, 173-182.
36. Elias, D. & Ferguson, M. (1974). Topological design of multipoint teleprocessing networks, *IEEE Transactions on Communications*, 22, 1753-1762.
37. Gavish, B. (1982). Topological design of centralized computer networks: formulation and algorithms, *Networks*, 12, 355–377.
38. Raidl, G. R. & Julstrom, B. A. (2003). Edge Sets: An Effective Evolutionary Coding of Spanning Trees, *IEEE Transaction on Evolutionary Computation*, 7(3), 225–239.
39. Xue, G. (2003). Minimum-cost QoS multicast & unicast routing in communication networks, *IEEE Transactions on Communication*, 51(5), 817–824.
40. Atiqullah, M. M. & Rao, S. S. (1993). Reliability optimization of communication networks using simulated annealing, *Microelectronics & Reliability*, 33, 1303–1319.
41. Baker, J. (1987). Adaptive selection methods for genetic algorithms, *Proceedings of 2th International Conference on Genetic Algorithms*, 100–111.
42. Ball. M. O. & Provan, J. S. (1983). Calculating bounds on reachability and connectedness in stochastic networks, *Networks*, 13, 253–278.
43. Beltran, H. F. D. & Skorin-Kapov, D, (1994). On minimum cost isolated failure immune networks, *Telecommunication Systems*, 3, 183–200.
44. Bertsekas, D. & Gallager, R. (1992). *Data Networks*, 2nd Edition, Prentice-Hall, New Jersey.
45. Charnes, A. & Cooper, W. W. (1959). Chance-constrained programming, M*anagement Science*, 6(1), 73–79.
46. Coan, B. A., Leland, W. E., Vecchi, M. P., Wwinrib, A. & Wu, L. T. (1991). Using distributed topology update and preplanned configurations to achieve trunk network survivability, *IEEE Transactions on Reliability*, 40(4), 404–416.
47. Dutta, A. & Mitra, S. (1993). Integrating heuristic knowledge and optimization models for communication network design, *IEEE Transactions on Knowledge Data Engineering*, 5(6), 999–1017.
48. Gavish, B. (1992). Topological design of computer communication network - The overall design problems, *European Journal of Operational Research*, 58, 149–172.
49. Gen, M. & Cheng, R. (2000). *Genetic Algorithms & Engineering Optimization*, John Wiley & Sons.
50. Gen, M., Ida. K. & Kim, J. R. (1998). A spanning tree-based genetic algorithm for bicriteria topological network design, *Proceedings of the 1998 Congress on Evolutionary Computation*, 15–20.
51. Glover, F., Lee, M. & Ryan, J. (1991). Least-cost network topology design for a new service: an application of a tabu search, *Annals of Operations Research*, 33, 351–362.
52. Jan, R. H. (1993). Design of reliable networks, *Computers & Operations Research*, 20, 25–34.
53. Jan, R. H., Hwang, F. J. & Chen, S. T. (1994). Topological optimization of a communication network subject to reliability constraint, *IEEE Transactions on Reliability*, 42, 63–70.
54. Kershenbaum, A., Kermani, P. & Grover, G. A. (1991). MENTOR: An algorithm for mesh network topological optimization & routing, *IEEE Transactions on Communications*, 39(4), 503–513.
55. Koh, S. J. & Lee, C. Y. (1995). A tabu search for the survivable fiber optic communication network design, *Computers & Industrial Engineering*, 28, 689–700.
56. Liu, B. (1997). Dependent-chance programming: a class of stochastic optimization, *Computers and Mathematics with Applications*, 34(12), 89–104.
57. Newport, K. T. & Varshney, P. K. (1991). Design of survival communications networks under performance constraints, *IEEE Transactions on Reliability*, 40(4), 443–440.

58. Pierre, S., Hyppolite, M.-A., Bourjolly, J.-M. & Dioume, O. (1995). Topological design of computer communication networks using simulated annealing, *engineering Applications of Artificial Intelligence*, 8, 61–69.
59. Pierre, S. & Elgibaoui, A. (1997). A tabu search approach for designing computer-network topologies with unreliable components, *IEEE Transactions on Reliability*, 46(2), 350–359.
60. Rose, C. (1992). Low mean internodal distance network topologies and simulated annealing, *IEEE Transactions on Communications*, 40(8), 1319–1326.
61. Schwartz, M. & Stern, T. E. (1980). Routing techniques used in computer communication networks, *IEEE Transactions on Communications*, COM-28(4), 539–552.
62. Sykes, E. A. & White, C. C. (1985). Specifications of a knowledge system for packet-switched data network topological design, *Proceedings of Expert Systems Government Symposium*, Mc Lean, VA, 102–110.
63. Tomy, M. J. & Hoang, D. B. (1987). Joint optimization of capacity and flow assignment in a packet-switched communications network, *IEEE Transactions on Communications*, COM-35(2), 202–209.
64. Venetsanopoulos, N. & Singh, I. (1986). Topological optimization of communication networks subject to reliability constraints, *Problem of Control & Information Theory*, 15, 63–78.
65. Yeh, M.-S., Lin, J.-S. & Yeh, W.-C. (1994). A new Monte Carlo method for estimating network reliability, *Proceedings of the 16th International Conference on Computers & Industrial Engineering*, 723–726.
66. Yun, Y., Gen, M. & Seo, S. (2003). Various hybrid methods based on genetic algorithm with fuzzy logic controller, *Journal of Intelligent Manufacturing*, 14, 401–419,
67. Lin, L. & Gen, M. (2007). *An evolutionary algorithm for improvement of QoS of next generation network in dynamic environment*, Artificial Neural Networks In Engineering, St. Louis, USA.
68. Blake, S. et. al., (1998). An Architecture for Differentiated Services, *RFC 2475*. [Online].Available :ftp://ftp.isi.edu/in-notes/rfc2475.txt
69. Braden, R., Clark, D. & Shenker, S. (1994). Integrated Services in the Internet Architecture: an Overview, *RFC 1633*. http://www.ietf.org/rfc/rfc1633.txt
70. Callon, R., et. al., (1999). *A Framework for Multiprotocol Label Switching*, draft-ietf-mpls-framework-05.txt.
71. ITU-T, [Online]. Available: http://www.itu.int/ITU-T/studygroups/com13/ngn2004/ working_definition.html
72. Sun, W. (1999). *QoS/Policy/Constraint Based Routing, Technical Report*, Ohio State University, [Online].Available: http://www.cse.wustl.edu/ jain/cis788-99/qos_routing/index.html.
73. Lin, L & Gen, M. (2006). A self-control genetic algorithm for reliable communication network design, *Proceeding of IEEE Congress on Evolutionary Computation*, Vancouver, Canada, 2006.
74. Aggarwal, K. K. Chopra, Y. C. & Bajwa, J. S. (1982). Reliability evaluation by network decomposition, *IEEE Transaction on Reliability*, R-31, 355–358.
75. Srivaree-ratana, C., Konak, A. & Smith, A. E. (2002). Estimation of All-terminal Netork Reliability Using an Artificial Neural Network, *Computers & Operations Research*, 29, 849–868.
76. Altiparmak, F., & Bulgak, A. A. (2002). Optimization of Buffer Sizes in Aaaembly Systems Using Intelligent Techniques, *proceedings of the 2002 Winter Simulation Conferene*, 1157–1162.
77. Kumar, A., Pathak, R. M. & Gupta, Y. P. (1995). Genetic algorithm based reliability optimization for computer network expansion, *IEEE Transaction on Reliability*, 44, 63–72.
78. Davis, L., Orvosh, D., Cox, A. & Qui, Y. (1993). A genetic algorithm for survivable network design, *Proceeding of 5th International Conference on Genetic Algorithms*, 408–415.
79. Abuali, F. N., Schoenefeld, D. A. & Wainwright, R. L. (1994). Terminal assignment in a communications network using genetic algorithms, *Proceeding of ACM Computer Science Conference*, 74–81.

80. Abuali, F. N., Schoenefeld, D. A. & Wainwright, R. L. (1994). Designing telecommunications networks using genetic algorithms and probabilistic minimum spanning trees, *Proceeding of 1994 ACM Symposium Applied Computing*, 242–246.
81. Walters, G. A. & Smith, D. K. (1995). Evolutionary design algorithm for optimal layout of tree networks, *Engineering Optimization*, 24, 261–281.
82. Deeter, D. L. & Smith, A. E. (1997). Heuristic optimization of network design considering all-terminal reliability, *in Proceeding of Reliability & Maintainability Symposium*, 194–199.
83. Colbourn, C. J. (1987). *The Communication of Network Reliability*, Oxford University Press.
84. Gen, M. & Cheng, R. W. (1997). *Genetic Algorithms and Engineering Design*, New York: John Wiley & Sons.
85. Altiparmak, F., Gen, M., Dengiz, B. & Smith, A. E. (2004). A network-based genetic algorithm for design of communication networks, *Journal of Society of Plant Engineers Japan*, 15, 4, 184-190.
86. Raidl, G. (2000). An efficient evolutionary algorithm for the degree-constrained minimum spanning tree problem, *Proceeding of Congress on Evolutionary Computation*, 1, 104-111.
87. Chou, H., Premkumar, G. & Chu, C. (2001). Genetic algorithms for communications network design - an empirical study of the factors that influence performance, *IEEE Transaction on Evolutionary Computation*, 5(3), 236-249.
88. Julstrom, B. (2003). A permutation-coded evolutionary algorithm for the bounded-diameter minimum spanning tree problem, *Proceeding of Genetic & Evolutionary Computation Conference*, 2-7.
89. Yun, Y. S. & Gen, M. (2003). Performance Analysis of Adaptive Genetic Algorithms with Fuzzy Logic & Heuristics, *Fuzzy Optimization and Decision Making*, 2(2), 161-175.
90. Kim, J. R. & Lee, J.U. (2007). Hierarchical spanning tree network design with nash genetic algorithm, *Computers & Industrial Engineering*, Submitted.
91. Sim, K. B., Ki, J. Y., & Lee, D. W. (2002). Optimization of Multi-objective Function based on the Game Theory and Co-evolutionary Algorithm, *Journal of Fuzzy Logic and Intelligent Systems*, 12(6), 491-496 (in Korean).
92. Sefrioui, M., & Periaux, J. (2000). Nash Genetic Algorithms: Examples and Applications, *Proceedings on IEEE Congress on Evolutionary Computation*, 509-516.

Chapter 5
Advanced Planning and Scheduling Models

Advanced planning and scheduling (APS) refers to a *manufacturing management process* by which raw materials and production capacity are optimally allocated to meet demand. APS is especially well-suited to environments where simpler planning methods cannot adequately address complex trade-offs between competing priorities. However, most scheduling problems of APS in the real world face inevitable constraints such as due date, capability, transportation cost, set up cost and available resources. Generally speaking, we should obtain an effective "flexibility" not only as a response to the real complex environment but also to satisfy all the combinatorial constraints. Thus, how to formulate the complex problems of APS and find satisfactory solutions play an important role in manufacturing systems.

5.1 Introduction

The global concepts of manufacturing systems throughout modern history were in close connection with the then principles and findings of the science, philosophy, and arts. It can be noticed that the manufacturing concepts reflected the principles, criteria, and values generally accepted by the society as the most important. For example, in the eighteenth, nineteenth, and the first half of the twentieth century, the scientific facts mainly exposed the unchangeability, determinism, rationality, exactness, and causality. That period can be characterized as the era of predominance of production. In the second half of the twentieth century, the advanced information technology assured the formal conditions for the expansion of various organizational forms. The second half of the twentieth century can thus be characterized as the era when organizational aspects were prevailing.

Future manufacturing concepts will have to be adapted to the needs of the modern society and, particularly, to the ecosystem more than ever. Unfortunately, the term adaptability is still bereft today of an important component; the adaptability still has a particularly mechanistic-technological connotation, whereas the component considering the adaptability to the ecosystem is almost completely neglected.

For this reason, in the existing manufacturing systems for optimization of production of goods, particularly the technological parameters are considered which are measurable or which we want to measure, whereas the global interaction of the goods with the ecosystem is much less considered. Therefore, today it is possible to speak particularly of the local mechanistic-technological optimum of the goods, whereas the global optimum, affected by several influencing parameters and mutual relations, is not reached in most cases. In present manufacturing systems, particularly the deterministic approaches are used for synchronization of material, energy, and information flows. It means that the methods based on the exact mathematical findings and the rules of logic are used for modeling, optimization, and functioning of systems. However, production is a very dynamic process with many unexpected events and continuously emerging new requirements. Therefore, exact description of a system is often not possible by conventional methods. Mathematical models are often derived by making simplifying assumptions. As a consequence, the model of the system is not in accordance with functioning of the real system. Such models of the systems are frequently not suitable and flexible enough to respond efficiently to fast changes in the environment.

The existing manufacturing concepts cannot successfully respond to the abovementioned problems to which modern production and society as a whole will be exposed in the future more than ever. However, in many different areas of science and technology it has been possible recently to notice the shift towards the conceiving of integrated systems capable of learning and efficiently responding to increasing complexity, unpredictability, and changeability of the environment. During the learning process, the system behavior gradually improves. Machine learning as area of artificial intelligence is increasingly gaining importance. Several successful integrated systems have been conceived by the methods of machine learning. Recently, some interesting manufacturing concepts and approaches to problem solving based on learning, self-organization, and on the bottom-up organizational principle were also proposed. The intelligence, order and consequent, efficiency of those systems develop step by step and emerge due to interactions of the basic components (entities) of the systems with the environment. However, more research will be required for the concepts to take roots in real environments to the full extent.

In an *integrated manufacturing system* (IMS), the functional areas involved in providing the finished products to customers are integrated in a single system. These areas vary from manufacture product to distribution and from product design to facility management as shown in Fig. 5.1

To find the optimal solutions in those fields gives rise to complex combinatorial optimization problems; unfortunately, most of them fall into the class of NP-hard problems. Hence to find a satisfactory solution in an acceptable time span is crucial for the performance of IMS. Genetic algorithms have turned out to be potent methods to solve such kinds of optimization problems. In an IMS the basic problems indicated in Fig. 5.1 are likely to use evolutionary techniques.

1. *Design*: Design problems generally have to be decided only once in IMS, and they form some kinds of fixed inputs for other subsequent problems, such as manufacturing and planning problems. Typically, design problems in an inte-

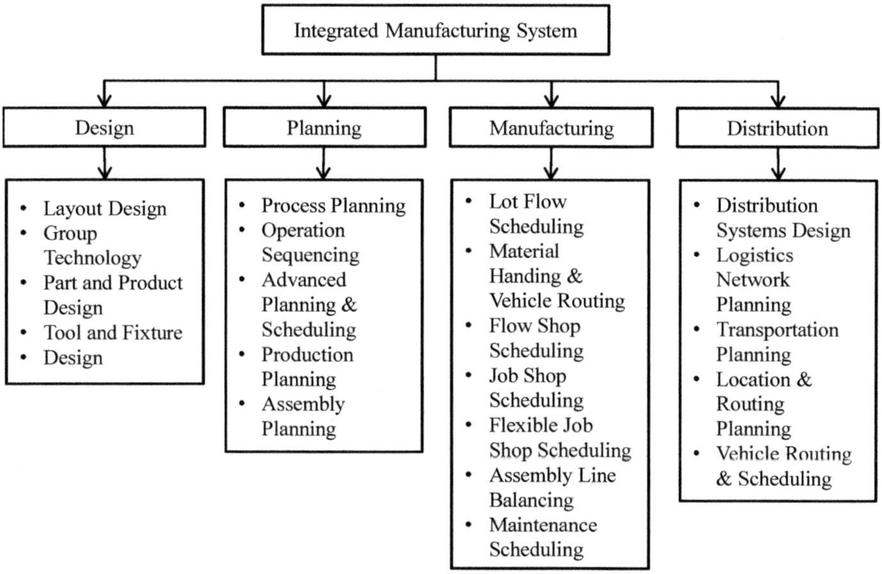

Fig. 5.1 Basic problems in an integrated manufacturing system

grated manufacturing system include layout design, assembly planning, group technology and so on.
2. *Planning*: Compared to scheduling problems, planning problems have a longer horizon. Hence the demand information needed to find the optimal solution for a planning problem comes from forecasting rather than arrived orders. Process planning, operation sequencing, production planning and assembly line balancing fall into the class of planning problems.
3. *Manufacturing*: In manufacturing, there are two kinds of essential issues: scheduling and routing. Machining, assembly, material handling and other manufacturing functions are performed to the best efficiency. Such kinds of problems are generally triggered by a new order.
4. *Distribution*: The efficient distribution of products is very important in IMS, as transportation costs become a nonnegligible part of the purchase price of products in competitive markets. This efficiency is achieved through sophisticated logistic network design and efficient traffic routing.

Advanced planning and scheduling (APS) includes a range of capabilities from finite-capacity scheduling at the shop floor level through to constraint-based planning in IMS. APS is a new revolutionary step in enterprise and inter-enterprise planning. It is revolutionary, due to the technology and because APS utilises planning and scheduling techniques that consider a wide range of constraints to produce an optimized plan [1]:

• Material availability

- Machine and labor capacity
- Customer service level requirements (due dates)
- Inventory safety stock levels
- Cost
- Distribution requirements
- Sequencing for set-up efficiency

As discussed in advanced manufacturing researches (Bermudez [2]) and (Eck [1]), the scope of APS is not limited to factory planning and scheduling, but has grown rapidly to include the full spectrum of enterprise and inter-enterprise planning and scheduling functions.

Strategic and Long-term Planning

This solution addresses issues like

- Which products should be made?
- What markets should the company pursue?
- How should conflicting goals be resolved?
- How should assets be deployed for the best ROI?

Supply Chain Network Design

This solution optimises the use of resources across the current network of suppliers, customers, manufacturing locations and DCs. What-if analyses can be performed to test the impact of decisions to open new or move existing facilities on profit and customer-service level. It can also be a helpful tool to determine where a new facility should be located to fulfil customer demand in the most optimal way. These supply chain network design tools are mostly applied to find the balance between holding more stock at a specific location or making more transportation costs.

Demand Planning and Forecasting

Both statistical and time-series mathematics are used in this solution to calculate a forecast based on sales history. A demand forecast is unconstrained because it considers only what customers want and not what can be produced. Based on the information from the forecast, it is possible to create more demand through promotions in periods where the demand is less than maximum production.

Sales and Operations Planning

This is the process which converts the demand forecast into a feasible operating plan which can be used by both sales and production. This process can include the use of a manufacturing planning and/or a supply chain network optimising solution to determine if the forecast demand can be met.

Inventory Planning

This solution determines the optimal levels and locations of finished goods inventory to achieve the desired customer service levels. In essence, this means that it calculates the optimal level of safety stock at each location.

Supply Chain Planning (SCP)

SCP compares the forecast with actual demand to develop a multi-plant constrained master schedule, based on aggregate-level resources and critical materials. The schedule spans multiple manufacturing and distribution sites to synchronise and optimise the use of manufacturing, distribution and transportation resources.

Manufacturing Planning

Develops a constrained master schedule for a single plant based on material availability, plant capacity and other business objectives. The manufacturing planning cycle is often only executed for critical materials, but that depends on the complexity of the bill of material. Also the desired replanning time is a factor that one must take into account when deciding which level of detail is used. For example, with a simple bill of material a complete MRP I/II explosion can be executed in a few minutes.

Distribution Planning

Based on actual transportation costs and material allocation requirements, a feasible plan on the distribution of finished goods inventory to different stocking point or customers is generated to meet forecast and actual demand. With this solution it is possible to support Vendor Managed Inventory.

Transportation Planning

A solution which uses current freight rates to minimise shipping costs. Also optimisation of outbound and inbound material flow is used to minimise transportation costs or to maximise the utilisation of the truck fleet. Another possibility is to consolidate shipments into full truckloads and to optimise transportation routes by sequencing the delivery/pickup locations.

Production Scheduling

Based on detailed product attributes, work centre capabilities and material flow, a schedule is determined that optimises the sequence and routings of production orders on the shop floor.

Shipment Scheduling

This solution determines a feasible shipment schedule to meet customer due dates. It determines the optimal method and time to ship the order taking customer due dates into account.

In this chapter, we introduce GA applications on job-shop scheduling (JSP) models, flexible JSP models, integrated operation sequence and resource selection (iOS/RS) models, integrated scheduling models for manufacturing and logistics, and manufacturing and logistics models with pickup and delivery. Depending on the common sense "from easy to difficult" and "from simple to complex", the core APS models are summarized in Fig. 5.2.

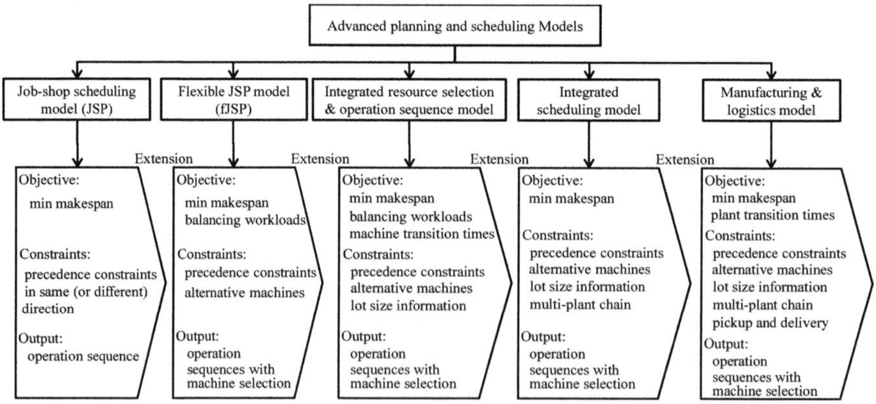

Fig. 5.2 The core models of advanced planning and scheduling models introduced in this chapter

5.2 Job-shop Scheduling Model

Following Blazewicz *et al.* [27], scheduling problems can be broadly defined as "the problems of the allocation of resources over time to perform a set of tasks". The scheduling literature is full of very diverse scheduling problems [28, 29]. The job-shop scheduling problem (JSP) concerns determination of the operation sequences on the machines so that the makespan is minimized. It consists of several assumptions as follows [30]:

A1. Each machine processes only one job at a time.
A2. Each job is processed on one machine at a time.
A3. A job does not visit the same machine twice.
A4. The processing time of each operation has been determined.
A5. There are no precedence constraints among operations of different jobs.
A6. Operations cannot be interrupted.
A7. Neither release times nor due dates are specified.

This problem has already been confirmed as one of the NP-hard problems. There are n jobs and m machines to be scheduled; furthermore each job is composed of a set of operations and the operation order on machines is prespecified, and each operation is characterized by the required machine and the fixed processing time [30, 51, 5].

Opinion in this field is mixed about who first proposed JSP in its current form. Roy and Sussmann were the first to propose the disjunctive graph representation [31], and Balas [32] was the first to apply an enumerative approach based on the disjunctive graph. Since then many researchers have tried various strategies for solving this problem[32]. Dispatching rules was adapted in Grabot's work [33], and Yang also used neural networks combined with a heuristics approach to solve the problem [34]; moreover some researchers especially Gen and Cheng using hybrid GA also obtained good solutions [5, 35].

Recently, many researchers tried to adapt different meta-heuristic approaches to obtain a near-optimal solution of JSP. Monch and Driessel considered distributed versions of a modified shifting bottleneck heuristic to solve complex job shops [**?**]. Nowicki and Smutnicki provide a new approximate Tabu Search (TS) algorithm that is based on the big valley phenomenon, and uses some elements of so-called path relinking technique as well as new theoretical properties of neighborhoods[36]. Tavakkoli-Moghaddam *et al.* used neural network (NN) approach to generate initial feasible solutions and adapted a simulated annealing (SA) algorithm to improve the quality and performance of the solution[37].

To improve the efficiency for finding better solutions in searching space, some special technical local searches have been adapted in JSP. Ida and Osawa [38] reformed the traditional left shift to short the idle time, and formulated an algorithm called Eshift [38]. Gonçalves *et al.* [39] proposed another technique based on the critical path to confirm the search space, and swapped the operations in critical block, and this approach can also improve the efficiency of algorithm for finding active schedule [39].

5.2.1 Mathematical Formulation of JSP

Notation

Indices

i, l: index of jobs, $i, l = 1,2,\cdots,n$
j, h: index of machines, $j, h = 1,2,\cdots,m$
k: index of operations, $k = 1,2,\cdots,m$

Parameters

n: total number of jobs
m: total number of machines
t_M: makespan
M_j: the j-th machine
J_i: the i-th job, $i = 1,2,\cdots,n$
o_{ikj}: the k-th operation of job J_i operated on machine M_j
p_{ikj}: processing time of operation o_{ikj}

Decision Variables

t_{ikj}: completion time of operation o_{ikj} on machine M_j for each job J_i

The Job-shop Scheduling Problem (JSP) we are treating is to minimize the makespan, so the problem could be described as an n-job m-machine JSP by simple equations as follows:

$$\min t_M = \max_{ikj}\{t_{ikj}\} \tag{5.1}$$

$$\text{s. t. } t_{i,k-1,h} + p_{ikj} \leq t_{ikj}, \forall i,k,h,j \tag{5.2}$$

$$t_{ikj} \geq 0, \forall i,k,j \tag{5.3}$$

The objective function at Eq. 5.1 is to minimize the makespan. The constraint at Eq. 5.2 is the operation precedence constraint, the $k-1$-th operation of job i should be processed before the k-th operation of the same job. The time chart for model is illustrated in Fig. 5.3.

Complexity of JSP

Most variants of the JSP except a few formulations with the number of machines or jobs limited to 1 or 2 are known to be NP-hard [28, 29]. In particular, JSPs with the

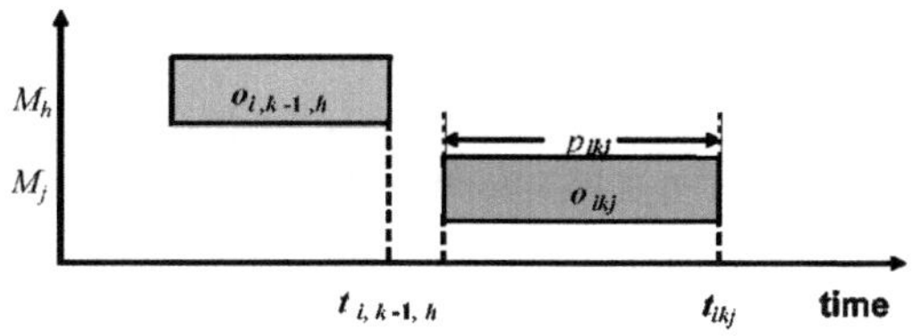

Fig. 5.3 Time chart for the constrain

number of machines m fixed ($m \geq 2$) using the C_{max} and F_{Σ} performance criteria are NP-hard in the strong sense [40, 41]. Since C_{max} problems can trivially be reduced to T_{max} and L_{max} problems, and F_{Σ} problems can trivially be reduced to T_{Σ} and L_{Σ} problems, these problems are NP-hard in the strong sense as well.

According to [29] theorem 11.6 and [40] theorem 1, strong NP-hardness of a problem implies that it is impossible to create a search heuristic which guarantees to find a solution for which the relative error is bounded by

$$\frac{\text{performance measure of found solution}}{\text{performance measure of optimum solution}} \leq 1+\varepsilon \tag{5.4}$$

and which runs in polynomial time both in the problem size and $1/\varepsilon$. This result is very important, since it indicates that efficient approximation algorithms with guaranteed performances should not be expected for these problems unless $P = NP$. For this reason, most research focused on finding (near) optimal schedules has been turned towards implicit enumeration algorithms (branch and bound techniques), local improvement methods (shifting bottleneck) and heuristic search methods such as genetic algorithms, tabu search and simulated annealing.

Experience has shown job shop problems not just to be NP-hard, but to be very difficult to solve heuristically even for problems in this complexity class. The 10x10 instance ft10 of [42] was first published in 1963, and remained open until the optimal solution was published by Adams *et al.* [43]. Recent results from [44], indicate that the open shop problem (which is not as well researched as the job shop problem) is even more difficult than the job shop problem. In [44] this extraordinary problem hardness is attributed in part to the larger searchspace of the open shop for problems of the same size, and in part to smaller gaps between the performance of optimal schedules and lower bounds.

5.2.2 Conventional Heuristics for JSP

Job-shop scheduling is one of the hardest combinatorial optimization problems. Since job-shop scheduling is a very important everyday practical problem, it is therefore natural to look for approximation methods that produce an acceptable schedule in useful time. The heuristic procedures for a job-shop problem can be roughly classified into two classes:

- One-pass heuristic

- Multi-pass heuristic

One-pass heuristic simply builds up a single complete solution by fixing one operation in schedule at a time based on *priority dispatching rules*. There are many rules for choosing an operation from a specified subset to be scheduled next. This heuristic is fast, and usually finds solutions that are not too bad. In addition, one-pass heuristic may be used repeatedly to build more sophisticated multi-pass heuristic in order to obtain better schedules at some extra computational cost.

5.2.2.1 Priority Dispatching Heuristics

Priority rules are probably the most frequently applied heuristics for solving scheduling problems in practice because of their ease of implementation and their low time complexity. The algorithms of Giffler and Thompson can be considered as the common basis of all priority rule-based heuristics [26]. Giffler and Thompson have proposed two algorithms to generate schedule: *active schedule* and *nondelay schedule* generation procedures [18]. A nondelay schedule has the property that no machine remains idle if a job is available for processing. An active schedule has the property that no operation can be started earlier without delaying another job. Active schedules form a much larger set and include nondelay schedules as a subset. The generation procedure of Giffler and Thompson is a tree-structured approach. The nodes in the tree correspond to partial schedules, the arcs represent the possible choices, and the leaves of the tree are the set of enumerated schedules. For a given partial schedule, the algorithm essentially identifies all processing conflicts, *i.e.*, operations competing for the same machine, and an enumeration procedure is used to resolve these conflicts in all possible ways at each stage. In contrast, heuristics resolve these conflicts with priority dispatching rules, *i.e.*, specify a priority rule for selecting one operation among the conflicting operations.

Generation procedures operate with a set of schedulable operations at each stage. Schedulable operations are operations unscheduled yet with immediately scheduled predecessors and this set can be simply determined from precedence structure. The number of stages for a one-pass procedure is equal to the number of operations $m \times n$. At each stage, one operation is selected to add into *partial schedule*. Conflicts among operations are solved by priority dispatching rules. Following the notations of Baker [7], we have

PS_t = a partial schedule containing t scheduled operations,
S_t = the set of schedulable operations at stage t, corresponding to a given PS_t,
σ_i = the earliest time at which operation $i \in S_t$ could be started,
ϕ_i = the earliest time at which operation $i \in S_t$ could be completed.

For a given active partial schedule, the potential start time σ_i is determined by the completion time of the direct predecessor of operation i and the latest completion time on the machine required by operation i. The larger of these two quantities is

σ_i. The potential finishing time ϕ_i is simply $\sigma_i + t_i$, where t_i is the processing time of operation i. The procedure to generate active schedule works as follows:

---

procedure priority dispatching heuristic (active schedule generation)
input: JSP data
output: a complete schedule
Step 1: Let $t = 0$ and begin with PS_t as the null partial schedule.
Initially S_t includes all operations with no predecessors.
Step 2: Determine $\phi_t^* = \min_{i \in S_t}\{\phi_i\}$ and the machine m^* on which ϕ_t^* could be realized.
Step 3: For each operation $i \in S_t$ that requires machine m^* and for which $\sigma_i < \phi_t^*$, calculate a priority index according to a specific priority rule. Find the operations with the smallest index and add this operation to PS_t as early as possible, thus creating new partial schedule PS_{t+1}.
Step 4: For PS_{t+1}, update the data set as follows:
1. remove operations i from S_t
2. form S_{t+1} by adding the direct successor of operation j to S_t
3. increment t by one

Step 5: Return to step 2 until a complete schedule is generated.

If we replace step 2 and step 3 of the algorithm with those given in following algorithm, the procedure can generate nondelay schedule, which is shown below:

---

procedure priority dispatching heuristic (nondelay schedule generation)
input: JSP data
output: a complete schedule
Step 1: Let $t = 0$ and begin with PS_t as the null partial schedule.
Initially S_t includes all operations with no predecessors.
Step 2: Determine $\sigma_t^* = \min_{i \in S_t}\{\sigma_i\}$ and the machine m^* on which $\sigma_t *$ could be realized.
Step 3: For each operation $i \in S_t$ that requires machine m^* and for which $\sigma_i = \sigma_t^*$, calculate a priority index according to a specific priority rule. Find the operations with the smallest index and add this operation to PS_t as early as possible, thus creating new partial schedule PS_{t+1}.
Step 4: For PS_{t+1}, update the data set as follows:
1. remove operations i from S_t
2. form S_{t+1} by adding the direct successor of operation j to S_t
3. increment t by one

Step 5: Return to step 2 until a complete schedule is generated.

The remaining problem is to identify an effective priority rule. For an extensive summary and discussion see Panwalkar and Iskander [25], Haupt [19], and Blackstone *et al.* [10]. Table 5.1 consists of some of the priority rules commonly used in practice.

Table 5.1 A list of job shop dispatch rules

Rule	Description
SPT (Shortest Processing Time)	Select the operation with the shortest processing time
LPT (Longest Processing Time)	Select an operation with longest processing time
LRT (Longest Remaining Processing Time)	Select the operation belonging to the job with the longest remaining processing time
SRT (Shortest Remaining Processing Time)	Select the operation belonging to the job with the shortest remaining processing time
LRM (LRT excluding the operation under consideration)	Select the operation belonging to the job with the longest remaining processing time excluding the operation under consideration

We shall discuss and compare different dispatch rules using a simple example given in Table 5.2, where p_{ikj} is the processing time of the k-th operation of job J_i operated on machine M_j.

Table 5.2 An example of a three-job three-machine problem

	p_{ikj}			M_j		
Operations	1	2	3	1	2	3
J_1	16	21	12	M_1	M_3	M_2
J_2	15	20	9	M_1	M_2	M_3
J_3	8	18	22	M_2	M_3	M_1

Shortest Processing Time: SPT

Using the trace table (in Table 5.3) for the operation selection process, the schedule can be constructed as follows, and Fig. 5.4 illustrates a Gantt chart showing the schedule for SPT dispatch rule.

Schedule $S = \{(o_{ikj}(t_{ikj} - t^F_{ikj})\}$:

$$\begin{aligned} S = \{ & o_{312}(t_{312} - t^F_{312}), o_{211}(t_{211} - t^F_{211}), o_{111}(t_{111} - t^F_{111}), o_{323}(t_{323} - t^F_{323}), \\ & o_{222}(t_{222} - t^F_{222}), o_{233}(t_{233} - t^F_{233}), o_{123}(t_{123} - t^F_{123}), o_{132}(t_{132} - t^F_{132}), \\ & o_{331}(t_{331} - t^F_{331})\} \\ = \{ & o_{312}(0-8), o_{211}(0-15), o_{111}(15-31), o_{323}(8-26), \\ & o_{222}(15-35), o_{233}(35-44), o_{123}(44-65), o_{132}(65-77), \\ & o_{331}(31-53)\} \end{aligned}$$

Table 5.3 An example of SPT

l	O_l	p_{ikj}	o_{ikj}^*	$S(l)$
1	$\{o_{111}, o_{211}, o_{312}\}$	p_{111}=16, p_{211}=15, p_{312}=8	o_{312}	$\{o_{312}\}$
2	$\{o_{111}, o_{211}, o_{323}\}$	p_{111}=16, p_{211}=15, p_{323}=18	o_{211}	$\{o_{312}, o_{211}\}$
3	$\{o_{111}, o_{222}, o_{323}\}$	p_{111}=16, p_{222}=20, p_{323}=18	o_{111}	$\{o_{312}, o_{211}, o_{111}\}$
4	$\{o_{123}, o_{222}, o_{323}\}$	p_{123}=21, p_{222}=20, p_{323}=18	o_{323}	$\{o_{312}, o_{211}, o_{111}, o_{323}\}$
5	$\{o_{123}, o_{222}, o_{331}\}$	p_{123}=21, p_{222}=20, p_{331}=22	o_{222}	$\{o_{312}, o_{211}, o_{111}\ o_{323}, o_{222}\}$
6	$\{o_{123}, o_{233}, o_{331}\}$	p_{123}=21, p_{233}=9, p_{331}=22	o_{233}	$\{o_{312}, o_{211}, o_{111}, o_{323}, o_{222}, o_{233}\}$
7	$\{o_{123}, o_{331}\}$	p_{123}=21, p_{331}=22	o_{123}	$\{o_{312}, o_{211}, o_{111}, o_{323}, o_{222}, o_{233}, o_{123}\}$
8	$\{o_{132}, o_{331}\}$	p_{132}=12, p_{331}=22	o_{132}	$\{o_{312}, o_{211}, o_{111}, o_{323}, o_{222}, o_{233}, o_{123}, o_{132}\}$
9	$\{o_{331}\}$	p_{331}=22	o_{331}	$\{o_{312}, o_{211}, o_{111}, o_{323}, o_{222}, o_{233}, o_{123}, o_{132}, o_{331}\}$

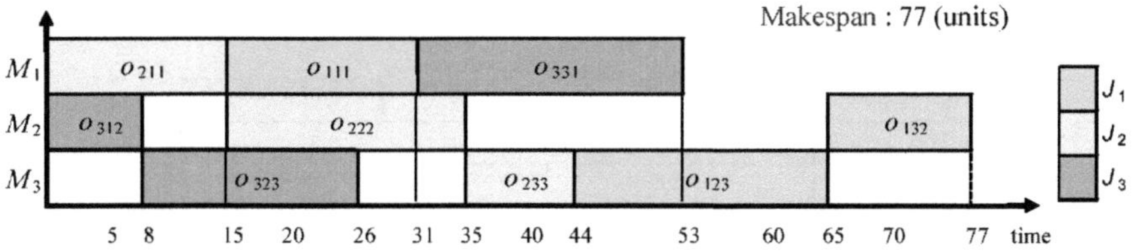

Fig. 5.4 Gantt chart for SPT dispatch rule

Longest Processing Time: LPT

Using the trace table (in Table 5.4) for the operation selection process, the schedule can be constructed as follows, and Fig. 5.5 illustrates a Gantt chart showing the schedule for LPT dispatch rule.

Schedule $S = \{(o_{ikj}(t_{ikj} - t^F_{ikj})\}$:

$$
\begin{aligned}
S = \{\ & o_{111}(t_{111} - t^F_{111}), o_{123}(t_{123} - t^F_{123}), o_{211}(t_{211} - t^F_{211}), o_{222}(t_{222} - t^F_{222}), \\
& o_{132}(t_{132} - t^F_{132}), o_{233}(t_{233} - t^F_{233}), o_{312}(t_{312} - t^F_{312}), o_{323}(t_{323} - t^F_{323}), \\
& o_{331}(t_{331} - t^F_{331})\} \\
= \{\ & o_{111}(0-16), o_{123}(16-37), o_{211}(16-31), o_{222}(31-51), \\
& o_{132}(51-63) o_{233}(51-60), o_{312}(0-8), o_{323}(60-78), o_{331}(78-100)\}
\end{aligned}
$$

Table 5.4 An example of LPT

l	O_l	p_{ikj}	o_{ikj}^*	$S(l)$
1	$\{o_{111}, o_{211}, o_{312}\}$	$p_{111}=16, p_{211}=15, p_{312}=8$	o_{111}	$\{o_{111}\}$
2	$\{o_{123}, o_{211}, o_{312}\}$	$p_{123}=21, p_{211}=15, p_{312}=8$	o_{123}	$\{o_{111}, o_{123}\}$
3	$\{o_{132}, o_{211}, o_{312}\}$	$p_{132}=12, p_{211}=15, p_{312}=8$	o_{211}	$\{o_{111}, o_{123}, o_{211}\}$
4	$\{o_{132}, o_{222}, o_{312}\}$	$p_{132}=12, p_{222}=20, p_{312}=8$	o_{222}	$\{o_{111}, o_{123}, o_{211}, o_{222}\}$
5	$\{o_{132}, o_{233}, o_{312}\}$	$p_{132}=12, p_{233}=9, p_{312}=8$	o_{132}	$\{o_{111}, o_{123}, o_{211}, o_{222}, o_{132}\}$
6	$\{o_{233}, o_{312}\}$	$p_{233}=9, p_{312}=8$	o_{233}	$\{o_{111}, o_{123}, o_{211}, o_{222}, o_{132}, o_{233}\}$
7	$\{o_{312}\}$	$p_{312}=8$	o_{312}	$\{o_{111}, o_{123}, o_{211}, o_{222}, o_{132}, o_{233}, o_{312}\}$
8	$\{o_{323}\}$	$p_{323}=18$	o_{323}	$\{o_{111}, o_{123}, o_{211}, o_{222}, o_{132}, o_{233}, o_{312}, o_{323}\}$
9	$\{o_{331}\}$	$p_{331}=22$	o_{331}	$\{o_{111}, o_{123}, o_{211}, o_{222}, o_{132}, o_{233}, o_{312}, o_{323}, o_{331}\}$

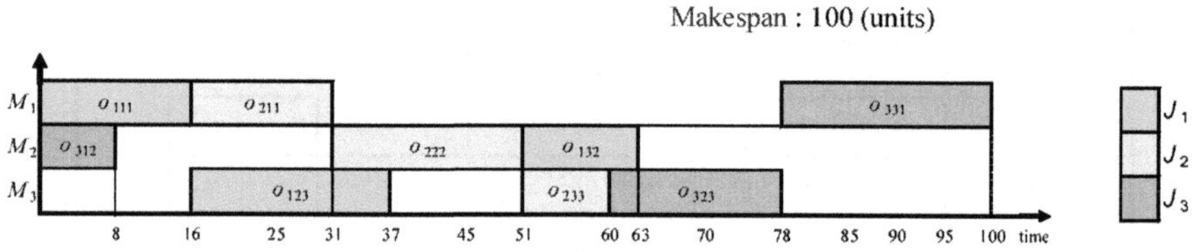

Fig. 5.5 Gantt chart for LPT dispatch rule

Longest Remaining Processing Time: LRT

LRT select the operation belonging to the job with the longest remaining processing time. Remaining processing time r_{ikj} of J_i corresponding to operation o_{ikj} is defined as follows:

$$r_{ikj} = p_{ikj} + p_{i,k+1,j} + \cdots + p_{i,N_i,j}$$

Using the trace table (in Table 5.5) for the operation selection process, the schedule can be constructed as follows, and Fig. 5.6 illustrates a Gantt chart showing the schedule for LRT dispatch rule.

Schedule $S = \{(o_{ikj}(t_{ikj} - t_{ikj}^F)\}$:

$$\begin{aligned}S = \{\ & o_{111}(t_{111} - t_{111}^F), o_{312}(t_{312} - t_{312}^F), o_{211}(t_{211} - t_{211}^F), o_{323}(t_{323} - t_{323}^F),\\ & o_{123}(t_{123} - t_{123}^F), o_{222}(t_{222} - t_{222}^F), o_{331}(t_{331} - t_{331}^F), o_{132}(t_{132} - t_{132}^F),\\ & o_{233}(t_{233} - t_{233}^F)\}\end{aligned}$$

$$= \{\, o_{111}(0-16), o_{312}(0-8), o_{211}(16-31), o_{323}(8-26),$$
$$o_{123}(26-47), o_{222}(31-51), o_{331}(31-53), o_{132}(51-63),$$
$$o_{233}(51-60)\}$$

Table 5.5 An example of LRT

l	O_l	r_{ikj}	o_{ikj}^*	$S(l)$
1	$\{o_{111}, o_{211}, o_{312}\}$	r_{111}=49, r_{211}=44, r_{312}=48	o_{111}	$\{o_{111}\}$
2	$\{o_{123}, o_{211}, o_{312}\}$	r_{123}=33, r_{211}=44, r_{312}=48	o_{312}	$\{o_{111}, o_{312}\}$
3	$\{o_{123}, o_{211}, o_{323}\}$	r_{123}=33, r_{211}=44, r_{323}=40	o_{211}	$\{o_{111}, o_{312}, o_{211}\}$
4	$\{o_{123}, o_{222}, o_{323}\}$	r_{123}=33, r_{222}=29, r_{323}=40	o_{323}	$\{o_{111}, o_{312}, o_{211}, o_{323}\}$
5	$\{o_{123}, o_{222}, o_{331}\}$	r_{123}=33, r_{222}=29, r_{331}=22	o_{123}	$\{o_{111}, o_{312}, o_{211}, o_{323}, o_{123}\}$
6	$\{o_{132}, o_{222}, o_{331}\}$	r_{122}=12, r_{222}=29, r_{331}=22	o_{222}	$\{o_{111}, o_{312}, o_{211}, o_{323}, o_{123}, o_{222}\}$
7	$\{o_{132}, o_{233}, o_{331}\}$	r_{133}=12, r_{233}=9, r_{331}=21	o_{331}	$\{o_{111}, o_{312}, o_{211}, o_{323}, o_{123}, o_{222}, o_{331}\}$
8	$\{o_{132}, o_{233}\}$	r_{132}=12, r_{233}=9	o_{132}	$\{o_{111}, o_{312}, o_{211}, o_{323}, o_{123}, o_{222}, o_{331}, o_{132}\}$
9	$\{o_{233}\}$	r_{233}=9	o_{233}	$\{o_{111}, o_{312}, o_{211}, o_{323}, o_{123}, o_{222}, o_{331}, o_{132}, o_{233}\}$

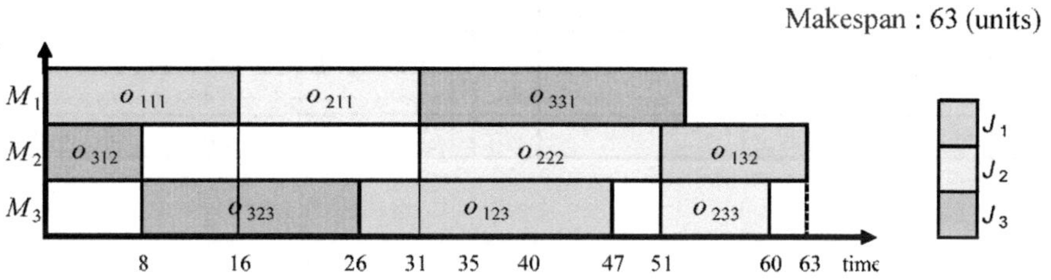

Fig. 5.6 Gantt chart for LRT dispatch rule

Shortest Remaining Processing Time: SRT

SRT select the operation belonging to the job with the shortest remaining processing time. Remaining processing time r_{ikj} of J_i corresponding to operation o_{ikj} is defined as follows:

$$r_{ikj} = p_{ikj} + p_{i,k+1,j} + \cdots + p_{i,N_i,j}$$

Using the trace table (in Table 5.6) for the operation selection process, the schedule can be constructed as follows, and Fig. 5.7 illustrates a Gantt chart showing the schedule for SRT dispatch rule.

Schedule $S = \{(o_{ikj}(t_{ikj} - t^F_{ikj})\}$:

$$\begin{aligned} S = \{ & o_{211}(t_{211} - t^F_{211}), o_{222}(t_{222} - t^F_{222}), o_{233}(t_{233} - t^F_{233}), o_{312}(t_{312} - t^F_{312}), \\ & o_{323}(t_{323} - t^F_{323}), o_{331}(t_{331} - t^F_{331}), o_{111}(t_{111} - t^F_{111}), o_{123}(t_{123} - t^F_{123}), \\ & o_{132}(t_{132} - t^F_{132})\} \\ = \{ & o_{211}(0-15), o_{222}(15-35), o_{233}(35-44), o_{312}(0-8), \\ & o_{323}(8-26), o_{331}(26-48), o_{111}(48-64), o_{123}(64-85), \\ & o_{132}(85-97)\} \end{aligned}$$

Table 5.6 An example of SRT

l	O_l	r_{ikj}	o_{ikj}^*	$S(l)$
1	$\{o_{111}, o_{211}, o_{312}\}$	r_{111}=49, r_{211}=44, r_{312}=48	o_{211}	$\{o_{211}\}$
2	$\{o_{111}, o_{222}, o_{312}\}$	r_{111}=49, r_{222}=29, r_{312}=48	o_{222}	$\{o_{211}, o_{222}\}$
3	$\{o_{111}, o_{233}, o_{312}\}$	r_{111}=49, r_{233}=9, r_{312}=48	o_{233}	$\{o_{211}, o_{222}, o_{233}\}$
4	$\{o_{111}, o_{312}\}$	r_{111}=49, r_{312}=48	o_{312}	$\{o_{211}, o_{222}, o_{233}, o_{312}\}$
5	$\{o_{111}, o_{323}\}$	r_{111}=49, r_{323}=40	o_{323}	$\{o_{211}, o_{222}, o_{233}, o_{312}, o_{323}\}$
6	$\{o_{111}, o_{331}\}$	r_{111}=49, r_{331}=22	o_{331}	$\{o_{211}, o_{222}, o_{233}, o_{312}, o_{323}, o_{331}\}$
7	$\{o_{111}\}$	r_{111}=49	o_{111}	$\{o_{211}, o_{222}, o_{233}, o_{312}, o_{323}, o_{331}, o_{111}\}$
8	$\{o_{123}\}$	r_{123}=33	o_{123}	$\{o_{211}, o_{222}, o_{233}, o_{312}, o_{323}, o_{331}, o_{111}, o_{123}\}$
9	$\{o_{132}\}$	r_{132}=12	o_{132}	$\{o_{211}, o_{222}, o_{233}, o_{312}, o_{323}, o_{331}, o_{111}, o_{123}, o_{132}\}$

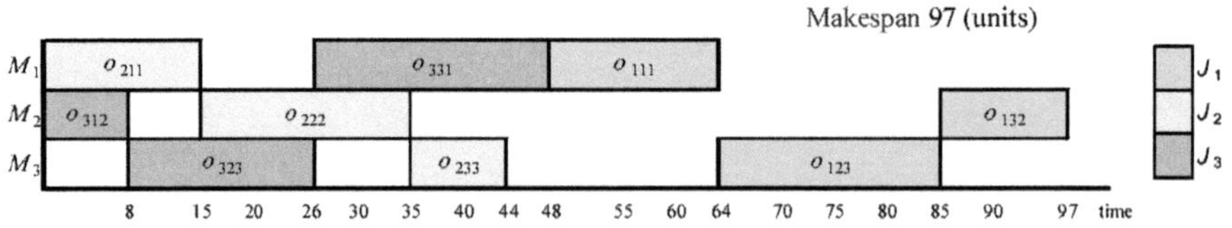

Fig. 5.7 Gantt chart for SRT dispatch rule

LPT Excluding the Operation Under Consideration: LRM

LRM selects the operation belonging to the job with the longest remaining processing time excluding the operation under consideration. Remaining processing time e_{ikj} of J_i corresponding to operation o_{ikj} is defined as follows:

$$e_{ikj} = p_{i,k+1,j} + p_{i,k+2,j} + \cdots + p_{i,N_i,j}$$

Using the trace table (in Table 5.7) for the operation selection process, the schedule can be constructed as follows, and Fig. 5.8 illustrates a Gantt chart showing the schedule for LRM dispatch rule.

Schedule $S = \{(o_{ikj}(t_{ikj} - t^F_{ikj})\}$:

$$\begin{aligned} S = \{ & o_{312}(t_{312} - t^F_{312}), o_{111}(t_{111} - t^F_{111}), o_{211}(t_{211} - t^F_{211}), o_{323}(t_{323} - t^F_{323}), \\ & o_{123}(t_{123} - t^F_{123}), o_{222}(t_{222} - t^F_{222}), o_{331}(t_{331} - t^F_{331}), o_{132}(t_{132} - t^F_{132}), \\ & o_{233}(t_{233} - t^F_{233})\} \\ = \{ & o_{312}(0-8), o_{111}(0-16), o_{211}(16-31), o_{323}(8-26), \\ & o_{123}(26-47), o_{222}(31-51), o_{331}(31-53), o_{132}(51-63), \\ & o_{233}(51-60)\} \end{aligned}$$

Table 5.7 An example of LRM

l	O_l	e_{ikj}	o_{ikj}^*	$S_{(l)}$
1	$\{o_{111}, o_{211}, o_{312}\}$	e_{111}=33, e_{211}=29, e_{312}=40	o_{312}	$\{o_{312}\}$
2	$\{o_{111}, o_{211}, o_{323}\}$	e_{111}=33, e_{211}=29, e_{323}=22	o_{111}	$\{o_{312}, o_{111}\}$
3	$\{o_{123}, o_{211}, o_{323}\}$	e_{123}=12, e_{211}=29, e_{323}=22	o_{211}	$\{o_{312}, o_{111}, o_{211}\}$
4	$\{o_{123}, o_{222}, o_{323}\}$	e_{123}=12, e_{222}=9, e_{323}=22	o_{323}	$\{o_{312}, o_{111}, o_{211}, o_{323}\}$
5	$\{o_{123}, o_{222}, o_{331}\}$	e_{123}=12, e_{222}=9, e_{331}=0	o_{123}	$\{o_{312}, o_{111}, o_{211}, o_{323}, o_{123}\}$
6	$\{o_{132}, o_{222}, o_{331}\}$	e_{132}=0, e_{222}=9, e_{331}=0	o_{222}	$\{o_{312}, o_{111}, o_{211}, o_{323}, o_{123}, o_{222}\}$
7	$\{o_{132}, o_{233}, o_{331}\}$	e_{132}=0, e_{233}=0, e_{331}=0	o_{331}	$\{o_{312}, o_{111}, o_{211}, o_{323}, o_{123}, o_{222}, o_{331}\}$
8	$\{o_{132}, o_{233}\}$	e_{132}=0, e_{233}=0	o_{132}	$\{o_{312}, o_{111}, o_{211}, o_{323}, o_{123}, o_{222}, o_{331}, o_{132}\}$
9	$\{o_{233}\}$	e_{233}=0	o_{233}	$\{o_{312}, o_{111}, o_{211}, o_{323}, o_{123}, o_{222}, o_{331}, o_{132}, o_{233}\}$

5.2.2.2 Randomized Dispatching Heuristic

While one-pass heuristics limit themselves to constructing a single solution, multi-pass heuristics (also called search heuristics) try to get much better solutions by generating many of them, usually at the expense of a much higher computation

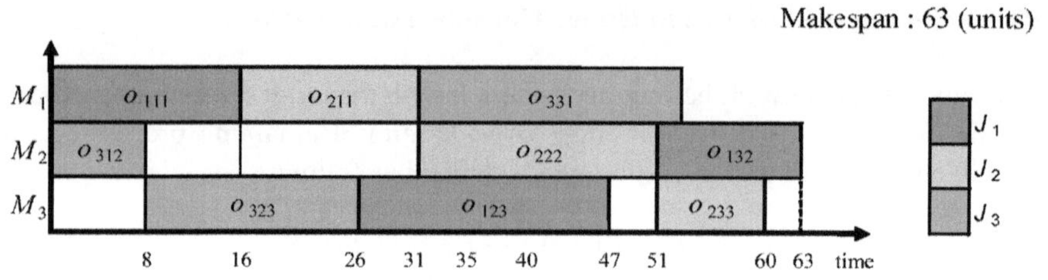

Fig. 5.8 Gantt chart for LRM dispatch rule

time. Techniques like branch-and-bound method and dynamic programming can guarantee an optimal solution, but are not practical for large-scale problems.

Randomized heuristics are an early attempt to provide more accurate solutions [7]. The idea of randomized dispatch is to start with a family of dispatching rules. At each selection of an operation to run, choose the dispatching rule randomly, repeated throughout an entire schedule generation. Repeat the entire process several times and choose the best result. The procedure to generate active schedule is given as follows:

---

procedure randomized dispatching heuristic(active schedule generation)
input: JSP data
output: a complete schedule
Step 0: Let the best schedule BS as a null schedule.
Step 1: Let $t = 0$ and begin with PS_t as the null partial schedule.
Initially S_t includes all operations with no predecessors.
Step 2: Determine $\phi_t^* = \min_{i \in S_t}\{\phi_i\}$ and the machine m^* on which ϕ_t^* could be realized.
Step 3: Select a dispatching rule randomly from the family of rules. For each operation $i \in S_t$ that requires machine m^* and for which $\sigma_i < \phi_t^*$, calculate a priority index according to the specified rule. Find the operations with the smallest index and add this operation to PS_t as early as possible, thus creating new partial schedule PS_{t+1}.
Step 4: For PS_{t+1}, update the data set as follows:
1. remove operations i from S_t
2. form S_{t+1} by adding the direct successor of operation j to S_t
3. increment t by one
Step 5: Return to step 2 until a complete schedule is generated.
Step 6: If the generated schedule in above step is better than the best one found so far, save it as BS. Return to step 1until iteration equals the predetermined times.

Various researchers tried to improve on the randomization approach. One change is to have a learning process so that more successful dispatching rules will have higher chances of being selected in the future. Morton and Pentico proposed a guided random approach. The *guided* means that an excellent heuristic is needed first, to "explore" the problem and provide good guidance as to where to search.

5.2.2.3 Shifting Bottleneck Procedure

The *shifting bottleneck heuristic* from Adams *et al.* [6] is probably the most powerful procedure known up to now among all heuristics for the job-shop scheduling problem. It sequences the machines one by one, successively, taking each time the machine identified as a bottleneck among the machines not yet sequenced. Every time after a new machine is sequenced, all previously established sequences are locally reoptimized. Both the bottleneck identification and the local reoptimization procedures are based on repeatedly solving a certain one-machine scheduling problem that is a relaxation of the original problem. The method of solving the one-machine problems is not new, although they have speeded up considerably the time required for generating these problems. Instead, the main contribution of their approach is the way to use this relaxation to decide upon the order in which the machines should be sequenced. This is based on the classic idea of giving priority to bottleneck machines. Let M_0 be the set of machines already sequenced ($M_0 = \emptyset$ at the start). A brief statement of the shifting bottleneck procedure is as follows:

---

procedure shifting bottleneck heuristic
input: JSP data
output: a complete schedule
Step 1: Identify a bottleneck machine m among the machines $k \in M \backslash M_0$ and sequence it optimally. Set $M_0 \leftarrow M_0 \cup \{m\}$.
Step 2: Reoptimize the sequence of each critical machine $k \in M_0$ in turn, while keeping the other sequences fixed. Then if $M_0 = M$, stop; otherwise go to step 1.

The details of this implementation can be found in Adams *et al.* [6] or in Applegate and Cook [92]. According to the computational experience of Adams *et al.*, the typical improvement of their procedure is somewhere between 4% and 10%. Storer, Wu *et al.* reported that with Applegate and Cook implementation of shifting bottleneck procedure, it was unable to solve two 10×50 problems [26]. Recently, Dauzere-Peres and Lasserre gave a modified version of shifting bottleneck heuristic [12].

5.2.3 Genetic Representations for JSP

Because of the existence of the precedence constraints of operations, JSP is not as easy as the traveling salesmen problem (TSP) to find a nature representation. There is no good representation with a system of inequalities for the precedence constraints. Therefore, the penalty approach is not easily applied to handle such kind of constraints. Orvosh and Davis [24] have shown that, for many combinatorial optimization problems, it is relatively easy to repair an infeasible or illegal chromosome and the repair strategy did indeed surpass other strategies such as rejecting strategy or penalizing strategy. Most GA and JSP researchers prefer to take repairing strategy to handle the infeasibility and illegality. A very important issue in building a genetic algorithm for a job-shop problem is to devise an appropriate representation of solutions together with problem-specific genetic operations in order that all chromosomes generated in either initial phase or evolutionary process will produce feasible schedules. This is a crucial phase that conditions all the subsequent steps of genetic algorithms. During the last few years, the following six representations for job-shop scheduling problem have been proposed:

- Operation-based representation
- Job-based representation
- Preference list-based representation
- Priority rule-based representation
- Completion time-based representation
- Random key-based representation

These representations can be classified into the following two basic encoding approaches:

- Direct approach
- Indirect approach

In the direct approach, a schedule (the solution of JSP) is encoded into a chromosome and genetic algorithms are used to evolve those chromosomes to find a better schedule. The representations, such as operation-based representation, job-based representation, job pair relation-based representation, completion time-based representation, and random keys representation belong to this class.

In the indirect approach, such as priority rule-based representation, a sequence of dispatching rules for job assignment, but not a schedule, is encoded into a chromosome and genetic algorithms are used to evolve those chromosomes to find out a better sequence of dispatching rules. A schedule then is constructed with the sequence of dispatching rules. Preference list-based representation, priority rule-based representation, disjunctive graph-based representation, and machine-based representation belong to this class.

We shall discuss them in turn using a simple example given in Table 5.2, where p_{ikj} is the processing time of the k-th operation of job J_i operated on machine M_j.

Operation-based Representation

This representation encodes a schedule as a sequence of operations and each gene stands for one operation. There are two possible ways to name each operation. One natural way is to use natural number to name each operation, like the permutation representation for TSP. Unfortunately because of the existence of the precedence constraints, not all the permutations of natural numbers define feasible schedules. Gen, Tsujimura and Kubota proposed an alternative: they name all operations for a job with the same symbol and then interpret them according to the order of occurrence in the sequence for a given chromosome [14, 20]. For an n-job m-machine problem, a chromosome contains $n \times m$ genes. Each job appears in the chromosome exactly m times and each repeating (each gene) does not indicate a concrete operation of a job but refers to an operation which is context-dependent. It is easy to see that any permutation of the chromosome always yields a feasible schedule.

chromosome v =

3	2	1	3	2	2	1	1	3

(1 : job 1, 2 : job 2, 3 : job 3)

Fig. 5.9 An example of operation-based representation

Consider the three-job three-machine problem given in Table 5.2. Suppose a chromosome is given as shown in Fig. 5.9, where 1 stands for job J_1, 2 for job J_2 and 3 for job J_3. Because each job has three operations, it occurs exactly three times in the chromosome. For example, there are three 2s in the chromosome which stands for the three operations of job J_2. The first 2 corresponds to the first operation of job J_2 which will be processed on machine 1, the second 2 corresponds to the second operation of job J_2 which will be processed on machine 2 and the third 2 corresponds to the third operation of job J_2 which will be processed on machine 3. We can see that all operations for job J_2 are named with the same symbol 2 and then interpreted according to their orders of occurrence in the sequence of this chromosome. The corresponding relations of the operations of jobs and processing machines are given in Fig. 5.10. Figure 5.11 illustrates a Gantt chart showing the schedule based on the example of operation-based representation.

Job-based Representation

This representation consists of a list of n jobs and a schedule is constructed according to the sequence of jobs. For a given sequence of jobs, all operations of the first job in the list are scheduled first, and then the operations of the second job in the list are considered. The first operation of the job under treatment is allocated in the best available processing time for the corresponding machine the operation required, and then the second operation, and so on until all operation of the job are scheduled. The process is repeated with each of the jobs in the list considered in the appropriate sequence. Any permutation of jobs corresponds to a feasible schedule.

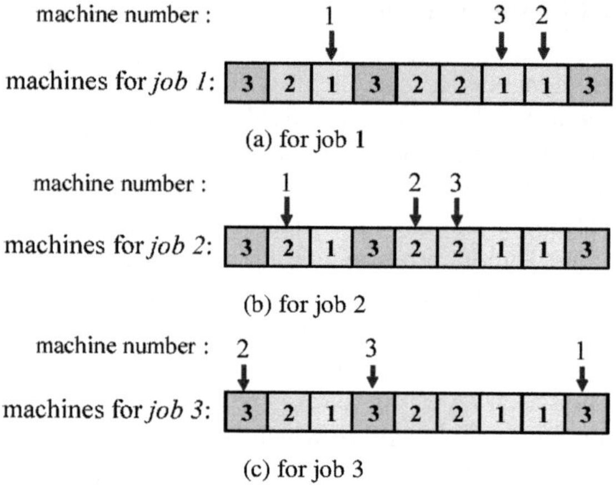

Fig. 5.10 Operations of jobs and corresponding machines

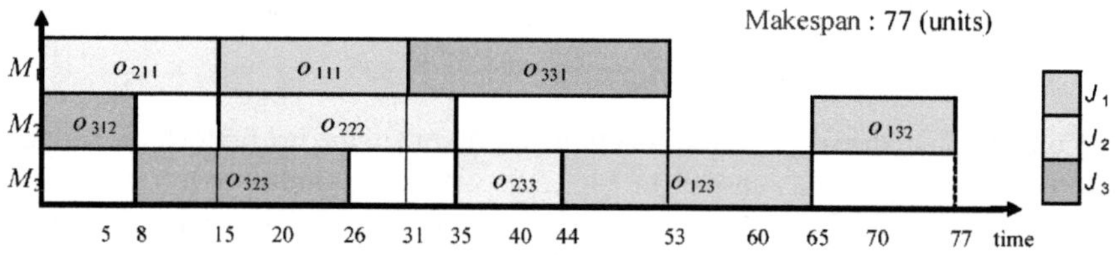

Fig. 5.11 Gantt chart for operation-based representation

Holsapple, Jacob, Pakath and Zaveri have used this representation to deal with static scheduling problem in flexible manufacturing context [8].

Fig. 5.12 An example of job-based representation

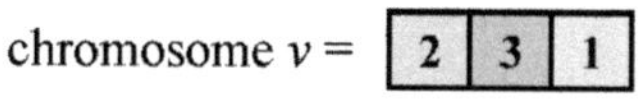

Consider the three-job three-machine problem given in Table 5.2. Suppose a chromosome is given as shown in Fig. 5.12, where 1 stands for job J_1, 2 for job J_2 and 3 for job J_3. The first job to be processed is job J_2.The operation precedence constraint for J_2 is $[M_1, M_2, M_3]$ and the corresponding processing time for each machine is $[15, 20, 9]$. First, job J_2 is scheduled as shown in Fig. 5.13a. Then job J_3 is processed. Its operation precedence among machines is $[M_2, M_3, M_1]$ and the corresponding processing time for each machine is $[8, 18, 22]$. Each of its operations is scheduled in the best available processing time for the corresponding machine the operation required as shown in Fig. 5.13b. Finally, job J_1 is scheduled as shown in Fig. 5.13c.

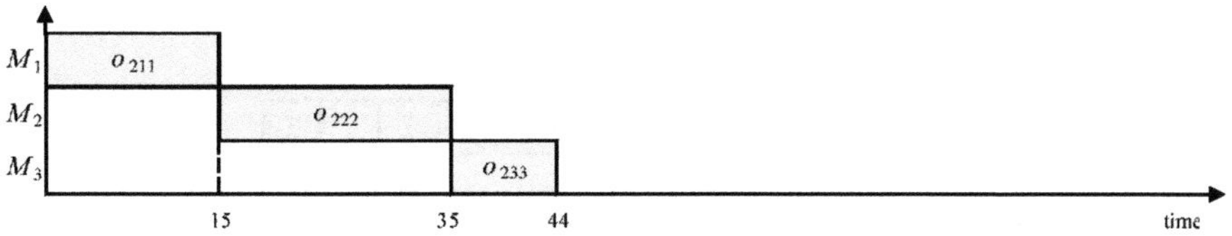

(a) Gantt chart for the first job J_2

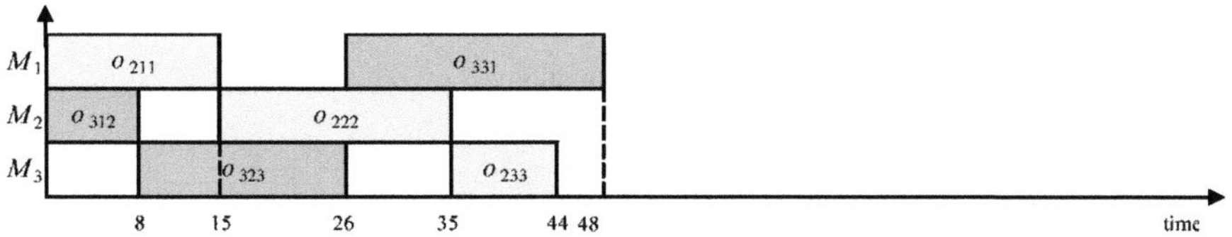

(b) Gantt chart for the second job J_3

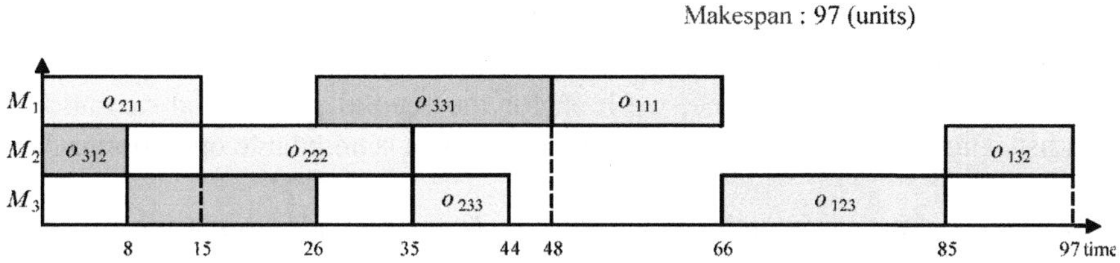

(c) Gantt chart for the third job J_1

Fig. 5.13 Gantt chart for job-based representation

Preference List-based Representation

This representation was first proposed by Davis for a kind of scheduling problem [13]. Falkenauer and Bouffouix used it for dealing with a job-shop scheduling problem with release times and due dates [16]. Croce *et al.* applied it for classical job-shop scheduling problem [11].

For an n-job m-machine job-shop scheduling problem, a chromosome is formed of m subchromosomes, each for one machine. Each subchromosome is a string of symbols with length n, and each symbol identifying an operation that has to be processed on the relevant machine. Subchromosomes do not describe the operation sequence on the machine, they are *preference lists*; each machine has its own preference list. The actual schedule is deduced from the chromosome through a simulation, which analyzes the state of the waiting queues in front of the machine and if necessary uses the preference lists to determine the schedule; that is, the operation which appears first in the preference list will be chosen.

Consider the three-job three-machine problem given in Table 5.2. Suppose a chromosome is given as shown in Fig. 5.14, The first gene $[2,3,1]$ is the preference list for machine M_1, $[1,3,2]$ for machine M_2 and $[2,1,3]$ for machine M_3. From these preference lists we can know that the first preferential operations are job J_2 on

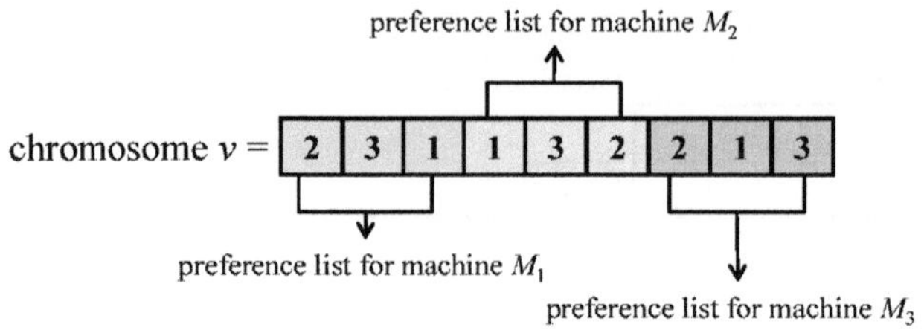

Fig. 5.14 An example of preference-based representation

machine M_1, J_1 on M_2, and J_2 on M_3. According to the given operation precedence constraints, only J_2 on M_1 is schedulable, so it is scheduled on M_1 first as shown in Fig. 5.15(a).

The next schedulable operation is J_3 on M_2 as shown in Fig. 5.15b. Now the current preferential operations are J_1 on M_2, and J_2 on M_3. Because all of them are not schedulable at current time, we look for the second preferential operation in each list. They are J_3 on M_1, J_3 on M_2, J_1 on M_3. The schedulable operation is J_3 on M_2.

Now the current preferential operations are J_1 on M_2, and J_2 on M_3, J_3 on M_1, J_1 on M_3, and the next preferential operations are J_1 on M_1, and J_2 on M_2 , J_3 on M_3. J_1 on M_1, and J_2 on M_2 , J_3 on M_3 are schedulable as shown in Fig. 5.15c.

After that the current preferential operations are J_1 on M_2, and J_2 on M_3, J_3 on M_1, J_1 on M_3. J_1 on M_3, and J_2 on M_3 , J_3 on M_1 are schedulable as shown in Fig. 5.15d. The last operation to be scheduled is J_1 on M_2 as shown in Fig. 5.15e.

Priority Rule-based Representation

Dorndorf and Pesch proposed a priority rule-based representation [15], where a chromosome is encoded as a sequence of dispatching rules for job assignment and a schedule is constructed with priority dispatching heuristics based on the sequence of dispatching rules. GAs here are used to evolve those chromosomes to find out a better sequence of dispatching rules.

Priority dispatching rules are probably the most frequently applied heuristics for solving scheduling problems in practice because of their ease of implementation and their low time complexity. The algorithms of Giffler and Thompson can be considered as the common basis of all priority rule-based heuristics [18, 26]. The problem is to identify an effective priority rule. For an extensive summary and discussion on priority dispatching rules see Panwalkar and Iskander [25], Haupt [19], and Blackstone *et al.* [10].

For an n-job m-machine problem, a chromosome is a string of $n \times m$ entries $[p_1, p_2, \cdots, p_i, \cdots, p_{nm}]$. An entry p_i represents one rule of the set of prespecified priority dispatching rules. The entry in the i-th position says that a conflict in the i-th iteration of the Giffler and Thompson algorithm should be resolved using priority

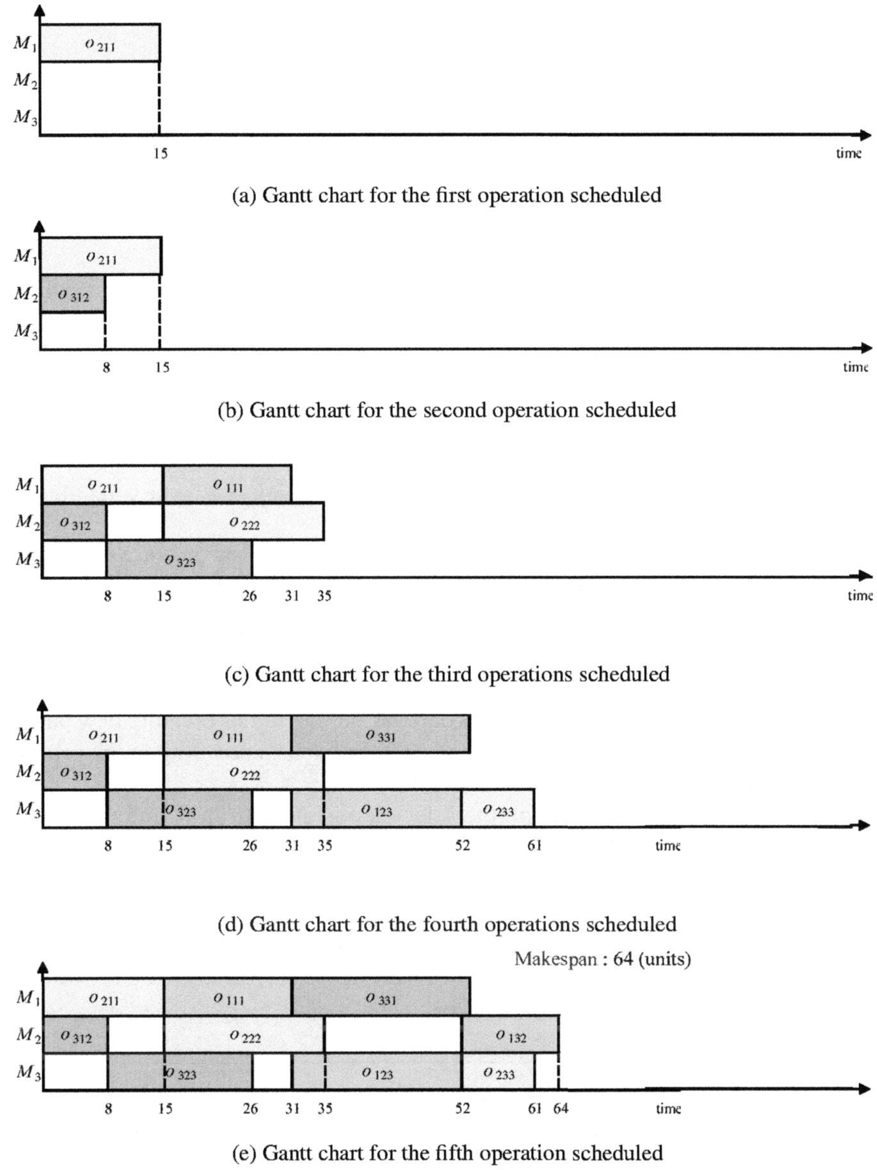

Fig. 5.15 Gantt chart for preference list-based representation

rule p_i. More precisely, an operation from the conflict set has to be selected by rule p_i; ties are broken by a random choice. Let

PS_t = a partial schedule containing t scheduled operations,
S_t = the set of schedulable operations at iteration t, corresponding to PS_t,
σ_i = the earliest time at which operation $i \in S_t$ could be started,
ϕ_i = the earliest time at which operation $i \in S_t$ could be completed,
C_t = the set of conflicting operations in iteration t.

The procedure to deduce a schedule from a given chromosome $(p_1, p_2, \ldots, p_{nm})$ works as follows:

---

procedure: deduce a schedule for priority rule-based representation
input: a chromosome
output: a schedule
Step 1: Let $t = 1$ and begin with PS_t as the null partial schedule and S_t include all operations with no predecessors.
Step 2: Determine $\phi_t^* = \min_{i \in S_t}\{\phi_i\}$ and the machine m^* on which ϕ_t^* could be realized. If more than one such machine exist, tie is broken by a random choice.
Step 3: Form conflicting set C_t which includes all operations $i \in S_t$ with $\sigma_i < \phi_t^*$ that requires machine m^*. Select one operation from C_t by priority rule p_t and add this operation to PS_t as early as possible, thus creating new partial schedule PS_{t+1}. If more than one operation exist according to the priority rule p_t, tie is broken by a random choice.
Step 4: Update $PS_{t+1}$by removing the selected operation from S_t and adding the direct successor of the operation to S_t. Increment t by one.
Step 5: Return to step 2 until a complete schedule is generated.

Table 5.8 Selected priority rules

No.	Rule	Description
1	SPT	Select the operation with the shortest processing time
2	LPT	Select an operation with longest processing time
3	LRT	Select the operation belonging to the job with the longest remaining processing time
4	SRT	Select the operation belonging to the job with the shortest remaining processing time
5	LRM	Select the operation belonging to the job with the longest remaining processing time excluding the operation under consideration

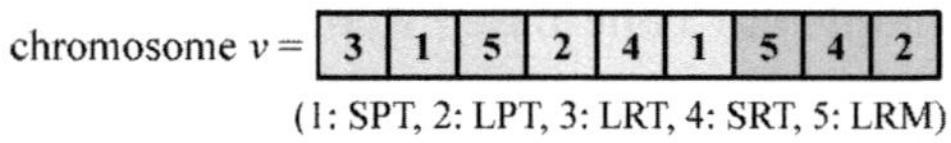

Fig. 5.16 An example of priority rule-based representation

Let us see an example. We use five priority rules given in Table 5.8. Suppose a chromosome is given as shown in Fig. 5.16. At the initial step, we have $S_1 = \{o_{111}, o_{211}, o_{312}\}$, determine $\phi_1^* = \min\{16, 15, 8\} = 8$ and the machine $m^* = 2$ on which ϕ_1^* could be realized. The set of conflicting operations $C_1 = \{o_{312}\}$. Operation o_{312} is scheduled on machine M_2.

Then we have updated $S_2 = \{o_{111}, o_{211}, o_{323}\}$, determine $\phi_2^* = \min\{16, 15, 26\} = 15$ and the machine $m^* = 1$ on which ϕ_2^* could be realized. The set of conflicting operations $C_2 = \{o_{111}, o_{211}\}$. Now operations o_{111} and o_{211} compete for machine M_1. Because the second gene in the given chromosome is 1 (which means SPT priority rule), operation o_{211} is scheduled on machine M_1. Repeat these steps until a complete schedule is deduced from the given chromosome as shown in the trace table (in Table 5.9). Figure 5.17 illustrates a Gantt chart showing the schedule by given chromosome.

Table 5.9 An example of priority rule-based representation

l	S_l	ϕ_l	ϕ_l^*	m^*	C_l	Rule	$S(l)$
1	$\{o_{111}, o_{211}, o_{312}\}$	{16, 15, 8}	8	M_2	$\{o_{312}\}$	*	$\{o_{312}\}$
2	$\{o_{111}, o_{211}, o_{323}\}$	{16, 15, 26}	15	M_1	$\{o_{111}, o_{211}\}$	SPT	$\{o_{312}, o_{211}\}$
3	$\{o_{111}, o_{222}, o_{323}\}$	{31, 35, 26}	26	M_3	$\{o_{323}\}$	*	$\{o_{312}, o_{211}, o_{323}\}$
4	$\{o_{111}, o_{222}, o_{331}\}$	{31, 35, 48}	31	M_1	$\{o_{111}\}$	*	$\{o_{312}, o_{211}, o_{323}, o_{111}\}$
5	$\{o_{123}, o_{222}, o_{331}\}$	{52, 35, 48}	35	M_2	$\{o_{222}\}$	*	$\{o_{312}, o_{211}, o_{323}, o_{111}, o_{222}\}$
6	$\{o_{123}, o_{233}, o_{331}\}$	{52, 44, 48}	44	M_3	$\{o_{123}, o_{233}\}$	SPT	$\{o_{312}, o_{211}, o_{323}, o_{111}, o_{222}, o_{233}\}$
7	$\{o_{123}, o_{331}\}$	{52, 48}	48	M_1	$\{o_{331}\}$	*	$\{o_{312}, o_{211}, o_{323}, o_{111}, o_{222}, o_{233}, o_{331}\}$
8	$\{o_{123}\}$	{52}	52	M_3	$\{o_{123}\}$	*	$\{o_{312}, o_{211}, o_{323}, o_{111}, o_{222}, o_{233}, o_{331}, o_{123}\}$
9	$\{o_{132}\}$	{77}	77	M_2	$\{o_{132}\}$	*	$\{o_{312}, o_{211}, o_{323}, o_{111}, o_{222}, o_{233}, o_{331}, o_{123}, o_{132}\}$

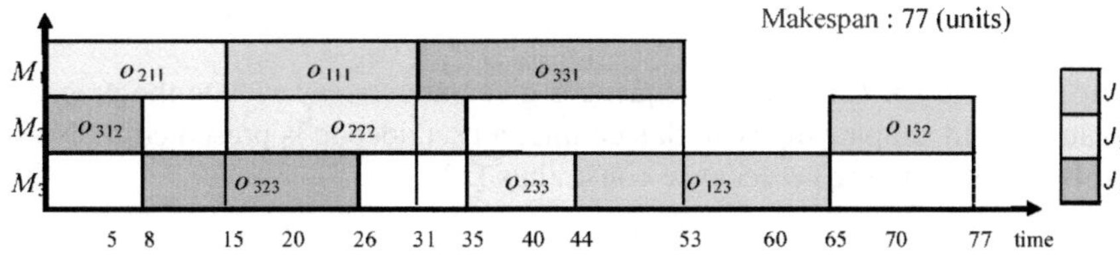

Fig. 5.17 Gantt chart for priority rule-based representation

Completion Time-based Representation

Yamada and Nakano proposed a completion time-based representation [93]. A chromosome is an ordered list of completion times of operations. For the same example given in Table 5.2, the chromosome can be represented as follows:

$$\text{chromosome } v = \left[\, c_{111} \;\; c_{122} \;\; c_{133} \;\; c_{211} \;\; c_{223} \;\; c_{232} \;\; c_{312} \;\; c_{321} \;\; c_{333} \,\right]$$

where c_{ikj} means the completion time for operation k of job i on machine j. It is easy to know that such representation is not suitable for most of genetic operators because it will yield an illegal schedule. Yamada and Nakano designed a special crossover operator for it.

Random Key-based Representation

Random key representation is first given by Bean [9]. With this technique, genetic operations can produce feasible offspring without creating addition overheads for a wide variety of sequencing and optimization problems. Norman and Bean further successfully generalized the approach to the job-shop scheduling problem [22, 23].

Random keys representation encodes a solution with *random number*. These values are used as sort *keys* to decode the solution. For n-job m-machine scheduling problem, each gene (a random key) consists of two parts: an integer in set $\{1, 2, \ldots, m\}$ and a fraction generated randomly from $(0, 1)$. The integer part of any random key is interpreted as the machine assignment for that job. Sorting the fractional parts provides the job sequence on each machine. Consider the same example given in Table 5.2. Suppose a chromosome be

$$\text{chromosome } v = [1.88\ 1.09\ 1.34 \mid 2.91\ 2.66\ 2.01 \mid 3.23\ 3.21\ 3.04\,]$$

Sort the keys for machine M_1 in ascending order results in the job sequence $2 \rightarrow 3 \rightarrow 1$, for machine M_2 the job sequence $3 \rightarrow 2 \rightarrow 1$, and for machine M_3 the job sequence $3 \rightarrow 2 \rightarrow 1$. Let o_{ikj} denote operation k of job i on machine j. The chromosome can be translated into a unique list of ordered operations as follows:

$$\left[\, o_{211}\ o_{331}\ o_{111}\ o_{312}\ o_{222}\ o_{132}\ o_{323}\ o_{233}\ o_{123} \,\right]$$

It is easy to see that the job sequences given above may violate the precedence constraints. Incompanying with this coding, a pseudocode is presented by Norman and Bean for handling precedence constraints [22].

5.2.4 Gen-Tsujimura-Kubota's Approach

Gen, Tsujimura and Kubota proposed an implementation of GA for solving job-shop scheduling problem [94]. Let $P(t)$ and $C(t)$ be parents and offspring in current generation t, respectively. The overall procedure of Gen-Tsujimura-Kubota's GA approach (gtkGA) for solving JSP is outlined as follows:

---

procedure: gtkGA for JSP model
input: JSP data, GA parameters ($popSize$, $maxGen$, p_M, p_C)
output: the best scheduling
begin
    $t \leftarrow 0$;
    initialize $P(t)$ by operation-based encoding routine;
    evaluate $P(t)$ by operation-based decoding routine;
    while (**not** terminating condition) **do**
        create $C(t)$ from $P(t)$ by partial schedule exchange crossover routine;
        create $C(t)$ from $P(t)$ by job-pair exchange mutation routine;
        evaluate $P'(t)$ and $C'(t)$ by operation-based decoding routine;
        select $P(t+1)$ from $P(t)$ and $C(t)$ by selection routine;
        $t \leftarrow t+1$;
    end
    output the best scheduling
end

Operation-based Encoding and Decoding

Gen, Tsujimura and Kubota proposed the operation-based representation for JSP. This representation encodes a schedule as a sequence of operations and each gene stands for one operation. The detailed description of operation-based representation was discussed at above subsection (on Page 317).

Partial Schedule Exchange Crossover

Gen, Tsujimura and Kubota designed a *partial schedule exchange* crossover operator for operation-based representation, which considers partial schedules to be the natural building blocks and intend to use such crossover to maintain building blocks in offspring in much the same manner as Holland described [3]. The partial schedule is identified with the same job in the first and last positions of the partial schedule. The procedure of partial schedule exchange crossover works as follows:

---

procedure: partial schedule exchange crossover
input: parents v_1 and v_2
output: offspring v_1' and v_2'
Step 1: Select partial schedules:

(1.1) Choose one position of job i in parent v_1 randomly.
(1.2) Find out the next nearest job i in the same parent v_1, and get *partial 1* as $[\,i \cdots i\,]$.
(1.3) The next partial schedule is not randomly generated from parent v_2. It must begin and end with job i as the same as the first partial schedule and get *partial 2*.

Step 2: Exchange partial schedules between parents v_1 and v_2.
Step 3: Check the missed genes and exceeded genes in each proto-children.
Step 4: Legalize offspring v_1' and v_2' by deleting exceeded genes and adding missed gens.

Figure 5.18 illustrates an example of partial schedule exchange crossover. First, we choose one position in parent v_1 randomly. Suppose that it is the second position at which is job 2, and find out the next nearest job 2 in the same parent v_1, which is in position 4. Now we get partial 1 as $[\,2\;\;1\;\;2\,]$. Then find partial 2 from parent v_2, that the two jobs 2 are the third and severth job 2 in parent v_2: $[\,2\;\;3\;\;3\;\;1\;\;2\,]$. After exchanging the partial schedules between parents v_1 and v_2, we can find the exceeded genes of proto-child 1 are $\{3,3\}$, and the missed genes of proto-child 2 are $\{3,3\}$. Last, legalize offspring v_1' and v_2' by deleting exceeded genes $\{3,3\}$ and adding missed genes $\{3,3\}$, respectively.

Job-pair Exchange Mutation

Gen, Tsujimura and Kubota use a job-pair exchange mutation, that is randomly pick up two nonidentical jobs and then exchange their positions as shown in Fig. 5.19. For operation-based representation, the mapping relation between chromosome and schedule is n-to-1 mapping and an offspring obtained by exchanging two nearest jobs may yield the same schedule as do its parents. So the separation between two nonidentical jobs is the further the better.

5.2.5 Cheng-Gen-Tsujimura's Approach

Cheng, Gen and Tsujimura further modified Gen, Tsujimura and Kubota's method in order to enhance its efficiency. Let $P(t)$ and $C(t)$ be parents and offspring in current generation t, respectively. The overall procedure of Cheng-Gen-Tsujimura's

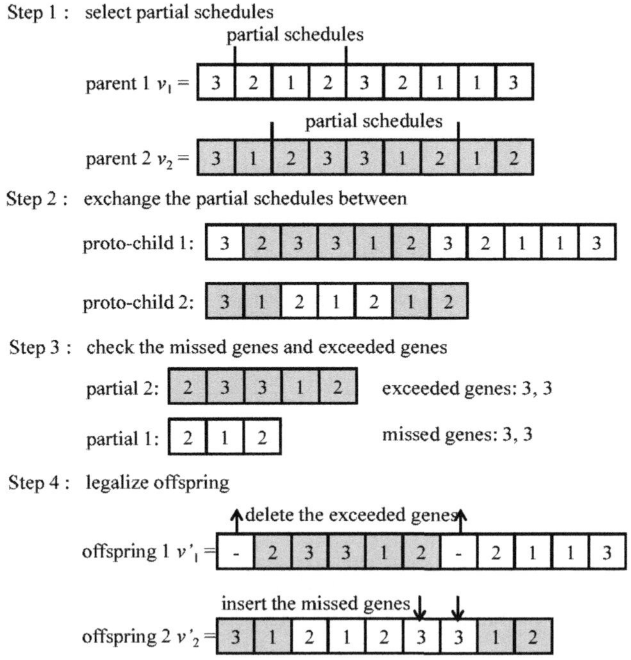

Fig. 5.18 Illustration of partial schedule exchange crossover

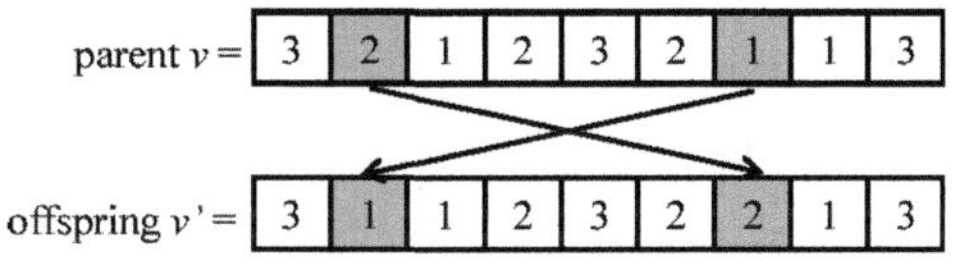

Fig. 5.19 Illustration of job-pair exchange mutation

GA approach (cgtGA) for solving JSP is outlined as follows:

Modified Operation-based Encoding and Decoding

In Gen, Tsujimura and Kubota's method, when decoding a chromosome to a schedule, they only consider two things: the operations order in the given chromosome and the precedence constraints of operations of jobs, and specify the order of the operations on each machine in schedule. The decoding procedure can just guarantee to generate a *semiactive schedule* but not an *active schedule*; that is, there may exist *permissible left-shift* (or *global left-shift*) within the schedule. Because optimal schedule is an active schedule, the decoding procedure will inhibit the efficiency of the proposed algorithm for large-scale job-shop problems.

---

procedure: cgtGA for JSP model
input: JSP data, GA parameters (*popSize*, *maxGen*, p_M, p_C)
output: the best scheduling
begin
    $t \leftarrow 0$;
    initialize $P(t)$ by modified operation-based encoding routine;
    evaluate $P(t)$ by modified operation-based decoding routine;
    climb C(t) by left-shifts routine;
    while (**not** terminating condition) **do**
        create $C(t)$ from $P(t)$ by extended order crossover routine;
        create $C(t)$ from $P(t)$ by job-pair exchange mutation routine;
        evaluate $P'(t)$ and $C'(t)$ by modified operation-based decoding routine;
        select $P(t+1)$ from $P(t)$ and $C(t)$ by selection routine;
        $t \leftarrow t+1$;
    end
    output the best scheduling
end

As an example, we still use the same three-job three-machine problem given in Table 5.2. Suppose we have a chromosome as

$$\text{chromosome } v = [\,2\ \ 1\ \ 1\ \ 1\ \ 2\ \ 3\ \ 2\ \ 3\ \ 3\,]$$

By the decoding procedure, we can get a corresponding machine list as

$$[\,1\ \ 1\ \ 3\ \ 2\ \ 2\ \ 2\ \ 3\ \ 3\ \ 1\,]$$

According to the machine list, we can make a schedule as shown in Table 5.10, which specifies the order of the operations on each machine.

Table 5.10 Schedule

Machine	Job sequence		
M_1	J_2	J_1	J_3
M_2	J_1	J_2	J_3
M_3	J_1	J_2	J_3

Figure 5.20a shows the Gantt chart of the semiactive schedule obtained by decoding procedure for the given chromosome and it has a makespan of 133. There are two permissible left-shifts in the semiactive schedule:

1. Job j_3 can start at time 0 on machine m_2.
2. Operation o_222 can start at time 15 on machine m_2.

By performing these left-shifts, we get an active schedule shown in Fig. 5.20b, which has a makespan of 64. Cheng, Gen and Tsujimura modified the decoding procedure to guarantee to generate an active schedule from a given chromosome.

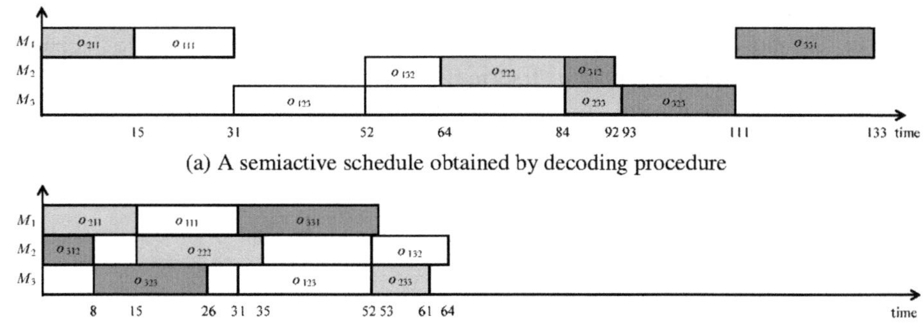

(a) A semiactive schedule obtained by decoding procedure

(b) An active schedule by performing left-shifts on the semiactive schedule

Fig. 5.20 The effect of left-shifts in altering a semiactive schedule

For the operation-based representation, each gene uniquely indicates an operation. The modified decoding procedure first translates the chromosome to a list of ordered operations. A schedule is generated by a one-pass heuristic based on the list. The first operation in the list is scheduled first, and then the second operation in the list is considered, and so on. Each operation under treatment is allocated in the best available processing time for the corresponding machine the operation required. The process is repeated until all operations in the list is scheduled into appropriate places. A schedule generated by this modified procedure can be guaranteed to be an active schedule. A chromosome here can be considered to assign a priority to each operation. Let us label the position from left to right of the list as from lowest to highest. An operation with lower position in the list has higher priority and will be scheduled first prior to those with a higher position.

Extended Order Crossover

The crossover operator given in the last section is rather complex because it tries to keep the orders of operations of a partial schedule in offspring the same as they are in the parent. The modified approach uses chromosome to determine the priority for each operation and a schedule is generated by one-pass priority dispatching heuristics. Therefore it is not necessary to keep the same orders for the operations of partial schedule in both parents, and offspring and the partial schedule exchange crossover can be then simplified as follows; Fig. 5.21 illustrates an example of extended order crossover.

---

procedure: extended order crossover
input: parents v_1 and v_2
output: offspring v'

Step 1: Pick up two partial schedule from parents v_1 and v_2 respectively. Two partial schedules contain the same number of genes. Exchange the partial schedules to generate offspring.

Step 2: Find the missed and exceeded genes for offspring v' by comparing two partial schedules.

Step 3: Legalizing offspring v' by deleting the exceeded genes and inserting the missed genes in a random way.

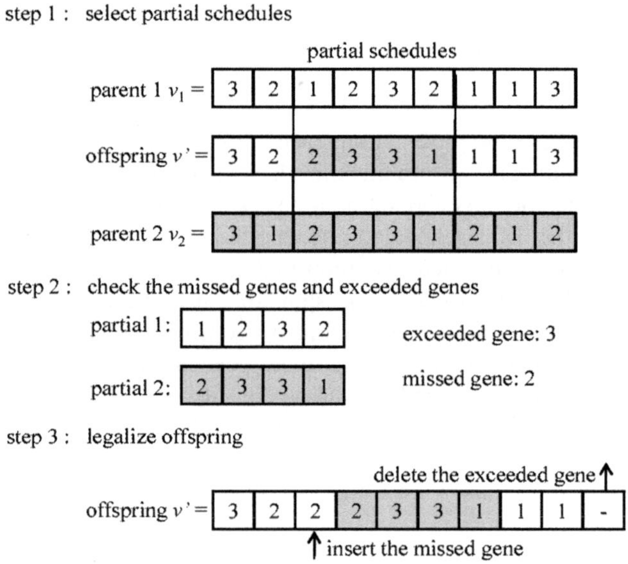

Fig. 5.21 Illustration of extended order crossover

5.2.6 Gonçalves-Magalhacs-Resende's Approach

Gonçalves *et al.* [39] proposed another technique based on the critical path to improve the efficiency of algorithm for finding active schedule [39]. Let $P(t)$ and $C(t)$ be parents and offspring in current generation t, respectively. The overall procedure of Goncalves-Magalhacs-Resende's GA approach (gmrGA) for solving JSP is outlined as follows.

---

procedure: gmrGA for JSP model
input: JSP data, GA parameters ($popSize$, $maxGen$, p_M, p_C)
output: the best scheduling
begin
  $t \leftarrow 0$;
  initialize $P(t)$ by random key-based encoding routine;
  evaluate $P(t)$ by random key-based decoding routine;
  climb C(t) by critical path-based local search routine;
  while (**not** terminating condition) **do**
    create $C(t)$ from $P(t)$ by uniform crossover routine;
    create $C(t)$ from $P(t)$ by random generation routine;
    climb C(t) by critical path-based local search routine;
    evaluate $P'(t)$ and $C'(t)$ by random key-based decoding routine;
    select $P(t+1)$ from $P(t)$ and $C(t)$ by selection routine;
    $t \leftarrow t+1$;
  end
  output the best scheduling
end

Random Key-based Encoding and Decoding

For n-job m-machine scheduling problems, each gene (a random key) consists of two parts: an integer in set $\{1,2,\ldots,m\}$ and a fraction generated randomly from $(0,1)$. The integer part of any random key is interpreted as the machine assignment for that job. Sorting the fractional parts provides the job sequence on each machine. The detailed description of random key-based representation was discussed above(on page 324).

Critical Path-based Local Search

Classification of schedules is the basic work to be done in order to attack scheduling problems. For the case of the JSP, thorough studies have been performed [45, 46]. Schedules for the JSP are classified as feasible, semi-active, active, and non-delay schedules. Using the schedule classification for the JSP proposed by Baker [7] as the stepping stone, Sprecher *et al.* [47] develop more formal and general definitions, that is, schedules for the resource-constrained project scheduling problem (rcPSP) are discriminated to be feasible, semi-active, active, or non-delay schedules.

In principle, there are an infinite number of feasible schedules for a JSP, because superfluous idle time can be inserted between two operations. We can shift the operation to the left as compact as possible. A shift is called local left-shift if some operations can be started earlier in time without altering the operation sequence. Wiest defines a left-justified schedule as a "feasible schedule in which...no job can be started at an earlier date by local left shifting of that job alone" [48]. A feasible schedule can be transformed into a semi-active schedule by a series of local left shifts. A shift is called global left-shift if some operations can be started earlier in time without delaying any other operation even though the shift has changed the operation sequence. Based on these two concepts, three kinds of schedules can be distinguished as follows [4, 46]:

- Semi-active schedule: a schedule is semi-active if no local left-shift exists.
- Active schedule: a schedule is active if no global left-shift exists.
- Non-delay schedule: a schedule is non-delay if no machine is kept idle at a time when it could begin processing some operations.

The relationship among active, semi-active, and non-delay schedules is shown by the Venn diagram of schedule relationship in Fig. 5.22. Optimal schedule is within the set of active schedules [4].

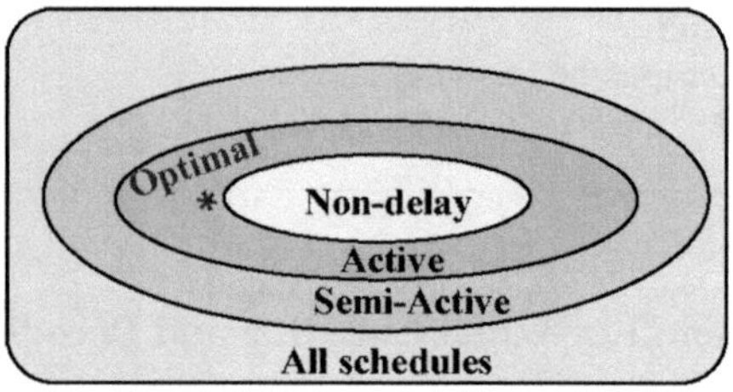

Fig. 5.22 Venn diagram of schedule relationship [4]

The makespan of a schedule is equal to the length of the critical path. A critical operation cannot be delayed without increasing the makespan of the schedule. For a neighbor solution, only when these two adjacent operations are on the critical path, the new solution is possibly superior to the previous one.

Gonçalves, Magalhacs and Resende employ a local search based on the disjunctive graph model of Roy and Sussmann [31] and the neighborhood of Nowicki Smutnicki [49] to improve solutions of the problem. The local search procedure begins by identifying the critical path in the solution obtained by the decoding procedure. Any operation on the critical path is called a critical operation. It is possible to break down the critical path into a number of blocks where a block is a maximal sequence of adjacent critical operations that require the same machine. Figure 5.23. give a procedure to create and confirm the critical path of JSP.

If a job predecessor and a machine predecessor of a critical operation are both critical, then choose the predecessor (from among these two alternatives) which has a higher priority in the current solution of the schedule. For JSP, a neighbor solution can be generated by exchanging the sequence of two adjacent critical operations that belong to the same block. In each block, we only swap the last two and first two operations. For the first (last) block, we only swap the last (first) two operations in the block. In cases where the first and/or the last block contains only two operations, these operations are swapped. If a block contains only one operation, then no swap is made. A current solution with possible operations swaps is shown in Fig. 5.24. If no swap of first or last operations in any block of critical path improves the makespan, the local search ends.

```
procedure: critical operations identification
input: a feasible schedule S with makespan t_M, data set of JSP
output: a set of critical operations crt-oper and its total number crt-length
begin
    crt-length ← 0;     //total number of critical operations of schedule S
    temp ← t_M
    for i ← 0 to n do
        for k ← 0 to m do
            if (temp == t_ik^F)
                insert o_ik in crt-oper ;
                  crt-length ← crt-length + 1;     //update the total number of critical operation
                temp ← temp − p_ikj;
        end
    end
    output crt-oper and crt-length
end
```

Fig. 5.23 Procedure of creating an critical path of JSP

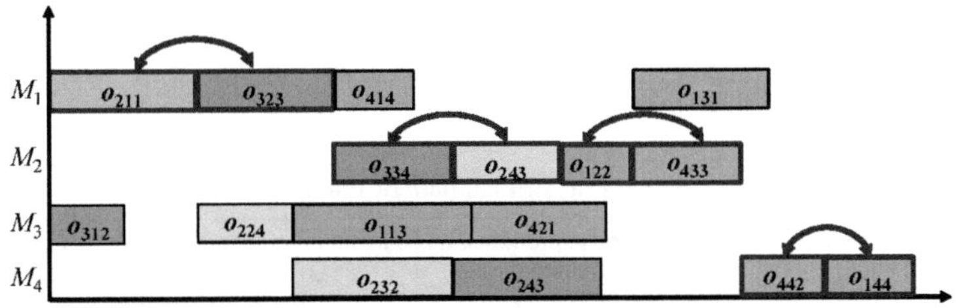

Fig. 5.24 Current solution with possible operations swaps [39]

Moreover, Gonçalves *et al.* proposed an effective local search procedure to reduce the idle time of solutions in Fig. 5.25. In this procedure, we first define: unprd-block (unprocessed blocks); first-crit-block (first critical block in the left side); last-crit-block (last critical block in the right side).

As an example in Fig. 5.26, we can get a critical path and find out possible operations swaps from the intial solution (makespan is 77) following the trace table at Table 5.11. It has only the critical block $o_{233} - o_{123}$ in M_3 for swapping. If we swap the two operations we obtain a better schedule with a makespan of 64, and then generate the new critical path.

Uniform Crossover

Parameterized uniform crossovers are employed in place of the traditional one-point or two-point crossover. After two parents are chosen randomly from the full, old population (including chromosomes copied to the next generation in the elitist pass),

```
procedure: critical path-based Local Search
input: crt-solution (Current Solution)
output: best solution
begin
    upd-flag=False;
    determine the critical path and all critical blocks of crt-solution;
    while (unprd-block && not upd-flag)
        if (not first-crit-block) then
            new-solution := swap first two operations of block in crt-solution;
            if (makespan (new-solution)<makespan (crt-solution)) then
                crt-solution := new-solution;
                upd-flag := true;
        if (not last-crit-block && not upd-flag) then
            new-solution := swap last two operations of block in crt-solution;
            if (makespan (new-solution)<makespan (crt-solution)) then
                crt-solution := new-solution;
                upd-flag := true;
    end while
    output best solution;
end
```

Fig. 5.25 Procedure of critical path local search

Table 5.11 Trace table for confirming the critical path of initial solution

crt-length	temp	last operation o_{ikj}	processing time p_{ikj}	current machine M_j	crt-oper
1	77	o_{132}	12	M_2	$\{o_{132}\}$
2	65	o_{123}	21	M_3	$\{o_{132}, o_{123}\}$
3	44	o_{233}	9	M_3	$\{o_{132}, o_{123}, o_{233}\}$
4	35	o_{222}	20	M_2	$\{o_{132}, o_{123}, o_{233}, o_{222}\}$
5	15	o_{211}	15	M_1	$\{o_{132}, o_{123}, o_{233}, o_{222}, o_{211}\}$

at each gene we toss a biased coin to select which parent will contribute the allele. Figure 5.27 presents an example of the crossover operator. It assumes that a coin toss of a head selects the gene from the first parent, a tail chooses the gene from the second parent, and that the probability of tossing a heads is for example 0.7 (this value is determined empirically). Figure 5.27 shows one potential crossover outcome:

Random Generation

Rather than the traditional gene-by-gene mutation with very small probability at each generation, one or more new members of the population are randomly gen-

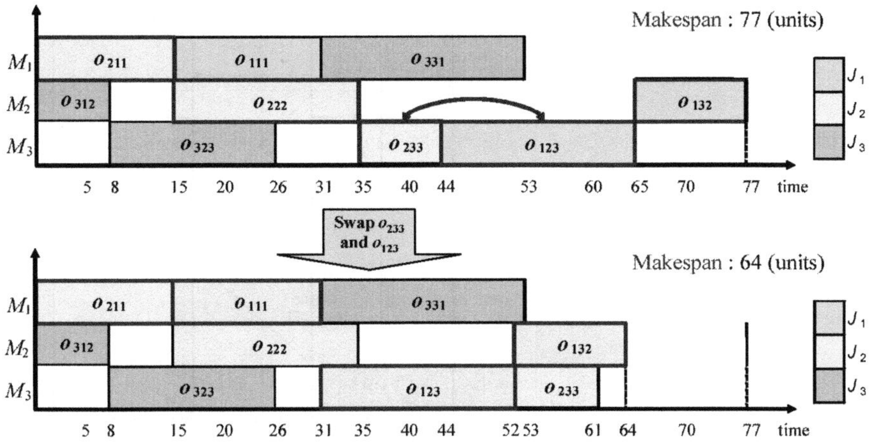

Fig. 5.26 Improvement after swap one critical block

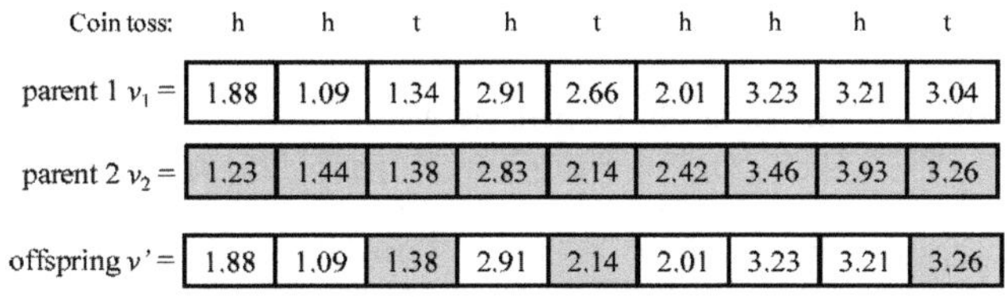

Fig. 5.27 Example of parameterized uniform crossover

erated from the same distribution as the original population. This process prevents premature convergence of the population, like in a mutation operator, and leads to a simple statement of convergence. The random generation is very similar operation to the immigration operation (on page 23). Figure 5.28 depicts the transitional process between two consecutive generations.

5.2.7 Experiment on Benchmark Problems

Fisher and Thompson proposed three well-known benchmarks for job-shop scheduling problems in 1963 [17]; since then, researchers in operations research have tested their algorithms on these problems. Most GA and JSP researchers used these benchmarks to test the performance of their algorithms. Table 5.12 summarizes the experimental results. It lists problem name, problem dimension (number of jobs × number of operations), the best known solution (BKS), the solution obtained by different algorithms, where Dorndorf1 stands for the hybrid approach of GA with Giffler and Thompson heuristic and Dorndorf2 for the hybrid one of GA with bottleneck shifting heuristic proposed by Dorndorf and Pesch.

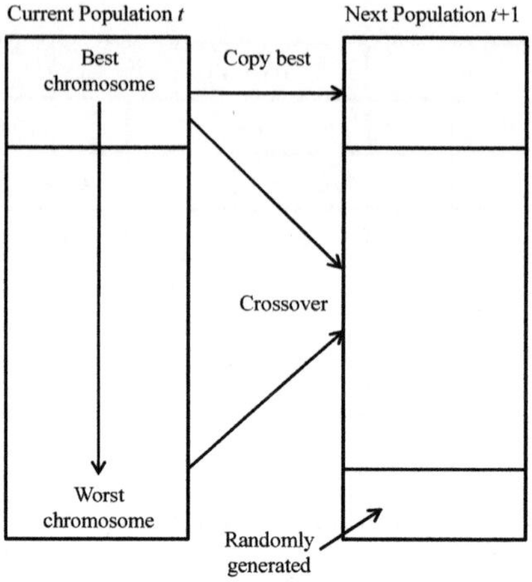

Fig. 5.28 Transitional process between consecutive generations

Table 5.12 Fisher and Thompson's benchmark problems

	FT06 (6 × 6)	FT10 (10 × 10)	FT20 (20 × 5)
BKS	55	930	1165
Gonçalves *et al.* [39]	55	930	1165
Aiex *et al.* [102]	55	930	1165
Binato *et al.* [103]	55	938	1169
Wang and Zheng [104]	55	930	1165
Gonçalves and Beirão [105]	55	936	1177
Nowicki and Smutnicki [106]	55	930	1165
Croce *et al.* [11]	55	946	1178
Cheng *et al.* [107]	55	948	1196
Dorndorf1 *et al.* [15]	55	960	1249
Dorndorf2 *et al.* [15]	55	938	1178
Gen *et al.* [94]	55	962	1175
Fang *et al.* [108]	–	949	1189
Yamada *et al.* [93]	55	930	1184
Paredis *et al.* [109]	–	1006	–
Nakano *et al.* [110]	55	965	1215

5.3 Flexible Job-shop Scheduling Model

The Flexible Job-shop Scheduling Problem is expanded from the traditional Job-shop Scheduling Problem, which possesses wider availability of machines for all the operations. The classic Job-shop Scheduling Problem (JSP) concerns determination of a set of jobs on a set of machines so that the makespan is minimized. It is obviously a single criterion combinational optimization and has been proved as an NP-hard problem with several assumptions as follows: each job has a specified processing order through the machines which is fixed and known in advance; processing times are also fixed and known corresponding to the operations for each job; set-up times between operations are either negligible or included in processing times (sequence-independent); each machine is continuously available from time zero; there are no precedence constraints among operations of different jobs; each operation cannot be interrupted; each machine can process at most one operation at a time.

The flexible Job-shop Scheduling Problem (fJSP) extends JSP by assuming that a machine may be capable of performing more than one type of operation [59]. That means, for any given operation, there must exist at least one machine capable of performing it. Two kinds of flexibility are considered to describe the performance of fJSP [57]:

- *Total flexibility*: in this case all operations are achievable on all the machines available;
- *Partial flexibility*: in this case some operations are only achievable on part of the available machines.

Most of the literature on the shop scheduling problem concentrates on the JSP case [4, 5]. The fJSP recently captured the interest of many researchers. The first paper that addresses the fJSP was given by Brucker and Schlie [53], who proposed a polynomial algorithm for solving the fJSP with two jobs, in which the machines able to perform an operation have the same processing time. For solving the general case with more than two jobs, two types of approaches have been used: hierarchical approaches and integrated approaches. The first was based on the idea of breaking down the original problem in order to reduce its complexity. Brandimarte [54] was the first to use this breaking down for the fJSP. He solved the assignment problem using some existing dispatching rules and then focused on the resulting job shop subproblems, which are solved using a tabu search heuristic. Mati proposed a greedy heuristic for simultaneously dealing with the assignment and the sequencing subproblems of the flexible job shop model [58]. The advantage of Mati's heuristic is its ability to take into account the assumption of identical machine. Kacem *et al.* [57] came to use GA to solve fJSP, and adapted two approaches to solve jointly the assignment and JSP (with total or partial flexibility). The first is the approach by localization (AL) . It makes it possible to solve the problem of resource allocation and build an ideal assignment mode (assignments schemata). The second is an evolutionary approach controlled by the assignment model, and applying GA to solve the fJSP. Wu and Weng [62] considered the problem with job earliness and tardiness objectives, and proposed a multiagent scheduling method. Xia and Wu [60] treated this problem with a hybrid of Particle Swarm Optimization (PSO) and simulated annealing as a local search algorithm. Zhang and Gen [61] proposed a multistage operation-based genetic algorithm to deal with the fJSP problem from the point view of dynamic programming.

5.3.1 Mathematical Formulation of fJSP

Basically, there are two versions of flexibility in flexible Job-shop Scheduling Problem (fJSP): total flexibility and partial flexibility. In the case of total flexibility, each operation is achievable on any machine. In the case of partial flexibility, some operations are only achievable on a part of the available machines. With loss of generality, we assume that each operation is achievable on any machine. This assumption is reasonable since in the case where a task cannot be processed on a machine, the processing time of the task on the machine is set to a very large number.

The flexible job shop scheduling problem is as follows: n jobs are to be scheduled on m machines. Each job i contains n_i ordered operations. The execution of each operation requires one machine, and will occupy that machine until the operation is completed. The fJSP problem is to assign operations on machines and to schedule operations assigned on each machine, subject to the constraints that:

1. The operation sequence for each job is prescribed;
2. Each machine can process only one operation at a time.

In this study, we consider to minimize the following three criteria:

1. Makespan (t_M) of the jobs;
2. Maximal machine workload (w_M), *i.e.*, the maximum working time spent at any machine;
3. Total workload (w_T), which represents the total working time over all machines.

The relationships between the three objectives are very complex. As in most multiobjective optimization problems, the three objectives considered conflict with one another. For example, the machines may have different efficiencies for each operation, so the total workload would always be minimized by using exclusively the most efficient machine. Yet, that may not minimize makespan since the less efficient machines would be idle. On the other hand, a small makespan requires a small maximal workload and a small maximal workload implies a small total workload. In this sense, the three objectives depend on each other.

In contrast to most multiobjective optimization problems in which there are no priorities among the objectives, makespan is given overwhelming importance in the fJSP problem. When two solutions with different makespans are compared, we always prefer the solution with a small makespan regardless of its performance on maximal machine workload and total workload. For an fJSP problem, there may be more than one solution that has the same makespan with different maximal workloads or total workloads. From this point of view, we first find the solutions with the minimum makespan, then minimize the maximal workload in the presence of the minimum makespan, and third minimize the total workload in the presence of the minimum makespan and minimum maximal workload. Makespan is given first importance, maximal machine workload given the second importance, and total workload is given least importance. Assumptions of this problem are summarized as follows:

A1. The precedence relationships among operations of a job is predetermined.
A2. Each operation can be implemented on any machine. In case a task cannot be processed on a machine, the processing time of the task on the machine is set to a very large number.
A3. Operation cannot be interrupted.
A4. Each machine processes only one operation at a time.
A5. The set-up time for the operations is machine-independent and is included in the processing time.

Before introducing the mathematical model some symbols and notations have been defined as follows.

Notation

Indices

i: index of jobs, $i, h = 1, 2, \ldots, n$;
j: index of machines, $j = 1, 2, \ldots, m$;
k: index of operations, $k = 1, 2, \ldots K_i$

Parameters

n: total number of jobs
m: total number of machines
K_i: total number of operations in job i (or J_i)
J_i: the i-th job
o_{ik}: the k-th operation of job i (or J_i)
M_j: the j-th machine
p_{ikj}: processing time of operation o_{ik}on machine j (or M_j)
U: a set of machines with the size m
U_{ik}: a set of available machines for the operation o_{ik}
W_j: workloads (total processing time) of machine M_j

Decision Variables

$$x_{ijk} = \begin{cases} 1 \text{ if machine } j \text{ is selected for the operation } o_{ik}; \\ 0 \text{ otherwise} \end{cases}$$

c_{ik} = completiom time of the operation o_{ik}

The fJSP model is then given as follows:

$$\min t_M = \max_i \max_k \{c_{ik}\} \tag{5.5}$$

$$\min W_M = \max_j \{W_j\} \tag{5.6}$$

$$\min W_T = \sum_{j=1}^{m} W_j \tag{5.7}$$

$$\text{s. t. } c_{ik} - t_{ikj} x_{ikj} - c_{i(k-1)} \geq 0, k = 2, \ldots, K_i; \forall\, i, j \tag{5.8}$$

$$\sum_{j=1}^{m} x_{ikj} = 1, \forall\, k, i \tag{5.9}$$

$$x_{ikj} \in 0, 1, \forall\, j, k, i \tag{5.10}$$

$$c_{ik} \geq 0, \forall\, k, i \tag{5.11}$$

The first objective function accounts for makespan, Eq. 5.6 combining with Eq. 5.7 giving a physical meaning to the fJSP, referring to reducing total processing time and dispatching the operations averagely for each machine. Considering both equations, our objective is to balancing the workloads of all machines. Equation 5.8 states that the successive operation has to be started after the completion of its precedent operation of the same job, which represents the operation precedence constraints. Equation 5.9 states that one machine must be selected for each operation.

5.3.2 Genetic Representations for fJSP

To demonstrate the fJSP model clearly, we first prepare a simple example. Table 5.13 gives the data set of an fJSP including three jobs operated on four machines. This is obviously a problem with total flexibility because all the machines are available for each operation ($U_{ik} = U$). There are several traditional heuristic methods that can be used to make a feasible schedule.

In this case, we use the SPT (select the operation with the shortest processing time) as selective strategy to find an optimal solution, and the algorithm is based on the procedure in Fig. 5.29. Before selection we firstly make some initialization:

- Starting from a table D presenting the processing times possibilities;

- On the various machines, create a new table D' whose size is the same as the table D;
- Create a table S whose size is the same as the table D (S is going to represent chosen assignments);
- Initialize all elements of S to 0 ($S_{ikj} = 0$);
- Recopy D in D' .

Table 5.13 Data set of a three-jobs four-machine problem [57]

		M_1	M_2	M_3	M_4
J_1	o_{11}	1	3	4	1
	o_{12}	3	8	2	1
	o_{13}	3	5	4	7
J_2	o_{21}	4	1	1	4
	o_{22}	2	3	9	3
	o_{23}	9	1	2	2
J_3	o_{31}	8	6	3	5
	o_{32}	4	5	8	1

```
procedure: SPT Assignment
input: dataset table D
output: best schedule S
begin
    for (i=1; i<=n)
        for (k=1; k<=K_i)
          min=+∞;
          pos=1;
          for (j=1;j<=m)
                  if (p'_ikj<min) then {min=p'_ikj; pos=j;}
          S_i,k,pos=1(assignment of o_ik to the machine M_pos);
          //updating of D';
          for (k'=k+1; k'<=K_i')
             p'_i',k,pos= p'_i',k,pos+ p_i,k,pos;
          for (i'= i +1; i'<=n)
             for (k'= 1; k'<=K_i')
                p'_i',k',pos= p'_i',k',pos+ p_i,k,pos;
    end
end
```

Fig. 5.29 SPT assignment procedure

Following this algorithm, we first assign o_{11} to M_1, and we add the processing time $p_{111} = 1$ to the elements of the first column of D' (shown in Table 5.14).

Table 5.14 D'(for i=1 and k=1)

		M_1	M_2	M_3	M_4
J_1	o_{11}	1	3	4	1
	o_{12}	4	8	2	1
	o_{13}	4	5	4	7
J_2	o_{21}	5	1	1	4
	o_{22}	3	3	9	3
	o_{23}	10	1	2	2
J_3	o_{31}	9	6	3	5
	o_{32}	5	5	8	1

Second, we assign o_{12} to M_4, and we add the processing time $p_{124} = 1$ to the elements of the fourth column of D' shown in Table 5.15 . By following the same method, we obtain assignment S shown in Table 5.16.

Table 5.15 D'(for i=1 and k=2)

		M_1	M_2	M_3	M_4
J_1	o_{11}	1	3	4	1
	o_{12}	4	8	2	1
	o_{13}	4	5	4	8
J_2	o_{21}	5	1	1	5
	o_{22}	3	3	9	4
	o_{23}	10	1	2	3
J_3	o_{31}	9	6	3	6
	o_{32}	5	5	8	2

Table 5.16 Assignment S

		M_1	M_2	M_3	M_4
J_1	o_{11}	1	0	0	0
	o_{12}	0	0	0	1
	o_{13}	1	0	0	0
J_2	o_{21}	0	1	0	0
	o_{22}	0	1	0	0
	o_{23}	0	0	1	0
J_3	o_{31}	0	0	1	0
	o_{32}	0	0	0	1

Furthermore, we can denote the schedule based on job sequence as:

$$S = (o_{11}, M_1), (o_{12}, M_4), (o_{13}, M_1), (o_{21}, M_2), (o_{22}, M_2), (o_{23}, M_1), (o_{31}, M_3), (o_{32}, M_4)$$
$$= (o_{11}, M_1 : 0-1), (o_{12}, M_4 : 1-2), (o_{13}, M_1 : 2-5), (o_{21}, M_2 : 0-1),$$
$$(o_{22}, M_2 : 1-4), (o_{23}, M_1 : 4-6), (o_{31}, M_3 : 1-3), (o_{32}, M_4 : 3-4)$$

Finally we can calculate the solution by Eqs. 5.5–5.7 as follows:

$$t_M = max 1, 2, 5, 1, 4, 6, 3, 4 = 6$$
$$W_M = max(1+3), (1+3), (3+2), (1+1) = 5$$
$$W_T = 4+4+5+2 = 15$$

Parallel Machine Representation (PM-R)

The chromosome is a list of machines placed in parallel (see Table 5.17). For each machine, we associate operations to execute. Each operation is coded by three elements: operation k, job J_i and (starting time of operation o_{ik} on the machine M_j).

Table 5.17 Parallel machine representation

M_1	(k, J_i, t_{ik1}^S)	...
M_2	(k, J_i, t_{ik2}^S)	...
M_3	(k, J_i, t_{ik3}^S)	...
...		
M_m	...	...

Table 5.18 Parallel jobs representation

J_1	(M_j, t_{1kj}^S)	...
J_2	(M_j, t_{2kj}^S)	...
J_3	(M_j, t_{3kj}^S)	...
...		
J_n	(M_j, t_{nkj}^S)	...

Parallel Jobs Representation (PJ-R)

The chromosome is represented by a list of jobs shown in Table 5.18. Information of each job is shown in the corresponding row where each case is constituted of two terms:

- Machine M_j which executes the operation
- Corresponding starting time t_{ikj}^S

Operations Machines Representation (OM-R)

Kacem *et al.* proposed the Operations Machines Representation (OM-R) approach [57], which was based on a traditional representation called Schemata Theorem Representation (ST-R) . It was first introduced in GAs by Holland [56].

In the case of a binary coding, a schemata is a chromosome model where some genes are fixed and the others are free (see Fig. 5.30). Positions 4 and 6 are occupied by the symbol: "$*$". This symbol indicates that the genes considered can take "0" or "1" as value. Thus, chromosomes C_1 and C_2 respect the model imposed by the schemata S.

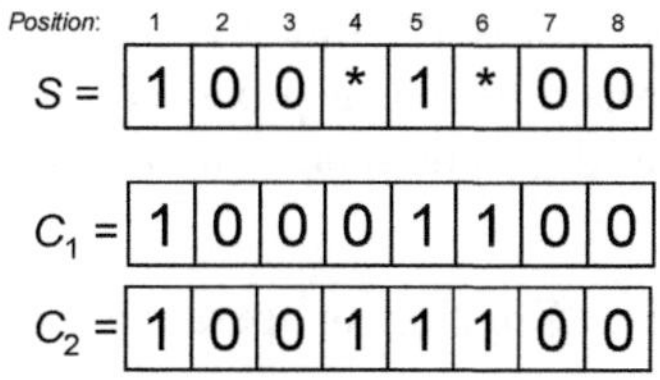

Fig. 5.30 Example for ST-R

Based on the ST-R approach, Kacem expanded it to Operations Machines Representation (OM-R). This consists in representing the schedule in the same assignment table S. We replace each case $S_{ikj} = 1$ by the couple (t^S_{ik}, t^F_{ik}), while the cases $S_{ikj} = 0$ are unchanged. To explain this coding, we present the same schedule introduced before (see Table 5.19). Furthermore, operation based crossover and the other two kinds of mutation (operator of mutation reducing the effective processing time, operator of mutation balancing work loads of machines) are included in this approach.

Table 5.19 Representation of OM-R

		M_1	M_2	M_3	M_4
J_1	o_{11}	0, 1	0	0	0
	o_{12}	0	0	0	1, 2
	o_{13}	2, 5	0	0	0
J_2	o_{21}	0	0, 1	0	0
	o_{22}	0	1, 4	0	0
	o_{23}	0	0	4, 6	0
J_3	o_{31}	0	0	0, 3	0
	o_{32}	0	0	0	3, 4

5.3.3 Multistage Operation-based GA for fJSP

Let $P(t)$ and $C(t)$ be parents and offspring in current generation t, respectively. The overall procedure of multistage operation-based GA (moGA) for solving fJSP is outlined as follows:

Two-vector Multistage Operation-based Representations

Considering the GA approach proposed by Imed Kacem, it is really a bit complex to take all the objectives into account, because all the crossovers and mutations are based on the chromosome which is described as a constructor of table. Therefore, it will spend more CPU-time for finding solutions; hence a multistage operation-based GA approach has been proposed. To develop a two-vector multistage operation-based-based genetic representation for the fJSP, there are three main phases:

Phase 1: Creating an operation sequence

Step 1.1: Generate a random priority to each operation in the model using priority-based encoding procedure for first vector v_1.

procedure: moGA for fJSP model
input: fJSP data, GA parameters $(popSize, maxGen, p_M, p_C)$
output: the best scheduling
begin
  $t \leftarrow 0$;
  initialize $P(t)$ by two-vector Multistage Operation-based encoding routine;
  evaluate $P(t)$ by priority-based decoding routine;
  reorder operation sequence according to operation starting time;
  while (**not** terminating condition) **do**
    create $C(t)$ from $P(t)$ by exchange crossover routine;
    create $C(t)$ from $P(t)$ by allele-based mutation routine;
    improve $P(t)$ and $C(t)$ to yield $P'(t)$ and $C'(t)$ by bottleneck shifting;
    evaluate $P'(t)$ and $C'(t)$ by priority-based decoding routine;
    select $P(t+1)$ from $P(t)$ and $C(t)$ by mixed sampling routine;
    reorder operation sequence according to operation starting time;
    $t \leftarrow t+1$;
  end
  output the best scheduling
end

Step 1.2: Decode a feasible operation sequence that satisfies the precedence constraints of fJSP.

Phase 2: Assigning operations to machines

Step 2.1: Input the operation sequence found in step 1.2.
Step 2.2: Generate a permutation encoding for machine assignment of each operation (second vector v_2).

Phase 3: Designing a schedule

Step 3.1: Create a schedule S using operation sequence and machine assignments.
Step 3.2: Draw a Gantt chart for this schedule.

Figure 5.31 present an fJSP which includes three jobs operated on four machines; we add another two nodes (starting node and terminal node) in the figure to make it a formal network presentation. Denoting each operation as one stage, and each machine as one state, the problem can be formulated into an eight-stage, four-state problem. Connected by the dashed arcs a feasible schedule can be obtained as Fig. 5.31:

Phase 1: Creating an Operation Sequence

Step 1.1: Generate a random priority to each operation in the model using priority-based encoding procedure for first vector v_1.

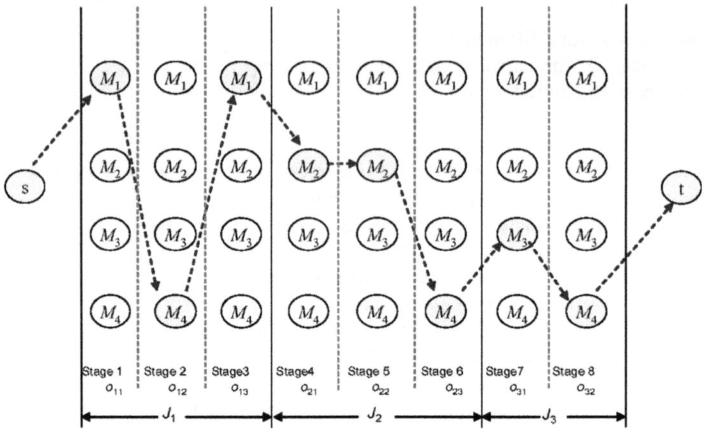

index j:	1	2	3	4	5	6	7	8
$v_1(j)$:	1	2	3	4	5	6	7	8
$v_2(j)$:	1	4	1	2	2	3	1	4

Fig. 5.31 Multistage operation-based representation of fJSP

Step 1.2: Decode a feasible operation sequence that satisfies the precedence constraints of fJSP

Priority-based encoding is obviously simpler than all the representations referred to before, and certainly can easily combing with almost all kinds of classic crossover and mutation method. Figures 5.32 and 5.33 separately give the encoding and decoding procedure for v_1.

Phase 2: Assigning Operations to Machines

Step 2.1: Input the operation sequence found in step 1.2.
Step 2.2: Generate a permutation encoding for machine assignment of each operation (second vector v_2).

The v_2 are represented by simple permutation encoding and decoding shown in Figs. 5.34 and 5.35.

Phase 3: Designing a Schedule

Step 3.1: Create a schedule S using operation sequence and machine assignments.

$$S = \{(o_{11}, M_1), (o_{12}, M_4), (o_{13}, M_1), (o_{21}, M_2), (o_{22}, M_2), (o_{23}, M_1), (o_{31}, M_3), (o_{32}, M_4)\}$$

```
procedure: priority-based encoding for sequencing operations
input: total number of operations K
output: chromosome v1()
begin
  for j=1 to K
    v1(j)←j;
    for i=1 to repeat
      k←random[1, K];
      l←random[1, K];
    until l≠k
    swap (v1(k), v1(l));
  output chromosome v1()
end
```

Fig. 5.32 Priority-based encoding for sequencing operation

```
procedure: priority-based decoding for sequencing operations
input: set of successive operations L_k after activity k;
       set of starting activities L_0;
       chromosome v1()
output: operation sequence
begin
  k←0;
  S←φ, S̄←φ;
  while ( S̄≠φ ) do
    S̄ ← S̄ ∪ L_k;
    k*←argmin{v1(j)| j ∈S̄};
    S̄← S̄\ k*;
    S← S∪ k*;
    k←k*;
  end
  output operation sequence
end
```

Fig. 5.33 Priority-based decoding for sequencing operations

$$= \{(o_{11}, M_1 : 0-1), (o_{12}, M_4 : 1-2), (o_{13}, M_1 : 2-5), (o_{21}, M_2 : 0-1), (o_{22}, M_2 : 1-4), (o_{23}, M_1 : 4-6), (o_{31}, M_3 : 1-3), (o_{32}, M_4 : 3-4)\}$$

```
procedure: permutation encoding for selecting machines
input:  number of stags K, number of states at stage j, n_j
output: chromosome v_2()
begin
    for j=1 to K
        v_2(j) ←0;
    for j=1 to J
        v_2(j) ← random[1, n_j];
output chromosome v_2()
end
```

Fig. 5.34 Machine permutation encoding for selecting machines

```
procedure: permutation decoding for selecting machines
input: chromosome v_2(), operation sequence S_1,
       number of stags K
output: schedule S
begin
    S ← φ;
    j←1;
    while (j ≤ J) do
        r ← v_2(j);
        o_j ← S_1(j);
        S ← S ∪ {(o_j, M_r)};
        j ←j+1;
    end
output schedule S
end
```

Fig. 5.35 Machine permutation decoding for selecting machines

Step 3.2: Draw a Gantt chart for this schedule.
Figure 5.36 illustrates a Gantt chart showing a feasible schedule for fJSP.

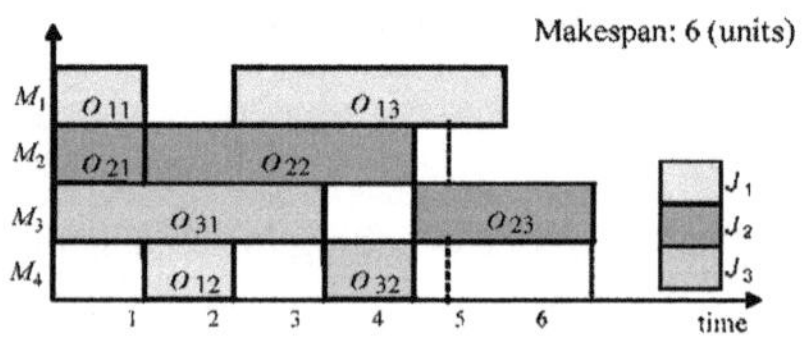

Fig. 5.36 Gantt chart for fJSP

5.3.3.1 Adjusting Operation Sequence for Schedule Improvement

Two individuals with high fitness values are likely to have dissimilar machine assignment and operation sequences, and the recombination may result in offspring of poor performance. This means that the genetic operations by themselves have limited ability in finding the global optimal solution. In this study, bottleneck shifting serves as one kind of local search method and is hybridized with the genetic algorithm.

Defining Neighborhood

A central problem of any local search procedure for combinatorial optimization problems is how to define the effective neighborhood around an incumbent solution. In this study, the effective neighborhood is based on the concept of critical path. To define neighborhood using critical path is not new for a job shop scheduling problem and has been employed by many researchers [43].

The feasible schedules of an fJSP problem can be represented with disjunctive graph $G = (N, A, E)$, with node set N, ordinary (conjunctive) arc set A, and disjunctive arc set E. The nodes of G correspond to operations, the real arcs (A) to precedence relations, and the dashed arc (E) to implementation sequence of operations to be performed on the same machine. For example, the following schedule of the 4×4 problem can be illustrated in the disjunctive graph shown in Fig. 5.37:

$$\begin{aligned} S = \{&(o_1,1,M_4:0-16),(o_1,2,M_3:21-33),(o_1,3,M_3:33-51),\\ &(o_1,4,M_1:51-69),(o_2,1,M_2:0-16),(o_2,2,M_4:94-112),\\ &(o_2,3,M_1:112-136),(o_2,4,M_4:136-148),(o_3,1,M_3:0-21),\\ &(o_3,2,M_1:21-45),(o_3,3,M_2:45-68),(o_3,4,M_1:69-105),\\ &(o_4,1,M_2:16-32),(o_4,2,M_4:32-62),(o_4,3,M_4:62-94),\\ &(o_4,4,M_3:94-118)\} \end{aligned}$$

In Fig. 5.37, S and T are dummy starting and terminating nodes respectively. The number above each node represents the processing time of that operation. The critical path is the longest path in a graph. For an fJSP schedule, the makespan is equal to the length of the critical path in the corresponding disjunctive graph. The critical path is highlighted with broad-brush arcs in Fig. 5.37. Any operation on the critical path is called a critical operation. A critical operation cannot be delayed without increasing the makespan of the schedule.

The job predecessor $P_J(r)$ of an operation r is the operation preceding r in the operation sequence of the job that r belongs to. The machine predecessor $P_M(r)$ of an operation r is the operation preceding r in the operation sequence on the machine that r is processed on. If an operation r is critical, then at least one of $P_J(r)$ and $P_M(r)$ must be critical, if they exist. In this study, if a job predecessor and a machine

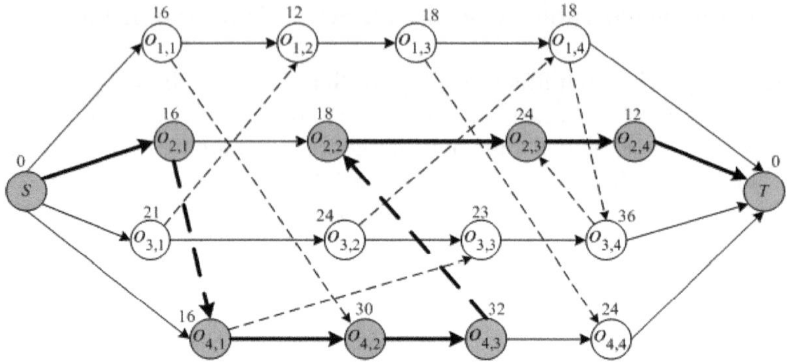

Fig. 5.37 Illustration of disjunctive graph

predecessor of a critical operation are both critical, then choose the predecessor (from these two alternatives) that appears first in the operation sequence.

A new schedule that is slightly different from the incumbent solution can be generated by changing the processing sequence of two adjacent operations performed on the same machine, *i.e.*, reversing the direction of the disjunctive arc that links the two operations. The neighborhood created in this way is called as type I here. Neighbor solutions can also be generated by assigning a different machine for one operation. This kind of neighborhood is called as type II.

The makespan of a schedule is defined by the length of its critical path, in other words, the makespan is no shorter than any possible path in the disjunctive graph. Hence, for a neighbor solution of type I, only when these two adjacent operations are on the critical path can the new solution be superior to the old one. Likewise, for a neighbor solution of type II, it cannot outperform the incumbent solution if the operation is not a critical one.

For the fJSP problem, we can only swap the operation sequence between a pair of operations that belong to different jobs. It is possible to break down the critical path into a number of blocks, each of which is a maximal sequence of adjacent critical operations that require the same machine. As a result, the possible swaps are further confined as follows:

1. In each block, we only swap the last two and first two operations;
2. In cases where the first (last) block contains only two operations, these operations are swapped;
3. If a block contains only one operation, then no swap is made.

Due to the strict restrictions above, possible swaps occur only on a few pairs of adjacent operations that belong to different jobs on the critical path. Neighbor solutions of type I are actually generated by implementing these possible swaps. Figure 5.38 shows the critical path, critical blocks and the possible swaps in a schedule. The total number of the neighbors of type I (N_{I}) is less than the total number of critical operations (N_C) since some critical operations cannot involve the possible swaps.

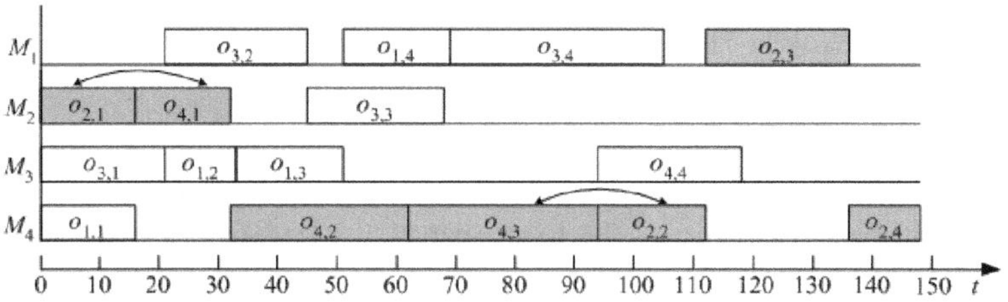

Fig. 5.38 Neighborhood of type I

A neighbor solution of type II can be created by assigning a different machine $j \in A_{ik}$ for a critical operation o_{ik}. Let n_{II_l} be the number of machines on which the l-th critical operation can be assigned. $n_{\mathrm{II}_l} - 1$ neighbors can be generated by assigning the operation on any of the other $n_{\mathrm{II}_l} - 1$ available machines. Hence, the total number of neighbors of type II (N_{II}):

$$N_{\mathrm{II}} = \sum_{l=1}^{N_C} n_{\mathrm{II}_l} - 1 \tag{5.12}$$

Since N_{I} is less than N_C, N_{II} generally represents a much larger number than N_{I}.

Local Search Transition Mechanism

During the local search, the original schedule will transit to a better neighbor solution, if it exists. This gives rise to a new problem: what is an improved solution? For the fJSP problem, there may be more than one critical path in a schedule, in which the makespan is determined by the length of the critical path. A solution with a smaller number of critical paths may provide more potential to find solutions with less makespan nearby because the makespan cannot be decreased without breaking all the current critical paths. An important problem of any local search method is how to guide to the most promising areas from an incumbent solution. In this study, a solution is taken to be an improved solution if it satisfies either of the two alternative requirements:

1. An improved solution has a larger fitness value than the incumbent solution;
2. The improved solution has the same fitness value as the incumbent solution, yet it has less critical paths.

Adjusting Neighborhood Structure

Let $N(i)$ denote the set of neighborhood of solution i. The enlarged two-pace neighborhood can be defined as the union of the neighborhood of each neighbor of the incumbent solution. Let $N_2(i)$ be the two-pace neighborhood of solution i; then,

$$N_2(i) = \cup_{j \in N(i)} N(j) \tag{5.13}$$

A larger neighborhood space size generally indicates a higher quality of the local optima because, in each step of the local search, the best solution among a larger number of neighbor solutions is selected as the incumbent solution for the next local search iteration. On the other hand, a larger neighborhood space size would bring a greater computational load because more neighbor solutions have to be evaluated and compared. That is, each step of the local search will take a longer time. Hence, the number of the local search iterations is decreased when the time spent on a local search is limited. As a result, the deep search ability is not fully utilized.

In order to enhance the search ability of the local search without incurring too much computational load during the search process over type II neighborhood, the local search procedure will implement over the enlarged two-pace neighborhood only when it reaches the local optimum of the one-pace neighborhood. The broad search process is illustrated in Fig. 5.39.

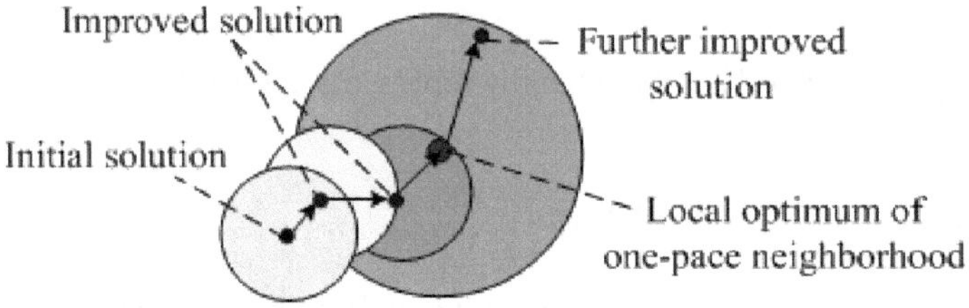

Fig. 5.39 Broad search over the enlarged two-pace neighborhood

Bottleneck Shifting Procedure

The main positive effect of hybridizing GAs with local search is the improvement in the convergence speed to local optima. On the other hand, the main negative effect is the increase in the computation time per generation. Thus, the number of generations is decreased when the available computation time is limited. As a result, the global search ability of genetic algorithms is not fully utilized [66]. One crucial problem of the hybrid genetic algorithm is how to allocate properly the available computation time wisely between genetic search and local search.

The depth of local search is among the main determinants of the computation time spent on local search. The depth refers to how many local search steps are implemented in each generation of the genetic search. The space size of type I neighborhood is much less than that of type II. Hence, more local search steps are implemented over type I neighborhood than over type II, and only one step is implemented for local search over type II neighborhood here.

Let L be the number of local search steps over type I neighborhood; the bottleneck shifting procedure is shown in Fig. 5.40.

```
procedure: Bottleneck Shifting
input: initial solution S0, local search parameters
output: improved solution S
begin
   S←S0;
   for i=1 to L
      S0←S;
      yield S by local search over type I neighborhood of S0;
      S0←S;
      yield S by local search over type II neighborhood of S0;
   for i=1 to L
      S0←S;
      yield S by local search over type I neighborhood of S0;
output the improved solution S
end
```

Fig. 5.40 Bottleneck shifting procedure

It is notable that the local search over both type I and type II neighborhoods does not guarantee to reach the local optima. That is, a newly generated solution evolves to the local optimum gradually along the evolution process of the whole population if the solution is preserved in the selection procedure. If a solution is replaced by its local optimum at the time when it is generated, the specific information contained in the chromosome may be lost in the genetic search because different chromosomes may come to the same local optimum. As a result, the diversity of the population is hindered and the risk of premature convergence increases. Moreover, finding the local optimum needs a great computational complexity because it needs too many steps for the local search to reach the local optima.

5.3.4 Experiment on Benchmark Problems

We have used random selection to generate the initial population. Then we applied the proposed hybrid multistage operation-based GA with the following parameters: *popSize*: 100; p_M=0.3; p_C=0.6. The results obtained by our method are compared with the results from Kacem *et al.* [57] and Xia and Wu [60].

Table 5.20 is an instance of partial flexibility. The symbol "-"in Table 5.20 shows that the machine is not available for the corresponding operation. In the flexible job shop, there are 8 jobs with 27 operations to be performed on 8 machines. One of the best solutions is shown in Fig. 5.41. This test instance seems to be oversimplified. It takes on average 16.4 generations for the h-moGA to converge to the solutions. The computation time averages 5 min. As shown in Table 5.21, it is easy to find the h-moGA outperform all the other approaches.

Table 5.20 Case of partial flexibility [57]

		M_1	M_2	M_3	M_4	M_5	M_6	M_7	M_8
J_1	o_{11}	5	3	5	3	3	-	10	9
	o_{12}	10	-	5	8	3	9	9	6
	o_{13}	-	10	-	5	6	2	4	5
J_2	o_{21}	5	7	3	9	8	-	9	-
	o_{22}	-	8	5	2	6	7	10	9
	o_{23}	-	10	-	5	6	4	1	7
	o_{24}	10	8	9	6	4	7	-	-
J_3	o_{31}	10	-	-	7	6	5	2	4
	o_{32}	-	10	6	4	8	9	10	-
	o_{33}	1	4	5	6	-	10	-	7
J_4	o_{41}	3	1	6	5	9	7	8	4
	o_{42}	12	11	7	8	10	5	6	9
	o_{43}	4	6	2	10	3	9	5	7
J_5	o_{51}	3	6	7	8	9	-	10	-
	o_{52}	10	-	7	4	9	8	6	-
	o_{53}	-	9	8	7	4	2	7	-
	o_{54}	11	9	-	6	7	5	3	6
J_6	o_{61}	6	7	1	4	6	9	-	10
	o_{62}	11	-	9	9	9	7	6	4
	o_{63}	10	5	9	10	11	-	10	-
J_7	o_{71}	5	4	2	6	7	-	10	-
	o_{72}	-	9	-	9	11	9	10	5
	o_{73}	-	8	9	3	8	6	-	10
J_8	o_{81}	2	8	5	9	-	4	-	10
	o_{82}	7	4	7	8	9	-	10	-
	o_{83}	9	9	-	8	5	6	7	1
	o_{84}	9	-	3	7	1	5	8	-

For another test example, there are 10 jobs with 30 operations to be performed on 10 machines shown in Table 5.22. One of the best solutions is shown in Fig. 5.42. On average, the h-moGA takes 26.50 generations and about 17 min to find the solutions. Table 5.23 gives the performance of the proposed method compared with other algorithms. The proposed h-moGA is also all the other approaches.

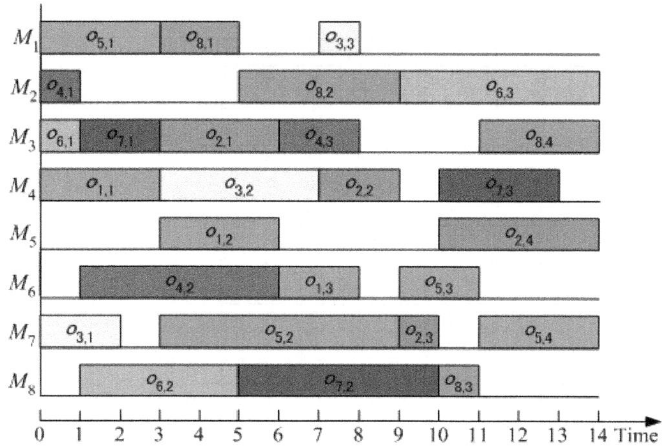

Fig. 5.41 Best solution 1 of problem $8 \times 8(t_M = 14, w_M = 12, w_T = 77)$

Table 5.21 Result comparison (8×8)

	Heuristic method (SPT)	Simple GA	AL+CGA by Kacem	PSO+SA by Xia and Wu	h-moGA
t_M	19	16	16	16	14
W_T	91	77	75	73	77
W_M	16	14	14	13	12

5.4 Integrated Operation Sequence and Resource Selection Model

In order to obtain greatest profit, manufacturing enterprises tend to improve the ability to respond to the rapidly changing market demands, which require the effective and efficient manufacturing of a variety of products with varying volume [76]. Based on the development of information technique and manufacturing conception, the Integrated Manufacturing System (IMS) is built as the most adaptive and available approach for modern requirements. To get an optimal operation sequence with flexible resource assignment is the main function of IMS; therefore several important issues come in fact of the managers for them, to find optimal production planning:

- How to get a minimum execution time (makespan) for responding to the emergency or forecasting orders?
- How to deal with different lot sizes of orders for minimizing the transportation cost by workers or robots?

Table 5.22 Case of a total flexibility [57]

		M_1	M_2	M_3	M_4	M_5	M_6	M_7	M_8	M_9	M_{10}
J_1	o_{11}	1	4	6	9	3	5	2	8	9	5
	o_{12}	4	1	1	3	4	8	10	4	11	4
	o_{13}	3	2	5	1	5	6	9	5	10	3
J_2	o_{21}	2	10	4	5	9	8	4	15	8	4
	o_{22}	4	8	7	1	9	6	1	10	7	1
	o_{23}	6	11	2	7	5	3	5	14	9	2
J_3	o_{31}	8	5	8	9	4	3	5	3	8	1
	o_{32}	9	3	6	1	2	6	4	1	7	2
	o_{33}	7	1	8	5	4	9	1	2	3	4
J_4	o_{41}	5	10	6	4	9	5	1	7	1	6
	o_{42}	4	2	3	8	7	4	6	9	8	4
	o_{43}	7	3	12	1	6	5	8	3	5	2
J_5	o_{51}	7	10	4	5	6	3	5	15	2	6
	o_{52}	5	6	3	9	8	2	8	6	1	7
	o_{53}	6	1	4	1	10	4	3	11	13	9
J_6	o_{61}	8	9	10	8	4	2	7	8	3	10
	o_{62}	7	3	12	5	4	3	6	9	2	15
	o_{63}	4	7	3	6	3	4	1	5	1	11
J_7	o_{71}	1	7	8	3	4	9	4	13	10	7
	o_{72}	3	8	1	2	3	6	11	2	13	3
	o_{73}	5	4	2	1	2	1	8	14	5	7
J_8	o_{81}	5	7	11	3	2	9	8	5	12	8
	o_{82}	8	3	10	7	5	13	4	6	8	4
	o_{83}	6	2	13	5	4	3	5	7	9	5
J_9	o_{91}	3	9	1	3	8	1	6	7	5	4
	o_{92}	4	6	2	5	7	3	1	9	6	7
	o_{93}	8	5	4	8	6	1	2	3	10	12
J_{10}	$o_{10,1}$	4	3	1	6	7	1	2	6	20	6
	$o_{10,2}$	3	1	8	1	9	4	1	4	17	15
	$o_{10,3}$	9	2	4	2	3	5	2	4	10	23

Table 5.23 Result comparisons (10×10)

	Heuristic method (SPT)	Classic GA	AL+CGA by Kacem	PSO+SA by Xia and Wu	h-moGA
t_M	16	7	7	7	7
W_T	59	53	45	44	43
W_M	16	7	6	6	5

- How to balance the workload of all the machines in our plants for reducing the work-in-process inventories and operation bottlenecks?
- How to reduce the complex transportation process between machines in a local plant?

The integrated Operation Sequences and Resource Selection (iOS/RS) problem is formulated in particular, for the reason that it was originally derived from the real

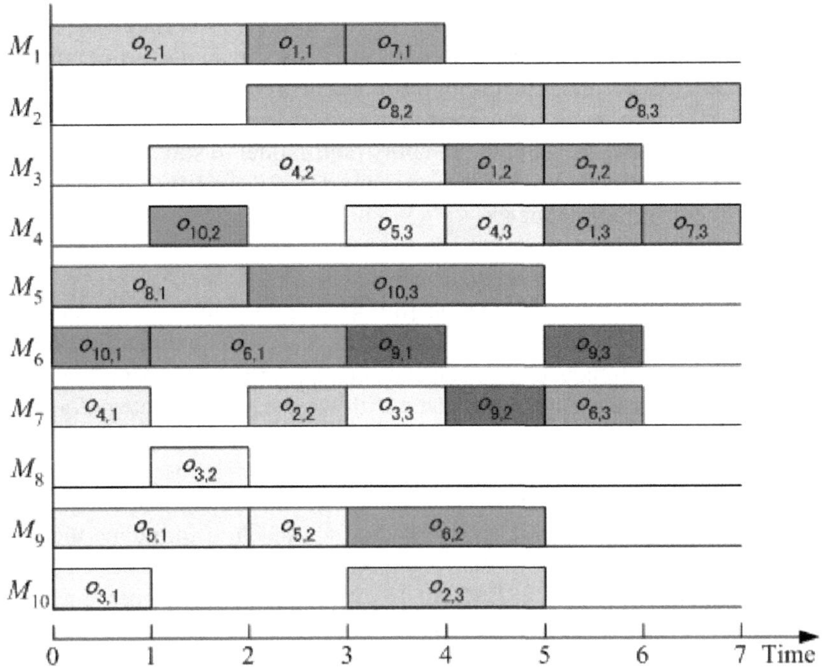

Fig. 5.42 Optimal solution of problem $10\times10(t_M = 7, w_M = 5, w_T = 43)$

production processes in manufacturing systems, and approximates to them. For instance, each order consists of some operations, while the sequences are not fixed, which in terms of several precedence constraints–lot size and unit load size for different orders–are considered in the scheduling process; this means the starting times of operations depend not only on the finishing time of preceding operations within the same order but also on their finishing time for unit load size.

During the past several years, many researchers have put great effort into the area of integrated process planning and scheduling problem. Tan [77] reported a brief review of the research in the process planning and scheduling area and discussed the extent of applicability of various approaches, and also proposed a linearized polynomial mixed integer programming model for this problem in recent research work [78]. Dellaert *et al.* [64] discussed multi-level lot-sizing problems in material requirements planning (MRP) systems. They developed a binary encoding genetic algorithm (GA) and designed five specific genetic operators to ensure that exploration takes place within the set of feasible solutions. Raa and Aghezzaf [75] also introduced a robust dynamic planning strategy for lot-sizing problems with stochastic demands in their recent research work. Recently, for improving the flexibility of machine assignment, Kacem *et al.* [57] proposed a genetic algorithm controlled by the assigned model which is generated by the approach of localization. Najid *et al.* [59] used simulated annealing (SA) for optimizing the flexible assignment of machines in flexible Job-shop Scheduling Problem. Lopez and Ramirez [68] newly describe the design and implementation of a step-based manufacturing information system to share flexible manufacturing resources data.

Anyway, all the research mentioned above considered the alternative machines for each operation, and they wanted to apply their model to solving the flexible assignment of various resources (machines). However, there exists a weakness which is fixing all the operation sequences or non-constraint operation sequences. That is, they ignore the flexibility especially for orders, which includes some precedence constraints and corresponding sequences.

Especially in recent work by Moon [69]–[73], a GA approach is proposed to solve such a kind of iOS/RS problem considering the orders with unfixed operation sequence; however, it encoded the chromosome by only considering the information of operation sequence, and hybridized with some heuristic strategy for resource selection (by minimum processing time). This approach actually improves the efficiency of convergence and the speed of calculation, but may lose some optimal solution, because it ignores the transition time between different machines. To avoid this limitation, we propose an effective coding approach to formulate the iOS/RS model into a two-dimensional format. The new idea is built on the basic concept of multistage decision making (MSDM) model. We separate all the operations as a set of stages, and in each stage (operation) several alternative states (machines) are offered for selection; hence our job is to make a decision in all the stages for choosing states and getting an optimal schedule. For this reason we formulate our multistage operation-based Genetic Algorithm (moGA), and define the chromosome with two vectors which contain information, operation sequence and machine selection.

5.4.1 Mathematical Formulation of iOS/RS

In the Integrated Manufacturing System, many hot topics have been addressed by many researchers, such as Flow-shop Scheduling Problem (FSP), Job-shop Scheduling Problem (JSP) , and flexible Job-shop Scheduling Problem (fJSP) . The iOS/RS problem is implemented as a further extension from these classical scheduling problems, and has been valuable for research because of its resemblance to the real manufacturing system.

We present the extension of the iOS/RS problem from the other classical scheduling problems (JSP, fJSP). It distinctly describe that the development of the iOS/RS problem is from not only some constraints considered close the real environment, but also the objectives treated for optimizing. To formulate the iOS/RS more precisely, several assumptions should be brought forward:

A1. All the orders for scheduling arrive simultaneously with no precedence.
A2. All the available machines are idle before scheduling.
A3. Unit load sizes are fixed for all the machines, and each machine cannot stop until finishing the operating load size of order.

A4. For the same order, unit load size can move to the next operation as soon as the preceding operation is finished, even if other leaving unit load sizes are still operating on the previous machine.
A5. Transition time between different machines will be considered.
A6. Setup time is included in processing time.

To demonstrate clearly the process planning problem in detail we give a simple example shown in Fig. 5.43, which presents the material to be machined in a modern manufacturing system by 2 orders for removing 7 volumes, the corresponding lot size for the two order are (60, 50), and the unit load size is 10. All the manufacturing plans of this example are offered in Table 5.24, which includes the type of all the operations and the alternative machine sets.

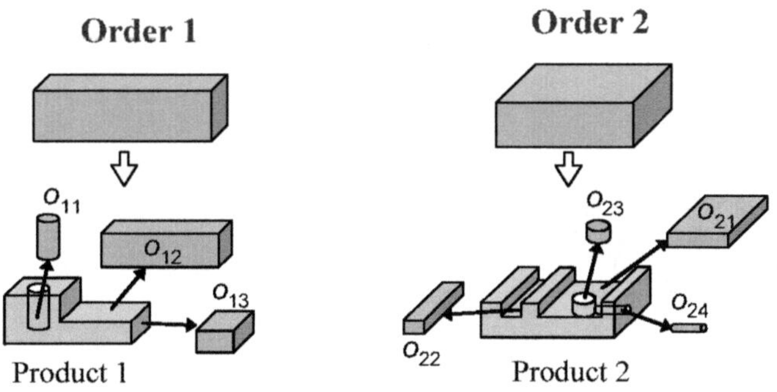

Fig. 5.43 Simple example for iOS/RS problem

Table 5.24 Manufacturing Plan

Operation index (order k, operation o_{ki})	Operation type	Machine selection M_m
1 $(1,o_{11})$	Milling	M_1, M_4
2 $(1,o_{12})$	Milling	M_2, M_5
3 $(1,o_{13})$	Milling	M_2, M_3, M_5
4 $(2,o_{21})$	Milling	M_2, M_3
5 $(2,o_{22})$	Milling	M_3, M_5
6 $(2,o_{23})$	Drilling	M_1, M_4, M_5
7 $(2,o_{24})$	Drilling	M_1, M_2

In Fig. 5.44, we describe operation sequence constraints of the two orders by using some kind of node graphs, so we can get the same precedence constraints data shown in Table 5.25. For the data offered we can acknowledge that the operation sequence is alternative and follows some precedence constraints for each order. For instance, in order 1: both $o_{11}, o_{12}, o_{13}$and o_{12}, o_{11}, o_{13} are legal, while o_{11}, o_{13}, o_{12} are illegal.

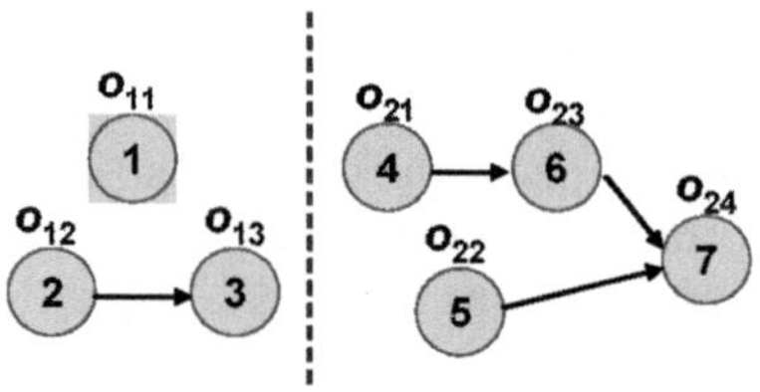

Fig. 5.44 Operation precedence constraints of two orders

Table 5.25 Operation precedence data

Order k	(i, j)
1	(2, 3)
2	(4, 6)
	(5, 7)
	(6, 7)

Tables 5.26 and 5.27 prepare the data set of processing time for each operation and transition time between different machines. They show that the five machines have different capability for each corresponding operation, although not available to some of them.

As shown in this example, the objective of the iOS/RS problem is to determine an optimal schedule with operation sequences for all the orders. Therefore, the iOS/RS problem can be defined as: given a set of K orders with lot size q_k which are to be processed on M machines with alternative operations sequences and alternative machines for operations, find an operations sequence for each job and a schedule in which jobs pass between machines and a schedule in which operations on the same jobs are processed such that it satisfies the precedence constraints and it is optimal with respect to minimize makespan and balance the workloads.

Before the formulation of mathematical model, we should first define all the related symbols and notation as following:

Table 5.26 Processing time of operations on different machines

Order k:	Order 1			Order 2			
Operation o_{ki} / Machine M_m	1 (o_{11})	2 (o_{12})	3 (o_{13})	4 (o_{21})	5 (o_{22})	6 (o_{23})	7 (o_{24})
M_1	5	-	-	-	-	6	8
M_2	-	7	6	6	-	-	12
M_3	-	-	5	7	9	-	-
M_4	7	-	-	-	-	5	-
M_5	-	6	8	-	5	6	-

Table 5.27 The transition time between different machines

	M_1	M_2	M_3	M_4	M_5	Available capacity L_m
M_1	-	7	27	26	17	1500
M_2	19	-	15	19	10	1500
M_3	5	17	-	36	27	1500
M_4	8	20	4	-	30	1500
M_5	18	18	14	5	-	1500

Notation

Indices

i, j: index of operation number, $i, j = 1, 2, \ldots, J_k$
k, l: index of orders, $k, l = 1, 2, \ldots, K$
m, n: index of machines, $m, n = 1, 2, \ldots, N$

Parameters

K: number of orders
N: number of machines
O_k : set of operations for order $k, i.e., O_k = o_{ki}|i = 1, 2, \ldots, J_k$
o_{ki} : the i-th operation for order k
J_k : number of operations for order k
M_m: the m-th machine
q_k: lot size of order k
A_m: set of operations that can be processed on machine m
p_{kim}: unit processing time of operation o_{ki} on machine m
L_m: capacity of machine m
r_{kij}: precedence constraints

$r_{kij} = 1$, if o_{ki} is predecessor of o_{kj} in order k; 0, otherwise.
u_{kij}: unit load size of order k from operation o_{ki} to operation o_{kj}
t_{mn} : unit shipping time between machine m to machine n
c_{ki} : completion time of operation o_{ki}
$c_{ki} = s_{ki} + q_k \sum_{m=1}^{N} p_{kim} x_{kim}$

Decision Variables

$$y_{kilj} = \begin{cases} 1 \text{ if operation } o_{ki} \text{ is performed immediately} \\ \quad \text{before operation } o_{lj}; \\ 0 \text{ otherwise} \end{cases}$$

s_{ki} : starting time of operation o_{ki}

$$x_{kim} = \begin{cases} 1 \text{ if machine } m \text{ is selected for operation } o_{ki}; \\ 0 \text{ otherwise} \end{cases}$$

$$v_{kmn} = \begin{cases} 1 \text{ for order } k, \text{ if machine } n \text{ is selected} \\ \quad \text{for transmitting from machine } m \text{ ,and } n \neq m \text{ ;} \\ 0 \text{ otherwise} \end{cases}$$

The overall model can be described as follows:
where:
c_M: makespan
w_B: workload balancing evaluation
w_m: workload of machine m, $w_m = \sum_{k=1}^{K} \sum_{i=1}^{J_k} q_k p_{kim} s_{kim}$
$\bar{w}$: average workload over all the machines, $w_m = \frac{1}{N} \sum_{m=1}^{N} w_m$
T_{mn}: total transition times between machine m and machine n for all orders:

$$\min\ f_1 = c_M = \max_{i,k,m} c_{kim} \tag{5.14}$$

$$\min\ f_2 = w_B = \frac{1}{N} \sum_{m=1}^{N} (w_m - \bar{w})^2 \tag{5.15}$$

$$\min\ f_3 = T_{mn} = \sum_{k=1}^{K} \sum_{m=1}^{N} \sum_{n=1}^{N} t_{mn} u_{kmn} \tag{5.16}$$

To define precisely the mathematical model, several constraints are presented as follows:

$$\text{s. t. } (s_{lj} - c_{ki}) x_{kim} x_{ljm} y_{kilj} \geq 0 \quad \forall\ (k,i),(l,j),m \tag{5.17}$$

$$\{s_{kj} - (s_{ki} + u_{kij} p_{kim} x_{kjn} + t_{mn})\} r_{kij} x_{kim} x_{kjn} \geq 0 \quad \forall\ i,j,k, \forall\ m,n \tag{5.18}$$

$$\{c_{kj} - u_{kij} p_{kjn} - (c_{ki} + t_{mn})\} r_{kij} x_{kim} x_{kjn} \geq 0 \quad \forall\ i,j,k, \forall\ m,n \tag{5.19}$$

$$\sum_{k=1}^{K}\sum_{i=1}^{J_k} q_k p_{kim} x_{kim} \le L_m \quad \forall\, m \tag{5.20}$$

$$r_{kij} y_{kjki} = 0 \quad \forall\, i,j,k \tag{5.21}$$

$$y_{kjki} = 0 \quad \forall\, i,k \tag{5.22}$$

$$y_{kilj} + y_{ljki} = 1 \quad \forall\, (k,i),(l,j),(k,i) \neq (l,j) \tag{5.23}$$

$$\sum_{m=1}^{N} x_{kim} = 1 \quad \forall\, i,k \tag{5.24}$$

$$x_{kim} = 0 \quad \forall\, (k,i) \notin A_m, \forall\, m \tag{5.25}$$

$$S_{ki} \ge 0 \quad \forall\, i,k \tag{5.26}$$

$$y_{kilj} \in 0,1 \quad \forall\, (k,i),(l,j) \tag{5.27}$$

$$x_{kim} \in 0,1 \quad \forall\, i,k,m \tag{5.28}$$

$$v_{kmn} \in 0,1 \quad \forall\, k,m,n \tag{5.29}$$

Equation 5.17 imposes that for any resource (machine), it cannot be selected for one operation until the predecessor is completed (see Fig. 5.45).

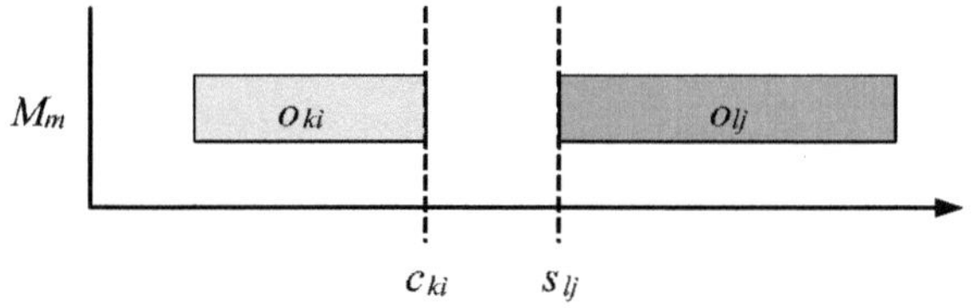

Fig. 5.45 Time chart for the constraints on the same machine

Equations 5.18 and 5.19 impose the intermediate transportation instances in a single plant. Both of the constraints must be satisfied simultaneously to ensure operations can be run uninterrupted on one machine (see Fig. 5.46).

Equation5.20 restricts the available capacity for each machine. Equation 5.21 ensures that the precedence constraints are not violated. Equations 5.22 and 5.23 ensure the feasible operation sequence. Equations 5.24 and 5.25 ensure the feasible resource selection. Equations 5.26–5.29 impose nonnegative condition.

5.4.2 Multistage Operation-based GA for iOS/RS

Let $P(t)$ and $C(t)$ be parents and offspring in current generation t, and use P and C represent the chromosome set P_1, P_2 and C_1, C_2. In overall procedure, we choose the appropriate function of crossover and mutation corresponding to the two vector of the chromosome, especially only once crossover for Phase 1 because after the

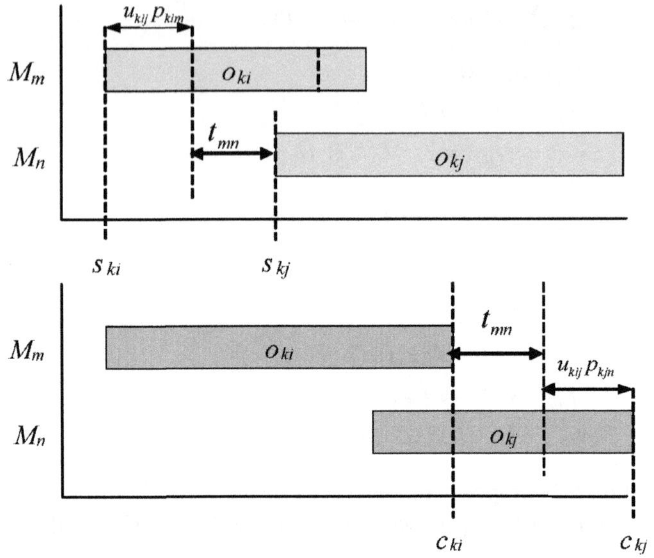

Fig. 5.46 Time chart for the constraints in the same plant with transportation

operation sequence is decided we do not need to change the stages (operations) position in Phase 2 for choosing states (machines).

Generally, Genetic Algorithms (GAs) can be thought of as a special kind of genetic search over the subspace of local optima [5], mainly adapting some local optimizer to Genetic Algorithms (GAs) for improving solutions toward local optimization. As is known, crossover and mutation can avoid the solution set trap in the local optimal result even if they probably overslip an opportunity to find a global optimal solution sometimes. Consequently, a GA which hybridizes a local search with a traditional genetic search can improve the convergence speed and outperform either method operating alone.

In Barretta's work [63], the GA has been used to get optimal lot sizing in a production planning problem, and they also consider the concept of multistage capacitated constraints. Moreover, Nishi and Konishi [74] also decentralized the supply chain planning system for multi-stage production processes.

5.4.2.1 Moon's Approach

To solve the multiobjective iOS/RS model, the GA has been proved to be an effective way for finding an optimal solution. Several related works by Moon [72] have referred a GA approach especially for solving such kinds of iOS/RS problems. However, to derive a feasible complete schedule, he only considered the optimal op-

```
procedure: moGA for iOS/RS model
input: iOS/RS data, GA parameters (popSize, maxGen, p_M, p_C)
output: Pareto optimal solutions E
begin
   t ← 0;
   initialize P_1(t) by priority-based encoding routine;
   initialize P_2(t) by machine permutation-based encoding routine;
   calculate objectives z_1(P) and z_3(P) by priority-based decoding routine;
   create Pareto E(P);
   evaluate eval(P) by adaptive-weight fitness assignment routine;
   while (not terminating condition) do
      create C_1(t) from P_1(t) by one-cut point crossover routine;
      create C_1(t) from P_1(t) by swap mutation routine;
      create C_2(t) from P_2(t) by neighbor search mutation routine;
      improve P_1(t) and C_1(t) to yield P'_1(t) and C'_1(t) by left-shift hillclimber;
      calculate objectives z_1(P) and z_3(P) by priority-based decoding routine;
      update Pareto E(P,C);
      evaluate eval(P,C) by adaptive-weight fitness assignment routine;
      select P(t+1) from P(t) and C(t) by roulette wheel selection routine;
      t ← t+1;
   end
   output Pareto optimal solutions E(P,C)
end
```

eration sequence, but selected the resources in terms of minimum processing time which is shown in Fig. 5.47.

```
procedure: a feasible solution generation
input: set of networks with precedence constraints for all orders and set of resources for each
        operation;
       generating a set of priorities on the nodes in the network;
begin
  while (any node remains) do
        if every node has a predecessor, then the network is infeasible: stop.
        else pick a node v with the highest priority among nodes with no predecessors;
                que ← v;
             select a resource with minimum processing time among available resources
               for an operation corresponds to the v, where the total machining time for
               selected resource must not exceed the maximum allowable time;
             delete v and all edges leading out of v from the network;
  end
end
```

Fig. 5.47 Feasible path generation algorithm by Moon

That means, for machines assignment we acknowledge that the minimum processing time assignment is the optimal choosing strategy for the solution, while the transition times between machines are ignored. Anyway, Moon-Kim-Gen's approach will inevitably lose some optimal solutions for some iOS/RS cases.

5.4.2.2 Effective Encoding Method in moGA

To avoid such kinds of loss and improve the convergence speed at the same time, we propose a multistage operation-based Genetic Algorithm (moGA) for solving this iOS/RS problem. Actually our idea of moGA comes from the basic concept of the multistage decision making model shown in Fig. 5.48, and has been used for solving some flexible job-shop scheduling problems [61]. There are several stages separating the route from the starting node to the terminal node, and in each stage several states are offered for choosing. After we make all the decisions for choosing states, we can draw a solution, and the fitness of the result is in terms of the different decisions made in the route.

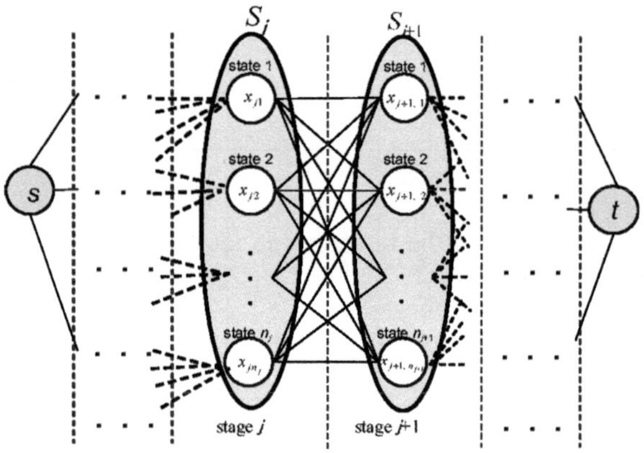

Fig. 5.48 Basic concept of multistage decision making model

Since we have accepted that both operation sequences and machine selection can affect the solution in iOS/RS problems, it is obvious that the chromosome presentation of multistage operation-based GA (moGA) for iOS/RS problems consists of two parts:

- Priority-based encoding for operation sequences;
- Machine permutation encoding for machine selection.

To develop a multistage operation-based genetic representation for the iOS/RS, there are three main phases:

Phase 1: Creating an operation sequence.

Step 1.1: Generate a random priority to each operation in the model using priority-based encoding procedure for first vector v_1.
Step 1.2: Decode a feasible operation sequence that satisfies the precedence constraints of iOS/RS

Phase 2: Assigning operations to machines

Step 2.1: Input the operation sequence found in step 1.2.
Step 2.2: Generate a permutation encoding for machine assignment of each operation (second vector v_2).

Phase 3: Designing a schedule

Step 3.1: Create a schedule S using operation sequence and machine assignments.
Step 3.2: Draw a Gantt chart for this schedule.

Phase 1: Creating an Operation Sequence

Step 1.1: Generate a random priority to each operation in the model using priority-based encoding procedure for first vector v_1.
Step 1.2: Decode a feasible operation sequence that satisfies the precedence constraints of iOS/RS

In this part, all the chromosomes are generated based on the simple iOS/RS example related in Fig. 5.43. First, we use a priority encoding procedure to formulate chromosome, with the procedure shown in Fig. 5.49, and we can draw a chromosome for the simple example as Fig. 5.50.

```
procedure: priority-based encoding for sequencing operations
input: total number of operations J
output: chromosome v1()
begin
  for j=1 to J
    v1(j)←j;
  for i=1 to
    repeat
      k←random[1, J];
      l←random[1, J];
    until l≠k
    swap (v1(k), v1(l));
  output chromosome v1()
end
```

Fig. 5.49 Priority-based encoding procedure

index j:	1	2	3	4	5	6	7
priority $v_1(j)$:	**5**	**1**	**7**	**2**	**4**	**6**	**3**

Fig. 5.50 A simple example of chromosome

By using the decoding procedure shown in Fig. 5.51, one feasible operation sequence will be obtained as follows:
operation sequences: $o_{12}, o_{21}, o_{22}, o_{11}, o_{23}, o_{13}, o_{24}$

```
procedure: priority-based decoding for sequencing operations
input: set of successive operations L_k after activity k;
       set of starting activities L_0;
       chromosome v_1()
output: operation sequence
begin
  k←0;
  S←φ, S̄←φ;
  while ( S̄≠φ ) do
     S̄ ← S̄ ∪ L_k;
     k*←argmin{v_1(j)| j ∈ S̄};
     S̄← S̄ \ k*;
     S← S ∪ k*;
     k←k*;
  end
  output operation sequence
end
```

Fig. 5.51 Priority-based decoding procedure

Phase 2: Assigning Operations to Machines

Step 2.1: Input the operation sequence found in step 1.2.
Step 2.2: Generate a permutation encoding for machine assignment of each operation (second vector v_2).

After finishing Phase 1 we draw a fixed operation sequence, which means the position of all stages (operations) has been decided. Hence in Phase 2 we input the fixed operation sequence, and transition time, and after states (machines) assignment, we will output the whole schedule and solution of makespan.

Since the iOS/RS problem is treating the flexible machine selection, we use machine permutation coding procedure to make another part of the chromosome. That is, we will first build a multistage operation-based frame shown in Fig. 5.52. It is obvious that the stage number is just the total operation number, and also in each stage, the machines available are treated as the state correspondingly, and then by the encoding procedure shown in Fig. 5.53, we can draw another part of moGA chromosome as follows. By using the decoding procedure shown in Fig. 5.54, we

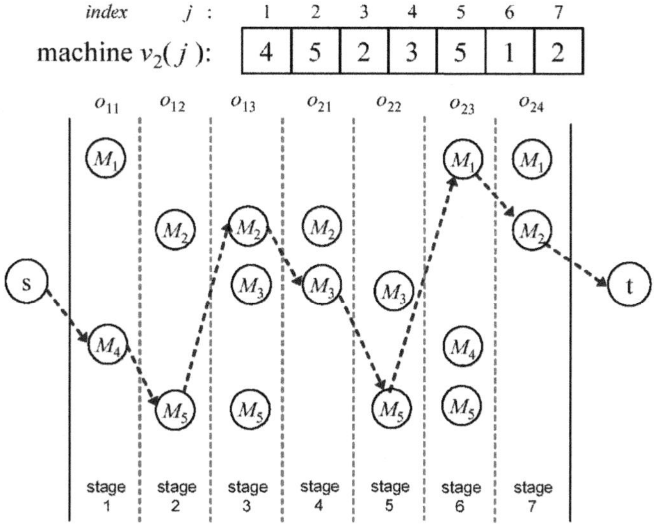

Fig. 5.52 Node graph of machine selection in moGA

```
procedure: permutation encoding for selecting machines
input: number of stages J, number of states at stage j, n_j
output: chromosome v_2()
begin
   for j=1 to J
     v_2(j) ←0;
     for j=1 to J
     v_2(j) ← random[1, n_j];
   output chromosome v_2()
end
```

Fig. 5.53 Machine permutation encoding

can assign the machine selection to the fixed stage sequence offered by Phase 1 shown in Fig. 5.52.

```
procedure: permutation decoding for selecting machines
input: chromosome v2(), operation sequence S1,
    number of stages J
output: schedule S
begin
  S← φ;
  j←1;
  while (j ≤J) do
    r ← v2(j);
    oj ← S1(j);
    S← S ∪ {(oj, Mr)};
    j ←j+1;
  end
  output schedule S
end
```

Fig. 5.54 Machine permutation decoding

Phase 3: Designing a Schedule

Step 3.1: Create a schedule S using operation sequence and machine assignments:

$$
\begin{aligned}
S = & \{Order1, Order2\} \\
& \{(o_{12}, M_5 : 0-360), (o_{21}, M_3 : 0-250), (o_{22}, M_5 : 360-610), \\
& (o_{11}, M_4 : 65-485), (o_{23}, M_1 : 428-728), (o_{13}, M_2 : 155-515), \\
& (o_{24}, M_2 : 515-1115)\}
\end{aligned}
$$

Step 3.2: Draw a Gantt chart for this schedule. After calculating the makespan by the data in Tables 5.26 and 5.27 , we can draw the Gantt chat shown in Fig. 5.55. We found the makespan for this simple example is 1115, and the evaluation of the workloads balancing is 6.65×10^4, while the total transition times is 4 (Order 1: $M_5 - M_4, M_4 - M_3$, Order 2: $M_3 - M_5, M_5 - M_1$).

5.4.2.3 Left-shift Hillclimber Local Search

A trend in genetic optimal scheduling practice is to incorporate local search techniques into the main loop of a GA to convert each offspring into an active schedule (Kusiak [67]; Gen and Cheng [4]). In general, feasible permutation of operations corresponds to a set of semiactive schedules. As the size of the problem increases, the size of the set of semiactive schedules will become larger and larger. Because traditional GAs were not accomplished in fine tuning, a search within the huge space of semiactive schedules will make GAs less effective.

We find that the optimal solution to an iOS/RS problem resides within the set of active schedules, which is much smaller than the set of semiactive schedules. By using a permutation encoding, we have no way to confine the genetic search within the space of the active schedules. One possible way would be to leave the genetic search in the entire search space of the semiactive schedules while we convert each

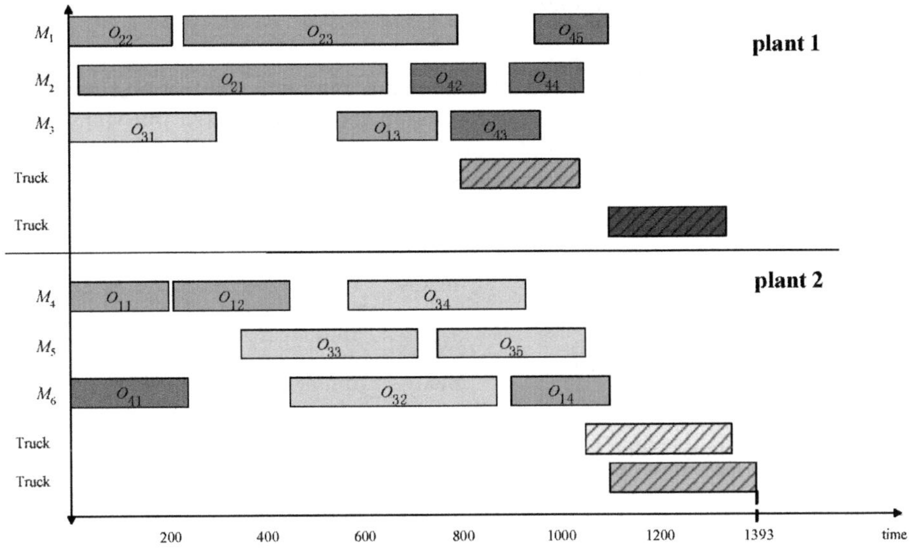

Fig. 5.55 Gantt chart for the best solution

chromosome into an active schedule by a left-shift hillclimber procedure shown as follows:

Step 1: Start with current solution;
Step 2: Get the neighborhoods of the current schedule as left-shift all the position of operation to climb the hill;
Step 3: If one of the operation can be moved left-side and obtain another better solution, then replace current schedule by the neighborhood with the better solution.

5.4.2.4 Adaptive moGA for Pareto Solutions

As the evaluation function for survival, the adapted-weighted sum method is used to construct the fitness function. Considering our mathematical model, makespan (f_1) and evaluation of workload (f_2) are not incompatible objectives, and swell each other. Hence, multiple objective functions f_1 and f_3 are combined into one overall objective function at hand to find the Pareto optimal solution. The fitness function is handled in the following way relates to Gen and Cheng's approach [5].

At generation t, choose the solution points which contain the minimum $z_1^{\min}$ (or $z_3^{\max}$) and $z_1^{\max}$ (or $z_3^{\max}$) corresponding to each objective function, then compare with the stored solution points at the previous generation and select the best points:

$$z_q^{\min(t)} = \min_k\{z_q^{\min(t-1)}, f_q^{(t)}(v_k)|k = 1,2,\ldots,iSize\}, q = 1,3 \qquad (5.30)$$

$$z_q^{\min(t)} = \max_k\{z_q^{\max(t-1)}, f_q^{(t)}(v_k)|k = 1,2,\ldots,iSize\}, q = 1,3 \quad (5.31)$$

where $z_q^{\max(t)}$ and $z_q^{\min(t)}$ are the maximum and minimum value of objective function q. v_k and *iSize* are the k-th feasible solution and number of feasible solutions on current generation, *i.e.*, the sum of population and generated offspring.

Solve the following equations to get weights for evaluation function:

$$w_1 = 1 - \frac{f_1^{(t)}(v_k) - z_1^{\min(t)}}{z_1^{\max(t)} - z_1^{\min(t)}}, \quad (5.32)$$

$$w_3 = 1 - \frac{f_3^{(t)}(v_k) - z_3^{\min(t)}}{z_3^{\max(t)} - z_3^{\min(t)}} \quad (5.33)$$

Calculate the fitness value for each feasible solution as follows:

$$eval(v_k) = w_1 f_1^{(t)}(v_k) + w_3 f_3^{(t)}(v_k) \quad (5.34)$$

5.4.3 Experiment and Discussions

5.4.3.1 Single Objective to Minimize Makespan

We first consider a relatively simple size problem with five machine centers to procedure five orders. Their lot sizes are $q_k = (40, 70, 60, 30, 60)$ and the available capacities for each machine are 1500. The unit load size for transportations are assumed to be 10 for all orders. The operations and their precedence constraints for the five orders are given in Fig. 5.56. The transportation times between resources are given in Table 5.28, and the processing times for each operation and their alternative resources are given in Table 5.29.

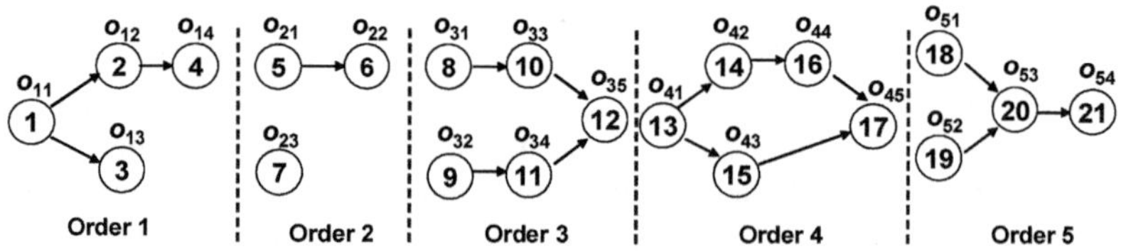

Fig. 5.56 Operation precedence constraints

To solve the problem using the moGA approach, the genetic parameters are set to maximum generation *maxGen* = 200; population size *popSize* = 100; crossover probability $p_C = 0.7$; and mutation probability $p_M = 0.3$.

Table 5.28 Processing time for all the operations in alternative machines

Order k :	Order 1				Order 2			Order 3					Order 4					Order 5			
Operation o_{kj} / Machine M_m	1 (o_{11})	2 (o_{12})	3 (o_{13})	4 (o_{14})	5 (o_{21})	6 (o_{22})	7 (o_{23})	8 (o_{31})	9 (o_{32})	10 (o_{33})	11 (o_{34})	12 (o_{35})	13 (o_{41})	14 (o_{42})	15 (o_{43})	16 (o_{44})	17 (o_{45})	18 (o_{51})	19 (o_{52})	20 (o_{53})	21 (o_{54})
M_1	7	7	-	6	-	3	8	-	10	6	15	-	-	-	-	-	5	5	-	13	-
M_2	-	-	6	-	9	5	-	-	-	5	-	-	-	-	-	5	-	8	7	-	-
M_3	-	-	5	-	-	-	12	5	-	-	7	-	6	-	6	-	-	-	10	-	7
M_4	5	6	-	-	-	-	-	-	10	-	-	6	-	-	4	3	-	-	-	-	-
M_5	-	-	8	-	-	-	-	8	7	-	-	-	8	9	-	-	4	-	8	8	-

Table 5.29 Transition time between machines

machine M_n / machine M_m	M_1	M_2	M_3	M_4	M_5	Available capacity L_m
M_1	-	7	27	26	17	1500
M_2	19	-	15	19	10	1500
M_3	5	17	-	36	27	1500
M_4	8	20	4	-	30	1500
M_5	18	18	14	5	-	1500

If only considering the objective of makespan, the best solution is 1500, with the corresponding chromosome shown in Fig. 5.57 and schedule shown as follows:

$$
\begin{aligned}
S = {} & \text{Order 1, Order 2, Order 3, Order 4, Order 5} \\
= {} & (o_{11}, M_4 : 0-200), (o_{12}, M_4 : 800-1040), (o_{13}, M_2 : 70-310), \\
& (o_{14}, M_1 : 1249-1489), (o_{21}, M_2 : 310-940), (o_{22}, M_1 : 1039-1249), \\
& (o_{23}, M_1 : 479-1039), (o_{31}, M_3 : 0-300), (o_{32}, M_4 : 200-800), \\
& (o_{33}, M_2 : 940-1240), (o_{34}, M_3 : 660-1080), (o_{35}, M_4 : 1040-1400), \\
& (o_{41}, M_3 : 300-480), (o_{42}, M_5 : 1027-1297), (o_{43}, M_3 : 480-660), \\
& (o_{44}, M_2 : 1240-1390), (o_{45}, M_5 : 1320-1440), (o_{51}, M_1 : 0-300), \\
& (o_{52}, M_5 : 67-547), (o_{53}, M_5 : 547-1027), (o_{54}, M_3 : 1080-1500)
\end{aligned}
$$

index j:	1	2	3	4	5	6	7	8	9	10	11	12	13	14	15	16	17	18	19	20	21
$v_1(j)$:	2	16	3	17	4	7	5	6	10	18	11	19	8	15	9	20	21	1	12	13	14
$v_2(j)$:	4	4	2	1	2	1	1	3	4	2	3	4	3	5	3	2	5	1	5	5	3

Fig. 5.57 The corresponding chromosome

The best schedule in detail is shown in Table 5.30; it clearly presents the assignment information of all the operations to each machine. Comparing all the finishing times we can draw that final operation which completed last is o_{54} operated on machine M_3. The concrete schedule is shown in form of Gantt chart in Fig. 5.58.

Table 5.30 Best schedule

Machine M_m	Operation o_{ki} (start time-finishing time)				
M_1	o_{51}(0-300)	o_{23}(479-1039)	o_{22}(1039-1249)	o_{14}(1249-1489)	
M_2	o_{13}(70-310)	o_{21}(310-940)	o_{33}(940-1240)	o_{44}(1240-1390)	
M_3	o_{31}(0-300)	o_{41}(300-480)	o_{43}(480-660)	o_{34}(660-1080)	o_{54}(1080-1500)
M_4	o_{11}(0-200)	o_{32}(200-800)	o_{12}(800-1040)	o_{35}(1040-1400)	
M_5	o_{52}(67-547)	o_{53}(547-1027)	o_{42}(1027-1297)	o_{45}(1320-1440)	

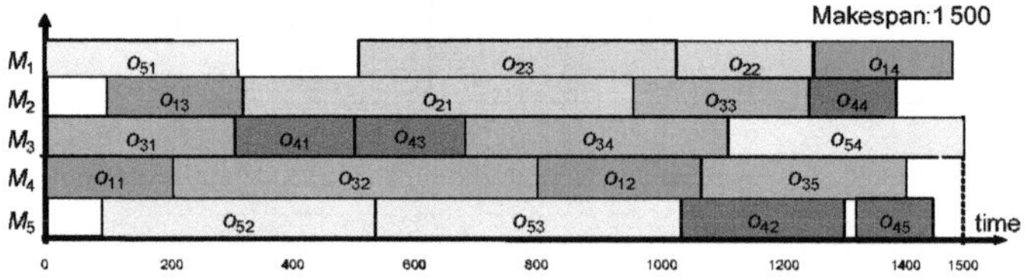

Fig. 5.58 Gant chart of the best schedule

Comparing the different workloads for the five available machines in Fig. 5.59, it presents the states for assignment for all the five machines. Workloads on machine 3 are obviously higher than the workload on other machines with the maximum workload 1500, which means the most plenary using for this schedule. Moreover, since the availability for these five machines are all set as 1500, we can draw the values of utilization factor, and give more concrete explanation to our manager. By all appearance, all the five machines have almost same higher utilization, and should take maintenance in the same time for the further scheduling.

Furthermore, in Fig. 5.60 we present the typical evolutionary process for the same experimental data (5 orders, 21 operations) with different parameter setting of p_C and p_M, and after generating 200 iterations, the result demonstrate that during we set p_C=0.7, p_M=0.3 for our proposed moGA the convergence situation outperform other parameter settings which means a faster convergence to the optimal solution 1500.

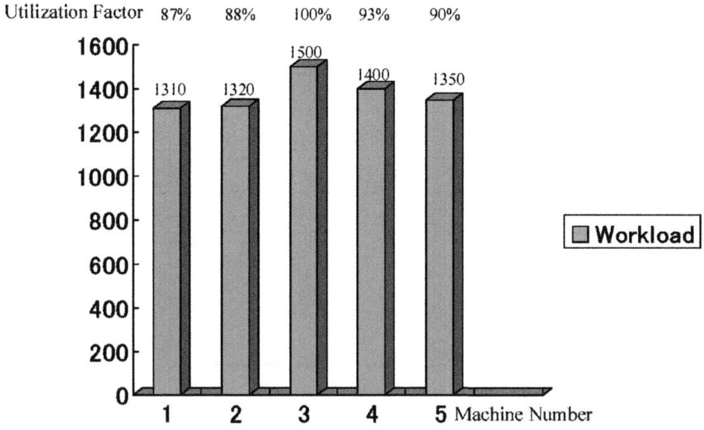

Fig. 5.59 Utilization information for machines

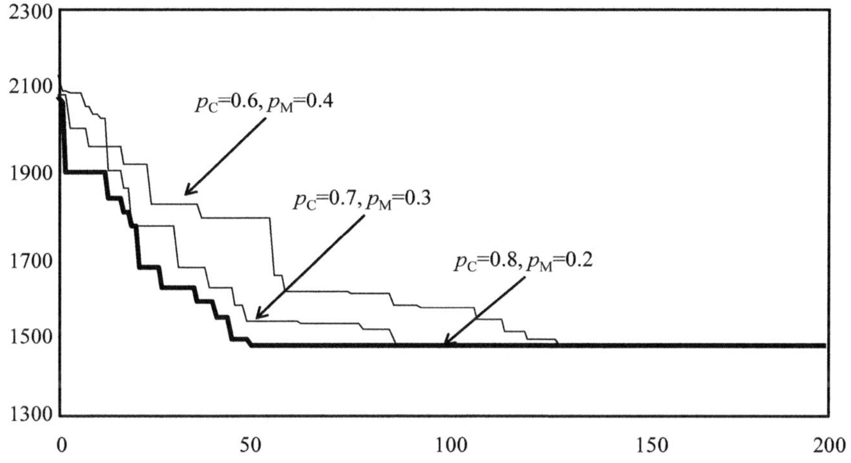

Fig. 5.60 Comparisons with different parameters setting

5.4.3.2 Bicriteria to Minimum the Makespan and Balance the Workload

In Table 5.31 we comparing our proposed moGA with the previous approach by Moon-Kim-Gen by using the same experimental data related in Moon's work. For the small case of the iOS/RS problem they can get the same optimal solution as the makespan 1500, while as the size of case become larger, our moGA obviously outperforms in finding the best solutions; furthermore the corresponding workload balancing evaluations remarkably outperform the previous approach. Consequently the GA combined with local search can effectively improve the capability for getting the active optimal solution.

Table 5.31 Comparisons of experimental result

No. of orders	No. of operations	No. of machines	Moon-Kim-Gen's approach		moGA	
			c_M	$w_B(\times 10^4)$	c_M	$w_B(\times 10^4)$
5	21	5	1500*	1.57	1500*	1.57
7	30	5	1965	1.49	1894*	1.21
10	43	5	2592	1.94	2487*	1.89
15	64	5	2810	1.03	2639*	0.95

5.4.3.3 Pareto Solution to Makespan and Total Transition Times

However, when we consider the total transition times for the schedule in Table 5.30, it will take 13 times for the robots in plant to making transportation between different machines. That means some additional cost in manufacturing system (especially in semi conduct manufacturing system). To reduce such kind of infection we consider the third object function f3 simultaneously for finding Pareto solution.

Furthermore, by using the all the cases of iOS/RS problems, we compare our moGA with the classical GA (Moon's approach) by using Pareto performance to f_1 and f_3 (makespan and total transition times). All the parameters setting are the same as the experiment as $p_C = 0.7, p_M = 0.3$, with the generation number as 200; Figs. 5.61–Fig. 5.64 show how our proposed moGA can obtain optimal Pareto solutions in all the cases. For the simple instance, the minimum value of makespan in this case is 1500; however the corresponding total transition times will be increased to 13, and it will waste too much cost for complex transporting processes, especially in some semiconductor manufacturing systems. Thereby we should find the most adaptive point, which take two other value (2058, 11). Moreover, in case 2, case 3, and case 4, we can clearly recognize that the simple GA have narrower solution spaces, which means it cannot improve the solution further for the combing of the heuristic machine selection.

5.5 Integrated Scheduling Model with Multi-plant

This section describes a scheduling system using integrated data structure and scheduling algorithms to combine manufacturing and transportation scheduling. In manufacturing industries, information systems become more important to apply in today's rapidly changing global business environment. Scheduling subsystem is a main module of a manufacturing information system. Until now, the main purpose of

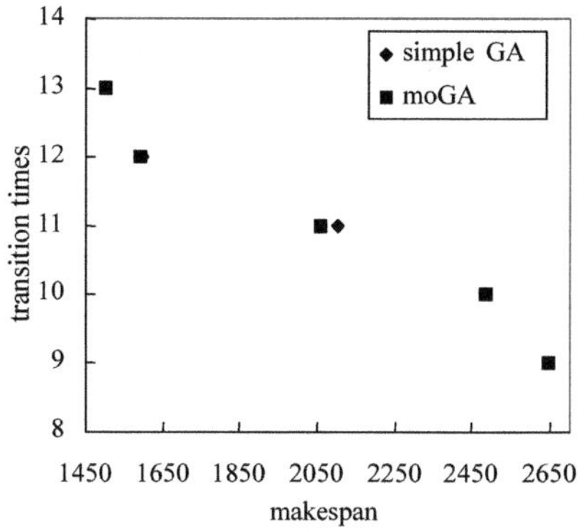

Fig. 5.61 Pareto solutions (case 1)

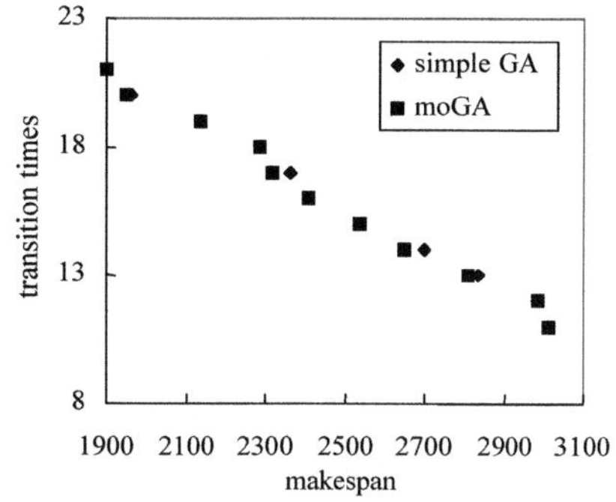

Fig. 5.62 Pareto solutions (case 2)

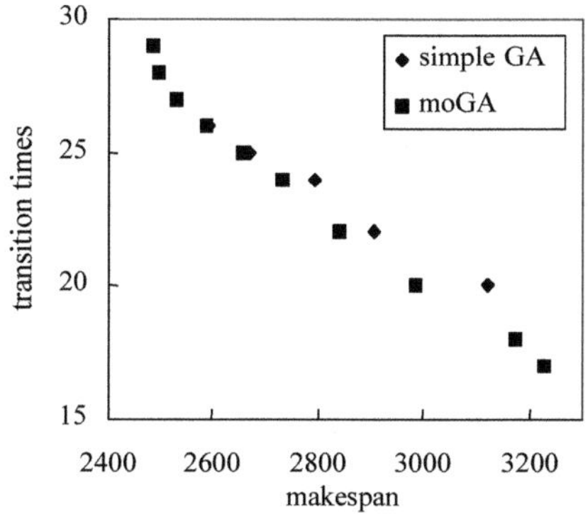

Fig. 5.63 Pareto solutions (case 3)

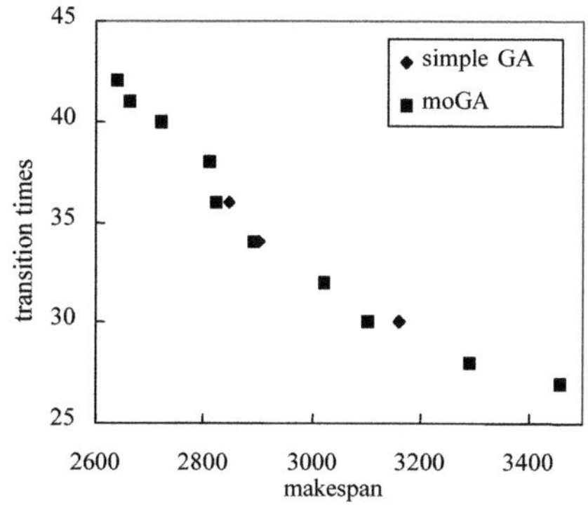

Fig. 5.64 Pareto solutions (case 4)

scheduling was improvement of equipment operation rate and reduction of cost by mass production. However, at present, manufacturers must make plans and schedules considering many kind of customer's demand. Therefore, it is necessary to consider various elements in the optimization of a schedule such as not only simple manufacturing scheduling but also logistics, inventory control, *etc*. Figure 5.65 shows a block diagram of the scheduling system.

Advanced planning and scheduling (APS) is an approach for these combined problems [79]. PSLX Technical Specifications [80] defined APS as "system to integrate decision making elements of organization such as planning or scheduling and to orient the whole optimization independently with synchronizing by each section over the boundary between organizations or enterprises". Generally, the APS problem is recognized as combined problem with whole or part of various elements; Moon *et al.* [72] proposed advanced planning and scheduling model which inte-

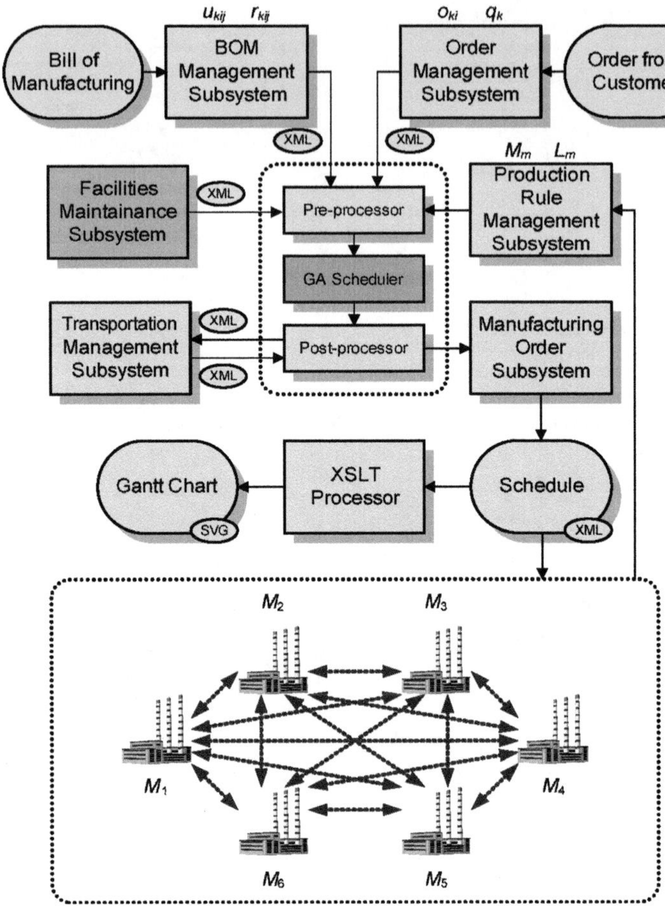

Fig. 5.65 Block diagram of experimental scheduling system

grated process planning and production scheduling; Okamoto *et al.* [81] integrated transportation scheduling in addition to Moon's APS model.

It is difficult to construct the information system including large numbers of elements as a single system. The enlarged system including various elements becomes difficult to read, change and improve. Moreover, the requirement specification of the system changes today, while the system has been designed by spending time. PSLX proposes the method to construct the APS system by small subsystems called agents, and functions of the system are realized by the cooperation of agents. The standardization of the data format for the communication is important to realize cooperation of agents. Recently, XML (Extensible Markup Language) [82] is widely utilized to exchange data on the Internet. For example, PSLX and MESX [83] developed the standards based on XML for industries. [84] proposed a GA-based scheduling agent for APS system using XML, and suggested manufacturing scheduling subsystem is the most important module on APS system. The system has extensible structure according to the merit of GA and XML.

5.5.1 Integrated Data Structure

In the APS, Bill of Manufacturing [85], which integrated Bill of Material (BOM) and process procedure information, is used at planning and scheduling. Generally, BOM is used at MRP (Material Requirements Planning), and process procedure information is used at CRP (Capacity Requirements Planning) and scheduling. The APS considers these planning elements at once using Bill of Manufacturing. However, required information for scheduling is not only process procedure information but also transportation, setup *etc*. Especially, transportation times cannot be disregarded in multi-plant chain environment.

Nishioka [86] proposed the method to integrate data structure for logistics and manufacturing scheduling. By adding the location attribute in the item as identifier, it is possible to handle the transportation as well as the process. However, there is a problem which causes increase in number of items and rules on this method. Therefore, in our approach, attributes are not embedded in the item name, and the integrated rule is dynamically generated from the static master data in Fig.5.66.

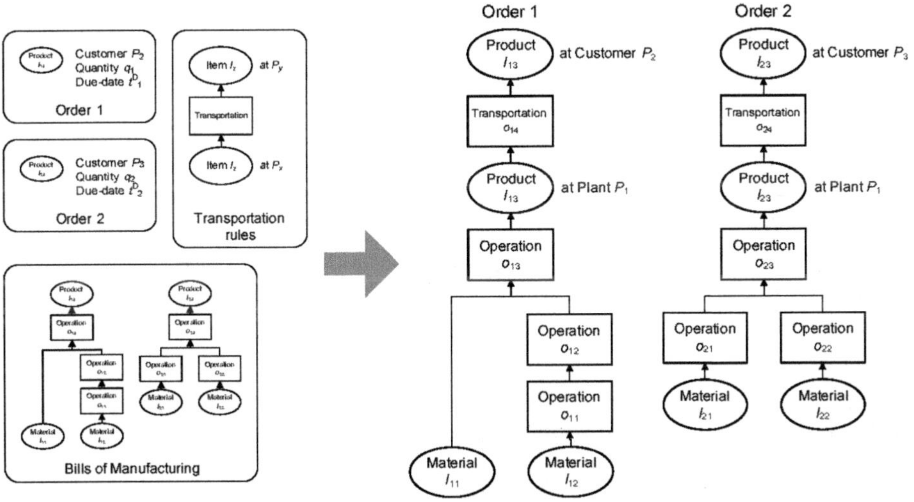

Fig. 5.66 Dynamic generation of integrated data model

In this model, it is necessary to add various attributes for items (materials, intermediate products and products), resources (machines and vehicles) and operations (processes, transportations and setups). The necessary transportations are different to the location of items, even if these items are in identical classes. However, the number of items explodes if the item in the different place is distinguished as a different item. The element such as item with attributes can be expressed in the tree structure, and using XML is suitable to exchange tree structured data (Fig. 5.67). The example of the each element which is not described in Fig. 5.66 is shown in Tables 5.32–5.34. Though these numerical values are shown in tabular format, these values are actually recorded as attributes. For example, quantity q_k is one of the attributes of order k.

```
<order code=”example.order.1” item=”product.I13”>
    <location code=”customer.P2”/>
    <qty value=”40”/>
</order>
<order code=”example.order.2” item=”product.I23”>
    <location code=”customer.P3” />
    <qty value=”30” />
</order>
```

Fig. 5.67 Order example in XML

Table 5.32 Parameters of simple example

k	q_k
1	40
2	30

M_m / o_{ki}	t^T_{kim} M_4
o_{14}	40
o_{24}	50

M_m / o_{ki}	p_{kim} M_1	M_2	M_3
o_{11}	3	5	4
o_{12}	3	4	-
o_{13}	2	-	4
o_{21}	-	3	2
o_{22}	3	4	-
o_{23}	3	2	4

Table 5.33 Setup time t^U_{kilj} of simple example

o_{ki} \ o_{lj}	o_{11}	o_{12}	o_{13}	o_{14}	o_{21}	o_{22}	o_{23}	o_{24}
o_{11}	0	5	5	-	5	5	5	-
o_{12}	5	0	5	-	5	5	5	-
o_{13}	5	5	0	-	5	5	5	-
o_{14}	-	-	-	40	-	-	-	40
o_{21}	5	5	5	-	0	5	5	-
o_{22}	5	5	5	-	5	0	5	-
o_{23}	5	5	5	-	5	5	0	-
o_{24}	-	-	-	50	-	-	-	50

Table 5.34 Transportation time in the same plant t^S_{mn} of simple example

M_m \ M_n	M_1	M_2	M_3
M_1	0	5	6
M_2	5	0	7
M_3	6	7	0

5.5.2 Mathematical Models

This model is based on Advanced Process Planning and Scheduling Model [72] and modified in order to include assembly process and transportations. This model consists of three elements of operation sequence, resources selection and scheduling. In comparison with the FMS (Flexible Manufacturing System) Problem and fJSP (Flexible Job-shop Scheduling Problem), this model includes precedence constraints got from Bill of Manufacturing, setup time, and transportation time.

Notation

Indices

i,J: index of operations $i,j=1,2,\ldots,J_k$
k,l: index of orders, $k,l=1,2,\ldots,K$
m,n: index of resources, $m,n=1,2,\ldots,N$
d,e: index of locations, $d,e=1,2,\ldots,D$

Parameters

K: number of orders
D: number of locations
N: number of resources
J_k: number of operations for order k
o_{ki}: i-th operation for order k
M_m: m-th resource
P_d: d-th location
q_k: lot size of order k
A_m: set of operations that can be processed on M_m

p_{kim}: unit processing time of o_{ki} on M_m
L_m: capacity of M_m
u_{kij}: unit load size from o_{ki} to o_{kj}
B_d: set of resources (machines) at plant P_d
V_d: set of resources (vehicles) that can use for transportation from P_d
r_{kij}: precedence constraints

$$r_{kij} = \begin{cases} 1, \text{ if } o_{ki} \text{ is predecessor of } o_{kj}, \\ 0, \text{ otherwise} \end{cases}$$

t^U_{kilj}: setup time from o_{ki} to o_{lj}
t^S_{mn}: transportation time from M_m to M_n in the same plant
t^T_{kim}: transportation time of o_{ki} on vehicle M_m
t^O_{kim}: operation time of o_{ki} on M_m

$$t^O_{kim} = \begin{cases} t^T_{kim}, & \text{if } o_{ki} \text{ is transportation}, \\ q_k p_{kim}, & \text{otherwise} \end{cases}$$

t^D_k: due date of order k
c_ki: completion time of o_{ki}

$$c_{ki} = s_{ki} + \sum_{m=1}^{N} t^O_{kim} x_{kim}$$

C_M: makespan
t_P: penalty of tardiness in due date

Decision Variables

s_{ki}: starting time of o_{ki}

$$x_{kim} = \begin{cases} 1, \text{ if } M_m \text{ is selected for } o_{ki}, \\ 0, \text{ otherwise} \end{cases}$$

$$y_{kilj} = \begin{cases} 1, \text{ if } o_{ki} \text{ precedes } o_{ij}, \\ 0, \text{ otherwise} \end{cases}$$

In the example of this scheduling problem, objectives of this model are to minimize makespan and to minimize penalty of tardiness in due date of each order. The overall model is described as follows:

$$\min c_M = \max_{I,K} \{c_{ki}\} \tag{5.35}$$

$$\min t_P = \sqrt{\sum_{k=1}^{K} \max\left\{0, \max_i\{c_{ki}\} - t_k^D\right\}^2} \tag{5.36}$$

$$\text{s. t.} \left\{s_{lj} - \left(c_{ki} + t_{kilj}^U\right)\right\} x_{kim} x_{ljm} y_{kilj} \geq 0 \quad \forall (k,i),(l,j),m \tag{5.37}$$

$$\left\{s_{kj} - \left(s_{ki} + u_{kij} p_{kim} + t_{mn}^s\right)\right\} r_{kij} x_{kim} x_{kjn} \geq 0 \quad \forall\, i,j,k,\ \forall m,n \in B_d,\ \forall d \tag{5.38}$$

$$\left\{c_{kj} - u_{kij} p_{kjn} - \left(c_{ki} + t_{mn}^s\right)\right\} r_{kij} x_{kim} x_{kjn} \geq 0 \quad \forall\, i,j,k,\ \forall m,n \in B_d,\ \forall d \tag{5.39}$$

$$\left(s_{kj} - c_{ki}\right) r_{kij} x_{kim} x_{kjn} \geq 0 \quad \forall\, i,j,k,\ \forall m \in B_d,\ \forall n \in V_d,\ \forall d \tag{5.40}$$

$$\left(s_{kj} - c_{ki}\right) r_{kij} x_{kim} x_{kjn} \geq 0 \quad \forall\, i,j,k,\ \forall m \in V_d,\ \forall n \in B_e,\ \forall d,e \tag{5.41}$$

$$\sum_{k=1}^{K} \sum_{i=1}^{J_k} q_k p_{kim} \leq L_m \quad \forall m \tag{5.42}$$

$$r_{kij} y_{kjki} = 0 \quad \forall i,j,k \tag{5.43}$$

$$y_{kilj} = 0 \quad \forall i,k \tag{5.44}$$

$$y_{kilj} + y_{ljki} = 1 \quad \forall (k,i),(l,j),(k,i) \neq (l,j) \tag{5.45}$$

$$\sum_{m=1}^{N} x_{kim} = 1 \quad \forall i,k \tag{5.46}$$

$$x_{kim} = 0 \quad \forall (k,i) \notin A_m, \quad \forall m \tag{5.47}$$

$$s_{ki} \geq 0 \quad \forall i,k \tag{5.48}$$

$$y_{kilj} \in \{0,1\} \quad \forall (k,i),(l,j) \tag{5.49}$$

$$x_{kim} \in \{0,1\} \quad \forall i,k,m \tag{5.50}$$

The objectives are to minimize makespan in Eq. 5.35 and to minimize tardiness in Eq. 5.36. The constraint at Eq. 5.37 ensures that a resource cannot be assigned to any operation until predecessors and setup operations are completed. The constraints at Eqs. 5.38 and 5.39 impose transportation time between two resources in the same plant. The constraints at Eq. 5.40 and 5.41 ensure that a lot cannot be processed and transported at the same time. The constraint at Eq. 5.42 restricts the available capacity for each machine. The constraint at Eq. 5.43 ensures that the precedence constraints are not violated. The constraints at Eq. 5.44 and 5.45 ensure the feasible operation sequence. The constraints at Eq. 5.46 and 5.47 ensure the feasible resource selection. The constraints at Eq. 5.48–5.50 impose nonnegative condition.

5.5.3 Multistage Operation-based GA

The Multistage Operation-based Genetic Algorithm (moGA) for scheduling agent [84] is based on the multistage decision model. The GA-based discrete dynamic programming (DDP) approach was proposed [87] for generating schedules in FMS environment. Network-based Hybrid Genetic Algorithm [88] was developed as an improved approach of GA-based DDP for FMS. The problem in applying this approach to our model is to fix operation sequence in each job for selecting alternative resources. Therefore, our moGA approach consists two parts, sequencing operations and selection resources.

Let *P(t)* and *C(t)* be parents and offspring in current generation t, respectively. The overall procedure of multistage operation-based GA is outlined as follows:

---

procedure: bicriteria moGA
input: data set, GA parameters ($popSize$, $maxGen$, p_M, p_C)
output: Pareto optimal solutions E
begin
  $t \leftarrow 0$;
  initialize $P1(t)$ by random key-based encoding routine;
  evalute $P2(t)$ by resource permutation encoding routine;
  calculate objectives $z_i(P)$, $i = 1,2$
      by random key-based decoding and resource permutation decoding routine;
  create Pareto $E(P)$;
  evaluate $eval(P)$ by adaptive-weight fitness assignment routine;
  while (**not** terminating condition) **do**
   create $C(t)$ from $P(t)$ by parameterized uniform crossover routine;
   create $C_1(t)$ from $P_1(t)$ by random key-based mutation routine;
   create $C_2(t)$ from $P_2(t)$ by inversion mutation routine;
   local search C(t) by bottleneck shifting;
   calculate objectives $z_i(C)$, $i = 1, 2$
      by random key-based decoding and resource permutation decoding routine;
   update Pareto $E(P,C)$;
   evaluate eval(P,C) by adaptive-weight fitness assignment routine;
   select $P(t+1)$ from $P(t)$ and $C(t)$ by elitist strategy routine;
   $t \leftarrow t+1$;
  end
  output Pareto optimal solutions $E(P,C)$
end

To develop a multistage operation-based genetic representation for the problem, there are three main phases:

Phase 1: Creating an operation sequence

Step 1.1: Generate a random priority to each operation using encoding procedure for first vector v_1.

Step 1.2: Decode a feasible operation sequence that satisfies the precedence constraints

Phase 2: Assigning operations to machine

Step 2.1: Input the operations sequence found in step 1.2.
Step 2.2: Generate a permutation encoding for machine assignment of each operation (second vector v_2).

Phase 3: Designing a schedule

Step 3.1: Create a schedule S using task sequence and processor assignments.
Step 3.2: Draw a Gantt chart for this schedule.

Phase 1: Creating an Operation Sequence

A random key-based representation [89, **?**] is used for operation sequence. By using this representation, any chromosome formed by crossover, mutation and random generation are feasible solutions. The encoding procedure is shown in Fig. 5.68, and we can draw chromosome shown in Fig. 5.69 as the result of example case.

```
procedure: Random key-based encoding
input: total number of operations J
output: chromosome v1()
begin
  for i =1 to J
    v1(i) ← random[0, 1);
  output chromosome v1();
end
```

Fig. 5.68 Random key-based encoding procedure

Operation ID i:	1	2	3	4	5	6	7	8
Priority $v_1(i)$:	0.08	0.12	0.20	0.54	0.43	0.29	0.58	0.79

Fig. 5.69 Chromosome v_1 drawn by random key-based encoding

By using the decoding procedure shown in Fig. 5.70, a feasible operation sequence will be obtained through trace table shown in Table 5.35. The result of this case is as follows:

$$
\begin{aligned}
Q &= \{1,2,6,5,4,7,8\} \\
&= \{o_{11},o_{12},o_{13},o_{22},o_{21},o_{14},o_{23},o_{24}\}
\end{aligned}
$$

```
procedure: Random key-based decoding
input: chromosome v_1(),
    set of predecessors R_i of operation i,
    total number of operations J
output: operation sequence Q()
begin
  P ← { i | i = 1, ..., J };
  for k =1 to J do
    i* ← argmin{v_1(i)|R_i ⊆{ Q }, i ∈P };
    Q(k) ← i*;
    P ←P ¥ i*;
  end
  output operation sequence Q();
end
```

Fig. 5.70 Random key-based decoding procedure for sequencing operations

Table 5.35 Trace table of random key-based decoding procedure

k	$i*$	R_{i*}	$v_1(i*)$	Q
1	1	F	0.08	{ 1 }
2	2	{ 1 }	0.12	{ 1, 2 }
3	3	{ 2 }	0.20	{ 1, 2, 3 }
4	6	F	0.29	{ 1, 2, 3, 6 }
5	5	F	0.43	{ 1, 2, 3, 6, 5 }
6	4	{ 3 }	0.54	{ 1, 2, 3, 6, 5, 4 }
7	7	{ 5, 6 }	0.58	{ 1, 2, 3, 6, 5, 4, 7 }
8	8	{ 7 }	0.79	{ 1, 2, 3, 6, 5, 4, 7, 8 }

Phase 2: Assigning Operations to Machine

After finishing sequencing operations, position of all stages (operations) has been decided. Hence it is possible to decide selecting resources as a multistage decision making problem (Fig. 5.71). The encoding procedure is shown in Fig. 5.72, and the trace table of this procedure is shown in Table 5.36. Moreover, we can draw chromosome shown in Fig. 5.73 as the result of example case.

By using the decoding procedure shown in Fig.5.74, a feasible operation sequence and resource selection will be obtained through the trace table shown in Table 5.37. The result of this case is as follows:

$$S = \{(o_{11}, M_3), (o_{22}, M_1), (o_{21}, M_2), (o_{23}, M_2), (o_{24}, M_4), (o_{13}, M_1), (o_{14}, M_4)\}$$

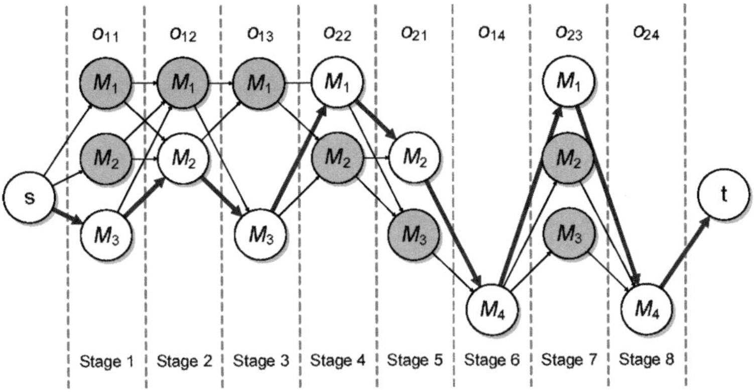

Fig. 5.71 Node graph of resource selection as multistage decision making model

procedure: Resource permutation encoding
input: total number of operations J,
total number of resources N,
set of all operations $O()$,
set of operations A_m
that can be processed on M_m
output: chromosome $v_2()$
begin
for $i = 1$ **to** J **do**
repeat
$m \leftarrow$ random$[1, N]$;
until $O(i) \in A_m$
$v_2(i) \leftarrow m$
end
output chromosome $v_2()$;
end

Fig. 5.72 Resource permutation encoding procedure

Operation ID i:	1	2	3	4	5	6	7	8
Resource $v_2(i)$:	3	2	3	4	2	1	1	4

Fig. 5.73 Chromosome v_2 drawn by resource permutation encoding procedure

Table 5.36 Trace table of resource permutation encoding procedure

i	$O(i)$	m	A_m	v_2
1	o_{11}	3	$\{\boldsymbol{o_{11}}, o_{13}, o_{21}, o_{23}\}$	{ 3 }
2	o_{12}	2	$\{o_{11}, \boldsymbol{o_{12}}, o_{21}, o_{22}, o_{23}\}$	{ 3, 2 }
3	o_{13}	3	$\{o_{11}, \boldsymbol{o_{13}}, o_{21}, o_{23}\}$	{ 3, 2, 3 }
4	o_{14}	4	$\{\boldsymbol{o_{14}}, o_{24}\}$	{ 3, 2, 3, 4 }
5	o_{21}	2	$\{o_{11}, o_{12}, \boldsymbol{o_{21}}, o_{22}, o_{23}\}$	{ 3, 2, 3, 4, 2 }
6	o_{22}	1	$\{o_{11}, o_{12}, o_{13}, \boldsymbol{o_{22}}, o_{23}\}$	{ 3, 2, 3, 4, 2, 1 }
7	o_{23}	1	$\{o_{11}, o_{12}, o_{21}, o_{22}, \boldsymbol{o_{23}}\}$	{ 3, 2, 3, 4, 2, 1, 1 }
8	o_{24}	4	$\{o_{14}, \boldsymbol{o_{24}}\}$	{ 3, 2, 3, 4, 2, 1, 1, 4, }

```
procedure: Resource permutation decoding
input: chromosome v2(),
   operation sequence Q(),
   set of all operations O(),
   total number of operations J
output: schedule S()
begin
   for k =1 to J do
      i = Q(k);
      m ←v2(i);
      S(k) ← { (O(i), Mm) };
   end
   output schedule S();
end
```

Fig. 5.74 Resource permutation decoding procedure

Table 5.37 Trace table of resource permutation decoding procedure

k	i	$O(i)$	$v_2(i)$	S
1	1	o_{11}	3	$\{(o_{11}, M_3)\}$
2	2	o_{12}	2	$\{(o_{11}, M_3), (o_{12}, M_2)\}$
3	3	o_{13}	3	$\{(o_{11}, M_3), (o_{12}, M_2), (o_{13}, M_3)\}$
4	6	o_{22}	1	$\{(o_{11}, M_3), (o_{12}, M_2), (o_{13}, M_3), (o_{22}, M_1)\}$
5	5	o_{21}	2	$\{(o_{11}, M_3), (o_{12}, M_2), (o_{13}, M_3), (o_{22}, M_1), (o_{21}, M_2)\}$
6	4	o_{14}	4	$\{(o_{11}, M_3), (o_{12}, M_2), (o_{13}, M_3), (o_{22}, M_1), (o_{21}, M_2), (o_{14}, M_4)\}$
7	7	o_{23}	1	$\{(o_{11}, M_3), (o_{12}, M_2), (o_{13}, M_3), (o_{22}, M_1), (o_{21}, M_2), (o_{14}, M_4), (o_{23}, M_1)\}$
8	8	o_{24}	4	$\{(o_{11}, M_3), (o_{12}, M_2), (o_{13}, M_3), (o_{22}, M_1), (o_{21}, M_2), (o_{14}, M_4), (o_{23}, M_1), (o_{24}, M_4)\}$

Phase 3: Designing a Schedule

After finishing sequencing operations and selecting resources, we can decide the starting time of each operation. In this example case, the start time of each operation will be assigned to earliest starting time. Finally, schedules are constructed as follows:

$$\begin{aligned} S = \{&(o_{11}, M_3 : 0 - 120), (o_{12}, M_2 : 47 - 167), (o_{13}, M_3 : 125 - 305), \\ &(o_{22}, M_1 : 0 - 200), (o_{21}, M_2 : 171 - 321), (o_{14}, M_4 : 305 - 365), \\ &(o_{23}, M_1 : 206 - 356), (o_{24}, M_2 : 425 - 475)\} \end{aligned}$$

The Gantt chart of this schedule is shown in Fig. 5.75.

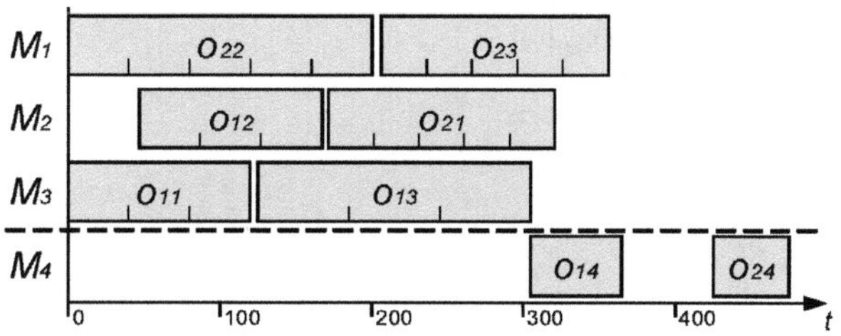

Fig. 5.75 Gantt chart of schedule from simple example

5.5.3.1 Local Search Technique

Hybrid genetic algorithms (hGA) [5] are technique which combines GA with other techniques such as local search heuristics in order to supplement the demerit of GA. To change operations on the critical path is effective to improve solution of scheduling problem. In our approach, following two types of local searches are used to improve two objectives (Fig. 5.76). Local search type is assigned randomly for each chromosome at first generation. When local search does not improve any more, another type of local search is assigned for the chromosome.

5.5.4 Numerical Experiment

This example problem is to schedule three orders in two plants, eight resources and three customers environment shown in Fig. 5.77. The experimental dataset consists of Bills of Manufacturing (Fig. 5.78), orders as XML document (Fig. 5.79) and other numerical parameters (Tables 5.38 and 5.39). Proposed moGA scheduling system

```
procedure: Bicriteria local search
input: schedule S, objective k, total number of resources N
output: schedule S', objective k
begin
  S'←S;
  if (k =1) then
    o*← argmax { completion time of o | oS' }; // Objective 1
  else
    o*← argmax { tardiness of o | oS' }; // Objective 2
  i ← false;
  o_s ← o*;
  while (pred(o_s) ≠ Φ) do
    if (i = false) then
      o_p← argmax { completion time of o | opred(o_s) };
    if (resourse of o_p = resource of o_s) then
      swap priority(o_p, o_s);
      reschedule(S');
      if (f_k(S') > f_k(S)) then
        i ← true;
    end
    o_s ←o_p;
   end
 end
 o_c ← o*;
 repeat
   if (i = false) then
     r_0 ← resource of o_c;
     for r = 1 to N do
       if (r ≠ r_0) then
         change resource of o_c to r;
         reschedule(S');
         if (f_k(S') > f_k(S)) then
           i ← true;
         else
           repair resource of o_c to r_0;
       end
     end
   end
   o_c ← argmax { completion time of o | opred(o_c) };
  until pred(o_c) ≠ Φ;
 if (i = false) then
   k ← k + 1;
   if (k > 2) then
     k ← 1;
 end
 output schedule S', objective k;
end
```

Fig. 5.76 Bicriteria local search procedure

outputs result of scheduling in XML form, and human readable Gantt charts in SVG (Scalable Vector Graphics) [90]. The genetic parameters used in this experiment are as follows:
$p_C = 0.7$, $p_M = 0.2$, $popSize = 100$, $maxGen = 100$, and the experimental result is shown in Fig. 5.80. The Gantt chart of experimental result in SVG is shown in Fig. 5.81. The experimental result by using other data with many orders is shown in Figs. 5.82 and 5.83.

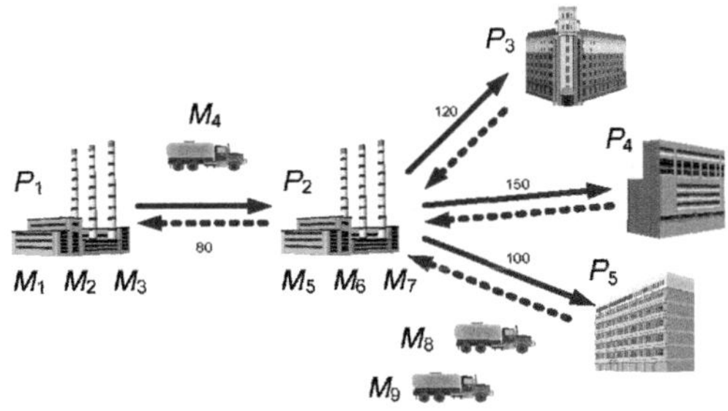

Fig. 5.77 Environment model for numerical experiment

Table 5.38 Unit processing time p_{kim}

o_{ki} \ M_m	M_1	M_2	M_3	M_4
o_{11}	7	5	4	-
o_{21}	-	5	4	7
o_{22}	4	-	8	-
o_{23}	6	5	-	3
o_{31}	-	6	3	5
o_{32}	4	-	7	6
o_{33}	-	6	-	4
o_{34}	8	3	6	-
o_{35}	-	5	6	8

o_{ki} \ M_m	M_6	M_7	M_8	M_9
o_{13}	6	5	-	3
o_{14}	8	7	3	6
o_{15}	-	8	3	6
o_{16}	3	5	8	7
o_{17}	7	3	-	3
o_{25}	-	4	6	-
o_{26}	6	-	4	5
o_{27}	6	4	5	5
o_{37}	3	-	6	8

Table 5.39 Transportation time in the same plant t^S_{mn}

M_m \ M_n	M_1	M_2	M_3	M_4
M_1	0	5	6	7
M_2	5	0	7	8
M_3	6	7	0	9
M_4	7	8	9	0

M_m \ M_n	M_6	M_7	M_8	M_9
M_6	0	5	6	7
M_7	5	0	7	8
M_8	6	7	0	9
M_9	7	8	9	0

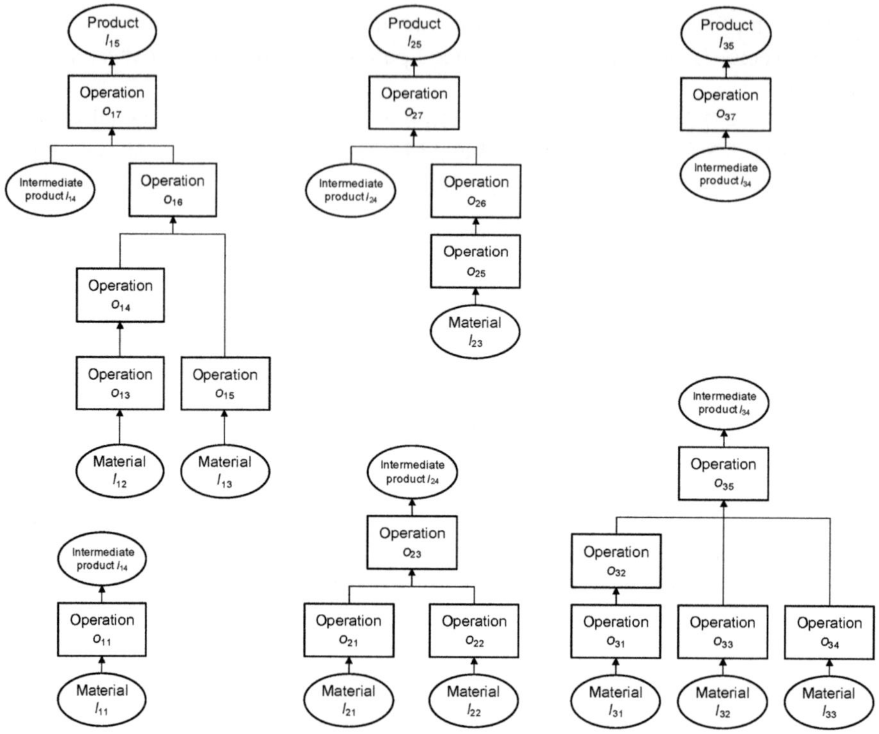

Fig. 5.78 Bills of Manufacturing for numerical experiment

Fig. 5.79 Order example in XML for numerical experiment

```
<order code=”exp.order.1“item=“exp.product.I15”>
    <location code=“exp.customer.P3”/>
    <qty value=“40”/>
    <due><time count=“1200”/></due>
</order>
<order code=“exp.order.2” item=“exp.product.I25”>
    <location code=“exp.customer.P4”/>
    <qty value=“60”/>
    <due><time count=“1000”/></due>
</order>
<order code=“exp.order.3” item=“exp.product.I35”>
    <location code=“exp.customer.P5”/>
    <qty value=“50”/>
    <due><time count=“800”/></due>
</order>
```

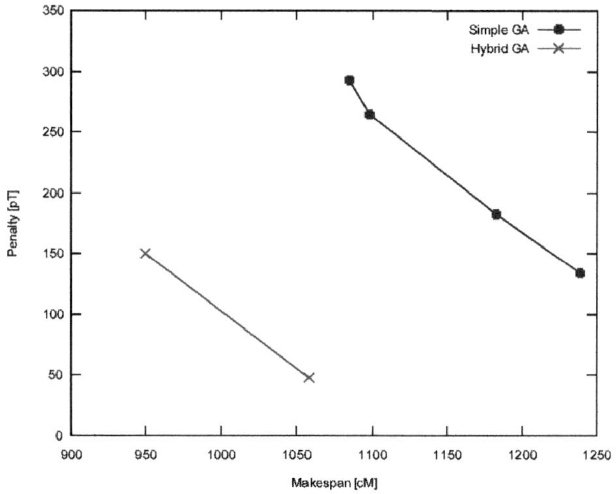

Fig. 5.80 Pareto optimal solutions of numerical experiment ($K = 3$)

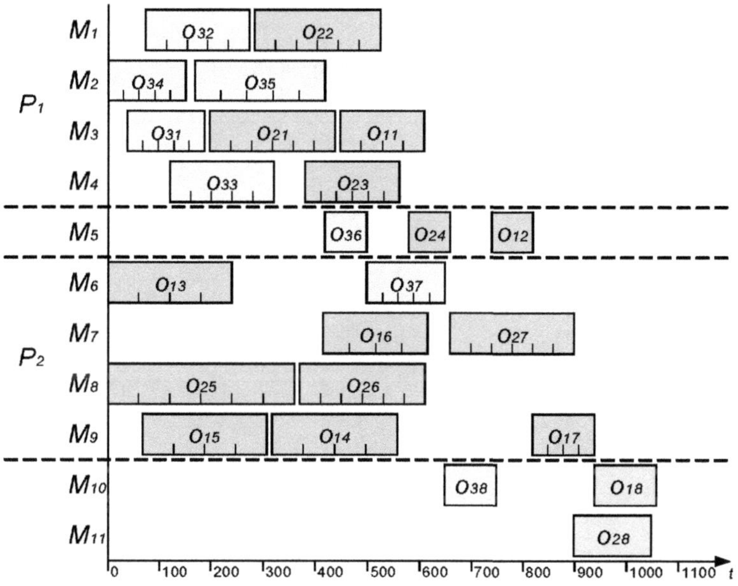

Fig. 5.81 Gantt chart of schedule ($t_P = 48$, $c_M = 1058$)

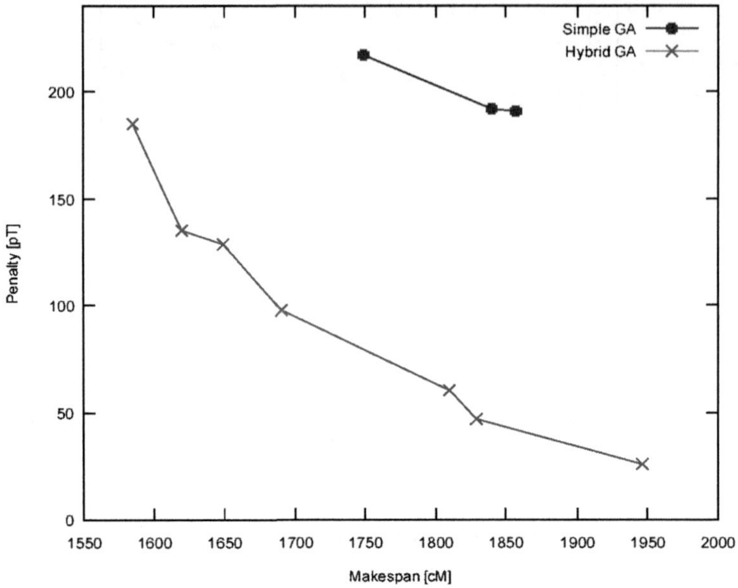

Fig. 5.82 Pareto optimal solutions of numerical experiment ($K = 6$)

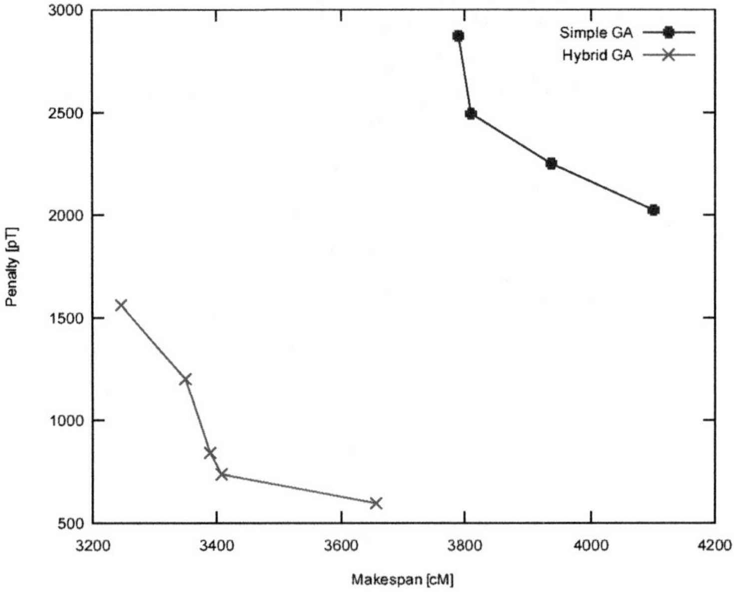

Fig. 5.83 Pareto optimal solutions of numerical experiment ($K = 9$)

5.6 Manufacturing and Logistics Model with Pickup and Delivery

In contemporary manufacturing, meeting promised delivery dates requires production schedules that take into account elements such as transportation. Previously scheduling focused on improvements in facility workloads and achieving cost reductions through mass production. Traditional manufacturing scheduling systems consider only manufacturing constraints while transportation scheduling systems typically consider only total vehicle mileage. These traditional systems cannot provide schedules that accommodate customer demands for just-in-time delivery [95].

It is difficult to develop a monolithic manufacturing scheduling system that includes large number of elements. Very large scheduling systems are difficult to develop and are inflexible. Furthermore, changes in system specifications often occur during system design. Okamoto *et al.* [96] proposed a method for constructing a manufacturing system using XML to exchange data among small subsystems including a scheduler based on a genetic algorithm (GA). This system responds even to a slight change in the constraint by adding a penalty to the adaptation value, which is performed by the genetic algorithm.

Moon *et al.* [72] integrated process planning and production scheduling in an Advanced Planning and Scheduling (APS) model that included the factor of transportation. Okamoto *et al.* [97, 98], expanded Moon's model by integrating the manufacturing process and transportation between plants and proposed a solution based on the genetic algorithm. Most of the studies concerning scheduling that consider both manufacturing and transportation [99, 100] have established a number of resource and transportation restrictions to systematize the problem from the viewpoint of complexity. These approaches offer only specialized solutions.

This section describes an integrated manufacturing and transportation model that incorporates pickup and delivery. In comparison with previous studies [97, 98], this model accommodates a single vehicle that transports multiple materials, intermediate, and finished products. We also developed a scheduler using a multiobjective genetic algorithm that minimizes both makespan and vehicle mileage.

5.6.1 Mathematical Formulation

When manufacturing and transportation schedules are created separately, the results of one become the constraints of the other. For example, the starting and completion times of the production schedule become time-window constraints in the transportation scheduling. In contrast, the transportation schedule defines the earliest start time and latest completion time of the manufacturing process. Some approaches can reduce the constraint violation by considering the scheduling of manufacturing and transportation separately. However, some of these find a solution by combining local optimum solutions, but not by finding a global optimum. In addition, some approaches arrive at a solution that does not satisfy all constraints.

Most manufacturing scheduling problems, such as the job-shop scheduling problem (JSP), are problems relating to process sequencing. For some problems, such as flexible job-shop scheduling (fJSP), machine selection is an issue. Transportation problems, such as the vehicle-routing problem (VRP), relate to routing (round-order sequencing) and vehicle-assignment. However, both manufacturing and transportation problems deal with two issues; sequencing and selection. Global optimization is possible only by deciding the sequence through integration, since it mutually and largely influences the result.

In our model, pickup and delivery services are considered as operations. For a production process using resources in different locations, service time and transportation time are considered. We then create different schedules by integrating pickup and delivery as operations. In this problem, we make the following assumptions:

A1. All resources including machines and vehicles are available at the same time ($t = 0$) and all operations can be started at $t = 0$.
A2. The predecessors of manufacturing processes are given for each order.
A3. At any given time, a resource can only execute one operation. A machine can execute a process and a vehicle can execute a pickup/delivery service or provide transportation. It becomes available for other operations once the operation currently assigned to it is completed.
A4. When a process is assigned a different machine (B) from the previous process (A), then pickup at (A), transportation from (A) to (B), and delivery at (B) must be scheduled.

Notation

Indices

i, J: index of operations i, j=1,2,...,J_k
k, l: index of orders, k, l=1,2,...,K
m, n: index of resources (locations),
m, n=1, 2, ..., N (0: location of the depot)
v: index of vehicles, v= 1, 2, ..., V
g, h: index of vehicle operations, g, h, $\in U$

Parameters

K: number of orders
N: number of resources (locations)
J_k: number of operations for order k
V: number of vehicles

E: set of out-of-service vehicle operations,
$E = \{o_S, o_F\}$
P: set of pickup operations,
$P = \left\{ o^P_{kij} \mid r_{kij} = 1, \forall i, j, k \right\}$
D: set of delivery operations,
$D = \left\{ o^D_{kij} \mid r_{kij} = 1, \forall i, j, k \right\}$
U: set of all vehicle operations, $U = E \cup P \cup D$
o_{ki}: i-th operation for order k
o^S: operation of starting from depot
o^F: operation of returning to depot
o^P_{kij}: pickup operation of items produced in o_{ki} to transport to o_{kj}
o^D_{kij}: delivery operation of items consumed in o_{kj} from o_{ki}
M_m: m-th resource (location)
A_m: set of operations that can be processed on M_m
B_v: set of locations that can be visited by vehicle v
p_{kim}: processing time for o_{ki} on M_m
d_{mn}: transportation time from M_m to M_n
L_m: capacity of M_m
W_v: capacity of vehicle v
r_{kij}: precedence constraints

$$r_{kij} = \begin{cases} 1, & \text{if } o_{ki} \text{ is predecessor of } o_{kj}, \\ 0, & \text{otherwise} \end{cases}$$

t^U_{kilj}: setup time from o_{ki} to o_{lj}
t^P_{kij}: service time for o^P_{kij}
t^D_{kij}: service time for o^D_{kij}
t_g: service time for service g
u_g: amount loaded into vehicle in service g
$u_g \geq 0, \quad \text{if } g \in P,$
$u_g < 0, \quad \text{if } g \in D,$
w_{gv}: total amount on vehicle v after service g
a_{gm}: location assignment

$$a_{gm} = \begin{cases} x_{kim}, & \text{if} g = o^P_{ki}, \text{ or } g = o^D_{k*i}, \\ 1, & \text{if } g \in E, m = 0, \\ 0, & \text{otherwise} \end{cases}$$

c^P_{kij}: completion time of o^P_{kij} (same as c_g, if $g = o^P_{kij}$)
c^D_{kij}: completion time of o^D_{kij} (same as c_g, if $g = o^D_{kij}$)
c_M: makespan
T: total mileage of all vehicles
s_{kilj}: operation sequence of processes

$$s_{kilj} = \begin{cases} 1, \text{ if } c_{ki} \le c_h, \\ 0, \text{ otherwise} \end{cases}$$

s_{gh}: operation sequence of pickup and delivery services

$$s_{gh} = \begin{cases} 1, \text{ if } c_g \le c_h, \\ 0, \text{ otherwise} \end{cases}$$

Decision Variables

c_{ki}: completion time for o_{ki}
c_g: completion time of pickup or delivery operation g

$$x_{kim} = \begin{cases} 1, \text{ if } M_m \text{ is selected for} o_{ki}, \\ 0, \text{ otherwise} \end{cases}$$

$$y_{ghv} = \begin{cases} 1, \text{ if } v \text{ is visited in the order of } g \text{ and } h, \\ 0, \text{ otherwise} \end{cases}$$

This model views the problem as a multiobjective optimization problem, and minimizes makespan and vehicle mileage. Objectives and constraints are as follows:

$$\min c_M = \max\{c_{ki}\} \tag{5.51}$$

$$\min T = \sum_{v=1}^{V} \sum_{m=0}^{N} \sum_{n=0}^{N} \sum_{g \in U} \sum_{h \in U} d_{mn} a_{gm} a_{hn} y_{ghv} \tag{5.52}$$

$$\text{s. t. } \left\{c_{lj} - \left(c_{ki} + t_{kilj}^{U} + p_{ljm}\right)\right\} s_{kilj} x_{kim} x_{ljm} \ge 0 \quad \forall (k,i), (l,j), m \tag{5.53}$$

$$\left\{c_{kij}^{P} - \left(c_{ki} + t_{kij}^{P}\right)\right\} r_{kij} x_{kim} x_{kjn} \ge 0 \quad \forall i,j,m,n; m \ne n \tag{5.54}$$

$$\left\{c_{kij}^{D} - \left(c_{kij}^{P} + d_{mn} + t_{kilj}^{D}\right)\right\} r_{kij} x_{kim} x_{kjn} \ge 0 \quad \forall i,j,m,n; m \ne n \tag{5.55}$$

$$\left\{c_{kij} - \left(c_{kij}^{D} + p_{kjn}\right)\right\} r_{kij} x_{kim} x_{kjn} \ge 0 \quad \forall i,j,m,n; m \ne n \tag{5.56}$$

$$\left\{c_h - \left(c_g + d_{mn} + t_h\right)\right\} a_{gm} a_{hn} y_{ghv} \ge 0 \quad \forall g,h,v,m,n; m \ne n \tag{5.57}$$

$$w_{gv} = 0 \quad \forall g \in E, \ \forall v \tag{5.58}$$

$$\left\{c_h - \left(c_g + d_{mn} + t_h\right)\right\} y_{ghv} = 0 \quad \forall g,h,v \tag{5.59}$$

$$w_{gv} \le W_v \quad \forall g, v \tag{5.60}$$

$$\sum_{k=1}^{K} \sum_{i=1}^{J_k} p_{kim} x_{kim} \le L_m \quad \forall m \tag{5.61}$$

$$r_{kij} s_{kjki} = 0 \quad \forall i,j,k \tag{5.62}$$

$$s_{kjki} = 0 \quad \forall i,k \tag{5.63}$$

$$s_{kilj} + s_{ljki} = 1 \quad \forall (k,i), (l,j); (k,i) \ne (l,j) \tag{5.64}$$

$$\sum_{m=1}^{N} x_{kim} = 1 \quad \forall i,k \tag{5.65}$$

$$x_{kim} = 0 \quad \forall (k,i) A_m, \ \forall m \tag{5.66}$$

$$\sum_{h \in P \cup D} y_{ghv} \leq 1 \quad \forall g = o^S, \ \forall v \tag{5.67}$$

$$\sum_{g \in P \cup D} y_{ghv} \leq 1 \quad \forall h = o^F, \ \forall v \tag{5.68}$$

$$\sum_{h \in U} y_{ghv} = \sum_{h \in U} y_{hgv} \leq 1 \quad \forall g \in P \cup D, \ \forall v \tag{5.69}$$

$$\sum_{g \in P \cup D} \sum_{h \cup U} a_{gm} y_{ghv} = 0 \quad \forall m \notin B_v, \ \forall v \tag{5.70}$$

$$c_{ki} \geq 0 \quad \forall i,k \tag{5.71}$$

$$c_g \geq 0 \quad \forall g \tag{5.72}$$

$$x_{kim} \in \{0,1\} \quad \forall i,k,m \tag{5.73}$$

$$y_{ghv} \in \{0,1\} \quad \forall g,h,v \tag{5.74}$$

Equation 5.51 is the objective function that minimizes the makespan. Equation 5.52 is the objective function that minimizes total mileage of all the vehicles. The constraint at Eq. 5.53 means that multiple operations cannot be processed by the same machine simultaneously. The constraints at Eqs. 5.54–5.56 ensure that the product pickup does not occur earlier than the end of the process, and the material delivery is completed before starting the process (Fig. 5.84). The constraint at Eq. 5.57 indicates that the transportation time is necessary when a vehicle services different places. The constraint at Eq. 5.58 ensures that nothing is loaded on the vehicles at starting time. The constraint at Eq. 5.59 shows the current loading of a vehicle from all services, and the constraint at Eq. 5.60 restricts the load limit of a vehicle and ensures that it is not exceeded. The constraint at Eq. 5.61 restricts the ability in each resource since it cannot be exceeded. The constraints at Eqs. 5.62–5.64 indicate precedence constraints and consistency of the operation sequence. The constraints at Eqs. 5.65 and 5.66 ensure that only one resource is assigned to a process. The constraints at Eqs. 5.67–5.69 show that any available vehicle can allocate all services. The constraint at Eq. 5.70 ensures that a vehicle does not remain in the place after completing the service, unless it is the depot. The constraints at Eqs. 5.71–5.74 impose nonnegative conditions.

5.6.2 Multiobjective Hybrid Genetic Algorithm

The Multistage Operation-based Genetic Algorithm (Scheduler moGA) for the scheduling agent [95]is based on a multistage decision model. This algorithm uses an enhanced GA-based discrete dynamic programming (DDP) approach, proposed by Yang [101] for generating schedules in FMS environments. The scheduler moGA

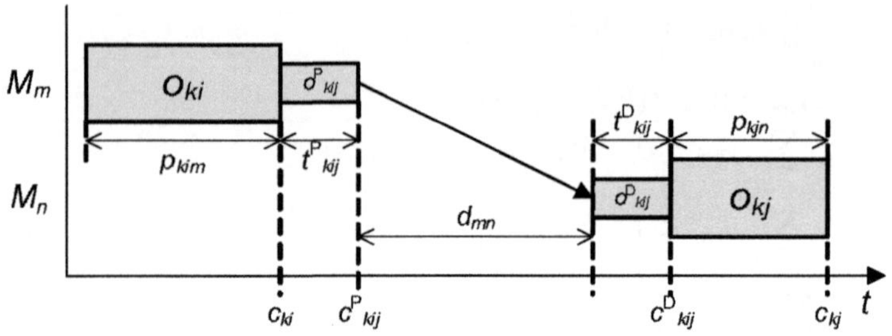

Fig. 5.84 Operation time considering pickup and delivery

approach consists of two parts: sequencing operations and selecting resources. To apply the scheduler moGA to a new integrated scheduling problem, we modified a few chromosome designs. Let $P(t)$ and $C(t)$ be parents and offspring in current generation t, respectively. The overall procedure of multiobjective hybrid GA is outlined as follows:

---

procedure: multiobjective hybrid GA
input: data set, GA parameters ($popSize$, $maxGen$, p_M, p_C)
output: Pareto optimal solutions E
begin
  $t \leftarrow 0$;
  initialize $P1(t)$ by random key-based encoding routine;
  evalute $P2(t)$ by resource permutation encoding routine;
  calculate objectives $z_i(P)$, $i = 1{,}2$ by random key-based decoding
    and resource permutation decoding routine;
  create Pareto $E(P)$;
  calculate $eval(P)$ by adaptive-weight fitness assignment routine;
  while (**not** terminating condition) **do**
    create $C(t)$ from $P(t)$ by parameterized uniform crossover routine;
    create $C_1(t)$ from $P(t)$ by random key-based mutation routine;
    create $C_2(t)$ from $P(t)$ by random resource mutation routine;
    local search C(t) by bottleneck shifting;
    calculate objectives $z_i(C)$, $i = 1, 2$ by random key-based decoding
      and resource permutation decoding routine;
    update Pareto $E(P,C)$;
    calculate eval(P,C) by adaptive-weight fitness assignment routine;
    select $P(t+1)$ from $P(t)$ and $C(t)$by elitist strategy and roulette wheel selection routine;
    $t \leftarrow t+1$;
  end
  output Pareto optimal solutions $E(P,C)$
end

To develop a multistage operation-based genetic representation for the problem, there are three main phases:

Phase 1: Creating an operation sequence

Step 1.1: Generate a random priority to each operation using encoding procedure for first vector v_1.
Step 1.2: Decode a feasible operation sequence that satisfies the precedence constraints.

Phase 2: Assigning operations to machine

Step 2.1: Input the operations sequence found in step 1.2.
Step 2.2: Generate a permutation encoding for machine assignment of each operation (second vector v_2).

Phase 3: Designing a schedule

Step 3.1: Create a schedule S using task sequence and processor assignments.
Step 3.2: Draw a Gantt chart for this schedule.

Phase 1: Creating an Operation Sequence

A random key-based representation [?, 5] is used for the operation sequence. Any chromosome formed by crossover, mutation, and random generation is a feasible solution. It contributes to the reduction of the complexity of genetic operations, since the complicated repair process is not required. The encoding procedure is shown in Fig. 5.85. As the result of an example case, a chromosome is shown in Fig.5.86.

Table 5.40 Simple example of processing time p_{kim}

	M_1	M_2	M_3
o_{11}	-	-	40
o_{12}	30	70	-
o_{13}	-	-	40
o_{21}	40	-	20
o_{22}	-	30	40
o_{23}	60	90	-

procedure: Random key-based encoding
input: total number of operations J
output: chromosome $v_1()$
begin
  for i =1 **to** J
    $v_1(i) \leftarrow$ random[0, 1);
  output chromosome $v_1()$;
end

Fig. 5.85 Random key-based encoding procedure

Operation ID i:	1	2	3	4	5	6	7	8	9	10	11	12	13	14
Priority $v_1(i)$:	0.71	0.95	0.67	0.48	0.83	0.34	0.42	0.80	0.75	0.17	0.40	0.78	0.27	0.13

Fig. 5.86 Chromosome v_1 drawn by random key-based encoding

Using the decoding procedure shown in Fig. 5.87, a feasible operation sequence can be obtained through a trace table, as shown in Table 5.41. The result of this case is as follows:

$$
\begin{aligned}
Q &= \{4,11,1,7,12,8,5,13,14,6,2,9,10,3\} \\
&= \left\{o_{21}, o^P_{212}, o_{11}, o^P_{112}, o^D_{212}, o^D_{112}, o_{22}, o^P_{223}, o^D_{223}, o_{23}, o_{12}, o^P_{123}, o^D_{123}, o_{13}\right\}
\end{aligned}
$$

The parameter R_i is a set of predecessors of operation i. If the i-th operation is a process, R_i is a set of deliveries and if the i-th operation is a delivery, R_i is a pickup. Otherwise, the i-th operation is a pickup and R_i is a process.

Phase 2: Assigning Operations to Machine

After completing the sequencing operations, the position of all stages (operations) is determined. Therefore, it is possible to select resources through a multistage decision making process. The encoding procedure is shown in Fig. 5.88, and the trace table of this procedure is shown in Table 5.42. The delivery gene is not used in order to ensure that the same vehicle is always allocated for pickup. For this reason, in this example, genes of ID 8, 10, 12, and 14 are empty. As the result of this example, a chromosome can be drawn as shown in Fig. 5.89.

Using the decoding procedure shown in Fig. 5.90 a feasible operation sequence and resource selection can be obtained through the trace table shown in Table 5.43. The result of this case is as follows:

$$
\begin{aligned}
S = \{&(o_{21},M_3), \left(o^P_{212},V_1\right), (o_{11},M_3), \left(o^P_{112},V_1\right), \left(o^D_{223},V_1\right), (o_{23},M_2), \\
&(o_{12},M_1), \left(o^P_{123},V_1\right), \left(o^D_{123},V_1\right), (o_{13},M_3)\}
\end{aligned}
$$

procedure: Random key-based decoding
input: chromosome $v_1()$,
  set of predecessors R_i of operation i,
  total number of operations J
output: operation sequence $Q()$
begin
  $P \leftarrow \{ i \mid i = 1, ..., J \}$;
  for $k = 1$ **to** J **do**
    $i^* \leftarrow \text{argmin}\{ v_1(i) \mid R_i \subseteq Q, i \in P \}$;
    $Q(k) \leftarrow i^*$;
    $P \leftarrow P \setminus i^*$;
  end
  output operation sequence $Q()$;
end

Fig. 5.87 Random key-based decoding procedure

Table 5.41 Trace table of random key based decoding

k	i^*	R_{i^*}	$v_1(i^*)$	Q
1	4	Φ	0.48	{ 4 }
2	11	{ 4 }	0.40	{ 4, 11 }
3	1	Φ	0.71	{ 4, 11, 1 }
4	7	{ 1 }	0.42	{ 4, 11, 1, 7 }
5	12	{ 11 }	0.78	{ 4, 11, 1, 7, 12 }
6	8	{ 7 }	0.80	{ 4, 11, 1, 7, 12, 8 }
7	5	{ 12 }	0.83	{ 4, 11, 1, 7, 12, 8, 5 }
8	13	{ 5 }	0.27	{ 4, 11, 1, 7, 12, 8, 5, 13 }
9	14	{ 13 }	0.13	{ 4, 11, 1, 7, 12, 8, 5, 13, 14 }
10	6	{ 14 }	0.34	{ 4, 11, 1, 7, 12, 8, 5, 13, 14, 6 }
11	2	{ 8 }	0.95	{ 4, 11, 1, 7, 12, 8, 5, 13, 14, 6, 2 }
12	9	{ 2 }	0.75	{ 4, 11, 1, 7, 12, 8, 5, 13, 14, 6, 2, 9 }
13	10	{ 9 }	0.17	{ 4, 11, 1, 7, 12, 8, 5, 13, 14, 6, 2, 9, 10 }
14	3	{ 10 }	0.67	{ 4, 11, 1, 7, 12, 8, 5, 13, 14, 6, 2, 9, 10, 3 }

```
Procedure: Resource permutation encoding
input: total number of operations J,
    total number of resources N,
    set of all operations O(),
    set of operations A_m that can be processed on M_m
output: chromosome v2()
begin
  for i = 1 to J do
    if (O(i) is not delivery) then
      do
        m ← random[1, N];
      until O(i) ∈ A_m;
      v2(i)←m;
    end
  end
  output chromosome v2();
end
```

Fig. 5.88 Resource permutation encoding procedure

Table 5.42 Trace table of resource permutation encoding

i	$O(i)$	m, v	A_m, B_v	v_2
1	o_{11}	3	$\{ o_{11}, o_{13}, o_{21}, o_{22} \}$	{ 3 }
2	o_{12}	1	$\{ o_{12}, o_{21}, o_{23} \}$	{ 3, 1 }
3	o_{13}	3	$\{ o_{11}, o_{13}, o_{21}, o_{22} \}$	{ 3, 1, 3 }
4	o_{21}	3	$\{ o_{11}, o_{13}, o_{21}, o_{22} \}$	{ 3, 1, 3, 3 }
5	o_{22}	2	$\{ o_{12}, o_{22}, o_{23} \}$	{ 3, 1, 3, 3, 2 }
6	o_{23}	2	$\{ o_{12}, o_{22}, o_{23} \}$	{ 3, 1, 3, 3, 2, 2 }
7	o^*_{112}	4	$\{ o^*_{112}, o^*_{123}, o^*_{212}, o^*_{223} \}$	{ 3, 1, 3, 3, 2, 2, 4 }
9	o^*_{123}	4	$\{ o^*_{112}, o^*_{123}, o^*_{212}, o^*_{223} \}$	{ 3, 1, 3, 3, 2, 2, 4, 4 }
11	o^*_{212}	4	$\{ o^*_{112}, o^*_{123}, o^*_{212}, o^*_{223} \}$	{ 3, 1, 3, 3, 2, 2, 4, 4, 4 }
13	o^*_{223}	4	$\{ o^*_{112}, o^*_{123}, o^*_{212}, o^*_{223} \}$	{ 3, 1, 3, 3, 2, 2, 4, 4, 4, 4 }

Operation ID i:	1	2	3	4	5	6	7	8	9	10	11	12	13	14
Resource $v_1(i)$:	3	1	3	3	2	2	4		4		4		4	

Fig. 5.89 Chromosome v_2 drawn by resource permutation encoding procedure

```
procedure: Resource permutation decoding
input: chromosome v2(), operation sequence Q(),
      set of all operations O(), total number of operations J
output: schedule S()
begin
   for k = 1 to J do
      i = Q(k);
      m ←v2(i);
      S(k) ← { (O(i), Mm) };
   end
   output schedule S();
end
```

Fig. 5.90 Resource permutation decoding procedure

Phase 3: Designing a Schedule

The schedule is created based on the decided operation sequence and resource selection. The starting time of each operation is allocated at the earliest possible time. It is best for the operation to end early for the two objective functions. But the service is deleted as there is no necessity for transportation in this step. In this example, o^P_{223} and o^D_{223} are deleted because o_{22} and o_{23} are assigned to the same machine. The scheduling result is as follows:

$$\begin{aligned} S = \{ & (o_{21}, M_3 : 0-20), \left(o^P_{212}, V_1 : 20-30\right), (o_{11}, M_3 : 35-75), \\ & \left(o^P_{112}, V_1 : 75-85\right), \left(o^D_{212}, V_1 : 40-50\right), \left(o^D_{112}, V_1 : 99-109\right), \\ & (o_{22}, M_2 : 50-80), \left(o^P_{223}, - : -\right), \left(o^D_{223}, - : -\right), \\ & (o_{23}, M_2 : 95-185), (o_{12}, M_1 : 109-139), \left(o^P_{123}, V_1 : 139-149\right), \\ & \left(o^D_{123}, V_1 : 163-173\right), (o_{13}, M_3 : 173-213)\} \end{aligned}$$

The Gantt chart in Fig. 5.91 shows this sample schedule.

Table 5.43 Trace table of resource permutation decoding

k	i	$O(i)$	$v_2(i)$	S
1	4	o_{21}	3	$\{(o_{21}, M_3)\}$
2	11	o^P_{212}	4	$\{(o_{21}, M_3), (o^P_{212}, V_1)\}$
3	1	o_{11}	3	$\{(o_{21}, M_3), (o^P_{212}, V_1), (o_{11}, M_3)\}$
4	7	o^P_{112}	4	$\{(o_{21}, M_3), (o^P_{212}, V_1), (o_{11}, M_3), (o^P_{112}, V_1)\}$
5	12	o^D_{212}	4	$\{(o_{21}, M_3), (o^P_{212}, V_1), (o_{11}, M_3), (o^P_{112}, V_1), (o^D_{212}, V_1)\}$
6	8	o^D_{112}	4	$\{(o_{21}, M_3), (o^P_{212}, V_1), (o_{11}, M_3), (o^P_{112}, V_1), (o^D_{212}, V_1), (o^D_{112}, V_1)\}$
7	5	o_{22}	2	$\{(o_{21}, M_3), (o^P_{212}, V_1), (o_{11}, M_3), (o^P_{112}, V_1), (o^D_{212}, V_1), (o^D_{112}, V_1), (o_{22}, M_2)\}$
8	13	o^P_{223}	(4)	$\{(o_{21}, M_3), (o^P_{212}, V_1), (o_{11}, M_3), (o^P_{112}, V_1), (o^D_{212}, V_1), (o^D_{112}, V_1), (o_{22}, M_2), (o^P_{223}, -)\}$
9	14	o^D_{223}	(4)	$\{(o_{21}, M_3), (o^P_{212}, V_1), (o_{11}, M_3), (o^P_{112}, V_1), (o^D_{212}, V_1), (o^D_{112}, V_1), (o_{22}, M_2), (o^P_{223}, -), (o^D_{223}, -)\}$
10	6	o_{23}	2	$\{(o_{21}, M_3), (o^P_{212}, V_1), (o_{11}, M_3), (o^P_{112}, V_1), (o^D_{212}, V_1), (o^D_{112}, V_1), (o_{22}, M_2), (o^P_{223}, -), (o^D_{223}, -), (o_{23}, M_2)\}$
11	2	o_{12}	1	$\{(o_{21}, M_3), (o^P_{212}, V_1), (o_{11}, M_3), (o^P_{112}, V_1), (o^D_{212}, V_1), (o^D_{112}, V_1), (o_{22}, M_2), (o^P_{223}, -), (o^D_{223}, -), (o_{23}, M_2), (o_{12}, M_1)\}$
12	9	o^P_{123}	4	$\{(o_{21}, M_3), (o^P_{212}, V_1), (o_{11}, M_3), (o^P_{112}, V_1), (o^D_{212}, V_1), (o^D_{112}, V_1), (o_{22}, M_2), (o^P_{223}, -), (o^D_{223}, -), (o_{23}, M_2), (o_{12}, M_1), (o^P_{123}, V_1)\}$
13	10	o^D_{123}	4	$\{(o_{21}, M_3), (o^P_{212}, V_1), (o_{11}, M_3), (o^P_{112}, V_1), (o^D_{212}, V_1), (o^D_{112}, V_1), (o_{22}, M_2), (o^P_{223}, -), (o^D_{223}, -), (o_{23}, M_2), (o_{12}, M_1), (o^P_{123}, V_1), (o^D_{123}, V_1)\}$
14	3	o_{13}	3	$\{(o_{21}, M_3), (o^P_{212}, V_1), (o_{11}, M_3), (o^P_{112}, V_1), (o^D_{212}, V_1), (o^D_{112}, V_1), (o_{22}, M_2), (o^P_{223}, -), (o^D_{223}, -), (o_{23}, M_2), (o_{12}, M_1), (o^P_{123}, V_1), (o^D_{123}, V_1), (o_{13}, M_3)\}$

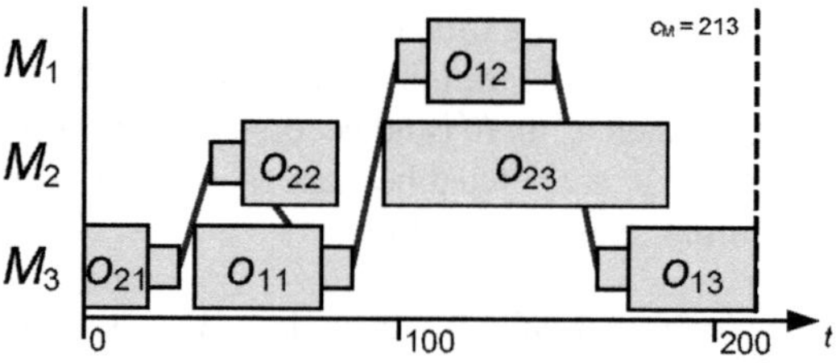

Fig. 5.91 Gantt chart of scheduling result from the example dataset

5.6.2.1 Schedule Improvement

Hybrid genetic algorithms (hGA) combine GA with other techniques, such as local search heuristics, in order to offset the problems inherent in GA. Changing operations in the critical path is an effective way of improving scheduling solutions. In this problem, we have two approaches to shift bottlenecks. One is to change the process order for each machine (Type-A). The other is to change the service order in each vehicle (Type-B).

In this study, a local search is applied only for the first rank, best-solution candidates. The solution candidates are divided into three areas, and different types of local searches apply to each area (Fig. 5.92). Generally, in a scheduling problem, the decoding procedure consumes a great deal of time. Therefore, if local searches repeat calls to the decoding procedure for all chromosomes, the algorithm takes an enormous amount of time.

Figure 5.93 shows the local search procedure, and Fig. 5.65 shows the complete procedure of mo-hGA which combined scheduler moGA with the local search.

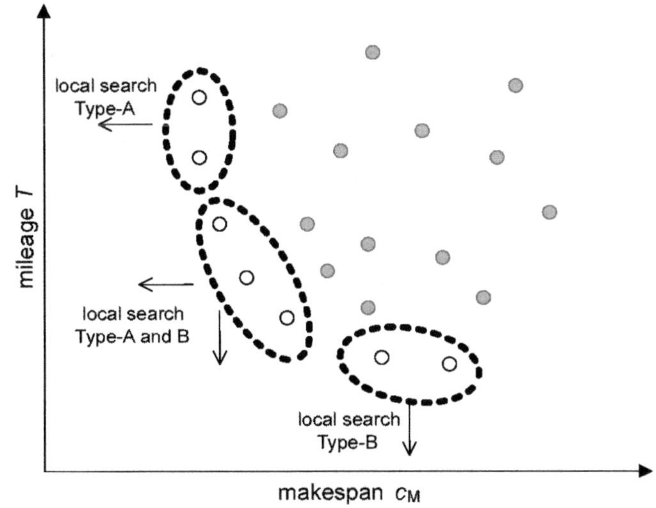

Fig. 5.92 Local search for multiobjective problem

5.6.3 Numerical Experiment

Figure 5.65 shows block diagram of the scheduling system used for this experiment. The main module of this system is GA Scheduler as an implementation of algorithms that was explained in previous section. This example problem requires scheduling eight orders, in a six-machine, four-vehicle environment. The experimental dataset consists of processing timetable (Table 5.44) and machine locations (Fig. 5.94), which are randomly generated. Other dataset and genetic parameters are:

$V = 1, 2, 3, 4$

t^U_{kilj}=30, $\forall (k,i),(l,j)$

$t^P_{kij}=t^D_{kij}$=20, $\forall\ i,j,k$

procedure: Bottleneck shifting
input: schedule $S()$, search type X
output: improved schedule $S'()$
begin
  Step 0. Copy $S'()$ from $S()$
  Step 1. Calculate earliest completion time and
      latest completion time for all operations in $S'()$
  Step 2. Mark an operation as critical in $S'()$
      if earliest completion time is same as
      latest completion time.
      (If $X = A$, mark only processes, and
      If $X = B$, mark only pickup and delivery services)
  Step 3. If there is no critical operation, go to step 8.
  Step 4. Pick and unmark a marked critical operation j
      that has the latest completion time
      among the marked operations.
  Step 5. Pick a critical operation i
      that was assigned the same resource as operation j.
  Step 6. If operation i is not found, go to Step 4.
  Step 7. Save $S'()$ as $S''()$.
      Swap priority of $i, j,$ in $S'()$ and reschedule.
      Repair $S'()$ from $S''()$ if schedule is not improved.
      Go to Step 4.
  Step 8. Output schedule $S'()$
end

Fig. 5.93 Bottleneck shifting local search procedure

$p_C = 0.7$, $p_M = 0.2$,
$popSize = 100$, $maxGen = 100$

Local search methods:

(1) Apply local search to all solution candidates
(2) Proposed method

The Gantt charts of the representative experimental results solved by the local search methods (1) and (2) are shown in Figs. 5.95 and 5.96, respectively. All best compromised solutions are shown as a pareto graph in Fig. 5.97. The final results show no difference between local search methods (1) and (2). However, the proposed method (2) required about 50–70% of the computation time of method (1). Then, its effectiveness was confirmed.

Table 5.44 Experimental dataset of processing time p_{kim}

	M_1	M_2	M_3	M_4	M_5	M_6
o_{11}	50	-	-	-	100	90
o_{12}	-	80	-	90	-	-
o_{13}	-	-	60	-	40	50
o_{21}	-	90	-	-	-	-
o_{22}	50	-	-	70	100	90
o_{23}	-	50	60	-	-	70
o_{24}	90	-	-	70	-	-
o_{31}	-	70	-	-	-	-
o_{32}	60	-	80	-	100	-
o_{33}	-	60	-	100	-	70
o_{41}	60	-	90	70	80	-
o_{42}	-	-	100	-	-	90
o_{43}	-	100	-	90	-	70
o_{51}	70	-	90	-	100	-
o_{52}	-	-	-	80	-	-
o_{53}	80	70	40	-	70	-
o_{54}	-	-	-	50	-	60
o_{61}	-	-	60	-	-	100
o_{62}	90	90	-	70	60	-
o_{63}	90	100	110	-	-	-
o_{71}	-	-	70	-	100	-
o_{72}	-	90	-	90	100	-
o_{73}	90	-	80	-	-	100
o_{81}	-	90	-	40	-	100
o_{82}	70	-	90	-	100	-
o_{83}	-	-	-	60	70	-
o_{84}	30	70	-	-	-	-

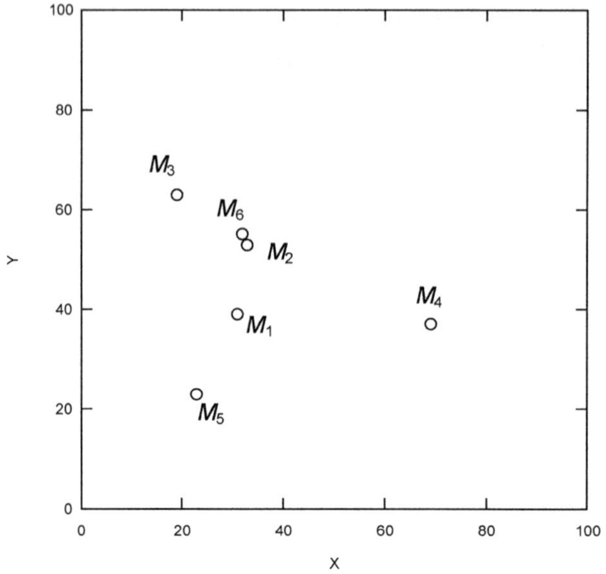

Fig. 5.94 Resource locations for the numerical experiment

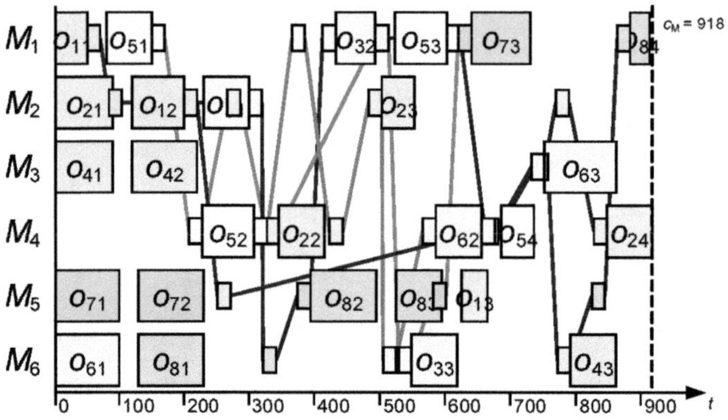

Fig. 5.95 Gantt chart of an experimental result

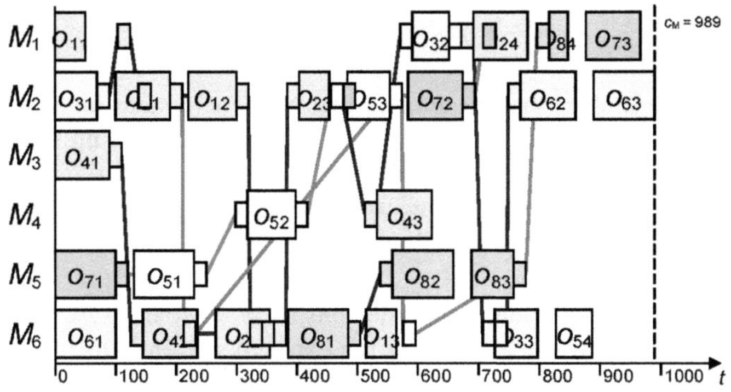

Fig. 5.96 Gantt chart of an experimental result

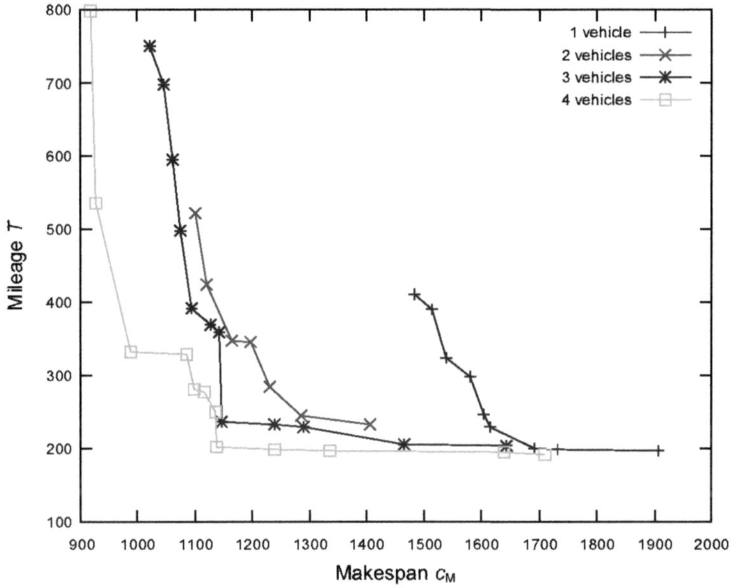

Fig. 5.97 Best compromised solutions of the experimental results

5.7 Summary

In this chapter, we analyze the structure of integrated manufacturing system, and extract the mathematical model in advanced planning and scheduling field. Typically, five kinds of scheduling problems are addressed, job-shop scheduling problem, flexible job-shop scheduling problem, integrated operation sequence and resource selection, integrated scheduling model with multi-plant, and manufacturing and logistics model with pickup and delivery.

We introduced conventional heuristics and several typical GA approaches for job-shop scheduling problem, and gave the experiments on benchmark problems. According to the flexibility in advanced planning and scheduling problems, a multistage operation-based GA is developed as an effective approach for representing the information of flexible resources assignment in combinatorial scheduling problem. In terms of characters of different problem, effective local search techniques are combined to obtain active schedule, *i.e.*, critical path local search, left shift local search. The effectiveness and efficiency of GA approaches were investigated with various scales of advanced planning and scheduling problems by comparing with recent related researches.

References

1. Eck, M. (2003). *Advanced Planning and Scheduling*, BWI paper: http://obp.math.vu.nl/logistics/papers/vaneck.doc
2. Bermudez, J. (1998). *Advanced Planning and Scheduling: Is it as good as it sounds?* The report on Supply Chain Management, March, 3–18.
3. Holland, J., (1975). *Adaptation in Natural and Artificial Systems*, University of Michigan Press, Ann Arbor.
4. Gen, M. and R., Cheng, (1997). *Genetic Algorithms and Engineering Design*, John Wiley Sons, New York.
5. Gen, M. and Cheng, R., (2002). *Genetic Algorithms and Engineering Optimization*, New York: John Wiley & Sons.
6. Adams, J., Balas, E., & Zawack, D. (1987). The shifting bottleneck procedure for job shop scheduling, *Internatioanl Journal of Flexible Maunfacturing Systems*, 34(3), 391–401.
7. Baker, K. (1974). *Introduction to sequencing & scheduling*, NewYork, John Wiley & Sons.
8. Baker, k. & Scudder, G. (1990). Sequencing with earliness & trainess penalties: A review, *Operations Research*, 38, 22–36.
9. Bean, J. (1994). Genetic algorithms & random keys for sequencing & optimizaion, *ORSA Journal on Computing*, 6(2), 154–160.
10. Blackstone, J., Phillips, D., & Hogg, G. (1982). A state of the art survey of dispatching rules for maunfacturing job shop operations, *International Journal of Production Research*, 20, 26–45.
11. Croce, F., Tadei, R., & Volta, G. (1995). A genetic algorithm for the job shop problem, *Computer & Operations Research*, 22, 15–24.
12. Dauzere-Pers, S. & Lasserre, J. (1993). A modified shifting bottleneck procedure for job-shop scheduling, *International Journal of Production Researches*, 31, 923–932.
13. Davis, L. (1985). Job shop scheduling with genetic algorithms, *Proceedings of the First International Conference on Genetic Algorithms*, 136–140.

14. De Jong, K. (1994). Genetic algorithms: a 25 year perspective, *Computational Intelligence: Imitating Life*, 125–134.
15. Dorndorf, W. & Pesch, E. (1995). Evolution based learning in a job shop scheduling enviorment, *Computer & Operations Research*, 22, 25–40.
16. Fralkenauer, E. & Bouffoix, S. (1991). A genetic algorithm for job shop, *Proceedings of IEEE International Conference on Robotics & Automation*, 824–829.
17. Fisher, H. & Thompson, G. (1963). Probabilistic learning combinations of job-shop scheduling rules, *Industrial Scheduling*, 15, 1225–1251.
18. Giffler, B. & Thompson, G. (1960). Algorithms for solving production scheduling problem, *Operations Research*, 8(4), 487–503.
19. Haupt, R. (1989). A survey of priority-rule based scheduling problem, *OR Spectrum*, 11, 3–16.
20. Kubota, A. (1995). Study on optimal scheduling for manufacturing system by genetic algorithms, *Master's thesis*, Asikaga Institute of Technology, Ashikaga, Japan.
21. Morton, T. & Pentico, D. (1993). *Heuristic scheduling systems-with applications to a production systems & project management*, New York, John Wiley & Sons.
22. Norman, B. & Bean, J. (1995). *Random keys genetic algorithm for job-shop scheduling:unabridged version*, Technical report, University of Michigan, Ann Arbor.
23. Norman, B. & Bean, J. (1995). *Random keys genetic algorithm for scheduling*, Technical report, Universuty of Michigan, Ann Arbor.
24. Orvosh, D. & Davis, L. (1994). Using a genetic algorithm to optimize problems with Feasibility constrains, *Proceedings of the First IEEE Conference on Evolutionary Computation*, 548–552.
25. Panwalkar, S. & Iskander, W. (1977). A survey of scheduling rules, *Operations Research*, 25, 45–61.
26. Storer, R., Wu, S., & Vaccari, R. (1992). New search spaces for sequencing problems with application to job shop scheduling, *Management Science*, 38(10), 1495–1510.
27. Blazewicz, J., Ecker, K. H., Schmidt, G. and Weglarz, J., (1994). *Scheduling in Computer and Manufacturing Systems*, Springer.
28. Brucker, P., (1998). *Scheduling Algorithms*, Springer.
29. French, S., (1982). *Sequencing and Scheduling. Mathematics and its applications*, Ellis Horwood Limited.
30. Cheng, R., Gen, M. and Tsujimura, Y. (1996). A Tutorial Sur-vey of Job-Shop Scheduling Problems Using Genetic Algorithm, Part I: Representation. *Computers and Industrial Engineering*, 30(4), 983–997,
31. Roy, B. and Sussmann, B. (1964). Les problems d'ordonnancement avec contraintes disjonctives, Note D.S. No. 9 bis, *SEMA*, Paris, France.
32. Balas, E., (1969). Machine Scheduling via disjunctive graphs : An implicit enumeration algorithm, *Operation Research*, 17, 941–957.
33. Grabot, B. and Geneste, L., (1994). Dispatching rules in scheduling: A fuzzy approach, *International Journal of Production Research*, 32(4), 903-915.
34. Yang, S. and Wang, D. (2000). Constraint Satisfaction Adaptive Neural Network and Heuristics Combined Approaches for Generalized Job-Shop Scheduling, *IEEE Trans. on Neural Networks*, 11(2), 474–486.
35. Cheng, R., Gen, M. and Tsujimura, Y. (1999). A Tutorial Survey of Job-shop Scheduling Problems using Genetic Algorithms, Part II: Hybrid Genetic Search Strategies, *Computers and Industrial Engineering*, 36(2), 343–364.
36. Nowicki, E. and Smutnicki, C., (2005). An advanced tabu search algorithm for the job-shop problem, *Journal of Scheduling*, 8(2), 145–159.
37. Tavakkoli-Moghaddam, R., Jolai, F., Vaziri, F., Ahmed, P.K. and Azaron, A., (2005). A hybrid method for solving stochastic job shop scheduling problems, *Applied Mathematics and Computation*, 170(1), 185–206.
38. Ida, k. and Osawa, A., (2005). proposal of algorithm for shortening idel time on job-shop scheduling problem and its numerical experiments, *J Jpn Ind manage Assoc*, 56(4), 294–301.

39. Gonçalves, J. F., Magalhães Mendes, J. J. and Resende, M. G. C., (2005). A hybrid genetic algorithm for the job shop scheduling problem, *European Journal of Operational Research*, 167(1), 77–95.
40. Garey, M. R. and Johnson,D. S., (1978). Strong NP-Completeness Results: Motivation, Examples and Implications, *Journal of the ACM*, 25, 499–508.
41. Garey, M. R., Johnson, D. S. and Sethi, R., (1976). The complexity of flowshop and jobshop scheduling, textitMathematics of Operations Research, 1(2), 117–129.
42. Fisher, H. and Thompson,G. L., (1963). Probabilistic learning combinations of local job-shop scheduling rules, In J.F. Muth and G.L. Thompson, *Industrial Scheduling*, Prentice Hall, 225-251.
43. Adams, J., Balas, E. and Zawack, D., (1988). procedure for The shifting bottleneck job shop scheduling, *Management Science*, 34(3), 391–401.
44. Prins, C., (2000). Competitive genetic algorithms for the open-shop scheduling problem, *Mathematical Methods of Operations Research*, 52, 389–411.
45. Conway, R.W., Maxwell, W.L., and Miller, L.W., (1967). *Theory of Scheduling*, Addison-Wesley, Reading, MA.
46. Baker, K.R., (1974). *Introduction to Sequencing and Scheduling*, Wiley, New York.
47. Sprecher, A., Kolisch R. and Drexl, A., (1995). Semi-active, active, and non-delay schedules for the resource-constrained project scheduling problem, *European Journal of Operational Research*, 80(1), 94–102.
48. Wiest, J. D., (1964). Some properties of schedules for large projects with limited resources, *Operations Research*, 12, 395–418.
49. Nowicki, E. and Smutnicki, C., (1996). A fast taboo search algorithm for the job-shop problem, *Management Science*, 42(6), 797-813.
50. Ying K. C. and Liao C. J. (2004). An Ant Colony Optimization for permutation flow-shop sequencing. *Computers & Operations Research*, 31(5), 791–801.
51. Jain A. S. and Meeran S. (1998). A State-of -the-art Review of Job-shop Scheduling Techniques, Technical Report, Department of Applied Physics, *Electronics and Mechanical Engineering, University of Dundee, Scotland.*
52. Balas, E. and Vazacopoulos, A., (1998). Guided local search with shifting bottleneck for job shop scheduling. *Management Science*, 44(2), 262–275.
53. Bruker, P. and Schlie, R., (1990). Job-shop scheduling with multi-purpose machines. *Computing*, 45, 369–375.
54. Brandimarte, P., (1993). Routing and scheduling in a flexible job shop by tabu search, *Annals of Operations Research*, 41, 157–183.
55. Chambers, J. B., (1996). Classical and Flexible Job Shop Scheduling by Tabu Search. PhD thesis, University of Texas at Austin, Austin, U.S.A.
56. Charon, I., (1996). Germinated, A. and Hudry, O., Methodes d'Optimization Combinatoires, Paris, France: Masson.
57. Kacem, I., Hammadi, S. and Borne, P., (2002). Approach by localization and multiobjective evolutionary optimization for flexible job-shop scheduling problems. *IEEE Trans. Systems, Man, and Cybernetics-Part C*, 32(1), 1-13.
58. Mati, Y., Rezg, N. and Xie, X., (2001). An Integrated Greedy Heuristic for a Flexible Job Shop Scheduling Problem, *IEEE International Conference on Systems, Man, and Cybernetics*, 4, 2534–2539.
59. Najid, N.M., Dauzere-Peres, S. and Zaidat, A., (2002). A modified simulated annealing method for flexible job shop scheduling problem, *IEEE International Conference on Systems, Man and Cybernetics*, 5(6).
60. Xia, W. and Wu, Z., (2005). An effective hybrid optimization approach for muti-objective flexible job-shop scheduling problem. *Computers & Industrial Engineering*, 48, 409–425.
61. Zhang, H. and Gen, M., (2005). Multistage-based genetic algorithm for flexible job-shop scheduling problem. *Journal of Complexity International,* 11, 223–232.
62. Wu, Z. and Weng, M. X., (2005). Multiagent scheduling method with earliness and tardiness objectives in flexible job shops. *IEEE Trans. System, Man, and Cybernetics-Part B*, 35(2), 293–301.

63. Berretta, R. and Rodrigues, L. F., (2004). A Genetic algorithm for a multistage capacitated lot-sizing problem, *International Journal of Production Economics*, 87(1), 67–81.
64. Dellaert, N., Jeunet, J. and Jornard, N., (2000). A genetic algorithm to solve the general multi-level lot-sizing problem with time-varying costs, *International Journal of Production Economy*, 68, 241–257.
65. Knowles, J. D. and Corne, D. W., (2000). M-PAES: A Genetic Algorithm for Multiobjective Optimization, *Proc. of the Congress on Evolutionary Computation,* 1, 325–332.
66. Krasnogor, N., (2002), *Studies on the theory and design space of Genetic algorithms*, Ph.D. dissertation, Univ. of the West of England, Bristol, U.K.
67. Kusiak, A., (2000). *Computational Integrated in Design & Manufacturing*, John Wiley & Sons, New York.
68. Lopez, O. and Ramirez, M., (2005).A STEP-based manufacturing information system to share flexible manufacturing resources data, *Journal of Intelligent Manufacturing*, 16(3), 287–301.
69. Moon, C., Lee, M., Seo, Y. and Lee, Y. H., (2002). Integrated machine tool selection and operation sequencing with capacity and precedence constraints using genetic algorithm, *Computers & Industrial Engineering*, 43(3), 605–621.
70. Moon, C., Lee, Y. H. and Gen, M., (2004). Evolutionary Algorithm for Process plan Selection with Multiple Objectives, *Journal of Industrial Engineering and Management Systems*, 3(2), 125–131.
71. Moon, C., (2004). *Evolutionary System Approach for Advanced Planning in Multi-Plant Chain*, PhD dissertation, Waseda University, Japan.
72. Moon, C., Kim, J. S. and Gen, M., (2004). Advanced planning and scheduling based on precedence and resource constraints for e-plant chains, *International Journal of Production Research*, 42(15), 2941–2955.
73. Moon, C. and Seo, Y., (2005). Advanced planning for minimizing makespan with load balancing in multi-plant chain, *International Journal of Production Research.* 43(20), 4381–4396.
74. Nishi, T. and Konishi, M., (2005). An autonomous decentralized supply chain planning system for multi-stage production processes, *Journal of Intelligent Manufacturing*, 16(3), 259–275.
75. Raa, B. and Aghezzaf, E. H., (2005). A robust dynamic planning strategy for lot-sizing problems with stochastic demands, *Journal of Intelligent Manufacturing*, 16(2), 207–213.
76. Su, P., Wu, N. and Yu, Z., (2003) Resource selection for distributed manufacturing in agile manufacturing, *Proc. of IEEE International Conference on Systems, Man and Cybernetics*, 2(1), 1578–1582.
77. Tan, W., (2000). Integration of process planning and scheduling-a review, *Journal of Intelligent Manufacturing*, 11(1), 51–63.
78. Tan, W., (2004). A linearized polynomial mixed integer programming model for the integration of process planning and scheduling, *Journal of Intelligent Manufacturing*, 15(5), 593–605.
79. Nishioka, Y. (2002). A New Turn of Manufacturing Enterprise Architecture with Advanced Planning and Scheduling, *Management Systems, Japan Industrial Management Association*, 12(1), 9–13 (in Japanese).
80. PSLX Consortium (2003). *PSLX Technical Specifications, Recommendation*, Version 1.0, [Online]. Available:http://www.pslx.org/.
81. Okamoto, A., Gen, M. & Sugawara, M. (2006). Integrated Data Structure and Scheduling Approach for Manufacturing & Transportation using Hybrid Multistage Operation-based Genetic Algorithm, *Journal of Intelligent Manufacturing*, 17, 411–421.
82. W3C (2004). Extensible Markup Language (XML) 1.0 (Third Edition), *W3C Recommendation*, [Online]. Available:http://www.w3.org/TR/2004/REC-xml-20040204.
83. MESX Joint Working Group. (2004). *MESX White Paper*, [Online]. Available:http://www.mstc.or.jp/faop/doc/informative/MESX-WP.pdf, (in Japanese).
84. Okamoto, A., Gen, M. & Sugawara, M. (2005). APS System based on Scheduler moGA & XML, *Journal of the Society of Plant Engineers Japan*, 17(2), 15–24 (in Japanese).
85. Hastings, N. A. J. & Yeh, C. -H. (1992). Bill of manufacture, *Production and Inventory Management Journal*, 4th Quarter, 27–31.

86. Nishioka, Y. (1999). Supply Chain Planning for Computer Optimized Manufacturing, *Management Systems, Japan Industrial Management Association*, 9(3), 132–136 (in Japanese).
87. Yang, J. B. (2001). GA-based discrete dynamic programming approach for scheduling in FMS environments, *IEEE Transactions on Systems, Man & Cybernetics*, Part B, 31(5), 824–835.
88. Kim, K., Yamazaki, G., Lin, L. & Gen, M. (2004). Network-based hybrid genetic algorithm for scheduling in FMS environments, *Artificial Life & Robotics*, 8, 67–76.
89. Bean, J. C. (1994). Genetics & random keys for sequencing & optimization, *ORSA Journal on Computing*, 6, 154–160.
90. W3C (2003). Scalable Vector Graphics (SVG) 1.1 Specification, *W3C Recommendation*, [Online]. Available:http://www.w3.org/TR/2003/REC-SVG11-20030114/.
91. Spears, W. M. & Dejong, K. A. (1991). On the virtues of parameterized uniform crossover, *Proceedings of the Fourth International Conference on Genetic Algorithms*, 230–236.
92. Applegate, D. & Cook, W. (1991). A computational study of the job shop scheduling problem, *ORSA Journal of Computing*, 3(2), 149–156.
93. Yamada, T. & Nakano, R. (1992). A genetic algorithm applicable to large-scale job-shop problems, *Parallel Problem Solving from Nature: PPSN II*, 281–290, Elsevier Science Publishers, North-Holland .
94. Gen, M.,Tsujimura, Y., & Kubota, E. (1994). Solving job-shop scheduling problem using genetic algorithms, *Proceedings of the 16th International Conference on Computers & Industrial Engineering*, 576–579, Ashikaga, Japan.
95. Okamoto, A., Gen, M. & Sugawara, M. (2005). Cooperation of Scheduling Agent & Transportation Agent in APS System, *Proceedings of the J.S.L.S. Kyushu Division Conference*, 1–11 (in Japanese).
96. Okamoto, A., Gen, M. & Sugawara, M. (2005). APS System based on Scheduler moGA & XML, *Journal of the Society of Plant Engineers Japan*, 17(2), 15–24 (in Japanese).
97. Okamoto, A., Gen, M. & Sugawara, M. (2006). Integrated Data Structure & Scheduling Approach for Manufacturing & Transportation using Hybrid Multistage Operation-based Genetic Algorithm, *Journal of Intelligent Manufacturing*, 17, 411–421.
98. Okamoto, A., Gen, M. & Sugawara, M. (2006). Integrated Scheduling Problem of Manufacturing & Transportation with Pickup & Delivery, *International Journal of Logistics and SCM Systems*, 1, 19–27.
99. Lee, C. Y. & Chen, Z. -L. (2001). Machine scheduling with transportation considerations, *Journal of Scheduling*, 4, 3–24.
100. Soukhal, A., Oulamara, A. & Martineau, P. (2005). Complexity of flow shop scheduling problems with transportation constraints, *European Journal of Operational Research*, 161, 32–41.
101. Yang, J. B. (2001). GA-based discrete dynamic programming approach for scheduling in FMS environments, *IEEE Transactions on Systems, Man & Cybernetics*, Part B, 31(5), 824–835.
102. Aiex, R. M. , Binato, S. & Resende, M. G. C. (2003). Parallel GRASP with path-relinking for job shop scheduling, *Parallel computing in numerical optimization*, 29(4), 393–430.
103. Binato, S., Hery, W.J., Loewenstern, D.M., & Resende, M.G.C., (2002). A GRASP for job shop scheduling. In: Ribeiro, C.C., Hansen, P. (Eds.), *Essays and Surveys in Metaheuristics*. Kluwer Academic Publishers.
104. Wang, L., & Zheng, D., (2001). An effective hybrid optimisation strategy for job-shop scheduling problems. *Computers & Operations Research*, 28, 585–596.
105. Gonçalves, J.F., & Beirão, N.C., (1999). Um Algoritmo Genètico Baseado em Chaves Aleatòrias para Sequenciamento de Operações. *Revista Associaç ao Portuguesa de Desenvolvimento e Investigaç ao Operacional*, 19, 123–137 (in Portuguese).
106. Nowicki, E. & Smutnicki, C. (1996). A fast taboo search algorithm for the job shop problem, *Management Science*, 42(6), 797–813.
107. Cheng, R., Gen, M. & Sasaki, M. (1995). Film-copy deliverer problem using genetic algorithms, *Computers and Industrial Engineering* , 29(1–4), 549–553.

108. Fang, H., Ross, P. & Corne, D. (1993). A promising genetic algorithm approach to job-shop scheduling re-scheduling and open-shop scheduling problems, *Proceedings of the fifth International Conference on Genetic Algorithms*, Morgan Kaufmann Publishers, San Mateo, CA, 375–382.
109. Paredis, J. (1992). Exploiting constraints as background knowledge for genetic algorithms: a case-study for scheduling, *Parallel Problem Solving from Nature 2, PPSN-II*, North-Holland. 231–240.
110. Nakano, R. & Yamada, T. (1992). Conventional genetic algorithms for job-shop problems, *Proceedings of the fourth International Conference on Genetic Algorithms*, Morgan Kaufmann Publishers, San Mateo, CA, 477–479.

Chapter 6
Project Scheduling Models

Project scheduling is a complex process involving many resource types and activities that require optimizing. The requirement of resource type may often influence the requirement of other types. Each activity in a project may be performed in one of a set of prescribed ways as the predecessor. The resource-constrained project scheduling problem (rc-PSP) is a classical well-known problem where activities of a project must be scheduled to minimize the project duration, *i.e.*, makespan. Nevertheless, the NP-hard nature of the problem which is difficult to use to solve realistic sized projects makes necessary the use of heuristic and meta-heuristics in practice.

6.1 Introduction

Today, in a highly competitive environment with rapidly changing operational requirements, manufacturing companies try to balance operational processes and improve their workload processes with the use of effective project management techniques. To balance operational processes, some of the practical scheduling problems have to be efficiently managed.

The problem of scheduling activities under resource and precedence restrictions with the objective of minimizing the project duration is referred to in the literature as *resource-constrained project scheduling problem* (rc-PSP) [1]–[3]. The resource constraints refer to limited renewable resources such as manpower, materials and machines which are necessary for carrying out the project activities. It can be clearly seen that the rc-PSP is a generalization of the static job shop, flow shop, assembly line balancing, related scheduling problem and hence belongs to the class of NP-hard problems [4]–[6].

Many variations of the rc-PSP model have been examined during the last three decade, and these tests can be divided into the various categories based on the problem characteristics, *i.e.*, the number of projects, execution mode, resource consumption, and interruption (see Fig. 6.1). According to the number of projects, the rc-PSP models can be divided into two groups: single project, where there is only one project to be scheduled and multiple project, where there are more than one project to be scheduled. According to the execution mode, the rc-PSP models can also be divided into two groups: single mode and multiple mode. In the multiple mode versions, each activity can be performed in one of several execution modes. According to resource consumption, the resources in rc-PSP models may be limited but renewable from period to period, non-renewable and of limited amount for the whole project life, or doubly constrained which means that both total amount over the project life and per period availability are limited. According to interruption, the rc-PSP models can also be divided into two groups such as non-preemptive and preemptive. In the non-preemptive case, an activity cannot be interrupted once it is started. However, interruption is allowed in the preemptive case.

Additionally, several versions of rc-PSP models arise by varying the objective function, such as minimization of project duration (makespan), minimization of project cost, and minimization of project net present value.

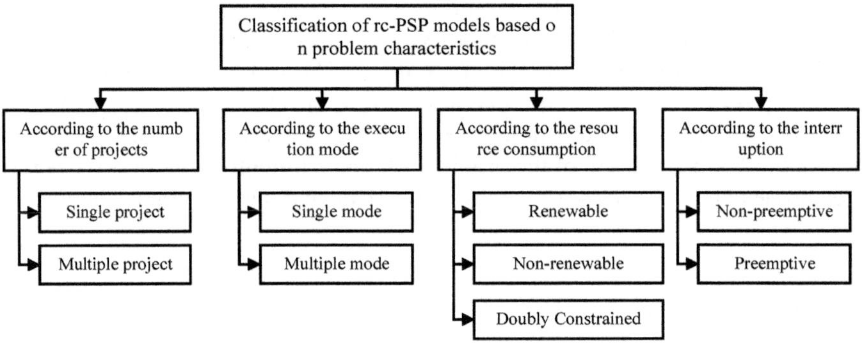

Fig. 6.1 Classification of resource-constrained project scheduling problems

Since the pioneering work of Kelley [7], the rc-PSP has been addressed by a great number of researchers [8]. The earliest attempts were made to find an exact optimal solution to the problem by using traditional optimum seeking methods such as zero-one programming [9]–[11] and implicit enumeration with branch and bound [12]–[18]. A great number of exact methods to solve the rc-PSP are proposed in the literature. Currently, the most competitive exact algorithms appear to be of Demeulemeester and Herroelen [16], Sprecher [19], Brucker *et al.* [17], Klein and Scholl [20, 21] and Mingozzi *et al.* [18]. The reader may refer to Hartmann and Drexl's study [22] for comparison of optimum seeking methods. As stated by Blazewicz *et al.* [4], the rc-PSP is NP-hard. Since for large projects, the size of the problem may render optimal seeking methods computationally impracticable, the researchers have focused their attention on solving rc-PSP by using heuristics with fairly simple scheduling rules capable of producing reasonably good suboptimal schedules. Over the past 40 years, a larger number of heuristic algorithms have been developed and tested. Most of the heuristic methods known so far can be viewed as priority dispatching rules, which assign activity priorities in making sequencing decisions for resolution of resource conflicts according to either of temporally related heuristic rules or resource related heuristic rules [3]. Several approaches of this type have been proposed in the literature, *e.g.*, Alvarez-Valdes and Tamarit [23], Boctor [24], Cooper [25, 26], Davis and Patterson [27], Lawrence [28], Kolisch Drexl [29], Kolisch and Hartmann [3], and Tormos and Lova [30]. For a survey and computational comparison of the solution approaches, the reader may refer to the paper of Alvarez-Valdes and Tamarit [23], Brucker *et al.* [1], Herroelen *et al.* [2], Icmeli *et al.*[31], Kolisch and Padman [32], and Ozdamar and Ulusoy [33]. Kolisch and Hartmann [3] present a classification and performance evaluation of existing heuristic procedures. An excellent review paper by Kolisch and Hartmann [34] discusses the different meta-heuristics for the rcPSP and is an update of the previously published paper of Hartmann and Kolisch [35]. This research revealed that very diverse meta-heuristic techniques have been applied to the rcPSP, but the best performing procedures all have some characteristics in common.

In recent years, the researchers started to develop meta-heuristics such as genetic algorithms [36]–[44], simulated annealing [45]–[47], tabu search [48]–[50] and local search-oriented approaches [51, 52]. Among these meta-heuristics, GAs have been used by many researchers since it has been found very useful for solving many practical scheduling problems. The GA was developed originally as a technique of function optimization derived from the principles of evolutionary theory. For realistic scheduling problems, GA can't be guaranteed the optimality and can sometimes suffer from the premature convergence situation of its solution because of their fundamental requirements that are not using *a priori* knowledge and not exploiting local search information. It has to take too much time to tune the unknown GA parameters such as population size, maximum generation, crossover and mutation probability, because they are affected by a balance between exploitation and exploration in the search space. Therefore, it is very important to regulate these parameters efficiently when applying GA successfully.

In this chapter, we will illustrate the successful applications of GAs to solve various rc-PSP models, *i.e*, with single project, multiple project and multiple modes.

6.2 Resource-constrained Project Scheduling Model

The basic version of the rc-PSP model consists a project of interrelated activities, each characterized by a known processing time and given resource requirements. The rc-PSP model can be defined by the following assumptions:

A1. A single project consists of a number of activities with known processing time.

A2. The processing times of activities are deterministic.

A3. The start time of each activity is dependent upon the completion of some other activities (precedence constraints).

A4. Resources are available in limited quantities and they are non-renewable during period.
A5. There is no substitution between resources.
A6. Activities can't be interrupted.
A7. There is only one execution mode for each activity.
A8. The managerial objective is to minimize a makespan of the project (project duration).

6.2.1 Mathematical Formulation of rc-PSP Models

In the rc-PSP model, we consider a single project which consists of j=1,$\cdots$,i ,$\cdots$, J activities performed by r=1,$\cdots$,R resources with a non-preemptive processing time p_j of periods. Additionally, the precedence relations between each pair of activities (i, j), where i immediately precedes j are taken into consideration.

In the model, the activities a_{rj} are interrelated by two kinds of constraints. In the first constraint, the precedence constraints which are known from traditional rc-PSP, force an activity no to be started before all its predecessors have been finished. The second constraint is the resource constrained where activity j requires l_{rj} units of resource during every period of its processing time p_j (resource r is only available with the constant period availability of b_r units for each period; activities might not be scheduled at their earliest precedence feasible start time but later).

Table 6.1 presents the data set for an example rc-PSP model, which contains 11 activities [53]. The duration, three kinds of resource consumptions (manpower, cost, material), and predecessors of each activity are given in Table 6.1. Using this data set, the precedence graph in Fig. 6.2 is constructed. In the precedence graph, the nodes denote the activities and direct arcs denote the precedence constraints. For example, the processing time of activity 11 is 3 time units and this activity consumes 5 units of manpower, 4 units of cost and 2 units of material.

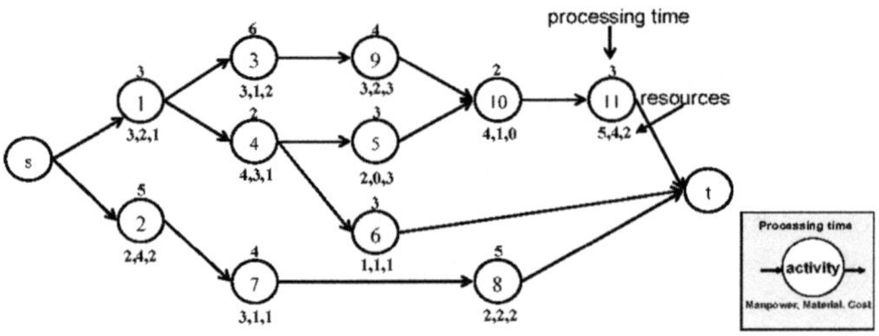

Fig. 6.2 Precedence graph of the rc-PSP model [53]

Table 6.1 Data set of a rc-PSP model [53]

Activity	Processing time	Resource consumptions			Predecessor of activities
		Manpower	Cost	Material	
S	(dummy activity)				s<1,2
1	3	3	2	1	1<3,4
2	5	2	4	2	2<7
3	6	3	1	2	3<9
4	2	4	3	1	4<5,6
5	3	2	0	3	5<10
6	3	1	1	1	6<t
7	4	3	1	1	7<8
8	5	2	2	2	8<t
9	4	3	2	3	9<10
10	2	4	1	0	10<11
11	3	5	4	2	11<t
T	(dummy activity)				

Using the precedence graph in Fig. 6.2, a feasible schedule $S = \{2, 1, 3, 7, 8, 9, 4, 6, 5, 10, 11\}$ where the maximum amount of manpower, cost and material resources are all 6 units (b_1=b_2=b_3=6) can be constructed as follows:

$$
\begin{aligned}
S = \{ & (a_{12}, a_{11}, a_{13}, a_{17}, a_{18}, a_{19}, a_{14}, a_{16}, a_{15}, a_{1,10}, a_{1,11}), \\
& (a_{22}, a_{21}, a_{23}, a_{27}, a_{28}, a_{29}, a_{24}, a_{16}, a_{25}, a_{2,10}, a_{2,11}), \\
& (a_{32}, a_{31}, a_{33}, a_{37}, a_{38}, a_{39}, a_{34}, a_{36}, a_{35}, a_{3,10}, a_{3,11})\} \\
S = \{ & (a_{12}, l_{12} : t_{12}^S - t_{12}^F), (a_{11}, l_{11} : t_{11}^S - t_{11}^F), (a_{13}, l_{13} : t_{13}^S - t_{13}^F), \\
& (a_{17}, l_{17} : t_{17}^S - t_{17}^F), (a_{18}, l_{18} : t_{18}^S - t_{18}^F), (a_{19}, l_{19} : t_{19}^S - t_{19}^F), \\
& (a_{14}, l_{14} : t_{14}^S - t_{14}^F), (a_{16}, l_{16} : t_{16}^S - t_{16}^F), (a_{15}, l_{15} : t_{15}^S - t_{15}^F), \\
& (a_{1,10}, l_{1,10} : t_{1,10}^S - t_{1,10}^F), (a_{1,11}, l_{1,11} : t_{1,11}^S - t_{1,11}^F), (a_{22}, l_{22} : t_{22}^S - t_{22}^F), \\
& (a_{21}, l_{21} : t_{21}^S - t_{21}^F), (a_{23}, l_{23} : t_{23}^S - t_{23}^F), (a_{27}, l_{27} : t_{27}^S - t_{27}^F), \\
& (a_{28}, l_{28} : t_{28}^S - t_{28}^F), (a_{29}, l_{29} : t_{29}^S - t_{29}^F), (a_{24}, l_{24} : t_{24}^S - t_{24}^F), \\
& (a_{26}, l_{26} : t_{26}^S - t_{26}^F), (a_{25}, l_{25} : t_{25}^S - t_{25}^F), (a_{2,10}, l_{2,10} : t_{2,10}^S - t_{2,10}^F), \\
& (a_{2,11}, l_{2,11} : t_{2,11}^S - t_{2,11}^F), (a_{32}, l_{32} : t_{32}^S - t_{32}^F), (a_{31}, l_{31} : t_{31}^S - t_{31}^F), \\
& (a_{33}, l_{33} : t_{33}^S - t_{33}^F), (a_{37}, l_{37} : t_{37}^S - t_{37}^F), (a_{38}, l_{38} : t_{38}^S - t_{38}^F), \\
& (a_{39}, l_{39} : t_{39}^S - t_{39}^F), (a_{34}, l_{34} : t_{34}^S - t_{34}^F), (a_{36}, l_{36} : t_{36}^S - t_{36}^F), \\
& (a_{35}, l_{35} : t_{35}^S - t_{35}^F), (a_{3,10}, l_{3,10} : t_{3,10}^S - t_{3,10}^F), (a_{3,11}, l_{3,11} : t_{3,11}^S - t_{3,11}^F)\}
\end{aligned}
$$

$$
\begin{aligned}
S = \{&(a_{12},2:0-5),(a_{11},3:0-3),(a_{13},3:3-9),(a_{17},3:5-9),\\
&(a_{18},2:9-14),(a_{19},3:9-13),(a_{14},4:13-15),(a_{16},1:15-18),\\
&(a_{15},2:15-18),(a_{1,10},4:18-20),(a_{1,11},5:20-23),(a_{22},4:0-5),\\
&(a_{21},2:0-3),(a_{23},1:3-9),(a_{27},1:5-9),(a_{28},2:9-14),\\
&(a_{29},2:9-13),(a_{24},3:13-15),(a_{26},1:15-18),(a_{25},0:15-18),\\
&(a_{2,10},1:18-20),(a_{2,11},4:20-23),(a_{32},2:0-5),(a_{31},1:0-3),\\
&(a_{33},2:3-9),(a_{37},1:5-9),(a_{38},2:9-14),(a_{39},3:9-13),\\
&(a_{34},1:13-15),(a_{36},1:15-18),(a_{35},3:15-18),(a_{3,10},0:18-20),\\
&(a_{3,11},2:20-23)\}
\end{aligned}
$$

Figure 6.3 illustrates the Gantt chart for this feasible schedule with makespan of 23 time units.

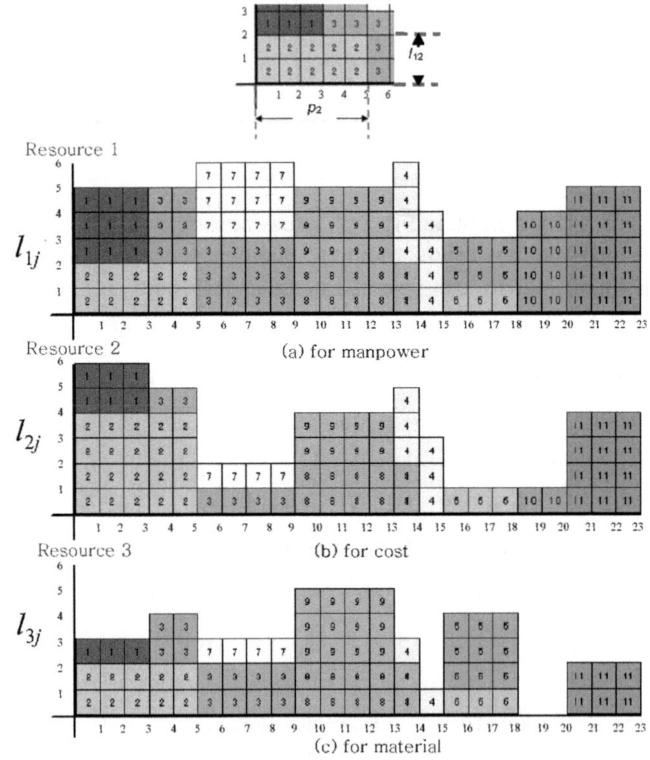

Fig. 6.3 A feasible schedule for the rc-PSP model

In order to formulate the mathematical model, the following indices, parameters and decision variable are introduced:

Notation

Indices

j: index of the activities $(i = 1, 2, \ldots, i, \ldots, J)$
r: index of the non-renewable resources $(j = 1, \ldots, R)$

Parameters

J: number of activities
R: number of non-renewable resources
p_j: the processing time of activity i
l_{rj}: the amount of resource r consumed by the activity j
b_r: the maximum-limited resource r only available with the constant period availability
Suc(i): the set of direct successors of activity i
Pre(i): the set of direct predecessors of activity i

Decision Variables

t_j: the finishing time of activity j

$$x_{jt} = \begin{cases} 1, & \text{if activity } j \text{ is scheduled in time } t \\ 0, & \text{otherwise} \end{cases}$$

Mathematical Model

The mathematical model for the rc-PSP can be stated as follows:

$$\min f_M = \max_j \{t_j\} \tag{6.1}$$

$$\text{s. t. } t_j - t_i \geq p_j, \quad \forall j \in Suc(i) \tag{6.2}$$

$$\sum_{j=1}^{J} l_{rj} x_{jt} \leq b_r, \quad \forall t, r \tag{6.3}$$

$$t_j \geq 0, \quad \forall j \tag{6.4}$$

$$x_{jt} = 0 \; or \; 1 \quad \forall j, t \tag{6.5}$$

In this mathematical model, the objective (Eq. 6.1) is to minimize the total makespan. The constraints given in Eqs. 6.2–6.5 are used to formulate the general feasibility of the problem. The constraint given in Eq. 6.2 ensures that none of the precedence constraints is violated. Constraint at Eq. 6.3 ensures that the amount of resource used by all activities does not exceed its limited quantity in any period. Constraints given in Eqs. 6.4 and 6.5 represent the usual integrity restriction.

6.2.2 Hybrid GA for rc-PSP Models

In this section of the chapter, we will introduce a hybrid genetic algorithm with fuzzy logic controller (flc-hGA) for solving the rc-PSP model. Particularly for the initialization of the population, a new scheme based on the serial method to find an optimal or near optimal solution is used [54, 55].

Let $P(t)$ and $C(t)$ be parents and offspring in current generation t. The overall procedure of flc-hGA for solving rc-PSP model is outlined as follows:

```
procedure: flc-hGA for rc-PSP
input: processing time, maximum-limited resource, GA parameters
output: the best schedule
begin
   t ← 0;
   initialize P(t) by priority-based encoding routine;
   evaluate P(t) by revised serial priority-based decoding routine;
   while ( not terminating condition) do
      create C(t) from P(t) by PMX or position-based crossover routine;
      create C(t) from P(t) by swap mutation or local search-based mutations routine;
      evaluate C(t) by priority-based decoding routine;
      if t > u then
          regulate adaptive GA parameters p_C and p_M by using FLC;
      select P(t+1) from P(t) and C(t) by elitist selection routine;
      t ← t+1;
   end
   output the best schedule;
end
```

The detailed information about the genetic representation, revised serial method, genetic operators and fuzzy logic controller are given in the following subsections.

6.2.2.1 Genetic Representation

For the representation of rc-PSP models, the priority-based encoding and priority-based decoding methods were used. To develop a priority-based genetic representation for the rc-PSP model, there are two main phases:

Phase 1: Creating an Activity Sequence

Step 1.1: Generate a random priority to each activity in the project using encoding procedure
Step 1.2: Decode a feasible activity sequence that satisfies the precedence constraints

Phase 2: Designing a Schedule

Step 2.1: Create a schedule *S* using the activity sequence found in step 1.2
Step 2.2: Draw a Gantt chart for this schedule

While introducing these methods, in order to illustrate the procedures for rc-PSP, we will be using the data set given in Table 6.1 and its corresponding precedence graph given in Fig. 6.2.

Phase 1: Creating an Activity Sequence

Step 1.1: Generate a random priority to each activity in the project using encoding procedure The first step in GAs is the generation of initial population that consists of chromosomes. In this step, an indirect representation scheme called priority-based encoding method is used. In this method, the position of a gene was used to represent a task node and the value of the gene was used to represent the priority of the task node for constructing a schedule among candidates. This encoding method verifies any permutation type representations, so that most of the existing genetic operators can be easily applied. Fig. 6.4 illustrates a priority-based chromosome obtained by using the priority-based encoding procedure.

activity ID j:	1	2	3	4	5	6	7	8	9	10	11
activity priority $v(j)$:	10	11	7	9	3	8	5	4	6	2	1

Fig. 6.4 A priority-based chromosome

Step 1.2: Decode a feasible activity sequence that satisfies the precedence constraints In order to decode the chromosomes generated by encoding procedure in step 1.1, the priorities of each activity are used to create a feasible activity sequence that satisfies the precedence constraints in the model. Figure 6.5 presents the priority-based decoding procedure for creating a schedule.

```
procedure: priority-based decoding (creating a schedule)
input: the set of direct successors of activity j chromosome v(j)
output: schedule S
begin
    s̄ ← ∅, S ← ∅;
    t ← 0, j ← 0;
    while (j ≠ t) do
        s̄ ← Suc(j);
        j* ← argmax{v(j) | j ∈ s̄};
        s̄ ← s̄ ¥ j* ;
        S ← S ∪ j*;
        j ← j*;
    end
    output schedule S
end
```

Fig. 6.5 Priority-based decoding procedure for creating a schedule

For the illustration of this decoding procedure, the chromosome found in Fig. 6.4 is used. Table 6.2 presents the trace table for the decoding procedure. In the example, we obtained a feasible schedule of S={2, 1, 4, 6, 3, 9, 7, 8, 5, 10, 11}.

Table 6.2 Trace table for activity sequence

j		$v(j)$	j^*	S
0	{1,2}	v(1)=10, v(2)=11	2	S={2}
2	{1,7}	v(1)=10, v(7)=5	1	S={2,1}
1	{3,4,7}	v(3)=7, v(4)=9, v(7)=5	4	S={2,1,4}
4	{3,5,6,7}	v(3)=7, v(5)=3, v(6)=8, v(7)=5	6	S={2,1,4,6}
6	{3,5,7}	v(3)=7, v(5)=3, v(7)=5	3	S={2,1,4,6,3}
3	{5,7,9}	v(5)=3, v(7)=5, v(9)=6	9	S={2,1,4,6,3,9}
9	{5,7,10}	v(5)=3, v(7)=5, v(10)=2	7	S={2,1,4,6,3,9,7}
7	{5,8,10}	v(5)=3, v(8)=4, v(10)=2	8	S={2,1,4,6,3,9,7,8}
8	{5,10}	v(5)=3, v(10)=2	5	S={2,1,4,6,3,9,7,8,5}
5	{10}	v(10)=2	10	S={2,1,4,6,3,9,7,8,5,10}
10	{11}	v(11)=1	11	S={2,1,4,6,3,9,7,8,5,10,11}

Phase 2: Designing a Schedule

Step 2.1: Create a *schedule S* using the activity sequence found in step 1.2. Using the trace table for the decoding procedure, the schedule can be constructed as S= { 2,1,4,6,3,9,7,8,5,10,11}.

Step 2.2: Draw a Gantt chart for this schedule. Using this schedule, Gantt chart for limited resources $(b_1, b_2, b_3)=(6, 6, 6)$ is constructed and shown in Fig. 6.6. The makespan of this solution is 25 time units.

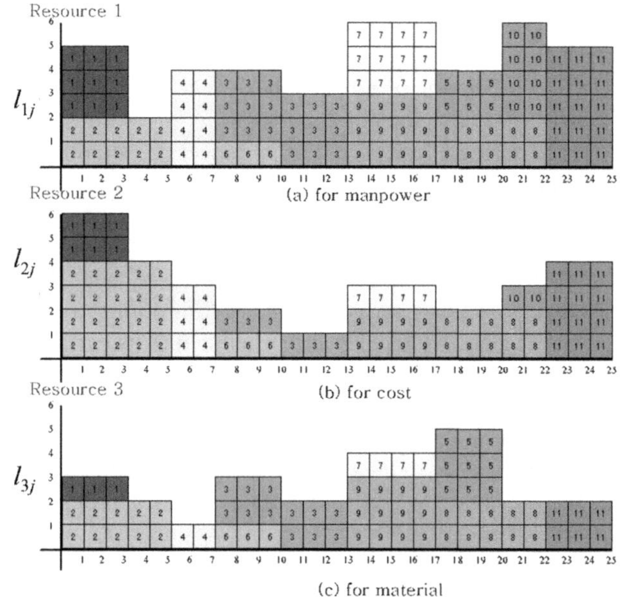

Fig. 6.6 Gantt chart for the rc-PSP model

6.2.2.2 Revised Serial Method for rc-PSP model

For project scheduling, serial and a parallel schedule generation schemes with minimum time lags have been discussed by many researchers [56, 3]. In Kolish [56] and Kolish and Hartmann [3], it has been shown that the parallel method does not generally perform better than the serial method. Consequently, a new revised serial method of priority-based decoding is introduced in this section.

The new revised serial method of priority rule based heuristic algorithms can be described as follows. The schedule set S_j are the activities where were already scheduled and thus belong to the partial schedule. The decision set $\bar{S}_j$ contains the unscheduled activities with every predecessor being in the scheduled set:

$$l_{rf}^{\pi} = b_r - \sum_j inA_t l_{rj}, \tag{6.6}$$

$$\bar{S_j} = \{j \mid j \notin S_j, Pre(j) \subseteq S_j\} \tag{6.7}$$

where

l_{rf}^{π}: the left over capacity of the non-renewable resource r in period f, and the decision set $\bar{S}_j$.

A_f: the set of activities being in progress in period f

K_j: the set of predecessor activities of activity j

l_{rj}: units of resource during every period of its processing time p_j

b_r: maximum-limited resource r only available with the constant period availability

The revised serial method always tries to schedule some activities and ensures for each activity that there exits an appropriate corresponding sequence. In this method, first one activity from the decision set is selected with a priority rule (the activity with the highest priority value is selected) and scheduled at its earliest precedence. If the resource feasible start time is satisfied, the selected activity is removed from the decision set and put into the schedule set. This, in turn, may replace a number of activities into the decision set, since all their predecessors are scheduled. This algorithm terminates when all activities are in the partial schedule set, *i.e.*, n=j. Figure 6.7 presents the procedure to generate initial schedules through revised serial method. where:

t_j^{EF}: the earliest finish time of activity j within the current partial schedule.

t_j^{LF}: the latest finish time of activity j as determined by backward recursion from the upper bound of the project's makespan.

---

procedure: revised serial method
input: the set of direct predecessors of activity j, chromosome $v(j)$
output: schedule S
begin
    $\bar{s} \leftarrow \varnothing$, $S \leftarrow \varnothing$;
    $t \leftarrow 0$, $j \leftarrow 0$;
    while ($|j| \neq t$) **do**
        $\bar{s} \leftarrow \mathrm{Suc}(j)$;
        $j^* \leftarrow \arg\max\{v(j) \mid j \in \bar{s}\}$;
        $\bar{s} \leftarrow \bar{s}$ ¥ j^* ;
        $t_{j^*}^{EF} \leftarrow \max\{t_i \mid i \in s\,Suc(j^*)\} + p_{j^*}$;
        $t_{j^*} \leftarrow \min\{f \mid t_{j^*}^{EF} \leq f \leq t_{j^*}^{LF},\ l_{rj^*} \leq l_{r\tau}^{\pi},\ \tau = f - d_{j^*} + 1, \ldots, f, r \in R\}$;
        $S \leftarrow S \cup j^*$;
        $j \leftarrow j+1$;
    end
    output schedule S
end

Fig. 6.7 Revised serial method

Using the same chromosome given in Fig. 6.4, we obtain the schedule S={ 2,1,3,7,8,9,4,6,5,10,11} where makespan is 23 time units by using revised serial method. In Fig. 6.8 is the Gantt chart for the limited resource (b_1, b_2, b_3) =(6, 6, 6).

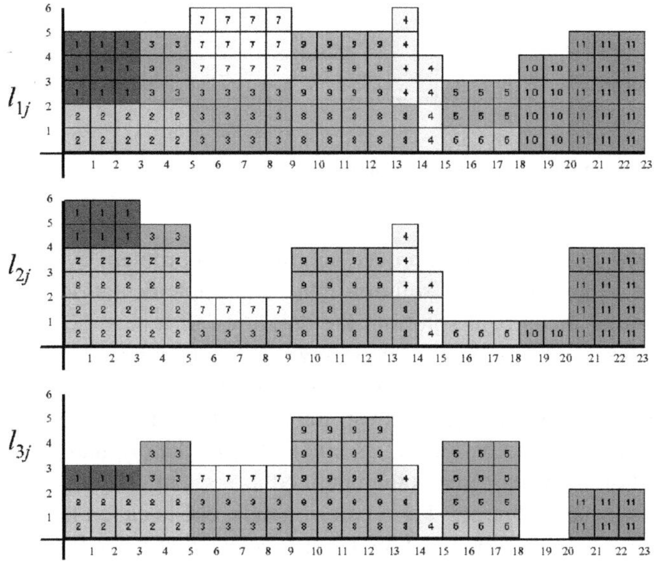

Fig. 6.8 Gantt chart for the rc-PSP model

6.2.2.3 Genetic Operators

In this subsection, the crossover, mutation and selection operators used in the flc-hGA approach are explained in detail.

Crossover Operator: As crossover operator, position-based crossover and partially mapped crossover (PMX) are used. The position-based crossover operator proposed by Syswerda [57] was adopted as shown in Fig. 6.9 Essentially, it takes some genes from one parent at random and fills vacuum position with genes from the other parent by a left-to-right scan. PMX, which was proposed by Goldberg and Lingle [58], and can be viewed as an extension of two-cut point crossover for binary string to permutation representation. It uses a special repairing procedure to resolve the illegitimacy caused by the simple two-cut point crossover. The reader may refer to Chapter 1 for more information.

Mutation Operator: As a mutation operator, swap mutation and local search-based mutation are used. In swap mutation, two positions are selected at random and their contents are swapped is used. For more detailed information about this mutation, the reader may refer to Chapter 1.

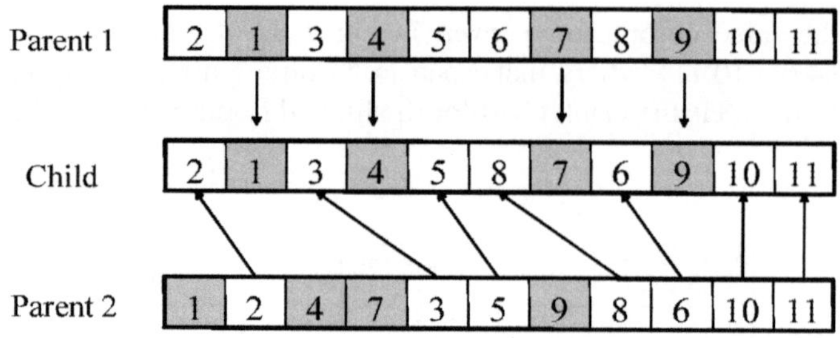

Fig. 6.9 The position-based crossover operator

Local search methods seek improved solutions to a problem by searching in the neighborhood of an incumbent solution [59]. The implementation of local search requires an initial incumbent solution, the definition of a neighborhood for an incumbent solution, and a method for choosing the next incumbent solution. The idea that hunting for an improved solution by making a small change can be used in mutation operators. Observing the swap mutation, the chromosome generated by pair-wise interchange can be viewed as a neighbor of the original chromosome. A neighborhood of a chromosome is then defined as a set of chromosomes generated by such pair-wise interchanges. For a pair of genes, one is called the pivot which is fixed for a given neighborhood and the other is selected at random as shown in Fig. 6.10. For a given neighborhood, a chromosome is called local optima if it is better than any other chromosomes according to the fitness value. The size of a neighborhood affects the quality of the local optima. There is a clear trade-off between small and large neighborhoods: if the number of neighbors is larger, the probability of finding a good neighbor may be higher, but looking for it takes more time. The scheme of the mutation operator is described in Fig. 6.11.

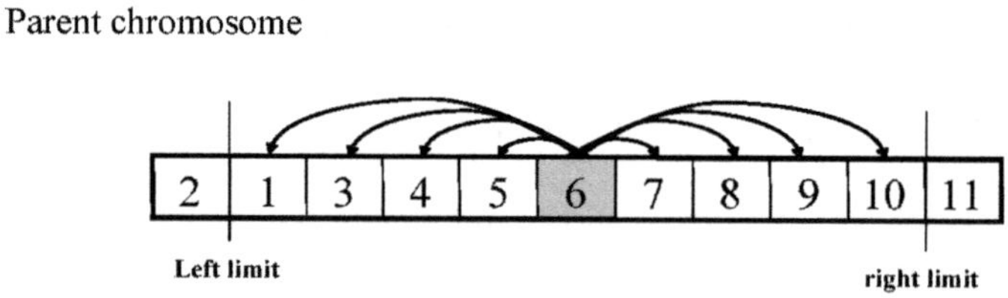

Fig. 6.10 The incumbent chromosome and its neighborhood

Selection Operator: As a selection operator, the elitist selection, which preserves the best chromosome in the next generation and overcome the stochastic errors of sampling, is used. With the elitist selection, if the best individual in the current generation is not reproduced into the new generation, one individual is randomly removed from the new population and the best one is added to the population.

```
procedure: Local Search-based Mutation
input: parents P
output: offspring C
begin
    i ← 1 ;
    let P be current incumbent chromosome;
    select a pivot gene randomly;
    while (i < specifiedNumber) do
          pick up the gene;
          evaluate the neighbors toward both sides
          until left and right limit;
          let the neighbor be current incumbent
          if it is better then
                          generate C_i;
          i ← i + 1;
      end
      output the incumbent in C;
end
```

Fig. 6.11 Local search-based mutation

6.2.2.4 Regulation of GA Parameters by Fuzzy Logic Controller

For the fine-tuning of GA parameters, fuzzy logic controller (FLC) has been proved to be very useful. Most research using FLC can adaptively regulate the GA parameters (generation number, population size, crossover ratio, mutation ratio and others). Thus by fine-tuning of these parameters, much computational time can be saved and the search ability of GA in finding global optimum can be improved more than the conventional GA without FLC.

In the proposed flc-hGA, we used the concept of Wang *et al.* [60]. The main idea of the concept consists of two FLCs; crossover FLC and mutation FLC that are implemented independently to adaptively regulate the crossover ratio and mutation ratio during the optimization process based on the FLC. The inputs of the mutation fuzzy controller are the same as the crossover fuzzy controller, the output of which is changed by the mutation ratio. The combination strategy used in the proposed approach between FLC and GA is shown in Fig. 6.12.

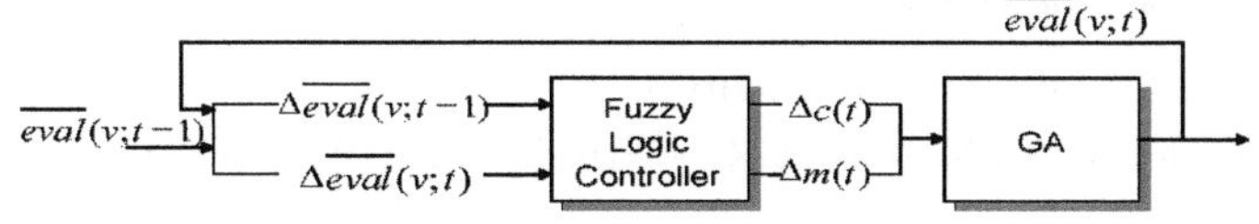

Fig. 6.12 Coordinated strategies of FLC and GA

6.2.3 Computational Experiments and Discussions

The computational experiments were conducted from two aspects: (1) comparison of flc-hGA using the revised serial initialization method with other methods and (2) evaluation of the several combinations of genetic operator that gives significant impact on the performance of the flc-hGA. The proposed algorithms are implemented in Visual Basic on a PC Pentium 1400 MHz clock-pulse and 255MB RAM as operation system. The values of parameters used in the computational experiments are as follows:

Population size: *popSize*= 20

Maximum number of generation: *maxGen*= 1000

Probability of crossover: $p_C = 0.3$

Probability of mutation: $p_M = 0.3$

The test problem consists of 27 activities including two dummy activities (start and end activity) [27]. Each activity has fixed multiple unit requirements of three different resources. The problem is shown in Fig. 6.13.

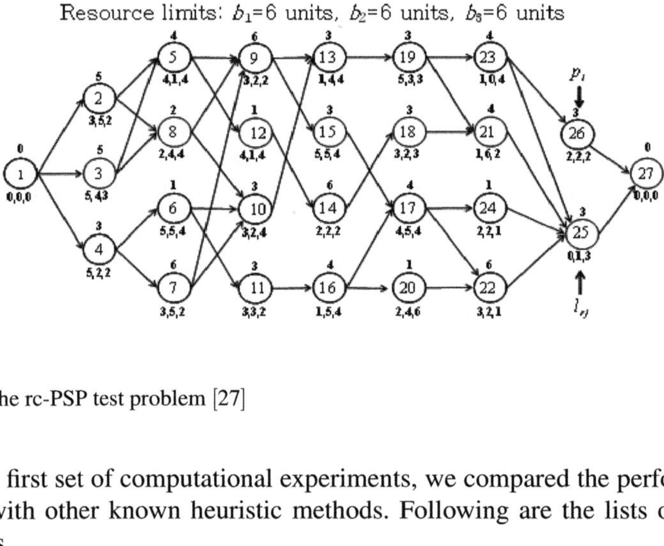

Fig. 6.13 The rc-PSP test problem [27]

For the first set of computational experiments, we compared the performance of flc-hGA with other known heuristic methods. Following are the lists of heuristic algorithms.

Minimum Late Finish Time (LFT): Priority $=\min_j \{t_j^{LF}\}$

t_j^{LF}=Late Finish time of activity j.

Greatest Resource Utilization (GRU): Priority $= \max_j \{l_{rj} \mid j \in J\}$

l_{rj}= resource utilization of activity j.

Shortest Imminent Operation (SIO): Priority $= \min_j \{p_j\}$

p_j=processing time of activity j.

This "shortest job first" select rule.

Minimum Job Slack (MINSLK): Priority $= \min_j \{T_j^{es} - t_j^{LS}\}$

t_j^{ES}=Early Start time of activity j.

p_j^{LS}=Late Start time of activity j.

Resource Scheduling Method (RSM): Priority= $\min_j \{\max[0, \max\{t_j^{EF} + d_j - t_j^{LS}\}]\}$

t_j^{EF}=Early Finish time of activity j.

t_j^{LS}=Late Start time of activity j.

Greatest Resource Demand (GRD): Priority $=P_J \sum_{r=1}^{m} l_{rj}$

p_i=processing time of activity j.

l_{rj}=per period requirement of resource type i by activity j.

M=number of different resource type.

Most Jobs Possible (MJP): Priority = $\max_j \{u_j\}$

u_j=the greatest number of activities being scheduled in any interval.

Select Jobs Randomly (RAN): This heuristic assigns priority among competing jobs on a purely random basis.

The results of first set of computational experiments are presented in Fig. 6.14. From these results, it can be clearly seen that the flc-hGA using serial method is superior to all existing heuristic methods.

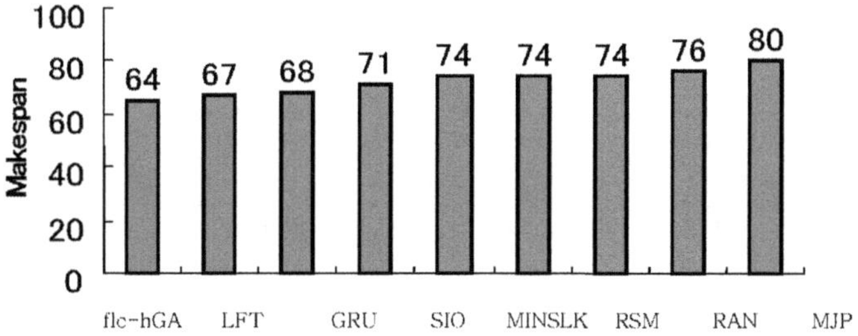

Fig. 6.14 Comparison with other heuristic results

Table 6.3 illustrates the 11 different of alternative optimal schedules found by solving the rc-PSP using flc-hGA with serial method. These alternative schedules are very important in rc-PSP model since sometimes there may be a need to change the schedules. Figure6.15 illustrates Gantt chart of alternative schedule 1 for the rc-PSP test problem.

For the second set of computational experiments, we used test problems with different number of activities to compare the performance of the proposed algorithm hGA and flc-hGA with various combinations of crossover and mutation operators using serial method. The combinations of genetic operators used in these computational experiments are as follows:

Position-based crossover + swap mutation (PB+SM)

Partially mapped crossover + swap mutation (PMX+SM)

Position-based crossover + local search-based mutation (PB+LSM)

Partially mapped crossover + local search-based mutation (PMX+LSM)

Table 6.3 Alternative schedules for the rc-PSP test problem

Alternative schedule 1	1	2	3	5	8	4	7	6	9	11	10	15	16	17	20	24	12	14	13	22	18	19	21	23	26	25	27
Start time	0	0	5	10	14	16	19	25	26	26	29	32	35	39	43	44	44	45	45	48	51	54	57	57	61	61	64
Finish time	0	5	10	14	16	19	25	26	32	29	32	35	39	43	44	45	45	51	48	54	54	57	61	61	64	64	64
Alternative schedule 2	1	2	3	5	8	4	7	6	9	11	10	15	16	20	17	24	12	14	13	22	18	19	21	23	26	25	27
Start time	0	0	5	10	14	16	19	25	26	26	29	32	35	39	40	44	44	45	45	48	51	54	57	57	61	61	64
Finish time	0	5	10	14	16	19	25	26	32	29	32	35	39	40	44	45	45	51	48	54	54	57	61	61	64	64	64
Alternative schedule 3	1	2	3	5	8	4	7	6	10	9	11	15	16	20	17	24	12	14	13	22	18	19	21	23	26	25	27
Start time	0	0	5	10	14	16	19	25	26	26	29	32	35	39	40	44	44	45	45	48	51	54	57	57	61	61	64
Finish time	0	5	10	14	16	19	25	26	29	32	32	35	39	40	44	45	45	51	48	54	54	57	61	61	64	64	64
Alternative schedule 4	1	2	3	8	4	7	5	6	9	11	10	15	16	20	17	24	12	14	13	22	18	19	21	23	26	25	27
Start time	0	0	5	10	12	15	21	25	26	26	29	32	35	39	40	44	44	45	45	48	51	54	57	57	61	61	64
Finish time	0	5	10	12	15	21	25	26	32	29	32	35	39	40	44	45	45	51	48	54	54	57	61	61	64	64	64
Alternative schedule 5	1	2	3	8	5	4	7	6	9	11	10	15	16	20	17	24	12	14	13	22	18	19	21	23	26	25	27
Start time	0	0	5	10	12	16	19	25	26	26	29	32	35	39	40	44	44	45	45	48	51	54	57	57	61	61	64
Finish time	0	5	10	12	16	19	25	26	32	29	32	35	39	40	44	45	45	51	48	54	54	57	61	61	64	64	64
Alternative schedule 6	1	3	2	8	5	4	6	7	9	11	10	16	20	15	17	24	12	14	13	22	18	19	21	23	26	25	27
Start time	0	0	5	10	12	16	19	20	26	26	29	32	36	37	40	44	44	45	45	48	51	54	57	57	61	61	64
Finish time	0	5	10	12	16	19	20	26	32	29	32	36	37	40	44	45	45	51	48	54	54	57	61	61	64	64	64
Alternative schedule 7	1	3	2	8	5	4	6	7	9	11	15	10	16	17	24	20	12	14	18	13	19	21	23	26	22	25	27
Start time	0	0	5	10	12	16	19	20	26	26	32	29	35	39	43	44	44	45	51	48	54	57	57	61	48	61	64
Finish time	0	5	10	12	16	19	20	26	32	29	35	32	39	43	44	45	45	51	54	51	57	61	61	64	54	64	64
Alternative schedule 8	1	3	4	2	8	6	5	7	9	11	10	15	16	20	17	24	12	14	13	22	18	19	23	21	26	25	27
Start time	0	0	5	8	13	15	16	20	26	26	29	32	35	39	40	44	44	45	45	48	51	54	57	57	61	61	64
Finish time	0	5	8	13	15	16	20	26	32	29	32	35	39	40	44	45	45	51	48	54	54	57	61	61	64	64	64
Alternative schedule 9	1	4	3	6	7	2	8	5	9	11	10	15	16	20	17	12	24	14	13	22	18	19	21	23	26	25	27
Start time	0	0	3	8	9	15	20	22	26	26	29	32	35	39	40	44	44	45	45	48	51	54	57	57	61	61	64
Finish time	0	3	8	9	15	20	22	26	32	29	32	35	39	40	44	45	45	51	48	54	54	57	61	61	64	64	64
Alternative schedule 10	1	4	3	7	6	2	5	8	9	11	10	15	16	20	17	12	24	14	13	22	18	19	21	23	26	25	27
Start time	0	0	3	8	14	15	20	24	26	26	29	32	35	39	40	44	44	45	45	48	51	54	57	57	61	61	64
Finish time	0	3	8	14	15	20	24	26	32	29	32	35	39	40	44	45	45	51	48	54	54	57	61	61	64	64	64
Alternative schedule 11	1	4	3	7	6	2	8	5	9	11	10	15	16	20	17	12	24	14	13	22	18	19	21	23	26	25	27
Start time	0	0	3	8	14	15	20	22	26	26	29	32	35	39	40	44	44	45	45	48	51	54	57	57	61	61	64
Finish time	0	3	8	14	15	20	22	26	32	29	32	35	39	40	44	45	45	51	48	54	54	57	61	61	64	64	64

The results are given in Tables 6.4–6.6. In the results, the best fitness states the percentage of instances for which either a feasible solution could be found or which could be shown to be unsolvable. The average value after 10 consecutive runs is used as the best fitness.

In the aspect of best fitness (see Table 6.4), flc-hGA has results better than hGA as the number of activities is increased. Methods with local search-based mutation out of the hGA and flc-hGA have the best result.

In Table 6.5, number of alternative schedules means an average value of the same best makespan after 10 runs. In the aspect of number of alternative schedules, numbers of alternative schedules of hGA and flc-hGA are increased as the number of activities is increased. It can be clearly seen that flc-hGA has results better than

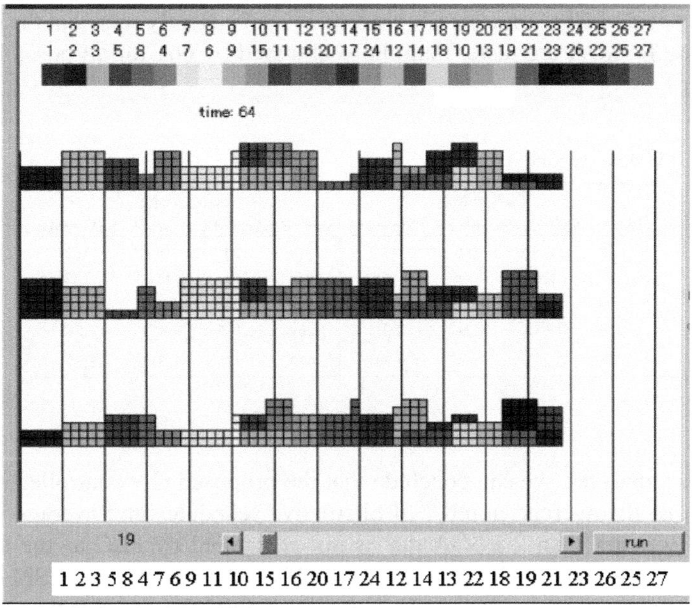

Fig. 6.15 The Gantt chart of alternative schedule 1

Table 6.4 The best fitness (percentage of solvable instances)

	hGA with				flc-hGA with			
Number of activities	PB+SM	PMX+SM	PB+LSM	PMX+LSM	PB+SM	PMX+SM	PB+LSM	PMX+LSM
10	90%	90%	100%	100%	100%	100%	100%	100%
20	80%	80%	80%	80%	90%	100%	100%	100%
50	80%	80%	100%	80%	100%	90%	100%	100%
70	70%	70%	80%	80%	90%	100%	100%	100%
100	60%	60%	70%	70%	80%	80%	100%	100%

hGA as the the number of activities is increased and methods which contain local search-based mutation out of the hGA and flc-hGA have the best result.

Table 6.5 The number of alternative schedules

	hGA with				flc-hGA with			
Number of activities	PB+SM	PMX+SM	PB+LSM	PMX+LSM	PB+SM	PMX+SM	PB+LSM	PMX+LSM
10	4	4	4	4	4	4	4	4
20	10	10	10.5	10.5	12	12	12	12
50	12.2	9.32	13.35	13.36	16.52	15.25	16.52	16.52
70	16.45	13.56	16.96	14.65	13.56	12.65	14.65	12.65
100	14.65	12.59	17.05	16.23	17.02	17.02	18	17.02

In the aspect of CPU time (see Table 6.6), it can be clearly seen that the results are obtained by flc-hGA are faster than those obtained by GA for all possible number of activities.

Table 6.6 CPU time (seconds)

	hGA with				flc-hGA with			
Number of activities	SM+PB	SM+PMX	LSM+PB	LSM+PMX	SM+PB	SM+PMX	LSM+PB	LSM+PMX
10	149	129	166	173	136	103	150	152
20	136	129	168	169	125	112	161	162
50	1,504	1,500	1,723	1,750	1,157	1,042	1,313	1,435
70	1,729	1,639	1,947	2,011	1,267	1,175	1,451	1,543
100	2,576	2,273	2,667	2,781	2,072	1,972	2,371	2,451

By these analyses, we can conclude that the proposed GA controlled by FLC in the aspect of the average number of alternative schedules and average CPU time is generally better than the hGA that is not controlled by FLC as the number of activities is increased. The hGA using LSM + PB and flc-hGA using LSM + PB have the best results out of the others. Therefore, we only compared their convergence behaviors in detail.

Figure 6.16 shows the evolutionary convergence behaviors of average fitness of flc-hGA using LSM + PB and hGA using LSM + PB, when there are 70 activities. The evolutionary environment was set as follows: population size was 10, crossover and mutation ratio were both 0.3, and maximum generation was 45. In the initial process of Fig. 6.16, hGA and flc-hGA have almost same convergence process. But flc-hGA has makespan better than hGA after generation number is 12 and products alternative schedules of the same project duration after generation number is 21. Figure 6.17 and Fig. 6.18 show the various behaviors of p_M and p_C for flc-hGA and hGA using LSM + PB. In hGA, the values of p_M and p_C are static as generation is increased. In flc-hGA, however, the values of p_M and p_C are adaptively regulated by FLC during the optimization process. These values make flc-hGA searching variously search space rather than GA.

In conclusion, by these computations, we can state that the flc-hGA yields better results than several heuristic procedures presented in the literature, the local search-based mutation has significant impact on the performance of the flc-hGA and also the convergence behavior of flc-hGA is better than the convergence behaviors of hGA without FLC.

6.3 Resource-constrained Multiple Project Scheduling Model

Most of the literature on project management has been dedicated to single projects. Nevertheless, in recent years there was been growing interest in problems related to the scheduling of multiple project environments. There is no doubt that trying

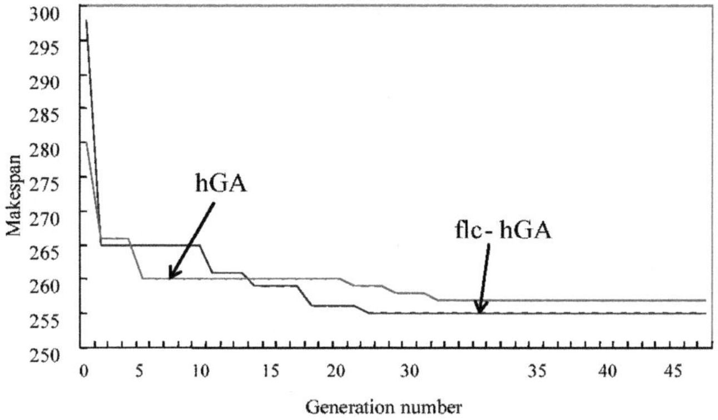

Fig. 6.16 The convergence behaviors of average fitness of flc-hGA and hGA

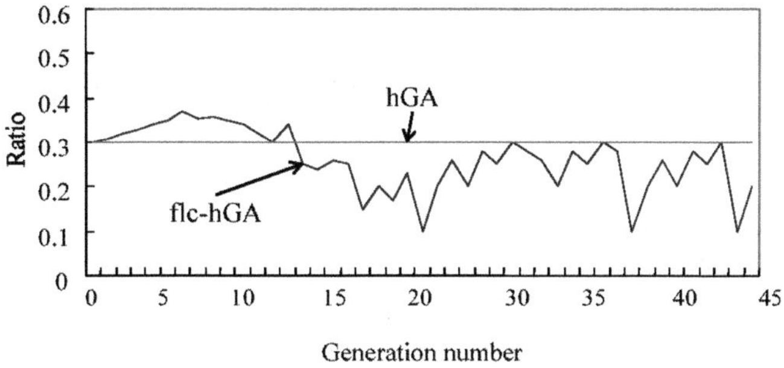

Fig. 6.17 The convergence behaviors of p_C in flc-hGA and hGA

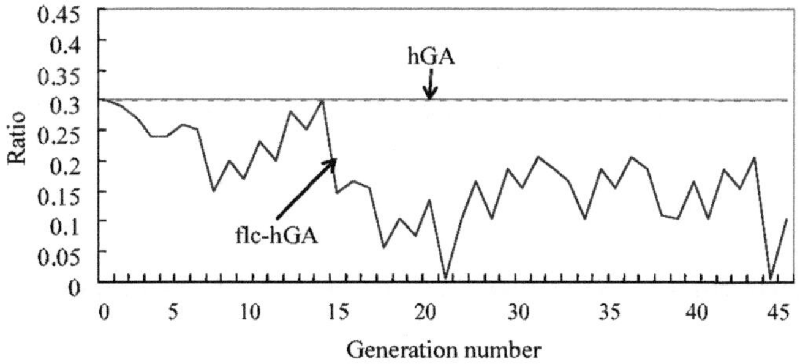

Fig. 6.18 The convergence behaviors of p_M in flc-hGA and hGA

to find "better" heuristics for computationally resource-constrained multiple project scheduling problems (rc-mPSP) with concepts of traditional single project management is hard since it is very difficulty to find major organizations managing in only one project. Recently, multiple projects have been addressed by a number of researchers [1], [61]–[67]. Their works have dealt with a variety of situations based on single-project in which one or both of these types of constraints are relaxed, or at least simplified, and a few studies on project management have started to explore the issue of how to manage an organization with multiple projects.

Fricke and Shendar [61] considered how the most important multiple project success factors in this environment differ from factors of success in traditional single project management, and are consistent with other emerging research in product development environment. Lova [64, 65] analyzed the effect of the schedule generation schemes such as serial or parallel, and priority rules that involved minimum latest finish time, minimum total slack, maximum total work content, shortest activity from shortest project, or first come first served in single project and multiple project environments.

The rc-mPSP consists of multiple projects which are a number of activities with known processing times and multiple resources. The general features of rc-mPSP model rely on some of the basic assumptions of rc-PSP model (refer to Section 6.2). Additionally, the rc-mPSP model can be defined by the following assumptions:

- A1. When a specific project is initiated, it must be finished without changing to another project, (precedence constraints of multiple projects).
- A2. The starting time of each activity is dependent upon the completion of some other activities (precedence constraints of activities).
- A3. The multiple resources are available in limited quantities but renewable from period to period.
- A4. Activities cannot be interrupted; there is only one execution mode for each activity.
- A5. The managerial objective is to minimize the total project time and the total tardiness penalty for all projects.

The most important multiple project success factors in this environment differ from factors of success in traditional single project management, and are consistent with other emerging research in product development environment.

6.3.1 Mathematical Formulation of rc-mPSP Models

In rc-mPSP model, we consider the models which consists of i=0,$\cdots$,m,$\cdots$, I+1 projects and j=0,$\cdots$, n,$\cdots$, J+1 activities with a non-preemptive processing time p_{ij} of periods. Additionally, the precedence relations between each project (i, m), where i immediately precedes m are taken into consideration.

In each model, the activities are interrelated by two kinds of constraints. In the first constraint, the precedence constraints which are known from traditional rc-PSP, force an activity must be not to be started before all its predecessors have been finished. In the second constraint, activity j in project i requires l_{ijr} units of resource $r \in R$ during every period of its processing time p_{ij} (resource r is only available with the constant period availability of b_r units for each period; activities might not be scheduled at their earliest precedence feasible start time but later).

In this rc-mPSP, the objectives are to minimize total project time and minimize total tardiness penalty of multiple projects. Total project time is the sum of the completion times for all projects which scheduled the activities such that precedence and resource constraints are obeyed in multiple projects. The total tardiness penalty is the sum of tardiness costs for all projects, where the tardiness cost for a project is the multiplication of its unit tardiness cost and the absolute difference between its completion time and its due date, given that the former is larger than the latter. The resource constraints refer to limited renewable resources such as manpower, material and machines which are necessary for carrying out the project activities.

Table 6.7 presents the data set for an rc-mPSP model, which contains five projects. The project-related data are duration of activity, resource consumption, predecessor of activities, duration of project, and predecessor of project for example rc-mPSP. Resources are available in limited quantities but are renewable at intervals. Using a network representation, in the former, every project is considered with its corresponding start and end dummy activities. Using this data set, the precedence graph in Fig. 6.19 is constructed. In the precedence graph, the nodes denote the activities and direct arcs denote the precedence constraints. In order to formulate

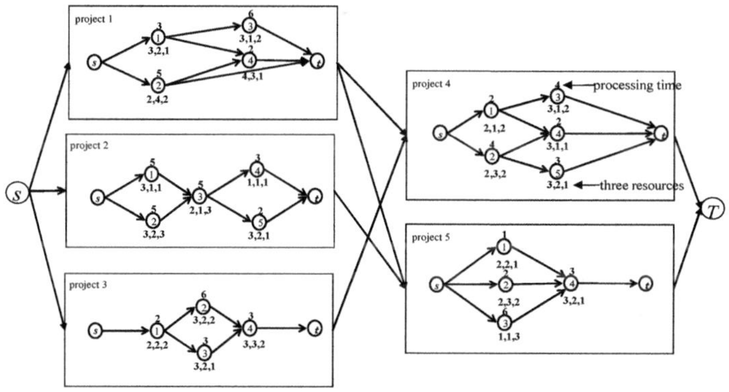

Fig. 6.19 Precedence relations for a rc-mPSP

the mathematical model, the following indices, parameters and decision variable are introduced:

Table 6.7 Project-related data for an rc-mPSP model

Project	Activity	Duration of activity	Resource consumption	Predecessor of activities	Duration of project	Predecessor of projects
S	(Dummy project)					S<1,2,3
1	s	(Dummy activity)		s<1,2	11	1<4,5
	1	3	3,2,1	1<3,4		
	2	5	2,4,2	2<4,t		
	3	6	3,1,2	3<t		
	4	2	4,3,1	4<t		
2	s	(Dummy activity)		s<1,2	11	2<4,5
	1	5	3,1,1	1<3		
	2	2	3,2,3	2<3		
	3	5	2,1,3	3<4,5		
	4	3	1,1,1	4<t		
	5	2	3,2,1	5<t		
3	s	(Dummy activity)		s<1	11	3<4,5
	1	2	2,2,2	1<2,3		
	2	6	3,2,2	2<4		
	3	3	3,2,1	3<4		
	4	3	3,3,2	4<t		
4	s	(Dummy activity)		s<1,2	11	4<T
	1	2	2,1,2	1<3,4		
	2	4	2,3,2	2<4,5		
	3	4	3,1,2	3<t		
	4	2	3,1,1	4<t		
	5	3	3,2,1	5<t		
5	s	(Dummy activity)		s<1,2,3	10	5<T
	1	1	2,2,1	1<4		
	2	2	2,3,2	2<4		
	3	6	1,1,3	3<4		
	4	3	3,2,1	4<t		
T	(Dummy project)					

Notation

Indices

i: project index, $i = 0, \cdots, m, \cdots, I+1$ $(orP_i, i = 1, \cdots, m, \cdots, I+1)$
(0 or = are dummy projects)
j: activity index in each project, $j = 0, \cdots, n, \cdots, J+1$ $(orA_j, j = 1, \cdots, n, \cdots, J+1)$
(We assume that J is the maximum number of activities in each project;J=J_i and j=0 or = J+1 are dummy activities)
r: resource index, $(j = 1, \ldots, R)$

Parameters

p_{ij}: the processing time of activity $j$in project i
l_{ijr}: the scheduling activity j in project i consumes resource units per period from

resource r

b_r: the maximum-limited resource r only available with the constant period availability

A_f: the set of activities being in progress in period f

$A_f = \{j \mid t_{ij}S \leq f < t_{ij}S + p_{ij}, = 0, \cdots, J+1, i = 0, \cdots, I+1\}$

t_i^D: the due date of the project i (the promised delivery time of project)

c_i^{TP}: the total penalty cost of the project i for unit time

t_{ij}^S: starting time of activity j in project i

t_{ij}^F: finish times of activity j in project i

t_{iJ}^F: finish times of last activity J in project i

Decision Variables

t_{ij}^S: starting time of activity j in project i

Mathematical Model

The mathematical model for the rc-PSP can be stated as follows:

$$\min t_F = \sum_{i=1}^{I} t_{iJ}^F \tag{6.8}$$

$$\min p_T = \sum_{i=1}^{I} c_i^{TP}(t_{iJ}^F - t_i^D) \tag{6.9}$$

$$\text{s. t. } t_{i-1}^S J + p_{j-1} \leq t_{ij}^S, \ \forall i, \ \forall j \tag{6.10}$$

$$\sum_{i \in A_f} \sum_{j \in A_f} l_{ijr} \leq b_r, \quad r \in R \tag{6.11}$$

$$t_{ij}^S \geq 0, \quad \forall i, \forall j \tag{6.12}$$

In this mathematical model, the first objective function (Eq. 6.8) minimizes the total project time that is the sum of the completion times for all projects. The second objective function (Eq. 6.9) minimizes the total tardiness penalty of multiple projects, which is the sum of penalty costs for all projects, where the penalty cost for a project is the multiplication of its unit penalty cost and the absolute difference between its completion time and its processing time, given that the former is larger than the latter. The constraint at Eq. 6.10 states precedence relations between related projects. Constraint at Eq. 6.11 corresponds to resource constraints regarding non-renewable resources. Constraints at Eq. 6.12 represent the usual integrality constraint.

6.3.2 Hybrid GA for rc-mPSP Models

In this section of the chapter, we will introduce a hybrid genetic algorithm (hGA) for solving the multi-objective rc-mPSP model [54, 68].

Let $P(t)$ and $C(t)$ be parents and offspring in current generation t. The overall procedure of hGA for solving rc-mPSP model is outlined as follows:

---

procedure: hGA for rc-mPSP
input: processing time, maximum-limited resource, GA parameters
output: the best schedule
begin
  $t \leftarrow 0$;
  iinitialize $P(t)$ by priority-based encoding routine;
  evaluate $P(t)$ by revised serial priority-based decoding routine;
  while (**not** terminating condition) **do**
    create $C(t)$ from $P(t)$ by swap mutation or local search-based mutations routine;
    evaluate $C(t)$ by priority-based decoding routine;
    if t > u **then**
      regulate adaptive GA parameters p_M by using FLC or adaptive method;
    select $P(t+1)$ from $P(t)$ and $C(t)$ by elitist selection routine;
    $t \leftarrow t+1$;
  end
  output the best schedule;
end

The detailed information about the genetic representation, genetic operators and the regulation of GA parameters are given in the following subsections.

6.3.2.1 Genetic Representation

For the representation of rc-mPSP models, the priority-based encoding and revised serial priority-based decoding methods were used. To develop a priority-based genetic representation for the rc-mPSP model, there are two main phases:

Phase 1: Creating an activity sequence

Step 1.1: Generate a random priority to each activity in each project using encoding procedure.

Step 1.2: Decode a feasible activity sequence that satisfies the precedence constraints.

Phase 2: Designing a schedule

Step 2.1: Create a schedule S using the activity sequence found in step 1.2.

Step 2.2: Draw a Gantt chart for this schedule.

While introducing these methods, in order to illustrate the procedures for rc-mPSP, we will be using the data set given in Table 6.7 and its corresponding precedence graph given in Fig. 6.19.

Phase 1: Creating an Activity Sequence

Step 1.1: Generate a random priority to each activity in each project using encoding procedure. In this step, an indirect representation scheme called priority-based encoding method is used. The encoding procedure is shown in Fig. 6.20. In this method, the position of a gene was used to represent a task node and the value of the gene was used to represent the priority of the task node for constructing a schedule among candidates. This encoding method verifies any permutation type representations, so that most of the existing genetic operators can be easily applied. The gene of priority-based chromosome presents a priority value of activity in all projects and is number of total activities in all projects. Figure 6.21 illustrates a priority-based chromosome obtained by using the priority-based encoding procedure. Here, we

```
procedure: priority-based encoding
input: number of activities m
output: chromosome v_k
begin
    for i = 1 to I
        for j = 1to J_i
            v_i(j) ← 0;
            ρ ← 1;
            while(ρ ≤ J_i) do
                j ← random(1, J_i);
                if (v_i(j) = 0) then
                    v_i(j) ← ρ;
                    ρ ← ρ+1;
            end
        end
    output chromosome v_k
end
```

Fig. 6.20 Priority-based encoding procedure for creating a schedule

use the position to denote an activity ID and the priority value to denote the priority associated with the activity as shown in Fig. 6.21. The value of a gene is an integer exclusively within $[1,n]$. The larger the integer value is the higher the priority becomes.

Step 1.2: Decode a feasible activity sequence that satisfies the precedence constraints In order to decode the chromosomes generated by encoding procedure in step 1.1, the priorities of each activity are used to create a feasible activity sequence that satisfies the precedence constraints in the model. In the decoding of the serial method by priority-based encoding, one activity in project from the decision set is

activity ID:	1	2	3	4	1	2	3	4	5	1	2	3	4	1	2	3	4	5	1	2	3	4
	7	14	4	11	2	17	20	12	9	1	15	21	5	6	18	3	13	8	19	16	10	22

Fig. 6.21 A priority-based chromosome

selected with a priority rule (the activity with the highest priority value is selected) and scheduled at the earliest precedence. We select activity obeyed feasible finish time and resource through decoding of serial method, and the selected activity is removed from the decision set and put into the schedule set. This, in turn, may replace a number of activities in the decision set, since all their predecessors are scheduled. The algorithm terminates at stage number n , when all activities of multiple projects are in the partial schedule set. Figure 6.22 presents the revised serial priority-based decoding procedure for creating a schedule. where

procedure: revised serial priority-based decoding (creating a schedule)
input: the set of successive activities Suc(i) after activity j, chromosome V_j
output: schedule S
begin
  $j \leftarrow 0$, $t \leftarrow 0$;
  $\bar{s} \leftarrow \varnothing$, $S \leftarrow \varnothing$;
  while ($|j| \neq t$) **do**
    $\bar{s} \leftarrow \mathrm{Suc}(i)$;
    $j^* \leftarrow \arg\max\{v(j) | j \in \bar{s}\}$;
    $\bar{s} \leftarrow \bar{s} \setminus j^*$;
    $t_{j^*}^{EF} \leftarrow \max\{t_i | i \in K_{j^*}\} + p_{j^*}$;
    $t_{j^*} \leftarrow \min\{f | t^{EF} \le f \le t_{j^*}^{LF}, l_{rj^*} \le l_{r\tau}^{\pi}, \tau = f - d_{j^*} + 1, \ldots, f, r \in R\}$;
    $S \leftarrow S \cup j^*$;
    $j \leftarrow j + 1$;
  end
  output schedule S
end

Fig. 6.22 Revised serial priority-based decoding procedure for creating a schedule

t_j^{EF} : the earliest finish time of activity j within the current partial schedule.

t_j^{LF}: the latest finish time of activity j as determined by backward recursion from the upper bound of the project's total flow time.

Suc(i): the set of successors of activity j.

For the illustration of this decoding procedure, the chromosome found in Fig. 6.21 is used and the schedule has been found as follows:

$$S = \{(2,1,4,3),(2,1,3,4,5),(1,3,2,4),(2,5,1,4,3),(3,2,1,4)\}$$

Phase 2: Designing a Schedule

Step 2.1: Create a schedule S using the activity sequence found in step 1.2
Using the the decoding procedure, the schedule can be constructed as

$$S = \{(2,1,4,3),(2,1,3,4,5),(1,3,2,4),(2,5,1,4,3),(3,2,1,4)\}$$

The total flow time of this schedule is 13+13+11+11+9=57 and total tardiness penalty is (2+2+0+0+0) *10=40 through serial method based on priority rule.
Step 2.2: Draw a Gantt chart for this schedule
Figure 6.24 shows the Gantt chart of this rc-mPSP model with limited resources $(b_1,b_2,b_3) = (6, 6, 6)$.

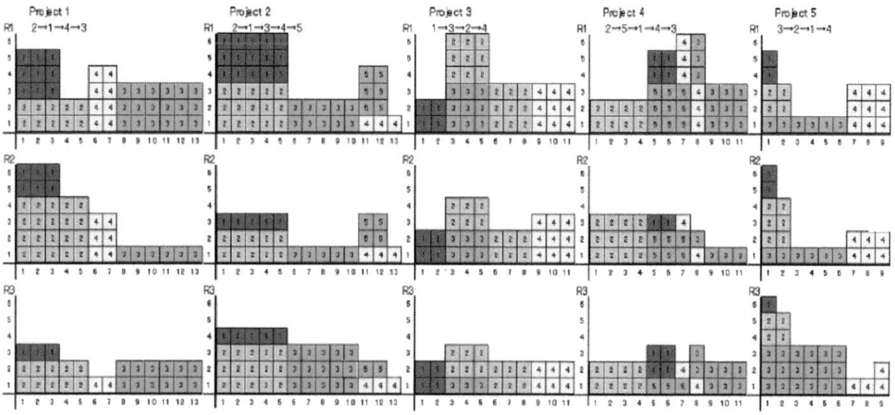

Fig. 6.23 Gantt chart for the rc-mPSP model

6.3.2.2 Genetic Operators

In this subsection, mutation and selection operators used in the hGA approach are explained in detail.
Mutation Operator: As a mutation operator, swap mutation (SM) and local search-based mutation (LSM) are used. In swap mutation [57], two positions are selected at random and their contents are swapped is used (see Fig. 6.24). In this example, project sequence changes to schedule S when using SM:

$S=\{P_1(2,1,4,3), P_2(2,1,3,4,5), P_3(1,3,2,4), P_4(2,5,1,4,3), P_5(3,2,1,4)\}$

to $S=\{P_3(1,3,2,4), P_2(2,1,3,4,5), P_1(2,1,4,3), P_4(2,5,1,4,3), P_5(3,2,1,4)\}$

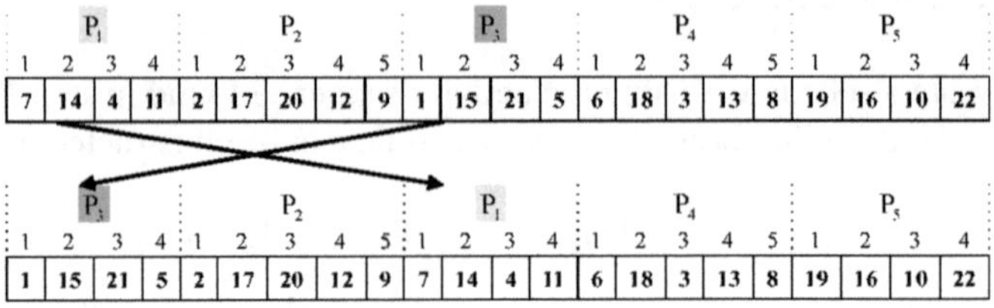

Fig. 6.24 The swap mutation operator

In local search based mutation, for a pair of genes, one is called pivot which is fixed for a given neighborhood and the other is selected at random as shown in Fig. 6.25. For a given neighborhood, a chromosome is called local optima if it

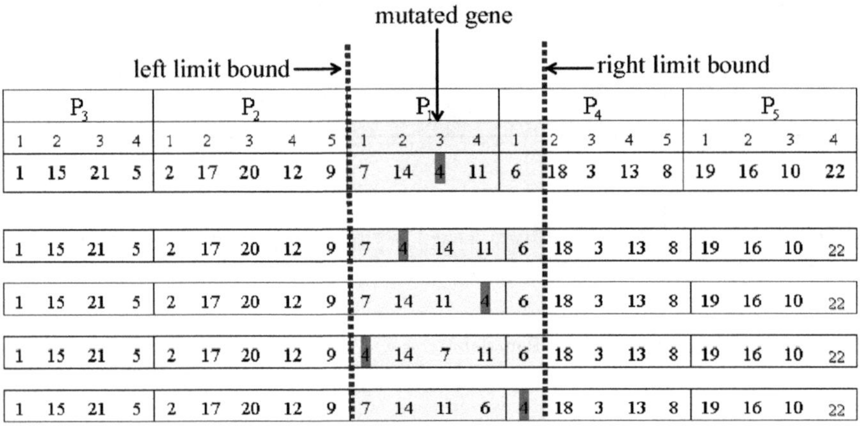

Fig. 6.25 The incumbent chromosome and its neighborhood

is better than any other chromosomes according to the fitness value. The size of a neighborhood affects the quality of the local optima. There is a clear trade-off between small and large neighborhoods: if the number of neighbors is larger, the probability of finding a good neighbor may be higher, but looking for it takes more time. The procedure of the local search-based mutation operator is shown in Fig. 6.26.

In the example, through LSM we adopt the schedule

$$S=\{P_3(1,3,2,4), P_2(1,2,3,4,5), P_1(1,3,2,4), P_4(1,2,3,4,5),$$
$$P_5(1,2,3,4)\}$$

in which total flow time is 11+13+11+11+9=55 and total tardiness penalty is $(0+2+0+0+0)\times 10=20$. Figure 6.27 illustrates the Gantt chart for this schedule.

```
procedure: revised serial priority-based decoding (creating a schedule)
procedure: local search-based mutation
inputs: P (parent)
outputs: C (offspring)
begin
        i ← 1 ;
        let P be current incumbent with the optimal t_F in pop_size;
        select a pivot gene randomly;
        while (i < specified_number) do
                pick up the gene;
                evaluate the neighbors toward both sides until left and right limit;
                let the neighbor be current incumbent if it is better;
                i ← i + 1;
        end
        save the incumbent in C
end
```

Fig. 6.26 Local search-based mutation

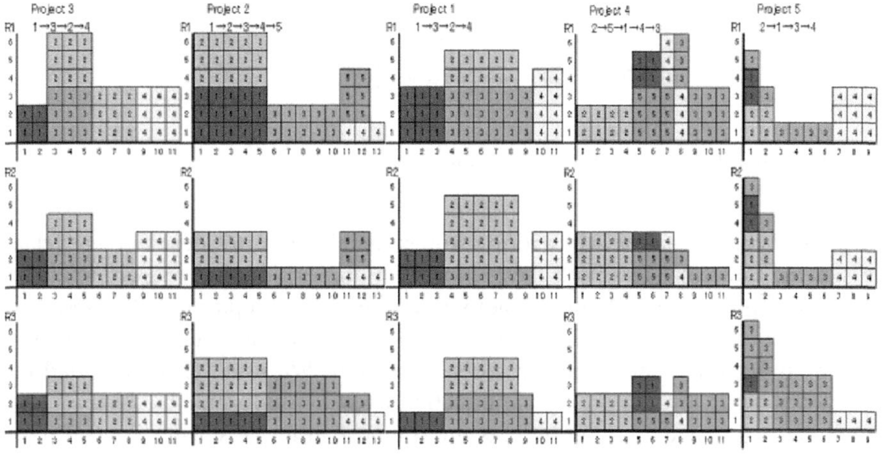

Fig. 6.27 Gantt chart for the schedule

Selection Operator: As a selection operator, the elitist selection, which preserves the best chromosome in the next generation and overcomes the stochastic errors of sampling, is used. With the elitist selection, if the best individual in the current generation is not reproduced into the new generation, one individual is randomly removed from the new population and the best one is added to the population.

6.3.2.3 Regulation of GA Parameters

In this section, we suggest two heuristic algorithms: hybrid genetic algorithm with fuzzy logic controller (flc-hGA) and adaptive hybrid genetic algorithm (a-hGA) which adaptively regulate the mutation ratio during the genetic search process.

Hybrid GA with Fuzzy Logic Controller: For the flc-hGA, we use the concept of Wang *et al.* [60]. The main idea of this concept is to consider two FLCs: crossover FLC and mutation FLC. However, we use the mutation FLC alone because crossover operator is not considered in the proposed hGA (refer to Section 6.3.2). The mutation FLC is implemented independently to regulate adaptively the mutation ratio during the genetic search process based on the FLC. The heuristic updating strategy for the mutation ratios is to consider the changes of the average fitness which occur at each population (parent and offspring populations) during the continuous two generations in Fig. 6.28.

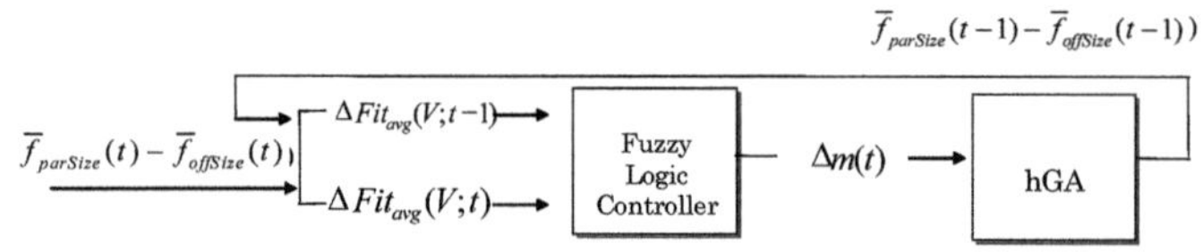

Fig. 6.28 Coordinated strategy of fuzzy logic control and hGA

Adaptive Hybrid Genetic Algorithm: Srinivas and Patnaik [69] developed an adaptive scheme that considers both the exploitation and exploration properties in the convergence process of GA; the capacity to converge to an optimum after locating the region containing the optimum, and the capacity to explore new regions of the solution space in the search of the global optimum. Wu *et al.* [70] suggested an improved scheme for reinforcing the mutation operator performed by Srinivas and Patnaik [69]. z_{max} and z_{avg} are the maximum fitness in the offspring size (offSize) and average fitness values of the offSize at each generation, respectively. z_{mut} is the best fitness value among the individuals to which the mutation with a rate p_M is applied. It has been observed that the difference between the z_{max} and z_{avg} of the population is likely to be less for a population that has converged to optimum solution than that for a population scattered in the solution space. Therefore, the rate of mutation operators should be varied depending on the value of the $z_{max} - z_{avg}$. On the other hand, if p_M has the same value for all the solutions of the population, this means that the solutions with high fitness values as well as the ones with low fitness values are subjected to the same level of mutation operation. This will certainly deteriorate the performance of hGAs. Therefore, Wu *et al.* [70] inserted an additional scheme into the original mutation scheme in order to prevent the deterioration of the performance as follows:

z_{max}: maximum fitness value in the *offSize*
z_{avg}: average fitness value in the *offSize*

z_{mut}: the best fitness value among the individuals to which the mutation with a probability p_M is applied

$f_j(\cdot)$: all the individuals except z_{mut}:

In this revised scheme, different values of α_1=0.05 and α_2= 0.5 are used.

6.3.2.4 Results of flc-hGA

We can find the optimal schedule

$S = \{(1,3,2,4),(1,2,3,5,4),(2,1,3,5,4),(1,2,3,4),(2,1,3,4)\}$

with total flow time: t_F= 11+13+11+8+9 = 52 and total tardiness penalty: p_T= (0+2+0+0+0)×10 = 20. Figure 6.29 illustrates the Gantt chart for this schedule.

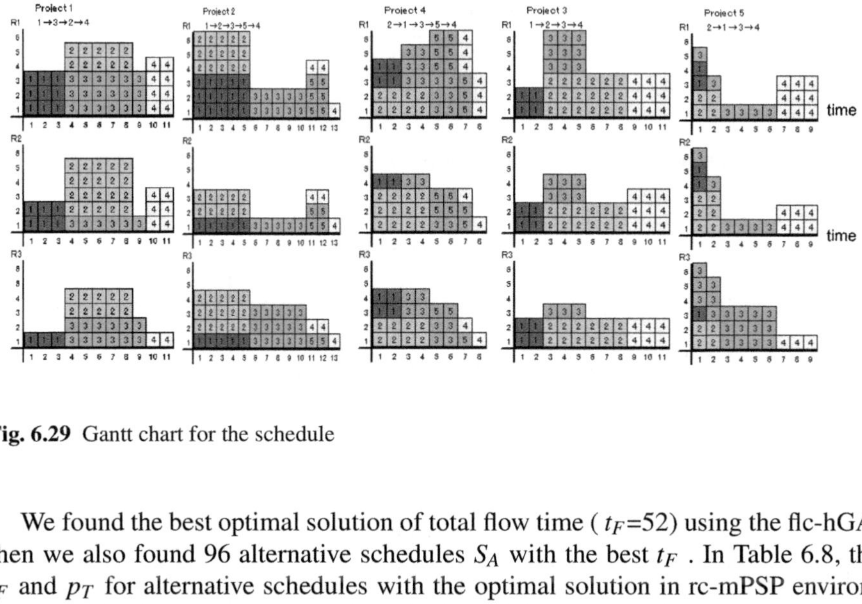

Fig. 6.29 Gantt chart for the schedule

We found the best optimal solution of total flow time (t_F=52) using the flc-hGA, then we also found 96 alternative schedules S_A with the best t_F . In Table 6.8, the t_F and p_T for alternative schedules with the optimal solution in rc-mPSP environment are listed. For the large sized rc-mPSP models, we could find many alternative schedules with the same t_F. Then t_F and p_T of many alternative schedules are different solutions. If a schedule is infeasible, we can substitute another alternative schedule for this infeasible schedule. Alternative schedules are very important in the rc-mPSP model.

6.3.3 Computational Experiments and Discussions

The computational experiments were conducted in two aspects: (1) comparison of hGA, flc-hGA and a-hGA using the revised serial initialization method and (2) per-

Table 6.8 Alternative schedules with optimal solution

	Alternative schedules with the optimal total flow time	t_F	p_T
1	P_1(1,3,2,4), P_2(1,2,3,4,5), P_3(1,2,3,4), P_4(1,2,3,5,4), P_5(1,2,3,4)	52	20
2	P_1(1,3,2,4), P_2(1,2,3,4,5), P_3(1,2,3,4), P_5(1,2,3,4), P_4(1,2,3,5,4)	52	20
3	P_1(1,3,2,4), P_2(1,2,3,4,5), P_5(1,2,3,4), P_3(1,2,3,4), P_4(1,2,3,5,4)	52	20
4	P_1(1,3,2,4), P_3(1,2,3,4), P_4(1,2,3,5,4), P_2(1,2,3,4,5), P_5(1,2,3,4)	52	20
5	P_1(1,3,2,4), P_3(1,2,3,4), P_2(1,2,3,4,5), P_4(1,2,3,5,4), P_5(1,2,3,4)	52	20
6	P_1(1,3,2,4), P_3(1,2,3,4), P_2(1,2,3,4,5), P_5(1,2,3,4), P_4(1,2,3,5,4)	52	20
7	P_1(1,3,2,4), P_2(2,1,3,4,5), P_3(1,2,3,4), P_4(1,2,3,5,4), P_5(1,2,3,4)	52	20
...	...	...	...
13	P_1(1,3,2,4), P_2(1,2,3,4,5), P_3(1,3,2,4), P_4(1,2,3,5,4), P_5(1,2,3,4)	52	20
...	...	...	...
19	P_1(1,3,2,4), P_2(1,2,3,4,5), P_3(1,2,3,4), P_4(2,1,3,5,4), P_5(1,2,3,4)	52	20
...	...	...	...
25	P_1(1,3,2,4), P_2(2,1,3,4,5), P_3(1,3,2,4), P_4(1,2,3,5,4), P_5(1,2,3,4)	52	20
...	...	...	...
31	P_1(1,3,2,4), P_2(2,1,3,4,5), P_3(1,2,3,4), P_4(2,1,3,5,4), P_5(1,2,3,4)	52	20
...	...	...	...
37	P_1(1,3,2,4), P_2(1,2,3,4,5), P_3(1,3,2,4), P_4(2,1,3,5,4), P_5(1,2,3,4)	52	20
...	...	...	...
43	P_1(1,3,2,4), P_2(2,1,3,4,5), P_3(1,3,2,4), P_4(2,1,3,5,4), P_5(1,2,3,4)	52	20
...	...	...	...
49	P_1(2,1,3,4), P_2(1,2,3,4,5), P_3(1,2,3,4), P_4(1,2,3,5,4), P_5(1,2,3,4)	52	20
...	...	...	...
55	P_1(2,1,3,4), P_2(2,1,3,4,5), P_3(1,2,3,4), P_4(1,2,3,5,4), P_5(1,2,3,4)	52	20
...	...	...	...
61	P_1(2,1,3,4), P_2(1,2,3,4,5), P_3(1,3,2,4), P_4(1,2,3,5,4), P_5(1,2,3,4)	52	20
...	...	...	...
66	P_1(2,1,3,4), P_2(1,2,3,4,5), P_3(1,2,3,4), P_4(2,1,3,5,4), P_5(1,2,3,4)	52	20
...	...	...	...
72	P_1(2,1,3,4), P_2(2,1,3,4,5), P_3(1,3,2,4), P_4(1,2,3,5,4), P_5(1,2,3,4)	52	20
...	...	...	...
78	P_1(2,1,3,4), P_2(2,1,3,4,5), P_3(1,2,3,4), P_4(2,1,3,5,4), P_5(1,2,3,4)	52	20
...	...	...	...
84	P_1(2,1,3,4), P_2(1,2,3,4,5), P_3(1,3,2,4), P_4(2,1,3,5,4), P_5(1,2,3,4)	52	20
...	...	...	...
90	P_1(2,1,3,4), P_2(2,1,3,4,5), P_3(1,3,2,4), P_4(2,1,3,5,4), P_5(1,2,3,4)	52	20
...	...	...	...
96	P_1(2,1,3,4), P_3(1,3,2,4), P_2(2,1,3,4,5), P_5(1,2,3,4), P_4(2,1,3,5,4)	52	20

formance evaluation of the the flc-hGA for various problem sizes. The proposed algorithms are programmed in Delphi on a PC Pentium 1400MHz clock-pulse and 256MB RAM as operation system. The values of parameters used in the computational experiments are as follows:

Population size: *popSize* = 150
Maximum number of generation:*maxGen* = 200
Probability of SM: $p_{M1} = 0.5$

Probability of LSM: p_{M2}= 0.5

For the first set of computational experiments, the rc-mPSP test problem consisting of eight projects including two dummy projects (start and end project) is used. Collectively, these eight projects can perform ten activities $A_1, \cdots, A_{10}$ including two dummy activities (start and end activity). Each activity has fixed multiple unit requirements of three difference resources and processing time. The due date t_i^D and total penalty cost c_i^{TP} of each project with project precedence Suc(i) requirements are listed in Table 6.9. The sucessor, processing time and resources data of activity for example of rc-mPSP model are listed in Table 6.10. Out of 149 alternative schedules, the optimal schedule of the rc-mPSP test problem with limited resources (b_1, b_2, b_3) = (6, 7, 6) is found as follows:

Table 6.9 Project data for the rc-mPSP test problem.

P_i	Suc(i)	t_i^D	c_i^{TP}
S	1,2	0	0
P_1	3,4	20	10
P_2	3,4,5	25	10
P_3	6	30	10
P_4	6,7	25	10
P_5	7	35	10
P_6	8	30	10
P_7	8, T	35	10
P_8	T	30	10

$$S=\{(1,2,4,3,5,6,7,8,9,10),(2,1,4,3,5,7,8,6,9,10),(1,5,3,2,4,7,6,8,10,9),$$
$$(1,2,4,5,8,3,7,10,6,9),(1,3,2,4,6,5,7,8,9,10),(4,2,3,5,7,1,6,10,9,8),$$
$$(1,2,4,5,8,3,7,10,6,9),(1,3,2,4,6,5,7,8,9,10)\}$$

The best optimal solution of total flow time of this schedule is calculated as

$$t_F = P_2(24)+P_1(18)+P_3(34)+P_4(28)+P_5(39)+P_6(27)+P_7(31)+P_8(28) = 229$$

Additionally, alternative schedules with the optimal total flow time are found =149 and the total tardiness penalty is calculated as

$$p_T = (0+0+4+3+4+0+0+0) \times 10 = 110$$

Table 6.11 depicts the computational results of the hGA, a-hGA, and flc-hGA for the rc-mPSP test problem. The computations were repeated 20 times. The flc-hGA performs with t_F, S_A, p_T and CPU times slightly better than the other GA approaches. In the t_F , the performances of the entire GA approaches can find the best solution, but the flc-hGA approach is considerabely superior to the other GA approaches in

Table 6.10 Successor, processing time and resources data of activity for the rc-mPSP test problem.

P_1

A_j	Suc(i)	p_{ij}	l_{ijr}
s	1,2	0	0
A_1	3,4	3	3,2,1
A_2	7	5	2,4,2
A_3	9	6	3,1,2
A_4	5,6	2	4,3,1
A_5	10	3	2,0,3
A_6	t	3	1,1,1
A_7	8	4	3,1,1
A_8	t	5	2,2,2
A_9	10	4	3,2,3
A_{10}	t	2	4,1,0

P_2

A_j	Suc(i)	p_{ij}	l_{ijr}
s	1	0	0
A_1	2	3	3,2,4
A_2	3,4,5	2	2,1,3
A_3	6	5	3,2,3
A_4	6	3	1,1,1
A_5	7	5	2,4,2
A_6	8,9	4	3,2,4
A_7	8,10	7	3,5,2
A_8	t	5	2,3,1
A_9	t	3	1,3,4
A_{10}	t	4	2,2,2

P_3

A_j	Suc(i)	p_{ij}	l_{ijr}
s	1,2,3	0	0
A_1	4,5	5	3,4,2
A_2	4,6	4	4,3,2
A_3	6	6	2,4,1
A_4	7,8	5	3,1,3
A_5	7,8	7	3,2,4
A_6	8	5	4,2,1
A_7	9,10	3	2,3,1
A_8	t	2	4,1,3
A_9	t	5	2,2,3
A_{10}	t	6	3,3,1

P_4

A_j	Suc(i)	p_{ij}	l_{ijr}
s	1,2	0	0
A_1	3,4	3	3,1,3
A_2	4,5	5	3,2,4
A_3	6	4	4,1,3
A_4	6	2	4,3,1
A_5	7,8	1	2,4,2
A_6	9	6	3,1,2
A_7	9,10	4	4,2,1
A_8	10	7	2,3,1
A_9	t	4	2,2,3
A_{10}	t	5	3,2,3

P_5

A_j	Suc(i)	p_{ij}	l_{ijr}
s	1	0	0
A_1	2,3	4	2,4,4
A_2	4	3	3,2,1
A_3	4	5	2,3,4
A_4	5,6	4	3,3,2
A_5	7	2	4,1,4
A_6	7	6	1,4,4
A_7	8,9	4	2,2,2
A_8	10	7	3,2,4
A_9	10	4	3,2,3
A_{10}	t	3	4,3,1

P_6

A_j	Suc(i)	p_{ij}	l_{ijr}
s	1,2,3,4	0	0
A_1	4	5	3,2,1
A_2	4,5	6	2,3,4
A_3	5,6	4	3,3,2
A_4	7,8	2	4,1,4
A_5	8,9	7	1,4,4
A_6	9,10	6	2,2,3
A_7	10	4	2,2,2
A_8	t	4	3,2,4
A_9	t	2	3,2,3
A_{10}	t	5	2,4,2

P_7

A_j	Suc(i)	p_{ij}	l_{ijr}
s	1	0	0
A_1	2,3	5	2,4,1
A_2	4	6	3,1,2
A_3	5	4	2,2,1
A_4	5	2	3,2,4
A_5	6,7	5	1,2,4
A_6	8,9	3	2,3,1
A_7	8,10	7	4,2,1
A_8	10	3	3,2,4
A_9	10	2	3,3,3
A_{10}	t	1	2,1,3

P_8

A_j	Suc(i)	p_{ij}	l_{ijr}
s	1,2,3	0	0
A_1	4,5	3	3,3,2
A_2	4,5	5	3,2,4
A_3	4	7	1,2,4
A_4	6,7,8	5	1,4,4
A_5	7,8	4	2,2,2
A_6	9	3	3,2,4
A_7	9,10	5	2,3,1
A_8	9,10	6	4,2,1
A_9	t	2	4,1,4
A_{10}	t	3	2,1,1

the average of t_F. The $p_T$is considerably related to the t_F . The flc-hGA approach is especially quick at the CPU time to stop condition than hGA and a-hGA.

Table 6.11 Computational results of the rc-mPSP test problem

	hGA			a-hGA			flc-hGA		
	Best	Worst	Average	Best	Worst	Average	Best	Worst	Average
t_F	229	235	234.21	229	235	233.45	229	233	231.75
S_A	149	124	137.24	130	121	125.75	125	123	121.42
p_T	110	170	160.21	110	170	150.45	110	150	130.75
CPU time	218	229	218.95	151	159	153.42	132	138	134.02

For more detailed comparison of the adaptive schemes used in each GA approaches, we analyzed the convergence behaviors of the mutation operator and average fitness values during the search process of each GA approach. Figures 6.30 and 6.31 show that the behaviors of average fitness of t_F and p_T. Figures 6.30 and 6.31 show that in the initial stage, a-hGA and flc-hGA have almost the same convergence behavior. But flc-hGA has the t_F better than a-hGA after generation number 5 and produces the alternative schedules of the same t_F after generation number 12.

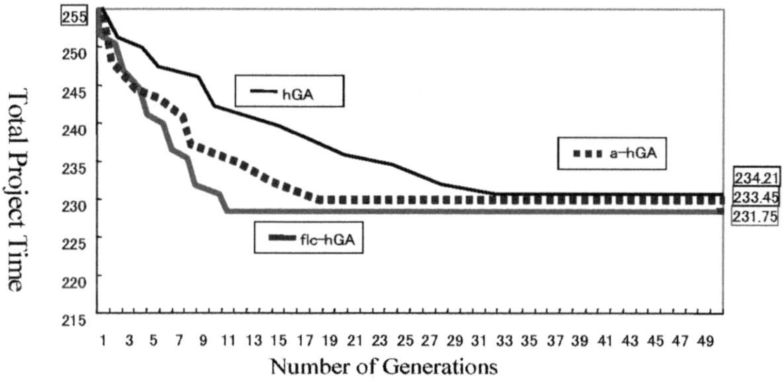

Fig. 6.30 Convergence behaviors of GA approaches for average fitness of total project time

In Fig. 6.32, the evolutionary behaviors of the mutation rate of the flc-hGA approach is decreased at generation number 12 after finding the optimal t_F. In the evolutionary behaviors of the mutation rate of the a-hGA approach, mutation probability changes suddenly to 0 when z_{mut} and $f_j(\cdot)$ in *offSize* are the same value of generation number 28. It is a problem of the a-hGA approach.

For the second set of computational experiments, several test problems with various number of projects and activities is considered. The rc-mPSP test problems of various sizes were solved by using flc-hGA which is found to be the best GA approach in the first set of experiments. Table 6.12 shows the average values of the

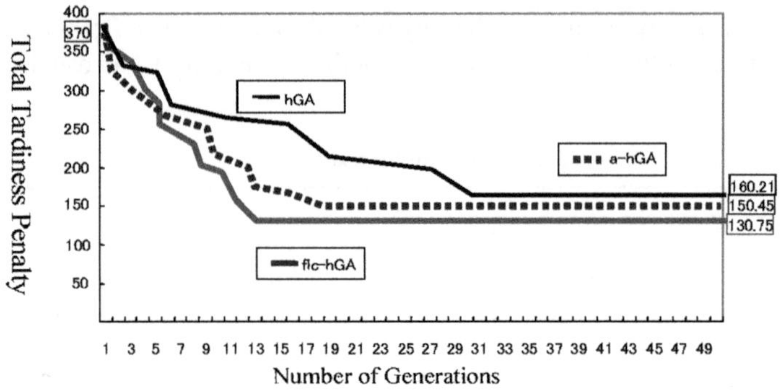

Fig. 6.31 Convergence behaviors of GA approaches for average fitness of total tardiness penalty

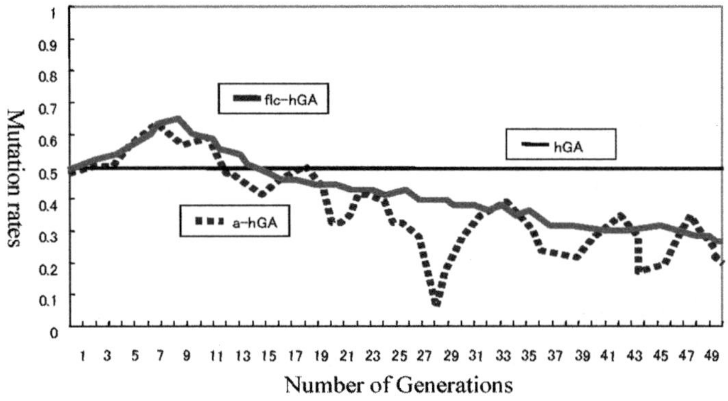

Fig. 6.32 Convergence behaviors of GA approaches for mutation rate of rc-mPSP test problem

objectives t_F, S_A, p_T, and computational CPU time over 20 runs for each parameter setting.

As seen from Table 6.12, the best solution in the first test problem using the heuristic algorithm is found without difficulty. In the other large size problems, we could obtain the best solution with a very high *maxGen* and *popSize* probability. From the results, we could see that the *popSize* has an influence on finding the best result more than *maxGen*.

In conclusion, the flc-hGA could find the best t_F and a lot of S_A with optimal t_F during minimizing CPU time. We could find them through primary computational experiments conducted in two aspects which compared the GA approaches of hGA, a-hGA, and flc-hGA with rc-mPSP to find the best algorithm and treated the rc-mPSP of various sizes using flc-hGA which is the best algorithm out of the others.

Table 6.12 Results found by flc-hGA for problems with different sizes

No. of projects	No. of total activities	Test No.	*maxGen*	*popSize*					CPU time (sec)
		1	150	100	0.3	139	62	0	105
10	90	2	100	150	0.5	139	67	0	158
		3	250	200	0.8	139	88	0	323
		1	250	150	0.3	233	75	0	411
15	150	2	150	250	0.5	233	115	0	504
		3	300	250	0.8	233	121	0	818
		1	300	250	0.3	340	116	110	1295
20	220	2	250	300	0.5	331	107	20	2153
		3	350	350	0.8	329	106	0	2,458
		1	400	350	0.3	456	87	80	3,955
25	300	2	350	400	0.5	451	60	20	4,855
		3	450	500	0.8	447	87	0	5,933

6.4 Resource-constrained Project Scheduling Model with Multiple Modes

When there are several different modes that may be selected for an activity, the problem is known as a resource-constrained project scheduling problem with multiple modes (rc-PSP/mM). In this problem, each mode corresponds to a different time-resource trade-off option for the activity under consideration. A feasible schedule specifies the implementation mode, as well as the start and finish times for each activity. The general features of the rc-mPSP/mM model relies on some of the basic assumptions of rc-PSP model (refer to Section 6.2). Additionally, the rc-PSP/mM model can be defined by the following assumptions:

A6. Each activity must be performed in a mode where each activity-mode combination has a fixed duration and requires a constant amount of one or more types of renewable resources when activity is being executed.

A7. Mode switching is not allowed when activities are resumed after splitting.

6.4.1 Mathematical Formulation of rc-PSP/mM Models

In the rc-PSP/mM model, we consider a single project which consists of i=1,$\cdots$,e,$\cdots$,I activities with the precedence relations between each pair of activities (e,i) where e immediately precedes i are taken into consideration. Additionally, each activity i must be performed in one of m_i possible modes, where each activity-mode combination has a fixed duration and requires a constant amount of one or more of R types of renewable resources when activity is being executed.

Table 6.13 presents the data set for an rc-PSP/mM model, which contains seven activities including dummy activities (starting and finishing). The modes, durations, three kinds of resource consumptions (manpower, cost, material), and predecessor of each activity are given in Table 6.13. Using this data set, the precedence graph in Fig. 6.33 is constructed.

Table 6.13 Data set of a rc-PSP/mM model

Activity	Mode	Duration of activity	Resource consumptions			Predecessor of activities
			Non-renewable resource 1	Non-renewable resource 2	Renewable resource 1	
1	1	(dummy activity)				
2	1	12	3	5	3	1
	2	15	4	4	2	
	3	18	3	3	4	
3	1	5	2	5	3	1
	2	11	5	2	4	
	3	13	4	3	2	
4	1	5	4	2	3	1
	2	14	5	4	2	
5	1	15	2	3	4	2,3
	2	12	5	2	3	
	3	8	4	3	2	
6	1	13	5	3	3	2,4
	2	12	6	4	2	
	3	15	2	3	3	
7	1	(dummy activity)				2,5,6

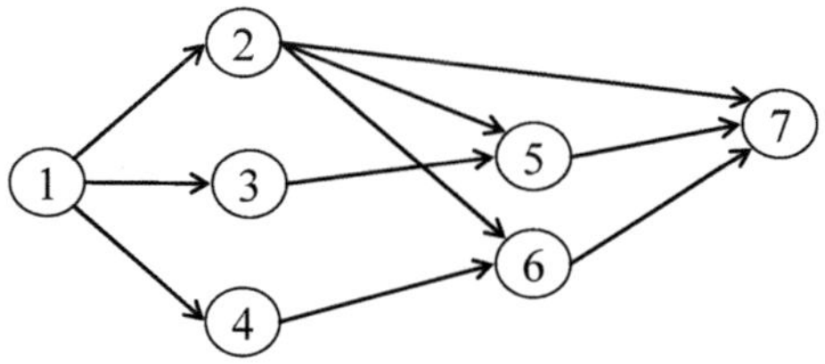

Fig. 6.33 Precedence graph of a rc-PSP/mM model

Notation

In order to formulate the mathematical model, the following indices, parameters and decision variable are introduced.

Indices

i: activity index, i=1,2,$\cdots$, e, $\cdots$, I
j: mode index, $j = 1, 2, \cdots, m_i$ (m_i is the number of possible modes for activity i)
k: renewable resource type index, $k = 1, 2, \cdots, K$.
n: non-renewable resource type index, $n = 1, 2, \cdots, N$.
t: period index, $t = t_i^{EF}, \cdots, t_i^{LF}$.

Variables

t_{ij}^F: finish processing time of activity i of selected mode j.
t_i^{EF}: early finish time of activity i.0
t_i^{LF}: lately finish time of activity i.

Parameters

$Pre(i)$: set of immediate predecessors of activity i.
r_{ijk}: amount of non-renewable resource k required to execute activity i when mode j is used.
i_k^M: maximum-limited renewable resource k only available with the constant period availability.
l_n^M: maximum-limited non-renewable resource k only available with the constant period availability.
p_{ij}: processing time of activity i of selected mode j.

Decision Variables

$$x_{ijt} = \begin{cases} 1, \text{ if activity } i \text{ executed in mode} ji \text{ scheduled to be finished in time } t \\ 0, \text{ otherwise} \end{cases}$$

Mathematical Model

The mathematical model for the rc-PSP can be stated as follows:

$$\min t_M = \sum_{j=1}^{m_I} \sum_{t=t_I^{EF}}^{t_I^{LF}} t x_{Ijt} \tag{6.13}$$

$$\text{s. t. } \sum_{j=1}^{m_i} \sum_{t=t_i^{EF}}^{t_i^{LF}} x_{Ijt} = 1, \quad i = 1, \cdots, I \tag{6.14}$$

$$\sum_{j=1}^{m_e}\sum_{t=t_e^{EF}}^{t_e^{EF}} tx_{ejt} + \sum_{j=1}^{m_i}\sum_{t=t_i^{EF}}^{t_i^{EF}} p_{ij}x_{ijt} \leq \sum_{j=1}^{m_i}\sum_{t=t_i^{EF}}^{t_i^{LF}} tx_{ijt},$$

$$e \in \text{Pre}(i),\ i = 1, \cdots, I \quad (6.15)$$

$$\sum_{i=1}^{I}\sum_{j=1}^{m_i} r_{ijk} \sum_{s=t}^{t+p_{ij}+1} x_{ijs} \leq i_k^M, \quad k \in K,\ t = 1, \cdots, T \quad (6.16)$$

$$\sum_{i=1}^{I}\sum_{j=1}^{m_i} r_{ijn} \sum_{t=t_i^{EF}}^{t_i^{LF}} x_{ijt} \leq l_n^M, \quad n \in N \quad (6.17)$$

$$t_{ij}^F \geq 0,\ \forall i,\ \forall j \quad (6.18)$$

$$t_{ij}^{EF} \geq 0,\ \forall i,\ \forall j \quad (6.19)$$

$$t_{ij}^{LF} \geq 0,\ \forall i,\ \forall j \quad (6.20)$$

$$x_{ijt} = 0\ or\ 1 \quad \forall i, j, t \quad (6.21)$$

In this mathematical model, the objective at Eq. 6.13 is to minimize the total makespan. The constraints given in Eqs. 6.14–6.21 are used to formulate the general feasibility of the problem. The constraint at Eq. 6.14 ensures that each activity is performed in one of its modes and finished within its time windows $\left[t_i^{EF}, t_i^{LF}\right]$. Constraint at Eq. 6.15 ensures that none of the precedence constraints is violated. Constraint at Eq. 6.16 ensures that the amount of renewable resource k used by all activities does not exceed its limited quantity in any time periods. The constraint at Eq. 6.17 limits the total resource consumption of non-renewable resource n to the available amount. Figure 6.34 defines these constrains used in this mathematical model for non-renewable resources. Constraints at Eq. 6.18 – 6.21 represent the usual integrity restriction.

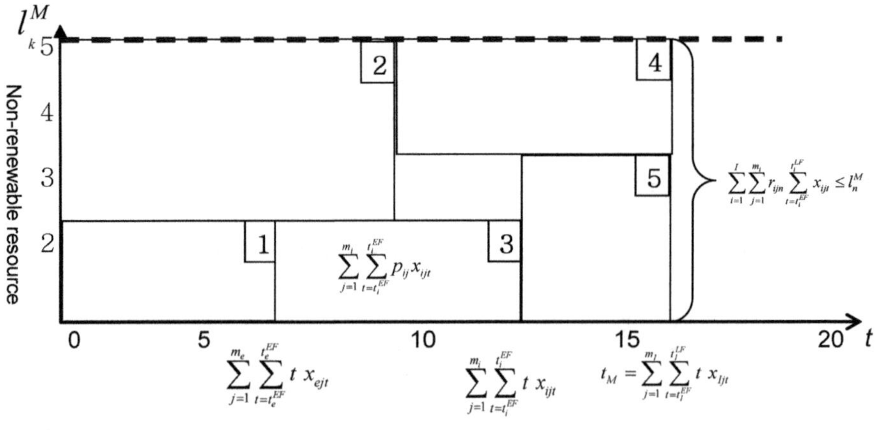

Fig. 6.34 The schedule of a non-renewable resource

6.4.2 Adaptive Hybrid GA for rc-PSP/mM Models

In this section of the chapter, we will introduce an adaptive hybrid genetic algorithm (a-hGA) for solving the rc-PSP/mM model. The a-hGA approach was composed of GA, iterative hill climbing routine and adaptive regulation mechanism. The main objective of this approach is to minimize the makespan while the resource and precedence constrained are satisfied [71].

Let *P*(*t*) and *C*(*t*) be parents and offspring in current generation *t*. The overall procedure of a-hGA for solving rc-PSP/mM model is outlined as follows:

---

procedure: a-hGA for rc-PSP/mM
input: problem data, GA parameters
output: the best schedule
begin
  $t \leftarrow 0$;
  iinitialize $P(t)$ by activity priority and multistage-based encoding routine;
  evaluate $P(t)$ by priority-based decoding routine;
  while (**not** terminating condition) **do**
    create $C(t)$ from $P(t)$ by order-based crossover routine;
    create $C(t)$ from $P(t)$ by neighborhood search mutation routine;
    climb $C(t)$ by iterative hill climbing routine;
    evaluate $C(t)$ by priority-based decoding routine;
    if t > u **then**
      regulate adaptive GA parameters p_C and p_M by using adaptive method;
    select $P(t+1)$ from $P(t)$ and $C(t)$ by elitist selection routine;
    $t \leftarrow t+1$;
  end
  output the best schedule;
end

The detailed information about the genetic representation, genetic operators, iterative hill climbing method and FLC are given in the following subsections.

6.4.2.1 Genetic Representation

For the representation of rc-PSP/mM models, priority-based encoding for activity sequences, multistage-based encoding for activity mode and priority-based decoding methods were used. The genetic representation of an individual solution is composed of two chromosomes where first chromosome shows the feasible activity sequence and the second chromosome consists of activity mode assignments. To develop this genetic representation for the rc-PSP/mM model, there are two main phases:

Phase 1: Creating an activity sequence and activity mode

Step 1.1: Generate a random priority to each activity in the project using encoding procedure.
Step 1.2: Generate an activity mode to each activity in the project using multistage-based encoding procedure.
Step 1.3: Decode a feasible activity sequence that satisfies the precedence constraints.
Step 1.4: Decide on the activity mode of each activity for the feasible activity sequence found in step 1.3.

Phase 2: Designing a schedule

Step 2.1: Create a schedule S using the activity sequence and activity modes found in step 1.4.
Step 2.2: Draw a Gantt chart for this schedule.

While introducing these methods, in order to illustrate the procedures for rc-PSP/mM, we will be using the data set given in Table 6.13 and its corresponding precedence graph given in Fig. 6.33.

Phase 1: Creating an Activity Sequence

Step 1.1: Generate a random priority to each activity in the project using encoding procedure. The first part of chromosome is obtained by using the priority-based encoding procedure.
Step 1.2: Generate an activity mode to each activity in the project using multistage-based encoding procedure. In this step, all activities are assigned with an activity mode by a multistage-based encoding procedure which is a permutation encoding procedure. The second part of Fig. 6.35 illustrates a multistage-based chromosome obtained by using the permutation encoding procedure.

activity ID j	1	2	3	4	5	6	7
activity priority $v(j)$	2	6	7	3	4	1	5
activity mode $m(j)$	1	1	2	1	3	1	1

Fig. 6.35 An individual solution composed of priority-based and a multistage-based chromosomes

Step 1.3: Decode a feasible activity sequence that satisfies the precedence constraints. In order to decode the chromosomes generated by priority-based encoding procedure, the priorities of each activity are used to create a feasible activity sequence that satisfies the precedence constraints in the model. Figure 6.36 presents the priority-based decoding procedure for creating a schedule.

procedure: priority-based decoding (creating a schedule)
input: the set of direct successors of activity *j* chromosome *v*(*j*)
output: schedule *S*
begin
    $\bar{s} \leftarrow \varnothing,\ S \leftarrow \varnothing;$
    $t \leftarrow 0,\ j \leftarrow 0;$
    while ($j \neq t$) **do**
        $\bar{s} \leftarrow \mathrm{Suc}(j);$
        $j^{*} \leftarrow \arg\max\{v(j) \mid j \in \bar{s}\};$
        $\bar{s} \leftarrow \bar{s}$ ¥ $j^{*};$
        $S \leftarrow S \cup j^{*};$
        $j \leftarrow j^{*};$
    end
    output schedule *S*
end

Fig. 6.36 Priority-based decoding procedure for creating a schedule

For the illustration of this decoding procedure, the chromosome found in Fig. 6.36 is used. Table 6.14 presents the trace table for the decoding procedure. In the example, we obtained a feasible schedule of $S = \{\ 1, 3, 2, 5, 4, 6, 7\ \}$.

Table 6.14 Trace table for activity sequence

j		$v(j)$	j^*	S
0	{1}	$v(1)=2$	1	$S=\{1\}$
1	{2,3,4}	$v(2)=6$, $v(3)=7$, $v(4)=3$	3	$S=\{1,3\}$
3	{2,4}	$v(2)=6$, $v(4)=3$	2	$S=\{1,3,2\}$
2	{4,5}	$v(4)=3$, $v(5)=4$	5	$S=\{1,3,2,5\}$
5	{4}	$v(4)=3$	4	$S=\{1,3,2,5,4\}$
4	{6}	$v(6)=1$	6	$S=\{1,3,2,5,4,6\}$
6	{7}	$v(7)=5$	7	$S=\{1,3,2,5,4,6,7\}$

Step 1.4: Decide on the activity mode of each activity for the feasible activity sequence found in step 1.3. Using the schedule from trace Table 6.14 and the multistage-based chromosome generated in step 1.2, the activity mode for the activity sequence is obtained as illustrated in Fig. 6.37.

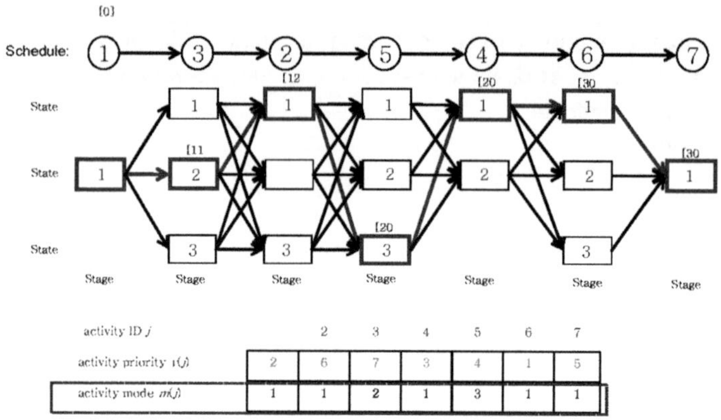

Fig. 6.37 Assignments of activity modes to activities

Phase 2: Designing a Schedule

Step 2.1: Create a schedule S using the activity sequence and activity modes found in step 1.4. The schedule using priority base and multistage-based encoding is as follows:

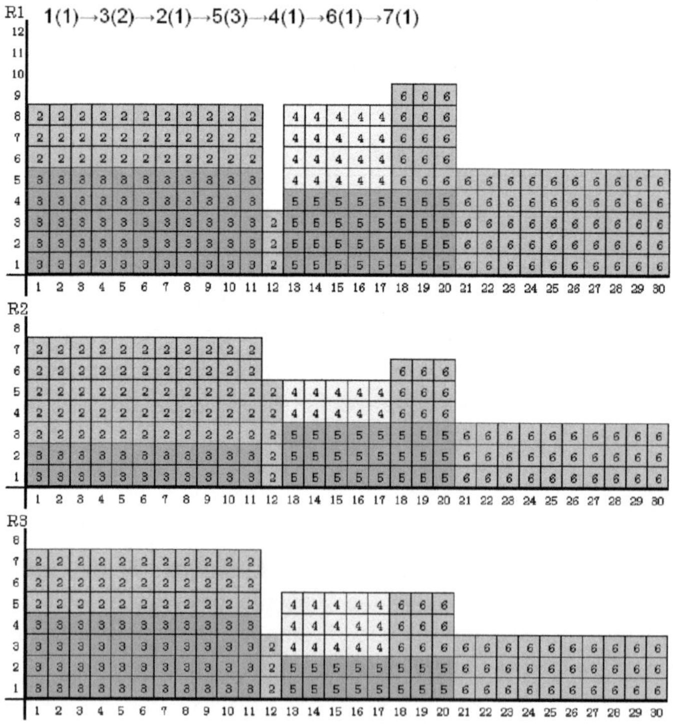

Fig. 6.38 Gantt chart for the rc-PSP/mM model

$$S=\{a_1(m_1),a_3(m_2),a_2(m_1),a_5(m_3),a_4(m_1),a_6(m_1),a_7(m_1)\}$$
$$=\{a_1(m_1):0-0,a_3(m_2):0-11,a_2(m_1):0-12,a_5(m_3):13-20,$$
$$a_4(m_1):13-17,a_6(m_1):18-30,a_7(m_1):30-30\}$$

Step 2.2: Draw a Gantt chart for this schedule. Using this schedule of activities and chromosome for activity mode, Gantt chart for limited resources (l_1^M, l_2^M, l_3^M)=(12, 8, 8) is constructed and shown in Fig. 6.38. The makespan of this solution is 30 time units.

6.4.2.2 Genetic Operators

In this subsection, the crossover, mutation and selection operators used in the flc-hGA approach are explained in detail.

Crossover Operator: As crossover operator, order-based crossover (OBX) is used for activity priority. The procedure of order-based crossover operator is as shown in Fig. 6.39; Fig.6.40 illustrates the activity priority chromosome generated by OBX.

---

procedure: Order-based crossover (OBX) for activity priority
input: parents *P*
output: offspring *C*
step 1: Select a set of positions from one parent of activity priority at random in activity priority.
step 2: Produce a proto-child by copying the cities on these positions into the corresponding positions of the proto-child.
step 3: Delete the cities which are already selected from the second parent. The resulting sequence of cities contains the cities the proto-child needs.
step 4: Place the cities into the unfixed positions of the proto-child from left to right according to the order of the sequence to produce one offspring *C*.
output the incumbent in *C*;
end

Fig. 6.39 Order-based crossover for activity priority

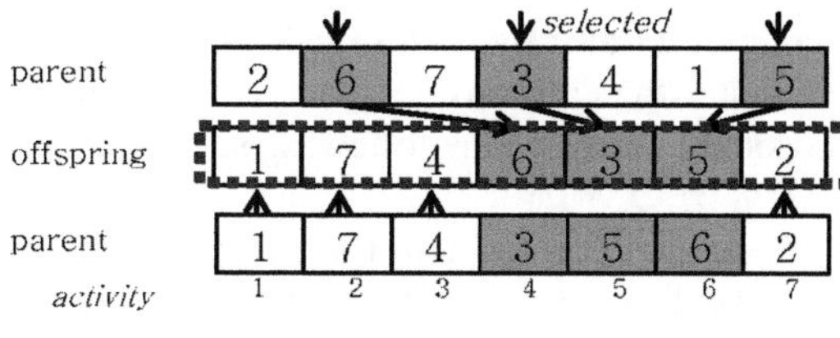

Fig. 6.40 Activity priority chromosome generated by OBX

Mutation Operator: As a mutation operator, local search-based mutation (LSM) is used for activity mode. The procedure of local search-based mutation is as shown in Fig. 6.41. Figure 6.42 illustrates the activity mode chromosome generated by local search-based mutation operator.

```
procedure: Local search-based mutation (LSM) for activity mode
input: parents P
output: offspring C
begin
   i ← 1 ;
   let P be current incumbent chromosome;
   select two pivot gene randomly;
    while (i < specifiedNumber) do
          pick up the gene;
          evaluate the neighbors until bound of activity mode;
          let the neighbor be current incumbent
               if it is better;
          i ← i + 1;
    end
    output the incumbent in C;
end
```

Fig. 6.41 Local search-based mutation for activity mode

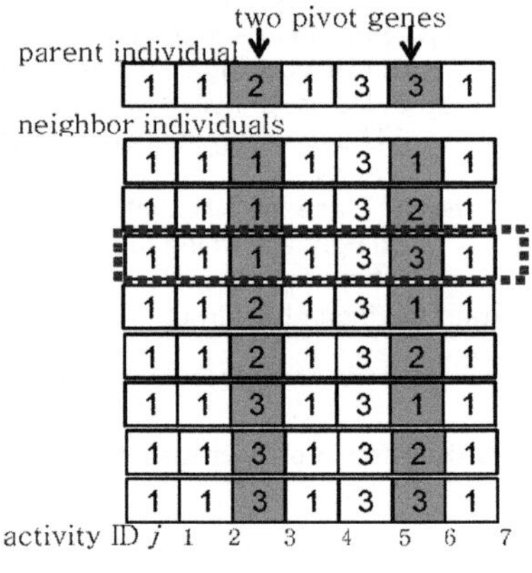

Fig. 6.42 Activity mode chromosome generated by LSM

Schedule Found by Using OBX and LSM for the mrc-PSP/mM model: Figure 6.43 illustrates the chromosome generated by OBX and LSM routines. Using this chromosome, we obtain the following schedule:

$$
\begin{aligned}
S=&\{a_1(m_1), a_4(m_1), a_2(m_1), a_3(m_2), a_6(m_2), a_5(m_1), a_7(m_1)\}\\
=&\{a_1(m_1):0-0, a_4(m_1):0-5, a_2(m_1):0-12, a_3(m_2):6-16,\\
&a_6(m_2):13-24, a_5(m_3):17-24, a_7(m_1):24-24\}
\end{aligned}
$$

Using this schedule, Gantt chart for limited resources (l_1^M, l_2^M, l_3^M)=(12,8,8) is constructed and shown in Fig. 6.44. The makespan of this solution is 27 time units.

activity ID *j*	1	2	3	4	5	6	7
activity priority *v*(*j*)	1	7	4	6	3	5	2
activity mode *m*(*j*)	1	1	1	1	3	3	1

Fig. 6.43 Chromosome generated by OBX and LSM routines

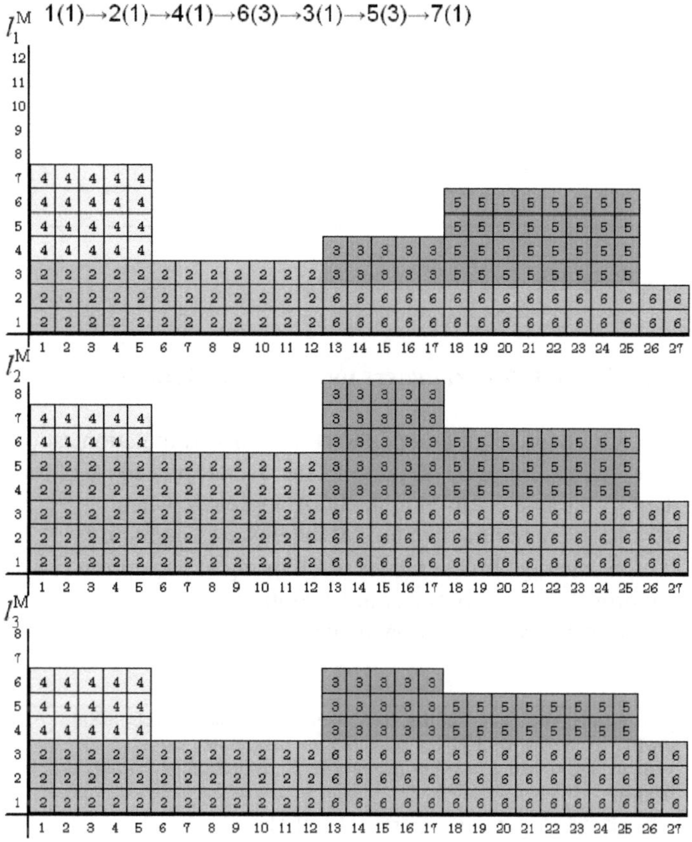

Fig. 6.44 Gantt chart for the rc-PSP/mM model

6.4.2.3 Iterative Hill Climbing Method

In this study, a local search method is proposed to enhance the search ability of GA. The local search method, *i.e.*, iterative hill climbing method, is based on critical stations in order to improve their effectiveness and efficiency. Figure 6.45 presents the overall procedure of the local search used in this proposed approach.

```
procedure: Iterative Hill-Climbing Method in GA Loop
input: string Vc
output: improved string Vc'
begin
    t←0;
    repeat
        local←FALSE;
        select a best string vC in current GA loop;
        repeat
            randomly generate a new string as many as popSize in the
                    neighborhood of vC;
            select a string with the best value of objective function tM
                    among the set of new strings;
            if f(vC) < f(vn) then
                    vC ←vn
            else  local ← TRUE
        until local
        t ← t+1;
    until t=popSize
    output  improved string Vc'
end
```

Fig. 6.45 Overall procedure of iterative hill climbing

6.4.2.4 Regulation of GA Parameters by Adaptive Method

In this section, we adaptively regulate the crossover and mutation ratios during the genetic search process. Figure 6.46 presents the overall procedure of this regulation routine of the GA parameters used in the hGA approach, where

$\overline{f_{parSize}}(t)$: the average fitness values of parents at generation t
$\overline{f_{offSize}}(t)$: the average fitness values of offspring at generation t
$p_C(t)$: the probability of crossover at generation t
$p_M(t)$: the probability of mutation at generation t

procedure: regulation of the crossover and mutation probabilities
input: crossover probability p_C, mutation probability p_M
output: improved crossover probability p_C, mutation probability p_M
begin
  if $(\overline{f_{par\_size}}(t)/\overline{f_{off\_size}}(t)) - 1 \geq 0.1$ **then**
    $p_C(t+1) = p_C(t) + 0.05, \quad p_M(t+1) = p_M(t) + 0.015;$
  if $(\overline{f_{par\_size}}(t)/\overline{f_{off\_size}}(t)) - 1 \leq 0.1$ **then**
    $p_C(t+1) = p_C(t) - 0.05, \quad p_M(t+1) = p_M(t) - 0.015;$
  if $-0.1 < (\overline{f_{par\_size}}(t)/\overline{f_{off\_size}}(t)) - 1 < 0.1$ **then**
    $p_C(t+1) = p_C(t), \quad p_M(t+1) = p_M(t);$
  output improved crossover probability p_C, mutation probability p_M
end

Fig. 6.46 Overall procedure of regulation of the GA parameters

activity ID i	1	2	3	4	5	6	7
activity priority $v(j)$	5	2	1	7	3	6	4
activity mode $m(j)$	1	1	2	1	3	2	1

Fig. 6.47 Chromosome generated after regulating the GA parameters

Figure 6.47 illustrates the chromosome. Using this chromosome, we obtain the following schedule:

$$S=\{a_1(m_1), a_4(m_1), a_2(m_1), a_3(m_2), a_6(m_2), a_5(m_3), a_7(m_1)\}$$
$$=\{a_1(m_1): 0-0, a_4(m_1): 0-5, a_2(m_1): 0-12, a_3(m_2): 6-16,$$
$$a_6(m_2): 13-24, a_5(m_3): 17-24, a_7(m_1): 24-24\}$$

Using this schedule, Gantt chart for limited resources (l_1^M, l_2^M, l_3^M)=(12,8,8) is constructed and shown in Fig. 6.48. The makespan of this solution is 24 time units.

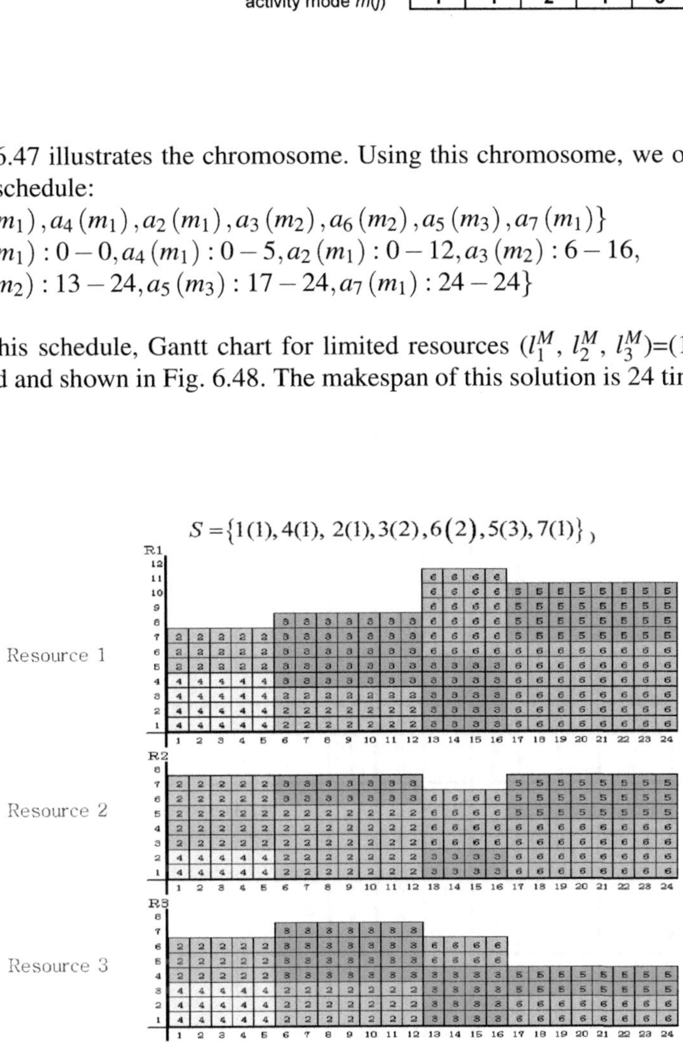

Fig. 6.48 Gantt chart for the schedule

6.4.3 Numerical Experiment

For the numerical experiment, the rc-mPSP/mM test problem consisting of 15 activities including two dummy activities (start and end) is used. Each activity has fixed multiple unit requirements of four difference resources and processing time. Table 6.15 presents the modes, durations, three kinds of resource consumptions (manpower, cost, material), and predecessor of each activity for the rc-PSP/mM test problem. Using this data set, the precedence graph in Fig. 6.49 is constructed. The values of parameters used in the computational experiments are as follows:

Population size: *popSize* = 20
Maximum number of generation: *maxGen* = 500
Probability of crossover: $p_C = 0.3$
Probability of mutation: $p_M = 0.025$

Table 6.15 Data set of the rc-PSP/mM test problem

Activity i	Mode j	Duration p_{ij}	l_1^M	l_2^M	l_3^M	l_4^M	Predecessor pre_i
1	1	(Dummy node)					
2	1	3	6	4	5	7	1
	2	9	5	3	3	6	
3	1	1	8	4	5	8	1
	2	1	7	3	4	8	
	3	5	6	4	3	5	
4	1	5	7	4	2	6	1
	2	8	6	3	5	7	
5	1	6	2	5	4	7	2
6	1	2	2	2	8	4	3
	2	6	2	3	7	1	
7	1	3	5	5	10	5	2,3
	2	8	5	4	7	10	
8	1	4	6	4	7	1	6
	2	10	3	2	10	3	
9	1	2	2	4	6	5	5
	2	7	1	3	4	8	
	3	10	1	2	3	7	
10	1	1	4	4	4	6	5,7
	2	1	5	2	6	8	
	3	9	4	6	3	5	
11	1	6	5	2	4	10	4,6,7
	2	9	3	1	3	9	
	3	10	6	1	6	7	
12	1	11	6	2	3	2	8,10
	2	8	7	4	5	6	
13	1	5	2	3	5	4	8,9,11
	2	6	3	3	4	3	
	3	7	4	2	3	4	
14	1	4	2	4	3	4	9
	2	3	3	5	2	3	
15	1	(Dummy site)					12,13,14

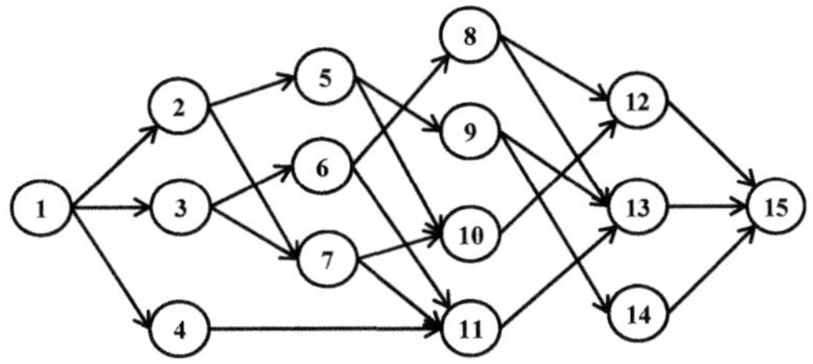

Fig. 6.49 Network representation of the rc-PSP/mM test problem

Table 6.16 illustrates the four different of alternative schedules with optimal duration found while solving the rc-PSP/mM by a-hGA. These alternative schedules are very important in the rc-PSP/mM problem since sometimes there may be a need to change the schedules.

Table 6.16 Four alternative schedules with minimized makespan

	Alternative schedules with the minimized makespan (t_M=27)
1	$a_1(m_1), a_2(m_1), a_3(m_2), a_6(m_1), a_5(m_1), a_4(m_1), a_8(m_1), a_9(m_1),$ $a_{14}(m_1), a_7(m_1), a_{10}(m_2), a_{11}(m_1), a_{12}(m_1), a_{13}(m_1), a_{15}(m_1)$
2	$a_1(m_1), a_3(m_2), a_2(m_1), a_6(m_1), a_5(m_1), a_4(m_1), a_8(m_1), a_9(m_1),$ $a_{14}(m_1), a_7(m_1), a_{10}(m_2), a_{11}(m_1), a_{12}(m_1), a_{13}(m_1), a_{15}(m_1)$
3	$a_1(m_1), a_4(m_1), a_2(m_1), a_3(m_2), a_5(m_1), a_6(m_1), a_8(m_1), a_9(m_1),$ $a_{14}(m_1), a_7(m_1), a_{10}(m_2), a_{11}(m_1), a_{12}(m_1), a_{13}(m_1), a_{15}(m_1)$
4	$a_1(m_1), a_3(m_2), a_2(m_1), a_4(m_1), a_5(m_1), a_6(m_1), a_8(m_1), a_9(m_1),$ $a_{14}(m_1), a_7(m_1), a_{10}(m_2), a_{11}(m_1), a_{12}(m_1), a_{13}(m_1), a_{15}(m_1)$

Using the chromosome illustrated in Fig. 6.50, we obtain the following schedule:

$$S=\{1,2,3,6,5,4,8,9,14,7,10,11,12,13,15\}$$
$$=\{a_1(m_1): 0-0, a_2(m_1): 0-3, a_3(m_2): 3-4, a_6(m_1): 4-6,$$
$$a_5(m_1): 4-10, a_4(m_1): 6-11, a_8(m_1): 6-10, a_9(m_1): 10-12,$$
$$a_{14}(m_1): 12-16, a_7(m_1): 12-15, a_{10}(m_2): 15-16, a_{11}(m_1): 16-22,$$
$$a_{12}(m_1): 16-27, a_{13}(m_1): 22-27, a_{15}(m_1): 27-27\}$$

activity ID j :	1	2	3	4	5	6	7	8	9	10	11	12	13	14	15
activity priority $v(j)$:	4	15	13	10	11	12	6	9	8	5	3	2	1	7	14
activity mode $m(j)$:	1	1	2	1	1	1	1	1	1	2	1	1	1	1	1

Fig. 6.50 Chromosome for rc-PSP/mM test problem

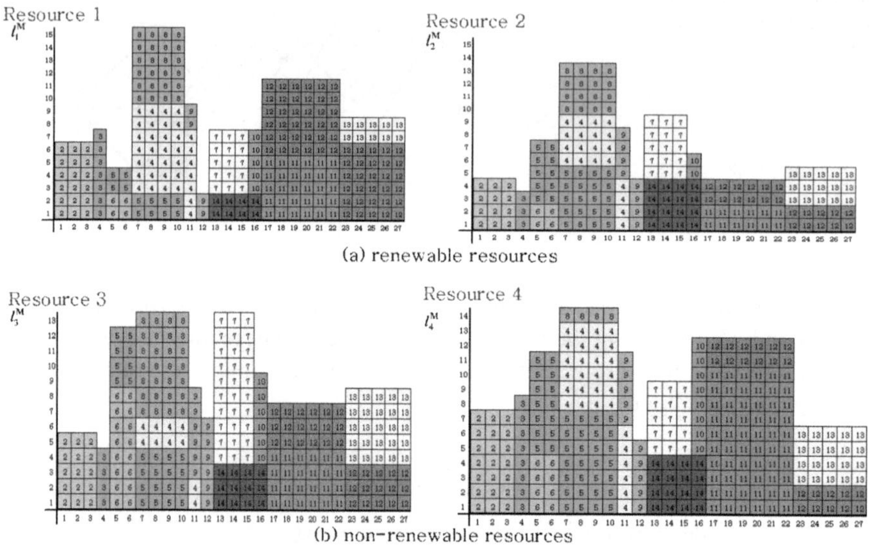

Fig. 6.51 Gantt chart for the schedule

Using this schedule, Gantt chart for limited resources $(l_1^M, l_2^M, l_3^M, l_4^M)$=(15,15,13,14) is constructed and shown in Fig. 6.51. The makespan of this solution is 27 time units.

6.5 Summary

In this chapter we discussed various project scheduling models as follows: (1) resource-constrained project scheduling model, which are to schedule the activities so that precedence and resource constraints are obeyed, (2) resource-constrained multiple project scheduling models, which are to manage the various projects with the activities so that precedence and resource constraints are obeyed, (3) resource-constrained project scheduling model with multiple modes, where by several different modes may be selected for an activity.

For solving each scheduling model, this chapter introduced effective hybrid GA approaches, used to improve the efficiency of genetic convergence. The revised serial method based on the priority-based decoding rule was to find an appropriate order of activities and problem-specific heuristics to determine the earliest start times and a fuzzy logic control to regulate automatically GA parameters and incorporation of the local search method performed for local exploitation around the near optimum solution for solving project scheduling problems.

References

1. Brucker, P., Drexl, A., Nohring, R., Neumann, K., & Pesch, E. (1999). Resource constrained project scheduling: Notation, models and methods, *European Journal of Operational Research*, 112(1), 3–41.
2. Herroelen, W., Demeulemeester, E., & De Reyck, B. (1998). Resource-constrained project scheluling - A survey of recent developments, *Computers and Operations Research*, 25(4), 279–302.
3. Kolisch, R. & Hartmann, S. (1999). Heuristic algorithms for solving the resource constrained project scheduling problem: Classification & computation analysis, In: J. Weglarz (Ed.), *Project Scheduling: Recent Models, Algorithms and Applications*, Kluwer Academic Publishers, Boston, 147–178.
4. Blazewicz, J., Lenstra, J., & Rinnooy Kan, A. (1983). Scheduling subject to resource constraints: Classification & complexity, *Discrete Applied Mathematics*, 5, 11–24.
5. Jozefowska, J., Mika, M., Rozycki, R., Waligora, G., & Weglarz, J. (2000). Solving the Discrete-continuous Project Scheduling Problem via Its Discretization, *Mathematical Methods of Operations Research*, 52, 489–499.
6. Yun, Y. S. & Gen, M. (2002). Advanced Scheduling Problem Using Constraint Programming Techniques in SCM Environment. *Computers & Industrial Engineering*, 43, 213–229.
7. Kelly, J. E. (1963). The critical path method: Resource planning and scheduling, in J.F. Muth, G.L. Thompson (Eds.), *Industrial Scheduling*, Prentice Hall, N.J., 347–365.
8. Gen, M. & Cheng, R.(1997). *Genetic Algorithms and Engineering Design*, New York, John Wiley & Sons.
9. Pritsker, A. A. B., Watters, L. J., & Wolfe, P. M. (1969). Multiproject scheduling with limited resources: a zero-one programming approach, *Management Science*, 16, 93–107.
10. Patterson, J. H. & Huber, W. D. (1974). A horizon-varying, zero-one approach to project scheduling. *Management Science*, 20, 990–998.
11. Patterson, J. H. & Roth, G. W. (1976). Scheduling a project under multiple resource constraints: a zero one programming approach, *AIIE Transactions*, 8, 449–55.
12. Davis, E. W. & Heidorn, G. E. (1971). An algorithm for optimal project scheduling under multiple resource constraints. *Management Science*, 17, 803–16.
13. Talbot, F. B. & Patterson, J. H. (1978). An efficient integer programming algorithm with network cuts for solving resource constrained scheduling problems, *Management Science*, 24(11), 1163–74.
14. Christofides, N., Alvarez-Valdes, R., & Tamarit, J. M. (1987). Project scheduling with resource constraints: A branch-and-bound approach, *European Journal of Operational Research*, 29(2), 262–273.
15. Demeulemeester, E. & Herroelen, W. (1992). A branch-and-bound procedure for the multiple resource-constrained project scheduling problem, *Management Science*, 38(12), 1803–1818.
16. Demeulemeester, E. & Herroelen, W. (1997). New benchmark results for the resource-constrained project scheduling problem, *Management Science*, 43, 1485–1492.
17. Brucker, P., Knust, S., Schoo, A., & Thiele, O. (1998). A branch-and-bound algorithm for the resources-constrained project scheduling problem, *European Journal of Operational Research*, 107, 272–288.
18. Mingozzi, A., Maniezzo, V., Ricciardelli, S., & Bianco, L. (1998). An exact Algorithm for project scheduling with resource constraints based on a new mathematical formulation, *Management Science*, 44, 714–729.
19. Sprecher, A. (1997). Solving the RCPSP efficiently at modest memory requirements, *Working paper University of Kiel*, Germany.
20. Klein, R. & Scholl, A. (1998). Progress: Optimally solving the generalized resource-constrained project scheduling problem, *Working paper, University of Technology*, Darmstadt.
21. Klein, R. & Scholl, A. (1998). Scattered branch & bound: An adaptative search strategy applied to resource-constrained project scheduling problem, *Working paper, University of Technology*, Darmstadt.

22. Hartmann, S. & Drexl, A. (1998). Project scheduling with multiple modes: A comparison of exact algorithms, *Networks*, 32(4), 283–298.
23. Alvarez-Valdez, R. & Tamarit, J. M. (1989). Heuristic algorithms for resource-constrained project scheduling: A review and empirical analysis, In: Slowinski, R., Weglarz, J. (Eds.), *Advances in Project Scheduling*, Elsevier, Amsterdam, 113–134.
24. Boctor, F. F. (1990). Some efficient multi-heuristic procedures for resource-constrained project scheduling, *European Journal of Operational Research*, 49, 3–13.
25. Cooper, D. F. (1976). Heuristics for scheduling resource-constrained projects: An experimental investigation, *Management Science*, 22(11), 1186–1194.
26. Cooper, D. F. (1977). A note on Serial & Parallel heuristics for resource-constrained project scheduling, *Foundations of Control Engineering*, 2(4), 131–133.
27. Davis, E. W. & Patterson, J. H. (1975). A comparison of heuristic & optimum solutions in resource constrained project scheduling, *Management Science*, 21(11), 944–955.
28. Lawrence, S. R. (1985). Resource constrained project scheduling - A computational comparison of heuristic scheduling techniques, *Technical report, Graduate School of Industrial Administration, Carnegie-Mellon University*, Pittsburgh.
29. Kolisch, R. & Drexl, A. (1996). Adaptative search for solving hard project scheduling problems, *Naval Research Logistics*, 43, 43–23.
30. Tormos, P. & Lova, A. (2003). Integrating heuristics for resource constrained project scheduling: One step forward, *Technical report, Department of Statistics & Operations Research, Universidad Politecnica de Valencia.*
31. Icmeli, O., Erenguc, S. S., & Zappe, C. J. (1993). Project scheduling problems: a survey. *International Journal of Operations & Productions Management*, 13(11), 80–91.
32. Kolisch, R. & Padman, R. (2001). An integrated survey of deterministic project scheduling, *Omega*, 49(3), 249–272.
33. Ozdamar, L. & Ulusoy, G. (1995). A survey on the resource-constrained project scheduling problem. *IIE Transactions*, 27, 574–586.
34. Kolisch, R. & Hartmann, S. (2004). Experimental investigation of heuristics for resource-constrained project scheduling: an update, *Working paper*, Technical University of Munich.
35. Hartmann, S. & Kolisch, R. (2000). Experimental evaluation of state-of-the-art heuristics for the resource-constrained project scheduling problem, *European Journal of Operational Research*, 127, 394–407.
36. Hartmann, S. (1998). A competitive genetic algorithm for the resource constrained project scheduling, *Naval Research Logistics*, 456, 733–750.
37. Kohlmorgen, U., Schmeck, H., & Haase, K. (1999). Experiences with fine-grained parallel genetic algorithms, *Annals of Operations Research*, 90, 203–219.
38. Alcaraz, J. & Maroto, C. (2001). A robust genetic algorithm for resource allocation in project scheduling, *Annals of Operations Research*, 102, 83–109.
39. Hartmann, S. (2002). A self-adapting genetic algorithm for project scheduling under resource constraints, *Naval Research Logistics*, 49, 433–448.
40. Kochetov, Y. & Stolyar, A. (2003). Evolutionary local search with variable neighborhood for the resource constrained project scheduling problem, *In Proceedings of the 3rd International Workshop of Computer Science & Information Technologies*, Russia.
41. Mendes, J. M. (2003). Sistema de apoio a decisao para planeamento de sistemas de producao do tipo projecto. Departamento de Engenharia Mecanica e Gestao Industrial. Faculdade de Engenharia da Universidade do Porto. Ph. D. Thesis (In Portuguese).
42. Valls, V., Ballestin, F., & Quintanilla, M. S. (2004). Justification and RCPSP: A technique that pays, *European Journal of Operational Research*, 165(2), 375–386.
43. Valls, V., Ballestin, F., & Quintanilla, M. S. (2004). A population-based approach to the resource constrained project scheduling problem, *Annals of Operations Research*, 131, 305–324.
44. Sakalauskas, L. & Felinskas, G. (2006). Optimization of resource constrained project schedules by genetic algorithm based on job priority list, *Information Technology & Control*, 35(4), 412–418.

45. Boctor, F. F. (1996). Resource-constrained project scheduling by simulated annealing, *International Journal of Production Research*, 34(8), 2335–2351.
46. Bouleimen, K. & Lecocq, H. (1998). A new efficient simulated annealing algorithm for the resource-constrained project scheduling problem, in: G. Barbarosoglu, S. Karabati, L. Ozdamar, C. Ulusoy (Eds), *Proceedings of the Sixth International Workshop on Project Management and Scheduling*, Bogazici University Printing Office, Istanbul, 19–22.
47. Bouleimen, K. & Lecocq, H. (2003). A new efficient simulated annealing algorithm for the resource-constrained project scheduling problem & its multiple mode version, *European Journal of Operational Research*, 149, 268–281.
48. Pinson, E., Prins, C., & Rullier, F. (1994). Using tabu search for solving the resource-constrained project scheduling problem, In :*Proceedings of the 4th International Workshop on Project Management & Scheduling*, Leuven, Belgium, 102–106.
49. Baar, T., Brucker, P., & Knust, S. (1997). Tabu search algorithms for resource constrained project scheduling problems, in S. Voss, S. Martello, I. Osman, C. Roucairol (Eds.), Metaheuristics : *Advances & Trends in Local Search Paradigms for Optimisation*, Kluwer, 1–18.
50. Nonobe, K. & Ibaraki, T. (2002). Formulation & tabu search algorithm for the resource constrained project scheduling problem. In C. C. Ribeiro & P. Hansen, editors, *Essays & Surveys in Metaheuristics*, 557–588. Kluwer Academic Publishers.
51. Fleszar, K. & Hindi, K. (2000). Solving the resource-constrained project scheduling problem by a variable neighbourhood search, *Technical report, Brunel University*, Department of Systems Engineering.
52. Palpant, M., Artigues, C., & Michelon, P. (2004). LSSPER: Solving the resource-constrained project scheduling problem with large neighbourhood search. *Annals of Operations Research*, 131, 237–257.
53. Patterson, J. (1984). A comparison of exact procedures for solving the multiple constrained resource projects scheduling problem, *Management Science*, 30, 854–867.
54. Kim, K. W. (2003). Optimum Design for Advanced Production Scheduling using Hybrid Genetic Algorithm with Fuzzy Logic, *PhD Thesis*, Waseda University, Graduate School of Information, Production & Systems, Kitakyushu, Japan.
55. Kim, K. W., Gen, M., & Yamazaki, G. (2002). Hybrid genetic algorithm with fuzzy logic for resource-constrained project scheduling, *Applied Soft Computing*, 44, 1–15.
56. Kolisch, R. (1996). Serial & Parallel Resource-constrained Project Scheduling Methods Revisited: Theory & Computation, *European Journal of Operational Research*, 90, 320–333.
57. Syswerda, G. (1991). Scheduling Optimization using Genetic Algorithms, 332–349 in Davis, L. Ed.: *Handbook of Genetic Algorithms*, New York, Van Nostrand Reinhold.
58. Goldberg, D. & Lingle R. (1985). Alleles: loci and the traveling salesman problem, *Proceedings of the 1st International Conference on GA*, 154–159.
59. Gen, M. & Cheng, R. (2000). Genetic Algorithm & Engineering Optimization, New York, John Wily & Sons.
60. Wang, P. Y., Wang, G. S., & Hu Z. G. (1997). Speeding up the search process of genetic algorithm by fuzzy logic, *European Congress on Intelligent Techniques & Soft Computing*, 665–671.
61. Fricke, S. E. & Shenhar, A. J. (2000). Managing Multiple Engineering Projects in a Manufacturing Support Environment, *IEEE Transactions on Engineering Management*, 47(2), 258–268.
62. Hendriks, M. H. A., Voeten, B., & Kroep, L. (1999). Human Resource Allocation in a Multi-project R&D Environment, *International Journal of Production Management*, 17(2), 181–188.
63. Isakow, S. A. & Golany, B. (2003). Managing Multiple Project Environments through Constant Work-in-process, *International Journal of Project Management*, 21, 9–18.
64. Lova, A. & Tormos, P. (2001). Analysis of scheduling schemes & heuristic rules performance in resource-constrained multiple project scheduling, *Annals of Operations Research*, 102, 263–286.

65. Lova, A., Maroto, C., & Tormos, P. (2000). A multicriteria heuristic method to improve resource allocation in multiproject scheduling, *European Journal of Operational Research*, 127, 408–424.
66. Scheiberg, M. & Stretton, J. (1994). Multiproject planning: turing portfolio indices, *International Journal of Project Management*, 12(2), 107–114
67. Wiley, V. D., Deckro, R. F., & Jackson Jr. J. A. (1998). Optimization analysis for design & planning of multiproject programs, *European Journal of Operational Research*, 107(2), 492–506.
68. Kim, K. W., Yun, Y. S., Yoon, J. M., Gen, M., & Yamazaki, G. (2005). Hybrid genetic algorithm with adaptive abilities for resource-constrained multiple project scheduling, *Computers in Industry*, 56(2), 143–160.
69. Srinivas, M. & Patnaik, L. M. (1994). Adaptive Probabilities of Crossover & Mutation in Genetic Algorithms, *IEEE Transaction on Systems, Man & Cybernetics*, 24(4), 656–667.
70. Wu, Q. H., Cao, Y. J., & Wen, J. Y. (1998). Optimal Reactive Power Dispatch Using an Adaptive Genetic Algorithm, *Electrical Power & Energy Systems*, 20(8), 563–569.
71. Kim, K. W., Gen, M., & Kim, M. H. (2006). Adaptive genetic algorithms for multi-resource constrained project scheduling problem with multiple modes, *International Journal of Innovative Computing, Information & Control*, 2(1), 41–49.

Chapter 7
Assembly Line Balancing Models

From ancient times to the modern day, the concept of assembly has naturally been changed a lot. The most important milestone in assembly is the invention of *assembly lines* (ALs). In 1913, Henry Ford completely changed the general concept of assembly by introducing ALs in automobile manufacturing for the first time. He was the first to introduce a moving belt in a factory, where the workers were able to build the famous model-T cars, one piece at a time instead of one car at a time. Since then, the AL concept revolutionized the way products were made while reducing the cost of production. Over the years, the design of efficient assembly lines received considerable attention from both companies and academicians. A well-known assembly design problem is *assembly line balancing* (ALB), which deals with the allocation of the tasks among workstations so that a given objective function is optimized.

7.1 Introduction

An *assembly line* (AL) is a manufacturing process consisting of various tasks in which interchangeable parts are added to a product in a sequential manner at a station to produce a finished product. Assembly lines are the most commonly used method in a mass production environment, because they allow the assembly of complex products by workers with limited training, by dedicated machines and/or by robots.

The installation of an assembly line is a long-term decision and usually requires large capital investments. Therefore, it is important that an AL is designed and balanced so that it works as efficiently as possible. Most of the work related to the ALs concentrate on the *assembly line balancing* (ALB). The ALB model deals with the allocation of the tasks among stations so that the precedence relations are not violated and a given objective function is optimized.

Besides balancing a newly designed assembly line, an existing assembly line has to be re-balanced periodically or after certain changes in the production process or the production plan. Because of the long-term effect of balancing decisions, the objective functions have to be carefully chosen while considering the strategic goals of the enterprise.

Based on the model structure, ALB models can be classified into two groups (see Fig. 7.1). While, the first group [1, 2] includes *single-model assembly line balancing* (smALB), *multi-model assembly line balancing* (muALB), and *mixed-model assembly line balancing* (mALB); the second group [3] includes *simple assembly line balancing* (sALB) and *general assembly line balancing* (gALB). The smALB model involves only one product. The muALB model involves more than one product produced in batches. The mALB refers to assembly lines which are capable of producing a variety of similar product models simultaneously and continuously (not in batches). Additionally, sALB, the simplest version of the ALB model and the special version of the smALB model, involves production of only one product with features such as paced line with fixed cycle time, deterministic independent processing times, no assignment restrictions, serial layout, one sided stations, equally equipped stations and fixed rate launching. The gALB model includes all of the models that are not sALB, such as balancing of mixed-model, parallel, u-shaped and two sided lines with stochastic dependent processing times; thereby more realistic ALB models can be formulated by gALB.

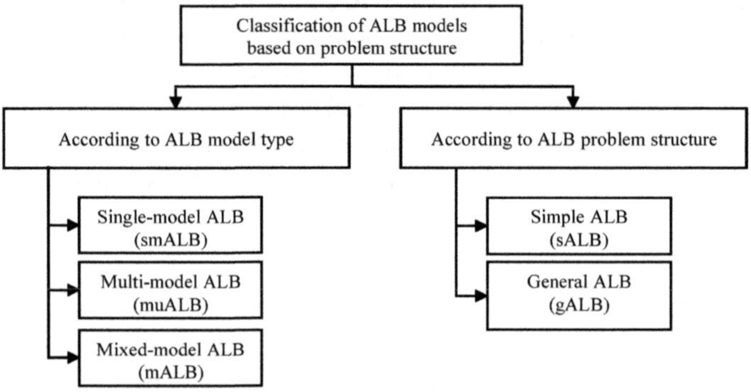

Fig. 7.1 Classification of assembly line balancing models

Additionally, several versions of ALB problems arise by varying the objective function [1]. Type-F is an objective independent problem which is to establish whether or not a feasible line balance exists. Type-1 and Type-2 have a dual relationship; the first tries to minimize the number of stations for a given cycle time, and the second tries to minimize the cycle time for a given number of stations. Type-E is the most general problem version, which tries to maximize the line efficiency by simultaneously minimizing the cycle time and a number of stations. Finally, Type-3, 4 and 5 correspond to maximization of workload smoothness, maximization of work relatedness and multiple objectives with Type-3 and Type-4, respectively [4].

Since the ALB model was first formulated by Helgeson *et al.* [5], many solution approaches have been proposed. Several optimum seeking methods, such as *linear programming* [6], *integer programming* [7], *dynamic programming* [8] and *branch-and-bound approaches* [9] have been employed to deal with ALB. However, none of these methods has proven to be of practical use for large problems due to their computational inefficiency. Since ALB models fall into the NP-hard class of combinatorial optimization problems [10], in recent years, to provide an alternative to traditional optimization techniques, numerous research efforts have been directed towards the development of heuristics [11] and meta-heuristics. While heuristic methods generating one or more feasible solutions were mostly developed until the mid 1990s, meta-heuristics such as tabu search [12], simulated annealing [13], genetic algorithms [14] and ant colony optimization [15] have been the focus of researchers in the last decade.

For more information, the reader can refer to several review studies, *i.e.* Baybars [3] that survey the exact (optimal) methods, Talbot *et al.* [16] that compare and evaluate the heuristic methods developed, Ghosh and Gagnon [17] that present a comprehensive review and analysis of the different methods for design, balancing and scheduling of assembly systems, Erel and Sarin [18] that present a comprehensive review of the procedures for smALB, muALB and mALB models, Rekiek *et al.* [19] that focus on optimization methods for the line balancing and resource planning steps of assembly line design, Scholl and Becker [20] that present a review and analysis of exact and heuristic solution procedures for sALB, Becker and Scholl [2] that present a survey on problems and methods for gALB with features such as cost/profit oriented objectives, equipment selection/process alternatives, parallel stations/tasks, u-shaped line layout, assignment restrictions, stochastic task processing times and mixed model assembly lines, Rekiek and Delchambre [21] that focus on solutions methods for solving sALB, and Ozmehmet Tasan and Tunali [22] that present a comprehensive review of GAs approaches used for solving various ALB models.

Among the meta-heuristics, the application of genetic algorithms (GAs) received considerable attention from the researchers, since it provides an alternative to traditional optimization techniques by using directed random searches to locate optimum solutions in complex landscapes and it is also proven to be effective in various combinatorial optimization problems. GAs are powerful and broadly applicable stochastic search and optimization techniques based on principles from evolutionary theory [23].

Falkenauer and Delchambre [14] were the first to solve ALB with GAs. Following Falkenauer and Delchambre [14], application of GAs for solving ALB models was studied by many researchers, *e.g.*, [4]–[24]-[51]. However, most of the researchers focused on the simplest version of the problem, with single objective and ignored the recent trends, *i.e.*, mixed-model production, u-shaped lines, and robotic lines *et.* in the complex assembly environments, where ALB models are multiobjective in nature [22].

In this chapter, we will illustrate the successful applications of multiobjective GAs (moGAs) to solve various ALB models, *i.e.*, sALB model and gALB models such as u-shaped ALB, robotic ALB and mALB.

7.2 Simple Assembly Line Balancing Model

The basic version of the ALB model is the simple assembly line balancing (sALB) model. The simple assembly line is a single-model assembly line that is capable of producing only one type of product. The simple assembly line can be defined by the following assumptions [1, 3]:

A1. The line is used to assemble one homogeneous product in mass quantities.
A2. The line is serial, paced line with fixed cycle time and there are no feeder or parallel subassembly lines.
A3. The processing times of tasks are deterministic.
A4. All stations are equally equipped with respect to machines and workers.
A5. A task cannot be split among two or more stations.
A6. There are no assignment restrictions besides the precedence constraints.
A7. All stations can process any one of the tasks and all have the same associated costs.
A8. The processing time of a task is independent of the station and furthermore, they are not sequence dependent.

Among the family of ALB models, the most well-known and well-studied is certainly the sALB model. Although it might be far too constrained to reflect the complexity of real-world line balancing, it nevertheless captures its main aspects and is rightfully regarded as the core model of ALB. In fact, vast varieties of more general problems are direct sALB model extensions or at least require the solution of sALB instances in some form. In any case, it is well suited to explain the basic principles of ALB and introduce its relevant terms.

7.2.1 Mathematical Formulation of sALB Models

A simple AL capable of producing only one type of product consists of *stations* $(i = 1,\ldots,m)$ arranged along a conveyor belt or similar mechanical material handling equipment. The workpieces (jobs) are consecutively launched down the line and are moved from station to station. At each station, certain tasks are repeatedly performed regarding the *cycle time* (maximum or average time available for each workcycle). The decision problem of optimally partitioning, *i.e., balancing*, the assembly work among the stations with respect to a given objective function is known as the sALB problem.

Manufacturing a product on an assembly line requires partitioning the total amount of work into a set of elementary operations named *tasks* $j = \{1,\ldots,n\}$. Performing a task j takes certain task time t_j and requires certain equipment and/or skills of workers. Due to technological and organizational conditions, *precedence constraints* between the tasks have to be observed. These elements can be summarized and visualized by a *precedence network graph*. It contains a node for each task, node weights for the task processing times and arcs for the precedence constraints.

Table 7.1 presents the data set for an example sALB model, which contains 12 tasks [52]. Using this data set, the precedence graph in Fig. 7.2 is constructed. The precedence graph contains 12 nodes for tasks, node weights for task processing times and arcs for orderings. For example, the processing time for task 6 is 5 time units. For the processing of task 6, tasks 3 and 5 (direct predecessors) and tasks 1, 2 and 4 (indirect predecessors) must be completed. Likewise, task 6 must be completed before tasks 7, 9 and 10 (direct successors), and tasks 8, 11, and 12 (indirect successors) start processing.

Table 7.1 Data set of the sALB model [52]

Task j	Suc(j)	Task time t_j
1	{2,4}	5
2	{3}	3
3	{6}	4
4	{5}	3
5	{6}	6
6	{7,9,10}	5
7	{8}	2
8	{12}	6
9	{12}	1
10	{11}	4
11	{12}	4
12	{}	7

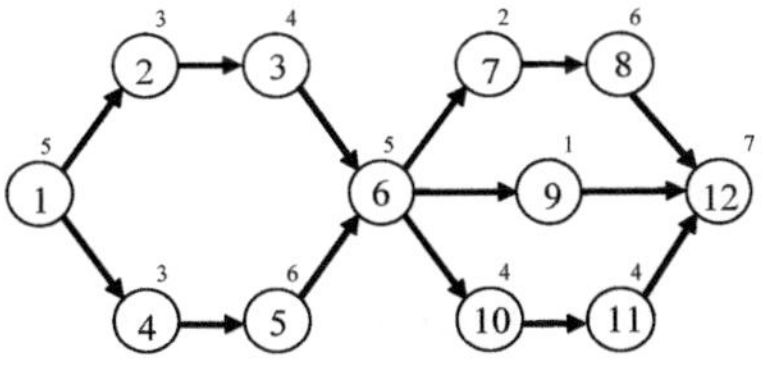

Fig. 7.2 Precedence graph of the sALB model [52]

Any type of sALB model consists in finding a feasible *line balance, i.e.*, an assignment of each task to a station such that the precedence constraints are fulfilled. The set of tasks S_i assigned to a station i $(= 1, \ldots, m)$ constitutes its *station load*, the cumulated task time:

$$t(S_i) = \sum_{j \in S_i} t_j \tag{7.1}$$

is called *station time*. When a fixed common *cycle time* c_T is given, a line balance is feasible only if the station time of neither station exceeds c_T. In case of $t(S_i) < c_T$, the station i has an idle time of ($c_T - t(S_i)$) time units in each cycle.

Using the precedence graph in Fig. 7.2, a feasible line balance with cycle time 11 time units and 6 stations can be constructed by the station loads $S_1 = \{1,2\}$, $S_2 = \{3,4\}$, $S_3 = \{5,6\}$, $S_4 = \{7,9,10,11\}$, $S_5 = \{8\}$, $S_6 = \{12\}$ (see Fig. 7.3) . While no idle time occurs in stations 3 and 4, stations 1, 2, 5 and 6 show idle times of 3, 4, 5 and 4 time units, respectively.

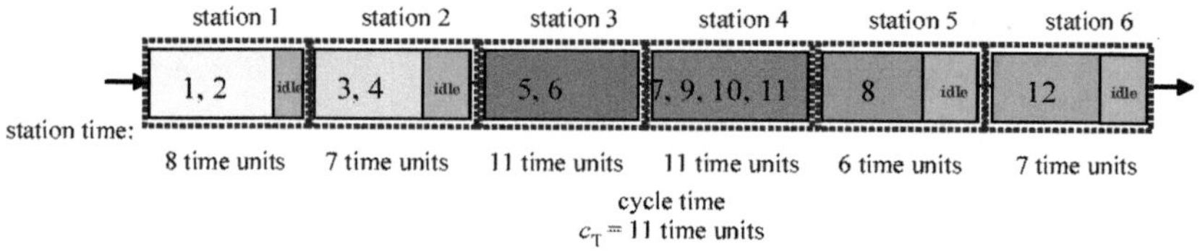

Fig. 7.3 A feasible line balance for a sALB model

Notation

In order to formulate the mathematical model, the following indices, parameters and decision variable are introduced.

Indices

i: index of the stations $(i = 1, \ldots, m)$
j,k: index of the tasks $(j,k = 1, \ldots, n)$

Parameters

m: number of stations actually employed
M: maximum number of stations available $(n \leq M)$
n: number of tasks
c_T: cycle time of the assembly line
t_j: processing time of task j
S_i: the set of tasks assigned to station i
T_s: a task sequence
Suc(j): the set of direct successors of task j

Pre(j): the set of direct predecessors of task j
u_i: *utilization* of the station i

$$u_i = \frac{1}{\max\limits_{1 \le i \le m} \{t(S_i)\}} t(S_i)$$

$\bar{u}$: *average utilization* of total stations

$$\bar{u} = \frac{1}{m} \sum_{i=1}^{m} u_i$$

Decision Variables

$$x_{ij} = \begin{cases} 1, \text{ if task } j \text{ is assigned to station } i \\ 0, \text{ otherwise} \end{cases}$$

Mathematical Model

The mathematical model for the sALB Type-1 can be stated as follows:

$$\max E = \frac{1}{m c_{\mathrm{T}}} \sum_{j \in S_i} t_j x_i j \tag{7.2}$$

$$\min m = \sum_{i=1}^{M} \max_{1 \le j \le n} \{x_{ij}\} \tag{7.3}$$

$$\min V = \sqrt{\frac{1}{m} \sum_{i=1}^{m} (u_i - \bar{u})^2} \tag{7.4}$$

$$\text{s.t. } \sum_{i=1}^{M} x_{ij} = 1, \forall\ j \tag{7.5}$$

$$\sum_{i=1}^{M} i x_{ik} \le \sum_{i=1}^{M} i x_{ij}, \ \forall\ j, \forall\ k \in \mathrm{Pre}(j) \tag{7.6}$$

$$t(S_j) = \sum_{j \in S_i} t_j = \sum_{j=1}^{n} t_j x_{ij} \le c_{\mathrm{T}}, \ \forall\ i \tag{7.7}$$

$$x_{ij} = 0 \text{ or } 1, \quad \forall\ i,\ j \tag{7.8}$$

In this mathematical model, the first objective (Eq. 7.2) of the model is to maximize the *line efficiency*. The second objective (Eq. 7.3) is to minimize the *number of stations* actually employed. The third objective (Eq. 7.4) of the model is to minimize the *variation of workload*. The constraints given in Eqs. 7.5–7.8 are used to formulate the general feasibility of the problem. The constraint given in Eq. (7.5) states that each task must be assign to one and only one station. The inequality at Eq. 7.6 represents the precedence constraints and it states that the direct predecessor of task j must be assigned to a station which is in front of or the same as the station that task j is assigned in. This constraint stresses that if a task is assigned to a station, then the predecessor of this task must be already assigned to a station. The inequality at Eq. 7.7 denotes that the available time at each station should be less than or equal to the given cycle time. Constraint given in Eq. 7.8 represents the usual integrity restriction.

7.2.2 Priority-based GA for sALB Models

In this section of the chapter, we will introduce a *priority-based genetic algorithm* (priGA) for solving the multi-objective sALB Type-1 model. The priGA approach was originally developed by Gen and Cheng [23] in order to handle the problem of creating encoding while treating the precedence constraints efficiently. Let $P(t)$ and $C(t)$ be parents and offspring in current generation t, respectively. The overall procedure of priGA for solving sALB model is outlined as follows:

---

procedure: priGA for sALB
input: problem data, GA parameters
output: the best solution
begin
  $t \leftarrow 0$;
  initialize $P(t)$ by *priority-based encoding routine*;
  evaluate $P(t)$ by *priority-based decoding routine*;
  while (**not** terminating condition) **do**
    create $C(t)$ from $P(t)$ by *weight mapping crossover routine*;
    create $C(t)$ from $P(t)$ by *insertion mutation routine*;
    evaluate $C(t)$ by *priority-based decoding routine*;
    select $P(t+1)$ from $P(t)$ and $C(t)$ by roulette wheel selection routine;
    $t \leftarrow t+1$;
  end
  output the best solution
end

The detailed information about the genetic representation, evaluation function, and genetic operators are given in the following subsections.

7.2.2.1 Genetic Representation

The primary issue in applying GA to problem is to convert the information of ALB model into a genetic representation form. Up to now, several genetic representations, *i.e.*, task-based, embryonic, workstation-based, grouping-based, and heuristic-based have been proposed, each having pros and cons concerning the type of applicable genetic operators.

1	2	4	3	5	6	7	10	9	11	8	12

(a) Task based representation

1	3	2	5	4	6

(b) Embryonic representation

1	1	2	2	3	3	4	5	4	4	4	6

(c) Workstation based representation

1	1	2	2	3	3	4	5	4	4	4	6	:	**1**	**2**	**3**	**4**	**5**	**6**

(d) Grouping based representation

H_1	H_2	H_5	H_4	H_7	H_6	H_3

(e) Heuristic based representation

Fig. 7.4 Chromosome representation schemes used in ALB models

The chromosome representation schemes are named in order to suit the characteristics of ALB model [22]. These representation schemes can be classified as follows:

- **Task-based Encoding**: The chromosomes are defined as feasible precedence sequences of tasks [42, 37]. The length of the chromosome is defined by the number of tasks. For example, the task based representation of the solution given in Fig. 7.3 is illustrated in Fig. 7.4a. In order to calculate the fitness of a task based chromosome, additional operations, which assign the tasks to workstations according to the task sequence in the chromosome, are needed. Task-based representation is the most appropriate representation for ALB Type-1 models, since Type-1 models consider the minimization of stations as an objective function.
- **Embryonic Encoding**: Embryonic chromosome representation that was proposed by Brudaru and Valmar [44] is actually a special version of the task based chromosome. The only difference between the two is that the embryonic repre-

sentation of a solution considers the subsets of solutions rather than the individual solutions. During the generations, the embryonic chromosome evolves through a full length solution. Therefore, the chromosome length varies throughout the generations. The length is initially defined by a random number and then increases until it reaches the number of tasks. Figure7.4b illustrates an example of embryonic representation of the solution given in Fig. 7.3

- **Workstation-based Encoding**: The chromosome is defined as a vector containing the labels of the stations to which the tasks are assigned [25, 35]. The chromosome length is defined by the number of tasks. For example, the workstation-based representation of the solution given in Fig. 7.3 is illustrated in Fig.7.4, where task 4 is assigned to station 3. This kind of chromosome representation scheme is generally used for ALB Type-2 models.
- **Grouping-based Encoding**: This type of representation was proposed by Falkenauer and Delchambre [14] especially for grouping problems,*i.e.*, ALB Type-1 models. The authors stated that the workstation-based representation, which is object oriented, is not suitable for ALB Type-1 models. In grouping-based representation, the stations are represented by augmenting the workstation-based chromosome with a group part. The group part of the chromosome is written after a semicolon to list all of the workstations in the current solution (see Fig. 7.4d). The length of the chromosome varies from solution to solution. As is seen in Fig. 7.4d, the first part is the same as in workstation-based chromosome. The difference comes from the grouping part, which list all the stations, *i.e.*, 1, 2, 3, 4, 5, and 6.
- **Heuristic-based (Indirect) Encoding**: This type of representation scheme represents the solutions in an indirect manner. In Goncalves and De Almeida [41], and Bautista *et al.* [34], the authors first coded the priority values of the tasks (or a sequence of priority rules), then they applied these rules to the problem to generate the solutions. The chromosome length is defined by the number of heuristics. For example, Fig. 7.4e shows an example chromosome having seven different heuristics, which are used in the sequence of H_1, H_2, H_5, H_4, H_7, H_6 and H_3 to assign the tasks to the workstations.

An appropriate chromosome representation scheme in conjunction with carefully designed genetic operators and fitness function is essential for GA design, since the application of standard crossovers or mutations to task-based, workstation-based and grouping-based chromosomes may result in infeasible solutions. This aspect must be dealt with by penalizing infeasibilities, rearranging the solution by certain heuristic strategies or constructing of special genetic operators which will force feasibility. In this context, heuristic-based chromosomes have the advantage to achieve the feasibility without such difficulties.

In this approach, a heuristic-based (indirect) representation, which is called the priority-based representation, is used. The priority-based representation scheme was proposed by Gen and Cheng [23] in order to handle the problem of creating encoding while treating the precedence constraints efficiently. To develop a priority-based genetic representation for the sALB model, there are three main phases:

Phase 1: Creating a task sequence

Step 1.1: Generate a random priority to each task in the model using encoding procedure.
Step 1.2: Decode a feasible task sequence T_S that satisfies the precedence constraints.

Phase 2: Assigning tasks to stations

Step 2.1: Input the task sequence found in step 1.2.
Step 2.2: Obtain a feasible solution set according to this task sequence.

Phase 3: Designing a schedule

Step 3.1: Create a schedule S using station assignments found in step 2.2.
Step 3.2: Draw a Gantt chart for this schedule.

While introducing these three phases, in order to illustrate the procedures for sALB, we will be using the data set given in Table 7.1and its corresponding precedence graph given in Fig. 7.2.

Phase 1: Creating a Task Sequence

Step 1.1: Generate a random priority to each task in the model using encoding procedure

The first step in GAs is the generation of initial population that consists of chromosomes. In this step, an indirect representation scheme called the priority-based encoding method is used. In this method, the position of a gene was used to represent a task node and the value of the gene was used to represent the priority of the task node for constructing a schedule among candidates. This encoding method verifies any permutation type representations, so that most of the existing genetic operators can be easily applied. Figure 7.5 illustrates the process of this encoding procedure on a chromosome.

Step 1.2: Decode a feasible task sequence T_S that satisfies the precedence constraints

In order to decode the chromosomes generated in step 1.1, the priorities of each task are used to create a feasible task sequence that satisfies the precedence constraints in the model. Figure 7.6 presents the priority-based decoding procedure for creating a task sequence.

To illustrate this decoding procedure, the precedence graph in Fig.7.2 is used. First, we try to find a node for the position next to node 1. Nodes 2 and 4 are eligible for that, which can be suitable for the next start node. Here we check their priority, that is 2 and 8 respectively. Next, the task 4 has the highest priority of 8 and is placed into the task sequence T_S. Then the next possible nodes are 2 and 5. They have priority of 2 and 5 respectively, and then we put 5 into task sequence T_S. Finally, we repeat these steps until we obtain a complete schedule T_S = 1, 4, 5, 2, 3, 6, 9, 7, 8, 10, 11, 12.

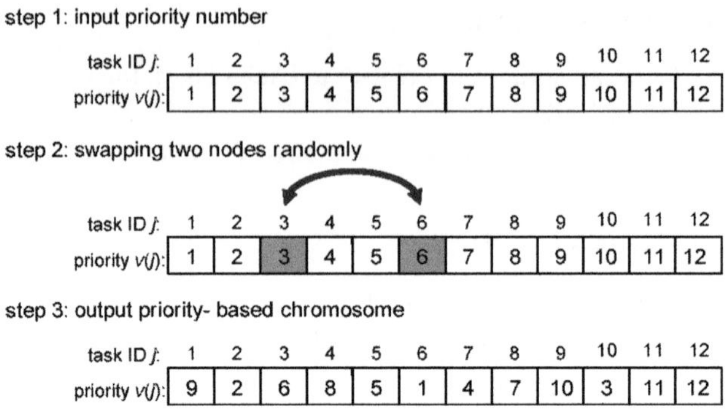

Fig. 7.5 Illustration of an example priority-based encoding

---

procedure: priority-based decoding (creating task sequence)
input: number of tasks n, chromosome $v(j)$
output: a task sequence T_S
begin

$\bar{s} \leftarrow \varnothing,\ T_S \leftarrow \varnothing;$

$n \leftarrow 0,\ j \leftarrow 0;$

while $(j \leq n)$ **do**

$\bar{s} \leftarrow \text{Suc}(j);$

$j^* \leftarrow \arg\max\{v(j) \mid j \in \bar{s}\};$

$\bar{s} \leftarrow \bar{s} \setminus j^*;$

$T_S \leftarrow T_S \cup j^*;$

$j \leftarrow j^*;$

end

output a task sequence T_S

end

Fig. 7.6 Priority-based decoding procedure for creating task sequence

Phase 2: Assigning Tasks to Stations

Step 2.1: Input the task sequence found in step 1.2
Step 2.2: Obtain a feasible solution set according to this task sequence

In this phase, the assignments of task to stations are formed using the task sequence found in step 1.2. Figure 7.7 presents the task to station assignment decoding procedure.

For the illustration of this decoding procedure, the task sequence T_S = 1, 4, 5, 2, 3, 6, 9, 7, 8, 10, 11, 12 found in step 1.2 is used. Table 7.2 presents the trace table

```
procedure: priority-based decoding (task to station assignment)
input: processing time t_j, chromosome v(j), the task sequence T_S
output: number of stations m, efficiency E, variation V, efficiency and variation EV
begin
  t_j ← 0, j=1, 2, ..., n;
  S_i ← 0, i=1, 2, ..., m;
  j ← 0, m ← 0, E ← 0, V ← 0, EV ← 0;
  for j = 1 to n
    while(t(S_i) ≤ c_T) do
        T_Sj ← T_S;
        S_i ← T_Sj;
        t(S_i) ← t(S_i) + t_j;
        T_S ← T_S \ T_Sj;
        m ← 1++;
        t_SUM ← T_Sj;
        E ← t_SUM / (m c_T);
        u_T ← u_i ++;
        ū ← u_T / m;
    end
  end
  for i = 1 to m
      W ← (u_i − ū)^2;
      V ← sqr(W / m);
      EV ← E + (1 − V);
  end
  output number of stations m, efficiency E, variation V, efficiency and variation EV
end
```

Fig. 7.7 Task to station assignment decoding procedure

for the decoding procedure. In the example, we obtained a feasible line balance with cycle time 10 time units and 6 stations represented by the station loads S_1 = 1, 4, S_2 = 2, 5, S_3 = 3, 6, 9, S_4 = 7, 8, S_5 = 10, 11, S_6 = 12. While no idle time occurs in stations 3, stations 1, 2, 4, 5 and 6 show idle times of 2, 1, 2, 2 and 3 time units, respectively.

Phase 3: Designing a Schedule

Step 3.1: Create a schedule S using station assignments found in step 2.2

Table 7.2 Trace table for task to station assignment decoding procedure

j	$\bar{S}$	$v(j)$	j^* (t_j)	S_i={}; $t(S_i)$ $(c_T$-$t(S_i))$
0	{1}	v(1)=9	1(5)	S_1={1}; 5 (5)
1	{2,4}	v(2)=2, v(4)=8	4(3)	**S_1={1,4}; 8 (2)**
4	{2,5}	v(2)=2, v(5)=5	5(6)	S_2={5}; 6 (4)
5	{2}	v(3)=6	2(3)	**S_2={5,2}; 9 (1)**
2	{3}	v(3)=6	3(4)	S_3={3}; 4 (6)
3	{6}	v(6)=1	6(5)	S_3={6}; 9 (1)
6	{7,9,10}	v(7)=4, v(9)=10, v(10)=3	9(1)	**S_3={3,6,9}; 10 (0)**
9	{7,10}	v(7)=4, v(10)=3	7(2)	S_4={7}; 2 (8)
7	{8,10}	v(8)=7, v(10)=3	8(6)	**S_4={7,8}; 8 (2)**
8	{10}	v(10)=3	10(4)	S_5={10}; 4 (6)
10	{11}	v(11)=11	11(4)	**S_5={10,11}; 8 (2)**
11	{12}	v(12)=12	12(7)	**S_6={12}; 7 (3)**

Using the trace table for the decoding procedure, the schedule can be constructed as follows:

Schedule $S = (j, S_i, t_j)$:

S=$\{(1, S_1, t_1), (4, S_1, t_4), (5, S_2, t_5), (2, S_2, t_2), (3, S_3, t_3),$
$(6, S_3, t_8), (9, S_3, t_9), (7, S_4, t_7), (8, S_4, t_8), (10, S_5, t_{10}),$
$(6, S_3, t_8), (9, S_3, t_9), (7, S_4, t_7), (8, S_4, t_8), (10, S_5, t_{10}),$
$(11, S_5, t_{11}), (12, S_6, t_{12})\}$
$=\{(1, S_1, 0-5), (4, S_1, 5-8), (5, S_2, 8-14), (2, S_2, 14-17),$
$(3, S_3, 17-21), (6, S_3, 21-26), (9, S_3, 26-27), (7, S_4, 27-29),$
$(8, S_4, 29-35), (10, S_5, 35-39), (11, S_5, 39-43), (12, S_6, 43-50)\}$

Step 3.2: Draw a Gantt chart for this schedule. Figure 7.8 illustrates a Gantt chart showing a feasible schedule for one unit of product.

However, since a single type of product is produced on a simple AL, the minimization of the makespan becomes an important production management issue. This time, Fig. 7.9 shows a Gantt chart for three units of product. Here, the makespan is 70 time units for three units of products.

7.2.2.2 Evaluation Function

The evaluation function is particularly related to the objective functions in the problem. It provides a means of evaluating the search nodes, and it also controls the selection process. A usual surrogate objective in ALB consists of maximization of the line utilization which is measured by the line efficiency as the productive frac-

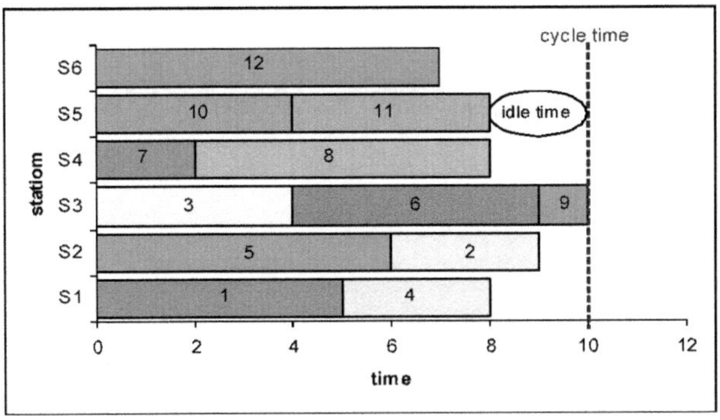

Fig. 7.8 Gantt chart for the sALB model (1 unit of product)

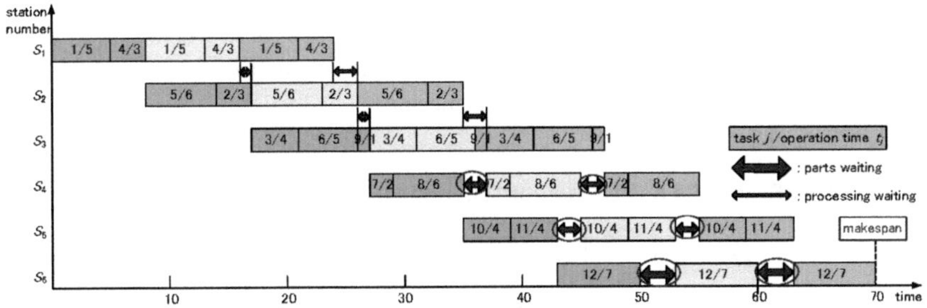

Fig. 7.9 Gantt chart for the sALB model (3 units of product)

tion of the line's total operating time and directly depends on the cycle time and the number of stations.

In the real-world, the ALB models consist of more than one objective. Further, in reality, success of an assembly line often requires an integrated consideration of many objectives simultaneously, *i.e.*, of a multiple objective ALB. In the past decade, researchers were mostly concentrating on using objectives, *i.e.*, the minimization of cycle time and the minimization of the number of stations. However, in practice, it is often more desirable to smooth out the workload assignments, and to assign related tasks to the same station if possible. Thus, in order to improve work efficiency and increase job satisfaction for workers, the objectives such as the maximization of the workload smoothness and the work relatedness became more important.

In this study, *adaptive weight approach* that utilizes some useful information from the current population to readjust weights for obtaining a search pressure toward a positive ideal point I is used. Two objective functions, *i.e.*, maximization of line efficiency (see Equation 7.2) and minimization of variation of workload (see Equation 7.4), are used in this study:

$$
\begin{aligned}
eval(v_k) &= w_1(f_1(v_k) - z_1^{\min}) + w_2(f_2(v_k) - z_2^{\min}) \\
&= w_1 \left\{ \frac{1}{m c_{\mathrm{T}}} \sum_{j=1}^{n} t_j - z_1^{\min} \right\} w_2 \left\{ 1 - \sqrt{\frac{1}{m} \sum_{i=1}^{m} (u_i - \overline{u})^2} - z_2^{\min} \right\}, \\
& \qquad k = 1, \cdots, popSize \qquad (7.9)
\end{aligned}
$$

7.2.2.3 Genetic Operators

In this subsection, the crossover, mutation and selection operators used in the proposed approach are explained in detail.

Crossover Operator: As a crossover operator, *weight mapping crossover*, which is similar to two-point crossover for binary strings and a remapping by the order of different binary strings, is used. The reader may refer to Chapter 1 for more information.

Mutation Operator: As a mutation operator, *insertion mutation*, in which one position is selected at random and inserts it in a random position, is used. Figure 7.10 illustrates the procedure of this mutation operator.

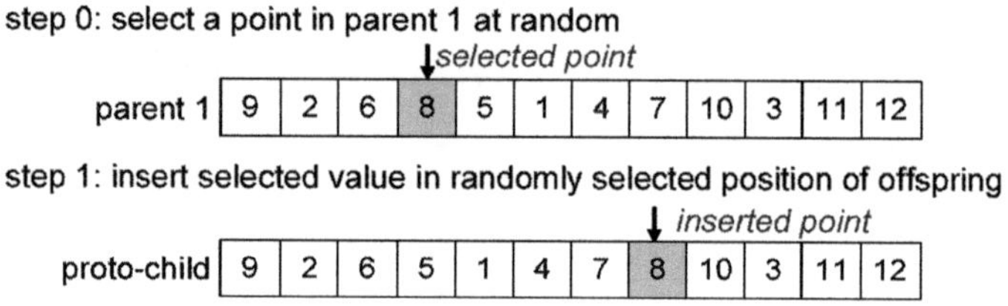

Fig. 7.10 Insertion mutation

Selection Operator: As a selection operator, the *roulette wheel selection*, which is a type of fitness-proportional selection, is adopted. It is used to reproduce a new generation proportional to the fitness of each individual. In this procedure, the solutions are placed on a roulette wheel where the section of the wheel for a better solution is larger than for a worse solution. The procedure of roulette wheel selection is shown in Fig. 7.11.

7.2.3 Computational Experiments and Discussions

Several well-know sALB test problems by Talbot *et al.* [16] and Hoffmann [53] are solved using the priGA approach. The values of parameters used in the computational experiments are as follows:

Population size: *popSize* = 100

procedure: roulette wheel selection
input: population $P(t\text{-}1)$, $C(t\text{-}1)$
output: population $P(t)$, $C(t)$
begin
step 0: calculate the total fitness of the population $F = \sum_{k=1}^{popSize} eval(v_k)$;
step 1: calculate selection probability p_k for each chromosome v_k
$p_k = \frac{eval(v_k)}{F}, \quad k = 1, 2..., popSize$
step 2: calculate cumulative probability q_k for each chromosome v_k
$q_k = \sum_{j=1}^{k} p_j, \quad k = 1, 2..., popSize$
step 3: calculate a random number r from the range [0, 1]
step 4: if $r \le q_1$, then select the first chromosome v_1; otherwise select the kth chromosome v_k ($2 \le k \le popSize$) such that $q_{k\text{-}1} < r < q_k$
output population $P(t)$, $C(t)$
end

Fig. 7.11 The roulette wheel selection procedure

Maximum number of generation: $maxGen = 1000$
Probability of crossover: $p_C = 0.7$
Probability of mutation: $p_M = 0.3$
Terminating condition: If the same fitness value is found for 100 generations, the algorithm is terminated.

The results of computational experiments are presented in Table 7.3. Two kinds of fitness functions are used, *i.e.*, Fitness Function (E): efficiency; Fitness Function (EV): efficiency + variation of workload. These results were compared with the optimal solutions reported in the literature.

As shown in Table 7.3, the performance of priGA is good when solving the test problems. And, we find better performance with a variation of workload. The better variation of workload means the better line balancing. The priGA algorithm confirms that our encoding and decoding procedures are efficient and can find an optimal solution rapidly. From the computational experiments, it is confirmed that the priGA can provide workable solutions, and also find good results rapidly.

7.3 U-shaped Assembly Line Balancing Model

Due to very different conditions in real manufacturing environments, assembly line systems show a great diversity. Particularly, ALs can be distinguished with regard to the line layout such as *serial* and *u-shaped lines* (see Fig. 7.12) [1, 21, 55].

Traditionally, an assembly line is organized as a serial line, where stations are arranged along a conveyor belt serially. Such serial lines are rather inflexible and have other disadvantages which might be overcome by a u-shaped assembly line. In

Table 7.3 The results of the computational experiments [54]

Problem	Number of tasks	Cycle time	Optimal number of stations	Fitness function (E)			Multiobjective function (EV)		
				Number of stations	E	V	Number of stations	E	V
Mitchell	21	14	8	8	93.8%	0.067	8	93.8%	0.032
		21	5	5	100.0%	0.000	5	100.0%	0.000
		35	3	3	100.0%	0.000	3	100.0%	0.000
Heskiaoff	28	138	8	8	92.8%	0.054	8	92.8%	0.009
		256	4	4	100.0%	0.000	4	100.0%	0.000
		342	3	3	99.8%	0.001	3	99.8%	0.001
Sawyer	30	30	12	12	90.0%	0.021	12	90.0%	0.011
		54	7	7	85.7%	0.035	7	85.7%	0.021
		75	5	5	86.4%	0.028	5	86.4%	0.006
Tonge	70	364	19	19	96.4%	0.009	19	96.4%	0.004
		410	9	9	95.1%	0.011	9	95.1%	0.004
		468	8	8	93.8%	0.014	8	93.8%	0.002
		527	7	7	95.1%	0.009	7	95.1%	0.002
Arcus2	111	8847	18	18	94.4%	0.026	18	94.4%	0.014
		10027	16	16	93.7%	0.033	16	93.7%	0.020
		10743	15	15	93.3%	0.035	15	93.3%	0.021
		11378	14	14	94.4%	0.024	14	94.4%	0.011
		17067	9	9	97.9%	0.006	9	97.9%	0.001

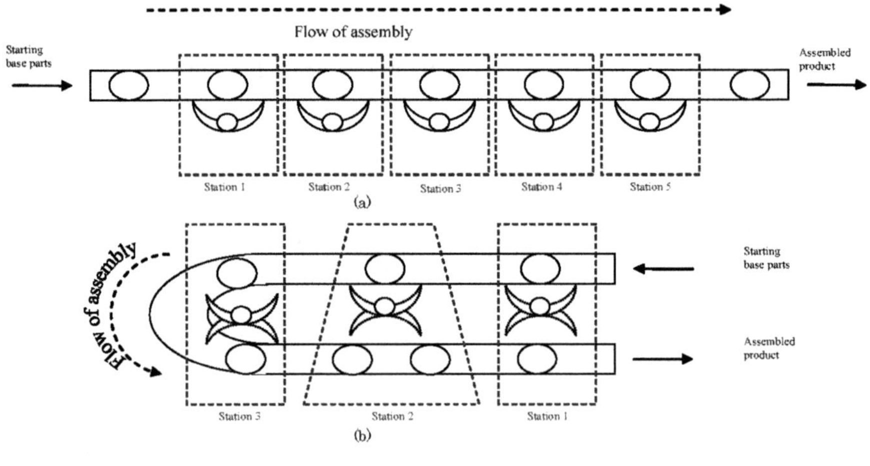

Fig. 7.12 Line layouts: **a** serial; **b** u-shaped lines

modern production lines with the implementation of *just-in-time* (JIT) production principles, u-shaped lines are becoming more preferred among other line layouts.

JIT is an "umbrella" term for a number of principles or techniques - typical u-shaped lines, pull production control, and quality management [56]. JIT production principles were introduced by Toyota in Japan as a means for reducing cost by lowering inventory levels; production inefficiencies are identified and targeted for improvement. The main purpose is to improve product quality and cost by eliminating all waste in the production system. While it is in a JIT production environment that the u-shaped line takes its most complete form, it becomes and optimal production principle when the u-shaped line is in place. The u-shaped line compliments the JIT principle by providing more alternatives. Namely, u-shaped lines provide more alternatives for assigning tasks to station (operators) than comparable serial lines because operators can handle not only adjacent tasks, but also tasks on both sides of the u-shaped line. Further, as an additional advantage, the u-shaped line is crowded with work places and less space is needed. Operators work together in u-shaped line and it can make communication easier and operator can trust each other.

In u-shaped lines, the stations are arranged along a rather narrow U, where both legs are close together, and the entrance and the exit of the line are in the same position. Stations in between those legs may work at two segments of the line facing each other simultaneously. This means that the workpieces can revisit the same station at a later stage in the production process without changing the flow direction of the line. This can result in better balance of station loads due to a larger number of *task-station combinations* where operators can handle adjacent tasks as well as tasks on both sides of the u-shaped line. Another advantage of u-shaped lines is that they simultaneously maximize both the use of *operational work space* and *operator communication* and *trust*, such that machines take up less space and workers are closer to one another. Besides improvements with respect to job enrichment and enlargement strategies, a u-shaped line design might result in a better balance of station loads due to the larger number of task-station combinations [57]–[59].

7.3.1 Mathematical Formulation of uALB Models

General features of the *u-shaped* ALB (uALB) model are the same as the sALB model (refer to Section 7.2.1). In addition to the assumptions of the sALB model (refer to Section 7.2), the uALB model can defined by the following distinguishing assumption:

A9. the uALB model allows for the forward and backward assignment of tasks to stations; for example, the first and last tasks of an assembly can be placed in the same station on a u-shaped line, but not on a serial line system.

Every solution feasible for sALB problems is feasible for uALB problems as well, because a u-shape does not need to include crossover stations. However, the optimal solutions on uALB problems may have an *improved line efficiency* compared to the optimal solution on sALB problems due to the increased possibilities of combining tasks to station loads.

For the illustration of the uALB model, the data set in Table 7.4 and precedence graph in Fig. 7.13 is used. Using the precedence graph in Fig. 7.13, a feasible line balance with cycle time 10 time units and 5 stations can be constructed by the station loads $S_1 = 1, 11$, $S_2 = 2, 4, 5$, $S_3 = 6, 7, 9$, $S_4 = 3, 10$, $S_5 = 8$ (see Fig. 7.14). The first station starts executing tasks 1 but gets back every task at the end of the production process to perform the task. While no idle time occurs in stations 1,2,3, and 4, and station 5 show idle times of 4 unit time.

Table 7.4 Data set for a uALB model [9]

Task j	Suc(j)	Task time t_j
1	{2,3,4,5}	6
2	{6}	2
3	{7}	5
4	{7}	7
5	{7}	1
6	{8}	2
7	{9}	3
8	{10}	6
9	{11}	5
10	{11}	5
11	{}	4

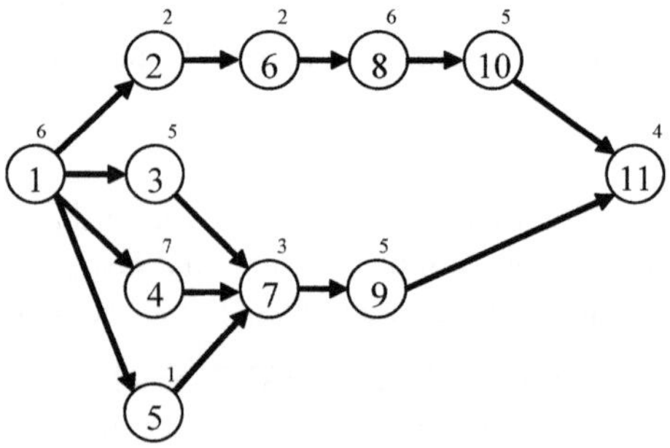

Fig. 7.13 Precedence graph of the uALB model [9]

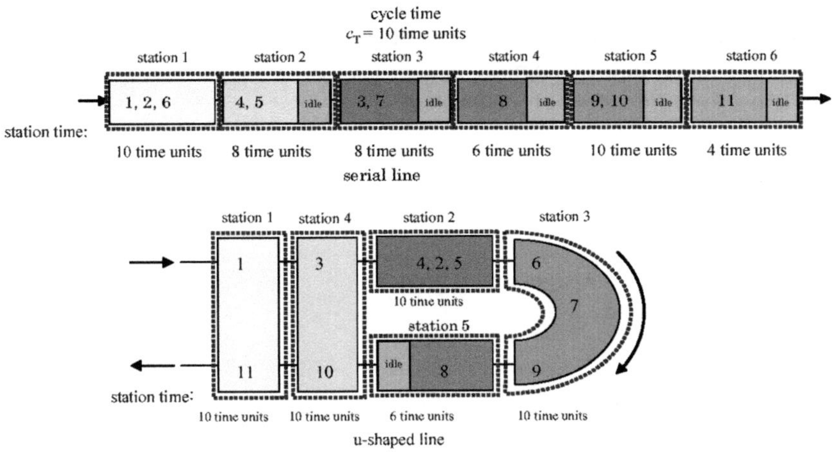

Fig. 7.14 A feasible line balance for the serial and u-shaped line

Notation

In order to formulate the mathematical model, the following indices, parameters and decision variable are introduced:

Indices

i: index of the stations $(i = 1,\ldots,m)$
j,k: index of the tasks $(j,k = 1,\ldots,n)$

Parameters

m: number of stations actually employed
M: maximum number of stations available $(n \leq M)$
n: number of tasks
c_T: cycle time of the assembly line
c_{aver}: average station time
c_{min}: lower bound of cycle time
t_j: processing time of task (j)
S_i: the set of tasks assigned to station i
T_s: a task sequence

Suc(j): the set of direct successors of task j
Pre(j): the set of direct predecessors of task j
u_i: *utilization* of the station i
$\overline{u}$: *average utilization* of total stations

Decision Variables

$$x_{ij} = \begin{cases} 1, & \text{if task } j \text{ is assigned to station } i \\ 0, & \text{otherwise} \end{cases}$$

Mathematical Model

The mathematical model for the uALB Type-1 can be stated as follows:

$$\max\ E = \frac{1}{m c_{\mathrm{T}}} \sum_{j=1}^{n} t_j \tag{7.10}$$

$$\min\ m = \sum_{i=1}^{M} \max_{1 \le j \le n} \{x_{ij}\} \tag{7.11}$$

$$\min\ V = \sqrt{\frac{1}{m} \sum_{i=1}^{m} (u_j - \overline{u})^2} \tag{7.12}$$

$$\text{s.t.}\ \sum_{i=1}^{M} x_{ij} = 1,\ \forall\ j \tag{7.13}$$

$$\max\left\{\max\{t_j\},\ c_{\mathrm{aver}}\right\} \le c_{\min},\ \forall\ j \tag{7.14}$$

$$t(S_i) = \sum_{j \in S_i} t_j = \sum_{j=1}^{n} t_j\, x_{ij} \le c_{\mathrm{T}},\ \forall\ i \tag{7.15}$$

$$x_{ij} = 0 \text{ or } 1,\ \forall\ i,\ j \tag{7.16}$$

In this mathematical model, the first objective (Eq. 7.10) of the model is to maximize the line efficiency. The second objective (Eq. 7.11) of the model is to minimize the number of stations. The third objective (Eq. 7.12) of the model is to minimize the variation of workload. The constraints given in each equation are used to formulate the general feasibility of the problem. Equation 7.13 states that each task must be assign to one and only one station. Inequality at Eq. 7.14 denotes lower bound of the cycle time, and inequality at Eq. 7.15 represents that the assembly time on each station must be no more than the cycle time. Constraint given in Eq. 7.16 represents the usual integrity restriction.

7.3.2 Priority-based GA for uALB Models

In this section, we will introduce the priGA for solving the multi-objective uALB model that was proposed by Hwang *et al.* [60]. The priGA approach was originally developed by Gen and Cheng [23] in order to handle the problem of creating encoding while treating the precedence constraints efficiently.

Let $P(t)$ and $C(t)$ be parents and offspring in current generation t, respectively. The overall procedure of priGA for solving uALB model is outlined as follows:

---

procedure: priGA for uALB
input: problem data, GA parameters
output: the best solution
begin
  $t \leftarrow 0$;
  initialize $P(t)$ by *priority-based encoding routine*;
  evaluate $P(t)$ by *priority-based decoding routine*;
  while (**not** terminating condition) **do**
    create $C(t)$ from $P(t)$ by *weight mapping crossover routine*;
    create $C(t)$ from $P(t)$ by *insertion mutation routine*;
    evaluate $C(t)$ by *priority-based decoding routine*;
    select $P(t+1)$ from $P(t)$ and $C(t)$ by roulette wheel selection routine;
    $t \leftarrow t+1$;
  end
  output the best solution
end

The detailed information about the genetic representation, evaluation function, and genetic operators are given in the following subsections.

7.3.2.1 Genetic Representation

For the representation of uALB models, the priority-based encoding method was used. To develop a priority-based genetic representation for the uALB model, there are three main phases:

Phase 1: Creating a task sequence

- Step 1.1: Generate a random priority to each task in the model using encoding procedure.
- Step 1.2: Decode a feasible task sequence T_S that satisfies the precedence constraints.

Phase 2: Assigning tasks to stations

Step 2.1: Input the task sequence found in step 1.2.
Step 2.2: Obtain a feasible solution set according to this task sequence.

Phase 3: Designing a schedule

Step 3.1: Create a schedule S using station assignments found in step 2.2.
Step 3.2: Draw a Gantt chart for this schedule.

While introducing these three phases, in order to illustrate the procedures, we will be using the data set given in Table 7.4 and its corresponding precedence graph given in Fig. 7.2.

Phase 1: Creating a Task Sequence

Step 1.1: Generate a random priority to each task in the model using encoding procedure. This step is the same as in sALB model.
Step 1.2: Decode a feasible task sequence T_S that satisfies the precedence constraints.

In order to decode the chromosomes generated in step 1.1, the priorities of each task are used to create a feasible task sequence that satisfies the precedence constraints in the model. Figure 7.15 presents the priority-based decoding procedure for creating a task sequence.

---

procedure: priority-based decoding (creating task sequence)
input: number of tasks n, chromosome $v(j)$
output: a task sequence T_S
begin

$\bar{s} \leftarrow \varnothing,\ T_S \leftarrow \varnothing$;
$n \leftarrow 0,\ j \leftarrow 0$;
while ($j \leq n$) **do**
$\bar{s} \leftarrow \mathrm{Suc}(j)$;
$\bar{s} \leftarrow \mathrm{Prec}(j)$;
$j^* \leftarrow \arg\max\{v(j) | j \in \bar{s}\}$;
$\bar{s} \leftarrow \bar{s} \setminus j^*$;
$T_S \leftarrow T_S \cup j^*$;
$j \leftarrow j^*$;
end
output a task sequence T_S

end

Fig. 7.15 Priority-based decoding procedure for creating task sequence

In order to show the application of this decoding procedure, the precedence graph in Fig. 7.13 is used. First, we try to find a 6 node for the first position. Nodes 1 and 11 are eligible for the first position, which can be suitable for the start node. Here we check their priority that is 7 and 3 respectively. Next, the task 1 has the highest priority of 7 and is placed into the task sequence T_S. Afterwards, the next possible nodes are 2, 3, 4, 5 and 11. They have 2, 10, 4, 5 and 3 priority respectively, and then we put 3 into task sequence T_S. Finally, we repeat these steps until we obtain a complete task sequence $T_S = 1, 3, 5, 4, 7, 9, 11, 2, 6, 8, 10$.

Phase 2: Assigning Tasks to Stations

Step 2.1: Input the task sequence found in step 1.2.
Step 2.2: Obtain a feasible solution set according to this task sequence.

In this phase, the assignments of task to stations are formed using the task sequence found in step 1.2. Figure 7.16 presents the task to station assignment decoding procedure.

For the illustration of this decoding procedure, the task sequence $T_S = 1, 3, 5, 4, 7, 9, 11, 2, 6, 8, 10$ found in step 1.2 is used. Table 7.5 presents the trace table for the decoding procedure. In the example, we obtained a feasible line balance with cycle time 7 time units and 7 stations given by the station loads $S_1 = 1, 5$, $S_2 = 2, 3$, $S_3 = 4$, $S_4 = 7, 11$, $S_5 = 6, 9$, $S_6 = 8$ $S_7 = 10$. While no idle time occurs in stations 1, 2, 3, 4 and 5, stations 6 and 7 show idle times of 1, and 2 time units, respectively.

Table 7.5 Trace table for task to station assignment decoding procedure

j	$\bar{s}$	$v(j)$	j^* (t_j)	S_i={}; $t(S_i)$ $(c_T - t(S_i))$
0	{1,11}	v(1)=7, v(11)=3	1(6)	S_1={1}; 6 (1)
11	{2,3,4,5,11}	v(2)=2, v(3)=10, v(4)=4, v(4)=46, v(11)=3	3(5)	S_2={3}; 5 (2)
1	{2,4,5,11}	v(2)=2, v(4)=4, v(5)=5, v(11)=3	5(1)	S_1={1,5}; 7 (0)
4	{2,4,11}	v(2)=2, v(4)=4, v(11)=3	4(7)	S_3={4}; 7 (0)
2	{2,7,11}	v(2)=2, v(7)=9, v(11)=3	7(3)	S_4={7}; 3 (4)
5	{2,9,11}	v(2)=2, v(9)=6, v(11)=3	9(5)	S_5={9}; 5 (2)
6	{2,11}	v(2)=2, v(11)=3	11(4)	S_4={7,11}; 7 (0)
9	{2,10}	v(2)=2, v(10)=1	2(2)	S_2={3,2}; 7 (0)
7	{6,10}	v(6)=8, v(10)=1	6(2)	S_5={9,6}; 7 (0)
10	{8,10}	v(8)=11,v(10)=1	8(6)	S_6={8}; 6 (1)
3	{10}	v(10)=1	10(5)	S_7={10}; 5 (2)

```
procedure: priority-based decoding (task to station assignment)
input: processing time t_j, chromosome v(j), the task sequence T_S
output: number of stations m, efficiency E, variation V, efficiency and variation EV
begin
    t_j ← 0, j=1, 2, ..., n;
    S_i ← 0, i=1, 2, ..., m;
    j ← 0, m ← 0, E ← 0, V ← 0, EV ← 0;
    for j = 1 to n
      while (t(S_i) ≤ c_T) do
              T_Sj ← T_S;
              S_i ← T_Sj;
              t(S_i) ← t(S_i) + t_j;
              T_S ← T_S \ T_Sj;
              m ← 1++;
              t_SUM ← T_Sj;
              E ← t_SUM / (m c_T);
              u_T ← u_i ++;
              ū ← u_T / m;
      end
    end
    for i = 1 to m
       W ← (u_i − ū)^2;
       V ← sqr(W / m);
       EV ← E + (1 − V);
    end
    output number of stations m, efficiency E, variation V, efficiency and variation EV
end
```

Fig. 7.16 Task to station assignment decoding procedure

Phase 3: Designing a Schedule

Step 3.1: Create a schedule using station assignments found in step 2.2

Using the trace table for the decoding procedure, the schedule can be constructed as follows:

Schedule $S = \{j, S_i, t_j)\}$:

$$S=\{(1,\ S_1,\ t_1),(5,\ S_1,\ t_5),(3,\ S_2,\ t_3),(2,\ S_2,\ t_2),(4,\ S_4,\ t_4),(11,\ S_4,\ t_{11}),(9,\ S_5,\ t_9),(6,\ S_5,\ t_6),(8,\ S_6,\ t_8),(10,\ S_7,\ t_{10})\}$$
$$=\{(1,\ S_1,\ 0-6),(5,\ S_1,\ 6-7),(3,\ S_2,\ 7-12),(2,\ S_2,\ 12-14),(4,\ S_4,\ 14-21),(11,\ S_4, 21-25),(9,\ S_5,\ 25-30),(6,\ S_5,\ 30-32),(8,\ S_6,\ 32-38),(10,\ S_7,\ 38-43)\}$$

Step 3.2: Draw a Gantt chart for this schedule. Figure 7.17 illustrates a Gantt chart showing a feasible schedule for one unit of product.

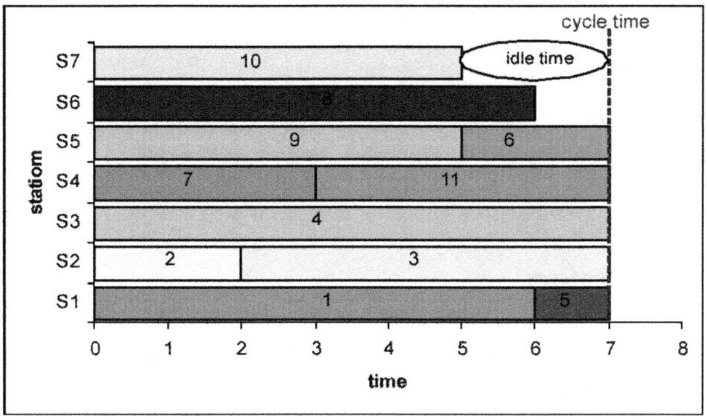

Fig. 7.17 Gantt chart for the uALB model (one unit of product)

However, since a single type of product is produced on a simple AL, the minimization of the makespan becomes an important production management issue. This time, Fig. 7.18 shows a Gantt chart for three units of product. Here, the makespan is 70 time units for 3 units of products.

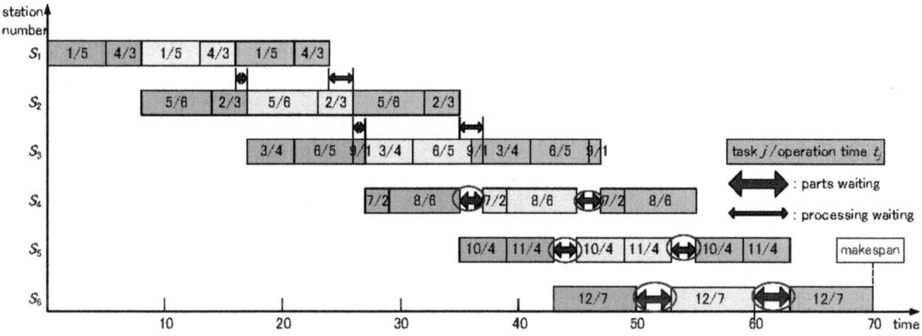

Fig. 7.18 Gantt chart for the uALB model (three units of product)

7.3.2.2 Evaluation Function

In this study, two objective functions, *i.e.*, maximization of line efficiency and minimization of variation of workload, are considered simultaneously. For the evaluation function, adaptive weight approach that utilizes some useful information from the current population to readjust weights for obtaining a search pressure toward a positive ideal point is used. For more information, please refer to Chapter 1.

$$eval(v_k) = w_1(f_1(v_k) - z_1^{\min}) + w_2(f_2(v_k) - z_2^{\min})$$
$$= w_1 \left\{ \frac{1}{m c_T} \sum_{j=1}^{n} t_j - z_1^{\min} \right\} + w_2 \left\{ 1 - \sqrt{\frac{1}{m} \sum_{i=1}^{m} (u_i - \bar{u})^2} - z_2^{\min} \right\},$$
$$k = 1, \cdots, popSize \quad (7.17)$$

7.3.2.3 Genetic Operators

In this subsection, the crossover, mutation and selection operators used in the proposed approach are explained in detail.
Crossover Operator: As a crossover operator, *weight mapping crossover*, which is similar to two-point crossover for binary strings and a remapping by the order of different binary strings, is used. The reader may refer to Chapter 1 for more information about the operator.
Mutation Operator: As a mutation operator, *swap mutation*, in which two positions are selected at random and their contents are swapped is used. Figure 7.19 illustrates the procedure of this mutation operator.

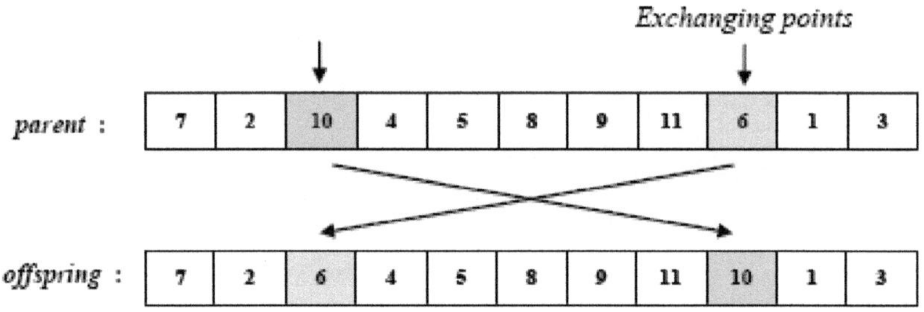

Fig. 7.19 Swap mutation

Selection Operator: As a selection operator, the *roulette wheel selection*, which is a type of fitness-proportional selection, is adopted. It is used to reproduce a new generation proportional to the fitness of each individual. In this procedure, the solutions are placed on a roulette wheel where the section of the wheel for a better solution is larger than for a worse solution. It is called the roulette wheel selection because the selection technique of the parent selection is that each individual is given a chance to become a parent in proportion to its fitness evaluation; the best chances of selecting a parent can be produced by a spinning roulette wheel with the size of its slots for each parent being proportional to their fitness. Obviously those with the largest fitness (and slot sizes) have more of a chance to be chosen, which is, needless to say, analogous to the notion of the survival of the fittest in natural evolution. Figure 7.11 presents the procedure of this selection operator.

7.3.3 Computational Experiments and Discussions

Several well-known test problems by Talbot *et al.* [16] are solved using the priGA. Both of the solutions for two line types–a serial line and a u-shaped line–are demonstrated. The values of parameters used in the computational experiments are as follows:

Population size: *popSize* = 100
Maximum number of generation: *maxGen* = 1000
Probability of crossover: $p_C = 0.7$
Probability of mutation: $p_M = 0.3$
Terminating condition: If the same fitness value is found for 100 generations, the algorithm is terminated

The results of computational experiments are presented in Table 7.6 for serial line and Table 7.7 for u-shaped line. By using line efficiency and the variation of workload fitness functions, the effectiveness of multi-objective functions is evaluated. Multi-objective (*EV*) is compared with the single objective (*E*). Additionally, the results of the proposed approach are compared with results of *integer programming* (IP) and *maximum ranked positional weight* (MRPW) methods.

As shown in Tables 7.6 and 7.7, the priGA performs test data thought out well. The bold characters indicate the results of improvement. In cases of small size problems, we obtained small improvement of line balance, but in many case of large size problems, the variation of workload improves. In particular, the u-shaped line performed line balancing (variation of workload) well because of the increased possibilities of combining tasks to station loads. In view of the results so far achieved, the priGA approach improved the variation of workload positively. Thus, the more confident and better variation of workload means an improved line balance has taken place. From the result it was confirmed that the genetic algorithm can provide workable solutions whereas the optimal uALB solution has an improved line efficiency compared to the optimal sALB solution in some cases. Again, this is due to the increased possibilities of combining tasks to station loads.

The average CPU time (20 times iterations) for the proposed approach is shown in Fig. 7.20. As it is seen from the figure, the solution time for uALB using the proposed approach requires longer execution time than serial AL. Since the execution time depends on the number of generations and population size, this can be attributed to the increased number of generations. Although to solve uALB with the proposed approach takes more CPU time, the CPU time requirement of the proposed approach is in reasonable range.

7.4 Robotic Assembly Line Balancing Model

In the past decades, robots have been extensively used in assembly lines called *robotic assembly lines* (rALs). An assembly robot can work 24 hours a day without worries or fatigue. Goals for implementation of robotic assembly lines include

Table 7.6 The results for test problems with serial line [60]

Problem	Number of tasks	Cycle time	Serial line						
			IP solutions	Proposed GA					
				Fitness function (E)			Multiobjective function (EV)		
			Number of stations	Number of stations	E	V	Number of stations	E	V
Mitchell	21	14	8	8	93.7%	0.057	8	93.7%	0.042
		15	8	8	87.5%	0.090	8	87.5%	0.090
		21	5	5	100.0%	0.000	5	100.0%	0.000
Heskiaoff	28	138	8	8	92.7%	0.054	8	92.7%	0.022
		205	5	5	99.9%	0.001	5	99.9%	0.001
		324	4	4	79.9%	0.329	4	79.0%	0.153
Sawyer	30	27	13	13	92.3%	0.107	13	92.3%	0.035
		33	11	11	89.2%	0.109	11	89.2%	0.046
		54	7	7	85.7%	0.233	7	85.7%	0.030
Kilbridge	45	79	7	7	99.8%	0.004	7	99.8%	0.004
		92	6	6	100.0%	0.000	6	100.0%	0.000
		184	3	3	100.0%	0.000	3	100.0%	0.000
Tonge	70	176	21	21	95.0%	0.052	21	95.0%	0.031
		364	10	10	96.4%	0.053	10	96.4%	0.039
		468	8	8	93.8%	0.129	8	93.8%	0.014
Arcus1	83	5853	13	13	92.4%	0.092	13	92.4%	0.019
		6842	12	12	92.2%	0.129	12	92.2%	0.043
		8412	10	10	90.0%	0.160	10	90.0%	0.038
		10816	8	8	87.5%	0.211	8	87.5%	0.160
Arcus2	111	5755	27	27	96.8%	0.049	27	96.8%	0.044
		10027	16	16	93.7%	0.093	16	93.7%	0.030
		10743	15	15	93.3%	0.123	15	93.3%	0.035
		17067	9	9	97.9%	0.033	9	97.9%	0.006

high productivity, quality of product, manufacturing flexibility, safety, decreasing demand of skilled labor, and so on. Different robot types may exist at the assembly facility. Each robot type may have different capabilities and efficiencies for various elements of the assembly tasks. Usually, specific tooling is developed to perform the activities needed at each station. Such tooling is attached to the robot at the station. In order to avoid the time waste required for tool change, the design of the tooling can take place only after the line has been balanced. Hence, allocating the best fitting robot for each station is critical for the performance of rALs.

Unlike manual assembly lines, where actual processing times for performing tasks vary considerably and optimal balance is rather of theoretical importance, the performance of rALs depends strictly on the quality of its balance. As extended from sALB, *robotic assembly line balancing* (rALB) is also NP-hard.

Table 7.7 The results for test problems with u-shape line [60]

			u-shaped line							
Problem	Number of tasks	Cycle time	IP solutions	MRPW	Proposed GA					
					Fitness function (*E*)			Multiobjective function (*EV*)		
			Number of stations	Number of stations	Number of stations	*E*	*V*	Number of stations	*E*	*V*
Mitchell	21	14	8	8	8	93.7%	0.055	8	93.7%	0.023
		15	8	8	8	87.5%	0.139	8	87.5%	0.023
		21	5	6	5	100.0%	0.000	5	100.0%	0.000
Heskiaoff	28	138	8	8	8	92.7%	0.112	8	92.7%	0.007
		205	5	6	5	99.9%	0.001	5	99.9%	0.001
		324	4	4	4	79.9%	0.332	4	79.0%	0.062
Sawyer	30	27	13	14	13	92.3%	0.071	13	92.3%	0.023
		33	10	11	10	98.1%	0.024	10	98.1%	0.014
		54	6	7	6	100.0%	0.000	6	100.0%	0.000
Kilbridge	45	79	7	7	7	99.8%	0.004	7	99.8%	0.004
		92	6	6	6	100.0%	0.000	6	100.0%	0.000
		184	3	3	3	100.0%	0.000	3	100.0%	0.000
Tonge	70	176	21	21	21	95.0%	0.043	21	95.0%	0.029
		364	10	10	10	96.4%	0.023	10	96.4%	0.009
		468	8	8	8	93.8%	0.063	8	93.8%	0.004
Arcus1	83	5853	13	14	13	92.4%	0.061	13	92.4%	0.014
		6842	12	12	12	92.2%	0.097	12	92.2%	0.039
		8412	10	10	10	90.0%	0.123	10	90.0%	0.038
		10816	8	8	8	87.5%	0.200	8	87.5%	0.089
Arcus2	111	5755	27	27	27	96.8%	0.036	27	96.8%	0.039
		10027	16	16	16	93.7%	0.074	16	93.7%	0.029
		10743	15	15	15	93.3%	0.063	15	93.3%	0.031
		17067	9	9	9	97.9%	0.019	9	97.9%	0.004

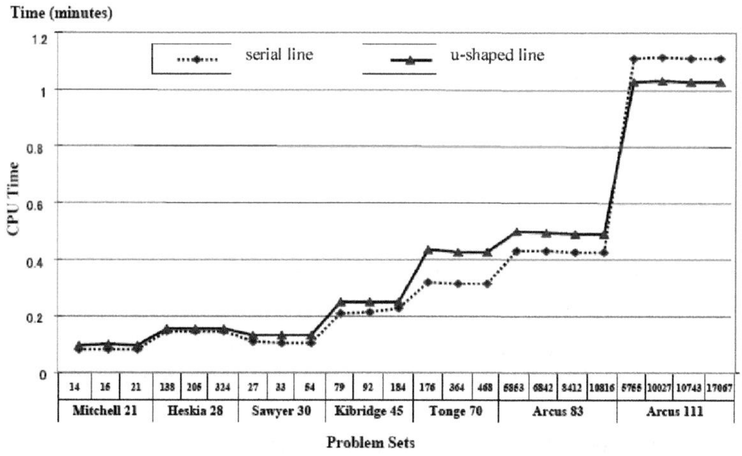

Fig. 7.20 CPU times required for solving serial and u-shaped ALB models

Rubinovitz and Bukchin [61] were the first to formulate the rALB model as one of allocating equal amounts of work to the stations on the line while assigning the most efficient robot type from the given set of available robots to each workstation. Their objective is to minimize the number of workstations for a given cycle time. Following this, those authors [62] presented a branch and bound algorithm for the problem. Bukchin and Tzur [63] treat the problem with the objective to minimize the total equipment cost, given a predetermined cycle time, where they developed an exact branch and bound algorithm; meanwhile a branch-and-bound-based heuristic procedure is suggested for large problems. Tsai and Yao [64] proposed a heuristic approach for the design of a flexible robotic assembly line which produces a family of products. Kim and Park [65] extended the problem by considering additional constraints, *i.e.*, due to limited space to store the parts and tools, restrictions for the joint assignment of tasks to stations are imposed. They proposed a mathematical formulation and a cutting plane procedure for this extension of the problem.

Khouja *et al.* [66] suggested statistical clustering procedures to design robotic assembly cells. Nicosia *et al.* [67] considered the problem of assigning operations to an ordered sequence of non-identical workstations under the constraints of precedence relationships and a given cycle time. The objective is to minimize the cost of the workstations. This formulation is very similar to the rALB problem.

The aforementioned rALB works have assumed that the cycle time is predetermined, and aimed at minimizing the number of workstations or the cost of the assembly systems. Hence, these works are of the rALB Type-1 model. Levitin *et al.* [50] dealt with an rALB Type-2 model, in which different robots may be assigned to the assembly line tasks, and each robot needs different assembly times to perform a given task due to its capabilities and specialization. The objective is to maximizing the production rate of the line. Two genetic algorithms are presented to solve the rALB Type-2 model.

Since the number of stations is determined by the number of robots in an rAL, in this section we will consider the rALB Type-2 model. This model is usually present when changes in the production process of a product take place. For example, a new product is introduced for assembly. In this case, the rAL has to be reconfigured using the present resources (such as robots) so as to improve its efficiency for the new production process. The model concerns how to assign the tasks to stations and how to allocate the available robots for each station in order to minimize the cycle time under the constraint of precedence relationships. In this case, the number of stations of the line and the available robots may remain fixed. The following assumptions are stated to clarify the setting in which the rALB model arises:

- A1. The precedence relations among assembly tasks are known and constant.
- A2. The processing times of tasks are deterministic and dependent on the assigned robot.
- A3. A task cannot be split among two or more stations.
- A4. There are no limitations on assignment of task or robots to any station besides the precedence constraints. In case a task can‘t be processed on a robot, the processing time of the task on the robot is set to very high.

A5. A single robot is assigned to each station.
A6. Material handling, loading and unloading times, as well as set-up and tool changing times are negligible, or are included in the activity times. This assumption is realistic on an sALB that works on the single product for which it is balanced. Tooling on such as rAL is usually designed such that tool changes are minimized within a station. If tool change or other type of set-up activity is necessary, it can be included in the activity time, since the transfer lot size on such a line is of a single product.
A7. The number of stations is determined by the number of robots, since the problem aims to maximize the productivity by using all robots at hand.
A8. The line is balanced for a single product.

7.4.1 Mathematical Formulation of rALB Models

The rALB Type-2 model focuses on the assignment of tasks to stations and to allocate a robot for each station with the objective of minimum cycle time given the number of stations as the available robots.

Table 7.8 presents the data set for an example rALB model, which contains 10 tasks, 4 robots assigned to 4 stations and the processing time of each task processed by each robot. Using this data set, the precedence graph in Fig. 7.21 is constructed. The precedence graph contains 10 nodes for tasks and arcs for orderings. For example, the processing time for task 7 by robot 3 is 23 time units. For the processing of task 8, tasks 6 and 7 (direct predecessors) and task 4 (indirect predecessor) must be completed. Likewise, task 8 must be completed before task 10 (direct successor) starts processing.

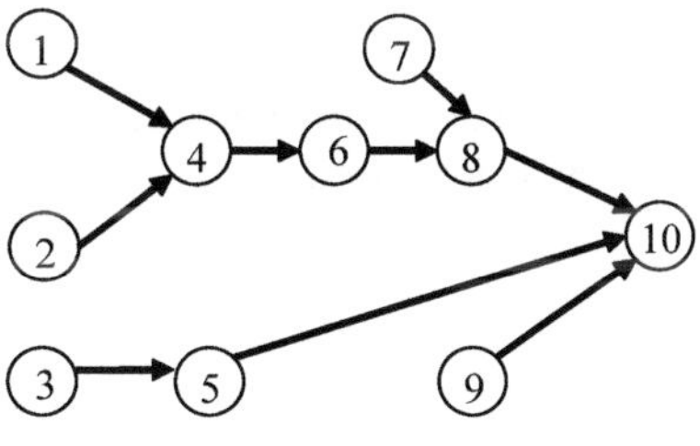

Fig. 7.21 Precedence graph of the rALB model

Using the precedence graph in Fig. 7.21, a feasible line balance with cycle time 85 time units and 4 stations can be constructed by the station loads S_1 = 2, 7, S_2 = 3, 5, S_3 = 1, 4, 6, S_4 = 9, 8, 10 (see Fig. 7.22). While no idle time occurs in station 4, stations 1, 2, and 3, 4 have very long idle times, which mean that the line is not effectively balanced.

Table 7.8 Data set of the rALB model

		Task time by robot l t_{il}			
Task i	Suc(i)	R_1	R_2	R_3	R_4
1	{4}	17	22	19	13
2	{4}	21	22	16	20
3	{5}	12	25	27	15
4	{6}	29	21	19	16
5	{10}	31	25	26	22
6	{8}	28	18	20	21
7	{8}	42	28	23	34
8	{10}	27	33	40	25
9	{10}	19	13	17	34
10	{}	26	27	35	26

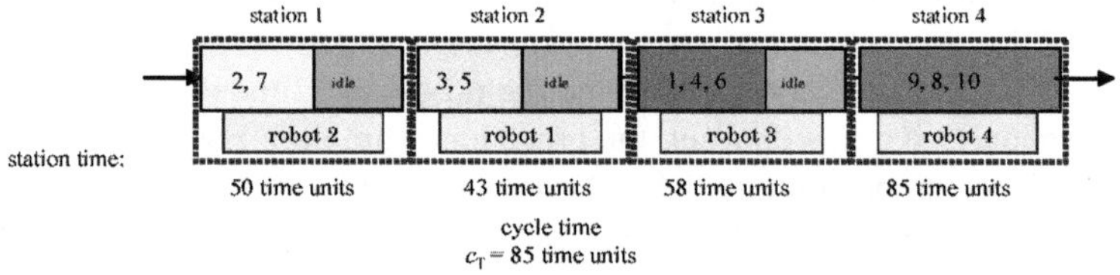

Fig. 7.22 A feasible line balance for a rALB model

Notation

In order to formulate the mathematical model, the following indices, parameters and decision variable are introduced:

Indices

i, j: index of the tasks $(i, j = 1, \ldots, n)$
k: index of the stations $(k = 1, \ldots, m)$
l: index of the robots$(l = 1, \ldots, m)$

Parameters

m: number of stations (robots)
n: number of tasks
t_{il}: processing time of task (i) by robot l
Pre(i): the set of predecessors of task i in the precedence diagram
Suc(i): the set of direct successors of task i

Decision Variables

$$x_{ik} = \begin{cases} 1, & \text{if task } j \text{ is assigned to station } k \\ 0, & \text{otherwise} \end{cases}$$

$$y_{lk} = \begin{cases} 1, & \text{if robot } l \text{ is assigned to station } k \\ 0, & \text{otherwise} \end{cases}$$

Mathematical Model

The mathematical model for the rALB Type-2 can be stated as follows:

$$\min\ c_{\mathrm{T}} = \max_{1 \le k \le m} \left\{ \sum_{i=1}^{n} \sum_{l=1}^{m} t_{il}\, x_{ik}\, y_{kl} \right\} \tag{7.18}$$

$$\text{s.t.} \quad \sum_{k=1}^{m} k x_{jk} - \sum_{k=1}^{m} k x_{ik} \ge 0, \quad \forall\ j,\ i \in \mathrm{Pre}(i) \tag{7.19}$$

$$\sum_{k=1}^{m} x_{ik} = 1, \quad \forall\ i \tag{7.20}$$

$$\sum_{l=1}^{m} y_{kl} = 1, \quad \forall\ k \tag{7.21}$$

$$\sum_{k=1}^{m} y_{kl} = 1, \quad \forall\ l \tag{7.22}$$

$$x_{ik} \in \{0,1\}, \quad \forall\ k,\ i \tag{7.23}$$

$$y_{kl} \in \{0,1\}, \quad \forall\ l,\ k \tag{7.24}$$

In this mathematical model, the first objective (**??**) is to minimize the cycle time (c_{T}). The constraints given in Eqs. 7.19–7.24 are used to formulate the general feasibility of the problem. Inequality at Eq. 7.19 represents the precedence constraints. It ensures that for each pair of assembly activities, the precedent cannot be assigned to a station behind the station of the successor, if there is precedence between the two activities. Equation 7.20 ensures that each task has to be assigned to one station.

Equation 7.21 ensures that each station is equipped with one robot. Equation 7.22 ensures that each robot can only be assigned to one station. Constraints given in Eqs. 7.23 and 7.24 represent the usual integrity restriction.

7.4.2 Hybrid GA for rALB Models

In this section, we will introduce a hybrid genetic algorithm (hGA) to solve the rALB Type-2 model. Let $P(t)$ and $C(t)$ be parents and offspring in current generation t, respectively. The overall procedure of hGA for solving rALB model is outlined as follows:

---

procedure: hybrid GA for rALB models
input: problem data, GA parameters
output: the best solution
begin
  $t \leftarrow 0$;
  initialize $P(t)$ by *task sequence and robot assignment encoding routine*;
  evaluate $P(t)$ by *breakpoint decoding routine*;
  while (**not** terminating condition) **do**
    create $C(t)$ from $P(t)$ by *mixed crossover routine*;
    create $C(t)$ from $P(t)$ by *allele-based mutation and immigration routine*;
    climb $C(t)$ from $P(t)$ by *robot assignment routine*;
    evaluate $C(t)$ by *breakpoint decoding routine*;
    select $P(t+1)$ from $P(t)$ and $C(t)$ by *mixed sampling selection routine*;
    $t \leftarrow t+1$;
  end
  output the best solution
end

The hGA uses a partial representation technique, that is, the coding space only contains a part of the feasible solutions, in which the optimal solution is included. New crossover and mutation operators are also developed to adapt to the specific chromosome structure and the nature of the problem. In order to strengthen the search ability, a *local search procedure*, which only investigates the neighbor solutions that have possibilities to improve the initial solution, is proposed under the framework of the GA.

The detailed information about the genetic representation, evaluation function, genetic operators, and local search procedure are given in the following subsections.

7.4.2.1 Genetic Representation

To develop a genetic representation for the rALB model, there are three main phases:

Phase 1: Creating a task sequence

Step 1.1: Order encoding for task sequence by randomly generating a list of tasks.
Step 1.2: Reordering the tasks to a feasible task sequence that satisfies the precedence constraints.
Step 1.3: Breakpoint decoding to assign the tasks into each station by breakpoint decoding.

Phase 2: Assigning robots to each station

Step 2.1: Order encoding for robot assignment by randomly assign the robots to each station.
Step 2.2: Breakpoint decoding to assign the robots into each station.

Phase 3: Designing a schedule

Step 3.1: Creating a schedule for the assembly line.
Step 3.2: Drawing a Gantt chart for this schedule.

7.4.2.2 Phase 1: Creating a Task Sequence

Step 1.1: Order encoding for task sequence by randomly generating a list of tasks.

A solution of the rALB model can be represented by two integer vectors, *i.e.*, v_1 and v_2. Task sequence vector (v_1), which contains a permutation of assembly tasks, ordered according to their technological precedence sequence, and robot assignment vector (v_2). The solution representation method can be visually illustrated as in Fig. 7.23.

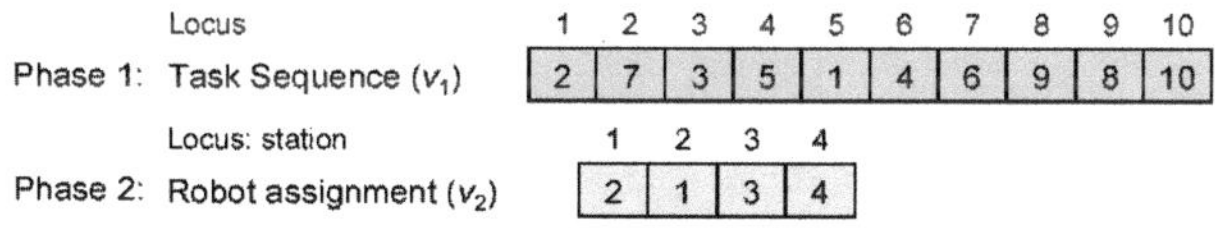

Fig. 7.23 Genetic representation of a rALB model

Step 1.2: Reordering the tasks to a feasible task sequence T_S that satisfies the precedence constraints.

Considering the precedence constraints, a task sequence may be infeasible since a precedent may appear before its successors. The reordering procedure will repair

procedure: reordering
input: set of predecessors for task *i* Pre(*i*), chromosome *v*(*i*)
output: reordered chromosome *v* '(*i*)
begin

$v'(i) \leftarrow \varnothing;$

$A \leftarrow \{i | \text{Pre}(i) = \varnothing\};$

while (|A| > 0) **do**

$i' \leftarrow \arg\min\{v(i) | v(i) \in A\};$

$v'(i) \leftarrow v'(i) \cup \{v(i')\};$

$A \leftarrow \{i | i \notin v'(i) \text{ and } \text{Pre}(i) \subseteq v'(i)\};$

end

output reordered chromosome *v* '(*i*)

end

Fig. 7.24 Reordering procedure

it into a feasible one before it is divided m parts to form a solution in the phase 1. The reordering procedure is shown in Fig. 7.24.

Step 1.3: Breakpoint decoding to assign the tasks into each station by breakpoint decoding.

The last decoding procedure in phase 1 is used to generate a feasible solution based on the task sequence and robot assignment schemes which are contained in the chromosomes. The breakpoint decoding procedure inserts m points along the reordered task sequence vector to divide it into m parts, each of which corresponds to a station. The breakpoint decoding procedure consists of four main steps:

A1. Calculate the lower bound of the cycle time (c_{LB}) for the solution represented by the task sequence vector and robot assignment vector:

$$c_{\text{LB}} = \frac{1}{m} \sum_{i=1}^{n} \min_{1 \le l \le m} \{t_{il}\} \tag{7.25}$$

A2. Find out a feasible cycle time as the upper bound of the cycle time (c_{UB});
A3. Find out the optimal cycle time *via* bisection method;
A4. Partition the task sequence into m parts with the optimal cycle time based on the robot assignment vector.

Figure 7.25 illustrates the process of breakpoint decoding procedure on a chromosome. Here, a cycle time is said to be feasible if all the tasks can be allocated to the stations by allowing as many tasks as possible for each station under the constraint of the cycle time. The procedure to calculate the upper bound of cycle time is illustrated in Fig. 7.26, and the bisection method to find out the optimal cycle time is shown in Fig. 7.27.

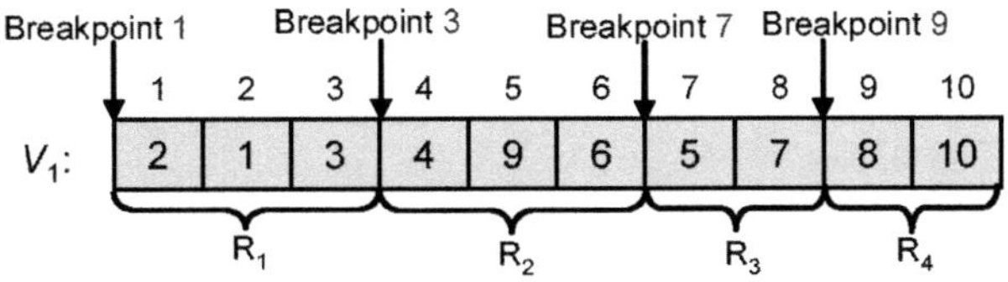

Fig. 7.25 Breakpoint decoding

```
procedure: calculating upper bound of the cycle time
input: m, n, v', a, c_LB
output: upper bound for cycle time c_UB
begin
    c_UB ← c_LB;   // set upper bound of the cycle time the same as lower bound
    repeat
        c_UB ← 2*c_UB;      // multiply the upper bound by "2"
        i ← 0;
        for l=1 to m do
            // assign as many tasks as possible to each station under the cycle time
            t ← 0;
            while (t < c_UB) and (i<n) do
                i ← i +1;
                t ← t + t_v'(i),a(l);     // add assembly time of the task to station time
            end;
            if t > c_T then
                i ← i -1;
        end
    until i ≥n
    output upper bound for cycle time c_UB
end
```

Fig. 7.26 Procedure for calculating the upper bound of cycle time

After calculating the optimal cycle time, it is easy to generate the breakpoints on the task sequence to divide it into m parts, each of which will correspond to a station, based on the robot assignment vector. Therefore, the tasks assigned for each station are determined according to a chromosome, and the cycle time can be calculated. Figure 7.28 presents the procedure for breakpoint decoding.

```
procedure: bisection method to calculate optimal cycle time
input: m, n, v1', v2, cUB
output: optimal cycle time cT
begin
    repeat
        c'LB←cUB/2;
        cT←(c'LB+ cUB)/2;
        i←0;
        for l=1 to m do
            // assign as many tasks as possible to each station under the cycle time
            T←0;
            while (t < cT) and (i<n) do
                i←i+1;
                T←T + tv'(i),a(l); // add assembly time of the task to station time
            end
            if t > cT then
                i ←i -1;
        end
        if i ≥n then
            cUB←cT; // update the lower and upper bounds
        else
            c'LB←cT;
    until cUB -cLB≤ 1; //until difference between upper and lower bounds is less than "1";
    cT←cUB;
    output optimal cycle time cT
end
```

Fig. 7.27 Bisection method to calculate the optimal cycle time

Phase 2: Assigning Robots to Each Station

Step 2.1: Order encoding for robot assignment by randomly assign the robots to each station.

In this phase, we code the robot assignment vector (v_2), which indicates the robot assigned for each station. The solution representation method can be visually illustrated as in Fig. 7.25. The coding space takes an exponential growth with the length of the chromosome; therefore, even one allele saved in the chromosome can decrease the coding space significantly. An obvious advantage to omit intentionally the breakpoint vector, which is used in recently research, in the chromosome is that the coding space is dramatically decreased. As a result, the speed to find the global optimal is accelerated.

Step 2.2: Breakpoint decoding to assign the robots into each station.

In this phase, this robot assignment vector (v_2) will decode into a solution with the task sequence vector (v_1), simultaneously by the breakpoint decoding procedure. It means when the tasks are assigned into each workstation, the robots will be

```
procedure: breakpoint decoding
input: m, n, v'_1, v_2, c_T
output: breakpoint vector v_3
begin
    i ← 0;
    for l to m do
    //assign as many tasks as possible to each station under the cycle time
     t ← 0;
    while (t ≤ c_T) and (I < n) do
          i ← i + 1;
          t ← t + t(v'_1(i), v_2(l));
    end
    if (t > c_T) then
          i ← i − 1;
          v_3(l) ← i;
    end
    output breakpoint vector v_3
end
```

Fig. 7.28 Procedure for breakpoint decoding

assigned into each workstation at the same time. The illustration of the solution is shown in Fig. 7.23.

Phase 3: Designing a Schedule

Step 3.1: Creating a schedule for the assembly line. Using the chromosomes, the schedule can be constructed as follows:

Schedule $S = \{j, R_i, t_j)\}$:

S=$\{(t_1,\ R_1,\ 0-17), (t_2,\ R_1,\ 17-38),\ (t_3,\ R_1,\ 38-50),$
$(t_4,\ R_2,\ 50-71), (t_5,\ R_3,\ 102-128), (t_6,\ R_2,\ 71-89),$
$(t_7,\ R_3,\ 128-151),\ (t_8,\ R_4,\ 151-176), t_9,\ R_2,\ 89-102),$
$(t_{10},\ R_4,\ 176-202)\}$

Step 3.2: Drawing a Gantt chart for this schedule

In the real-world, the assembly line is not just for producing one unit of the product. It should produce several units. Figures 7.29 and 7.30 show the Gantt charts for one unit and three units of product, respectively.

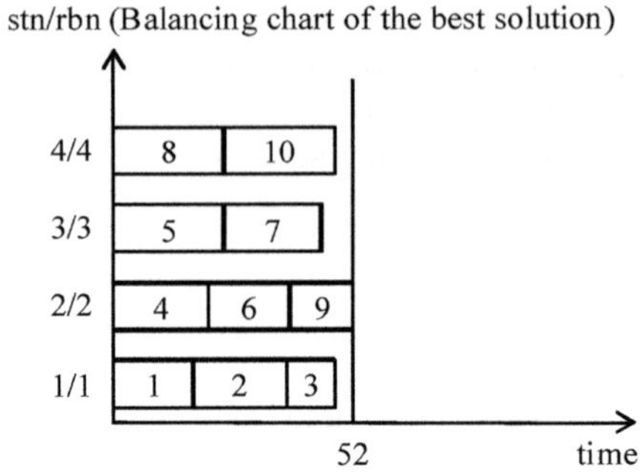

Fig. 7.29 The balance chart of the best solution

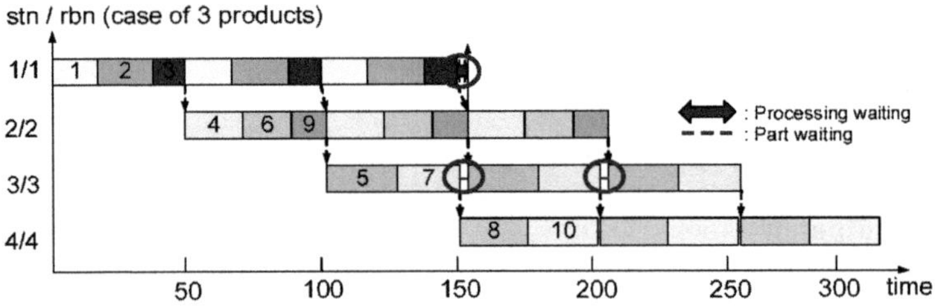

Fig. 7.30 Gantt chart for the rALB model (three units of product)

7.4.2.3 Evaluation Function

In a rAL, the station whose total assembly time is equal to the cycle time is called a critical station in this study (Fig. 7.31). Any task in the critical station is called a critical task.

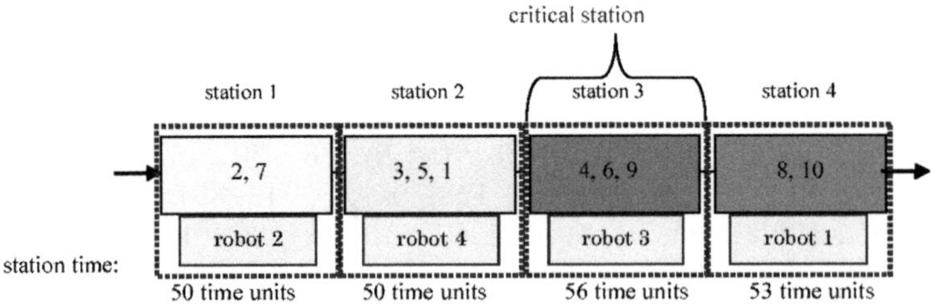

Fig. 7.31 Illustration of the critical station

In a solution candidate of rALB, the number of critical stations may be more than one. For the rALB model, a preferable solution will have a small cycle time. Because the cycle time is determined by the assembly time on the critical station, a small number of critical stations may facilitate the GA to find solutions with a shorter cycle time. In this study, when two individuals with the same cycle time are compared, the individual with less critical stations is given a larger fitness value than the one with more critical stations. And we also consider the workload of each station by standard deviation:

$$\begin{aligned} eval(v_k) &= w_1(f_1(v_k) - z_1^{\min}) + w_2(f_2(v_k) - z_2^{\min}) + w_3(f_3(v_k) - z_3^{\min}) \\ &= w_1(M\mathrm{c} - z_1^{\min}) + w_2(n_c - z_2^{\min}) + w_3 \left\{ \frac{1}{m-1} \sum_{k=1}^{m} ((c_\mathrm{T} - t_k^\mathrm{W}) - \bar{t})^2 - z_3^{\min} \right\}, \\ &\qquad k = 1, \cdots, popSize \end{aligned} \tag{7.26}$$

where

M is a very large number,
c is the cycle time,
n_c is the number of critical stations,
t_k^W is the total processing time of each station,

$$\bar{t} = \frac{1}{m} \sum_{k=1}^{m} (c_\mathrm{T} - t_k^\mathrm{W}) \tag{7.27}$$

We use the standard deviation by the idle time of each station for balancing the workload of each station.

7.4.2.4 Genetic Operators

In this study, two kinds of crossover methods, *i.e.*, exchange crossover and mixed crossover are used.

Crossover Operator

1. *Exchange crossover.* A good schedule could be expected by exchanging task sequence and robot assignment schemes between a pair of parents. This kind of crossover is accomplished by selecting two parents and exchanging the task sequence vectors of the two parents to generate offspring.
2. *Mixed crossover.* The mixed order crossover consists of two crossover methods, *i.e.*, order crossover (OX) and partial-mapped crossover (PMX). For the rALB chromosome, the task sequence vector and the robot assignment vector are both of the style of permutation representation. Yet, the useful features of the task sequence vector and the robot assignment vector are different.

3. The network does not contain parallel arcs (*i.e.*, two or more arcs with the same tail and head nodes). This assumption is essentially for the notational convenience:

   - For the task sequence vector, the useful feature of permutation representation to represent a rALB solution is the order information among the tasks.
   - For the robot assignment vector, the acting feature is the number at each allele which indicates the robot no. assigned for a specific station.

For the task sequence, order crossover is implemented to create the offspring task sequence vector by combining the structure of the parents. The steps of OX operator are as follows:

Step 1. Select a subsection of task sequence from one parent at random.
Step 2. Produce a proto-child by copying the substring of task sequence into the corresponding positions.
Step 3. Delete the tasks that are already in the substring from the second parent. The resulted sequence of operations contains tasks that the proto-child needs.
Step 4. Place the tasks into the unfixed positions of the proto-child from left to right according to the order of the sequence in the second parent.

The procedure is illustrated in Fig. 7.32. With the same steps, we can produce the second offspring as [13, 9, 1, 10, 5, 14, 2, 15, 11, 6, 12, 7, 3, 4, 16, 8] from the same parents.

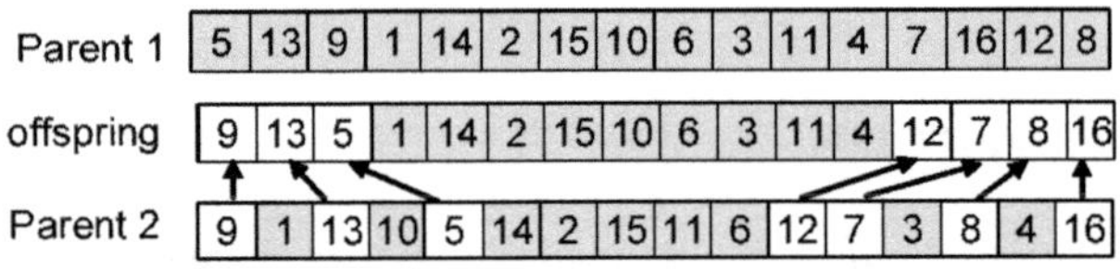

Fig. 7.32 Illustration of the OX operator

Partial-mapped crossover is used to crossover robot assignment vectors, due to its high heritability in passing the number at each allele of the parent vectors to the offsprings. The steps of PMX operator are as follows:

Step 1. Select two positions along robot assignment vector uniformly at random. The substrings defined by the two positions are called the mapping sections.
Step 2. Exchange two substrings between parents to produce proto-children.
Step 3. Determine the mapping relationship between two mapping sections.
Step 4. Legalize offspring with the mapping relationship.

The procedure is illustrated in Fig. 7.33. It shows that robots 7, 12, 14 are duplicated while robots 3, 4, 10 are missed in the proto-child 1. According to the mapping relationship determined in step 3, the excessive robots are replaced by the missed robots, while keeping the swapped substrings unchanged.

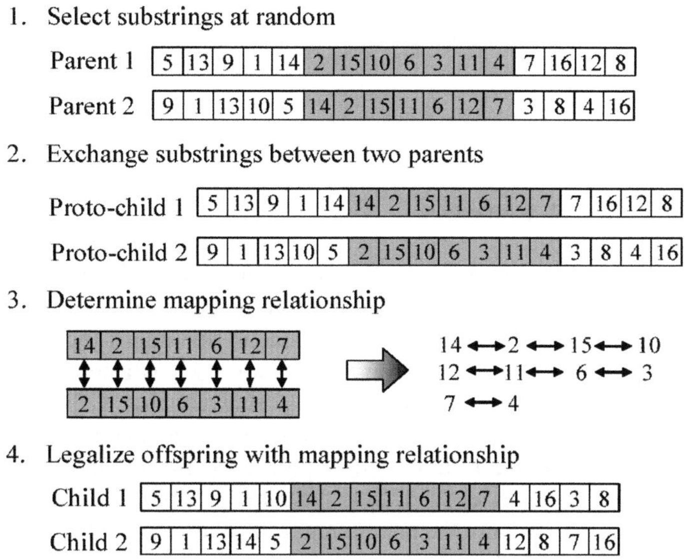

Fig. 7.33 Illustration of the PMX operator

Mutation Operator

In this study, two kinds of mutation operators are implemented, *i.e.*, allele-based mutation and immigration mutation. For both the robot task sequence and robot assignment vectors, allele-based mutation randomly decides whether an allele should be mutated in a certain probability. Then, another position will be generated at random to perform exchange values with the mutated allele.

In contrast to the canonical gene-by-gene mutation with very small probability at each generation, immigration mutation randomly generates one or more new members of the population from the same distribution as the initial population. This process prevents premature convergence of the population, and leads to a simple statement of convergence. Figure 7.34 illustrates the procedure of the mutation, which is used in this proposed algorithm.

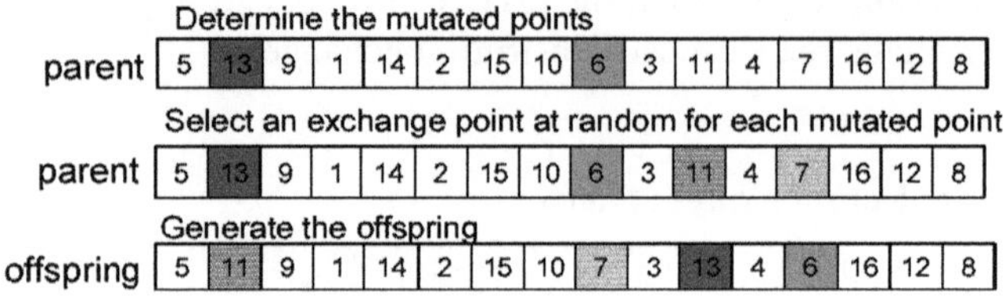

Fig. 7.34 Illustration of gene-by-gene Mutation

7.4.2.5 Local Search Procedure

Two individuals with high fitness values are likely to have dissimilar machine assignment and operation sequences, and the recombination may result in offsprings of poor performance. This means that the genetic operations by themselves have limited ability in finding the global optimal solution. Many researchers have found that the convergence speed of simple GAs is relatively slow [23]. One promising approach for improving the convergence speed to the global optimum is the use of local search in genetic algorithms [68]. Within a hybrid of genetic algorithm and local search, a local optimizer is added to the genetic algorithm and applied to every child before it is inserted into the population [69]. Within the hybrid approach, GAs are used to perform global exploration among populations, while heuristic methods are used to perform local exploitation around chromosomes. Because of the complementary properties of GAs and traditional heuristics, the hybrid approach often outperforms either method operating alone.

In this study, a local search method is proposed to enhance the search ability of GA. The local search method, *i.e.*, *Robot Assignment Search*, is based on critical stations in order to improve their effectiveness and efficiency.

Robot Assignment Search: The neighborhood can be defined as the set of solutions obtainable from an initial solution by some specified perturbation. A robot assignment neighbor solution is generated by exchanging the robot assigned on the critical station with another robot. The robot assignment search is the local search which works over robot assignment neighborhood.

Let $N(i)$ denote the set of machine assignment neighborhood of solution i. The enlarged two-pace machine assignment neighborhood is defined as the union of the neighborhood of each robot assignment neighbor of solution i (see Fig. 7.35).

$$N^2(i) = \cup_{j \in N(i)} N(j)$$

During the robot assignment search, the local search will implement over two-pace neighborhood when it reaches the local optima of one-pace neighborhood, and is called two-pace robot assignment search. During the robot assignment search, when the local optima of two-pace robot assignment neighborhood is reached, the neighbors of the two-pace local optima help the robot assignment search escape

$$N^2(i) = \cup_{j \in N(i)} N(j)$$

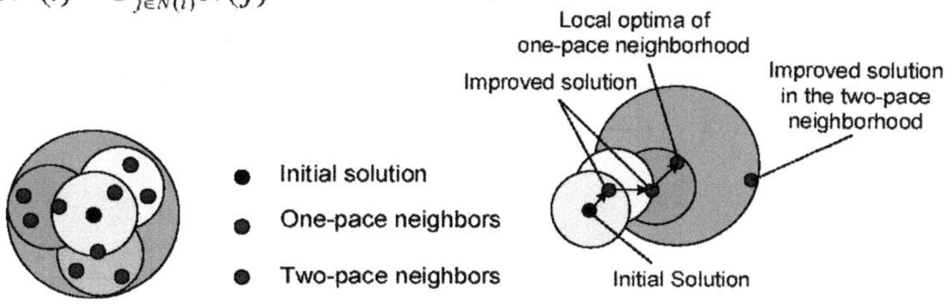

Fig. 7.35 Illustration of the neighborhood

from the local optima. Figure 7.36 presents the overall procedure of the local search used in this proposed approach.

procedure: overall local search
input: initial solution S_0, local search parameters
output: improved solution S
begin
    $S \leftarrow S_0$
    yield S by breakpoint search from S_0;
    $S_0 \leftarrow S$
    yield S by task sequence search from S_0;
    output improved solution S
end

Fig. 7.36 Overall procedure of robot assignment search

7.4.3 Computational Experiments and Discussions

Since there is no benchmark data set available for rALB in literature, eight representative precedence graphs, which are widely used in the sALB Type-1 models [70], are used. These precedence graphs contain 25–297 tasks. From each precedence graph, four different rALB Type-2 models are generated by using different WEST ratios: 3, 5, 7, 10, 15. WEST ratio, as defined by Dar-El [11], measures the average number of activities per station. This measure indicates the expected quality of achievable solutions and complexity of the problem. For each problem, the number of station is equal to the number of robots, and each task can be processed on any robot. The task time data are generated at random, while two statistical dependences are maintained:

- Statistical dependence of task times on the task type,
- Statistical dependence of task times on the robot on which the task is processed.

The computational experiments are performed on a Pentium 4 processor (2.6-GHz clock). The values of parameters used in the computational experiments are as follows:

Population size: $popSize = 100$
Maximum number of generation: $maxGen = 1000$
Probability of exchange crossover: $p_{C1} = 0.2$
Probability of mixed crossover: $p_{C2} = 0.8$
Probability of task sequence mutation: $p_{M1} = 0.05$
Probability of robot assignment mutation: $p_{M2} = 0.10$
Probability of immigration: $p_I = 0.20$

Terminating condition: If the same fitness value is found for 100 generations, the algorithm is terminated.

To evaluate the performance of the hGA, these 32 test instances are solved using the hGA approach. Additionally, the two algorithms proposed by Levitin *et al.* [50] are also used to solve the 32 problems. To solve the rALB, Levitin *et al.* [50] proposed two algorithms named as the recursive assignment method and the consecutive assignment method. If we look at each robot as one type and delete this type from the total types of robots when the robot is allocated for a station, these two algorithms can be adapted to solve the 32 rALB Type-2 models here. Levitin *et al.* [50] have assumed that all types of robots are available without limitations, while we assume that only the robots at hands can be used. Our assumption is reasonable since the AL has to be redesigned under the constraint of resources that currently exist. Table 7.9 presents the performance of the hGA approach for rALB Type-2 model.

From the result, it can be states that the hGA approach performs better than Levitin *et al.*'s two algorithms along with increasing the scale for the problems. The computational experiments show that this algorithm is computationally efficient and effective to find the optimal solution. Additionally, Fig. 7.37 illustrates the evolutionary process of the three algorithms on problem 148-21.

For 25 tasks, 6 robots and 6 stations problem, the precedence graph and data of the test is shown in Table 7.10 and Fig. 7.38.

The best solution found for this example problem is illustrated in Fig. 7.39. Moreover, the balancing chart for this solution is illustrated in Fig. 7.40, which shows that the idle time of each station is very small.

The schedule generated for this solution is as follows:

Schedule $S = \{j, S_i, t_j)\}$:

S=$\{(t_1,\ R_1,\ 0-44), (t_2,\ R_5,\ 44-97), (t_3,\ R_5,\ 97-158),$
$(t_4,\ R_5,\ 158-213), (t_5,\ R_3,\ 213-241), (t_6,\ R_3,\ 241-292),$
$(t_7,\ R_3,\ 292-336), (t_8,\ R_3,\ 336-369), (t_9,\ R_1,\ 419-479),$
$(t_{10},\ R_2,\ 836-903), (t_{11},\ R_1,\ 479-530), (t_{12},\ R_3,\ 369-419),$
$(t_{13},\ R_6,\ 625-674), (t_{14},\ R_6,\ 674-779), (t_{15},\ R_1,\ 530-625),$
$(t_{16},\ R_2,\ 903-951), (t_{17},\ R_2,\ 951-979), (t_{18},\ R_2,\ 979-1008),$
$(t_{19},\ R_4,\ 1044-1081), (t_{20},\ R_6,\ 779-836), (t_{21},\ R_4,\ 1081-1160),$

Table 7.9 The results of the computational experiments [71, 72]

The Problem			Cycle time (C_T)		
No. of tasks	No. of stations	WEST ratio	Levitin *et al.*'s Recursive	Levitin *et al.*'s Consecutive	Hybrid GA approach
25	3	8.33	518	503	503
	4	6.25	351	330	327
	6	4.17	343	234	213
	9	2.78	138	125	123
35	4	8.75	551	450	449
	5	7.00	385	352	344
	7	5.00	250	222	222
	12	2.92	178	120	113
53	5	10.60	903	565	554
	7	7.57	390	342	320
	10	5.30	35	251	230
	14	3.79	243	166	162
70	7	10.00	546	490	449
	10	7.00	313	287	272
	14	5.00	231	213	204
	19	3.68	198	167	154
89	8	11.13	638	505	494
	12	7.42	455	371	370
	16	5.56	292	246	236
	21	4.24	277	209	205
111	9	12.33	695	586	557
	13	8.54	401	339	319
	17	6.53	322	257	257
	22	5.05	265	209	192
148	10	14.80	708	638	600
	14	10.57	537	441	427
	21	7.05	404	325	300
	29	5.10	249	210	202
297	19	15.63	1129	674	646
	29	10.24	571	444	430
	38	7.82	442	348	344
	50	5.94	363	275	256

$(t_{22},\ R_4,\ 1160-1189), (t_{23},\ R_2,\ 1008-1044), (t_{24},\ R_2,\ 1189-1210),$
$(t_{25},\ R_4,\ 1210-1245)\}$

Finally, the Gantt chart for this solution for three units of the products is shown in Fig. 7.41.

The performance of the hGA method was validated through simulation experiments. The simulation showed that this algorithm is computationally efficient and effective to find the optimal solution. The solution obtained by this algorithm outperformed the results from previous work.

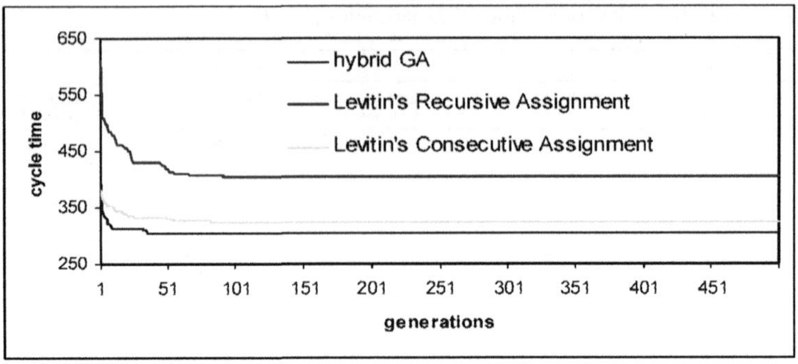

Fig. 7.37 Evolutionary process of problem 148-21 [71, 72]

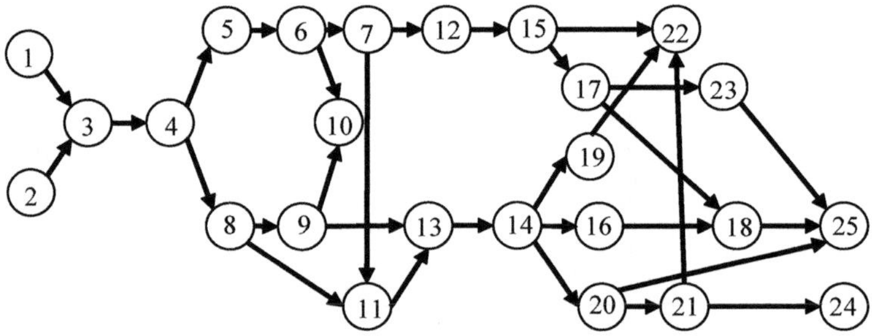

Fig. 7.38 Precedence graph of the rALB model

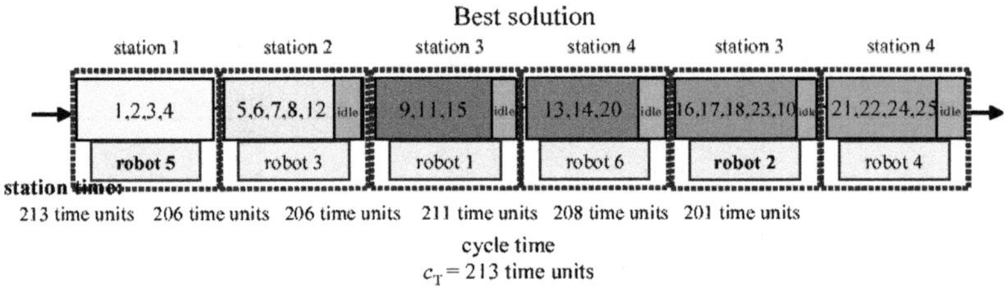

Fig. 7.39 Illustration of the best solution

7.5 Mixed-model Assembly Line Balancing Model

Recent market trends show that there is a growing demand for customized products, increasing the pressure for manufacturing flexibility. With these increasing and varied demands of customers, manufacturers are forced to produce different models of the same product on an assembly line so that the potential market is not lost. In such environments, *mixed-model assembly lines* (mALs) appear to be the most

Table 7.10 Data set of the rALB model (25 tasks, 6 robots, 6 stations)

		Processing time of tasks by each robot					
i	Suc(*i*)	R1	R2	R3	R4	R5	R6
1	3	87	62	42	60	44	76
2	3	67	47	42	45	53	100
3	4	82	58	54	40	61	60
4	5, 8	182	58	62	60	55	100
5	6	71	47	28	57	62	76
6	7, 10	139	48	51	73	61	117
7	11, 12	98	99	44	49	59	82
8	9, 11	70	40	33	29	36	52
9	10, 13	60	114	47	72	63	93
10	-	112	67	85	63	49	86
11	13	51	35	41	44	85	69
12	15	79	39	50	80	67	95
13	14	57	47	56	85	41	49
14	16, 19, 20	139	65	40	38	87	105
15	17, 22	95	63	42	65	61	167
16	18	54	48	51	34	71	133
17	18, 23	71	28	35	29	32	41
18	25	112	29	49	58	84	69
19	22	109	47	38	37	52	69
20	21, 25	63	45	39	43	36	57
21	22, 24	75	68	45	79	84	83
22	-	87	36	74	29	82	109
23	25	58	36	55	38	42	107
24	-	44	54	23	21	36	71
25	-	79	64	48	35	48	97

appropriate. In a mAL, a set of similar models of a product, *e.g.*, versions of a car, which may differ from each other with respect to size, colour *etc.*, can be assembled simultaneously. Nowadays, they are often adopted in the automobile industry, *e.g.* the German car manufacturer BMW offers a catalog of optional features which, theoretically, results in 10^{32} different models [73], and similar industries in order to avoid unnecessary inventories and increase manufacturing flexibility to respond to the ever changing demands of the customers.

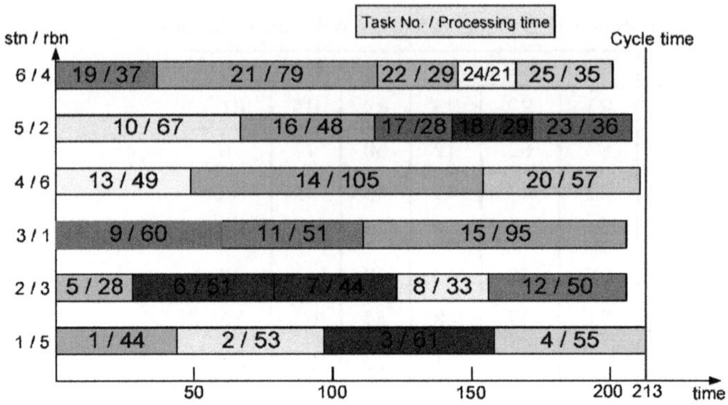

Fig. 7.40 The balancing chart of the best solution

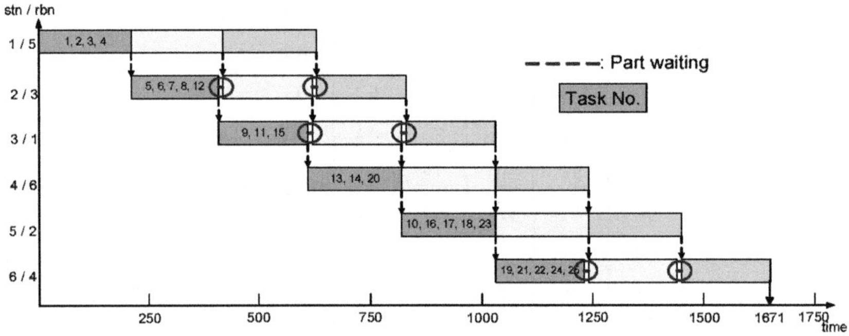

Fig. 7.41 Gantt chart for producing three units

In a market environment characterized by a growing trend for higher product variability, mALs are steadily replacing the traditional single-model assembly lines, as they allow for different versions of a product to be assembled in the same assembly line. Moreover, mALs are becoming popular in recent years as an integral part of JIT production systems, which require producing only the necessary products in the necessary quantities at the necessary time. Production in small quantities is important because it enables the firm to respond quickly to changes in market conditions while simultaneously reducing inventories of specific product models. This approach is feasible when the different models can be assembled without a significant changeover delay between them.

Most research addresses two key problems in the design of mALs, *i.e.*, the *mixed-model assembly line balancing* (mALB) which involves the assignment of tasks to stations and mixed-model sequencing which involves the sequencing of models on the line [74]. In this subsection, we will be dealing with the mALB model. mALB arises when designing (or redesigning) a mAL. The mAL is a more complex environment and consists in finding a feasible assignment of the tasks to the stations in such a way that the assembly costs are minimized, the demand of the products to be assembled is met and the constraints of the assembly process are satisfied. This problem is much more complex since it entails the additional considerations of the interactions between the assembled models. The mALB model reflects modern assembly lines more realistically, where demand is characterized by high variability and relatively small volume for each model.

7.5.1 Mathematical Formulation of mALB Models

Generally, the mALB model relies on same basic assumptions of sALB model (refer to Section 7.2.1), *i.e.*, deterministic processing times, no assignment restrictions, serial line layout, fixed rate launching. In addition to the assumptions of the sALB model (refer to Section 7.2), the mixed-model assembly line can be defined by the following assumptions [1]:

A1. The assembly line is capable of producing more than one type of product simultaneously, not in batches.

A2. The assembly of each model requires performing a set of tasks which are connected by precedence relations (*i.e.*, precedence graph for each model).

A3. A subset of tasks is common to all models; the precedence graphs of all models can be combined into a non-cyclical joint precedence graph.

A4. Tasks which are common to several models are performed by the same station but they may have different processing times (*i.e.*, zero processing times indicate that the task is not required for the model).

A5. The total time available for production is fixed and known (given by the number of shifts and the shift durations).

A6. The demands for all models (expected model mix) during the planning period are fixed and known.

Table 7.11 presents the data sets for an example mALB model, which contains nine tasks and two models. Using these data sets, the precedence graphs of each model are constructed as shown in Fig. 7.42. The precedence graphs contain nine nodes for tasks.

Using the precedence graphs in Fig. 7.42, a feasible line balance with cycle time 14 time units and 4 stations can be constructed by the station loads (see Fig. 7.43). To analyze this solution, the balancing chart is drawn in Fig. 7.44, which shows that the idle time of the station 2, 3, 4 for each model is very large, and it also means this line did not get the balancing for producing.

Table 7.11 Data set of the mALB model

	Suc(i) of model j		Task time of model j t_{ij}	
Task i	model I	model II	model I	model II
1	{3}	{3}	6	2
2	{3}	{3}	4	5
3	{5}	{7}	6	4
4	{6}	{8}	4	5
5	{7}	{7}	2	1
6	{8}	{8}	5	4
7	{9}	{9}	5	6
8	{9}	{9}	8	7
9	{}	{0}	4	8

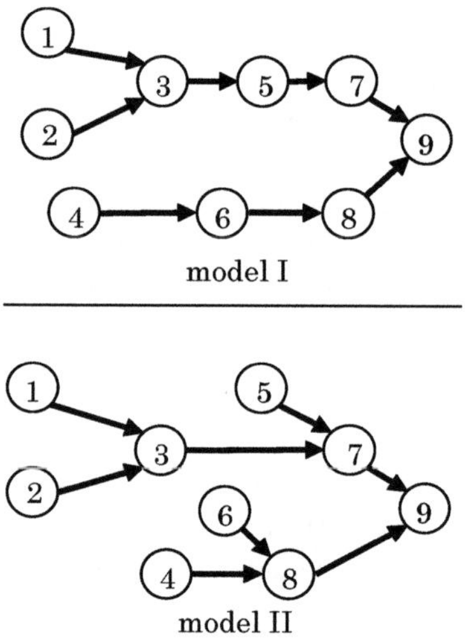

Fig. 7.42 Precedence diagrams for each model

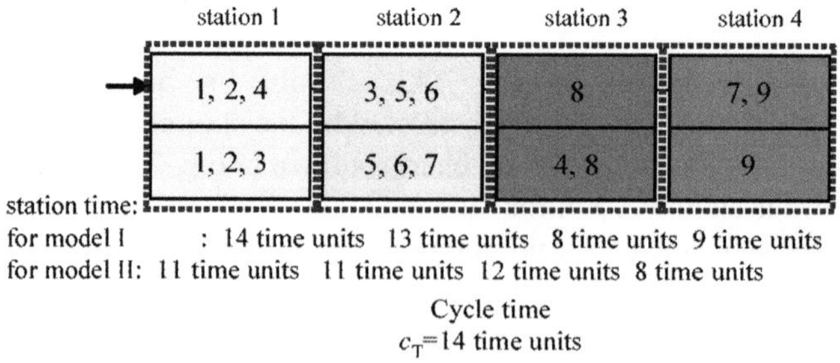

Fig. 7.43 The feasible line balance for mALB model

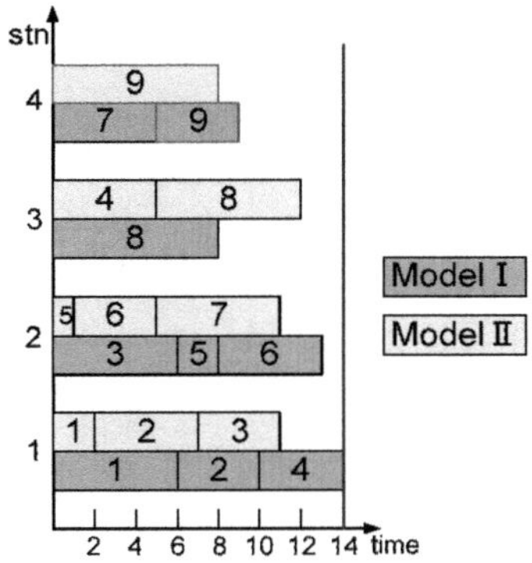

Fig. 7.44 Balancing chart of the feasible solution

Notation

In order to formulate the mathematical model, the following indices, parameters and decision variable are introduced:

Indices

i, g: index of the assembly tasks $(i, j = 1, \dots, n)$
j: index of the models $(j = 1, \dots, m)$
k: index of the stations $(k = 1, \dots, w)$

Parameters

n: number of tasks
m: number of models
w: number of stations
t_{ij}: processing time of task i for model j
c_T: cycle time
c_{UB}: upper bound of cycle time
Pre(i): the set of direct predecessors of task i
Suc(i): the set of direct successors of task i

Decision Variables

$$x_{ijk} = \begin{cases} 1, & \text{if task } i \text{ of model } j \text{ is assigned to station } k \\ 0, & \text{otherwise} \end{cases}$$

Mathematical Model

The mathematical model for the mALB Type-2 can be stated as follows:

$$\min\ c_{\mathrm{T}} = \max_{1\le k\le m}\left\{\sum_{i=1}^{n}\sum_{l=1}^{m} t_{il}\, x_{ik}\, y_{kl}\right\} \tag{7.28}$$

$$\text{s.t.}\quad \sum_{k=1}^{w} k x_{ijk} - \sum_{k=1}^{w} k x_{gik} \ge 0,\quad \forall\ i,\ j,\ g \in \mathrm{Suc}(i) \tag{7.29}$$

$$\sum_{i=1}^{n} t_{ij} x_{ijk} = c_{\mathrm{UB}},\quad \forall\ j, k \tag{7.30}$$

$$\sum_{k=1}^{n} x_{ijk} = 1,\quad \forall i, j \tag{7.31}$$

$$x_{ijk} \in \{0,1\},\quad \forall i,\ j,\ k \tag{7.32}$$

In this mathematical model, the objective (Eq. 7.28) is to minimize the cycle time. The constraints given in Eqs. 7.29–7.32 are used to formulate the general feasibility of the problem. Inequality at Eq. 7.29 represents the precedence constraints. It ensures that for each pair of assembly activities, the precedent cannot be assigned to a station behind the station of the successor, if there is precedence between the two activities. Inequality at Eq. 7.30 restricts the total process time of each model at each station. Equation 7.31 ensures that every task of each model is assigned to exactly one station. Constraint given in Eq. 7.32 represents the usual integrity restriction.

7.5.2 Hybrid GA for mALB Models

In this section, a new representation method adapting the GA to the mALB Type-2 model will be proposed. Advanced genetic operators adapted to the specific chromosome structure and the characteristics of the mALB model will be used. In order to strengthen the search ability, we also combine some local search techniques with GA.

Let $P(t)$ and $C(t)$ be parents and offspring in current generation t, respectively. The overall procedure of hGA for solving mALB model is outlined as follows:

The detailed information about the genetic representation, evaluation function, genetic operators and local search procedure are given in the following subsections.

procedure: hybrid GA for mALB
input: problem data, GA parameters
output: the best solution
begin
  $t \leftarrow 0$;
  initialize $P(t)$ by *sequence encoding routine*;
  evaluate $P(t)$ by *breakpoint decoding routine*;
  while (**not** terminating condition) **do**
    create $C(t)$ from $P(t)$ by *weight mapping crossover routine*;
    create $C(t)$ from $P(t)$ by *swap mutation and immigration routine*;
    climb $C(t)$ from $P(t)$ by *task sequence search routine*;
    evaluate $C(t)$ by *breakpoint decoding routine*;
    select $P(t+1)$ from $P(t)$ and $C(t)$ by *mixed sampling selection routine*;
    $t \leftarrow t+1$;
  end
  output the best solution
end

7.5.2.1 Genetic Representation

To develop a genetic representation for the mALB model, there are three main phases:

Phase 1: Creating a task sequence

- Step 1.1: Order encoding for task sequence T_S by randomly generating a list of tasks.
- Step 1.2: Reordering the tasks to a feasible task sequence that satisfies the precedence constraints.
- Step 1.3: Breakpoint decoding to assign the tasks into each station by breakpoint decoding.

Phase 2: Assigning tasks to each station

- Step 2.1: Order encoding for task to station assignment by randomly assign the task to each station.
- Step 2.2: Breakpoint decoding to assign the tasks into each station.

Phase 3: Designing a schedule

- Step 3.1: Creating a schedule for the assembly line.
- Step 3.2: Drawing a Gantt chart for this schedule.

Phase 1: Creating a Task Sequence

Step 1.1: Order encoding for task sequence by randomly generating a list of tasks.

A solution of the mALB model can be represented by the vectors, in which task sequence vector (v_j) contains a permutation of assembly tasks for each model, ordered according to their technological precedence sequence. The solution representation method can be visually illustrated as in Fig. 7.45.

	Locus	1	2	3	4	5	6	7	8	9
Model I :	Task Sequence (v_1)	2	1	4	3	6	5	8	7	9
	Locus	1	2	3	4	5	6	7	8	9
Model II :	Task Sequence (v_2)	1	2	3	6	5	7	4	8	9

Fig. 7.45 Solution representation of a mALB model

Step 1.2: Reordering the tasks to a feasible task sequence that satisfies the precedence constraints

Considering the precedence constraints, a task sequence may be infeasible since a precedent may appear before its successors. The reordering procedure will repair it into a feasible one before it is divided *m* parts to form a solution. The reordering procedure is shown in Fig. 7.46.

procedure: reordering
input: (n, Pre(k), Suc(k), A, v) for each model
output: reordered chromosome v'
begin

$v' \leftarrow \varnothing$;
while ($|A| > 0$) **do**
$i' \leftarrow \arg\min\{v(i) | v(i) \in A\}$;
$v' \leftarrow v' \cup \{v(i')\}$;
$A \leftarrow A / v(i')$;
$A \leftarrow A \cup \{i | i \in \text{Suc}(v(i')) \text{ and } \text{Pre}(i) \subseteq v'\}$;
end
output reordered chromosome v'

end

Fig. 7.46 Reordering procedure

Step 1.3: Breakpoint decoding to assign the tasks into each station by breakpoint decoding.

The breakpoint decoding procedure inserts m points along the reordered task sequence vector to divide it into m parts, each of which corresponds to a station. The breakpoint decoding procedure consists of four main steps:

A1. Calculate the lower bound of the cycle time (c_{LB}) for the solution represented by the task sequence vector:

$$c_{\mathrm{LB}} = \frac{1}{w} \sum_{i=1}^{n} \min\{t_{ij}\}, \ \forall \ j \tag{7.33}$$

A2. Find a feasible cycle time as the upper bound of the cycle time (c_{UB});
A3. Find the optimal cycle time *via* bisection method;
A4. Partition the task sequence into m parts with the optimal cycle time.

Figure 7.47 illustrates the process of breakpoint decoding procedure on a chromosome. Here, a cycle time is said to be feasible if all the tasks can be allocated to the stations by allowing as many tasks as possible for each station under the constraint of the cycle time. The procedure to calculate the upper bound of cycle time is illustrated in Fig. 7.48, and the bisection method to find out the optimal cycle time is shown in Fig. 7.49.

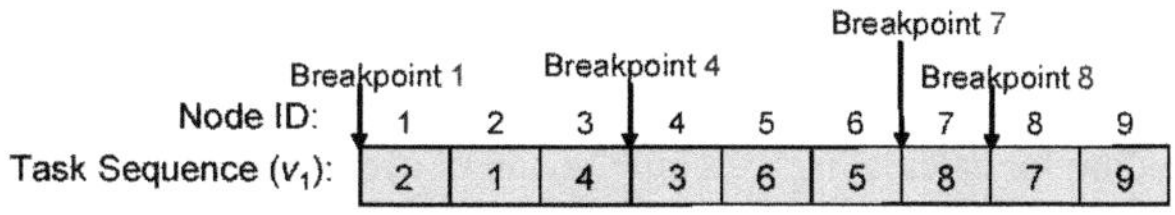

Fig. 7.47 Breakpoint decoding

Phase 2: Assigning Tasks to Each Station

Step 2.1: Order encoding for task to station assignment by randomly assign the task to each station
Step 2.2: Breakpoint decoding to assign the tasks into each station

After calculating the optimal cycle time, it is easy to generate the breakpoints on the task sequence to divide it into m parts, each of which will correspond to a station. Therefore, the tasks assigned for each station are determined according to a chromosome, and the cycle time can be calculated. Figure 7.50 presents the procedure for breakpoint decoding.

```
procedure: calculating upper bound of the cycle time
input: (w, n, v', c_LB) for each model
output: c_UB, c'_LB
begin
    c_UB ← c_LB;   // set upper bound of the cycle time the same as lower bound
    repeat
        c_UB ← 2*c_UB;  // multiply the upper bound by "2"
        i ← 0;
        for l=1 to w do
            // assign as many tasks as possible to each station under the cycle time
            t ← 0;
            while (t < c_UB) and (i<n) do
                i ← i +1;
                t ← t + t(v'(i));  // add assembly time of the task to station time
            end
            if t > c then
            i ← i - 1;
        end
    until i ≥n
    c'_LB ← c_UB/2;
    output c_UB, c'_LB
end
```

Fig. 7.48 Procedure for calculating the upper bound of cycle time

Phase 3: Designing a Schedule

Step 3.1: Creating a schedule for the assembly line. Using the chromosomes, the schedule for each product can be constructed as $S_m = (j, S_i, t_j)$
Step 3.2: Drawing a Gantt chart for this schedule. In the real-world, the assembly line is not just for producing one unit of the product. It should produce several units. Therefore, in this section a Gantt chart that shows the schedule for one unit of each model and later three units of each model needs to be constructed.

7.5.2.2 Evaluation Function

A central problem of any search method is how to guide to the most promising areas from a set of initial solutions. So we also consider the workload of each station by standard deviation of the idle time for each station. The small deviation of the idle time for each station may facilitate the genetic algorithm to find solutions with good balancing for the problem. In this study, when two individuals with the same cycle time are compared, the individual with less deviation is given a larger fitness value than the one with larger deviation:

```
procedure: calculating optimal cycle time
input: (w, n, v', c'_LB, c_UB) for each model
output: optimal cycle time c_T
begin
    repeat
        c_T ← (c'_LB + c_UB)/2;
        i ← 0;
        for l=1 to w do
            // assign as many tasks as possible to each station under the cycle time
            t ← 0;
            while (t < c_T) and (i<n) do
                i ← i +1;
                t ← t + t(v'(i));   // add assembly time of the task to station time
            end
            if t > c_T then
                i ← i - 1;
        end
        if i ≥ n then
            c_UB ← c_T;
        else
            c'_LB ← c_T;
    until c_UB-c'_LB ≤ 1;
    c_T ← c_UB ;
    output optimal cycle time c_T
end
```

Fig. 7.49 Bisection method to calculate the optimal cycle time

$$\begin{aligned} eval(v_k) &= w_1(f_1(v_k) - z_1^{\min}) + w_2(f_2(v_k) - z_2^{\min}) \\ &= w_1(Mc_\mathrm{T} - z_1^{\min}) + w_2\left\{\frac{1}{w-1}\sum_{k=1}^{w}((c_\mathrm{T} - t_k^\mathrm{W}) - \bar{t})^2 - z_2^{\min}\right\}, \\ & \qquad k = 1, \cdots, popSize \end{aligned} \tag{7.34}$$

where

M is a very large number
c is the cycle time
t_k^W is the total processing time of each station,

$$\bar{t} = \frac{1}{w}\sum_{k=1}^{w}(c_\mathrm{T} - t_k^\mathrm{W}) \tag{7.35}$$

```
procedure: breakpoint decoding
input: (w, n, v', c_T) for each model
output: breakpoint vector v'
begin
    i ←0;
    for l to w do
        //assign as many tasks as possible to each station under the cycle time
        t ←0;
        while (t ≤ c_T) and (i < n) do
            i ←i + 1;
            t ←t + t(v'(i));
        end
        if (t > c_T) then
            i ←i - 1;
            v'(l) ← i;
    end
    output breakpoint vector v'
end
```

Fig. 7.50 Procedure for breakpoint decoding

7.5.2.3 Genetic Operators

In this subsection, the crossover, mutation, immigration and selection operators used in the hGA approach are explained in detail.

Crossover Operator: As a crossover operator, we will be using weight mapping crossover. This crossover operator can be viewed as an extension of one-cut point crossover for permutation representation. As one-cut point crossover, two chromosomes (parents) would be chose a random-cut point and generate the offspring by using segment of own parent to the left of the cut point, then remapping the right segment based on the weight of other parent of right segment. The reader may refer to Chapter 1 for more information.

Mutation Operators: As a mutation operator, we will use swap mutation, which selects two positions at random, and exchange the two genes. Figure 7.51 gives an example of swap mutation.

Immigration Operator: The hGA algorithm is modified to include immigration routine, in each generation, generate and evaluate $popSize \cdot p_I$ random members, and replace the $popSize \cdot p_I$ worst members of the population with the $popSize \cdot p_I$ random members (p_I , called the immigration probability).

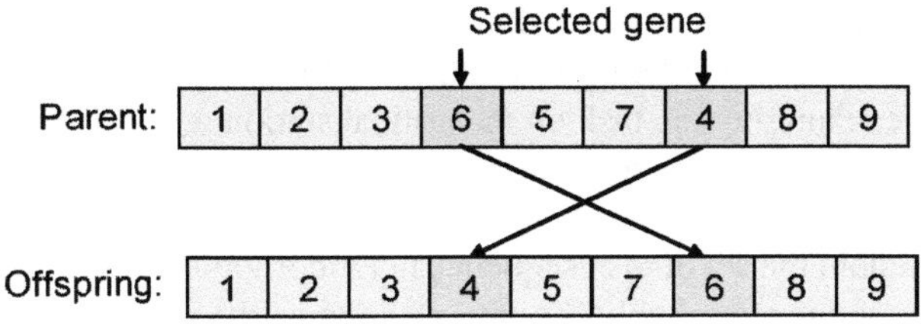

Fig. 7.51 Illustration of swap mutation

7.5.2.4 Local Search Procedure

In this study, a local search method is proposed to enhance the search ability of GA. The local search method is based on critical stations, which are the stations whose total assembly time are equal to the cycle time are called critical stations, in order to improve their effectivecannot ness and efficiency.

Task Sequence Search: Let $s(i)$ be the station to which task i is assigned. First we introduce the logical function $W_i(k)$ which returns false if task i cannot be transferred from station $s(i)$ to station k, and true otherwise:

$$\begin{aligned} W_i(k) &= \text{false if} \\ &\quad s(i) > k \text{ and } \exists\, g \in \text{Suc}(i),\ s(g) > k, \text{ or} \\ &\quad s(i) < k \text{ and } \exists\, g : i \in \text{Suc}(g),\ s(g) < k \\ W_i(k) &= \text{true, otherwise.} \end{aligned}$$

For a pair of tasks i, g, if

$$W_i(s(g)) = W_g(s(i)) = \text{true} \ \wedge\ i \notin \text{Pre}(i) \ \wedge\ g \notin \text{Pre}(i)$$

then, the two tasks are said to be exchangeable. Let T_l be the total assembly time of station l. For a critical task i and task g which is not in $s(i)$, if

$$t_{g,s(i)} < t_{i,s(i)} \text{ and } T_{s(g)} - t_{g,s(g)} + t_{i,s(g)} < c_T$$

then the exchange between tasks i and g is called worthwhile. The task sequence search is as follows:

Step 1. Decode the chromosome and identify a critical station l^*,

Step 2. For a task i that assigned to the critical station l^*, look for a task g that is not assigned to station l^*, and is exchangeable with task i.

Step 3. Repeat step 2 until the exchange between tasks i and g is worthwhile or all the tasks in the critical station are tried.

Step 4. Repeatedly perform the above search steps until the chromosome cannot be improved any more.

The neighborhood can be defined as the set of solutions obtainable from an initial solution by some specified perturbation. A task sequence neighbor solution is generated by exchanging the task on the critical station with another task of another station under the precedence constrains. The task sequence search is the local search which works over task sequence neighborhood.

Let $N(i)$ denote the set of task sequence neighborhood of solution i. The enlarged two-pace task sequence neighborhood is defined as the union of the neighborhood of each task sequence neighbor of solution I. During the task sequence search, the local search will implement over two-pace neighborhood when it reaches the local optima of one-pace neighborhood, and is called a two-pace task sequence search. During the task sequence search, when the local optima of two-pace task sequence neighborhood is reached, the neighbors of the two-pace local optima help the task sequence search escape from the local optima. The overall procedure of the local search used in this local search algorithm is shown in Fig. 7.52.

procedure: overall local search
input: initial solution S_0, local search parameters
output: improved solution S
begin
    $S \leftarrow S_0$;
    yield S by breakpoint search from S_0;
    $S_0 \leftarrow S$;
    yield S by task sequence search from S_0;
    output improved solution S
end

Fig. 7.52 Overall procedure of task sequence search

7.5.2.5 Computational Experiments and Discussions

In order to evaluate the performance of the hGA approach, eight representative precedence graphs [70] which are widely used in the sALB Type-1 models are selected. These precedence graphs contain with 25-297 tasks. From each precedence graph, three different mALB Type-2 models are generated. For each problem, the number of stations and the number of models are given, and each task of each model can be processed on any station. The task time data for each model are generated at random, while a statistical dependence is maintained, which is statistical dependence of task times on the task type for each model.

The computational experiments are performed on a Pentium 4 processor (2.6-GHz clock). The values of parameters used in the computational experiments are as follows:

Population size: $popSize = 100$

Maximum number of generation: *maxGen* = 1000
Probability of exchange crossover: $p_{C1} = 0.2$
Probability of mixed crossover: $p_{C2} = 0.8$
Probability of task sequence mutation: $p_{M1} = 0.05$
Probability of robot assignment mutation: $p_{M2} = 0.10$
Probability of immigration: $p_I = 0.20$
Terminating condition: If the same fitness value is found for 100 generations, the algorithm is terminated.

To evaluate the performance of the hGA, the 25 test instances are solved using the hGA approach with or without the local search. Our assumption is reasonable since the assembly line has to be redesigned under the constraint of resources that currently exist. In this research, we focused on creating the new genetic operators and local searches, which are fitting the mALB models. So, we compared this hGA approach with the algorithm which is without the local searches for solving the mALB models. Table 7.12 presents the performance of the approach for the mALB Type-2 model.

Table 7.12 The results of the computational experiments [?]

Problem information			Cycle time (c_T)	
Number of tasks	Number of stations	Number of models	GA without local searches	Hybrid GA approach
25	8	2	170	170
		3	215	215
		4	234	234
35	12	3	170	168
		4	200	168
		5	237	236
53	14	4	253	253
		5	259	257
		6	314	308
70	19	4	272	264
		5	273	267
		6	281	269
89	21	4	249	244
		5	294	292
		6	343	341
111	22	4	317	308
		6	339	332
		8	353	346
148	29	4	310	300
		6	328	316
		8	361	353
297	50	6	352	342
		8	361	349
		10	362	349

From the result, it can be stated that the hGA approach performs better than the GA approach without the local searches along with the increasing of the scale for the problems. The computational experiments show that the hGA approach is computationally efficient and effective to find the optimal solution.

7.5.3 Rekiek and Delchambre's Approach

Rekiek and Delchambre [21] investigated the multiobjective assembly line design (ALD), which have a complex structure due to multiple decisions, *e.g.*, tooling, balancing decisions. In this book, the authors presented a new moGA using *grouping genetic algorithms* (gGAs), *i.e.*, *multiobjective grouping* GA (mogGA) and a *multi criteria decision analysis* (MCDA) method, *i.e.*, PROMETHEE II. In this study, PROMETHEE II (Preference Ranking Organisation METHod of Enrichment Evaluation II) [75] was used as a technique to rank the set of potential alternatives using different evaluation criteria. PROMETHEE-2 belongs to the 'family' of outranking techniques. The method uses preference function which is a function of the difference between several alternatives for any criterion [75].

The gGA was first proposed by Falkenauer [76] especially for solving grouping optimization problems, where the aim was to group members of a set into a small number of families in order to optimize objective function under given constraints. The gGA has a special chromosome representation scheme and genetic operators, which are used to suit the representation scheme. Later, Falkenauer and Delchambre [14] implemented the gGA to two grouping optimization problems; *i.e.*, bin packing problem and sALB Type-1 model. This study was the first attempt to balance an ALB model with GA. Other implementations of gGAs for solving ALB models can be found in Falkenauer [29], Rekiek *et al.* [33], Rekiek [77] and Brown and Sumichrast [49]. For more information about gGA, the reader may refer to Falkenauer [78], where the author described the drawbacks, *i.e.* problems with the standard representation scheme and standard genetic operators, which occur when applying the typical GA to grouping problems.

In order to deal with the mALB in ALD decisions, the authors introduced a gGA based on Equal Piles approach. They tried to assign tasks to fixed a number of stations in such a way that the workload of each station was nearly equal by leveling the average size of each station (minimizing the standard deviation of sizes). Therefore, the method warranted obtaining the desired number of stations and trying to equalize the workloads of stations as much as possible. The authors used grouping-based genetic representation where the stations are represented by augmenting the workstation-based chromosome with a group part. The group part of the chromosome is written after a semicolon to list all the stations in the current solution. The length of the chromosome varies from solution to solution. For this kind of chromosome, the authors proposed gGA crossover operator and used a mutation operator. The chromosomes for mALB model are constructed by using the equal piles approach [80]. The objectives for this problem include minimization of standard deviation of AL, and minimization of the line imbalance. For the ALD decision, mogGA iteratively searches for Pareto solutions while PROMETHEE II narrows the search area.

Additionally, in the case of resource planning, the authors also proposed a new method which is based on mogGA and the branch-and-cut algorithm followed by the MCDA method. To deal with the changes during the operation of ALs, a new method that treats two problems partially at the same time was also proposed. In this method, first tasks that perform similar activities are grouped together in a work center and then for each work center tasks are assigned to stations. The main aim of such a method is to solve effectively balancing and material flow requirements of the production system.

The authors finally used commercial software, *i.e. Optiline* [80], developed by Universite Libre de Bruxelles in Belgium to solve a real-world problem, which consists of a European car manufacturer producing two different types of cars. The gGA used in OptiLine has a fundamental advantage over other methods in handling this kind of complex problems, which may explain its success.

7.5.4 Ozmehmet Tasan and Tunali's Approach

Ozmehmet Tasan and Tunali [22] proposed four moGA approaches for solving the mALB Type-1 models. Their study mainly focuses on solving mALB model, which concerns the allocation of the assembly tasks for each model equally among workstations so that a new multi-objective function including the number of workstations and the horizontal imbalances is minimized. In the proposed multi-objective function, the achievement of horizontal balancing will be evaluated more realistically by considering both cycle time violations and slack times in different priority levels. Additionally, the demand ratios will be included in the objective function in order to reflect the importance of each model produced in the assembly line.

For the evaluation of the solutions, the following aggregated multi-objective function (see Eq. 7.36) was used:

$$eval(v_k) = w_1(f_1(v_k) - z_1^{\min}) + w_2(f_2(v_k) - z_2^{\min}),\ k = 1, \cdots, popSize \quad (7.36)$$

$$f_1(v_k) = smc_{\mathrm{T}} \quad (7.37)$$

$$\begin{aligned} f_2(v_k) &= w_3 \sum\sum (d_m / d) \max\{1, \tau_{ki} - c_{\mathrm{T}}\} \\ &\quad + w_4 \sum\sum (d_m / d) \max\{1, c_{\mathrm{T}} - \tau_{ki}\} \end{aligned} \quad (7.38)$$

where

- s number of stations, index; i=1,..., s
- n number of tasks, index; j=1,...,n
- m number of models, index; k=1,..., m
- d_m demand for model m during planning horizon

d total demand during planning horizon

$$d = \sum_{k=1}^{m} d_k \tag{7.39}$$

c_T (average) cycle time, launch interval
t_{jk} processing time of task j for one unit of model k
S_i station load, the set of tasks assigned to station i
τ_{ki} processing time per unit of model k in station i

$$\tau_{ki} = \sum_{j \in WS_i} t_{jk} \tag{7.40}$$

The first part of this objective function will help to minimize the number of workstations, s employed in the line, which is considered as a primary objective of mALB model, vertical balancing. The second part of the objective function will contribute minimizing the horizontal imbalances to eliminate the inefficiencies across the models. In the new multi-objective function, the achievement of horizontal balancing is evaluated more realistically by considering both cycle time violations and slack times in different priority levels. Additionally, the demand ratios are included in the objective function in order to reflect the importance of each model produced in the assembly line. In this objective function, the symbols w_1, w_2, w_3, and w_4 represent the weights given to minimizing the number of workstations, horizontal imbalances, cycle time violations, and slack time, respectively. Changing the values of these weights, results in different solutions [81].

The proposed approaches are hybrid approaches, which combine GA with *tabu search* (TS). It is well known that pure GAs are able to locate the promising regions for global optima in a search space, but sometimes they have difficulty in finding the exact minimum of these optima [82, 83]. The motivation behind the hybridization concept is usually to obtain better performing systems that exploit and unite advantages of the individual pure strategies, *i.e.*, such hybrids are believed to benefit from synergy. Since it is likely that the solutions found by GAs can still be improved by using other solution methods, the proposed hybridization which incorporates TS [84, 85] into GA is likely to facilitate local convergence.

Since mALB models are direct sALB extensions or at least require the solution of simple AL instances in general, the authors initially introduced hybrid metaheuristic approaches to solve the simplest form of the model, *i.e.* sALB model [86, 87], and eventually by modifying the assumptions, more complex form of the model, *i.e.*, mALB model is solved [81]. The hybrid approaches for the mALB model are composed of two major steps, *i.e.*, construction of the combined problem and the solution of this combined problem. First, the authors integrated the precedence relations of each model to form a combined model and, following, they solved this combined model using one of these hybrid approaches using the aggregated multi-objective function.

The first three approaches, *i.e.*, sequential, parallel synchronous with TS refinement, and parallel synchronous with TS mutation, use the hybrid GA approach, which combines GA and TS to solve mALB Type-1 models. In the first hybrid approach, TS and GA are used sequentially to solve the problem, where the resulting solutions of TS are included into the initial population of GA. In the second hybrid approach, TS is embedded into GA within a refinement operation. Likewise, in the third hybrid approach, TS is embedded into GA where TS is used as a mutation operator. Unlike the previous three hybrid approaches that generate the initial population randomly, the fourth hybrid approach employs a new initialization scheme to generate the initial population. Considering the fact that the mALB models involves several sALB models, to generate the initial population, in this study, a new scheme which aims at reducing the search space within the global search space is introduced. In implementing this new scheme, they first solved the line balancing problem for each model independently by using pure GA with small sized populations and following, incorporated the results that satisfy the precedence relations of the combined model into GA's randomly generated initial population.

Some features of the hybrid approaches can be summarized as follows:

- For the genetic representation of individuals, task-based natural encoding scheme, which identifies the feasible precedence sequences of tasks for the combined model, is used. The length of the chromosome is defined by the number of tasks. Infeasible solutions, which violate the precedence constraints, are not allowed in the population.
- To select the parents, roulette wheel selection scheme is used.
- In order to incorporate some characteristics of assembly line balancing into the proposed approaches, Low Level Heuristic Based crossover operators, which help to transfer problem specific information genetically from parents to offsprings is used.
- Modified Inversion mutation operator, which is a modification of a standard mutation operator, is proposed.
- For survival of individuals, modified elitist strategy, where the best offspring and the best parents are survived to next generation, is used.
- The methodology is terminated if the fitness function of the best solution does not improve more than 1% after a predetermined number of generations, *i.e.* T_G=50, or the total number of generations exceeds a maximum number, *i.e.* T_{max}=200.

To evaluate the performance of the hybrid approaches, experiments on a number of benchmark problem sets reported in the literature were carried out. The results of this computational experiments stated that the hybrid approaches outperformed the pure GA and pure TS with respect to all measures provided and all data sets. In comparison to the hybrid approaches, the performances of the pure GA and pure TS are rather poor. Among hybridization schemes, the parallel synchronous scheme outperformed the sequential scheme and also the new initialization scheme outperformed the random initialization. Regarding these findings, the authors stated that the parallel synchronous hybridization of GA with TS using the new initialization scheme has the potential to improve the performance of the pure methods.

7.6 Summary

In this chapter, we have illustrated the successful applications of multiobjective GAs to solve various assembly line balancing models, *i.e.* simple assembly line balancing model and general assembly line balancing models such as balancing of u-shaped assembly line, robotic assembly line and mixed-model assembly line.

For solving the multi-objective simple assembly line balancing Type-1 model and multi-objective u-shaped assembly line balancing model, we introduced a priority-based GA, which developed by Gen and Cheng [23] in order to handle the problem of creating encoding while treating the precedence constraints efficiently. We also introduced a hybrid GA to solve the rALB Type-2 model. This hGA used a partial representation technique, that is, the coding space only contains part of the feasible solutions, in which the optimal solution was included. A critical path local search was developed to improve the chromosomes. The effectiveness and efficiency of GA approaches were investigated with various scales of assembly line balancing problems by comparing with recent related research.

References

1. Scholl, A. (1999). *Balancing and Sequencing of Assembly Lines*. Physica-Verlag, Heidelberg.
2. Becker, C., & Scholl, A. (2006). A survey on problems and methods in generalized assembly line balancing. *European Journal of Operational Research*, 168, 694-715.
3. Baybars, I. (1986). A Survey of Exact Algorithms for the Simple Assembly Line Balancing Problem. *Management Science*, 32, 909-932.
4. Kim, Y. K., Kim, Y. J., & Kim, Y. H. (1996). Genetic algorithms for assembly line balancing with various objectives. *Computers & Industrial Engineering*, 30(3), 397-409.
5. Helgeson, W. B., Salveson, M. E., & Smith, W. W. (1954). How to balance an assembly line, Technical Report, Carr Press, New Caraan, Conn.
6. Salveson, M. E. (1955). The assembly line balancing problem. *Journal of Industrial Engineering*, 6, 18-25.
7. Bowman, E. H. (1960). Assembly line balancing by linear programming. *Operations Research*, 8(3), 385-389.
8. Held, M., Karp, R. M., & Shareshian, R. (1963). Assembly line balancing-Dynamic programming with precedence constraints. *Operations Research*, 11, 442-459.
9. Jackson, J. R. (1956). A Computing Procedure for a Line Balancing Problem. *Management Science*, 2, 261-272.
10. Karp, R. M. (1972). Reducibility among combinatorial problems. In Miller R.E & Thatcher J.W. Editors: *Complexity of Computer Applications*, 85-104, New York: Plenum Press.
11. Dar-El, E. M. (1973). MALB-A heuristic technique for balancing large single-model assembly lines. *AIIE Transactions*, 5(4), 343-356.
12. Scholl, A. & Voss, S. (1996). Simple assembly line balancing-heuristic approaches. *Journal of Heuristics*, 2, 217-244.
13. Suresh, G., & Sahu, S. (1994). Stochastic assembly line balancing using simulated annealing. International Journal of Production Research, 32(8), 1801-1810.
14. Falkenauer, E., & Delchambre, A. (1992). A genetic algorithm for bin packing and line balancing. *Proceedings of the 1992 IEEE International Conference on Robotics and Automation*, Nice, France, 1189-1192.

15. Bautista, J., & Pereira, J. (2002). Ant algorithms for assembly line balancing, *Lecture Notes in Computer Science*, 2463, 65-75.
16. Talbot, F. B., Patterson, J. H., & Gehrlein, W. V. (1986). A comparative evaluation of heuristic line balancing techniques. *Management Science*, 32, 430 - 454.
17. Ghosh, S., & Gagnon, R. J. (1989). A comprehensive literature review and analysis of the design, balancing and scheduling of assembly systems. *International Journal of Production Research*, 27, 637-670.
18. Erel, E., & Sarin, S. C. (1998). A survey of the assembly line balancing procedures. *Production Planning and Control*, 9, 414-434.
19. Rekiek, B., Dolgui, A., Delchambre, A., & Bratcu, A. (2002). State of art of optimization methods for assembly line design. *Annual Reviews in Control*, 26, 163-174.
20. Scholl, A. & Becker, C. (2006). State-of-the-art exact and heuristic solution procedures for simple assembly line balancing. *European Journal of Operational Research*, 168, 666-693.
21. Rekiek, B. & Delchambre, A. (2006). *Assembly line design: The balancing of mixed-model hybrid assembly lines with genetic algorithms*. Springer Series in Advanced Manufacturing, London.
22. Ozmehmet Tasan, S. & Tunali, S. (2007). A review of the current applications of genetic algorithms in assembly line balancing. *Journal of Intelligent Manufacturing*, DOI: 10.1007/s10845-007-0045-5.
23. Gen, M., & Cheng, R. (2000). *Genetic Algorithms and Engineering Optimization*, New York: John Wiley & Sons.
24. Leu, Y. Y., Matheson, L. A., & Rees, L. P. (1994). Assembly line balancing using genetic algorithms with heuristic generated initial populations and multiple criteria. *Decision Sciences*, 15, 581-606.
25. Anderson, E. J., & Ferris, M. C. (1994). Genetic Algorithms for Combinatorial Optimization: The Assembly Line Balancing Problem. *ORSA Journal on Computing*, 6, 161-173.
26. Rubinovitz, J., & Levitin, G. (1995). Genetic algorithm for assembly line balancing. *International Journal of Production Economics*, 41, 343-354.
27. Tsujimura, Y., Gen, M., & Kubota, E. (1995). Solving fuzzy assembly line balancing using genetic algorithms. *Computers & Industrial Engineering*, 29(1-4), 543-547.
28. Suresh, G., Vinod, V. V., & Sahu, S. (1996). A genetic algorithm for assembly line balancing. *Production Planning and Control*, 7(1), 38-46.
29. Falkenauer, E. (1997). A grouping genetic algorithm for line balancing with resource dependent task times. *Proceedings of the Fourth International Conference on Neural Information Processing*, New Zealand, 464-468.
30. Ajenblit, D. A., & Wainwright, R. L. (1998). Applying genetic algorithms to the U-shaped assembly line balancing problem. *Proceedings of the 1998 IEEE International Conference on Evolutionary Computation*, Anchorage, Alaska, USA, 96-101.
31. Chan, C. C. K., Hui, P. C. L., Yeung, K. W., & Ng, F. S. F. (1998). Handling the assembly line balancing problem in the clothing industry using a genetic algorithm. *International Journal of Clothing Science and Technology*, 10(1), 21-37.
32. Kim, Y. J., Kim, Y. K., & Cho, Y. (1998). A heuristic-based genetic algorithms for workload smoothing in assembly lines. *Computers & Operations Research*, 25(2), 99-111.
33. Rekiek, B., de Lit, P., Pellichero, F., Falkenauer, E., & Delchambre, A. (1999). Applying the equal piles problem to balance assembly lines. *Proceedings of the ISATP 1999*, Porto, Portugal, 399-404.
34. Bautista, J., Suarez, R., Mateo, M., & Companys, R. (2000). Local search heuristics for the assembly line balancing problem with incompatibilities between tasks. *Proceedings of the 2000 IEEE International Conference on Robotics and Automation*, San Francisco, CA, 2404-2409.
35. Kim, Y. K., Kim, Y., & Kim, Y. J. (2000). Two-sided assembly line balancing: a genetic algorithm approach. *Production Planning and Control*, 11(1), 44-53.
36. Ponnambalam, S. G., Aravindan, P., Naidu, G., & Mogileeswar, G. (2000). Multi-objective genetic algorithm for solving assembly line balancing problem. *International Journal of Advanced Manufacturing Technology*, 16(5), 341-352.

37. Sabuncuoglu, I., Erel, E., & Tanyer, M. (2000). Assembly line balancing using genetic algorithms. *Journal of Intelligent Manufacturing*, 11(3) 295-310.
38. Carnahan, B. J., Norman, B. A., & Redfern, M. S. (2001). Incorporating physical demand criteria into assembly line balancing. *IIE Transactions*, 33, 875-887.
39. Simaria, A. S., & Vilarinho, P. M. (2001a). A genetic algorithm approach for balancing mixed model assembly lines with parallel workstations. *Proceedings of The 6th Annual International Conference on Industrial Engineering Theory, Applications and Practice*, November 18-20, 2001, San Francisco, USA.
40. Chen, R. S., Lu, K. Y., & Yu, S. C. (2002). A hybrid genetic algorithm approach on multi-objective of assembly planning problem. *Engineering Applications of Artificial Intelligence*, 15, 447-457.
41. Goncalves, J. F., & De Almedia, J. R. (2002). A hybrid genetic algorithm for assembly line balancing. *Journal of Heuristic*, 8, 629-642.
42. Miltenburg, J. (2002). Balancing and sequencing mixed-model U-shaped production lines. *International Journal of Flexible Manufacturing Systems*, 14, 119-151.
43. Valente, S. A., Lopes, H. S., & Arruda, L. V. R. (2002). Genetic algorithms for the assembly line balancing problem: a real-world automotive application. In: Roy, R., Kppen, M., Ovaska, S., Fukuhashi, T., Hoffman, F. *Soft Computing in Industry - Recent Applications*. Berlin: Springer-Verlag, 319-328.
44. Brudaru, O., & Valmar, B. (2004). Genetic algorithm with embryonic chromosomes for assembly line balancing with fuzzy processing times. *The 8th International Research/Expert Conference Trends in the Development of Machinery and Associated Technology*, TMT 2004, Neum, Bosnia and Herzegovina.
45. Martinez, U., & Duff, W. S. (2004). Heuristic approaches to solve the U-shaped line balancing problem augmented by Genetic Algorithms. *In the Proceedings of the 2004 Systems and Information Engineering Design Symposium*, 287-293.
46. Simaria, A. S., & Vilarinho, P. M. (2004). A genetic algorithm based approach to mixed model assembly line balancing problem of type II. *Computers and Industrial Engineering*, 47, 391-407.
47. Stockton, D. J., Quinn, L., & Khalil, R. A. (2004a). Use of genetic algorithms in operations management Part 1: applications. *Proceeding of the Institution of Mechanical Engineers-Part B: Journal of Engineering Manufacture*, 218(3), 315-327.
48. Stockton, D. J., Quinn, L., & Khalil, R. A. (2004b). Use of genetic algorithms in operations management Part 2: results. *Proceeding of the Institution of Mechanical Engineers-Part B: Journal of Engineering Manufacture*, 218(3), 329-343.
49. Brown, E. C., & Sumichrast, R. T. (2005). Evaluating performance advantages of grouping genetic algorithms. *Engineering Applications of Artificial Intelligence*, 18, 1-12.
50. Levitin, G., Rubinovitz, J., & Shnits, B. (2006). A genetic algorithm for robotic assembly line balancing. *European Journal of Operational Research*, 168, 811-825.
51. Noorul Haq, A., Jayaprakash, J., & Rengarajan, K. (2006). A hybrid genetic algorithm approach to mixed-model assembly line balancing. *International Journal of Advanced Manufacturing Technology*, 28, 337-341.
52. Runarsson, T. P., & Jonsson, M.T. (1999). Genetic production systems for intelligent problem solving. *Journal of Intelligent Manufacturing*, 10, 181-186.
53. Hoffmann T. R. (1992). EUREKA A hybrid system for assembly line balancing, *Management Science*, 38, 39-47
54. Chen, Y. Q. (2007). *Study on Multi-objective Assembly Line Balancing Problem by Hybrid Evolutionary Algorithm*. Ms Thesis, Waseda University, Graduate School of Information, Production and Systems, Kitakyushu, Japan.
55. Baudin, M. (2002). *Lean Assembly: The nuts and bolts of making assembly operations flow*. Productivity, New York.
56. Wemmerlov, U., & Hyer, N. L. (1989). Cellular manufacturing in the U.S. industry: A survey of users, *International Journal of Production Research*, 27(9), 1511-1530.
57. Miltenburg, J., & Wijngaard, J. (1994). The U-line line balancing problem. *Management Science*, 40(10), 1378-1388.

58. Monden, Y. (1998). *Toyota production system–An integrated approach to just-in-time*, 3rd ed. Dordrecht: Kluwer.
59. Scholl, A. & Klein, R. (1999). Balancing assembly lines effectively-a computational comparison. *European Journal of Operational Research*, 114, 50-58.
60. Hwang, R. K., Katayama, H., & Gen, M. (2007). U-shaped Assembly Line Balancing Problem with Genetic Algorithm. *International Journal of Production Research*, DOI: 10.1080/00207540701247906.
61. Rubinovitz, J. & Bukchin, J. (1993). RALB-a heuristic algorithm for design and balancing of robotic assembly line. *Annals of the CIRP*, 42, 497-500.
62. Rubinovitz, J. & Bukchin, J. (1991). Design and balancing of robotic assembly lines. *Proceedings of the 4th World Conference on Robotics Research*, Pittsburgh, PA.
63. Bukchin, J. & Tzur, M. (2000). Design of flexible assembly line to minimize equipment cost. *IIE Transactions*, 32, 585-598.
64. Tsai, D. M. &. Yao, M. J (1993). A line-balanced-base capacity planning procedure for series-type robotic assembly line. *International Journal of Production Research*, 31, 1901-1920.
65. Kim, H. & Park, S. (1995). Strong cutting plane algorithm for the robotic assembly line balancing. *International Journal of Production Research*, 33, 2311-2323.
66. Khouja, M., Booth, D. E., Suh, M., & Mahaney, Jr. J. K. (2000). Statistical procedures for task assignment and robot selection in assembly cells. *International Journal of Computer Integrated manufacturing*, 13, 95-106.
67. Nicosia, G., Paccarelli, D. & Pacifici, A. (2002). Optimally balancing assembly lines with different workstations. *Discrete Applied Mathematics*, 118, 99-113.
68. Krasnogor, N., & Smith, J. (2000). A memetic algorithm with self-adaptive local search: TSP as a case study. *Proceedings of Genetic and Evolutionary Computation Conference*, July 10-12, Las Vegas, NV, 987-994, 2000.
69. Moscato, P., & Norman, M. (1992). A memetic approach for the traveling salesman problem: implementation of a computational ecology for combinatorial optimization on message-passing systems. *Proceedings of the International Conference on Parallel Computing and Transputer Applications*, Amsterdam.
70. Scholl, A. (1993). *Data of Assembly Line Balancing Problems*. Schriften zur Quantitativen Betriebswirtschaftslehre 16/93, Th Darmstadt.
71. Gao, J. (2007). *A study on Hybrid Genetic Algorithms for Manufacturing Optimization*. PhD Thesis, Xi'an Jiaotong University, Xi'an, China.
72. Lin, L., Gen, M. & Gao, J. (2008) Optimization and improvement in robot-based assembly line system by hybrid genetic algorithm, *IEEJ Transactions on Electronics, Information and Systems*, in reviewing.
73. Meyr, H. (2004). Supply chain planning in the German automotive industry. *OR Spectrum*, 26(4), 447-470.
74. Miltenburg, J., & Sinnamon, G. (1989). Scheduling mixed model multi-level just-in-time production systems. *International Journal of Production Research*, 27, 1487-1509.
75. Brans, J.P., Vincke, P., & Mareschal, B. (1986). How to select and how to rank projects: The PROMETHEE method. *European Journal of Operational Research*, 24, 228-238.
76. Falkenauer, E. (1991). A genetic algorithm for grouping. *Proceedings of the Fifth International Symposium on Applied Stochastic Models and Data Analysis*, Granada, Spain.
77. Rekiek, B. (2000). *Assembly Line Design (multiple objective grouping genetic algorithm and the balancing of mixed-model hybrid assembly line)*. PhD Thesis, Free University of Brussels, CAD/CAM Department, Brussels, Belgium.
78. Falkenauer, E. (1998). *Genetic Algorithms for Grouping Problems*. Wiley, New York.
79. Falkenauer, E. (1995). Solving equal piles with the Grouping Genetic Algorithm. In L. J. Eshelman (ed.), *Proceedings of the Sixth International Conference on Genetic Algorithms*, 492-497. Morgan Kaufmann Publ., San Francisco, CA.
80. Optiline. www.optimaldesign.com/OptiLine/OptiLine.htm.
81. Ozmehmet Tasan, S. (2007). *Solving Simple and Mixed-Model Assembly Line Balancing Problems Using Hybrid Meta-Heuristic Approaches*. PhD Thesis, Dokuz Eylul University, Graduate School of Natural and Applied Sciences, Izmir, Turkey.

82. Goldberg, D. E. (2002). *Design of innovation: Lessons from and for competent genetic algorithms*. Boston, MA: Kluwer Acadamic Publishers.
83. Haupt, R. L. & Haupt, S. E. (2004). *Practical Genetic Algorithms*. Second edition with CD, Wiley, New York, NY.
84. Chiang, W. C. (1998). The application of a tabu search metaheuristic to the assembly line balancing problem. *Annals of Operations Research*, 77, 209-227.
85. Glover, F., & Laguna, M. (1998). *Tabu Search*. Dordrecht: Kluwer Academic Publishers.
86. Ozmehmet Tasan, S. & Tunali, S. (2006). Improving the genetic algorithms performance in simple assembly line balancing. *Lecture Notes in Computer Science*, 3984, 78-87.
87. Ozmehmet Tasan, S. & Tunali, S. (2007). Hybrid meta-heuristic approaches for solving SALBP. *Computers & Industrial Engineering*, in reviewing.

Chapter 8
Tasks Scheduling Models

Recent advances in computing, storage, and communication technologies have made a wide variety of multimedia applications possible. Multimedia systems combine a various information sources, such as audio, video, text, graphics and image, into the wide range of application (Chen, 1996). Video and audio data are referred to as continuous media due to their real-time delivery requirements, whereas text, graphics and image data are referred to as discrete media. Since continuous media are displayed within a certain time constraint, their computation and manipulation of continuous media should be handled under more limited condition than that of discrete media. The objective of the task scheduling models is to minimize total tardiness and the scheduling these tasks on multiprocessor system is NP-hard problem. A continuous task scheduling, real-time task scheduling on homogeneous system and real-time task scheduling on heterogeneous system are introduced in this chapter.

8.1 Introduction

Real-time tasks can be classified to many kinds. Some real-time tasks are invoked repetitively. For example, one may wish to monitor the speed, altitude, and attitude of an aircraft every 100ms. This sensor information will be used by periodic tasks that control the control surfaces of the aircraft, in order to maintain stability and other desired characteristics. In contrast, there are many other tasks that are aperiodic, that occur only occasionally. Aperiodic tasks with a bounded interarrival time are called sporadic tasks. Real-time tasks can also be classified according to the consequences of their not being executed on time. Critical (or hard real-time) tasks are those whose timely execution is critical. If deadline are missed, catastrophes occur. Noncritical (or soft real-time) tasks are, as name implies, not critical to the application [1].

For task scheduling, the purpose of general task scheduling is fairness which means that the computer's resources must be shared out equitably among users.

However, the purpose of hard real-time task scheduling is to execute, by the appropriate deadlines, its critical control tasks and the objective of the scheduling soft real-time tasks is to minimize total tardiness [2].

There are some traditional scheduling algorithm for hard real-time tasks on uniprocessor, such as Rate Monotonic (RM) and Earliest Deadline First (EDF) [3] scheduling algorithms. They guarantee the optimality in somewhat restricted environments. Several derived algorithms from RM, EDF is used for soft real-time tasks. However, these algorithms have some drawbacks in resource utilization and pattern of degradation under the overloaded situation. With the growth of soft real time applications, the necessity of scheduling algorithms for soft real-time tasks is on the increase. Rate Regulating Proportional Share (rrPS) [4] scheduling algorithm and Modified Proportional Share (mPS) [5] scheduling algorithm are designed for soft real-time tasks. However, these algorithms also cannot show the graceful degradation of performance under an overloaded situation and are restricted in a uniprocessor system.

Furthermore, the scheduling on multiprocessor system is an NP-hard problem. According to Yalaoui and Chu [6], the problem of scheduling tasks on identical parallel processors to minimize the total tardiness is at least an NP-hard problem since Du and Leung showed that the problem is an NP-hard problem for a single processor case [7]. Lenstra *et al.* also showed that the problem with two processors is NP-hard problem [8]. Nevertheless the exact complexity of this problem remains open for more than two processors. Consequently various modern heuristics based algorithms have been proposed for practical reasons.

Real-time tasks are characterized by computational activities with timing constraints. The tasks can be classified to *periodic*, *sporadic* and *aperiodic tasks* .
Periodic: As shown in Fig. 8.1, the task needs to be cyclically executed at constant activation rates.

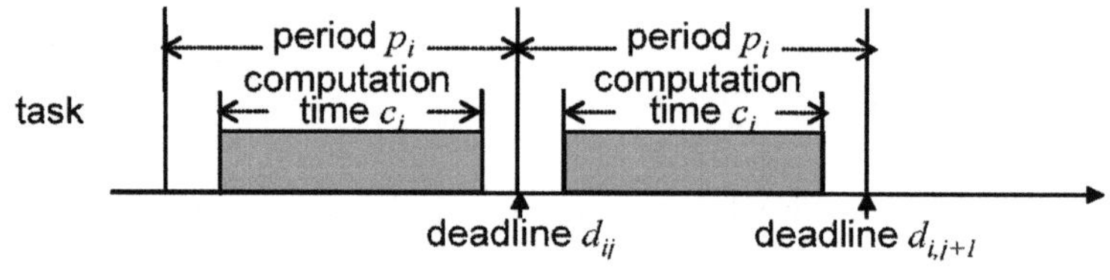

Fig. 8.1 Illustration of periodic task

Sporadic: Task is not periodic, but may be invoked at irregular intervals.
Aperiodic: Task is not periodic and has no upper bound on their invocation rate.

Also depending on the task attributes, tasks can be classified into two categories: *hard real-time tasks* and *soft real-time tasks*.

During the last decade, there has been a growing interest in using Genetic Algorithms (GA) to solve the task scheduling problems that are NP hard [35]. Based on the model structure, the core models of task scheduling are summarized in Fig. 8.2.

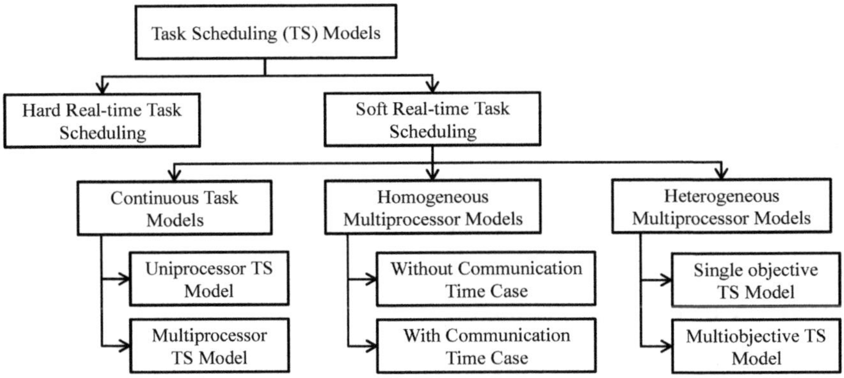

Fig. 8.2 The core models of task scheduling introduced in this Chapter

8.1.1 Hard Real-time Task Scheduling

In hard real-time tasks, the violation of timing constraints of certain task should not be acceptable [9, 10].The consequences of not executing a task before its deadline may lead to catastrophic consequences in certain environments, *i.e.*, in patient monitoring system, nuclear plant control, *etc*. In hard real-time task, the performance of scheduling algorithm is measured by its ability to generate a feasible schedule for a set of real-time tasks. If all the tasks start after their release time and complete before their deadlines, the scheduling algorithm is feasible. Typically, there is Rate Monotonic (RM) and Earliest Deadline First (EDF) derived scheduling algorithms for hard real-time tasks [11, 12]. They guarantee the optimality in somewhat restricted environments.

8.1.1.1 Rate Monotonic (RM) Scheduling Algorithm

The RM Scheduling is proposed by Liu and Layland [3]. This Scheduling Algorithm is based on the uniprocessor static-priority preemptive scheme and optimal among all fixed priority scheduling algorithms. This algorithm assigns static priority to the tasks such that tasks with shorter period get higher priorities. This algorithm assumes as follows:

A1. No task has any nonpreemptable section and the cost of preemption is negligible.
A2. Only processing requirements are significant; memory, I/O, and other resource requirements are negligible.
A3. All tasks are independent; there are no precedence constraints.
A4. All tasks in the tasks set are periodic.
A5. The relative deadline of a task is equal to its period.

The algorithm of RM scheduling is shown in Fig. 8.3.

---

procedure: RM Scheduling Algorithm
input: task set
output: schedule
step 1: Set up all tasks' priority inversely to their period.
Set up all tasks' start time, end time, remaining time and deadline.
step 2: If system is idle, then add task to schedule and go to step 4.
Otherwise go to step 3.
step 3: If new task's priority is higher than processing task's priority,
then update processing task's remaining time and exchange tasks.
Otherwise update new task's start time.
step 4: If all the tasks have not been scheduled, then go to step 2. Otherwise stop.

Fig. 8.3 Procedure of RM Scheduling Algorithm

Figure 8.4 represents the example of RM scheduling. Where τ_i is the i-th task, i is the number of task, c_i is the computation time of task i, r_i is the release time of task i and p_i is the period of task i. In this figure, τ_3's period is longer then τ_2's period. So, τ_3 can't preempt τ_2 at time 3. On the other hand, τ_1 preempts τ_3 at time 4.5.

Table 8.1 The example data of RM scheduling

i	Computation time c_i	Release time r_i	Deadline p_i
1	0.50	0	2
2	2.00	1	6
3	1.75	3	10

If the total utilization of the tasks is no greater than $n(2^{1/n} - 1)$, where n is the number of tasks to be scheduled, then RM scheduling algorithm will schedule all the tasks to meet their respective deadlines. The goal of RM scheduling algorithm is to meet all task deadlines and to keep the feasibility of scheduling through admission control $n(2^{1/n} - 1)$. Strict admission control prevents unpredictable behaviors when the overloaded condition occurs. But it may cause low utilization of resources.

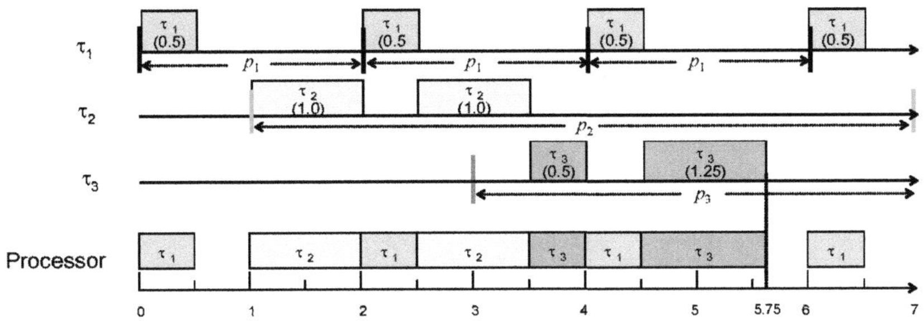

Fig. 8.4 Illustration of RM scheduling

8.1.1.2 Earliest Deadline First (EDF) Scheduling Algorithm

EDF Scheduling is proposed by Liu and Layland [3]. This scheduling algorithm is based on uniprocessor dynamic-priority preemptive scheme and optimal among all dynamic priority scheduling algorithms. This algorithm schedules the task with the earliest deadline first. This algorithm assumes as follows;

A1. No task has any nonpreemptable section and the cost of preemption is negligible.
A2. Only processing requirements are significant; memory, I/O, and other resource requirements are negligible.
A3. All tasks are independent; there are no precedence constraints.
A4. All tasks in the tasks set are periodic.
A5. The relative deadline of a task is equal to its period.

The algorithm of EDF scheduling is shown in Fig. 8.5.

---

procedure: EDF Scheduling Algorithm
input: task set
output: schedule
step 1: Set up all tasks' start time, end time, remaining time and deadline.
step 2: If system is idle, then add task to schedule and go to step 4.
Otherwise go to step 3.
step 3: If new task's deadline is earlier than processing task's deadline,
then update processing task's remaining time and exchange tasks.
Otherwise update new task's start time.
Step 4: If all the tasks have not been scheduled, then go to step 2. Otherwise stop.

Fig. 8.5 Procedure of EDF scheduling algorithm

Figure 8.6 represents the example of EDF scheduling. Where τ_i is the i-th task, i is the number of task, c_i is the computation time of task i and p_i is the period of task i.

Table 8.2 The example data of EDF scheduling

i	Computation time c_i	Deadline p_i
1	1.0	3
2	0.5	2
3	2.0	5

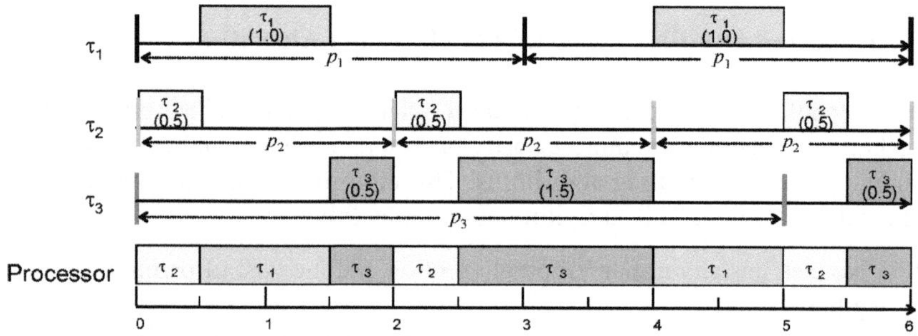

Fig. 8.6 Illustration of EDF scheduling

If the total utilization of the task set is no greater than 1, the task set can be feasibly scheduled on a single processor by the EDF scheduling algorithm. The goal of EDF scheduling algorithm is to meet all task deadlines and to keep the feasibility of scheduling through admission control 1. Strict admission control prevents unpredictable behaviors when the overloaded condition occurs. But it may cause low utilization of resources.

8.1.1.3 Analysis for Scheduling Algorithms Derived from RM and EDF

There are several RM and EDF derived algorithms for soft real-time tasks. But these algorithms have some drawbacks to cope with continuous tasks in soft real-time tasks related resource utilization and pattern of degradation under the overloaded situation. First, in continuous tasks, it is not necessary for every instance of a repetitive task to meet its deadline. For soft real-time tasks, slight violation of time limits is not so critical. Second, RM and EDF scheduling algorithms require strict admission control to prevent unpredictable behavior when the overloaded situation occurs. The strict admission control may cause low utilization of resources [31].

8.1.2 Soft Real-time Task Scheduling

The usefulness of results produced by a task decreases over time after the deadline expires without causing any damage to the controlled environment [32]. Recently, Rate Regulating Proportional Share (rrPS) and Modified Proportional Share (mPS) Scheduling algorithms were been designed for continuous tasks in soft real-time tasks.

8.1.2.1 Rate Regulating Proportional Share (rrPS) Scheduling Algorithm

The rrPS scheduling algorithm was proposed by Kim and Lee [31]. The rrPS scheduling algorithm is based on the stride scheduler and is proposed to schedule continuous tasks. To specify timing requirements of continuous media, a pair of parameters (p_i, c_i) is introduced, where p_i means the period of a task i and c_i means the computation time of task i needed over each period.

Three parameters, which are tc_i, sd_i and ps_i, are used in this scheduling algorithm. The tc_i is the period of a task and the computation time during the period. It represents the relative amount of resources that should receive. The sd_i is inversely proportional to the ticket and represents the interval between selections. The ps_i represents the virtual time index for next selection and has the same value as the stride at the first stage. The task with the minimum pass is selected, and its pass is advanced by its stride.

The rate regulator, the key concept of the scheduling algorithm, prevents certain tasks from taking more resources than its share for a given period. It can be represented as Eq. 8.1.

$$\frac{1}{t}\sum_{j=s_i}^{t} a_i \leq \frac{c_i}{p_i} \tag{8.1}$$

Notation

In above equation, notations are defined as follows:

Indices

t: time index
i: task index

Parameters

c_i: the computation time of task i
p_i: the period of task i

DecisionVariables

s_i: the start time of task i
$\sum a_i$: total number of allocations that task i has received from start to current time

The algorithm of rrPS scheduling is shown in Fig. 8.7.

---

procedure: rrPS Scheduling Algorithm
input: task set
output: schedule
step 1: Set up all tasks' tc_i, sd_i and ps_i.
step 2: Select task with minimum ps_i.
step 3: If task satisfies rate regulator, then update ps_i and go to step 5.
Otherwise go to step 4.
step 4: Select a task with the minimum ps_i in remains and go to step 3.
step 5: if all the tasks have not been scheduled, then go to step 2. Otherwise stop.

Fig. 8.7 Procedure of rrPS scheduling algorithm

Figure 8.8 represents the example of rrPS scheduling. Where τ_i is the i-th task, i is the number of tasks, c_i is the computation time of task i and p_i is the period of task i. rrPS scheduling algorithm contributes to process the scheduling of continuous media.

Table 8.3 The example data of rrPS scheduling

i	Computation time c_i	Period p_i	tc_i	sd_i
1	1	5	0.20	5
2	2	4	0.50	2
3	3	6	0.33	3

This algorithm considers time dependency of continuous media and it keeps fairness of resource allocation under normal scheduling condition. Even though the rrPS

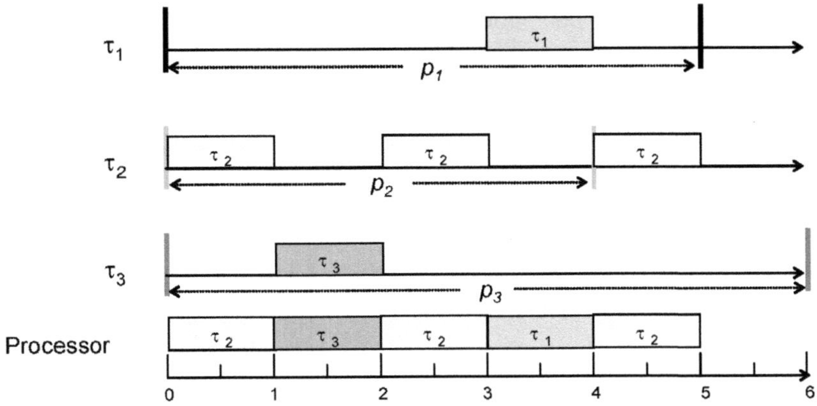

Fig. 8.8 Illustration of rrPS scheduling

scheduling algorithm has several the advantages, it has some difficulties in adapting continuous media as follows.

First, this algorithm does not show gradual degradation of performance under the overloaded condition. It works as if it kept fairness for the number of time quanta were not served from the scheduler. However it loses fairness for the ratio of computation time that is not served from the scheduler. In other words, weights which do not serve a given time quantum are different in the length of the computation time of tasks.

Second, this algorithm also has the possibility of avoidable context switching overhead. It means that the rate regulator can cause unnecessary context switching in task scheduling. Overheads of the context switching are machine dependent factors, which are page size, memory cycle time, number of registers, types of instruction format, *etc.* Since the portion of context switching overhead can be relatively small, it is seldom considered in many scheduling algorithm designs. When systems schedule many tasks, excessive context switching induces the trashing due to hot conflicts for systems and resources among tasks can cause serious degradation of system performance.

To avoid the above two drawbacks, the following points are considered [33].

1. Graceful degradation of performance under the overloaded condition.
2. Elimination of unnecessary context switching overhead.
3. Protection of the admission control policy that may cause deteriorating resources.

8.1.2.2 Modified Proportional Share (mPS) Scheduling Algorithm

The mPS scheduling algorithm is proposed by author [36]. In this scheduling algorithm, tc_i' of mPS scheduling algorithm represents the relative amount of resources that it should receive. It can be defined as Eq. 8.2. It considers the ratio of resource allocation in both normal condition and overloaded conditions.

$$tc_i' = \begin{cases} \frac{tc_i}{\sum\limits_{j=1}^{n} tc_j}, & if \ \sum\limits_{j=1}^{n} tc_j > 1, \ \forall i \\ tc_i, & if \ \sum\limits_{j=1}^{n} tc_j \leq 1, \ \forall i \end{cases} \tag{8.2}$$

$$tc_i = \frac{c_i}{p_i}, \ \forall i \tag{8.3}$$

Notation

In the above equations, notations are defined as follows:

Indices

i: task index

Parameters

n: the total number of tasks
c_i: the computation time of task i
p_i: the period of task i

The ps_i represents the virtual time index for next selection. It is 0 at the first stage and is updated as Eq. 8.4 in every time quantum:

$$ps_i(t) = tc_i' - \frac{1}{i} \sum_{j=s_i}^{i} a_i(j), \ \forall i, t \tag{8.4}$$

Notation

In the above equations, notations are defined as follows:

Indices

t: time index
i: task index

Parameters

c_i: the computation time of task i
p_i: the period of task i

DecisionVariables

s_i: the start time of task i
$\sum a_i$: total number of allocations that task i has received from start to current time

In the previous time quantum, the selected task has preferential right to schedule and if it did not violate the upper bound of rate regulator, it is selected again. If not, another task with the maximum pass will be selected.

By Eqs. 8.2 and 8.4, the rate regulator of mPS scheduling algorithm can be defined as Eq. 8.5:

$$ps_i(t) > 0, \ \forall i, t \tag{8.5}$$

The amount of context switching using this policy of preferential right is decreased.

The tc_i' and ps_i of a task are determined by its period and computation time. And the mPS scheduling algorithm selects a task by comparing the tc_i' and ps_i of the task. During scheduling, the number of context switching is decreased. The mPS scheduling algorithm shows graceful degradation of performance using tc_i', which is considered the rate regulating overloaded condition as shown in Eq. 8.5.

The algorithm of mPS scheduling is shown in Fig. 8.9.

---

procedure: mPS Scheduling Algorithm
input: task set
output: schedule
step 1: Set up all tasks' tc'_i and ps_i.
step 2: Select task with maximum tc'_i.
step 3: Update all tasks' ps_i.
step 4: If task satisfies rate regulator, then go to step 5. Otherwise go to step 6.
step 5: Select a task with the maximum ps_i in remains and go to step 4.
step 6: if all the tasks have not been scheduled, then go to step 3. Otherwise stop.

Fig. 8.9 Procedure of mPS scheduling algorithm

Table. 8.4 represents the example of mPS scheduling. Where τ_i is the ith task, i is the number of task, c_i is the computation time of task i and p_i is the period of task i.

The mPS scheduling algorithm shows better performance than the rrPS scheduling algorithm for graceful degradation of performance under the overloaded condition and less context switching as shown in Fig. 8.10. However, computational burden and solution accuracy of mPS could be improved by new algorithm based on Genetic Algorithm (GA).

Table 8.4 The example data of mPS scheduling

i	Computation time c_i	Period p_i	tc_i
1	1	5	0.19
2	2	4	0.48
3	3	6	0.32

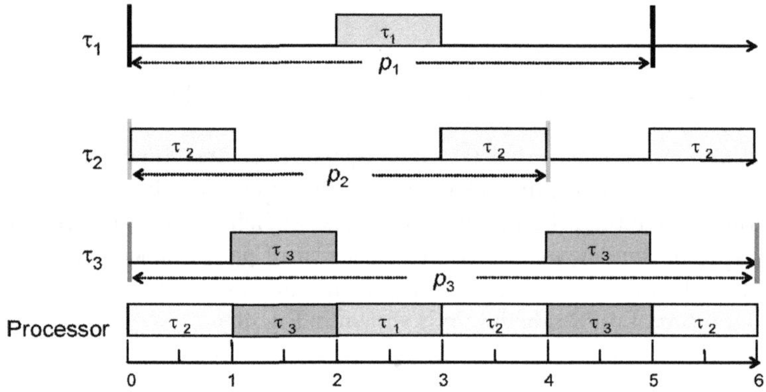

Fig. 8.10 Illustration of mPS scheduling

8.2 Continuous Task Scheduling

The availability of inexpensive high-performance processors has made it attractive to use multiprocessor systems for real-time applications. The programming of such multiprocessor systems presents a rather formidable problem. In particular, real-time tasks must be serviced within certain preassigned deadlines dictated by the physical environment in which the multiprocessor systems operates [13].

In this section, soft real-time task scheduling models on uniprocessor system and multiprocessor systems [14] are introduced, respectively. In particular, the algorithms are focused on the scheduling for continuous tasks that are periodic and nonpreemptive, and the objective is to minimize the total tardiness.

8.2.1 Continuous Task Scheduling Model on Uniprocessor System

Rate Regulating Proportional Share (rrPS) scheduling algorithm and Modified Proportional Share (mPS) Scheduling algorithm are designed for continuous soft real-time task. However, computational burden and solution accuracy of these algorithms could be improved by a new algorithm based on GA.

The proportion-based GA is proposed for this task scheduling problem. The objective is to minimize the weighted sum of variance of tardiness and the total number of context switching among tasks. Some drawbacks (*i.e.*, low resource utilization and avoidable context switching overhead) of Rate Monotonic (RM) and Earliest Deadline First (EDF) derived algorithms for soft real-time tasks could be fixed in the algorithm. Not only advantages of RM and EDF approaches but also the plus side of GA, such as, high speed, parallel searching and high adaptability is kept.

The following assumptions describe the property of this continuous task scheduling model on uniprocessor system:

A1. All tasks are periodic.
A2. All tasks are assigned to same processor (uniprocessor system).
A3. All tasks are preemptive.
A4. Only processing requirements are significant; memory, I/O and other resource requirements are negligible.
A5. All tasks are independent. This means that there are no precedence constraints.
A6. The deadline of a task is equal to its period.
A7. The system is a soft real-time system.

A task may be periodic, sporadic, or aperiodic. If the task needs to be cyclically executed at constant activation rates, the task is periodic. Sporadic task is not periodic, but may be invoked at irregular intervals and aperiodic task is not periodic and have no upper bound on their invocation rate [15].

Periodic tasks are characterized by their period and computation time. We focused on continuous media, that is, periodic tasks. In hard real-time tasks, these periodic tasks must finish their computation before their deadlines. However, since we focused on soft real-time tasks, rare and slight violation of deadline could be admitted. If a task cannot be finished until its next invocation, it is considered as tardiness. Actually, in case of playing video frames, 30 frames are played during (PER) for 1s to feel continued motion without halted gap. However, error within quality of service (QoS) of media is not perceived. QoS of video is 10^{-1} and QoS of audio is 10^{-3} [28, 29].

Table 8.5 represents the example data of continuous task scheduling and Fig. 8.11 represents the example of scheduling for soft real-time task, graphically. In

Fig. 8.11, task τ_1 needs τ_3 time units as computation time. However, task τ_1 is not executed really in its first period and is executed during two time units in its second period. Task τ_1's executed time in each period is smaller than its computation time. Therefore, tardiness has occurred in task τ_1's first period and second period. However, the other tasks τ_2 and τ_3 are executed during its computation time. Tasks τ_2 and τ_3 keep their deadlines.

Table 8.5 The example data of continuous task scheduling

i	Computation time c_i	Period p_i
1	3	6
2	1	3
3	2	4

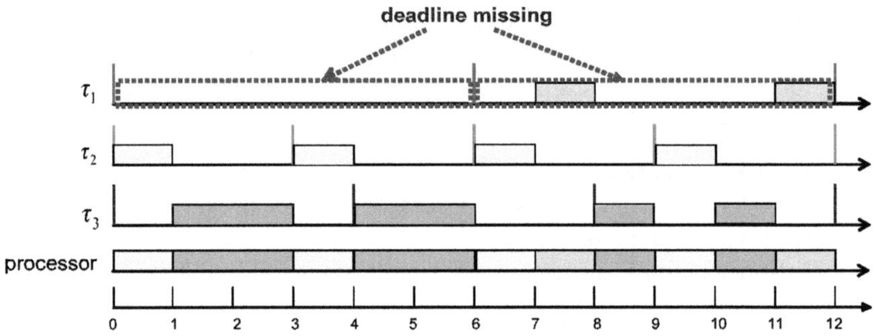

Fig. 8.11 Example of continuous soft real-time task

8.2.1.1 Mathematical Formulation

The problem is formulated under the following assumption:

A1. Computation time and period of each task is known a *priori.*
A2. A time unit is artificial time unit which is hypothetic measurement unit, not real measurement unit such as nanosecond, millisecond, *etc.*
A3. The context switching overheads among tasks are all same.

Task Constraints

Continuous task can be formulated as follows:

$$\gamma_i = \frac{c_i}{p_i},\ i = 1,2,\ldots,N \tag{8.6}$$

$$r_{ij} = \begin{cases} 0, & j = 1 \\ d_{ij-1}, & j = 2,3,\ldots,n_i \end{cases},\ \forall i \tag{8.7}$$

$$d_{ij} = r_{ij} + p_i,\ i = 1,2,\ldots,N, j = 1,2,\ldots,n_i \tag{8.8}$$

$$s_{ij} < s_{i'j'} < f_{ij},$$
$$\text{if priority } (\tau_{i'}) > \text{ priority } (t_i) \text{ and } r_{i'j'} \geq r_{ij},\ \forall\, i,j,i',j' \tag{8.9}$$

Notation

In the above equations, notations are defined as follows:

Indices

i': task index $i' \neq i, i' = 1,2,\ldots,N$,
j': index of j'-th executed task τ_i, $j' = 1,2,\ldots,n_i$

Parameters

N: total number of tasks
T: scheduled time
n_i: total number of executed task τ_i
r_{ij}: j-th release time of task τ_i
s_{ij}: j-th start time of task τ_i
f_{ij}: j-th finish time of task τ_i
i: proportional amount of executing time of processor for task τ_i
priority(τ_i): priority of task τ_i

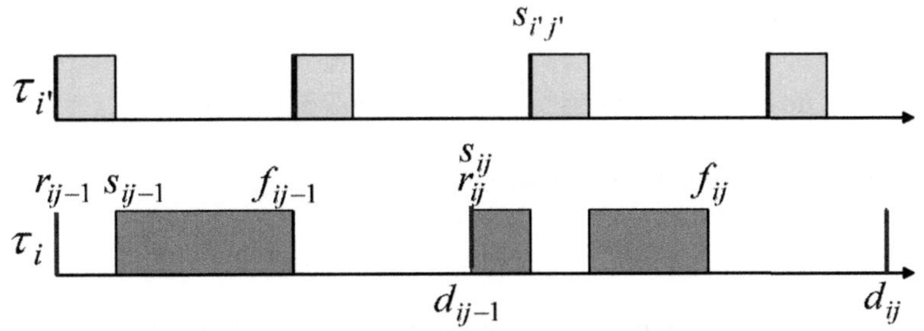

Fig. 8.12 Characteristics of continuous tasks II

Equation 8.6 means that all tasks have their share executing time of processor based on their period and computation time. The release time of task is defined in Eq. 8.7. Equation 8.8 means that the deadline of a task is equal to its period as shown in Fig. 8.12. All tasks are cyclically executed at constant activation rates as their periods. In the first stage, the release time of task is 0. From the next stage, the release time of the j-th executed task is equal to the deadline of the j-1- th executed task. Equation 8.9 means that all tasks are preemptive. If the processor is currently occupied by a task and a higher priority task is invoked at the moment, the higher priority task occupies the processor immediately as shown in Fig. 8.12.

System Constraints

Soft real-time system with uniprocessor can be formulated as follows:

$$\sum_{i=1}^{N} x_{it} \in \{0,\ 1\},\ \forall t \tag{8.10}$$

$$y_{ij} = \begin{cases} 1,\ \text{if } \sum_{t=r_{ij}}^{d_{ij}} x_{it} < c_i \\ 0,\ \text{otherwise} \end{cases},\ j = 1,2,\ldots,n_i,\ i = 1,2,\ldots,N \tag{8.11}$$

Notation

In the above equations, notations are defined as follows:

Indices

t: time index $t = 1,2,\ldots,T$,

DecisionVariables

y_{ij}: occurrence of tardiness at the j-th executed time of task τ_i

$$x_{ij} = \begin{cases} 1, \text{if } \tau_i \text{ is assigned to processor in time } t \\ 0, \text{otherwise} \end{cases}$$

$$i = 1,2,\ldots,N,\ t = 1,2,\ldots,T \tag{8.12}$$

In Eq. 8.10, the total number of assigned tasks in time t is 1 or 0. It means that all tasks are assigned to same processor. In other word, system is uniprocessor system. Equation 8.11 means system is soft real-time system and slight violence of deadline could be admitted. The value y_{ij}=1 means that executed time of task during one period is small than its computation time and that is tardiness.

Objective Function

The problem of scheduling for continuous tasks is to determine execution sequence of tasks with the objective of minimizing the weighted sum of variance of tardiness and the total number of context switching among tasks:

$$\min F(x) = \alpha \frac{\sigma_m^2}{\sigma_W^2} + \beta \frac{\mu_s}{\mu_W} \tag{8.13}$$

In the above objective function, parameters are defined as follows:

Parameters

α, β: coefficients

$$\alpha + \beta = 1,\ \alpha, \beta \in [0..1] \tag{8.14}$$

n_s: total number of context switching
m_i: average tardiness of τ_i

$$m_i = \frac{1}{n_i} \sum_{j=1}^{n_i} y_{ij},\ i = 1,2,\ldots,N \tag{8.15}$$

$\overline{m}$: average of m_i

$$\overline{m} = \frac{1}{N} \sum_{i=1}^{N} m_i,\ i = 1,2,\ldots,N \tag{8.16}$$

σ_m^2: variance of m_i

$$\sigma_m^2 = \frac{1}{N} \sum_{i=1}^{N} (m_i - \overline{m})^2,\ i = 1,2,\ldots,N \tag{8.17}$$

σ_W^2: worst value of σ_m^2
μ_s: average of n_s

$$\mu_s = \frac{n_s}{T} \tag{8.18}$$

μ_W: worst value of μ_s

8.2.1.2 Proportion-based GA

Let $P(t)$ and $C(t)$ be parents and offspring in current generation t, respectively. The overall procedure of proportion-based GA for solving continuous task scheduling is outlined as follows:

---

procedure: proportion-based GA (pGA)
input: task set, scheduling time T, GA parameters ($popSize$, $maxGen$, p_M, p_C)
output: best schedule set
begin
  $t \leftarrow 0$;
  initialize $P(t)$ by proportion-based encoding;;
  evaluate $P(t)$ by proportion-based decoding;
  while (**not** terminating condition) **do**
    create $C(t)$ from $P(t)$ by by period unit crossover;
    create $C(t)$ from $P(t)$ by altering mutation;
    evaluate $C(t)$ by proportion -based decoding;
    select $P(t+1)$ from $P(t)$ and $C(t)$ by roulette wheel selection;
    $t \leftarrow t+1$;
  end
  output best schedule set on processors;
end

Proportion-based Genetic Representation

To develop a proportion-based genetic representation for continuous task scheduling model, there are three main phases:

Phase 1: Generate tasks sequence which is assigned to a processor using proportion-based encoding procedure.
Phase 2: Create a schedule S using the task sequence by proportion-based decoding procedure.
Phase 3: Draw a Gantt chart and calculate the fitness for this schedule.

Phase 1: A chromosome V_k=v_t, $k = 1, 2, \ldots, popSize$, represents tasks which are assigned to a processor and the t-th gene v_t represents the task which is assigned to a processor in time t, where $popSize$ is total number of chromosomes in each generation. The length of a chromosome is T. It means the minimum term of scheduling. Figure 8.13 represents the structure of chromosome for proposed genetic algorithm during eight time units. The proportion-based encoding procedure is shown in Fig. 8.14 and an example of the proportion-based encoding is shown in Fig. 8.15.

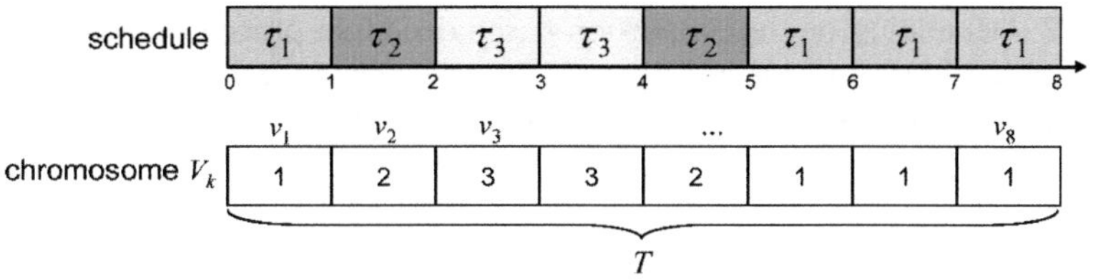

Fig. 8.13 Chromosome representation

---

procedure: proportion-based encoding
input: number of tasks N, number of time periods T
output: chromosome $V_k=\{v_t\}$
step 1: Calculate γ_i and γ_s. γ_s is total sum of γ_i.
step 2: Calculate q_i. q_i is the cumulative rate of share for executing time of processor and the following equation.

$$q_i = \begin{cases} \dfrac{\gamma_i}{\gamma_s}, & i=1 \\ q_{i-1} + \dfrac{\gamma_i}{\gamma_s}, & i=2,3,\dots,N \end{cases}$$

step 3: Generate a random number r from range [0..1].
if $r \le q_1$ then assign 1 to v_t, else if $q_{i-1} < r \le q_i$, i=2,3,…,N, then assign i to v_t.
step 4: Increase t by 1 and repeat steps 3-4 until $t = T$.

Fig. 8.14 Proportion-based encoding procedure

date set

i	c_i	p_i	q_i
1	2	6	0.27
2	2	3	0.80
3	1	4	1.00

q_i: the cumulative rate of share for processor service time

Gantt chart (N=3, T=12)

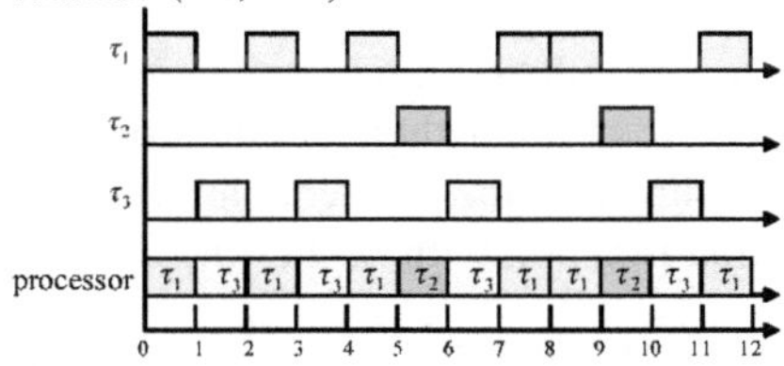

trace table

t	r	v_t	V
1	0.26	1	1
2	0.88	3	1 3
3	0.08	1	1 3 1
4	0.98	3	1 3 1 3
5	0.25	1	1 3 1 3 1
6	0.45	2	1 3 1 3 1 2
7	0.89	3	1 3 1 3 1 2 3
8	0.13	1	1 3 1 3 1 2 3 1
9	0.03	1	1 3 1 3 1 2 3 1 1
10	0.55	2	1 3 1 3 1 2 3 1 1 2
11	0.99	3	1 3 1 3 1 2 3 1 1 2 3
12	0.18	1	1 3 1 3 1 2 3 1 1 2 3 1

r: randomly selected number

Fig. 8.15 Example of the proportion-based encoding

Phase 2: Depending on the proportion-based decoding procedure as shown in Fig. 8.16, we can create a schedule S using the task sequence of chromosome V_k. An example of the decoding process is shown in Fig. 8.17.

---

procedure: proportion-based decoding
input: chromosome $v(\cdot)$
output: schedule S
step 1: Derive x_{it} from V_k.
step 2: Calculate n_{ij}^c. n_{ij}^c is executed time of task τ_i from r_{ij} to d_{ij}.
step 3: If $n_{ij}^c > c_i$ then regulate executed time by changing the value 1 of randomly selected locus to 0.
step 4: If $n_{ij}^c < c_i$ then assign i to processor in idle time.
step 5: Generate schedule S by appending task number i with value 1 at time t.

Fig. 8.16 Proportion-based decoding procedure

Phase 3:Draw a Gantt chart and calculate the fitness for this schedule S generated in Phase 2. The fitness function is essentially the objective function in the problem. It provides a means of evaluating the search node and it also controls the selection process (Gen and Cheng, [18]). The fitness evaluation method options considered were "Fitness is Evaluation" (FE), Windowing (WD) and Linear Normalization (LN) (Tracy and Douglass, [37]; Goldberg, [38]). We compare with the results of these evaluation methods.

FE is the most basic fitness evaluation method. Because we use the roulette wheel selection, we convert the minimization problem to maximization problem (Emilda *et al.* [39]; Hou *et al.* [40]), that is, the used evaluation function with FE $eval_F$ is then

$$eval_F(V_k) = 1/F(x), \; \forall\, k \tag{8.19}$$

WD is a technique for assigning fitness to a population of chromosomes to boost the fitness of the weaker members, in order to prevent their elimination and the resulting dominance by a small number of chromosomes. Parents are selected by roulette wheel selection. We define evaluation function with WD $eval_W$ as follows:

$$eval_W(V_k) = eval_F(V_k) - \min\{eval_F(V_k)\}, \; \forall\, k \tag{8.20}$$

LN is a fitness evaluation method intended to increase competition between similar individuals. It works in the following manner: Each individual's fitness is calculated with the fitness function. The individuals are ordered from least to most fit. Each individual's fitness becomes its order number. Parents are selected by roulette wheel selection. Chromosomes are ordered in increasing order of their $F(x)$. The evaluation function with LN $eval_L$ for each individual is defined as follows:

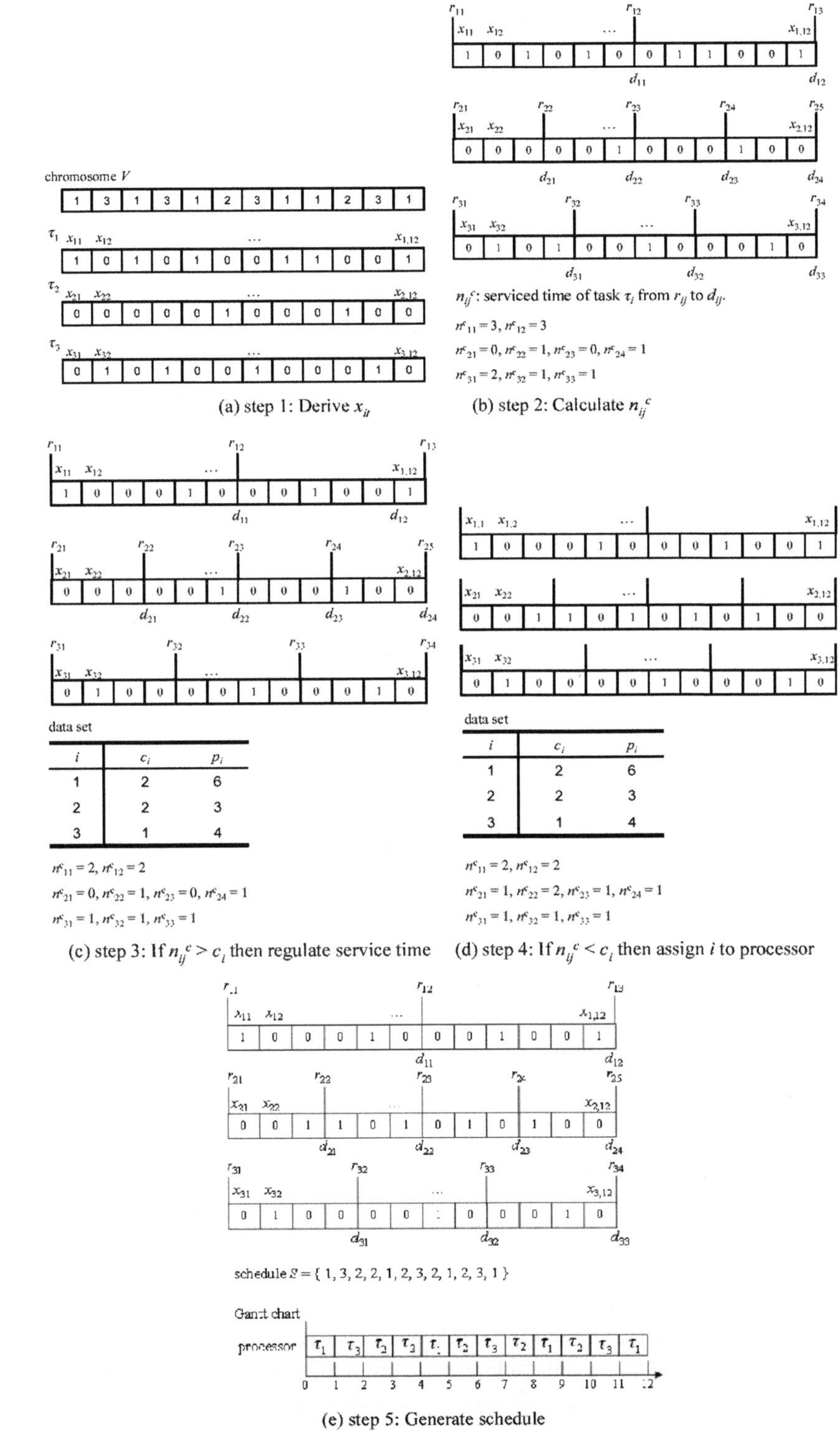

Fig. 8.17 Example of the proportion-based decoding

$$eval_L(V_{k'}) = \begin{cases} popSize, & k' = 1 \\ eval_L(V_{k'-1}), & k' = 1,2,\ldots,popSize \end{cases} \tag{8.21}$$

where k' is the ordered number of chromosome V_k.

Genetic Operators

Yoo and Gen [30] propose Period Unit Crossover (PUX). This operator creates two new chromosomes (the offspring) by mating two chromosomes (the parents), which are combined as shown in Fig. 8.18; the period (from r_{ij} to d_{ij}) is selected by i, j randomly chosen, and offspring chromosomes are built by exchanging the substrings of selected period between parents. V_1' and V_2' mean offspring 1 and offspring 2. The procedure of period unit crossover is shown in Fig. 8.19.

i	c_i	p_i
1	2	6
2	2	3
3	1	4

random number i = 2, j = 3
r_{23} = 6
d_{23} = 9

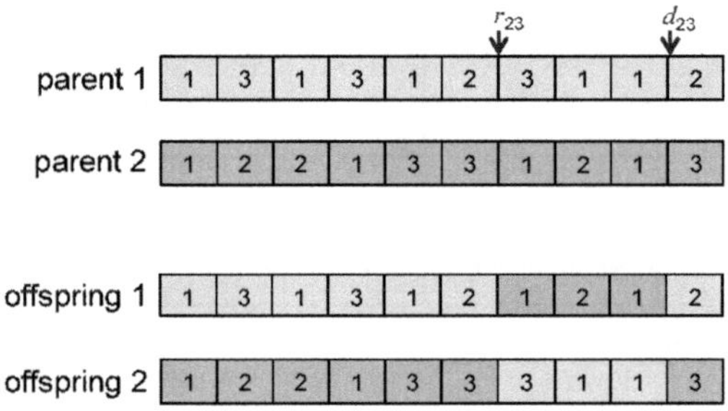

Fig. 8.18 Period unit crossover (PUX)

---

procedure: period unit crossover (PUX)
input: parents v_1 and v_2
output: offspring v_1' and v_2'
step 1: Generate random number i from range [1..N] and j from range [1..n_i].
step 2: Exchange substrings of period (from r_{ij} to d_{ij}) between parents v_1 and v_2, generate two offspring v_1' and v_2'.

Fig. 8.19 Procedure of period unit crossover

The purpose of mutation is a little change in chromosome. Yoo and Gen [30] use the classical one-bit altering mutation (Jackson and Rouskas [15]; Anagnostou *et al.* [28]). Figure 8.20 illustrates this mutation operator. Where V' means offspring, the procedure of altering mutation is shown in Fig. 8.21.

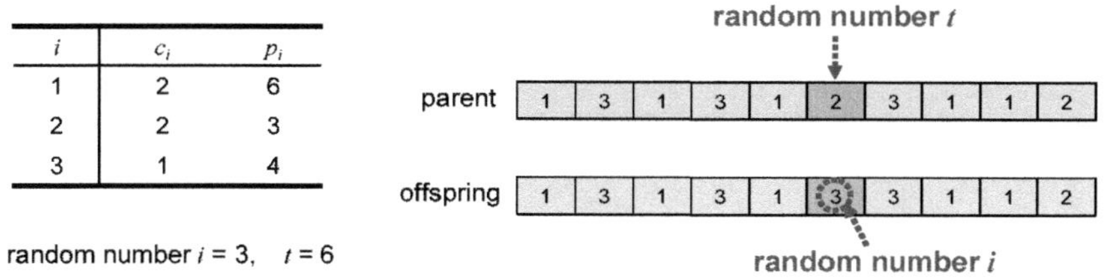

Fig. 8.20 Altering mutation

---

procedure: altering mutation
input: parents v
output: offspring v'
step 1: Generate random number i from range $[1..N]$, t from range $[1..T]$.
step 2: Change the value of v_t to i and generate offspring v'.

Fig. 8.21 Procedure of altering mutation

8.2.1.3 Numerical Experiments

In the experiment, there are several numerical tests performed. Proportion-based GA (pGA) is compared with traditional EDF and mPS scheduling algorithms. The EDF scheduling algorithm is a dynamic algorithm which schedules the task with the earliest deadline first and guarantee the optimality for hard real-time tasks. The mPS scheduling algorithm is designed for soft real-time tasks. Numerical tests are performed with three task sets. Tables 8.6–8.8 represent small scale, medium scale and large scale tasks sets respectively. For tasks' computation time and deadline, we use random number based on exponential distribution. They are different in total number of tasks and the degree of overloaded state.

Table 8.6 Task set 1

i	c_i	p_i
1	4	6
2	2	3
3	3	4

Table 8.7 Task set 2

i	c_i	p_i	i	c_i	p_i
1	1	4	6	2	12
2	2	7	7	4	9
3	1	5	8	1	8
4	2	8	9	2	15
5	1	6	10	3	17

Table 8.8 Task set 3

i	c_i	p_i	i	c_i	p_i
1	2	9	16	2	13
2	2	6	17	3	14
3	1	7	18	4	19
4	4	12	19	5	21
5	2	15	20	2	8
6	1	5	21	2	15
7	2	8	22	3	17
8	5	17	23	3	9
9	2	9	24	1	5
10	1	7	25	1	4
11	2	12	26	1	9
12	1	7	27	2	7
13	2	9	28	1	8
14	4	17	29	3	17
15	3	15	30	2	13

Table 8.9 represents the results of numerical tests with the task set 8.6, 8.7 and 8.8. The parameters were set as probabilities for crossover are tested from 0.5 to 0.8, from 0.001 to 0.4 for mutation, with the increments 0.05 and 0.001 respectively. For population size, individuals from 20 to 200 are tested. Table 8.9 shows the best computation result selected from number of computation results with various combinations of parameters.

Table 8.9 Comparison with results of scheduling algorithms for $F(x)$

Task set	EDF	mPS	pGA		
			FE	WD	LN
1	3.75	2.25	1.99	1.06	1.05
2	1.22	1.03	0.76	0.44	0.27
3	0.48	0.31	0.19	0.26	0.06

Table 8.9 shows that the value of objective function $F(x)$ value in proposed algorithm (pGA) is smaller than those of EDF and mPS algorithms. The objective of proposed scheduling algorithm is to minimize the weighted sum of variance of tardiness and the total number of context switching among tasks. Therefore, we can see that the performance of proposed pGA is better than those of other two algorithms in our test cases. Also, α and β can be adjusted for the important points. α and β are

set equally. In pGA, results are different according to the evaluation method. From the result of simulation, evaluation method with LN is found to be better than FE and WD for stability. The evaluation method with LN fitted can evaluate individuals with similar fitness values. In this continuous tasks scheduling problem, the fitness values of individuals are very similar. It means that the variance of tardiness is very small.

8.2.2 Continuous Task Scheduling Model on Multiprocessor System

The continuous task scheduling problem is defined as determining the execution schedule of continuous media tasks with minimizing the total tardiness under the following conditions:

A1. All tasks are periodic.
A2. All tasks are nonpreemptive.
A3. Only processing requirements are significant; memory, I/O and other resource requirements are negligible.
A4. All tasks are independent. This means that there are no precedence constrains.
A5. The deadline of a task is equal to its period.
A6. Systems are multiprocessor soft real-time systems.

Figure 8.22 represents the example of a scheduling for soft real-time tasks on multiprocessor systems, graphically. Where, i is task index, c_i is computation time of the i-th task, p_i is period of the i-th task and τ_{ij} is the j-th executed task of the i-th task. In Fig. 8.22, the serviced unit time of tau_{31} is 2 and smaller then the computation time of τ_{31}. It means that a tardiness has occurred in τ_{31} and the tardiness is 1. However, the other tasks keep their deadlines.

Table 8.10 The example data of continuous task scheduling

i	Computation time c_i	Period p_i
1	4	6
2	2	3
3	3	4

8.2.2.1 Mathematical Formulation

The continuous soft real-time tasks scheduling problem on multiprocessor systems can be formulated as follows:

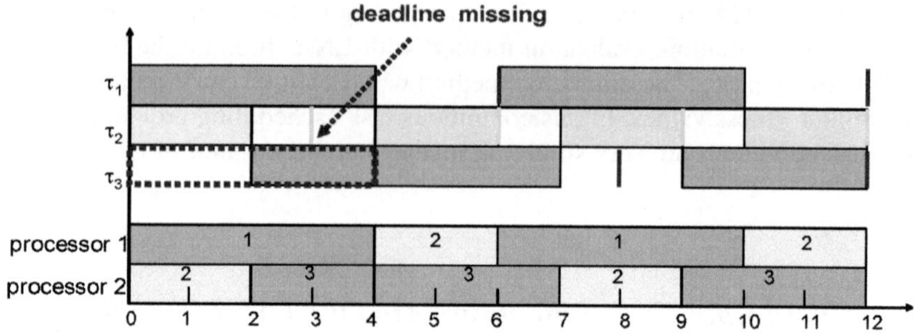

Fig. 8.22 Example of continuous soft real-time tasks scheduling on multiprocessor system

$$\min F(s) = \sum_{i=1}^{N} \sum_{j=1}^{n_i} \max\{0, \ (s_{ij} + c_i - d_{ij})\} \tag{8.22}$$

$$\text{s.t.} \ \ r_{ij} \le s_{ij} < d_{ij}, \ \forall \, i, j \tag{8.23}$$

Notation

In the above equations, notations are defined as follows:

Indices

m: processor index, $(m = 1, \ldots, M)$
i: task index, $(n = 1, \ldots, N)$
j: j-th executed task, $(j = 1, \ldots, n_j)$

Parameters

M: total number of processors
N: total number of tasks
τ_{ij}: j-th executed task of i-th task
c_i: computation time of i-th task
p_i: period of i-th task
T: scheduled time
n_i: total number of executed times for i-th task

$$n_i = \lfloor \frac{T}{P_i} \rfloor, i = 1, 2, \ldots, N \tag{8.24}$$

r_{ij}: j-th release time of i-th task

$$r_{ij} = \begin{cases} 0, & j = 1 \\ d_{ij-1}, & j = 2,3,\ldots,n_i \end{cases}, \quad \forall i \tag{8.25}$$

d_{ij}: j-th deadline time of i-th task

$$d_{ij} = r_{ij} + p_i, \ i = 1,2,\ldots,N, \ j = 1,2,\ldots,n_i \tag{8.26}$$

DecisionVariables

s_{ij}: j-th start time of i-th task

Equation 8.22 is the objective function and means to minimize the total tardiness as shown in Fig. 8.23. Equation 8.23 is the constraint of this problem and means that all tasks can start their computation between their release time and deadline.

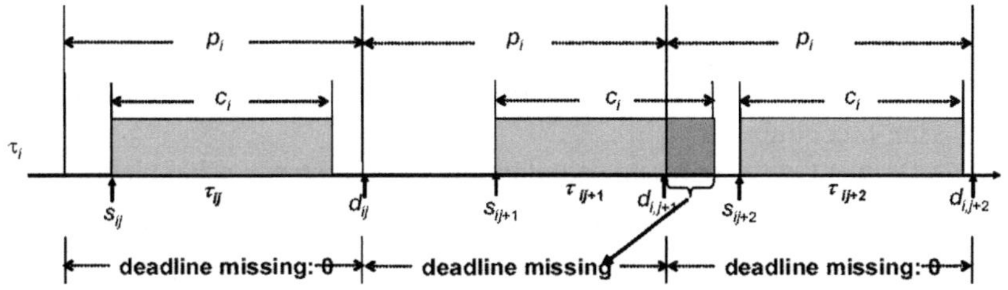

Fig. 8.23 Occurrence of tardiness

8.2.2.2 Period-based GA for Continuous Task Scheduling Model

Let $P(t)$ and $C(t)$ be parents and offspring in current generation t, respectively. The overall procedure of period-based GA for solving continuous task scheduling is outlined as follows:

Period-based Genetic Representation

To develop a period-based genetic representation for continuous task scheduling model, there are three main phases:

---

procedure: Period-based GA (pd-GA)
input: task set, total number of processors M, scheduling time T,
GA parameters $(popSize, maxGen, p_M, p_C)$
output: best schedule set
begin
$t \leftarrow 0$;
initialize $P(t)$ by period-based encoding;
evaluate $P(t)$ by period-based decoding;
while (**not** terminating condition) **do**
create $C(t)$ from $P(t)$ by multi period unit crossover;
create $C(t)$ from $P(t)$ by altering mutation;
evaluate $C(t)$ by period-based decoding;
select $P(t+1)$ from $P(t)$ and $C(t)$ by roulette wheel selection;
$t \leftarrow t+1$;
end
output best schedule set on processors;
end

Phase 1: Generate tasks sequence which is assigned to a processor using proportion-based encoding procedure.
Phase 2: Create a schedule S using the task sequence by proportion-based decoding procedure.
Phase 3: Draw a Gantt chart and calculate the fitness for this schedule.

Phase 1: A chromosome V_k=v_l, $k = 1, 2, \ldots, popSize$, represents the relation of tasks and processors. Where *popSize* is total number of chromosomes in each generation. The locus of l-th gene represents the order of tasks and the executed task and the value of gene v_l represents the number of the assigned processor. The length of a chromosome L can be calculated as follows:

$$L = \sum_{i=1}^{N} n_i \tag{8.27}$$

Figure 8.24 represents the structure of a chromosome for the proposed genetic algorithm. The task τ_{11}, τ_{21} and τ_{N1} are assigned to processor 1, 3 and 1 respectively.

Figure 8.25 represents the example of the period-based encoding. In this example, four tasks are assigned to two processors. And the total scheduling time T is 12. The length of a chromosome L is 9 by Eq. 8.27. The chromosome is generated by nine random numbers.
Phase 2: Depending on the following period-based decoding procedure, we can create a schedule S using the task sequence of chromosome V_k. Figure 8.26 represents example of the period based decoding with the chromosome in Fig. 8.25. Figure 8.26a represents the creating the scheduling set on each processor. The scheduling task set on processor 1 S_1 is generated by abstracting tasks with the value of gene

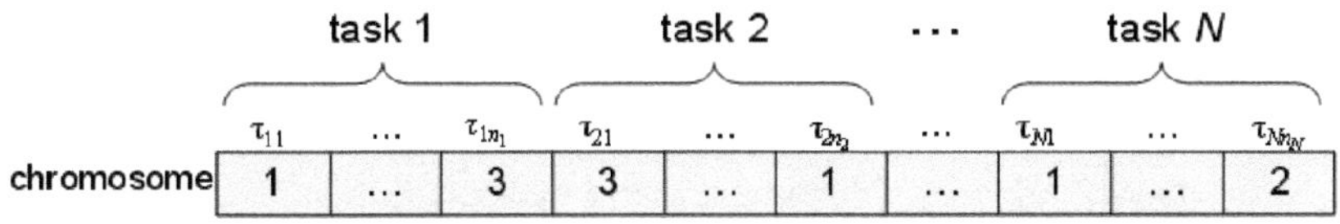

Fig. 8.24 Structure of a chromosome

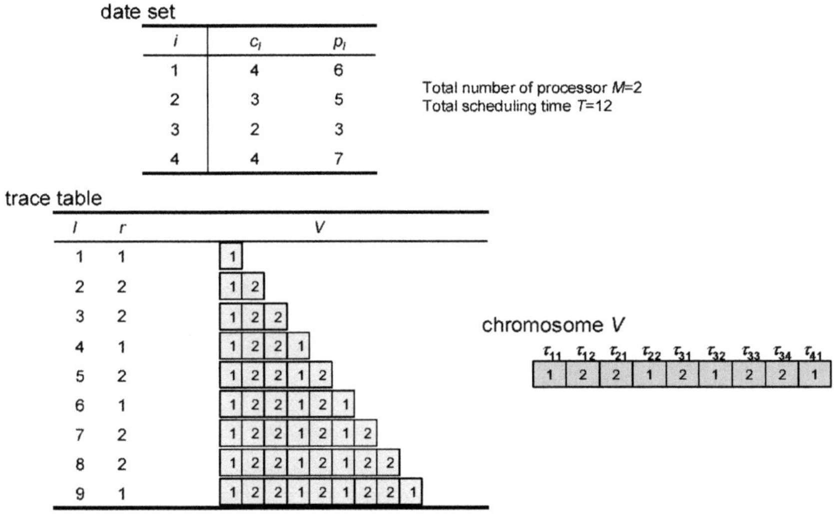

Fig. 8.25 Example of the period-based encoding

1, and S_2 is also generated in the same way. Figure 8.26b represents the sorting the scheduling set on each processor by task's release time. 8.26c represents the creating schedule set and calculating of the total tardiness. The tasks τ_{32} and τ_{33} cannot be executed on the processor 1 and 2 respectively. Therefore, the total tardiness is 5. **Phase 3:** Draw a Gantt chart and calculate the fitness for this schedule S generated in Phase 2. The fitness function is essentially the objective function in the problem. The fitness function is essentially the objective function for the problem. It provides the means of evaluating the search node and it also controls the selection process [15]. The fitness function used for this GA is based on the $F(s)$ of the schedule. Because the roulette wheel selection is used, the minimization problem is converted to the maximization problem, that is, the used evaluation function is then

$$eval(V_k) = 1/F(s),\ \forall k \tag{8.28}$$

Selection is the main way GA mimics evolution in natural systems. The common strategies called roulette wheel selection [16, 17] have been used.

S_1

task	release time r_{ij}
τ_{11}	0
τ_{22}	5
τ_{32}	3
τ_{41}	0

S_2

task	release time r_{ij}
τ_{12}	6
τ_{21}	0
τ_{31}	0
τ_{33}	6
τ_{34}	9

$S_1=\{\tau_{11}, \tau_{41}, \tau_{32}, \tau_{22}\}$

$S_2=\{\tau_{31}, \tau_{21}, \tau_{12}, \tau_{33}, \tau_{34}\}$

(a) step 1

data set

i	c_i	p_i
1	4	6
2	3	5
3	2	3
4	4	7

chromosome V

τ_{11}	τ_{12}	τ_{21}	τ_{22}	τ_{31}	τ_{32}	τ_{33}	τ_{34}	τ_{41}
1	2	2	1	2	1	2	2	1

$n_1=2$, $n_2=2$, $n_3=4$, $n_4=1$

schedule set in processor 1

$S_1=\{\tau_{11}, \tau_{22}, \tau_{32}, \tau_{41}\}$

schedule set in processor 2

$S_2=\{\tau_{12}, \tau_{21}, \tau_{31}, \tau_{33}, \tau_{34}\}$

(b) step 2

schedule set in processor 1 S_1

task	release time r_{ij}	computation time c_i	deadline d_{ij}
τ_{11}	0	4	6
τ_{41}	0	4	7
τ_{32}	3	2	6
τ_{22}	5	3	10

$S_1=\{\tau_{11}, \tau_{41}, \tau_{32}, \tau_{22}\}$

processor 1 : deadline missing 3

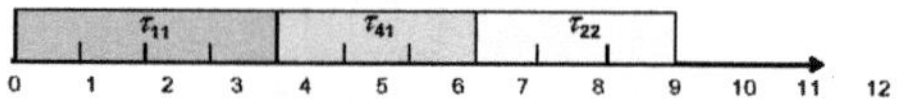

schedule set in processor 2 S_2

task	release time r_{ij}	computation time c_i	deadline d_{ij}
τ_{31}	0	2	3
τ_{21}	0	3	5
τ_{12}	6	4	12
τ_{33}	6	2	9
τ_{34}	9	2	12

$S_2=\{\tau_{31}, \tau_{21}, \tau_{12}, \tau_{33}, \tau_{34}\}$

processor 2 : deadline missing 2

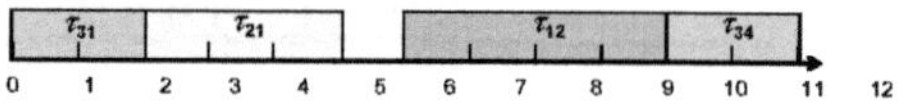

total deadline missing time $F(x)$=5

(c) step 3

Fig. 8.26 Example of the period-based decoding

Genetic Operators

The period unit crossover is proposed in this algorithm. This operator creates two new chromosomes (the offspring) by mating two chromosomes (the parents), which are combined as shown in Fig. 8.27. The periods of each task are selected by random number j and each offspring chromosome is built by exchanging selected periods between parents, where V_1' and V_2' means offspring 1 and 2, respectively.

For another GA operator, mutation, the classical one-bit altering mutation [18] is used.

date set

i	c_i	p_i
1	4	6
2	3	5
3	2	3
4	4	7

random number j

i	j
1	2
2	1
3	3
4	1

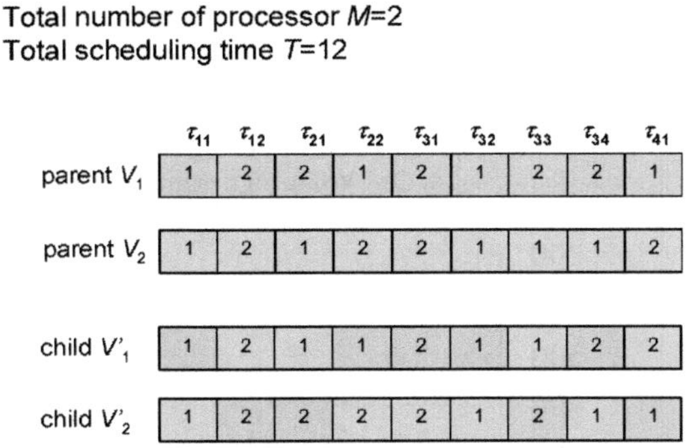

Fig. 8.27 Example of the mPUX

8.2.2.3 Numerical Experiments

In the experiment, there are several numerical tests performed. The period-based GA (pdGA) is compared with Oh-Wu's algorithm [19] by Oh and Wu and Monnier's algorithm by Monnier *et al.* [17]. Oh-Wu's algorithm and Monnier's algorithm use GA. However, these algorithms are designed for discrete tasks and use two-dimensional chromosomes.

For the numerical test, tasks are generated randomly based on exponential distribution and normal distribution as follows. Random tasks have been used by several researchers in the past [17]:

$$c_i^E = \text{random value based on exponential distribution with mean 5}$$
$$c_i^N = \text{random value based on normal distribution with mean 5}$$
$$r^E = \text{random value based on exponential distribution with mean } c_i^E$$
$$r^N = \text{random value based on normal distribution with mean } c_i^N$$
$$p_i^E = c_i^E + r^E$$
$$p_i^N = c_i^N + r^N$$

where c_i^E and c_i^N is the computation time of ith task based on exponential distribution and normal distribution, respectively. p_i^E and p_i^N is the period of the$i$-th task based on exponential distribution and normal distribution respectively.

The parameters were set to 0.7 for crossover (p_C), 0.3 for mutation (p_M), and 30 for population size (*popSize*). Probabilities for crossover are tested from 0.5 to 0.8, from 0.001 to 0.4 for mutation, with the increments 0.05 and 0.001 respectively. For population size, individuals from 20 to 200 are tested. Each combination of parameters is tested 20 times, respectively. The best combination of parameters is selected by average performance of 20 runs.

Numerical tests are performed with 100 tasks. Figures 8.28 and 8.29 show the comparisons of results by three different scheduling algorithms. In these figures, the total tardiness of the pd-GA is smaller than that of other algorithms.

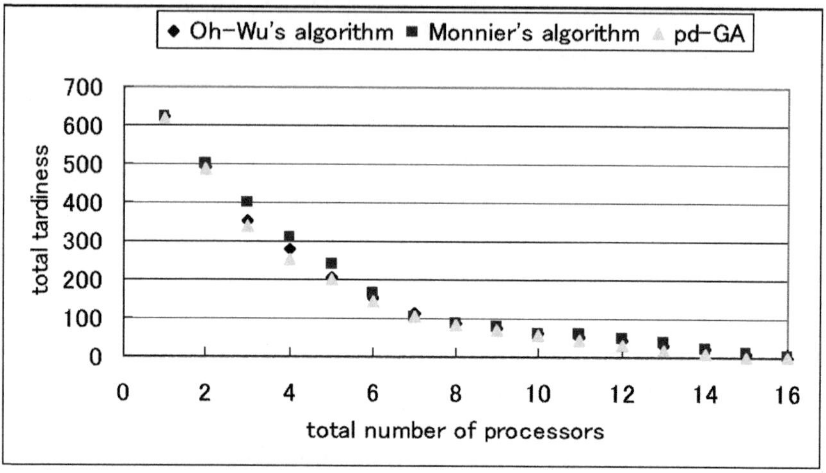

Fig. 8.28 Comparison of results (exponential)

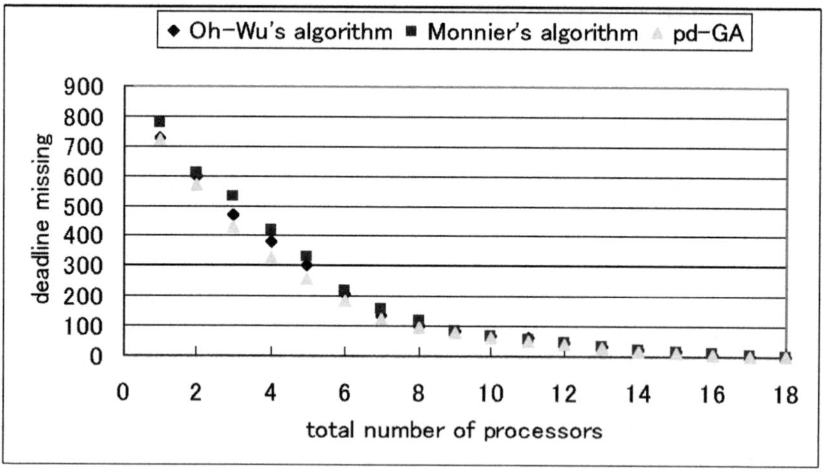

Fig. 8.29 Comparison of results (normal)

Table 8.11 shows the numerical data of Figs. 8.28 and 8.29. Tables 8.12 and 8.13 are the comparison of results in terms of better, worse and equal performance. In

Table 8.12, pd-GA performed better than Oh-Wu's algorithm in 9 cases and Monnier's algorithm in 11 cases. In Table 8.13, pd-GA performed better than Oh-Wu's algorithm in 10 cases and Monnier's algorithm in 12 cases.

Table 8.11 Numerical data (total tardiness) of Figs. 8.28 and 8.29

Algorithm	Total number of processors			
	Exponential		Normal	
	8	15	8	17
Oh-Wu's algorithm	86	7	103	2
Monnier's algorithm	85	12	117	8
pd-GA	81	0	97	0

Table 8.12 Comparison of other algorithms in terms of better, worse and equal performance (exponential)

Algorithm	pd-GA			total
	<	=	>	
Oh-Wu's algorithm	2	9	9	20
Monnier's algorithm	1	8	11	20

Table 8.13 Comparison of other algorithms in terms of better, worse and equal performance (normal)

Algorithm	pd-GA			total
	<	=	>	
Oh-Wu's algorithm	2	8	10	20
Monnier's algorithm	0	8	12	20

8.3 Real-time Task Scheduling in Homogeneous Multiprocessor

In this section, a new scheduling algorithm for nonpreemptive soft real-time tasks on multiprocessor without communication time using multiobjective genetic algorithm (moGA) is introduced. The objective of this scheduling algorithm is to minimize the total tardiness and total number of processors used. For these objectives, this algorithm is combined with adaptive weight approach (AWA) that utilizes some useful information from the current population to readjust weights for obtaining a search pressure toward a positive ideal point [18].

8.3.1 *Soft Real-time Task Scheduling Problem (sr-TSP) and Mathematical Model*

The problem of scheduling the tasks of precedence and timing constrained task graph on a set of homogeneous processors is considered in a way that simultaneously minimizes the number of processors used and the total tardiness under the following conditions:

A1. All tasks are nonpreemptive.
A2. Every processor processes only one task at a time.
A3. Every task is processed on one processor at a time.
A4. Only processing requirements are significant; memory, I/O, and other resource requirements are negligible.

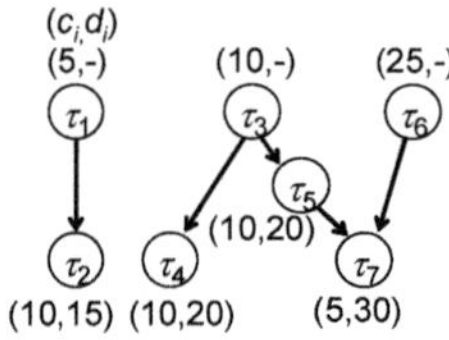

time constraint

i	suc(τ_i)	c_i	d_i
1	2	5	-
2	-	10	15
3	4,5	10	-
4	-	10	20
5	7	10	20
6	7	25	-
7	-	5	30

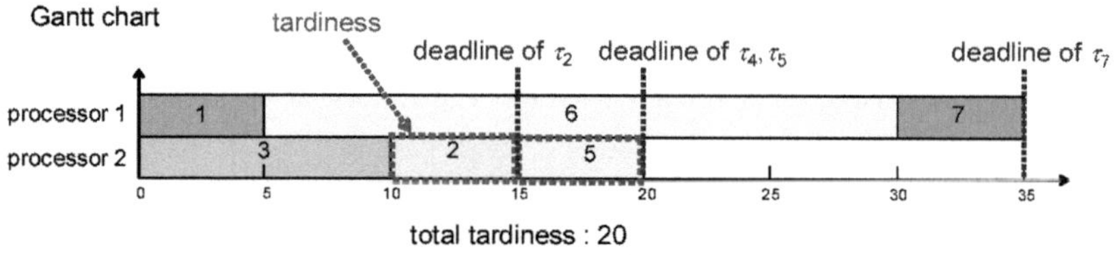

Fig. 8.30 Example of soft real-time tasks scheduling on multiprocessor

Figure 8.30 represents the example of scheduling for soft real-time tasks, graphically. In this example, the number of tasks is seven and the number of processors is two.

Where i and j are indices of tasks, τ_i is the i-th task, c_i is the computation time of the i-th task, d_i is the deadline of the i-th task, e_{ij} is the edge from τ_i to τ_j and suc(τ_i) is set of immediate successors of task τ_i. In Fig. 8.30, tardiness occurs in τ_2 and τ_5 and τ_4 is not assigned any processor. Total tardiness is 20.

The problem is formulated under the following assumptions. Computation time and deadline of each task is known. A time unit is artificial time unit. Soft real-time tasks scheduling problem (sr-TSP) is formulated as follows:

$$\min f_1 = M \tag{8.29}$$

$$\min f_2 = \sum_{i=1}^{N} \max\{0,\ t_i^S + c_i - d_i\} \tag{8.30}$$

$$\text{s.t. } t_i^E \le t_i^S \le d_i,\ \forall i \tag{8.31}$$

$$t_i^E \ge t_j^E + c_j,\ \tau_j \in \text{pre}(\tau_i),\ \forall i \tag{8.32}$$

$$1 \le M \le N \tag{8.33}$$

Notation

In the above equations, notations are defined as follows:

Indices

i, j: task index, $(i, j = 1, \ldots, N)$
m: processor index, $(m = 1, \ldots, M)$

Parameters

$G = (T, E)$: task graph
$T = \{\tau_1, \tau_2, \ldots, \tau_N\}$: a set of N tasks
$E = \{e_{ij}\}, i, j = 1, 2, \ldots, N, i \ne j$: a set of directed edges among the tasks representing precedence
τ_i: i-th task, $i = 1, 2, \ldots, N$
p_m: m-th processor, $m = 1, 2, \ldots, M$
c_i: computation time of task τ_i
$\text{pre}^*(\tau_i)$: set of all predecessors of task τ_i
$\text{suc}^*(\tau_i)$: set of all successors of task τ_i
$\text{pre}(\tau_i)$: set of immediate predecessors of task τ_i
$\text{suc}(\tau_i)$: set of immediate successors of task τ_i
t_i^E: earliest start time of i-th task

$$t_i^E = \begin{cases} 0, & if\ \neg\exists\tau_j : e_{ji} \in E \\ \max\limits_{\tau_j \in \text{pre}^*(\tau_i)} \{t_j^E + c_j\}, & \text{otherwise} \end{cases},\quad \forall i \tag{8.34}$$

t_i^L: latest start time of i-th task

$$t_i^L = \begin{cases} d_i - c_i, & if\ \neg\exists\tau_j : e_{ij} \in E \\ \min\{\min\limits_{\tau_j \in \text{suc}^*(\tau_i)} \{t_j^L - c_j\}, d_i - c_i\}, & \text{otherwise} \end{cases},\quad \forall i \tag{8.35}$$

DecisionVariables

t_i^S: real start time of i-th task
M: total number of processors used

Equations 8.29 and 8.30 are the objective function in this scheduling problem. In Eq. 8.29 there is the means to minimize the total number of processors used and in Eq. 8.30 the means to minimize total tardiness of tasks. Constraints conditions are shown in Eqs. 8.31–8.33. Equation 8.31 means that the task can be started after its earliest start time, begin its deadline. Equation 8.32 defines the earliest start time of task based on precedence constraints. Equation 8.33 is the nonnegative condition for the number of processors.

8.3.2 Multiobjective GA for srTSP

Let $P(t)$ and $C(t)$ be parents and offspring in current generation t, respectively. The overall procedure of multiobjective GA for solving soft real-time task scheduling problem (srTSP) is outlined as follows:

---

procedure: moGA for sr-TSP
input: task graph data set
output: the best schedule set S
begin
  $t \leftarrow 0$;
  initialize $P(t)$ by encoding strategy I and strategy II;
  calculate f_1, f_2 by decoding routine;
  create Pareto $E(P)$;
  evaluate $eval(P)$ by adaptive weight approach (AWA);
  while (**not** terminating condition) **do**
    create $C(t)$ from $P(t)$ by one-cut crossover;
    create $C(t)$ from $P(t)$ by altering mutation;
    calculate f_1, f_2 by decoding routine;
    update Pareto $E(P,C)$;
    evaluate $eval(P,C)$ by AWA;
    select $P(t+1)$ from $P(t)$ and $C(t)$;
    $t \leftarrow t+1$;
  end
  output the best schedule set S;
end

8.3.2.1 Encoding and Decoding

A chromosome V_k, $k = 1, 2, \ldots, popSize$, represents one of all the possible mappings of all the tasks into the processors, where *popSize* is the total number of chromosomes in a generation. A chromosome V_k is partitioned into two parts $u(\cdot)$, $v(\cdot)$; $u(\cdot)$ means scheduling order and $v(\cdot)$ means allocation information. The length of each part is the total number of tasks. The scheduling order part should be a topological order with respect to the given task graph that satisfies precedence relations. The allocation information part denote the processor to which the task is allocated. To develop a genetic representation for sr-TSP model, there are three main phases:

Phase 1: Creating a task sequence: generate a random priority to each task in the model using priority-based encoding procedure for the first vector $u(\cdot)$.
Phase 2: Assigning tasks to processors: generate a permutation encoding for processor assignment of each task (second vector $v(\cdot)$).
Phase 3: Designing a schedule

Step 3.1: Create a schedule S using task sequence and processor assignments.
Step 3.2: Draw a Gantt chart for this schedule.

Phase 1: Creating a task sequence: generate a random priority to each task in the model using priority-based encoding procedure for the first vector $u(\cdot)$. Figure 8.31 illustrates the process of this encoding procedure on a chromosome.

```
procedure: Encoding Strategy I for sr-TSP
input: task graph data set
output: u(·)
begin
  l←1, w←φ;
  while (T ≠ φ)
    w ← w ∪arg{τ_i |pre*(τ_i) = φ, ∀i };
    T ← T- {τ_i}, i∈w;
    while (w ≠ φ)
      j ← random(w);
      u(l) ← j;
      l ← l+1;
      w ← w – {j};
      pre*(τ_i) ← pre*(τ_i) – {τ_j}, ∀i;
  output u(·);
end
```

Fig. 8.31 Encoding Strategy I for sr-TSP

Figure 8.32 represents the example of encoding strategy I procedure.

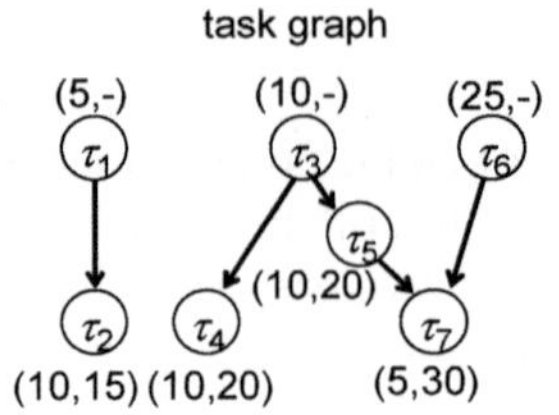

data set

i	suc(τ_i)	c_i	d_i		
1	2	5	-	0	7
2	-	10	15	5	5
3	4,5	10	-	0	0
4	-	10	20	10	10
5	7	10	20	10	10
6	7	25	-	0	0
7	-	5	30	25	25

W	j	T
ϕ		$\{\tau_1, \tau_2, \tau_3, \tau_4, \tau_5, \tau_6, \tau_7\}$
{1,3,6}	3	$\{\tau_2, \tau_4, \tau_5, \tau_7\}$
{1,6}	6	$\{\tau_2, \tau_4, \tau_5, \tau_7\}$
{1}	1	$\{\tau_2, \tau_4, \tau_5, \tau_7\}$
ϕ		$\{\tau_2, \tau_4, \tau_5, \tau_7\}$
{2,4,5}	4	$\{\tau_7\}$
{2,5}	5	$\{\tau_7\}$
{2}	2	$\{\tau_7\}$
ϕ		$\{\tau_7\}$
{7}	7	ϕ
ϕ		ϕ

chromosome V

l	1	2	3	4	5	6	7
task no. $u(\cdot)$	3	6	1	4	5	2	7

Fig. 8.32 Example of encoding strategy I procedure

Step 1.2: Decode a feasible task sequence that satisfies the precedence constraints of rtTSP.

Phase 2: Assigning tasks to processors: generate a permutation encoding for processor assignment of each task (second vector $v(\cdot)$). Figure 8.33 illustrates the process of this encoding procedure on a chromosome. In encoding strategy II procedure, α, β is the boundary constant to decide on increasing the number of processor and decreasing the number of processor respectively.

Phase 3: Designing a schedule, step 3.1: Create a schedule S using task sequence and processor assignments. Figure 8.34 represents the example of encoding strategy II procedure. Decoding procedure is shown in Fig. 8.35.

In Fig. 8.36, insert(i) means to insert τ_i at idle time if τ_i is computable in idle time, start(i) means to assign τ_i to maximum finish time of all assigned task to p_M, add_idle means to add idle time to idle time list if idle time is occurred, I^S means the start time of idle duration, I^F means the end time of idle duration, $idle_m$ means the list of idle time and t_m means the maximum finish time of all assigned task to p_M.

Step 3.2: Draw a Gantt chart for this schedule: Figure 8.36 represents the example of drawing process for a Gantt chart with chromosome in Figs. 8.32 and 8.34.

procedure: Encoding Strategy II for sr-TSP

input: task graph data set, $u(\cdot)$, $M = \begin{cases} M(k-1), & \text{if } 1 < k \le popSize \\ |\text{subgraph}|, & \text{if } k = 1 \end{cases}$, α, β

output: $v(\cdot)$, Mk

```
begin
    l←1, tm ← 0, ∀m, idle ← 0;
    while (l ≤ N)
    m ← random[1, M];
    i ← u(l);
    if (tm < tiE) then
        tiS ← tiE;
        idle ← idle + (tiS - tm);
    else
        tiS ← tm;
        if ( (di is not defined && tiS >tiL) || tiS > di) then
            if (idle/ci<α) then
                M ← M+1;
                m ← M;
                idle ← idle + tiE;
                tm ← tiE + ci;
            else
                idle ←max{0, (idle -ci)} ;
        else
            tm ← tiS+ci;
            v(l) ← m;
            l ← l+1;
            idle ← idle + Σ(max{ tm } - tm);
    while (idle/ΣM x max{tm}> β)
        M ← M-1;
        idle ←idle- idle/ΣM x max{tm};
    output v(·), Mk;
end
```

Fig. 8.33 Procedure of encoding strategy II for sr-TSP

8.3.2.2 Evolution Function and Selection

The multi-objective optimization problems have been receiving growing interest from researchers with various backgrounds since early 1960. Recently, GAs have received considerable attention as a novel approach to multiobjective optimization problems, resulting in a fresh body of research and applications known as genetic multi-objective optimizations [20].

Adaptive weight approach (AWA) [18] that utilizes some useful information from the current population to readjust weights for obtaining a search pressure toward a positive ideal point is combined in this scheduling algorithm.

The evaluation function is designed as follows:

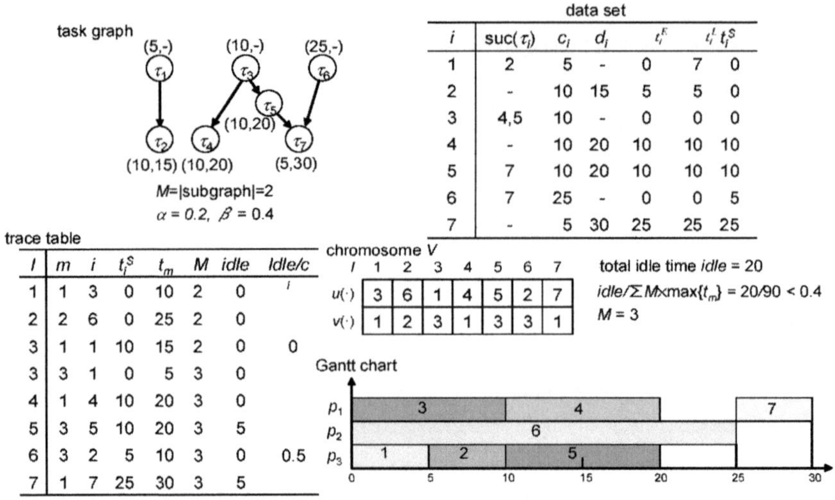

data set

i	suc(τ_i)	c_i	d_i	t_i^E	t_i^L	t_i^S
1	2	5	-	0	7	0
2	-	10	15	5	5	0
3	4,5	10	-	0	0	0
4	-	10	20	10	10	10
5	7	10	20	10	10	10
6	7	25	-	0	0	5
7	-	5	30	25	25	25

trace table

l	m	i	t_i^S	t_m	M	$idle$	$Idle/c$
1	1	3	0	10	2	0	
2	2	6	0	25	2	0	
3	1	1	10	15	2	0	0
3	3	1	0	5	3	0	
4	1	4	10	20	3	0	
5	3	5	10	20	3	5	
6	3	2	5	10	3	0	0.5
7	1	7	25	30	3	5	

Fig. 8.34 Example of encoding strategy II procedure

---

procedure: Decoding for sr-TSP
input: task graph data set, chromosome $u(\cdot)$, $v(\cdot)$
output: schedule set S, the total number of processor used $f1$, total tardiness of tasks $f2$
begin
    $l \leftarrow 1$, $tm \leftarrow 0$, $\forall m$, $idlem \leftarrow \phi$, $\forall m, f1 \leftarrow 0, f2 \leftarrow 0, S \leftarrow \phi$;
    while $(l \le N)$ **do**
        $i \leftarrow u(l)$;
        $m \leftarrow v(l)$;
        if $(tm = 0)$ **then** $f1 \leftarrow f1 + 1$;
            $IS^*, IF^* \leftarrow$ find $\{IS, IF | (IS, IF) \in idlem, IS \le di\}$;
        if (IS^* is exist && $tm > tiL$) **then** *insert(i)*;
        else *start(i)*;
        add_idle();
        $f2 \leftarrow f2 + \max\{0, (tiS + ci - di)\}$;
        $S \leftarrow S \cup \{(i, m: tiS - tiF)\}$;
        $l \leftarrow l + 1$;
    output $S, f1, f2$;
end

Fig. 8.35 Procedure of decoding for sr-TSP

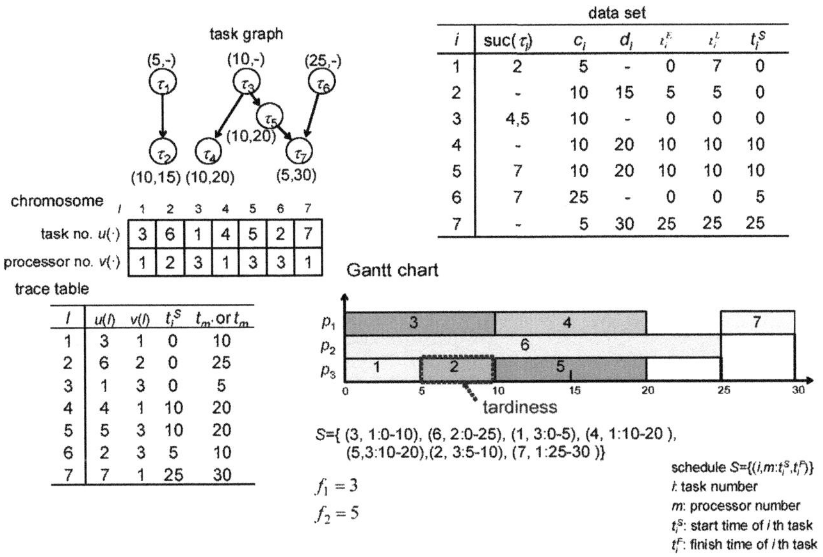

Fig. 8.36 Example of decoding procedure

$$
\begin{aligned}
eval(V_k) &= 1/F(V_k) \\
&= \frac{1}{\sum_{q=1}^{2} \frac{f_q(V_k)}{f_q^{\max} - f_q^{\min}}}
\end{aligned}
\tag{8.36}
$$

For selection, the commonly strategy called roulette wheel selection [16, 17] has been used.

8.3.2.3 GA Operators

The one-cut crossover is used. This operator creates two new chromosomes (the offspring) by mating two chromosomes (the parent). The one-cut crossover procedure will be written as in Fig. 8.37.

In Fig. 8.37, $u'(\cdot)$, $v'(\cdot)$ are proto-offspring chromosome. Figure 8.38 represents the example of one-cut crossover procedure.

For another GA operator, mutation, the classical one-bit altering mutation [15] is used.

procedure: One-cut Crossover
input: parents $u_1(\cdot)$, $v_1(\cdot)$, $u_2(\cdot)$, $v_2(\cdot)$
output: offspring $u_1(\cdot)'$, $v_1(\cdot)'$, $u_2(\cdot)'$, $v_2(\cdot)'$
begin
  $r \leftarrow$ random[1,N];
  $u_1(\cdot)' \leftarrow u_1(\cdot)$;
  $v_1(\cdot)' \leftarrow v_1[1{:}r]\ //\ v_2[r+1{:}N]$;
  $u_2(\cdot)' \leftarrow u_2(\cdot)$;
  $v_2(\cdot)' \leftarrow v_2[1{:}r]\ //\ v_1[r+1{:}N]$;
  output offspring chromosomes $u_1(\cdot)'$, $v_1(\cdot)'$, $u_2(\cdot)'$, $v_2(\cdot)'$;
end

Fig. 8.37 Procedure of one-cut crossover

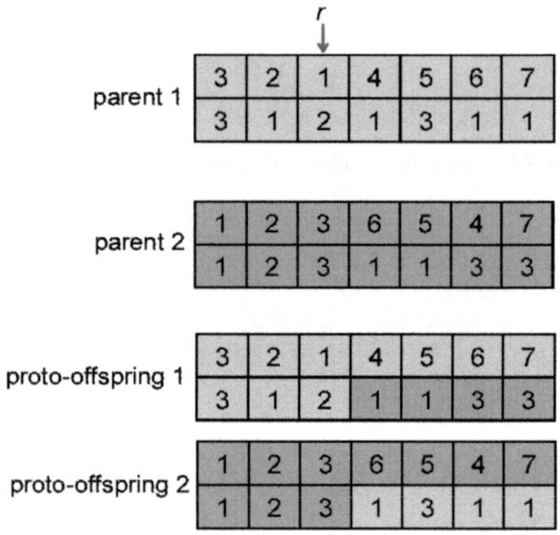

Fig. 8.38 Example of one-cut crossover

8.3.3 Numerical Experiments

In the experiment, the introduced moGA is compared with Monnier-GA by Monnier *et. al.* [17] and Oh-Wu's algorithm by Oh and Wu [19]. Numerical tests are performed with a randomly generated task graph. P-Method [21] is used for generation of the task graph. The P-Method of generating a random task graph is based on the probabilistic construction of an adjacency matrix of a task graph. Element a_{ij} of the matrix is defined as 1 if there is a precedence relation from τ_i to τ_j; otherwise, a_{ij} is zero. An adjacency matrix is constructed with all its lower triangular and diagonal elements set to zero. Each of the remaining upper triangular elements of the matrix is examined individually as part of a Bernoulli process with parameter ε, which represents the probability of a success. For each element, when the

Bernoulli trial is a success, then the element is assigned a value of one; for a failure the element is given a value of zero. The parameter ε can be considered to be the sparsity of the task graph. With this method, a probability parameter of ε=1 creates a totally sequential task graph, and ε=0 creates an inherently parallel one. Values of ε that lie in between these two extremes generally produce task graphs that possess intermediate structures.

Tasks' computation time and deadline are generated randomly based on exponential distribution and normal distribution and the parameters set as probabilities for crossover are tested from 0.5 to 0.8, from 0.001 to 0.4 for mutation, with the increments 0.05 and 0.001 respectively. For population size, individuals from 20 to 200 are tested.

Numerical tests are performed with 100 tasks. Tables 8.14 and 8.15 show the comparisons of results by three different scheduling algorithms. There is no tardiness inclusively. The computing time of proposed moGA is a little bit longer than those of the other two. However, the number of utilized processors is fewer than those of the other two algorithms. The variance of processor utilization rate by moGA is more desirable than those of the others.

Table 8.14 Computation results three algorithms (exponential)

Terms	Monnier-GA	Oh-Wu's algorithm	moGA
# of processors M	38	37	32
Makespan	149	157	163
Computing times (*msec*)	497	511	518
Average utilization of processors	0.447582	0.453392	0.567352

Table 8.15 Computation results three algorithms (normal)

Terms	Monnier-GA	Oh-Wu's algorithm	moGA
# of processors M	38	34	30
Makespan	189	198	196
Computing times (*msec*)	498	517	519
Average utilization of processors	0.483363	0.507823	0.627490

Figures 8.39 and 8.40 represents the Pareto solution of moGA and those of Oh-Wu's algorithm. In these figures, the Pareto solution curve by moGA is closer to the ideal point than that of Oh-Wu's algorithm.

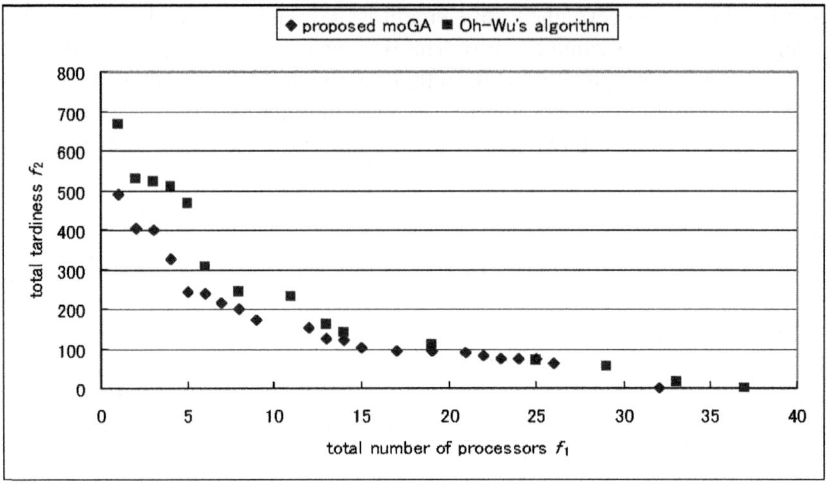

Fig. 8.39 Pareto solution (exponential)

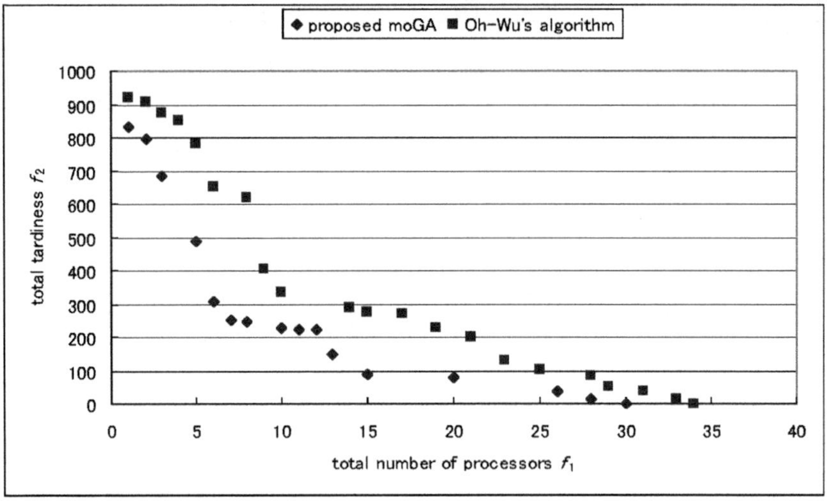

Fig. 8.40 Pareto solution (normal)

8.4 Real-time Task Scheduling in Heterogeneous Multiprocessor System

In a heterogeneous multiprocessor system, task scheduling is more difficult than that in a homogeneous multiprocessor system. Recently, several approaches of the genetic algorithm (GA) are proposed. Theys *et al.* presented a static scheduling algorithm using GA on a heterogeneous system [22], and, Page and Naughton. presented a dynamic scheduling algorithm using GA on a heterogeneous system [23]. Dhodhi *et al.* presented a new encoding method of GA for task scheduling on a heterogeneous system [24]. However, these algorithms are designed for general tasks without time constraints.

In this section, a new scheduling algorithm for nonpreemptive tasks with a precedence relationship in a soft real-time heterogeneous multiprocessor system [25] is introduced.

8.4.1 Soft Real-time Task Scheduling Problem (sr-TSP) and Mathematical Model

The problem of scheduling the tasks with precedence and timing constrained task graph on a set of heterogeneous processors is considered in a way that minimizes the total tardiness $F(x,t^S)$. Conditions are same as those of Section 8.3.

Soft real-time tasks scheduling problems on heterogeneous multiprocessor systems to minimize the total tardiness are formulated as follows:

$$\min f(x,t^S) = \sum_{i=1}^{N} \max\{0, \sum_{m=1}^{M} (t_i^S + c_{im} - d_i) \cdot x_{im}\} \tag{8.37}$$

$$\text{s.t. } t_i^E \le t_i^S \le d_i, \ \forall i \tag{8.38}$$

$$t_i^E \ge t_j^E + \sum_{m=1}^{M} \cdot x_{im}, \tau_j \in \text{pre}(\tau_i), \ \forall i \tag{8.39}$$

$$\sum_{m=1}^{M} x_{im} = 1, \ \forall i \tag{8.40}$$

$$x_{im} \in \{0, 1\}, \ \forall i, m \tag{8.41}$$

Notation

In the above equations, notations are defined as follows:

Indices

i, j: task index, $(i, j = 1, \ldots, N)$
m: processor index, $(m = 1, \ldots, M)$

Parameters

$G = (T, E)$: task graph
$T = \{\tau_1, \tau_2, \ldots, \tau_N\}$: a set of N tasks
$E = \{e_{ij}\}, i, j = 1, 2, \ldots, N, i \neq j$: a set of directed edges among the tasks representing precedence
τ_i: i-th task, $i = 1, 2, \ldots, N$
p_m: m-th processor, $m = 1, 2, \ldots, M$
c_i: computation time of task τ_i
$\text{pre}^*(\tau_i)$: set of all predecessors of task τ_i
$\text{suc}^*(\tau_i)$: set of all successors of task τ_i
$\text{pre}(\tau_i)$: set of immediate predecessors of task τ_i
$\text{suc}(\tau_i)$: set of immediate successors of task τ_i
t_i^E: earliest start time of i-th task

$$t_i^E = \begin{cases} 0, & if\ \neg\exists\tau_j : e_{ji} \in E \\ \max\limits_{\tau_j \in \text{pre}^*(\tau_i)} \{t_j^E + \sum\limits_{m=1}^{M} c_{jm} \cdot x_{jm}\}, & \text{otherwise} \end{cases}, \forall i \tag{8.42}$$

t_i^F: finish time of task τ_i

$$t_i^F = \min\{t_i^S + \sum_{m=1}^{M} c_{im} \cdot x_{im}, d_i\}, \ \forall i \tag{8.43}$$

DecisionVariables

t_i^S: real start time of i-th task

$$x_{im} = \begin{cases} 1, & \text{if processor } p_m \text{ is selected for task } \tau_i \\ 0, & \text{otherwise} \end{cases} \tag{8.44}$$

Equation 8.37 is the objective function in this scheduling problem. Equation 8.37 covers how to minimize total tardiness of tasks. Constraints conditions are shown in Eq. 8.38–8.41. Equation 8.38 shows that a task can be started after its earliest start time, to begin its deadline. Equation 8.39 defines the earliest start time of a task based on precedence constraints. Equation 8.40 means that every task is processed on one processor at a time. Figure 8.41 represents the time chart of sr-TSP.

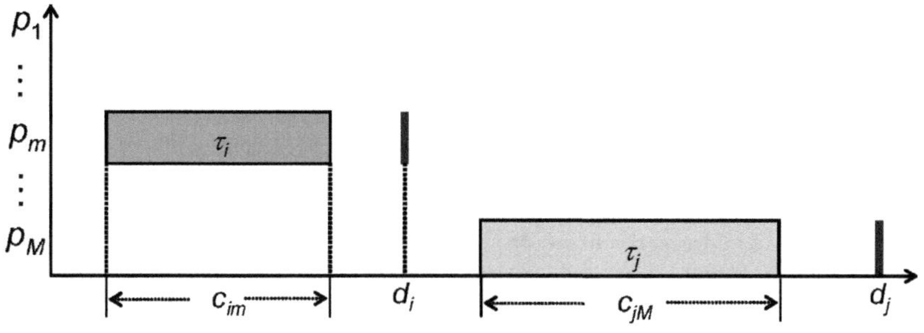

Fig. 8.41 Time chart of sr-TSP

8.4.2 SA-based Hybrid GA Approach

During reproduction and replacement steps, proto-offspring chromosomes are created by mating, with probability p_C, pairs of parents selected in the current population. Then chromosomes are mutated with probability p_M, randomly using one of the mutation operators [18]. The offspring chromosomes are produced by the probability of simulated annealing (SA). Then a new population is built through evaluating chromosomes and selecting.

This iterative evolution process is stopped as soon as one solution is found. However, this algorithm limits the number of offspring produced to *maxGen*, in order to avoid prohibitive calculation time, and to ensure that the GA will stop when treating an infeasible problem. Consequently, introduced hybrid genetic algorithm combined simulated annealing (hGA+SA) obeys the following algorithm:

8.4.2.1 Encoding and Decoding

A chromosome V_k, $k = 1, 2, \ldots, popSize$, represents one of all the possible mappings of all the tasks in the processors, where *popSize* is the total number of chromosomes in a generation. A chromosome V_k is partitioned into two parts $u(\cdot)$, $v(\cdot)$. The $u(\cdot)$ means scheduling order and the $v(\cdot)$ means allocation information. The length of each part is the total number of tasks. The scheduling order part should be a topological order with respect to the given task graph that satisfies the precedence relationship. The allocation information part denotes the processor to which task is allocated. To develop a multistage operation-based genetic representation for the srTSP, there are three main phases:

Phase 1: Creating a task sequence: generate a random priority to each task in the model using priority-based encoding procedure for the first vector $u(\cdot)$.

```
procedure: sr-TSP by hGA+SA
input: task graph data set
output: best schedule set S
begin
   t ← 0;
   initialize P(t) by encoding routine;
   evaluate eval(P) by decoding routine;
   while (not terminating condition) do
      create C(t) from P(t) by one-cut crossover;
      create C(t) from P(t) by altering mutation;
      improve C(t) and P(t) by the probability of SA;
      evaluate eval(P,C) by decoding routine;
      select P(t+1) from P(t) and C(t);
      t ← t+1;
   end
   output best schedule set S;
end
```

Phase 2: Assigning tasks to processors: generate a permutation encoding for processor assignment of each task (second vector $v(\cdot)$).

Phase 3: Designing a schedule

Step 3.1: Create a schedule S using task sequence and processor assignments.
Step 3.2: Draw a Gantt chart for this schedule.

Phase 1: Creating a task sequence: generate a random priority to each task in the model using priority-based encoding procedure for the first vector $u(\cdot)$. Encoding procedure for sr-TSP will be written as in Fig. 8.42.

In Fig. 8.42, W is temporary defined working data set for tasks without predecessors. In encoding procedure, feasible solutions are generated by respecting the precedence relationship of task and allocated processor is selected randomly.

Phase 2: Assigning tasks to processors: generate a permutation encoding for processor assignment of each task (2nd vector $v(\cdot)$). Figure 8.43 represents the example of this encoding procedure.

Phase 3: Designing a schedule.

Step 3.1: Create a schedule S using task sequence and processor assignments. The decoding procedure is will be written as in Fig. 8.44.

Fig. 8.44, insert(i) means to insert τ_i at idle time if τ_i is computable in idle time. At start(i), the real start time of the i-th task t_i^S and the finish time of the i-th task t_i^F can be calculated. update_idle means that the list of idle time is updated if new idle time duration is occurred. The objective value $F(x,t^S)$ and schedule set S is generated through this procedure.

Step 3.2: Draw a Gantt chart for this schedule. Figure 8.45 represents the example of drawing process for a Gantt chart with chromosome as in Fig. 8.43.

```
procedure: Encoding for sr-TSP
input: task graph data set, total number of processors M
output: u(·), v(·)
begin
    l←1, W←ϕ;
    while (T ≠ ϕ)
        W ← W ∪arg{τ_i |pre*(τ_i) = ϕ, ∀i };
        T ← T- {τ_i}, i∈W;
        while (W ≠ ϕ)
            j ← random(W);
            u(l) ← j ;
            W ← W – {j };
            pre*(τ_i) ← pre*(τ_i) – {τ_j}, i;
            m← random[1:M];
            v(l) ← m ;
            l ← l+1;
    output u(·), v(·) ;
end
```

Fig. 8.42 Procedure of encoding for sr-TSP

task graph

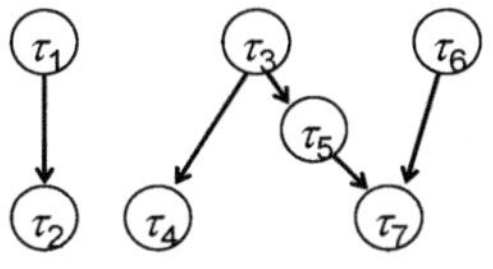

total number of processors M=3

chromosome V_k

l	1	2	3	4	5	6	7
$u(\cdot)$	3	6	1	4	5	2	7
$v(\cdot)$	1	2	3	1	3	3	1

data set

i	suc(τ_i)	c_{im}			d_i
		c_{i1}	c_{i2}	c_{i3}	
1	2	2	3	5	-
2	-	12	11	10	17
3	4,5	10	12	13	-
4	-	10	8	9	23
5	7	6	12	10	25
6	7	22	25	24	-
7	-	5	4	7	32

trace table

l	m	j	W	T
			ϕ	$\{\tau_1, \tau_2, \tau_3, \tau_4, \tau_5, \tau_6, \tau_7\}$
1	1	3	{1,3,6}	$\{\tau_2, \tau_4, \tau_5, \tau_7\}$
2	2	6	{1,6}	$\{\tau_2, \tau_4, \tau_5, \tau_7\}$
3	3	1	{1}	$\{\tau_2, \tau_4, \tau_5, \tau_7\}$
			ϕ	$\{\tau_2, \tau_4, \tau_5, \tau_7\}$
4	1	4	{2,4,5}	$\{\tau_7\}$
5	3	5	{2,5}	$\{\tau_7\}$
6	3	2	{2}	$\{\tau_7\}$
			ϕ	$\{\tau_7\}$
7	1	7	{7}	ϕ
			ϕ	ϕ

Fig. 8.43 Example of encoding procedure

```
procedure: Decoding for sr-TSP
input: task graph data set, chromosome u(·), v(·)
output: schedule set S, total tardiness of tasks F
begin
    l←1, F ← 0, S ← φ;
    while (l ≤ N)
        i ← u(l);
        m← v(l);
        if (exist suitable idle time) then
                insert(i);
        start(i);
        update_idle();
        F ← F +max{0,(tiS+cim-di )} ;
        S ← S ∪ {(i, m: tiS – tiF)};
        l ← l+1;
    end
    output S, F
end
```

Fig. 8.44 Procedure of decoding for sr-TSP

task graph

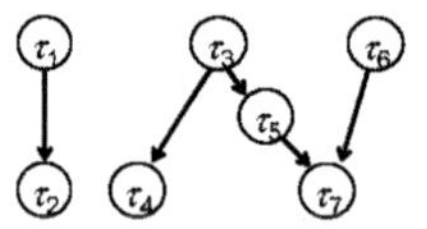

data set

i	suc(τ_i)	c_{i1}	c_{i2}	c_{i3}	d_i
1	2	2	3	5	-
2	-	12	11	10	17
3	4,5	10	12	13	-
4	-	10	8	9	23
5	7	6	12	10	25
6	7	22	25	24	-
7	-	5	4	7	32

(c_{i1}, c_{i2}, c_{i3} under c_{im})

trace table

l	i	m	t_i^S	t_i^F
1	3	1	0	10
2	6	2	0	25
3	1	3	0	5
4	4	1	10	20
5	5	3	10	20
6	2	3	5	10
7	7	1	25	30

$F(x) = 5$

S={ (3, 1:0-10), (6, 2:0-25), (1, 3:0-5), (4, 1:10-20), (5,3:10-20),(2, 3:5-10), (7, 1:25-30)}

schedule S={(i,m:t_i^S-t_i^F)}
i: task number
m: processor number
t_i^S: start time of i th task
t_i^F: finish time of i th task

chromosome V_k

l	1	2	3	4	5	6	7
$u(\cdot)$	3	6	1	4	5	2	7
$v(\cdot)$	1	2	3	1	3	3	1

Gantt chart

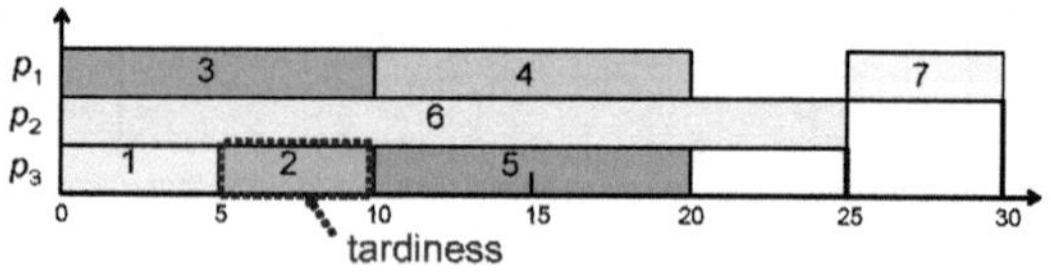

Fig. 8.45 Example of decoding procedure

8.4.2.2 Evolution Function and Selection

The fitness function is essentially the objective function for the problem. It provides a means of evaluating the search node and it also controls the selection process [18, 20].

The fitness function used for our algorithm is based on the $F(x,t^S)$ of the schedule. The used evaluation function is then

$$eval(V_k) = 1/F(x,t^S),\ \forall k \tag{8.45}$$

Selection is the main way GA mimics evolution in natural systems: the fitter an individual is, the highest is its probability of being selected. For selection, the commonly strategy called roulette wheel selection [16, 17] has been used.

8.4.2.3 GA Operators

For crossover, the one-cut crossover in Section 8.3 is used. For another GA operator, mutation, the classical one-bit altering mutation [15] is used.

8.4.2.4 Improving of Convergence by the Probability of SA

The convergence speed to the local optimum of the GA can be improved by adopting the probability of simulated annealing (SA). SA means the simulation of the annealing process of metal. If the temperature is lowered carefully from a high temperature in the annealing process, the melted metal will produce the crystal at 0 K. Kirkpatrick developed an algorithm that finds the optimal solution by substituting the random movement of the solution for the fluctuation of a particle in the system in the annealing process and making the objective function value correspond to the energy of the system, which decreases (involving the temporary increase by Boltzman's probability) with the descent of temperature [25, 26, 34]. Even though the fitness function value of newly produced strings is lower than those of current strings, the newly produced ones are fully accepted in the early stages of the searching process. However, in later stages, a string with a lower fitness function value is seldom accepted. The procedure of improved GA by the probability of SA will be written as in Fig. 8.46.

8.4.3 Numerical Experiments

In the experiment, this hybrid GA combined Simulated Annealing (hGA+SA) is compared with Monnier's GA and proposed simple GA which is not combined with SA. The Monnier's GA is concerned with the homogeneous multiprocessor system

```
procedure: Improving of GA chromosome by the probability of SA
input: parent chromosome V, proto-offspring chromosomes V', temperature T, cooling rate of SA ρ
output: offspring chromosomes V''
begin
        r ← random[0,1];
        ΔE ← eval(V')-eval(V);
        if ( ΔE >0 || r <Exp(ΔE/T) )   then
                V'' ← V';
        else
                V'' ← V;
        T ← T x ρ;
        output offspring chromosomes V''
end
```

Fig. 8.46 Procedure of improving of GA chromosome by the probability of SA

and the hGA+SA is designed for heterogeneous multiprocessor system. As there are no algorithms which are concerned with the heterogeneous multiprocessor system, the hGA+SA is compared with Monnier's GA on heterogeneous multiprocessor system. The Monnier's GA is proposed by Monnier *et al.* [17]. This algorithm based on simple GA uses linear fitness normalization techniques for evaluating chromosomes. The linear fitness normalization technique is effective to increase competition between similar chromosomes. However this method is limited in special problem with similar chromosomes, and in this algorithm, the insertion method is not used. In other words, although there is idle time, the task can't be executed in idle time.

Numerical tests are performed with randomly generated task graph. P-Method [21] for generation task graph is used. Tasks' computation times and deadlines are generated randomly based on exponential distribution. The parameters of GA is the same as those of Section 8.3.

Numerical tests are performed with 100 tasks. Figure 8.47 and 8.48 show the comparison of three algorithms for $F(x,t^S)$. In Figs. 8.47 and 8.48, $F(x,t^S)$ of hGA+SA is smaller that that of each algorithms.

In Table 8.16 and 8.17, some terms such as makespan, computing time and the utilization of processors are compared on the total number of processors without tardiness. Total number of processors without tardiness of hGA+SA is smaller than that of other algorithms and the average utilization of processors of hGA+SA is more desirable than those of the others.

8.5 Summary

In this chapter, we introduced GA approaches for soft real-time task scheduling models. Several derived algorithms from Rate Monotonic, Earliest Deadline First

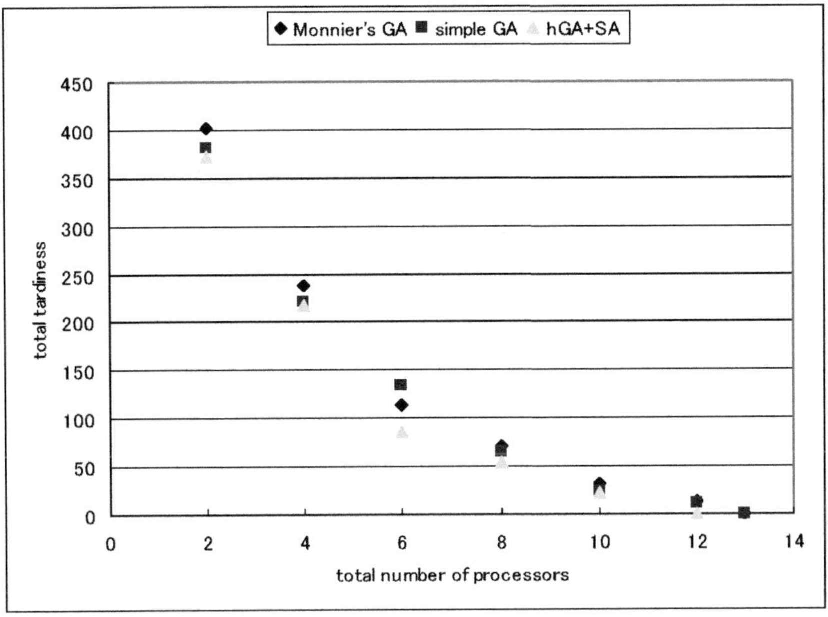

Fig. 8.47 Comparison with three algorithms for $F(x,t^S)$ (exponential)

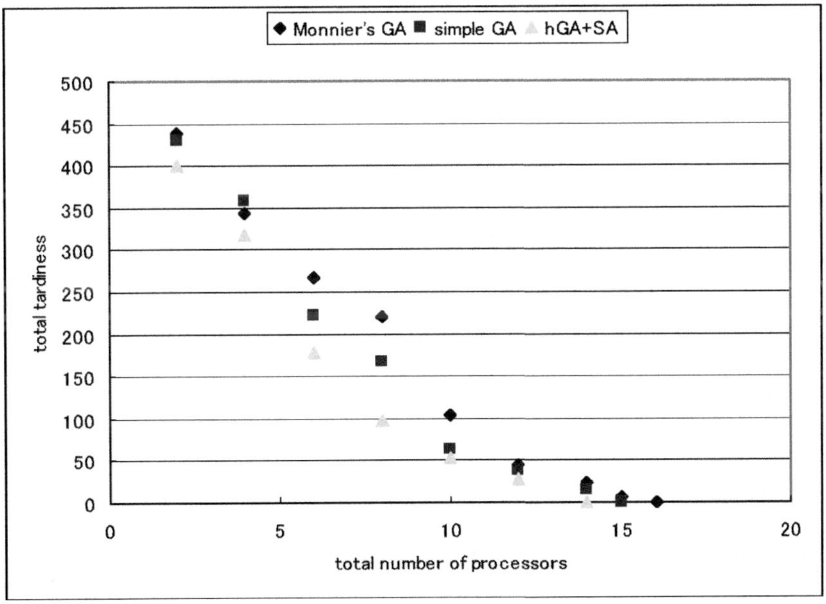

Fig. 8.48 Comparison with three algorithms for $F(x,t^S)$ (normal)

Table 8.16 Comparison with three algorithms (exponential)

Terms	Monnier's GA	Simple GA	hGA+SA
# of processors *M*	13	13	12
Makespan	123	120	132
Computing times (*msec*)	243	245	338
Average utilization of processors	0.4334	0.4375	0.5702

Table 8.17 Comparison with three algorithms (normal)

Terms	Monnier's GA	Simple GA	hGA+SA
# of processors *M*	16	15	14
Makespan	153	149	132
Computing times (*msec*)	248	249	342
Average utilization of processors	0.5332	0.5413	0.5598

for hard real-time tasks or some scheduling algorithms such as Rate Regulating Proportional Share and Modified Proportional Share have been used for soft real-time tasks. However, these algorithms have some drawbacks in resource utilization and pattern of degradation under the overloaded situation. Furthermore, the scheduling on multiprocessor system is an NP-hard problem.

For solving task scheduling models, a simulated annealing algorithm was combined with GA. The convergence of GA is improved by introducing the probability of SA as the criterion for acceptance of the new trial solution. This hybridization does not hurt own advantages of GA but finds more accurate solutions in the later stage of a searching process. The multiobjective GA for soft real-time task scheduling was also introduced. Not only minimization of the total tardiness but also minimization of the total number of processor was used and the makespan taken into considerations. The effectiveness and efficiency of GA approaches were investigated with various task scheduling problems by comparing with related research.

References

1. Krishna, C. M. & Kang, G. S. (1997). *Real-Time System*, McGraw-Hill.
2. Yoo, M. R., Ahn, B. C., Lee, D. H. & Kim, H. C. (2001). A New Real-Time Scheduling Algorithm for Continuous Media Tasks. *Proceedings of Computers and Signal Processing*, 26-28.
3. Liu, C. L. & Layland, J. W. (1973). Scheduling Algorithm for Multiprogramming in a Hard Real-Time Environment. *Journal of the ACM*, 20(1), 46-59.
4. Kim, M. H., Lee, H. G. & Lee, J. W. (1997). A Proportional-Share Scheduler for Multimedia Applications. *Proceedings of Multimedia Computing and Systems*, 484-491.
5. Yoo, M. R. (2002). A Scheduling Algorithm for Multimedia Process. Ph.D. dissertation, University of YeoungNam.

6. Yalaoui, F. & Chu, C. (2002). Parallel Machine Scheduling to Minimize Total Tardiness. *International Journal of Production Economics*, 76(3), 265-279.
7. Du, J. & Leung, J. (1990). Minimizing Total Tardiness on One Machine is NP-hard. *Mathematics of Operational Research*, 15, 483-495.
8. Lenstra, J. K., Kan, R. & Brucker, P. (1997). Complexity of Machine Scheduling Problems. *Annals of Discrete Mathematics*, 343-362.
9. Zhu, K., Zhuang, Y. & Viniotis, Y. (2001). Achieving End-to-End Delay Bounds by EDF Scheduling without Traffic Shaping. *Proceedings of 20th Annual Joint Conference on the IEEE Communications Societies*. 1493-1501.
10. Budin, L., Domagoj, J. & Marin, G. (1999). Genetic Algorithm in Real-Time Imprecise Computing. *Proceedings of Industrial Electronics*. 84-89.
11. Diaz, J. L., Garcia, D. F. & Lopez, J. M. (2004). Minimum and Maximum Utilization Bounds for Multiprocessor Rate Monotonic Scheduling. *IEEE Transactions on Parallel and Distributed Systems*, 15(7), 642-653.
12. Bernat, G., Burns, A. & Liamosi, A. (2001). Weakly Hard Real-Time Systems. *Transactions on Computer Systems*. 50(4), 308-321.
13. Denouzos, M. L. & Mok, A. K. (1989). Multiprocessor on-line scheduling of hard-real-time tasks. *IEEE Transactions on Software Engineering*, 15(12), 392-399.
14. Yoo, M. R. & Gen, M. (2005). Multimedia Tasks Scheduling using Genetic Algorithm. *Asia Pacific Management Review*. 10(6), 373-380.
15. Jackson, L. E. & Rouskas, G. N. (2003). Optimal Quantization of Periodic Task Requests on Multiple Identical Processors. *IEEE Transactions on Parallel and Distributed Systems*, 14(8), 795-806.
16. Gen, M. & Cheng, R. (1997). *Genetic Algorithms & Engineering Design*, John Wiley & Sons.
17. Monnier, Y., Beauvais, J. P. & Deplanche, A. M. (1998). A Genetic Algorithm for Scheduling Tasks in a Real-Time Distributed System. *Proceedings of 24th Euromicro Conference*, 708-714.
18. Gen, M. & Cheng, R. (2000). *Genetic Algorithms & Engineering Optimization*, John Wiley & Sons.
19. Oh, J. & Wu, C. (2004). Genetic-algorithm-based Real-time Task Scheduling with Multiple Goals. *Journal of Systems and Software*, 71(3), 245-258.
20. Deb, K. (2001). *Multi-objective Optimization using Evolutionary Algorithms*, John Wiley & Sons.
21. Al-Sharaeh, S. & Wells, B. E. (1996). A Comparison of Heuristics for List Schedules using The Box-method and P-method for Random Digraph Generation. *Proceedings of the 28th Southeastern Symposium on System Theory*, 467-471.
22. Theys, M. D., Braun, T. D., Siegal, H. J., Maciejewski A. A. & Kwok, Y. K. (2001). Mapping Tasks onto Distributed Heterogeneous Computing Systems Using a Genetic Algorithm Approach. Zomaya, A. Y., Ercal, F. & Olariu, S. editors, *Solutions to Parallel and Distributed Computing Problems*, 6, 135-178, John Wiley & Sons.
23. Page, A. J. & Naughton, T. J. (2005). Dynamic Task Scheduling using Genetic Algorithm for Heterogeneous Distributed Computing. *Proceedings of 19th IEEE International Parallel and Distributed Processing Symposium*, 189.1.
24. Dhodhi, M. K., Ahmad, I. , Yatama, A. & Ahmad, I.(2002). An Integrated Technique for Task Matching and Scheduling onto Distributed Heterogeneous Computing Systems. *Journal of Parallel and Distributed Computing*, 62, 1338-1361.
25. Yoo, M. R. & Gen, M. (2007). Scheduling algorithm for real-time tasks using multiobjective hybrid genetic algorithm in heterogeneous multiprocessors system. *Computers and Operations Research*, 34(10), 3084-3098
26. Ishii, H., Shiode, H. & Murata, T. (1998). A Multiobjective Genetic Local Search Algorithm and Its Application to Flowshop Scheduling. *IEEE Transactions. on Systems, Man and Cybernetics*, 28(3), 392-403.
27. Kim, H. C, Hayashi, Y. & Nara, K. (1997). An Algorithm for Thermal Unit Maintenance Scheduling through combined use of GA, SA and TS. *IEEE Transactions on Power Systems*, 12(1), 329-335.

28. Anagnostou, M. E., Theologou, M. E., Vlakos, K. M. & Tournis, D. (1991). Quality of service requirements in ATM-based B-ISDNs, *Computer Commun*, 14(4), 197-204.
29. Hehmann, D. B., Salmony, M. G. & Stuttgen, H. J. (1989). High-Speed Transport Systems for Multimedia Applications, *Protocols for High-Speed Networks*.
30. Yoo, M.R. & Gen, M. (2006). Multimedia Task Scheduling using Proportion-Based Genetic Algorithm, *IEEJ Transactions on Electronics, Information and Systems*, 126C(3), 347-352.
31. Kim, M. H., Lee, H. G. & Lee, J. W. (1997). A Proportional-Share Scheduler for Multimedia Applications, *Proceedings of Multimedia Computing & Systems*, 484–491.
32. Krishna, C. M. & Kang, G. S. (1997). *Real-Time System*, McGraw-Hill.
33. Yoo, M. R. (2002). *A Scheduling Algorithm for Multimedia Process*, Ph.D. dissertation, University of YeoungNam, in Korean.
34. Yoo, M. R. (2006). *Study on Real-time Task Scheduling by Hybrid Multiobjective Genetic Algorithm*, Ph.D dissertation, Waseda University.
35. Gen, M. & Yoo, M. R. (2008). Real Time Tasks Scheduling Using Hybrid Genetic Algorithm, in Abo Ala-Ola Ed.: *Computational Intelligence in Multimedia Processing: Recent Advances*, 319-350, Springer Berlin.
36. Liu, C. L. & Layland, J. W. (1973). Scheduling Algorithm for Multiprogramming in a Hard Real-Time Environment, *Journal of the ACM*, 20(1), 46-59.
37. Tracy, M., & Douglass, S. (2001). *Genetic Algorithms and their Application to Continuum Generation*, The Ohio State University, REU.
38. Goldberg, D.E. (1989). *Genetic Algorithms in Search, Optimization and Machine Learning*, Addison-Wesley, Reading, MA.
39. Emilda, S., JacobL., Ovidiu, D. & Prabhakaran B. (2002). Flexible Disk Scheduling for Multimedia Presentation Servers, *IEEE International Conference on Networks*, 151-155.
40. Hou, E. S. H., Hong, R., & Ansari, N. (1990). Efficient Multiprocessor Scheduling Based on Genetic Algorithms, *Proceedings of 16th Annual Conference of IEEE Industrial Electronics Society*, 2, 1239-1243.

Chapter 9
Advanced Network Models

Everywhere we look in our daily lives, networks are apparent. National highway systems, rail networks, and airline service networks provide us with the means to cross great geographical distances to accomplish our work, to see our loved ones, and to visit new places and enjoy new experiences. Manufacturing and logistics networks give us access to life's essential foodstock and to consumer products. And computer networks, such as airline reservation systems, have changed the way we share information and conduct our business and personal lives.

In all of these problem domains, and in many more, we wish to move some entity (electricity, a product, a person or a vehicle, a message) from one point to another in an underlying network, and to do so as efficiently as possible, both to provide good service to the users of the network and to use the underlying (and typically expensive) transmission facilities effectively. In the most general sense, we want to learn how to model application settings as mathematical objects known as network design models and to study various ways (algorithms) to solve the resulting models [1]. In this chapter, the following advanced network models are introduced as shown in Fig. 9.1.

9.1 Airline Fleet Assignment Models

Fleet assignment problem (FAP) is a logistics network problem. Given the flight schedule of an airline, the fleet assignment problem consists of determining the aircraft type to assign to each flight leg in order to minimize the total costs while satisfying aircraft routing and availability constraints. The domestic airport network from ANA in Japan as shown in Fig. 9.2.

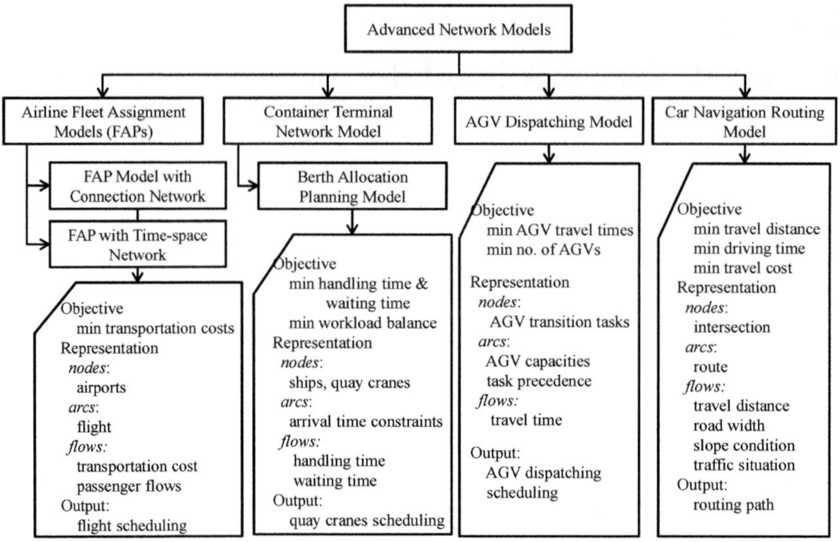

Fig. 9.1 The advanced network models introduced in this chapter

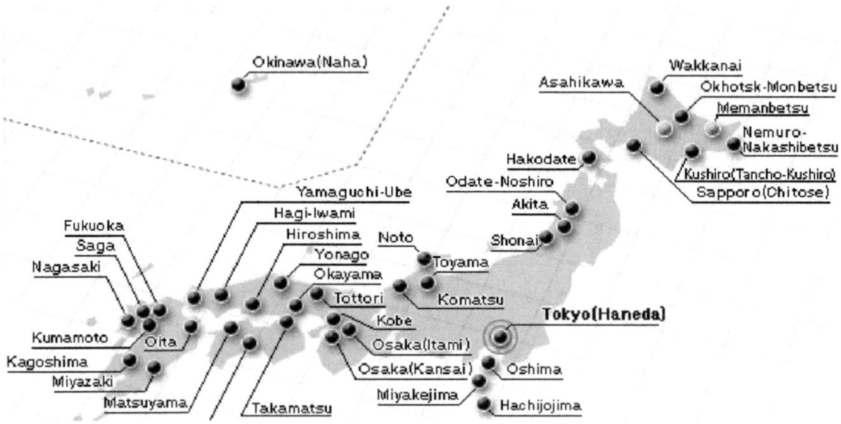

Fig. 9.2 The ANA domestic airport map in Japan

Connection Network Models

Abara [2] publishes the first significant fleet assignment problem (FAP) application based on work done at American Airlines. Abara's model could either maximize operating profit or reduce operating cost by maximizing the utilization of the most efficient fleets. He using a connection-based network structure to track individual aircraft in the air and on the ground in order to ensure that the assignment of aircraft was physically feasible and to control their turns or routings. Abara defines a turn as the successive assignment of one aircraft to two flights; the aircraft "turns" from the arrival of the first flight to the departure of the second. Minimum turn times are required to allow for passenger and baggage unloading, cleaning, refueling, inspection, and passenger and baggage loading. To make the FAP/routing problem practical to solve, he limits the number of possible aircraft turns for each flight. Revenue is based on a stochastic model of demand, but Abara assumes that each leg is independent. Abara first solves the FAP. LP relaxation, fixes variables and then solves the MIP. He reports that the solutions to the LP relaxation are largely integer. As a measure of solution quality, Abara observes that the number of high demand legs covered by large equipment increased from 76% to 90% and that the net profit impact of FAP is 1.4 margin points or $105 million per year.

In the connection model, the cover constraint requires that each flight is preceded by an arrival or an originating arc that is covered by a fleet type. The *balance constraint* assures the flow balance at each leg in the network for each fleet type. In addition, the schedule balance constraint ensures that the same number of aircraft of each type remain at each station every night so that the same assignment can repeat daily. The availability constraint limits the number of aircraft, the number aircraft available of different types. Other side-constraints such as a limit on the number of aircraft that stay overnight at each station can be added as needed.

In this network, the nodes represent the points of time when flights arrive or depart. There are three types of arcs representing the different types of connections: the *flight connection arcs* link the arrival nodes to the departure nodes, the *terminating* (connection) arcs link the arrival nodes to the master sink node to represent aircraft arriving and remaining at the station for the rest of the day, and the *originating* (connection) arcs link the master source node to the departure nodes to represent the aircraft that are present at the station at the beginning of the day. All flight connections have to be feasible with respect to flight arrival and departure times; that is, the minimum turn-time has to be observed between the arrival flight and the following departure flight to allow for the connection. An illustration of connection network in Fig. 9.3: (1) F1–F5–F8, (2) F3–F6– F2, (3) F4–F7. F1 represents flight 1, and so on, OR is a sequence origination, and TE is sequence termination.

Other literature on the FAP with a fixed schedule spans over several decades, the research papers being those of Abara [2], Subramanian *et al.* [3], Hane *et al.* [4], and Rushmeier and Kontogiorgis [5]. Formulations and solution approaches are very similar. They rely on mixed integer multi-commodity network flow formulations based on a space-time graph representations that are solved by branch-and-bound. The optimization problem is formulated as mixed-integer multi-commodity flow, in which commodities correspond to fleets and coupling constraints capture the coverage of flights as well as the operational requirements. It has been proven by Gu *et al.* [19] that a closely related fleet assignment problem, including only flight coverage and fleet count constraints, is NP-hard.

Barnhart *et al.* [6] presented a single model and solution approach to solve simultaneously the fleet assignment and aircraft routing problems. The solution approach is robust in that it can capture costs associated with aircraft connections and complicating constraints such as maintenance requirements. By setting the number of

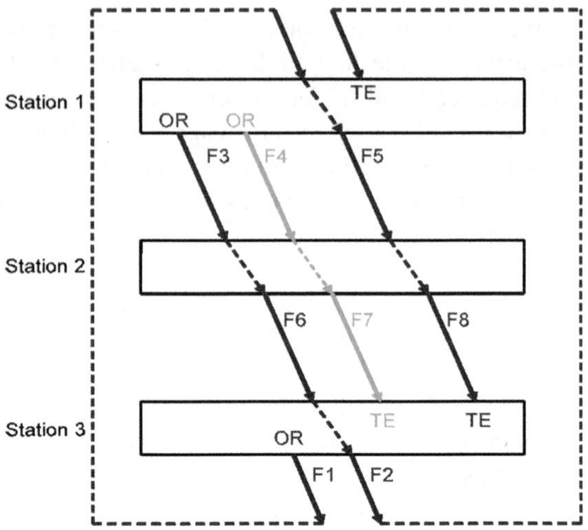

Fig. 9.3 Connection network

fleets to one, this approach can be used to solve the aircraft routing problem alone. In Yan and Tseng [7], the research develops a model and a solution algorithm to help carriers simultaneously solve for better fleet routes and appropriate timetables. The model is formulated as an integer multiple commodity network flow problem. An algorithm based on Lagrangian relaxation, a sub-gradient method, the network simplex method, cost flow augmenting algorithm and the flow decomposition algorithm is developed to efficiently solve the problem. Lohatepanont and Barnhart [8] focus on the steps of the airline schedule planning process involving schedule design and fleet assignment, and presented integrated models and solution algorithms that simultaneously optimize the selection of flight legs and the assignment of aircraft types to the selected flight legs. Sherali *et al.* [20] present a tutorial on the basic and enhanced models and approaches that have been developed for the FAP.

Time-Space Network Model

Hane *et al.* [4] develop a formulation similar to Abara's but make significant computational improvements. The Hane model includes many of the fundamental elements common to subsequent FAP formulations. Their model minimizes operating costs, which includes the cost of spilling a potential passenger due to lack of capacity, subject to:

- Cover constraint–every flight leg must be assigned exactly one fleet type
- Balance constraint–aircraft cannot appear or disappear in the network
- Aircraft available constraint–for each fleet, the total number of aircraft on the ground or in the air at any point in time cannot exceed the total available.

There is a balance constraint associated with each node to ensure that aircraft do not appear or disappear from the network. Arc flows are nonnegative to ensure that we don't assign more aircraft to departing flights than are present at this airport. There is a time specified in the network when we count aircraft either on the ground or in the air. This count is used to ensure that total aircraft counts are not exceeded.

The time-space network representation superimposes a set of networks, one for each fleet type. This allows for fleet type-dependent flight-times and turn-times. If the flight- and turn-times are not significantly different among some particular fleet types, then a composite network representation can be constructed for these types. In each type's network, each event of flight arrival or departure at a specific time is associated with a node. In order to allow for feasible aircraft connections, an arrival node is placed at the flight's ready-time, given by its arrival time plus the necessary turn-time, thus representing the time the aircraft is ready to takeoff. There are three types of arcs in the network for each fleet type: ground arcs representing aircraft staying at the same station for a given period of time, flight arcs representing flight legs, and wraparound arcs (or overnight arcs) connecting the last events of the day with the first events of the day, which, due to the same daily schedule, replicate the first events of the following day.

There is a time-space network for each airport, fleet type combination. Nodes indicate arrival and departure events. Arcs correspond to the assignment of equipment to a flight arrival, departure, or for some period on the ground. There is a return arc to ensure that the number of aircraft on the ground at the end of the time horizon (typically a day or week) is the same as at the beginning of the next time period. Figure 9.4 illustrates a sample time-space network with four arrivals (B, D, E, H), four departures (A, C, F, G), six ground arcs, and one overnight arc.

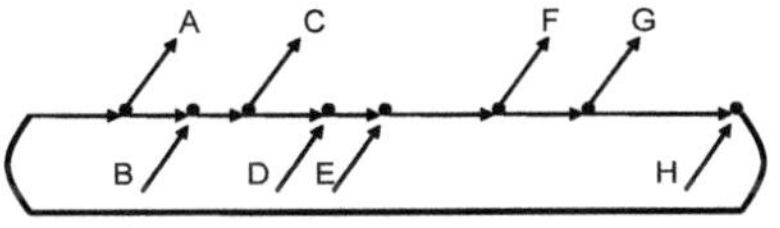

Fig. 9.4 Time-space network

The FAP problem is often formulated as a mixed-integer, linear, multi-commodity flow problem with side constraints defined on a time-space network as Abara [2]; Subramanian *et al.* [3], and Hane *et al.* [4]. In Desaulniers *et al.* [9] and Rexing *et al.* [10], a variant of the problem is tackled where some flexibility on the flight departure times is allowed. These departure times must fall within given time intervals called time windows. Such flexibility opens up new feasible flight connection opportunities and, thus, can yield a more profitable fleet assignment. Bandet [11] and Desaulniers *et al.* [12] presented formulations for the fleet assignment and aircraft routing problem with time windows. Desaulniers *et al.* solved their models by breaking down the problem and generating favorable flight paths with time-constrained shortest path sub-problems. The models are demonstrated to be effective on problems with fewer than 400 flights per day.

The time-space network technique is very helpful in modeling the feet routing and fight scheduling problems, as indicated by Yan and Young's [13] studies. Hence, this popular technique is adopted for the same application in this research. In summary, this study applies a time-space network technique to formulate an integrated model for feet routing and fight scheduling problems. An algorithm based on Lagrangian relaxation, Yan and Young's sub-gradient method, the network simplex method, the least cost flow augmenting algorithm and the flow decomposition algorithm is developed for solving the model. In particular, developing a good heuristic for finding a feasible upper bound (for minimization problems) and determining a proper sub-gradient method for modifying the Lagrangian multipliers to provide a good lower bound are two significant matters that directly affect the solution algorithm's performance. Introduce time window flexibility for the daily fleet assignment problem. Others research on routing and scheduling problems with time constraints are presented in Desrochers and Soumis [14].

The problem resulting from this decomposition is a shortest path problem with time windows and time costs, as recently discussed in Ioachim *et al.* [15]. This research proposed an efficient dynamic programming algorithm to solve that type of time constrained path problem. It has already been used in several applications, among which are aircraft routing with schedule synchronization (Ioachim *et al.*[15]) and simultaneous optimization of flight and pilot schedules in a recovery environment (Stojkovic and Soumis, 2001 [21]). Based on the time windows, Rexing *et al.* [10] formulate the problem as an integer linear program similar to the one proposed for the case with fixed departure times and solve it using a preprocessor, an LP solver and a branch-and-bound scheme.

Current fleet assignment methods represent the schedule as a space-time network, in which a directed flight arc corresponds to the movement of an aircraft of a particular type along a flight leg. Another alternative, introduced in Barnhart *et al.* [16], would be to consider simultaneous fleet assignment and passenger routing. Bélanger *et al.* [22] address the fleet assignment problem for a weekly flight schedule where it is desirable to assign the same type of aircraft to the legs operating with the same flight number on different days of the week. Even thought it may reduce schedule profitability, aircraft type homogeneity is sought in order to improve the planning and implementation of the operations. Bélanger *et al.* [23] proposed a model for periodic fleet assignment problem with time windows in which departure times are also determined, and develop new branch-and-bound strategies which are embedded in their branch-and-price solution strategy.

Lee *et al.* [17] improve the robustness of a flight schedule by re-timing its departure times. The problem is modeled as a multi-objective optimization problem, and a Multi-Objective Genetic Algorithm (MOGA) is developed to solve the problem. Yan *et al.* [18] developed a short-term flight scheduling model with variable market shares in order to help a Taiwan airline to solve for better fleet routes and flight schedules in today's competitive markets, and developed a heuristic method to efficiently solve the model.

9.1.1 Fleet Assignment Model with Connection Network

The connection network considers a set of passenger flow demands and a set of airports. From each airport to airport, they also have different passenger flow demand. We must assign aircraft to passenger flow demand satisfaction. For formulating this fleet assignment model with connection network, the assumptions are shown as follows:

A1. Each airport to airport requires the passenger flow demand to be a certain amount.
A2. The flow is provided by different aircraft types with maximum capacity.
A3. The airline company has to spend the transportation charges when the passenger is transported to the destination.
A4. An aircraft departing any airport must pay the preparing cost.
A5. Each aircraft can take a passenger to any connection routes until the aircraft capacity limit is reached.
A6. We don't consider aircraft amounts; there is adequate fleet.
A7. In connection network we don't consider the element of time.

First, a simple example for FAP with connection network in Japan is given in Fig. 9.5. In the example, we choose five airports (Fukuoka, Osaka, Nagoya, Tokyo, and Sapporo) that can connect with each other.

In this problem, the airline company has two types of aircraft: B747 and A330. B747 have 400 seats, and A330 have 250 seats. In Fig. 9.6, there are five airports can connection to each other. The d_{ij} is passenger flow demand from nodes i to j and the e_{fi} is the preparing cost of the aircraft type f in origin nodes i.

In this case, the objective is to minimize the total costs when assigning aircrafts to satisfied passenger demand. Hence, we have to formulate the transportation cost and preparing cost of each flight leg and airports.

9.1.1.1 Mathematical Formulation

Let $G = (N,A)$ be a network, consisting of a finite set of nodes $N = \{0,1,2,\ldots,n\}$, and a set of arcs $A = \{(i,j),(k,l),\ldots,(s,t)\}$ joining n pairs of nodes in N. Arc (i, j) is said to be incident with nodes i and j, and is directed for node i to node j. Suppose that each arc (i, j) has assigned to it nonnegative numbers c_{fij}.

Notation

Indices

i, j: the index of airport, $i, j = 0,1,\ldots,N$, where index 0 is a dummy node
f: index of aircraft type, $f = 1,2,\ldots,F$

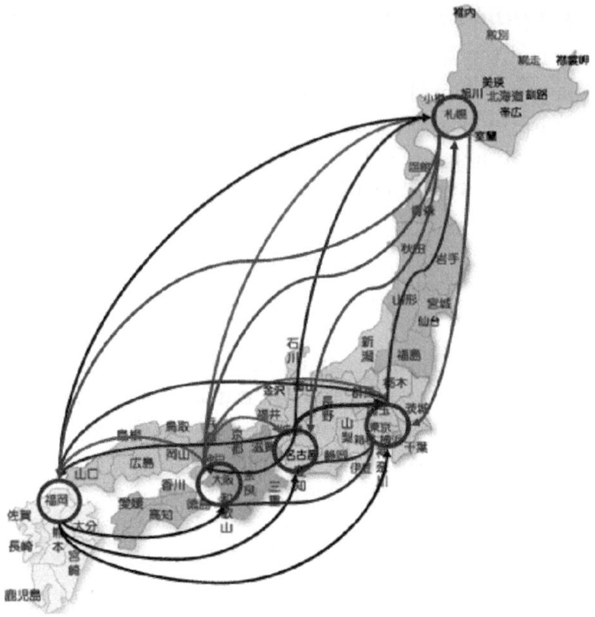

Fig. 9.5 Connection network for five airports in Japan

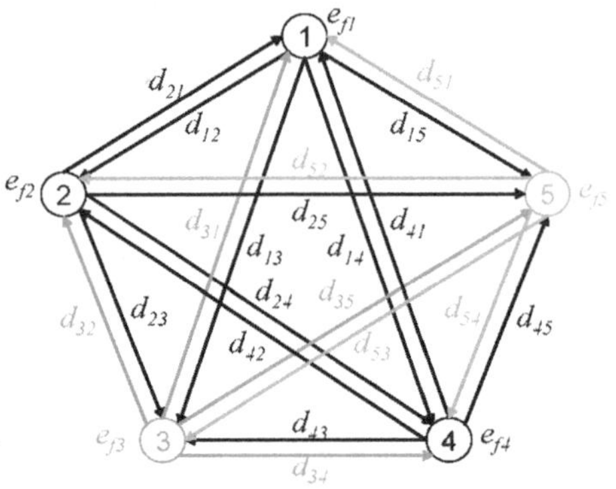

Fig. 9.6 The representation of connection network for five airports

Parameters

c_{fij}: unit transportation cost of the aircraft type f from node i to j
d_{ij}: passenger flow demand from nodes i to j
e_{fi}: the preparing cost of the aircraft type f in origin nodes i
F: set of all aircrafts
N: set of all nodes

Decision Variables

$$x_{fij} = \begin{cases} 1, \text{ aircraft type } f \text{ flies from nodes } i \text{ to } j \\ 0, \text{ otherwise} \end{cases}$$

$$y_{fi} = \begin{cases} 1, \text{ if aircraft type } f \text{ departure at origin nodes } i \\ 0, \text{ otherwise} \end{cases}$$

The fleet assignment model with connection network (cnFAP) is formulated as follows, in which the objective is to minimize the total transportation cost.

$$\min \sum_{f=1}^{F} \sum_{i=0}^{N} \sum_{j=0}^{N} c_{fij} x_{fij} + \sum_{f=1}^{F} \sum_{i=0}^{N} e_{fi} y_{fi} \tag{9.1}$$

$$\text{s.t.} \sum_{j=0}^{N} x_{fij} - \sum_{k=0}^{N} x_{fki} = 0, \ \forall i, \ f \tag{9.2}$$

$$\sum_{f=1}^{F} x_{fij} \geq d_{ij}, \ \forall i, \ j \tag{9.3}$$

$$x_{f0j} = x_{fj0}, \ \forall f, \ j \tag{9.4}$$

$$x_{fij} \geq 0, \ \forall f, \ i, \ j \tag{9.5}$$

$$y_{fi} = \begin{cases} 1, \text{ if } x_{f0i} > 0, \ \forall f, \ i \\ 0, \text{ otherwise} \end{cases} \tag{9.6}$$

The constraint at Eq. 9.2 is flow balance constraint that assures the flow balance at each leg in the network for each fleet type. Constraint at Eq. 9.3 demands satisfaction that assure all the transport flow must satisfy the passenger flow demand in the network for each fleet type. Constraint at Eq. 9.4) is a schedule balance constraint that ensures that the same number of aircraft of each type remain at each station every night. Equations 9.5 and 9.6 define the nature of the decision variables.

9.1.1.2 Priority-based GA for FAP with Connection Network

Let $P(t)$ and $C(t)$ be parents and offspring in current generation t, respectively. The overall procedure of priority-based GA (priGA) for solving this FAP model is outlined as follows:

Priority-based Representation

To develop a priority-based genetic representation for the FAP with connection network, there are three main phases:

---

procedure: priGA for FAP with connection network
input: FAP data, GA parameters ($popSize$, $maxGen$, p_M, p_C)
output: the best flight schedule S
begin
    $t \leftarrow 0$;
    initialize $P(t)$ by priority-based encoding routine;
    evaluate $P(t)$ by priority and locus mapping-based decoding routine;
    while (**not** terminating condition) **do**
        create $C(t)$ from $P(t)$ by position-based crossover routine;
        create $C(t)$ from $P(t)$ by swap mutation routine;
        evaluate $C(t)$ by priority and locus mapping-based decoding routine;
        select $P(t+1)$ from $P(t)$ and $C(t)$ by roulette wheel selection routine;
        $t \leftarrow t+1$;
    end
    output the best flight schedule S;
end

Phase 1: Creating a flight route

Step 1.1: Generate a random list of connection relationship of airport
Step 1.2: Generate a random priority to each flight route

Phase 2: Assigning aircraft to each flight route

Step 2.1: Assign aircraft to satisfied with customer demand
Step 2.2: Complete all the connection flights
Step 2.3: Obtain flight routes

Phase 3: Making a flight schedule

Step 3.1: Achieve the aircraft connection relationship
Step 3.2: Create a flight schedule S

Phase 1: Creating a Flight Route

A gene in a chromosome is characterized by two factors: locus (*i.e.*, the position of the gene located within the structure of chromosome), and allele (*i.e.*, the value the gene takes). In the priority-based encoding method, the position of the gene is used to represent the task ID and its value is used to represent the priority of the task for constructing a sequence among candidates. A feasible sequence can be uniquely determined from this encoding with considering operation precedence constraint. This encoding method easily verifies that any permutation of the encoding corresponds to the service sequences, so that most existing genetic operators can easily be applied to the encoding.

As a priority-based encoding method, depending on the following two steps, the initial chromosomes are randomly generated first:

Step 1.1: Generate a random list of connection relationship of airport
Step 1.2: Generate a random priority to each flight route

This priority-based encoding procedure is shown in Fig. 9.7. As an example, we generated a list of chromosome, the locus is meaning of airport, therefore, the ID 1 to 4 is showing from airport 1 to other airports (2, 3, 4, and 5). Base on this rule considering five airports, an initial chromosome is illustrated in Fig. 9.8.

```
procedure 3.1: priority-based encoding for flight route schedule
input: number of flight legs N
output: chromosome v_k(i)
begin
  for i=1 to N
    v_k(i)← i;
  for i=1 to ⌈N/2⌉
    repeat
      l ←random[1, N];
      j ←random[1, N];
    until l≠j
    swap (v_k(l), v_k(j));
  output the chromosome v_k(i)
end
```

Fig. 9.7 The procedure for priority-based encoding method

From station:	st. 1				st. 2				st. 3				st. 4				st. 5			
i:	1	2	3	4	5	6	7	8	9	10	11	12	13	14	15	16	17	18	19	20
To Station :	2	3	4	5	1	3	4	5	1	2	4	5	1	2	3	5	1	2	3	4
chromosome $v_1(i)$:	1	10	9	3	4	2	11	12	14	5	16	6	7	8	15	13	19	20	17	18

Fig. 9.8 The chromosome by priority-based encoding for five airports

Phase 2: Assigning Aircraft to Each Flight Route

In order to explain the aircraft route developing effectively, a sample example is illustrated as follows. We consider five airports for connection to each other. The passenger flow demand is as Table 9.1. The transportation cost by type 1 (A330) and type 2 (B747) is as Tables 9.2 and 9.3. The loading capacity of aircraft types is as Table 9.4. Preparing cost for each airport by aircraft types is as Table 9.5.

Table 9.1 The passenger flow demand for five airports

i \ j	1	2	3	4	5
1	0	650	950	620	540
2	720	0	650	850	550
3	850	740	0	630	720
4	600	780	540	0	890
5	550	650	750	920	0

Table 9.2 Unit transportation cost by aircraft type 1 (A330). (unit: thousand)

i \ j	1	2	3	4	5
1	0	40	60	80	60
2	30	0	50	80	30
3	60	40	0	60	30
4	90	40	60	0	50
5	60	30	40	50	0

Table 9.3 Unit transportation cost by aircraft type 2 (B747). (unit: thousand)

i \ j	1	2	3	4	5
1	0	30	50	70	50
2	20	0	40	70	20
3	50	30	0	50	20
4	80	30	50	0	40
5	50	20	30	40	0

Table 9.4 Loading capacity of aircraft types

f	seats
1	250
2	400

Table 9.5 Preparing cost for aircraft types

f \ i	1	2	3	4	5
1	5000	3000	3500	4000	4500
2	6000	4000	4500	5000	5500

As priority-based decoding method, depending on the following three steps, assign the aircrafts to each flight route:

Step 2.1: Assign aircrafts to satisfied with customer demand
Step 2.2: Complete all the connection flights
Step 2.3: Obtain flight routes

The priority-based decoding procedure is shown in Fig. 9.9 to generate the aircraft flight route. And the aircraft route is development from the example of chromosome *via* priority-based decoding is shown as Fig. 9.8. For example, the highest priority value is 20 (see Fig. 9.8), its mean an aircraft from airport 5 to airport 2, then the higher priority value in airport 2 is 12; therefore, this aircraft come back to airport 5; after that we generated the one aircraft flight route. We can use this priority-based decoding method to solve this problem.

```
procedure 2: priority and locus mapping-based decoding for flight route schedule
input: total number of nodes N; chromosome v(); m(*) is the ID set belong to
       locus* in the chromosome.
output: route schedule F()
begin
       i=1
       S ←v
       F(i) ← Ø
       while (S ≠ Ø)  do  // ensure all the leg are completed
             j* ←argmax {v(i) | j ∈ S}
             S ← S| j*
             k0(i) ←⌈j*/N-1⌉  // get the origin node point
             F(i) ←F(i) U k0(i)
             F(i) ←F(i) U L(j*)
             while (L(j*) ≠k0(i))  do  // check the connection nodes are return
                   j* ←argmax {v(i) | j ∈m(L(j*))}
                   S ← S| j*
                   F(i) ←F(i) U L(j*)
             end
             i++
       end
       output route schedule F(i)
end
```

Fig. 9.9 The procedure of priority and locus mapping-based decoding

Phase 3: Making a Flight Schedule

Depending on the following two steps, create a flight schedule S.

Step 3.1: Achieve the aircraft connection relationship
Step 3.2: Create a flight schedule *S*

For the five airports simple example, first we draw the passenger demand network in Fig. 9.10; second we can calculate the total costs by this chromosome; other information such as aircrafts route, max flow, aircraft type and route cost, we show in Table 9.6.

Genetic Operators

Position-based Crossover: As crossover operator, we adopt the position-based crossover introduced in Subsection 2.2.2.3. This method random select a set of positions from one parent, then produce a proto-child by copying the elements on these

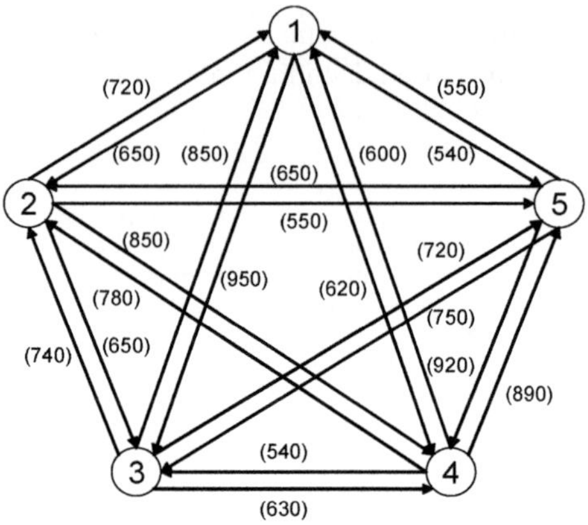

Fig. 9.10 Passenger flow demand network by five airports

Table 9.6 The sample case result by five airports

	Routs	Max flow	Aircraft type	Route cost	Flight leg set
1	{ (5,2) (2,5) }	650	T1:1 T2:1	41500	{ (1,2) (1,3) (1,4) (1,5) (2,1) (2,3) (2,4) (3,1) (3,2) (3,4) (3,5) (4,1) (4,2) (4,3) (4,5) (5,1) (5,3) (5,4) }
2	{ (5,1) (1,3) (3,4) (4,3) (3,1) (1,4) (4,5) }	950	T1:1 T2:2	411000	{ (1,2) (1,5) (2,1) (2,3) (2,4) (3,2) (3,5) (4,1) (4,2) (5,3) (5,4) }
3	{ (5,4) (4,2) (2,4) (4,1) (1,5) }	920	T1:1 T2:2	311500	{ (1,2) (2,1) (2,3) (3,2) (3,5) (5,3) }
4	{ (5,3) (3,5) }	750	T2:2	51000	{ (1,2) (2,1) (2,3) (3,2) }
5	{ (3,2) (2,1) (1,2) (2,3) }	740	T2:2	107000	{ Ø }

Total costs: 922000

positions into the corresponding positions of the proto-child. After that, produce the offspring by filing unfixed positions of the proto-child from another parent. An example of position-based crossover is shown in Fig. 9.11.

Swap Mutation: As mutation operator, we adopt the swap mutation. It selects two elements at random and then swaps the elements on these positions.

Base on this priority-based GA routine to calculate this five airports example, we got the best chromosome, and experiments result for this sample case. The GA parameter setting are *popSize* =20, *maxGen*=1000, p_C=0.5, p_M=0.5. The best chromosome for five airports shown in Figs. 9.12 and 9.13 is the experiments result for aircrafts routes. The final objective value of minimum total cost is 765,000, better than the previous random generated case showed in Table 9.6.

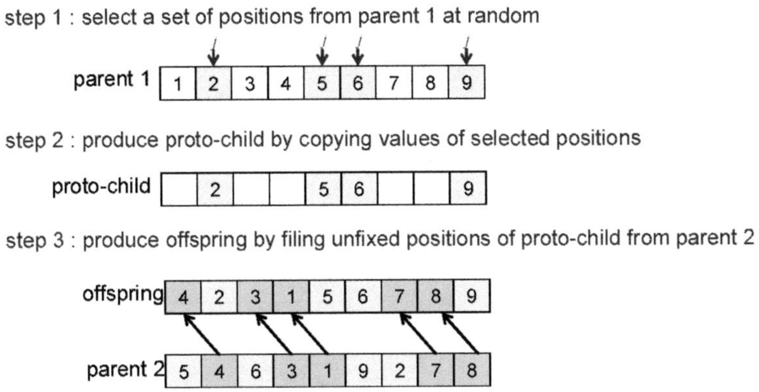

Fig. 9.11 The procedure for position-based crossover

From station:	st. 1				st. 2				st. 3				st. 4				st. 5			
i:	1	2	3	4	5	6	7	8	9	10	11	12	13	14	15	16	17	18	19	20
To Station :	2	3	4	5	1	3	4	5	1	2	4	5	1	2	3	5	1	2	3	4
chromosome $v_1(i)$:	12	16	11	10	19	17	9	18	20	2	8	3	7	1	14	5	15	13	6	4

Fig. 9.12 The best chromosome for five airports

	Routs	Max flow	Aircraft type
1	{ (3,1) (1,3) }	950	T1:1 T2:2
2	{ (2,1) (1,2) }	720	T2:2
3	{ (2,5) (5,1) (1,4) (4,3) (3,4) (4,1) }	650	T1:1 T2:1
4	{ (2,3) (3,5) (5,3) (3,2) }	750	T2:2
5	{ (2,4) (4,5) (5,4) (4,2) }	920	T1:1 T2:2

Total costs: 765000

Fig. 9.13 The best solution and aircrafts routes for five airports

9.1.1.3 Experiments and Discussion

For the numerical example, we consider the real problem of the airline company, four types of aircraft and ten airports from the domestic network in Japan. Figure 9.14 shows the name of the ten airports for this numerical example. These are Okinawa, Fukuoka, Kagoshima, Hiroshima, Kochi, Osaka, Nagoya, Sendai, Tokyo, and Sapporo in this case.

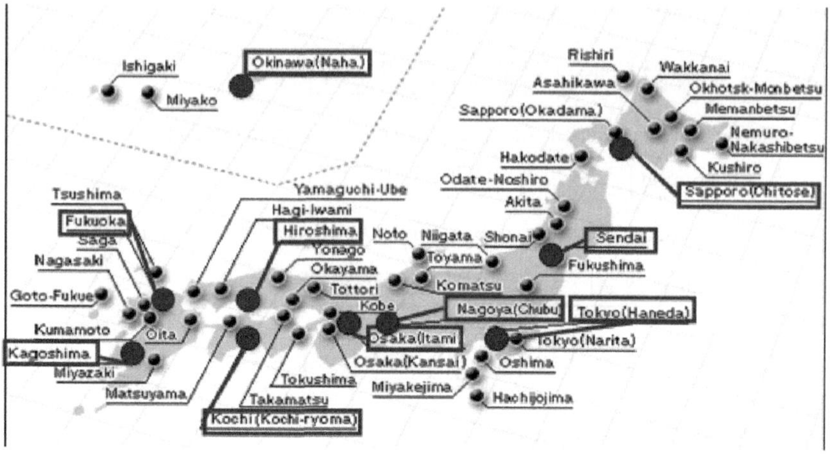

Fig. 9.14 Numerical example of ten airports in Japan

In this example, we have passenger flow demand for each flight between ten airports, and aircrafts flying costs for four types (737, 747, 767 and 777). We also have seats capacity for four types of aircraft, and the preparation cost from original starting node for each airport. Our numerical data is from the internet site of ANA, and ANA GROUP Flight Schedule 2006. Overall numerical data are shown in Tables 9.7–9.13.

Table 9.7 Passenger flow demand for ten airports

Demand d_{ij} i	j	Fukuoka 1	Kagoshima 2	Kochi 3	Hiroshima 4	Osaka 5	Nagoya 6	Tokyo 7	Sendai 8	Sapporo 9	Okinawa 10
Fukuoka	1	0	0	0	0	1000	1960	4500	400	1250	1600
Kagoshima	2	0	0	0	0	1000	800	1500	200	500	500
Kochi	3	0	0	0	0	700	0	800	0	0	0
Hiroshima	4	0	0	0	0	0	0	2000	0	600	300
Osaka	5	1000	1000	700	0	0	0	4000	1200	500	300
Nagoya	6	1960	800	0	0	0	0	0	500	1000	900
Tokyo	7	4500	1500	800	2000	4000	0	0	0	6000	3000
Sendai	8	400	200	0	0	1200	500		0	200	900
Sapporo	9	1250	500	0	600	500	1000	6000	200	0	0
Okinawa	10	1600	500	0	300	300	900	3000	900	0	0

The values of parameters used in the computational experiments are as follows:

Population size: $popSize = 20$
Maximum number of generation: $maxGen = 500$
Probability of crossover: $p_C = 0.5$
Probability of mutation: $p_M = 0.5$

Table 9.8 Transportation cost for aircraft type 747

747 cost c_{fij} i	j	Fukuoka 1	Kagoshima 2	Kochi 3	Hiroshima 4	Osaka 5	Nagoya 6	Tokyo 7	Sendai 8	Sapporo 9	Okinawa 10
Fukuoka	1	0	20	70	30	60	75	110	130	160	75
Kagoshima	2	20	0	85	45	70	85	120	145	175	95
Kochi	3	70	85	0	15	30	45	80	105	140	85
Hiroshima	4	30	45	15	0	30	45	80	100	135	70
Osaka	5	60	70	30	30	0	15	45	75	110	85
Nagoya	6	75	85	45	45	15	0	35	60	100	95
Tokyo	7	110	120	80	80	45	35	0	40	75	115
Sendai	8	130	145	105	100	75	60	40	0	40	120
Sapporo	9	160	175	140	135	110	100	75	40	0	125
Okinawa	10	75	95	85	70	85	95	115	120	125	0

Table 9.9 Transportation cost for aircraft type 777

777 cost c_{fij} i	j	Fukuoka 1	Kagoshima 2	Kochi 3	Hiroshima 4	Osaka 5	Nagoya 6	Tokyo 7	Sendai 8	Sapporo 9	Okinawa 10
Fukuoka	1	0	30	80	40	70	85	120	140	170	85
Kagoshima	2	30	0	95	55	80	95	130	155	185	105
Kochi	3	80	95	0	25	40	55	90	115	150	95
Hiroshima	4	40	55	25	0	40	55	90	110	145	80
Osaka	5	70	80	40	40	0	25	55	85	120	95
Nagoya	6	85	95	55	55	25	0	45	70	110	105
Tokyo	7	120	130	90	90	55	45	0	50	85	125
Sendai	8	140	155	110	110	85	70	50	0	50	130
Sapporo	9	170	185	145	145	120	110	85	50	0	135
Okinawa	10	85	105	80	80	95	105	125	130	135	0

Table 9.10 Transportation cost for aircraft type 767

767 cost c_{fij} i	j	Fukuoka 1	Kagoshima 2	Kochi 3	Hiroshima 4	Osaka 5	Nagoya 6	Tokyo 7	Sendai 8	Sapporo 9	Okinawa 10
Fukuoka	1	0	40	90	50	80	95	130	150	180	95
Kagoshima	2	40	0	105	65	90	105	140	165	195	115
Kochi	3	90	105	0	35	50	65	100	125	160	105
Hiroshima	4	50	65	35	0	50	65	100	120	155	90
Osaka	5	80	90	50	50	0	35	65	95	130	105
Nagoya	6	95	105	65	65	35	0	55	80	120	115
Tokyo	7	130	140	100	100	65	55	0	60	95	135
Sendai	8	150	165	120	120	95	80	60	0	60	140
Sapporo	9	180	195	155	155	130	120	95	60	0	145
Okinawa	10	95	115	90	90	105	115	135	140	145	0

Table 9.11 Transportation cost for aircraft type 737

737 cost c_{fij} i	j	Fukuoka 1	Kagoshima 2	Kochi 3	Hiroshima 4	Osaka 5	Nagoya 6	Tokyo 7	Sendai 8	Sapporo 9	Okinawa 10
Fukuoka	1	0	50	100	60	90	105	140	160	190	105
Kagoshima	2	50	0	115	75	100	115	150	175	205	125
Kochi	3	100	115	0	45	60	75	110	135	170	115
Hiroshima	4	60	75	45	0	60	75	110	130	165	100
Osaka	5	90	100	60	60	0	45	75	105	140	115
Nagoya	6	105	115	75	75	45	0	65	90	130	125
Tokyo	7	140	150	110	110	75	65	0	70	105	145
Sendai	8	160	175	135	130	105	90	70	0	70	150
Sapporo	9	190	205	170	165	140	130	105	70	0	155
Okinawa	10	105	125	115	100	115	125	145	150	155	0

Table 9.12 Preparing cost from original airport

orig. cost e_{fi} f	i	Fukuoka 1	Kagoshima 2	Kochi 3	Hiroshima 4	Osaka 5	Nagoya 6	Tokyo 7	Sendai 8	Sapporo 9	Okinawa 10
747	1	7000	4000	3500	4000	9000	8000	9000	5000	7000	3000
777	2	6000	3000	2500	3000	8000	7000	8000	4000	6000	2000
767	3	5000	2000	2000	2500	7000	6000	7000	3500	5000	1500
737	4	4000	1000	1500	2000	6000	5500	6500	2500	3000	1000

Table 9.13 Aircraft seats capacity for four types of aircraft

Air. Type f	Seats
747	350
777	250
767	200
737	150

After the simulation, the final objective value of minimum total cost is 11110000, and the best chromosome is shown in Fig. 9.15.

From station: st. 1

i:	1	2	3	4	5	6	7	8	9
To Station	2	3	4	5	6	7	8	9	10
chromosome $v_1(i)$	12	48	71	21	70	11	69	1	61

From station: st. 2

i:	10	11	12	13	14	15	16	17	18
To Station	1	3	4	5	6	7	8	9	10
chromosome $v_1(i)$	22	49	34	35	46	14	82	3	55

From station: st. 3

i:	19	20	21	22	23	24	25	26	27
To Station	1	2	4	5	6	7	8	9	10
chromosome $v_1(i)$	66	30	83	16	75	84	54	90	89

From station: st. 4

i:	28	29	30	31	32	33	34	35	36
To Station	1	2	3	5	6	7	8	9	10
chromosome $v_1(i)$	80	36	81	79	10	57	5	39	72

From station: st. 5

i:	37	38	39	40	41	42	43	44	45
To Station	1	2	3	4	6	7	8	9	10
chromosome $v_1(i)$	23	13	40	2	47	41	28	6	65

From station: st. 6

i:	46	47	48	49	50	51	52	53	54
To Station	1	2	3	4	5	7	8	9	10
chromosome $v_1(i)$	29	74	31	88	17	76	52	77	53

From station: st. 7

i:	55	56	57	58	59	60	61	62	63
To Station	1	2	3	4	5	6	8	9	10
chromosome $v_1(i)$	58	9	37	19	4	38	25	60	73

From station: st. 8

i:	64	65	66	67	68	69	70	71	72
To Station	1	2	3	4	5	6	7	9	10
chromosome $v_1(i)$	62	26	56	87	27	63	42	7	64

From station: st. 9

i:	73	74	75	76	77	78	79	80	81
To Station	1	2	3	4	5	6	7	9	10
chromosome $v_1(i)$	15	86	32	33	78	18	51	85	50

From station: st. 10

i:	82	83	84	85	86	87	88	89	90
To Station	1	2	3	4	5	6	7	8	9
chromosome $v_1(i)$	67	24	59	44	45	8	43	68	20

Fig. 9.15 The best chromosome for 10 airports

9.1.2 Fleet Assignment Model with Time-space Network

Fleet assignment problem (FAP)–given a flight schedule and set of aircraft–is to determine which type of aircraft should fly each flight segment. In contrast with the connection network, the "time-space network" focuses on representing flight legs, and leaves it to the model to decide on the connections, as long as these are feasible to the time and space considerations.

The objective of the time-space network is–given a flight schedule with fixed departure times and costs according to each aircraft type on each flight leg–to find the least cost assignment of aircraft types to flights, such that each flight is covered by exactly one fleet type, flow of aircraft by type is balanced at each airport,and only the available number aircraft of each type are used. The characteristic of the time-space network must satisfy balance constraints that force the aircraft to circulate through the network of flights. The balance of aircraft is enforced by the conservation of flow equations for a time-expanded multi-commodity network. We make the time line cycle, which forces the solution to be a circulation through the network.

As a simple example for FAP with time-space network in Japan, we choose five airports (Tokyo, Osaka, Fukuoka, Sendai, and Sapporo) that can connect with each other. The airline company has four types of aircraft–B747, B777, A320 and A340. In Fig. 9.16, there are five airports can connect to each other. The a_{fijt} is arrival time from node i to node j. The d_{fijt} is departure time from node i to node j. The x_{fijt} is aircraft type f which flies from node i to node j at time t. The y_{fi} is aircraft type f starting at origin node i. The assumptions of the FAP with time-space network are shown as following:

A1. In time-space network, our costs are including flying cost and ground cost for each aircraft.
A2. An aircraft first departing for service flights must pay the starting cost.
A3. We have enough aircraft to satisfy all the flights.
A4. Before aircraft departs, it must have 1h for preparation time.
A5. Only consider completed flights for one day.

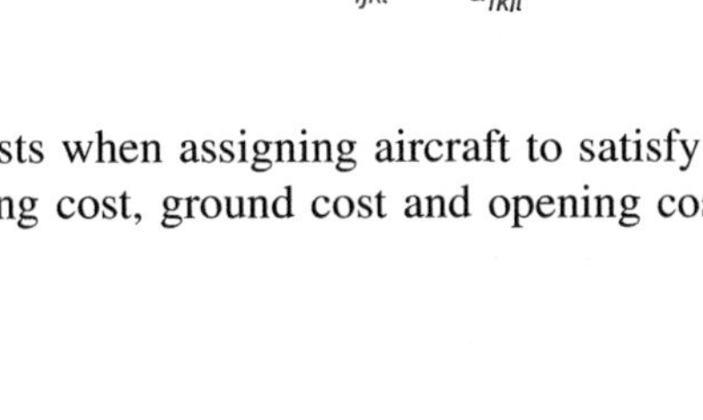

Fig. 9.16 The symbol of time-space network connection for five airports

The objective is to minimize total costs when assigning aircraft to satisfy each flight. Hence, we have to formulate flying cost, ground cost and opening cost for each aircraft.

9.1.2.1 Mathematical Formulation

Let $G = (N,A)$ be a network, consisting of a finite set of nodes $N = \{0,1,2,\ldots,n\}$, and a set of arcs $A = \{(i,j),(k,l),\ldots,(s,t)\}$ joining n pairs of nodes in N. Arc (i,j)

is said to be incident with nodes i and j, and is directed for node i to node j. Suppose that each arc (i, j) has assigned to it nonnegative numbers c_{fij}.

Notation

Indices

i, j, k: the index of airport, $i, j, k = 0, 1, \ldots, N$, where index 0 is a dummy node
f: aircraft no., $f = 1, 2, \ldots, F$

Parameters

c^L_{fij}: unit flight leg cost of the aircraft type f from node i to node j
c_{fjG}: unit ground cost of the aircraft type f in node j
d_{fjkt}: departure time form node j to node k
a_{fijt}: arrival time form node i to node j
e_{fi}: using the aircraft no. f in origin node i
s_{ijt}: all the required flight schedule
t^G_f: limited preparing time in the ground for next departure flight
F: set of all aircrafts
N: set of all nodes
T: set of all times

Decision Variables

$$x_{fijt} = \begin{cases} 1, & \text{aircraft no. } f \text{ departure from node } i \text{ to node } j \text{ at time } t \\ 0, & \text{otherwise} \end{cases}$$

$$y_{fi} = \begin{cases} 1, & \text{if aircraft no. } f \text{ opening at origin node } i \\ 0, & \text{otherwise} \end{cases}$$

The fleet assignment model with time-space network (tsFAP) is formulated as follows, in which the objective is to minimize the total cost:

$$\min \sum_{f=1}^{F}\sum_{i=1}^{N}\sum_{j=1}^{N}\sum_{t=0}^{T} c^L_{fij} x_{fijt} + \sum_{f=1}^{F}\sum_{j=1}^{N}\sum_{t=0}^{T} c^G_{fj}\Big(\sum_{k=1}^{N} d_{fjkt} x_{fijt} - \sum_{i=1}^{N} a_{fjkt} x_{fijt}\Big) + \sum_{f=1}^{F}\sum_{i=0}^{N} e_{fi} y_{fi} \tag{9.7}$$

$$\text{s.t.} \quad \sum_{f=1}^{F} x_{fijt} \geq s_{ijt}, \ \forall i,\ j,\ t \tag{9.8}$$

$$\sum_{i=1}^{N}\sum_{t=0}^{T} x_{fijt} - \sum_{k=1}^{N}\sum_{t=0}^{T} x_{fjkt} = 0,\ \forall f,\ j \tag{9.9}$$

$$\sum_{k=1}^{N} d_{fjkt}\, x_{fjkt} - \sum_{i=1}^{N} a_{fijt}\, x_{fijt} \geq t_f^G,\ \forall f,\ j,\ t \tag{9.10}$$

$$\sum_{j=1}^{N} x_{fjkt},\ \forall f,\ i,\ t \tag{9.11}$$

$$\sum_{t=0}^{N} x_{f0jt} = \sum_{t=0}^{T} x_{fj0t},\ \forall f,\ j \tag{9.12}$$

$$y_{fj} = \begin{cases} 1, \text{if } \sum_{t=0}^{T} x_{f0jt} = 1,\ \forall f,\ j \\ 0, \text{otherwise} \end{cases} \tag{9.13}$$

$$x_{fijt} = \{0,\ 1\},\ \forall f,\ i,\ j,\ t \tag{9.14}$$

Constraint at Eq. 9.8 is all the flight legs must be satisfied the fixed schedule. Constraint at Eq. 9.9 is flow balance for each fleet type in the flight network. Constraint at Eq. 9.10 is limited preparing time for each fleet type in the ground for the next departure flight. Constraint at Eq. 9.11 is the aircraft must be choice the one flight leg to work. Constraint at Eq. 9.12 is the dummy node from node j. Equations 9.13 and 9.14 define the nature of decision variables.

9.1.2.2 Extended Priority-based GA for FAP with Time-space Network

Let $P(t)$ and $C(t)$ be parents and offspring in current generation t, respectively. The overall procedure of priority-based GA (priGA) for solving this FAP model is outlined as follows:

Genetic Representation

To develop a priority-based genetic representation for the FAP with time-space network, there are four main phases:

Phase 1: Deciding flight precedence relationship

Step 1.1: Sort starting time for each flight
Step 1.2: Decide precedence relationship of flight sequence

Phase 2: Determining flight sequence

Step 2.1: Generate a random priority to each flight using encoding procedure
Step 2.2: Determine the flight sequence using decoding procedure

---

procedure: priGA for FAP with connection network
input: FAP data, GA parameters ($popSize$, $maxGen$, p_M, p_C)
output: the best flight schedule S
begin
  $t \leftarrow 0$;
  initialize $P(t)$ by extended priority-based encoding routine;
  evaluate $P(t)$ by priority-based decoding routine;
  while (**not** terminating condition) **do**
    create $C(t)$ from $P(t)$ by partially mapped crossover routine;
    create $C(t)$ from $P(t)$ by insertion mutation routine;
    evaluate $C(t)$ by priority-based decoding routine;
    select $P(t+1)$ from $P(t)$ and $C(t)$ by roulette wheel selection routine;
    $t \leftarrow t+1$;
  end
  output the best flight schedule S;
end

Phase 3: Assigning aircrafts to each flight

Step 3.1: Generate a random aircraft number to each flight using encoding procedure
Step 3.2: Assign aircraft using decoding procedure

Phase 4: Designing a flight schedule

Step 4.1: Create a flight schedule S
Step 4.2: Draw a Gantt chart for schedule S

Phase 1: Deciding Flight Precedence Relationship

Considering a 10 flights simple example as shown in Table 9.14, the following 2 steps decide the precedence relationship for each flight with departure times:

Step 1.1: Sort starting time for each flight as shown in Fig. 9.17.
Step 1.2: Decide precedence relationship of flight sequence as follows; the flight precedence relationship of 56 flights is shown in Fig. 9.18.

Flight sequence: 1, 21, 22, 23, 31, 2, 24, 32, 3, 33

k:	1	2	3	4	5	6	7	8	9	10
Flight no.: $f[k]$:	001	021	022	023	031	002	024	032	003	033
Starting time: $s[k]$:	0700	1000	1000	1200	1400	1600	1600	1600	1800	2000

Fig. 9.17 Sorted flight data of a simple example with 10 flights.

Table 9.14 Simple data

k	Flight no.	$s[k]$	dep. sta.	arr. t.	arr. sta.
1	001	0700	1	0800	2
2	002	1600	1	1700	2
3	003	1800	1	1900	3
4	021	1000	2	1100	1
5	022	1000	2	1130	3
6	023	1200	2	1300	1
7	024	1600	2	1730	3
8	031	1400	3	1530	2
9	032	1600	3	1700	1
10	033	2000	3	2130	2

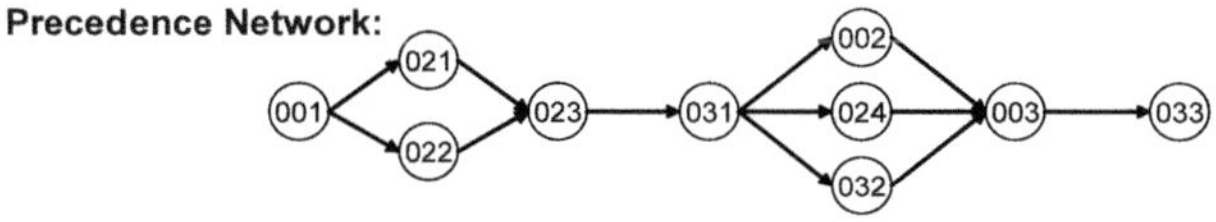

Fig. 9.18 The precedence network for a simple example.

Phase 2: Determining Flight Sequence

To determine a list of flight sequences satisfying the precedence relationship, we consider the following two steps using priority-based encoding and decoding procedures:

Step 2.1: Generate a random priority to each flight using encoding procedure.
Step 2.2: Determine the flight sequence using decoding procedure as shown in Fig. 9.19.

For considering the 56 flights example, the priority-based chromosome and the flight sequence is shown in Fig. 9.20.

Phase 3: Assigning Aircraft to Each Flight

To assign aircrafts to each flight based on the flight sequence by flight-based encoding and decoding procedures:

Step 3.1: Generate a random aircraft number to each flight using encoding procedure as shown in Fig. 9.21.

```
procedure 4: priority-based decoding for determine flight sequence
input: (S_j), (f[k]), (v_1)
output: determined flight sequence g[i], ∀i
begin
  i=1;
  for j=1 to J
     while S_j≠∅ do
         k* ←argmax {v_k |k∈S_j}
         g[i] ←f[k*];
         S_j ←S_j / {f[k*]};
         i ←i+1;
  output determined flight sequence g[i], ∀i
end
```

Fig. 9.19 Decoding procedure for determining flight sequence

k:	1	2	3	4	5	6	7	8	9	10
Flight no.:	001	021	022	023	031	002	024	032	003	033
Priority $v_1(i)$:	9	1	2	8	5	3	7	6	10	4

Flight sequence {001, 022, 021, 023, 031, 024, 032, 002, 003, 033}

Fig. 9.20 Determining flight sequence for a simple example

Step 3.2: Assign aircraft using decoding procedure as shown in Fig. 9.22. A simple example of assigning aircraft is shown in Fig. 9.23.

```
procedure 5: flight-based encoding for assigning aircrafts
input: max no. of available aircrafts L, no. of flights F
output: chromosome (v_2[k])
begin
  for k=1 to F
    v_2(k) · random[a, q];
  output the chromosome (v_2[k])
end
```

Fig. 9.21 Encoding a chromosome for assigning aircraft

```
procedure 6: priority-based decoding for assigning aircrafts
input: Sj, chromosome v2, flight information (e[u], s[u], O, d[u])
output: aircrafts route (Gl)
begin
    for l=1 to S1
        u ←v1[l]; // flight no.
        Gl ←{u};
        h[l] ←e[u]; // destination of flight u
        t[l] ←s[u]+d[u] // starting time t[l] on flight l
    for k=|S1 |+1 to F
        times ←1;
        a ←1;
        u ←v1[k];
        flag ←0;
        for l=1 to L
            if (s[u] ≥ Gp+O) and (h[e]=p[u]) then
                times ←times+1;
                a ←k;
            if (times=v2[k]) then
                Gp ← Gp U {u};
                h[l] ←e[u];
                t[l] ←t[l]+O+d[u]
                flag ←1;
        if flag=0 then
            L ←L+1;
            GL ←{u};
            h[L] ←e[u];
            t[L] ←s[u];
    output aircrafts route (Gl)
end
```

Fig. 9.22 Flight-based decoding for assigning aircraft

k:	1	2	3	4	5	6	7	8	9	10
Flight no.:	001	021	022	023	031	002	024	032	003	033
Aircraft $v_2(i)$:	a	b	d	b	c	a	e	c	d	e

Aircraft a: {001, 002} Aircraft b: {021, 023} Aircraft c: {031, 032}

Aircraft d: {022, 003} Aircraft e: {024, 033}

Fig. 9.23 Assigning aircraft to each flight by simple data

Phase 4: Designing a Flight Schedule

Depending on the following two steps, make a flight schedule by the aircraft assigned with the data set and draw a Gantt chart based on the flight schedule.

Step 4.1: Create a flight schedule *S*. For the simple example, the flight schedule can be shown as follows:

$$S = \{\text{aircraft a, aircraft b, aircraft c, aircraft d, aircraft e}\}$$

(aircraft no., flight no., departure time-arrival time,
departure station-arrival station)
= {[(a, 1, 0700-0800, 1-2), (a, 6, 1200-1300, 2-1)],
[(b, 4, 1000-1100, 2-1), (b, 2, 1000-1100, 1-2)],
[(c, 5, 1000-1130, 2-3), (c, 8, 1400-1530, 3-2)],
[(d, 9, 1600-1700, 3-1), (d, 3, 1800-1900, 1-3)],
[(e, 7, 1600-1730, 2-3), (e, 10, 2000-2130, 3-2)] }

Step 4.2: Draw a Gantt chart for schedule S as shown in Fig. 9.24.

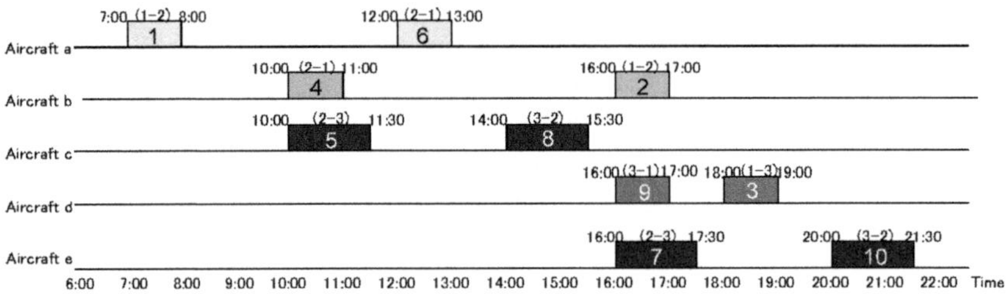

Fig. 9.24 Gantt chart for simple data

Genetic Operators

Partially Mapped Crossover: As crossover operator, we adopt the partially mapped crossover (PMX) introduced in Subsection 2.2.2.3. PMX can be viewed as an extension of two-point crossover for binary string to permutation representation. It uses a special repairing procedure to resolve the illegitimacy caused by the simple two-point crossover.
Insertion Mutation: As mutation operator, we adopt the insertion mutation. It selects a gene at random and inserts it in another random position.

9.1.2.3 Experiments and Discussion

For the numerical example, we consider the real problem of the airline company, 58 flights connecting with 5 airports (Tokyo, Osaka, Fukuoka, Sendai, and Sapporo) which is shown in Fig. 9.25, and 36 aircraft of 4 types (B747, B777, A320 and A340) from domestic networks in Japan. Also we have several original data for fly cost, ground cost, and open cost for different types of aircraft. The objective function is minimum cost of aircraft route. The original data is shown in Tables 9.15–9.25

Finally, the best chromosome and aircrafts route is show in Fig. 9.26. As a result, we need 19 aircraft to serve 58 flights, and we get the minimum cost. The best flight

Table 9.15 Tokyo airport

Flight no.	Dep. - Arr.	Arr. Sta.
001	07:00 - 08:00	2
002	07:00 - 08:30	5
003	07:30 - 08:30	3
004	09:30 - 11:00	5
005	10:30 - 11:30	3
006	10:30 - 11:30	4
007	11:00 - 12:00	2
008	11:30 - 13:00	5
009	14:30 - 16:00	5
010	15:00 - 16:00	2
011	15:30 - 16:30	3
012	16:00 - 17:00	3
013	17:00 - 18:30	5
014	17:00 - 18:00	4
015	19:00 - 20:00	2
016	19:30 - 21:00	5

Table 9.16 Fukuoka airport

Flight no.	Dep. - Arr.	Arr. Sta.
040	07:00 - 08:15	2
041	08:00 - 09:00	1
042	10:00 - 11:00	4
043	11:30 - 12:30	1
044	12:30 - 13:45	2
045	13:00 - 15:30	5
046	15:30 - 16:30	1
047	17:30 - 18:45	2
048	17:30 - 20:00	5
049	19:00 - 20:00	1
050	19:15 - 20:15	4

Table 9.17 Osaka airport

Flight no.	Dep. - Arr.	Arr. Sta.
020	07:45 - 09:00	3
021	08:00 - 09:15	4
022	09:00 - 10:00	1
023	09:00 - 10:45	5
024	11:00 - 12:15	3
025	13:00 - 14:00	1
026	13:30 - 15:15	5
027	16:00 - 17:00	1
028	17:00 - 18:15	4
029	19:00 - 20:00	1

Table 9.18 Sendai airport

Flight no.	Dep. - Arr.	Arr. Sta.
060	07:45 - 08:45	3
061	08:15 - 09:15	1
062	09:00 - 10:15	5
063	10:00 - 11:15	2
064	16:00 - 17:00	1
065	17:30 - 18:45	5
066	19:00 - 20:15	2
067	19:15 - 20:15	3

Table 9.19 Sapporo airport

Flight no.	Dep. - Arr.	Arr. Sta.
080	08:00 - 09:30	1
081	08:00 - 10:30	3
082	09:30 - 11:00	1
083	11:30 - 12:45	4
084	12:30 - 14:00	1
085	14:30 - 16:00	1
086	16:00 - 17:45	2
087	16:30 - 17:45	4
088	16:30 - 18:00	1
089	17:45 - 19:15	2
090	19:30 - 21:00	1

Table 9.20 Fly cost by B747 aircraft (x100Yen/min)

i \ j	Tokyo	Osaka	Fukuoka	Sendai	Sapporo
Tokyo	0	3	4	3	4
Osaka	3	0	3	4	5
Fukuoka	4	3	0	5	6
Sendai	3	4	5	0	4
Sapporo	4	5	6	4	0

Table 9.21 Fly cost by B777 aircraft (x100Yen/min)

i \ j	Tokyo	Osaka	Fukuoka	Sendai	Sapporo
Tokyo	0	2	3	2	3
Osaka	2	0	2	3	4
Fukuoka	3	2	0	4	5
Sendai	2	3	4	0	3
Sapporo	3	4	5	3	0

Table 9.22 Fly cost by A320 aircraft (x100Yen/min)

i \ j	Tokyo	Osaka	Fukuoka	Sendai	Sapporo
Tokyo	0	3	3	3	3
Osaka	3	0	3	3	4
Fukuoka	3	3	0	4	4
Sendai	3	3	4	0	3
Sapporo	4	4	3	3	0

Table 9.23 Fly cost by A340 aircraft (x100Yen/min)

i	Tokyo	Osaka	Fukuoka	Sendai	Sapporo
Tokyo	0	4	4	4	4
Osaka	3	0	3	4	4
Fukuoka	4	4	0	5	6
Sendai	3	4	4	0	4
Sapporo	4	5	5	5	0

Table 9.24 Ground cost by aircraft type (x100Yen/min)

Ground cost	B747	B777	A320	A340
Tokyo	10	8	7	9
Osaka	10	8	7	9
Fukuoka	8	7	6	8
Sendai	6	5	6	5
Sapporo	8	7	7	5

Table 9.25 Total number of aircraft (x100Yen/min)

Type no.	Type	Total no. of aircrafts
1	B747	10
2	B777	8
3	A320	10
4	A340	8
		Total : 36

Table 9.26 Open cost by aircraft type (x100Yen/min)

Type	Open cost
B747	3000
B777	2500
A320	2000
A340	2800

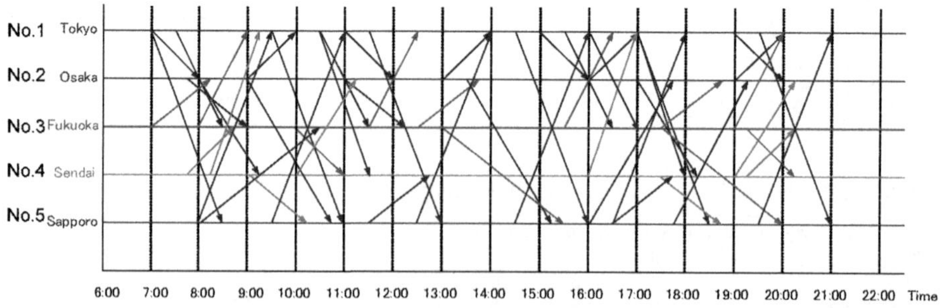

Fig. 9.25 The map of 58 flights for 5 airports

schedule are shown, and Gantt chart of aircraft and aircraft route for five airports are show in Figs. 9.27 and 9.28.

S = {aircraft a, aircraft b, aircraft c, aircraft d, aircraft e, . . . , aircraft s}
(aircraft no., flight no., departure time-arrival time,
departure station-arrival station)
= [(a, 27, 0700–0815, 3–2), (a, 21, 1100–1215, 2–3), (a, 33, 1530–1630, 3–1), (a, 15, 1900–2000, 1–2)], [(b, 2, 0700–0830, 1–5), (b, 48, 0930–1100, 5–1), (b, 9, 1430–1600, 1–5), (b, 55, 1745–1915, 5–2)], [(c, 1, 0700–0800, 1–2), (c, 19, 0900–1000, 2–1), (c, 7, 1100–1200, 1–2), (c, 22, 1300–1400, 2–1), (c, 10, 1500–1600, 1–2), (c, 25, 1700–1815, 2–4), (c, 45, 1915–2015, 4–3)], [(d, 3, 0730–0830, 1–3), (d, 29, 1000–1100, 3–4), (d, 42, 1600–1700, 4–1), (d, 16, 1930–2100, 1–5)], [(e, 38, 0745–0845, 4–3), (e, 31, 1230–1345, 3–2), (e, 24, 1600–1700, 2–1)], [(f, 17, 0745–0900, 2–3), (f, 34, 1730–1845, 3–2)], [(g, 28, 0800–0900, 3–1), (g, 5, 1030–1130, 1–3), (g, 35, 1730–2000, 3–5)], [(h, 46, 0800–0930, 5–1), (h, 6, 1030–1130, 1–4), (h, 43, 1730–1845, 4–5)], [(i, 18, 0800–0915, 2–4), (i, 44, 1900–2015, 4–2)], [(j, 47, 0800–1030, 5–3), (j, 30, 1130–1230, 3–1), (j, 13, 1700–1830, 1–5), (j, 56, 1930–2100, 5–1)], [(k, 39, 0815–0915, 4–1), (k, 8, 1130–1300, 1–5), (k, 51, 1430–1600, 5–1), (k, 14, 1700–1800, 1–4)], [(l, 20, 0900–1045, 2–5), (l, 50, 1230–1400, 5–1), (l, 12, 1600–1700, 1–3), (l, 36, 1900–2000, 3–1)], [(m, 40, 0900–1015, 4–5), (m,49, 1130–1245, 5–4)], [(n, 4, 0930–1100, 1–5), (n, 52, 1600–1745, 5–2), (n, 26, 1900–2000, 2–1)], [(o, 41, 1000–1115, 4–2), (o, 23, 1330–1515, 2–5), (o, 53, 1630–1745, 5–4)], [(p, 32, 1300–1530, 3–5), (p, 54, 1630–1800, 5–1)], [(q, 11, 1530–1630, 1–3), (q, 37, 1915–2015, 3–4)]

i:	1	2	3	4	5	6	7	8	9	10	11	12	13	14	15	16	17	18	19	20	21	22	23	24	25	26	27	28
Aircraft $v_2(i)$:	c	b	d	n	g	h	c	k	b	c	q	l	j	k	a	d	f	i	c	l	a	c	o	e	c	n	a	g

i:	29	30	31	32	33	34	35	36	37	38	39	40	41	42	43	44	45	46	47	48	49	50	51	52	53	54	55	56
Aircraft $v_2(i)$:	d	j	e	p	a	f	g	l	q	e	k	m	o	d	h	i	c	h	j	b	m	l	k	n	o	p	b	j

17 aircrafts routes:

Aircraft no.	Routs	Type	Cost
a	{ 27, 21, 33, 15 }	3	10385
b	{ 2, 48, 9, 55 }	3	10820
c	{ 1, 19, 7, 22, 10, 25, 45 }	3	10115
d	{ 3, 29, 42, 16 }	3	10670
e	{ 38, 31, 24}	3	10670
f	{ 17, 34 }	3	10970
g	{ 28, 5, 35 }	3	10310
h	{ 46, 6, 43}	4	10015
i	{ 18, 44 }	3	10895

Aircraft no.	Routs	Type	Cost
j	{ 47, 30, 13, 56 }	3	10550
k	{ 39, 8, 51, 14 }	2	9745
l	{ 20, 50, 12, 36 }	3	10895
m	{ 40, 49 }	2	9550
n	{ 4, 52, 26}	2	12490
o	{ 41, 23, 53 }	2	9850
p	{ 32, 54 }	2	12280
q	{ 11, 37 }	2	12640

Minimum total costs: 182850

Fig. 9.26 The best chromosome and 17 aircraft routes

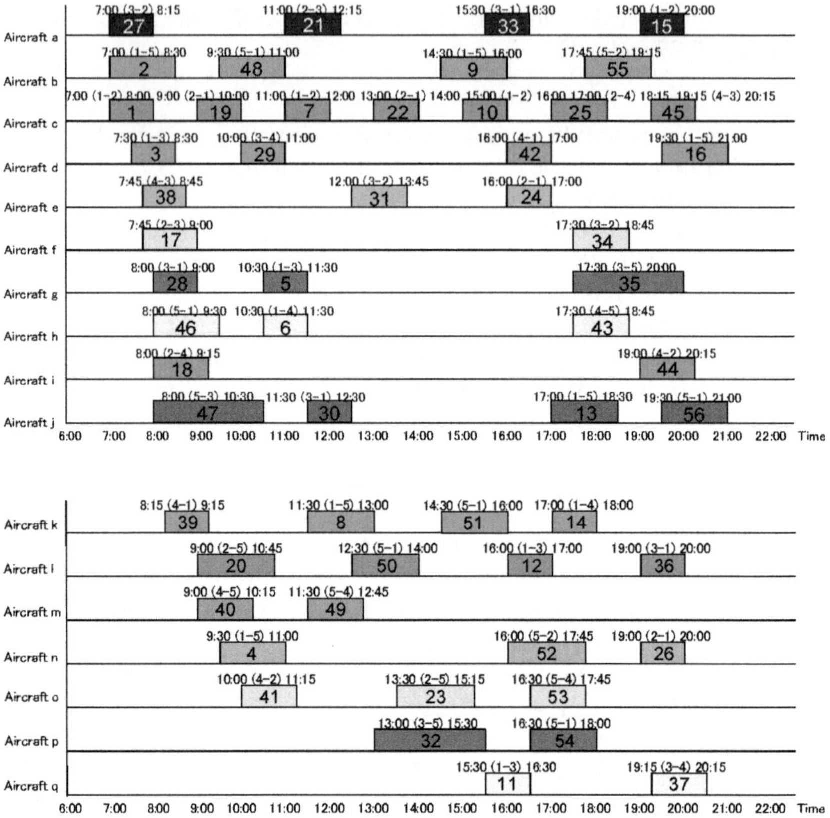

Fig. 9.27 The Gantt chart of aircraft routes

9.2 Container Terminal Network Model

In recent years, container transportation has become most important along with the increase of container business. With the rapid development of container transportation, container terminals (CT) have also been constructed in large numbers. So how to improve the efficiency in CT systems and how to heighten the service level becomes the most important research issue in CT systems. It has a deep and revolutionary significance for the modernization process of CT systems.

To improve the efficiency of CT systems is to finish the set of jobs using the limited resource in such a way as to minimize the total processing/ serving time or the total cost. After the container ships arrive at the seaport, the process will be

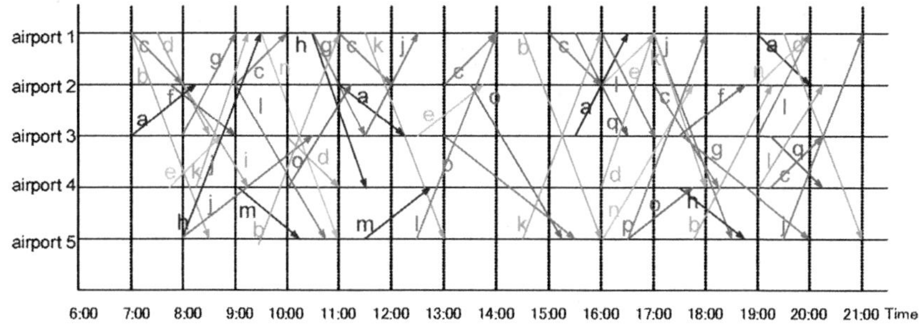

Fig. 9.28 The 17 aircraft routes for five airports

conduct as entrance process taken following steps:

1. Container ship enters the berth.
2. Containers are unloaded by the quay cranes (QCs).
3. Containers are transported by the vehicles.
4. Containers are unloaded by the yard crane.
5. Containers stay in the storage locations.
6. Containers depart.

For the different processes, there are several optimization problems of CT systems occurring in real-world applications:

For *yard crane problems*, Kim *et al.* [24] discussed how to determine the optimal amount of storage space and the optimal number of transfer cranes for import containers. A cost model was developed for decision making. Bose *et al.* [25] described the main logistics processes in seaport container terminals and presented methods for their optimization. They investigated different dispatching strategies for straddle carriers to gantry cranes and showed the potential of evolutionary algorithms to improve the solutions. Zhang *et al.* [26] addressed the crane deployment problem and formulated it as a mixed integer programming (MIP) model and solved by Lagrangean relaxation. They augmented the Lagrangean relaxation model by adding additional constraints and modified the solution procedure accordingly. Chung *et al.* [27] considered the problem of scheduling the movements of cranes in a container storage yard. The problem was formulated as a mixed-integer linear program. A new solution approach, called the successive piecewise-linear approximation method, was also developed. Linn *et al.* [28] presented an algorithm and a mathematical model for optimal yard crane deployment. Kim and Kim [29] discussed a method of routing yard-side equipment during loading operations in container terminals. A GA approach and a beam search algorithm were suggested to solve the problem. Ng and Mak [30] studied the problem of scheduling a yard crane to perform a given set of loading/unloading jobs with different ready times. A branch and bound algorithm was proposed to solve the scheduling problem optimally. Ng [31] examined the problem of scheduling multiple yard cranes to perform a given set of jobs with different ready times in a yard crane zone with only one bi-directional traveling lane. This research developed a dynamic programming-based heuristic to solve the scheduling problem and an algorithm to find lower bounds for benchmarking the schedules found by the heuristic.

For the problems of *container transfer and storage locations*, Kim and Kim [32] focused on how to route transfer cranes optimally in a container yard during loading operations of export containers at port terminals. This routing problem was formulated as a mixed integer program. Based on the mixed integer program, an optimizing algorithm was developed. Kozan and Preston [33] considered GA approaches to reduce container handling/transfer times and ships' time at the port by speeding up handling operations. Also discussed, is the application of the model to assess the consequences of increased scheduled throughput time as well as different strategies such as the alternative plant layouts, storage policies and number of yard machines. Preston and Kozoan [34] modeled the seaport system with the objective of determining the optimal storage strategy for various container-handling schedules. A container location model (CLM) was developed, with an objective to function designed to minimize the turnaround time of container ships, and solved using GA. Kozan and Preston [35] modeled the seaport system with the objective of determining the optimal storage strategy and container-handling schedule. A GA, a tabu search (TS) and a tabu search/GA hybrid were used to solve the problem. Imai *et al.* [36] were concerned with the ship's container stowage and loading plans that satisfy these two criteria. The problem was formulated as a multi-objective integer programming. In order to obtain a set of noninferior solutions of the problem, the weighting method was employed.

For the *berth and quay crane problems*, Kim and Park [37] discussed the problem of scheduling quay cranes (QCs). A mixed-integer programming model, which considers various constraints related to the operation of QCs, was formulated. This study proposed a branch and bound (B & B) method to obtain the optimal solution of the QC scheduling problem and a heuristic search algorithm, called greedy randomized adaptive search procedure (GRASP), to overcome the computational difficulty of the B & B method. Han *et al.* [38] described a nonlinear model for the berth scheduling problem and solved this model by the genetic algorithm GA and the hybrid optimization strategy GASA respectively. Jung and Kim [39] proposed a method to schedule loading operations when multiple yard cranes are operating in the same block. The loading scheduling methods in this paper are based on a genetic algorithm and a simulated annealing method. An encoding method considering the special properties of the optimal solution of the problem was suggested. Zhou *et al.* [40] proposed a dynamic stochastic berth allocation model to simulate the real decision-making. A heuristics algorithm was developed based on genetic algorithm, and a reduced basic search space for the algorithm was discussed by considering approximate optimal solution properties. Imai *et al.* [41] first constructed a new integer linear programming formulation for easier calculation and then the for easier calculation and then the formulation was extended to model the berth allocation problem at a terminal with indented berths. The berth allocation problem at the indented berths was solved by GAs. Hansen *et al.* [42] proposed a dynamic stochastic berth allocation model to simulate the real decision making. A heuristics algorithm was developed based on GA, and a reduced basic search space for the algorithm was discussed by considering approximate optimal solution properties. Imai *et al.* [43] addressed efficient berth and crane allocation scheduling at a multi-user container terminal. They developed a heuristic to find an approximate solution for the problem by employing GA. Lee *et al.* [44] considered a quay crane scheduling problem for determining a handling sequence of holds for quay cranes assigned to a container vessel considering interference between quay cranes. This paper provided a mixed integer programming model for the considered quay crane scheduling problem that is NP-complete in nature. A GA approach was proposed to obtain near optimal solutions.

For other optimization problems, such as Wen and Zhou [45] proposed a GA approach to gain an efficient and feasible solution by improving solution encoding for vehicle routing problem, a GA operators and involution strategy. For integrated scheduling problems in CT systems, Zeng and Yang [46] developed a bilevel programming model for container terminal scheduling to improve the integration efficiency of container terminals. A GA approach was designed to solve the model and numerical tests were provided to illustrate the validity of the model and the algorithm. Zhou *et al.* [47] discussed a systemic fuzzy model with including loading container sequence determination, yard-crane allocation and container vehicle assignation. An improved GA was proposed. Lau and Zhao [48] formulated a mixed-integer programming model, which considers various constraints related to the integrated operations between different types of handling equipment. This study proposed a heuristic method, called multi-layer genetic algorithm (MLGA) to obtain the near-optimal solution of the integrated scheduling problem and an improved heuristic algorithm, called genetic algorithm plus maximum matching (GAPM), to reduce the computation complexity of the MLGA method.

9.2.1 Berth Allocation Planning Model

Berth allocation planning (BAP) problem is how to allocate the berths to container ships when it entered the port. Usually, the berth allocation plan should be drawn up at first according to the date of arrival reported by the ship companies and then the one daily berth allocation plan should be decided concretely in two to three days before the ships arrival. If the resources are not busy, the ship will be served just-in-time; otherwise, we need to consider the suitable allocation of the container ship to berths. Up to now, this kind of allocation planning of container ships was conducted by human experience, and actually, the plans can‘t be executed normally in many container terminals. It is not only a waste of the handling cost; transportation cost *etc.*, but also brings big loss to the shipping company. Therefore, to allocate the container ships to a berth reasonably is the first step and also the key step in container terminal management optimization. In this section, we analyze the factors which need to be considered in berth allocation planning problems.

Ship Time Factor

The arrival time of container ships is determined by the container ship company in advance. If the port is beside the sea, the entering time of container ship is irregular, in other words, it can enter the port at at any time during the day. But if it is an inland port close to the sea, the entering time is affected by the tide, and the arrival time is concentrated at certain times. The departure times have the same conditions in the port: the ship can depart as soon as the work is over in the port beside the sea, but at an inland port beside the sea, it has to wait for the tide for departure. After the ships enter the berth, they maybe can't be served at once. Some ports serve using a First Come First Service (FCFS) rule, but other ports serve ships according to the priority of the ships. And there is the same aim on berth; it is to make the ships depart on time. The departure time is decided based on the equipment processing capacity of the port. But more and more container ships only execute the sailing time recently decided by the ship company, *i.e.*, the departure time of ships is decided by the ship company, and if the departure time is delayed, it needs to be compensated by the container terminal company. Therefore, to reduce the waiting time and departure time delay of container ships in the berth is very important to increase the service level of the container terminal company. Not only to allocate the berths to container ships in the suitable location, but also to arrange the quay cranes reasonably to load and unload container more efficiently are key points to realize that aim.

Berth Location Factor

Berth allocation planning problem is to determine the berth location to served container ships at some time. "Berth" is the place where the container ships can be stopped by the sea-line. The form of the sea-line, the water-depth of the sea *etc.* can influence the berth allocation planning. And, the ships which need to be served by the fixed equipment have to be arranged at the special location. In this section, we introduce how to arrange ships with different lengths by a given sea-line port. It must consider the safety distance between the ships, that between the ships and the berths, and the water-depth of the berths *etc*. Besides the above geographical factors, location of quay cranes should also be considered because the handling time of unloading and loading are affected by the number and the performance of the quay cranes. And the workload balance of QCs is also considered, not to be done by only one QC while other QCs are free. In this way the problem should be combined with the quay crane scheduling problem, AGV dispatching problem, and the storage spaces allocation problem *etc*. Therefore, when formulating the mathematical model for the BAP problem, we have to consider factors: ship time, berth location, and port layout.

9.2.1.1 Mathematical Formulation

First, there are some assumptions to formulate a mathematical model in this problem:

A1. Every ship must be serviced once and exactly once at any berth.
A2. Ships serviced at any berth with acceptable physical conditions.
A3. The handling time of each ship is dependent on the berth where it is serviced.
A4. The total number of QCs on the berth is fixed.
A5. The working time of each QC on the same berth is the same.

Notations

Indices

i: ship number $1,2,\ldots,n$
j: berth number $1,2,\ldots,m$
k: the number of service orders $1,2,\ldots,n$

Parameters

n: number of ships
m: number of berths
a_i: arrival time of ship i
d_i: departure time of ship i
c_i: number od container required for loading/unloading of ship i
H: number of QCs
v: working speed of QCs
U_i: working times of QCs on berth j
A: average working times of QCs

Decision variables

$$x_{ijk} = \begin{cases} 1, & \text{if } i \text{ is serviced as the } k\text{-th ship at berth } j \\ 0, & \text{otherwise} \end{cases}$$

s_i : starting time of ship i
h_j : number of QCs assigned to berth j

The handling time and waiting time of berths is the serviced time of container ships, the delay time of container ships departure is the first objective function, and the standard deviation of working time of QCs is the second objective function. Although this simple sum has some incidental factors, it is directly perceived through the senses, so it is used here. The mathematical model for the BAP problem is as follows:

$$\min z_1 = \sum_{i=1}^{n}\sum_{j=1}^{m}\sum_{k=1}^{n}\frac{c_i}{v\cdot h_j}x_{ijk} + \sum_{i=1}^{n}\sum_{j=1}^{m}\sum_{k=1}^{n}(s_i - a_i)x_{ijk}$$

$$+\sum_{i=1}^{n}\sum_{j=1}^{m}\sum_{k=1}^{n}(s_i + \frac{c_i}{v\cdot h_j} - d_i)\,x_{ijk} \tag{9.15}$$

$$\min z_2 = \sqrt{\frac{1}{H}\sum_{j-1}^{m} h_j(U_j - A)^2} \tag{9.16}$$

$$\text{s.t.} \quad \sum_{j=1}^{m}\sum_{k=1}^{n} x_{ijk} = 1,\ \forall i \tag{9.17}$$

$$\sum_{i=1}^{n} x_{ijk},\ \forall j,\ k \tag{9.18}$$

$$(s_i + \frac{c_i}{v\cdot h_j})x_{ij,k-1} \le s_l x_{ljk},\ \forall i,\ k,\ l,\ j \tag{9.19}$$

$$\sum_{j=1}^{m} h_j = H \tag{9.20}$$

$$U_j = \sum_{i=1}^{n}\sum_{k=1}^{n}\frac{c_i}{v\cdot h_j}x_{ijk},\ \forall j \tag{9.21}$$

$$A = \frac{1}{H}\sum_{j=1}^{m} h_j \cdot U_j \tag{9.22}$$

$$x_{ijk} = 0 \text{ or },\ \forall i,\ k,\ j \tag{9.23}$$

$$s_i \ge a_i,\ \forall i \tag{9.24}$$

$$h_j : \text{integer},\ \forall j \tag{9.25}$$

The constraint at Eq. 9.17 indicates that ship must be serviced once and exactly once at any berth. Constraint at Eq. 9.18 indicates that each berth can only service one ship at the moment. In other words, the ship can be serviced on berths only once. Constraint at Eq. 9.19 indicates that at each berth, the next ship can be serviced after the previous ship is serviced finally. Constraint at Eq. 9.20 indicates that each QC on berth could be assigned. Equation 9.21 is the working time of QC on the berth. Equation 9.22 indicates the average working time of all the QCs on berth.

9.2.2 Multi-stage Decision-based GA

Let $P(t)$ and $C(t)$ be parents and offspring in current generation t, respectively. The overall procedure of multi-stage decision-based GA (mdGA) for solving BAP model is outlined as follows.

---

procedure: mdGA for BAP model
input: BAP data, GA parameters ($popSize$, $maxGen$, p_M, p_C)
output: Pareto optimal solutions E
begin
  $t \leftarrow 0$;
  initialize $P(t)$ by priority and multistage-based encoding routine;
  calculate objectives $z_1(P)$ and $z_2(P)$ by priority and multistage based decoding routine;
  create Pareto $E(P)$;   calculate $eval(P)$ by adaptive-weight fitness assignment routine;
  while (**not** terminating condition) **do**
    create $C(t)$ from $P(t)$ by PMX crossover and one cut-point crossover routines;
    create $C(t)$ from $P(t)$ by swap mutation routine;
    calculate objectives $z_1(Z)$ and $z_2(Z)$ by priority and multistage-based decoding routine;
    update Pareto $E(P,C)$;
    calculate $eval(P,C)$ by adaptive-weight fitness assignment routine;
    select $P(t+1)$ from $P(t)$ and $C(t)$ by roulette wheel selection routine;
    $t \leftarrow t+1$;
  end
  output Pareto optimal solutions $E(P,C)$
end

9.2.2.1 Genetic Representation

To develop a priority-based genetic representation for the FAP with time-space network, there are four main phases:

Phase 1: Deciding ship precedence relationship

Step 1.1: Sort arrival time of ships
Step 1.2: Decide ships precedence relationship

Phase 2: Determining ship sequence

Step 2.1: Generate a random priority to each ship using encoding procedure
Step 2.2: Determine the ship sequence using decoding procedure

Phase 3: Assigning QCs to berths

Step 3.1: Generate a random berth number to each ship using encoding procedure

Step 3.2: Generate a random QC number to each berth using encoding procedure

Step 3.3: Assign QCs to berths based on the ship sequence using decoding procedure

Phase 4: Designing a ship schedule

Step 4.1: Create a ship schedule S

Step 4.2: Draw a Gantt chart for schedule S

Phase 1: Deciding Ship Precedence Relationship

Considering the five ships simple example as shown in Table 9.27 and Fig. 9.29, the following two steps decide the precedence relationship for each ship with arrival times:

Step 1.1: Sort arrival time of ships

Step 1.2: Decide ships precedence relationship

Table 9.27 Ship arrival time

	Arrival time
Ship 1	9:00
Ship 2	00:00
Ship 3	12:00
Ship 4	7:00
Ship 5	9:00

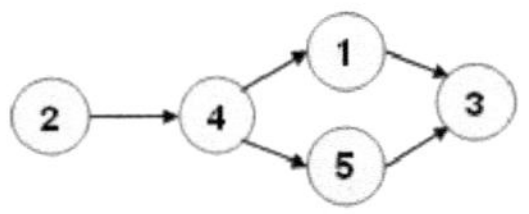

Fig. 9.29 Ship precedence relationship

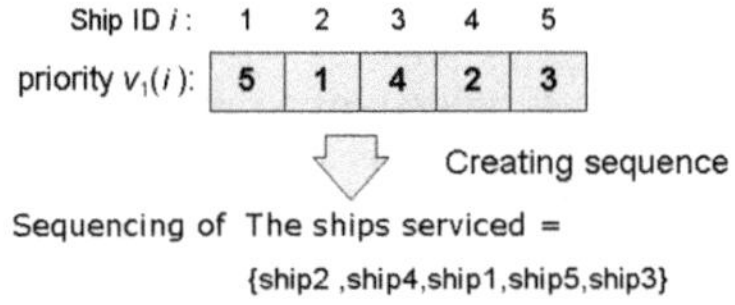

Fig. 9.30 An example of priority-based decoding

Phase 2: Determining Ship Sequence

To determine a list of ship sequences satisfying the precedence relationship, we consider the following two steps using priority-based encoding and decoding procedures:

Step 2.1: Generate a random priority to each ship using encoding procedure
Step 2.2: Determine the ship sequence using decoding procedure

For considering the five ships example, the priority-based chromosome and the ship sequence is shown in Fig. 9.30.

Phase 3: Assigning QCs to Berths

As shown in Tables 9.28 and 9.29, there are data about loading/unloading containers and cranes. Assign QC to each berth based on the ship sequence by multistage-based encoding and decoding procedures.

Table 9.28 A simple example data of loading/unloading containers

	Total number of Loading/unloading container (TEU)
Ship 1	428
Ship 2	455
Ship 3	259
Ship 4	172
Ship 5	684

Table 9.29 A simple example data of cranes

	Total number of crane
Berth 1	3
Berth 2	1
Berth 3	2
Berth 4	1

Step 3.1: Generate a random berth number to each ship using encoding procedure as shown in Fig. 9.31.

Fig. 9.31 Multistage-based chromosome for ship assignment

Ship ID l:	1	2	3	4	5
Berth $v_2(i)$:	2	1	4	2	3

Step 3.2: Generate a random QC number to each berth using encoding procedure as shown in Fig. 9.32.

Step 3.3: Assign QCs to berths based on the ship sequence using decoding procedure as shown follows.

Assignment = Ship (QC-Berth) = 1(2-3), 2(1-2), 3(4-1), 4(2-3), 5(3-1)

Berth ID j:	1	2	3	4
Number of QCs $v_3(j)$:	2	3	1	1

Fig. 9.32 Multistage-based chromosome for QC assignment

Phase 4: Designing a Ship Schedule

Depending on the following two steps, make a ship schedule by the QC and berth assigned in step 3.2, and draw a Gantt chart based on the ship schedule.

Step 4.1: Create a ship schedule S

$$S = \{\text{berth number } j\}$$
$$= \{\text{ship number } i, h_j, a_i - d_i, s_i - \text{operation time, waiting time, delay\_depart time}\}$$

Step 4.2: Draw a Gantt chart for schedule S

9.2.2.2 Genetic Operators

Partially Mapped Crossover: As crossover operator for priority-based encoding, we adopt the partially mapped crossover (PMX) introduced in Subsection 2.2.2.3. PMX can be viewed as an extension of two-point crossover for binary string to permutation representation. It uses a special repairing procedure to resolve the illegitimacy caused by the simple two-point crossover.
One Cut-point Crossover: As the second crossover operator, we adopt the simple one cut-point crossover for multistage-based encoding.
Swap Mutation: As mutation operator, we adopt the swap mutation. It selects two elements at random and then swaps the elements on these positions.

9.2.3 Numerical Experiment

The test problem is dependent on the real world project data which comes from one of the Shanghai container terminal companies in China. For this system simulation, C# programmes are performed on a Pentium 4 processor (2.6-GHz clock), 1GB RAM. The test data are shown as follows.

Experiment 1

There are four berths and seven QCs on the seaport in this numerical example. And according to the arrival time for each container ship given in Table 9.30, we can create a precedence network shown in Fig. 9.33.

The values of parameters used in the computational experiments are as follows:

Population size: $popSize = 150$

Table 9.30 Ship schedule1

	Ship name	Arrive time	Departure time	Total number of loading/unloading container (TEU)
1	MSG	9:00	20:00	428
2	NTD	9:00	21:00	455
3	CG	0:30	13:00	259
4	NT	21:00	23:50	172
5	LZ	0:30	23:50	684
6	XY	8:30	21:00	356
7	LZI	7:00	20:30	435
8	GC	11:30	23:50	350
9	LP	21:30	23:50	150
10	LYQ	22:00	23:50	150
11	CCG	9:00	23:50	333

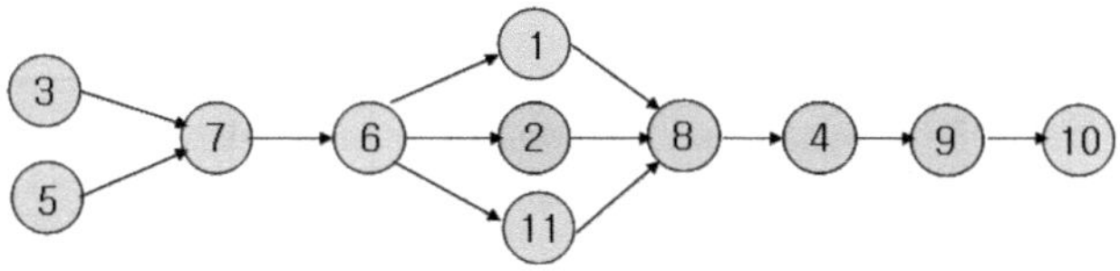

Fig. 9.33 The precedence network

Maximum number of generation: $maxGen = 500$
Probability of crossover: $p_C = 0.7$
Probability of mutation: $p_M = 0.3$

The nondominated solutions are shown in Table 9.31.

The solution 1 in Table 9.31 represents the nondominated solution which minimizes service time of berth. Total service time is 3080 [min], waiting time is 286 [min], delay time is 0 [min], workload standard deviation of QCs is 4.6. The chromosome and Gantt chart are shown in Figs. 9.34 and 9.35. The relative chromosome and computational result are shown as follows:
S={berth1, berth2, berth3, berth4}

={[(ship5, 3 crane, 0:30–20:00, 0:30–342,0,0), (ship7,3 crane, 7:00–20:30, 7:00–218, 0, 0), (ship1, 3 crane, 9:00–20:00, 10:38–218, 98, 0), (ship8, 3 crane, 11:30–23:00, 14:12–175, 162, 0), (ship4, 3 crane, 21:00–23:50, 21:00–86, 0, 0), (ship10, 3 crane, 22:00–23:50, 22:26–75, 26, 0)], [(ship3, 2 crane, 0:30–13:00, 0:30–195, 0, 0), (ship2, 2 crane, 9:00–21:00, 9:00–342, 0, 0), (ship9, 2 crane, 21:30–

Table 9.31 Calculated nondominated solutions

	Total serve time (h)	Workload balance	Distance from ideal point
1	51.33	4.60	1.00
2	51.90	4.30	0.90
3	53.15	3.12	0.55
4	55.13	2.50	0.46
5	57.68	1.80	0.57
6	60.45	1.52	0.80
7	62.73	1.50	1.00

23:50, 21:30–113, 0, 0)], [(ship11, 1 crane, 9:00–23:50, 9:00–500, 0, 0)], [(ship6, 1 crane, 8:30–21:00, 8:30–534, 0 ,0)]}

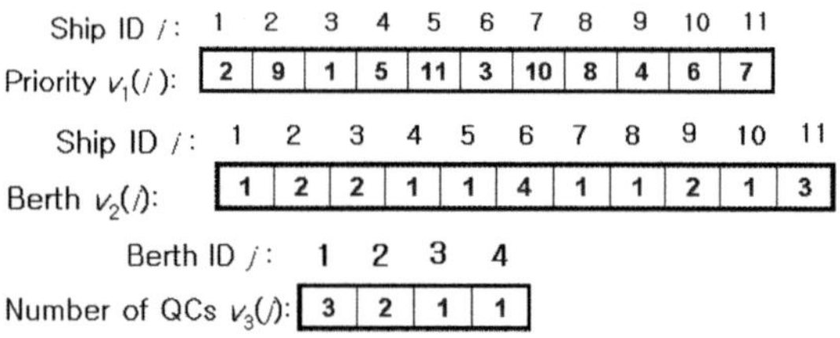

Fig. 9.34 The chromosome with minimizes service time of berth

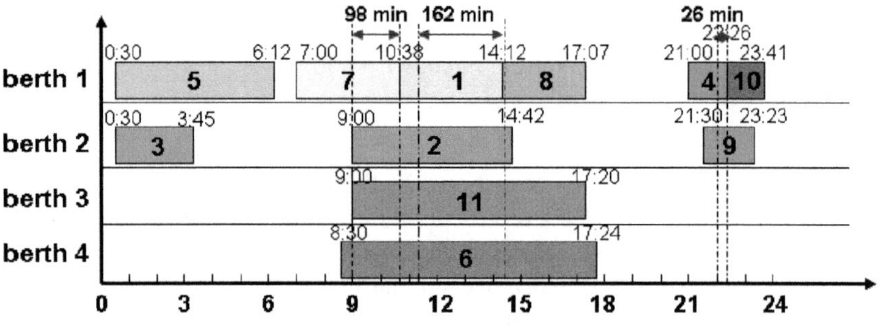

Fig. 9.35 Gantt chart of nondominated solution 1

The solution 4 in Table 9.31 represents the nondominated solution which balances the workload of QCs. Total service time is 3764 [min], waiting time is 566 [min], delay time is 115 [min], workload standard deviation of QCs is 1.49. The chromosome and Gantt chart are shown in Figs. 9.36 and 9.37. The relative chromosome and computational result are shown as follows:

S={berth1, berth2, berth3, berth4}

={[(ship5, 3 crane, 0:30–20:00, 0:30–342,0,0), (ship3,3 crane, 0:30–13:00, 6:12–130, 342, 0), (ship11, 3 crane, 9:00–23:50, 9:00–167, 0, 0), (ship8, 3 crane, 11:30–23:50, 11:47–175, 17, 0), (ship4, 3 crane, 21:00–23:50, 21:00–86, 0, 0)], [(ship6, 1 crane, 8:30–21:00, 8:30–534, 0, 0), (ship10, 1 crane, 22:00–23:50, 22:00–225, 0, 115)], [(ship7, 2 crane, 7:00–20:30, 7:00–327, 0, 0),(ship2, 2 crane, 9:00–21:00, 12:27–342, 207, 0), (ship9, 2 crane, 21:30–23:50, 21:30–113, 0, 0)], [(ship1, 1 crane, 9:00–20:00, 9:00–642, 0, 0)]}

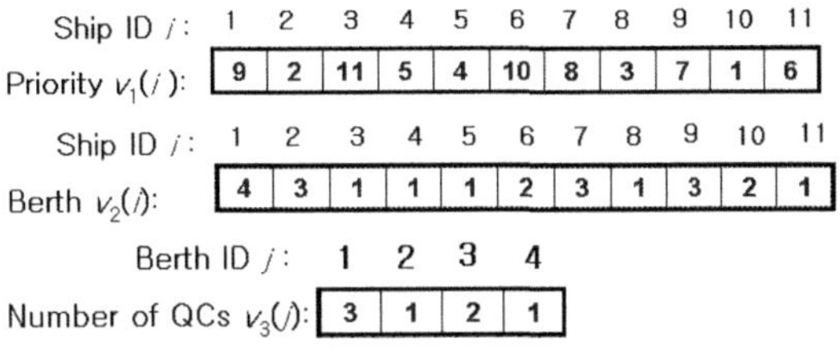

Fig. 9.36 The chromosome with balances the workload of QCs

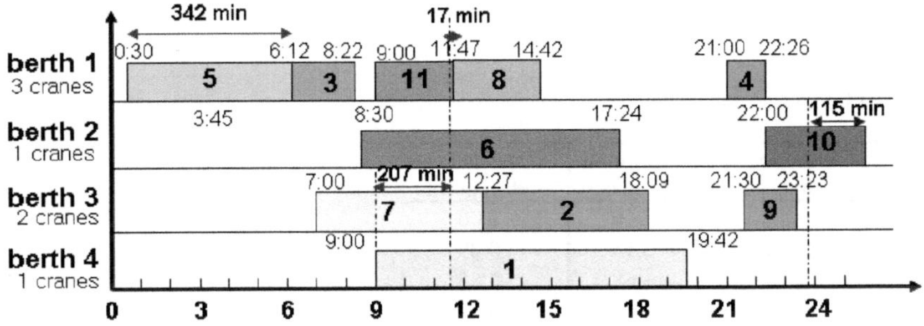

Fig. 9.37 Gantt chart of nondominated solution 4

Experiment 2

There are four berths and seven QCs on the seaport in this numerical example. And according to the arrival time for each container ship given in Table 9.32.

The nondominated solutions are shown in Table 9.33.

The solution 7 in Table 9.33 represents the nondominated solution which minimizes service time of berth. Total service time is 5712 [min], waiting time is 1371[min], delay time is 243 [min], workload standard deviation of QCs is 2.44. The chromosome and Gantt chart are shown in Figs. 9.38 and 9.39. The relative

Table 9.32 Ship schedule2

	Ship name	Arrive time	Departure time	Total number of loading/unloading container (TEU)
1	ZHE	1:00	17:00	525
2	ZHW	1:00	17:00	515
3	ZYE	1:00	15:00	722
4	ZYW	1:00	15:00	741
5	ZX	0:00	14:00	400
6	JWH	5:30	17:30	664
7	JYD	1:30	12:00	227
8	XNT	5:30	22:00	795
9	DY	7:54	10:25	34
10	MZ	13:54	14:59	31
11	ZH	0:00	14:30	149
12	XY	15:00	22:00	236
13	YL	20:06	23:50	105

Table 9.33 Pareto solution values

	Total serve time (h)	Workload balance	Distance from ideal point
1	90.32	5.60	1.00
2	91.67	5.13	0.89
3	91.80	3.68	0.55
4	91.87	4.62	0.77
5	93.33	3.98	0.62
6	94.17	3.00	0.39
7	95.20	2.44	0.26
8	103.47	2.34	0.28
9	114.75	1.98	0.33
10	130.23	1.59	0.49
11	134.70	1.48	0.55
12	142.57	1.47	0.64
13	152.82	1.60	0.77
14	171.48	1.36	1.00

chromosome and computational result are shown as follows:
r S={berth1, berth2, berth3, berth4}

={[(ship7, 1 crane, 1:30–12:00, 1:30–341,0,0), (ship9,1 crane, 7:54–10:25, 7:54–51, 0, 0), (ship10, 1 crane, 13:54–14:59, 13:54–47, 0, 0), (ship12, 1 crane, 15:00–22:00, 15:00–354, 0, 0)](ship5, 4 crane, 0:00–14:00, 00:00–150, 0, 0), (ship2, 4 crane, 1:00–17:00, 2:30–194, 90, 0), (ship4, 4 crane, 1:00–15:00, 5:44–278, 284, 0), (ship6, 4 crane, 5:30–14:30, 10:22–249, 292, 0), (ship8, 4 crane, 5:30–22:00, 14:31–299, 541, 0), (ship13, 4 crane, 20:06–23:50, 20:06–40, 0, 0)], [(ship3, 1 crane, 1:00–15:00, 1:00–1083, 0, 243)], [(ship11, 1 crane, 0:00–14:30, 0:00–224, 0, 0),(ship1, 1 crane, 1:00–17:00, 3:44–788, 164, 0)]}

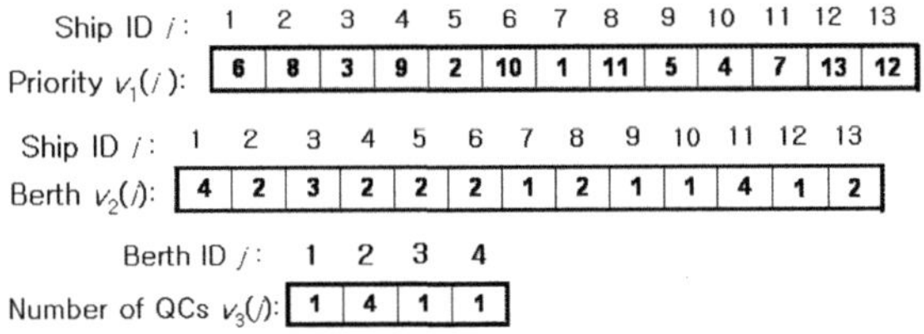

Fig. 9.38 The chromosome with minimizes service time of berth

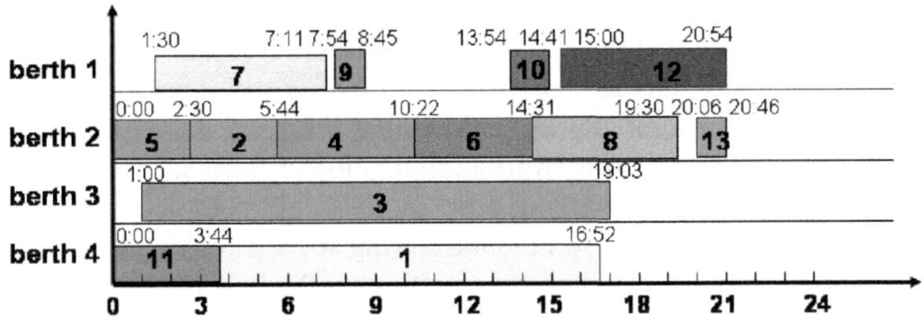

Fig. 9.39 Gantt chart of nondominated solution 7

9.3 AGV Dispatching Model

Automated material handling has been called the key to integrated manufacturing. An integrated system is useless without a fully integrated, automated material handling system. In the manufacturing environment, there are many automated material handling possibilities. Currently, automated guided vehicles systems (AGV systems), which include automated guided vehicles (AGVs), are state-of-the-art, and are often used to facilitate automatic storage and retrieval systems (AS/RS).

For a recent review on AGV problems and issues, the reader is referred to Qiu *et al.* [61]; Vis [62]; Le-Anh and Koster [56]; Lin [57]. An AGV is a driverless transport system used for horizontal movement of materials. AGVs were introduced in 1955. The use of AGVs has grown enormously since their introduction. AGV systems are implemented in various industrial contexts: container terminals, part transportation in heavy industry, manufacturing systems (Kim and Hwang [53]). In fact, new analytical and simulation models need to be developed for large AGV systems to overcome: large computation times, NP-completeness, congestion, deadlocks and delays in the system and finite planning horizons (Moon and Hwang [58]; Kim and Hwang [54]; Hwang *et al.* [52]).

In this chapter, we focus on the simultaneous scheduling and routing of AGVs in a flexible manufacturing system (FMS). For solving the AGV dispatching problem in FMS, the special difficulty arises from (1) the task sequencing is an NP-complete problem, and (2) a random sequence of AGV dispatching usually does not correspond to the operation precedence constraint and routing constraint.

The key point of this paper is that we model an AGV system by using network structure. We can present all of the decision variables and system constraints on the network for treating this AGVs dispatching problem. Let $G = (V, E)$ be a connected directed network with $n = |V|$ nodes and $m = |E|$ edges, nodes representing the tasks that transport the material from a pickup point to a delivery point. On the edges, we can consider most AGV problem's constraints, for example capacity of AGVs, precedence constraints among the tasks, deadlock control. After the network structure is generated, all of the system constraints are included. When we treat the problems depending on this network, we can overleap these system constraints, so the decisions become simple. Furthermore, these problems can be solved by using a lot of heuristic algorithms as network optimization problems.

For solving this AGV dispatching problem as an NP-complete problem, we introduce a random key-based GA. For considering several AGVs routing paths, as a different form of general genetic representation methods, a random key-based encoding method can represent various efficient routing paths with different vehicles by each chromosome.

9.3.1 Network Modeling and Mathematical Formulation

The problem is to dispatch AGVs for transporting the product between different machines in an FMS. At the first stage, we model the problem by using network structure. Assumptions considered are as follows:

A1. AGVs only carry one kind of products at the same time.
A2. A network of guide paths is defined in advance, and the guide paths have to pass through all the pickup/delivery points.
A3. The vehicles are assumed to travel at a constant speed.
A4. The vehicles can only travel forward, not backward.

A5. As many vehicles travel on the guide path simultaneously, collisions must be avoided by hardware, but are not considered in this chapter.

A6. On each working station, there are pickup spaces to store the operated material and delivery space to store the material for the next operation.

A7. The operation can be started any time after an AGV takes the material to come. And also the AGV can transport the operated material from the pickup point to the next delivery point at any time.

For scheduling, n jobs are to be scheduled on m machines in FMS. The i-th job has n_i operations that have to be processed. Each machine processes only one operation at a time. The set-up time for the operations is sequence-independent and is included in the processing time. For AGV dispatching, each machine is connected to the guide path network by a pick-up/delivery (P/D) station where pallets are transferred from/to the AGVs. The guide path is composed of aisle segments on which the vehicles are assumed to travel at constant speed. As many vehicles travel on the guide path simultaneously, collisions must be avoided by hardware, but this is not considered in this chapter. We are subject to the constraints that, for scheduling, the operation sequence for each job is prescribed, and each machine can process only one operation at a time; each AGV can transport only one kind of products at a time. For AGV dispatching, AGVs only carry one kind of product at same time. The vehicles can only travel forward, not backward. The objective function is minimizing the time required to complete all jobs (*i.e.*, makespan) t_{MS}.

Notation

Indices

i, i': index of jobs, $(i, i' = 1, 2, \ldots, n)$;
j, j': index of processed, $(j, j' = 1, 2, \ldots, n_i)$;

Parameters

n: number of jobs;
m: number of machines;
n_i: number of operations of job i;
n_{AGV}: number of AGVs;
o_{ij}: the j-th operation of job i;
p_{ij}: processing time of operation o_{ij};
M_{ij}: machine assigned for operation o_{ij};
T_{ij}: transition task for operation o_{ij};
t_{ij}: transition time from $M_{i,j-1}$ to M_{ij};

Example 9.1: An FMS has four machines 1, 2, 3 and 4.

Three job types J_1, J_2 and J_3 are to be carried out and Table 9.34 shows the requirements for each job (Proth *et al.* [60]). The first process of J_1 is carried out at machine 1 with p_{11} (60 processing times). The second process of J_1 is carried out at machine 2 with p_{12} (80 processing times); this table also gives the precedence constraints among the operations O_{ij} in each job J_i. For instance, the second process of J_1 can be carried out only after the first process of J_1 is complete. Note that J_2 has only two processes to be completed. Figure 9.40 shows the Gantt chart of the schedule of Example 9.1 without considering AGVs routing.

Table 9.34 Job requirements of Example 9.1

O_{ij}	M_{ij}				p_{ij}			
J_i \ P_i	P_1	P_2	P_3	P_4	P_1	P_2	P_3	P_4
J_1	1	2	3	4	60	80	100	70
J_2	4	3	-	-	100	40	-	-
J_3	1	4	2	-	70	20	50	-

M_{ij}: machine # of assigned P_j-th operation O_{ij} of job J_i
p_{ij}: processing time of operation O_{ij}

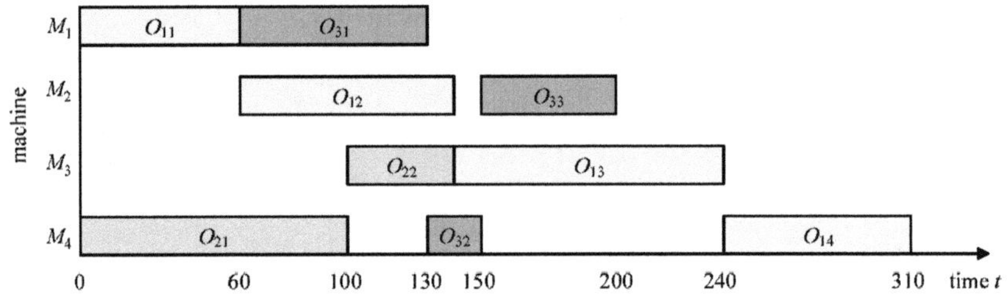

Fig. 9.40 Gantt chart of the schedule of Example 9.1 without considering AGVs routing

For designing an AGV system in a manufacturing system, the transition time t_{uv} / c_{vu} between pickup point on machine u and delivery point on machine v are defined in Table 9.35, and depend on Naso and Turchiano (2005) [59], we give a layout of facility for Example 9.1 in Fig. 9.41. Note, although the network of guide paths is unidirectional, it has to take a very large transition time from pickup point (P) to delivery point (D) on the same machine. It is unnecessary in the real application. So we defined an inside cycle for each machine, that is the transition time is the same with P to D and D to P. We give a routing example for carrying out job J_1 by using one AGV in Fig. 9.42.

Definition 1: A node is defined as task T_{ij} that presents a transition task of j-th process of job J_i for moving pickup point of machine $M_{i,j-1}$ to delivering point of machine M_{ij}.

Table 9.35 Transition time between pickup point on machine u and delivery point on machine v

t_{uv} / c_{uv}		Loading / unloading	Machine number 1	2	3	4
Loading / unloading		1 / 1	1 / 7	8 / 13	16 / 23	18 / 20
Machine number	1	13 / 18	3 / 3	2 / 9	10 / 19	13 / 18
	2	18 / 22	22 / 28	2 / 2	4 / 13	12 / 18
	3	8 / 14	12 / 20	18 / 26	3 / 3	2 / 10
	4	5 / 7	9 / 12	15 / 18	23 / 28	2 / 2

t_{uv}: transition time from pickup point on machine u to delivery point on machine v

c_{uv}: transition time from delivery point on machine u to pickup point on machine v

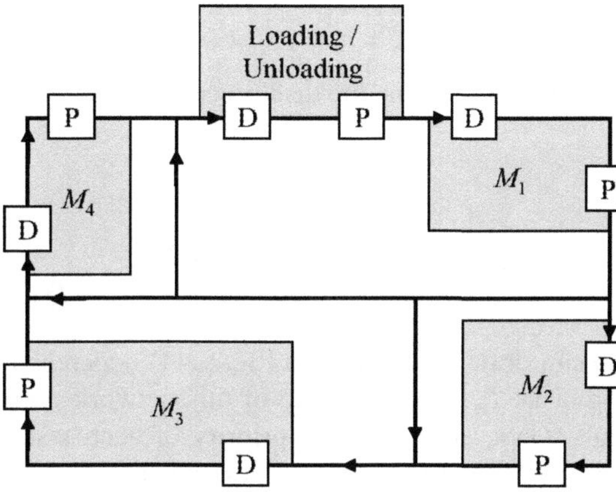

Fig. 9.41 Layout of facility (P: pickup point, D: delivery point)

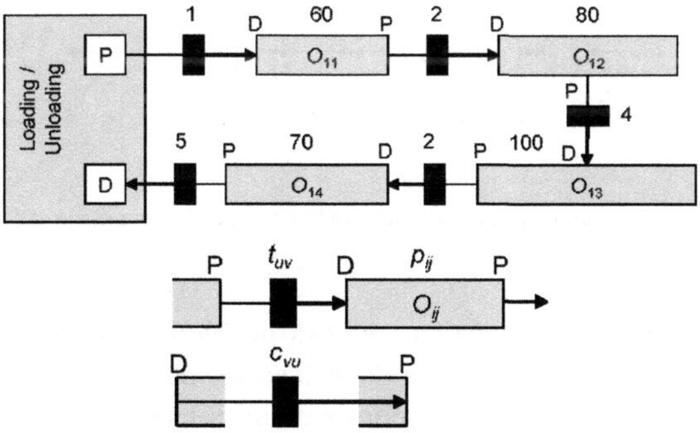

Fig. 9.42 A routing example for carry out job J_1 by using one AGV

Definition 2: An arc can be defined as many decision variables arise. For example, capacity of AGVs, precedence constraints among the tasks, costs of movement. In this chapter, we define an arc as a precedence constraint, and give a transition time $c_{jj'}$ from delivery point of machine M_{ij} to pickup point of machine $M_{i'j'}$ on the arc.

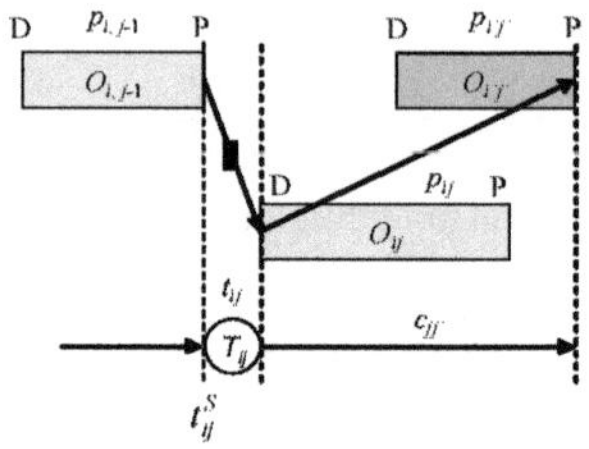

Fig. 9.43 Illustration of problem representations

Definition 3: We define the task precedence for each job:

$$\text{Job } J_1 : T_{11} \succ T_{12} \succ T_{13} \succ T_{14}$$
$$\text{Job } J_2 : T_{21} \succ T_{22}$$
$$\text{Job } J_3 : T_{31} \succ T_{32} \succ T_{33}$$

We can draw a network (as Fig. 9.44) dependent on the precedence constraints among tasks T_{ij}. The objective of this network problem assigns all of the tasks to several AGVs, and gives the priority of each task to make the AGV routing sequence. A result of Example 9.1 is shown as follows, final time required to complete all jobs (*i.e.* makespan) is 321 and 3 AGVs are used. Figure 9.45 shows the result on Gantt chart.

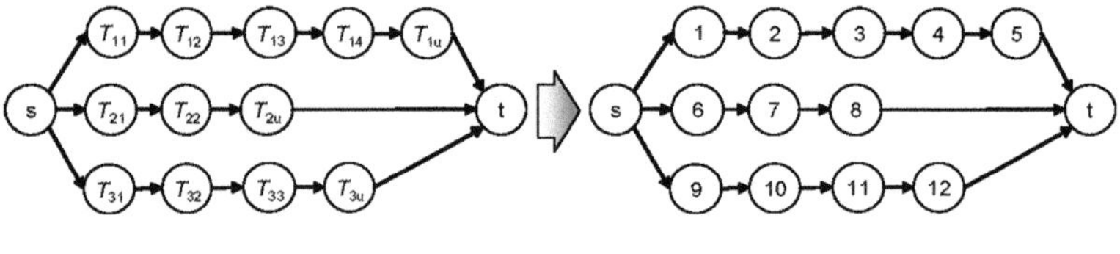

Fig. 9.44 Illustration of the network structure of Example 9.1

The objective of this network problem assigns all of the tasks to several AGVs, and gives the priority of each task to make the AGV routing sequence with minimizing time required to complete all jobs (*i.e.*, makespan).

Decision Variables

x_{ij}: assigned AGV number for task T_{ij};
t^S_{ij}: starting time of task T_{ij};
c^S_{ij}: starting time of operation o_{ij};

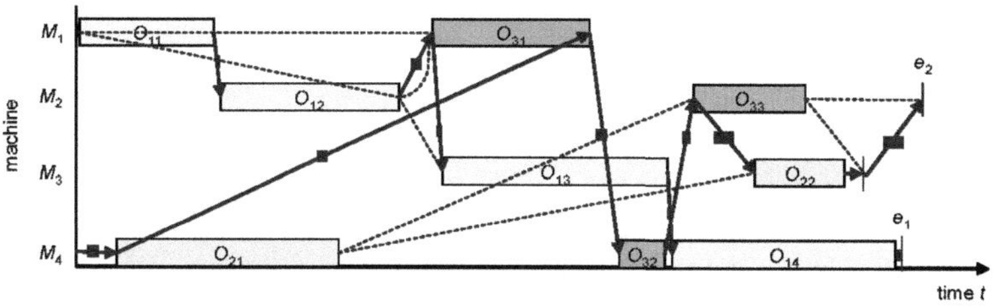

Fig. 9.45 Gantt chart of the schedule of Example 9.1 considering two AGVs routing (e_l: AGV)

The problem can be formulated as follows:

$$\min \quad t_{\rm MS} = \max_i \left\{ t^S_{i,n_i} + t_{M_i,n_i,0} \right\} \tag{9.26}$$

$$\text{s.t.} \quad c^S_{ij} - c^S_{i,j-1} \geq p_{i,j-1} + t_{ij}, \ \forall i, \ j = 2, \ldots, n_i \tag{9.27}$$

$$\left(c^S_{ij} - c^S_{i'j'} - p_{i'j'} + \Gamma \mid M_{ij} - M_{i'j'} \mid \geq 0 \right) \vee$$
$$\left(c^S_{i'j'} - c^S_{ij} - p_{ij} + \Gamma \mid M_{ij} - M_{i'j'} \mid \geq 0 \right), \ \forall (i,j), \ (i',j') \tag{9.28}$$

$$\left(t^S_{ij} - t^S_{i'j'} - t_{i'j'} + \Gamma \mid x_{ij} - x_{i'j'} \mid \geq 0 \right) \vee$$
$$\left(t^S_{i'j'} - t^S_{ij} - p_{ij} + \Gamma \mid x_{ij} - x_{i'j'} \mid \geq 0 \right), \ \forall (i,j), \ (i',j') \tag{9.29}$$

$$\left(t^S_{i,n_i} - t^S_{i'j'} - t_{i'j'} + \Gamma \mid x_{ij} - x_{i'j'} \mid \geq 0 \right) \vee$$
$$\left(t^S_{i'j'} - t^S_{i,n_i} - t_i + \Gamma \mid x_{ij} - x_{i'j'} \mid \geq 0 \right), \ \forall (i,n_i), \ (i',j') \tag{9.30}$$

$$c^S_{ij} \geq t^S_{i,j+1} - p_{ij} \tag{9.31}$$

$$x_{ij} \geq 0, \ \forall i, \ j \tag{9.32}$$

$$t^S_{ij} \geq 0, \ \forall i, \ j \tag{9.33}$$

where Γ is a very large number, and t_i is the transition time for pickup point of machine M_{i,n_i} to delivery point of loading/unloading. Constraint at Eq. 9.28 describes the operation precedence constraints. In constraints at Eq. 9.29–9.31, since one or other constraint must hold, they are called disjunctive constraints. This represents the operation un-overlapping constraint (Eq. 9.29) and the AGV non-overlapping constraint (Eqs. 9.30 and 9.31).

9.3.2 Random Key-based GA

Let $P(t)$ and $C(t)$ be parents and offspring in current generation t, respectively. The overall procedure of random key-based GA (mdGA) for solving AGV dispatching model is outlined as follows:

---

procedure: rkGA for AGV dispatching model
input: AGV dispatching data,GA parameters($popSize$, $maxGen$, p_M, p_C, p_I)
output: the best dispatching schedule S
begin
  $t \leftarrow 0$;
  initialize $P(t)$ by random key-based encoding routine;
  evaluate $P(t)$ by random key-based decoding routine;
  while (**not** terminating condition) **do**
    create $C(t)$ from $P(t)$ by arithmetical crossover routine;
    create $C(t)$ from $P(t)$ by swap mutation routine;
    create $C(t)$ from $P(t)$ by immigration routine;
    evaluate $C(t)$ by random key-based decoding routine;
    select $P(t+1)$ from $P(t)$ and $C(t)$ by roulette wheel selection routine;
    $t \leftarrow t+1$;
  end
  output the best dispatching schedule S
end

9.3.2.1 Genetic Representation

The AGV dispatching problem is a combination of AGV assignment and task sequencing. A solution can be described by the assignment of task on AGVs, and the task sequences that transport the material from a pickup point to a delivery point. In this chapter, the chromosome is presented as a random key-based encoding, *i.e.*, the locus of each gene represent task ID and the gene are presented as real number. As we know, real numbers can be separated with integer and decimal. We consider the part of integer be AGV assignment, and consider the part of decimal the priority of each task. The detailed encoding and decoding process are separate into following phases:

Phase 1: Random Key-based Encoding Process
Phase 2: Grouping Process (AGV Assignment)

  Step 2.1: Extraction of Integer Part
  Step 2.2: AGV Assignment

Phase 3: Task Sequencing Growth Process

Step 3.1: Extraction of Decimal Part
Step 3.2: Task Sequencing

Phase 1: Random Key-based Encoding Process

In general, there are two ways to generate the initial population by encoding procedure, heuristic initialization and random initialization. The mean fitness of the heuristic initialization is already high so that it may help the GAs to find solutions faster. Unfortunately, in most large scale problems, for example, network design problems, it may just explore a small part of the solution space and it is difficult to find global optimal solutions because of the lack of diversity in the population. Therefore, random initialization is effected In this chapter so that the initial population is generated with the priority-based encoding method. Figure 9.46 shows the pseudocode of the encoding method and an example of generated random key-based chromosome considering Example 9.1 is shown in Fig. 9.47.

```
procedure: randomkey-based encodingmethod
input: numberof jobs n, numberof operations n_i of job i, numberof AGV n_AGV,
output: kth initialchromosome v_k
begin
    for i = 1 to n
        for j = 1 to n_i + 1
            l ← n*(i-1) + j;
            v_k[l] ← random[1 : n_AGV];
            v_k[l] ← v_k[l] + random(0 : 1);
    output v_k;
end
```

Fig. 9.46 Pseudocode of random key-based encoding method

Node ID :	1	2	3	4	5	6	7	8	9	10	11	12
Task ID :	T_{11}	T_{12}	T_{13}	T_{14}	T_{1u}	T_{21}	T_{22}	T_{2u}	T_{31}	T_{32}	T_{33}	T_{3u}
Random Key :	1.92	1.86	1.64	1.59	1.47	2.82	2.57	2.41	1.71	2.79	2.66	2.31

Fig. 9.47 An example of random key-based encoding method

Phase 2: Grouping Process (AGV Assignment)

Step 2.1: Extraction of Integer Part. As the first stage of Phase 2, we extract the integer part from random key-based chromosome; Figs. 9.48 and 9.49 show the pseudocode and an example of the integer extraction.

procedure : AGV assignment
input : number of jobs n, number of operations n_i of job i, integer part I
output : group G_l, $\forall l \in n_{\text{AGV}}$
begin
  $G_l \leftarrow \phi$, $\forall l \in n_{\text{AGV}}$
  for $i = 1$ **to** n
    for $j = 1$ **to** $n_i + 1$
      $k \leftarrow n * (i-1) + j$;
      $l \leftarrow I[k]$;
      $G_l \leftarrow G_l \cup \{k\}$;
  output G_l, $\forall l \in n_{\text{AGV}}$;
end

Fig. 9.48 Pseudocode of integer extraction

Node ID :	1	2	3	4	5	6	7	8	9	10	11	12
Task ID :	T_{11}	T_{12}	T_{13}	T_{14}	T_{1u}	T_{21}	T_{22}	T_{2u}	T_{31}	T_{32}	T_{33}	T_{3u}
Integer Part I:	1	1	1	1	1	2	2	2	1	2	2	2

Fig. 9.49 An example of integer extraction

Step 2.2: AGV Assignment. As the second stage of Phase 2, we give the grouping process by pseudocode of AGV assignment as shown in Fig. 9.50. Depending on the grouping procedure, we assign the tasks to the same AGV with the same number in integer part I. The result of grouping Gl, (l=1 and 2) considering the above example in Fig. 9.49 is shown as follows:

$$G_1 : \{T_{11}, T_{12}, T_{13}, T_{14}, T_{1u}, T_{31}\}$$
$$G_2 : \{T_{21}, T_{22}, T_{2u}, T_{32}, T_{33}, T_{3u}\}$$

Phase 3: Task Sequencing Growth Process

Step 3.1: Extraction of Decimal Part. As first stage of Phase 3, we extract the decimal part from random key-based chromosome; Figures 9.51 and 9.52 show the pseudocode and an example of the decimal extraction.

---

procedure : AGV assignment
input : number of jobs n, number of operations n_i of job i, integer part I
output : group G_l, $\forall l \in n_{\text{AGV}}$
begin
    $G_l \leftarrow \phi$, $\forall l \in n_{\text{AGV}}$
    for $i = 1$ **to** n
        for $j = 1$ **to** $n_i + 1$
            $k \leftarrow n * (i-1) + j$;
            $l \leftarrow I[k]$;
            $G_l \leftarrow G_l \cup \{k\}$;
    output G_l, $\forall l \in n_{\text{AGV}}$;
end

Fig. 9.50 Pseudocode of AGV assignment

Step 3.2: Task Sequencing. As second stage of Phase 3, first, we separate into n_{AGV} sets T_l. The tasks will include in the same set T_l the tasks assigned in same group set G_l. Second, we calculate the task sequence S_l depend on the decimal part D, which the task will be assigned firstly with highest priority, *i.e.*, highest value in D. The pseudocode of task sequencing is shown in Fig. 9.53. The result of task sequencing S_l, (l=1 and 2) considering above example of decimal extraction D and group set G_l, (l=1 and 2) is:

$$T_1 : \{0.92, 0.86, 0.64, 0.59, 0.47, 0.71\}$$
$$T_2 : \{0.82, 0.57, 0.41, 0.79, 0.66, 0.31\}$$

and

$$S_1 : \{T_{11}, T_{12}, T_{31}, T_{13}, T_{14}, T_{1u}\}$$
$$S_2 : \{T_{21}, T_{32}, T_{33}, T_{22}, T_{2u}, T_{3u}\}$$

Based on the S_1 and S_2, we can calculate the time required to complete all jobs (*i.e.*, makespan). The final result is 389 as shown in Fig. 9.45.

```
procedure : task sequencing
input : number of jobs n, number of operations n_i of job i, decimal part D, group G_l, ∀l ∈ n_AGV
output : set of task sequence S_l, ∀l ∈ n_AGV
begin
    T_l ← φ, ∀l ∈ n_AGV
    L_l ← 0, ∀l ∈ n_AGV
    for i = 1 to n
        for j = 1 to n_i + 1
            k ← n * (i − 1) + j;
            l ← {l' | k ∈ G_l'};
            T_l ← T_l ∪ {D[k]};
            L_l ← L_l + 1;
    S_l ← G_l, ∀l ∈ n_AGV
    for l = 1 to n_AGV
        for i = 1 to L_l − 1
            for j = i to L_l
                if T_l[j] > T_l[i] then
                    exchange(T_l[j], T_l[i]);
                    exchange(S_l[j], S_l[i]);
    output S_l, ∀l ∈ n_AGV;
end
```

Fig. 9.51 Pseudocode of decimal extraction

Node ID :	1	2	3	4	5	6	7	8	9	10	11	12
Task ID :	T_{11}	T_{12}	T_{13}	T_{14}	T_{1u}	T_{21}	T_{22}	T_{2u}	T_{31}	T_{32}	T_{33}	T_{3u}
Decimal Part D :	0.92	0.86	0.64	0.59	0.47	0.82	0.57	0.41	0.71	0.79	0.66	0.31

Fig. 9.52 An example of decimal extraction

9.3.2.2 Genetic Operators

Arithmetical Crossover: The basic concept of this kind of operator is borrowed from the convex set theory (Gen and Cheng [49]). Generally, the weighted average of two vectors $v_j(i)$ and $v_k(i)$ of the j-th chromosome and the k-th chromosome is calculated as follows:

$$\lambda_1 \cdot v_j(i) + \lambda_2 \cdot v_k(i)$$

if the multipliers are restricted as

$$\lambda_1 + \lambda_2 = 1, \ \ \lambda_1 > 0 \text{ and } \lambda_2 > 0$$

```
procedure : task sequencing
input : number of jobs n, number of operations n_i of job i, decimal part D, group G_l, ∀l ∈ n_AGV
output : set of task sequence S_l, ∀l ∈ n_AGV
begin
    T_l ← φ,  ∀l ∈ n_AGV
    L_l ← 0,  ∀l ∈ n_AGV
    for i = 1 to n
        for j = 1 to n_i + 1
            k ← n*(i − 1) + j;
            l ← {l' | k ∈ G_l'};
            T_l ← T_l ∪ {D[k]};
            L_l ← L_l + 1;
    S_l ← G_l,  ∀l ∈ n_AGV
    for l = 1 to n_AGV
        for i = 1 to L_l − 1
            for j = i to L_l
                if T_l[j] > T_l[i] then
                    exchange(T_l[j], T_l[i]);
                    exchange(S_l[j], S_l[i]);
    output S_l,  ∀l ∈ n_AGV ;
end
```

Fig. 9.53 Pseudocode of task sequencing

The weighted form is known as convex combination. If the nonnegative condition on the multipliers is dropped, the combination is known as an affine combination. Finally, if the multipliers are simply required to be in real space E, the combination is known as a linear combination. Similarly, arithmetic operators are defined as the combination of two chromosomes v_j and v_k as follows:

$$v'_j(i) = \lambda_1 v_j(i) + \lambda_2 v_k(i),\ \forall i$$
$$v'_k(i) = \lambda_1 v_k(i) + \lambda_2 v_j(i),\ \forall i$$

Swap Mutation: It selects two positions at random and then swaps the gene on these positions. In this chapter, we adopt swap mutation for generating various offspring. **Immigration**: The algorithm is modified to (1) include immigration routine, in each generation, (2) generate and (3) evaluate $popSize \cdot p_{\mathrm{I}}$ random members, and (4) replace the $popSize \cdot p_{\mathrm{I}}$ worst members of the population with the $popSize \cdot p_{\mathrm{I}}$ random members (p_{I}, called the immigration probability).

9.3.3 Numerical Experiment

For evaluating the efficiency of the AGV dispatching algorithm suggested in a case study, a simulation program was developed using Java on a Pentium 4 processor (3.2-GHz clock). GA parameter settings were taken as follows: population size, *popSize* =20; crossover probability, p_C=0.70; mutation probability, p_M =0.50; immigration rate, $\mu = 0.15$.

First case study, we consider Example p.1 that was shown in Section 9.3.1. The best chromosome is shown in Fig. 9.54 and the comparison results are shown in Fig. 9.55. As shown in Fig. 9.55a, by a simple idea we assign the AGV to tasks depending on the processing sequence of the jobs there are intolerable ideal times. Compared with results in Fig. 9.55a, b, we can see the effectiveness of the proposed approach, and how to decide perfect AGV dispatching has a tremendous impact in the manufacturing system:

$$AGV1 : T_{11} \rightarrow T_{12} \rightarrow T_{13} \rightarrow T_{14} \rightarrow T_{1u}$$
$$AGV2 : T_{21} \rightarrow T_{31} \rightarrow T_{32} \rightarrow T_{33} \rightarrow T_{22} \rightarrow T_{2u} \rightarrow T_{3u}$$

Node ID :	1	2	3	4	5	6	7	8	9	10	11	12
Task ID :	T_{11}	T_{12}	T_{13}	T_{14}	T_{1u}	T_{21}	T_{22}	T_{2u}	T_{31}	T_{32}	T_{33}	T_{3u}
Random Key :	1.92	1.88	1.73	1.65	1.59	2.90	2.51	2.47	2.82	2.74	2.64	2.37

Fig. 9.54 Illustration of the best chromosome for Example 9.1

The second case study was used by Yang [63] and Kim *et. al.* [55]. In this case study, 10 jobs are to be scheduled on 5 machines. The maximum number of processes for the operations is 4. Table 9.36 gives the assigned machine numbers and process time and Table 9.37 gives the transition times between pickup points and delivery points. Depending on Naso and Turchiano [59], we give a layout of facility for the experiment in Fig. 9.56.

The best result of Example 9.2 is shown as follows; final time required to complete all jobs (*i.e.* makespan) is 574 and 4 AGVs are used. Figure 9.57 shows the result on a Gantt chart:

$$AGV1 : T_{11} \rightarrow T_{22} \rightarrow T_{41} \rightarrow T_{81} \rightarrow T_{91} \rightarrow T_{82} \rightarrow T_{92} \rightarrow T_{83} \rightarrow T_{84}$$
$$AGV2 : T_{21} \rightarrow T_{41} \rightarrow T_{12} \rightarrow T_{15} \rightarrow T_{10,2} \rightarrow T_{52} \rightarrow T_{71} \rightarrow T_{44}$$
$$AGV3 : T_{61} \rightarrow T_{62} \rightarrow T_{63} \rightarrow T_{64} \rightarrow T_{43} \rightarrow T_{72}$$
$$AGV4 : T_{31} \rightarrow T_{32} \rightarrow T_{10,1} \rightarrow T_{33} \rightarrow T_{13} \rightarrow T_{10,3} \rightarrow T_{93}$$

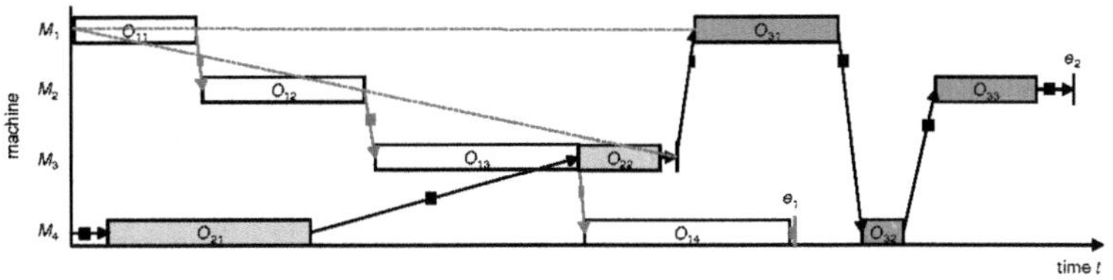

a Gantt chart of the schedule of Example 9.1 with simple idea

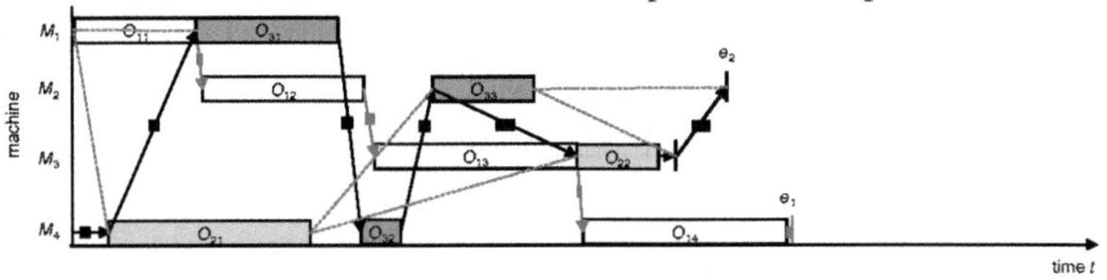

b Gantt chart of the schedule of Example 9.1 by proposed random key-based GA

Fig. 9.55 Comparison of different approaches

Table 9.36 Job requirements of Example 9.2

O_{ij}	M_{ij}				p_{ij}			
J_i \ P_i	P_1	P_2	P_3	P_4	P_1	P_2	P_3	P_4
J_1	1	2	1	-	80	120	60	-
J_2	2	1	-	-	100	60	-	-
J_3	5	3	3	-	70	100	70	-
J_4	5	3	2	2	70	100	100	40
J_5	4	2	-	-	90	40	-	-
J_6	4	4	1	2	90	70	60	40
J_7	1	3	-	-	80	70	-	-
J_8	5	4	5	4	70	70	70	80
J_9	5	4	1	-	70	70	60	-
J_{10}	5	1	3	-	70	60	70	-

Table 9.37 Transition time between pickup point u and delivery point v

t_{uv} / c_{uv}	Loading / unloading	M_1	M_2	M_3	M_4	M_5
Loading / unloading	1 / 1	1 / 7	8 / 13	14 / 18	16 / 23	18 / 20
M_1	13 / 18	3 / 3	2 / 9	8 / 14	10 / 19	13 / 18
M_2	18 / 22	22 / 28	2 / 2	2 / 7	4 / 12	12 / 18
M_3	13 / 11	17 / 22	24 / 29	1 / 1	1 / 6	7 / 11
M_4	8 / 14	12 / 20	18 / 26	24 / 29	3 / 3	2 / 10
M_5	5 / 7	9 / 12	15 / 18	19 / 23	23 / 28	2 / 2

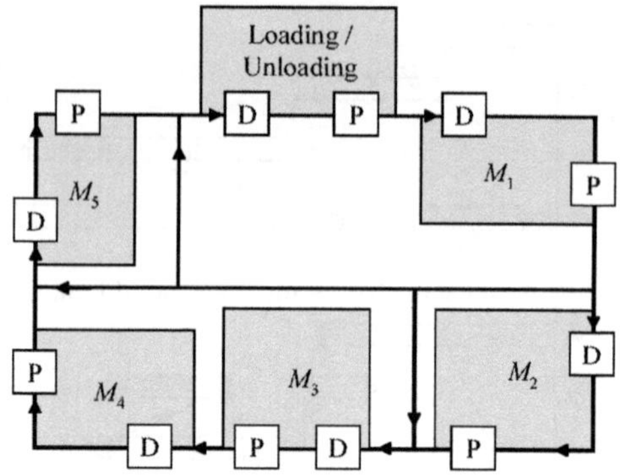

Fig. 9.56 Layout of facility (P: pickup point, D: delivery point)

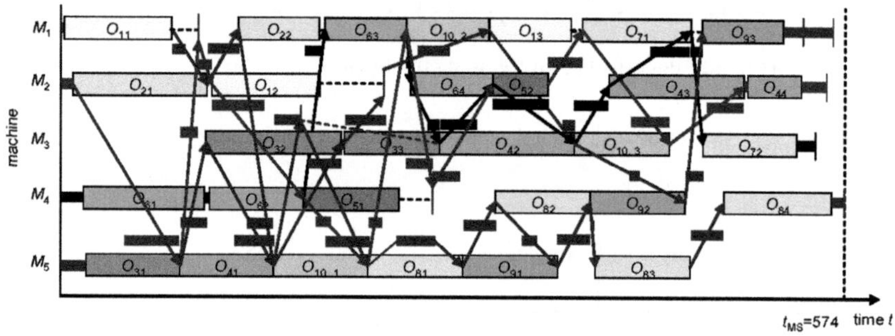

Fig. 9.57 Gantt chart of the schedule of Example 9.2 considering four AGVs dispaching

9.4 Car Navigation Routing Model

Intelligent Transport Systems (ITS) is a new transport system which is comprised of an advanced information and telecommunications network for users, roads and vehicles. ITS offers a number of solutions for modern transportation problems. Car navigation system (CNS) is the most widely used form of information terminals for the ITS. Route planning is the most important problem for a car navigation system. It guides a car driver through a complicated route in road networks, while displaying geographical information along the route. It typically uses Global Positioning Service (GPS) to acquire position data to locate the user on a road in the unit's map database. In the real life environment, some IT skills such as Vehicle Information and Communication System (VICS) and Transport Protocol Experts Group (TPEG) are used in the ITS area to provide dynamic traffic information. VICS is a technology used in Japan for delivering traffic and travel information to road vehicle drivers. VICS information (information on traffic congestion, road regulations, *etc.*) is edited and processed at the VICS Center. It is then transmitted in real time to car navigation systems where it is displayed in text and graphic form. TPEG technology is already standardized world-wide and recognized as providing the "tool-kit" for delivering all types of traffic and travel information content by any required delivery channel, *e.g.*, Digital Radio (DAB), Internet, Digital TV (DVB), GPRS and Wi-Fi *etc.* These techniques have been used in car navigation system devices to improve convenience in travel.

Route selection problems in car navigation systems are search problems for finding an optimal route from a starting point to a destination point on a road map. As commonly thought, this problem is a shortest path problem and the most efficient one-to-many nodes shortest path algorithm is Dijkstra's algorithm [65]. For the one-to-one node shortest path problem, the leading algorithm is A^* [66]. Moreover, in the case of car navigation systems, the shortest path may not be the best from other considerations such as, traffic time, costs, environmental problems and many other criteria. It is a multiple criteria aspect of route planning that we wish to tackle.

The problem is proven to be NP-complete that cannot be exactly solved in a polynomial time. Genetic algorithms (GA) have received considerable attention regarding their potential as a novel approach to multiobjective optimization problems, known as evolutionary or genetic multiobjective optimization [64, 50, 70]. In Chakraborty's paper [67], he proposed a multi Multiobjective Route Selection based on a genetic algorithm. But there is no clearly procedure for solving this kind of problem. Dynamic route planning for car navigation systems by genetic algorithm has been proposed by Kanoh [68]. Although road class, number of lanes and some other constraints are considered to calculate penalty value, the genetic operator is the basic way. It is hard to get the best compromised result without considering a set of selected route. So, an effective MOGA for multiple criteria route selection should be proposed to solve this problem.

9.4.1 Data Analyzing

In a real road map, we should confirm the elements for multiobjective route selection. They are listed bellow.

Node information:

1. Position of node
2. Height of intersection node

Link information:

1. Signal information
2. Link composed (from node, to node)
3. Distance of link composed by two connected nodes
4. Number of lanes in the road represented by link
5. Road class
6. Slope condition
7. Traffic situation (in real time)

After getting the data set from a real road map, we should design a data structure to get a bridge from the real road map to car navigation systems. Usually the structure of each digital map is different. For easily exchanging or integrating data from other systems, we proposed an XML-based structure to describe the navigation information such as road link and node (road crossing) information. When we get the vector of nodes and links, we can convert them into an XML file. Here we show a simple example of an XML file. The example data set list is given in Tables 9.38 and 9.39.

Table 9.38 The example data set of node information

Index	Node id	Position	Height (m)
1	030	(447, 435)	15
2	031	(507, 468)	14
3	032	(607, 509)	12
4	034	(423, 483)	14
5	035	(480,512)	15
...	...	...	...
...	...	...	...

```
<?xml version="1.0"?>
<Nodes>
   <Node ID="030">
      <position>447,435</position>
      <height>15</height>
   </Node>
   <Node ID="031">
      <position>507,468</position>
      <height>14</height>
   </Node>
   <Node ID="032">
      <position>607,509</position>
      <height>12</height>
   </Node>
   <Node ID="034">
      <position>423,483</position>
      <height>14</height>
   </Node>
   <Node ID="035">
      <position>480,512</position>
      <height>15</height>
   </Node>
   ...
   ...
</Nodes>
<?xml version="1.0"?>
   <Links>
```

Table 9.39 The example data set of link information

Link id.	From node	To node	Distance	No. of lanes	Traffic congestion level	Toll	Speed limit
030-031	030	031	63	1	1	0	30
031-032	031	032	139	1	2	0	30
032-036	032	036	63	1	2	0	30
033-044	033	044	51	1	2	0	30
034-035	034	035	63	1	1	0	30
…	…	…	…	…	…	…	…
…	…	…	…	…	…	…	…


```
< Link ID="030-031" >
    <from_node>030</from_node>
    <to_node>031</to_node>
    <distance>63</distance>
    <lane_number>1</lane_number>
    <traffic_congestion>1</traffic_congestion>
    <toll>0</toll>
    <speed_limit>30</speed_limit>
</Link>
<Link ID="031-032">
    <from_node>031</from_node>
    <to_node>032</to_node>
    <distance>139</distance>
    <lane_number>1</lane_number>
    <traffic_congestion>2</traffic_congestion>
    <toll>0</toll>
    <speed_limit>30</speed_limit>
</Link >
<Link id="032-036">
    <from_node>032</from_node>
    <to_node>036</to_node>
    <distance>63</distance>
    <lane_number>1</ lane_number>
    <traffic_congestion>2</traffic_congestion>
    <toll>0</toll>
    <speed_limit>30</speed_limit>
</Link>
<Link id="033-044">
    <from_node>033</from_node>
    <to_node>044</to_node>
    <distance>51</distance>
    <lane_number>1</ lane_number>
```

<traffic_congestion>2</traffic_congestion>
<toll>0</toll>
<speed_limit>30</speed_limit>
</Link>
<Link ID="034-035">
<from_node>034</from_node>
<to_node>035</to_node>
<distance>63</distance>
<lane_number>1</lane_number>
<traffic_congestion>1</traffic_congestion>
<toll>0</toll>
<speed_limit>30</speed_limit>
</Link>
...
...
</Links>

9.4.2 Mathematical Formulation

In this problem, to find a compromised optimal route, the road map is first converted into an undirected graph $G = (V, A)$ comprises a set of nodes $V = \{1, 2, \cdots, n\}$ and a set of edges $A \in V \times V$ connecting nodes in V. A path from node O to node D is a sequence of edges (O, l), (l, m), $\cdots$, (k, D) from A in which no node appears more than once; in this problem O means a original node of a route, D is destination node. A path can also be equivalently represented as a sequence of nodes $(O, l, m, \cdots, k, D)$. In Fig. 9.58, a network model represents a simple graphical representation of a road map and the possible routes from original node 1 to the destination node 10.

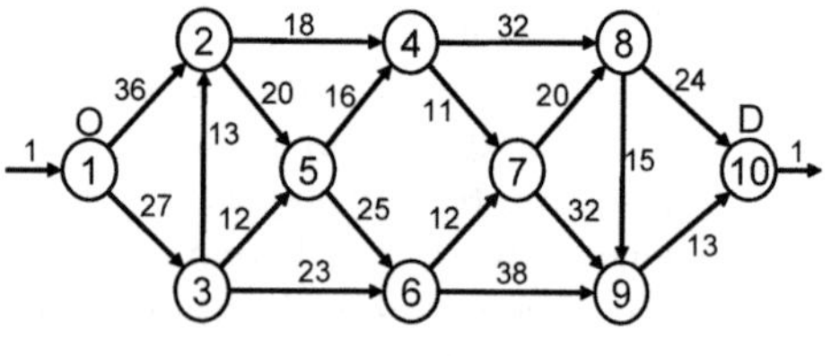

Fig. 9.58 A simple model of network

Notation

Indices

i, j, k: =1, 2,..., n index of node

Parameters

n: number of nodes
c_{ij}: cost of arc (i, j)
O: original node
D: destination node
P: a selected route from O to D

Decision Variables

$$x_{ij} = \begin{cases} 1, \text{ if link } (i, j) \text{ is included in the path} \\ 0, \text{ otherwise} \end{cases}$$

The multiobjective model is formulated as follows, which the objectives are extensive from the three objectives: distance, driving time and cost.

The Objective Function for Distance on a Route *P*

$$\min f_1(P) = \sum_{(i,j)\in P} d_{ij}x_{ij} \tag{9.34}$$

where (i,j) is links from node i to node j in link set A, d_{ij} is the length of a link

The Function for Driving Time on a Route *P*

$$\min f_2(P) = \sum_{(i,j)\in P} h(\beta u_{ij}(t) + (1-\beta)v_{ij}(t+\tau_{ij}))x_{ij} \tag{9.35}$$

where t is enter time of link (i, j) and $t+\tau_{ij}$ is exit time, $u_{ij}(t)$ is inflow rate at time t while enter in link(i, j), $v_{ij}(t+\tau_{ij})$ is outflow rate at time $t+\tau_{ij}$ when exit from link(i, j). β is a weight constraint and $0 \leq \beta < 1$. The inflow and outflow can be predicted by traffic data.

The Function for Driving Cost on a Route *P*

$$\min f_3(P) = \sum_{(i,j)\in P} g(s_{ij}, a_{ij}, m_{ij})x_{ij} \tag{9.36}$$

where s_{ij} is slope condition of link (i,j), a_{ij} is average fuel consumption cost when car driving and m_{ij} is toll fee.

The constraints of these objective functions are defined as follows:

$$\text{s.t.} \sum_{(j,k)\in A} x_{kj} - \sum_{(k,i)\in A} x_{ik} = \begin{cases} 1 & (k=O) \\ 0 & (k \in V\backslash(O,D)) \\ -1 & (k=D) \end{cases} \tag{9.37}$$

$$x_{ij} = 0 \text{ or } 1 \ \ (i,j = 1,2,\cdots,n) \tag{9.38}$$

9.4.3 Improved Fixed Length-based GA

Let $P(t)$ and $C(t)$ be parents and offspring in current generation t. The overall procedure for solving multi-criteria route selection problem is outlined as follows:

---

procedure: MOGA for Route selection
input: road map data (V, A, C), GA parameters ($popSize$, $maxGen$, p_M, p_C)
output: Pareto optimal solutions E, the best compromised route in $E(t)$;
begin
  $t \leftarrow 0$;
  initialize $P(t)$ by Improved Fixed-length based encoding routine;
  calculate objectives $z_i(P)$, $i = 1, \cdots ,q$ by Improved Fixed-length based decoding routine;
  create Pareto $E(P)$;
  calculate eval(P) by adaptive-weight fitness assignment routine;
  while (**not** terminating condition) **do**
    create $C(t)$ from $P(t)$ by weight mapping crossover routine;
    create $C(t)$ from $P(t)$ by insertion mutation routine;
    calculate objectives $z_i(C)$, $i = 1,\cdots,q$ by Improved Fixed-length based decoding routine;
    update Pareto $E(P,C)$;
    calculate eval(P,C) by adaptive-weight fitness assignment routine;
    select $P(t+1)$ from $P(t)$ and $C(t)$ by mixed roulette wheel selection routine;
    $t \leftarrow t+1$;
  end
  output Pareto optimal solutions $E(P,C)$
end

9.4.3.1 Genetic Representation

Depending on the following phases and steps, give the presentation of genetic representation.

Phase 1 : Analyzing real road map data (VICS real time data)

Step 1.1: Confirm the elements in required area on real road map
Step 1.2: Get node and link data from the selected area
Step 1.3: Translate node and link data into XML file which will be used in whole system with unified structure
Step 1.4: Draw network structure of road map

Phase 2 : Designing a chromosome and solving by moGA (GA engine)

Step 2.1: Input the task sequence found in Step 1.2
Step 2.2: Obtain a feasible solution set according to this task sequence

Phase 3 : Drawing a route on the display of car navigation systems

In the decoding method, the position of a gene is used to represent the node index in route and its value is used to decide which node will selected in successive sets of current node. When the destination node occurred in successive sets, the encoding process is ending. A path can be easily determined from this encoding. An example of generated chromosome and its decoded path is shown in Fig. 9.59, for the network shown in Fig. 9.58. The max value of successive sets of each node in the example network is in Fig. 9.59. All values in the gene locus are randomly created from 1 to 3. At the beginning, we try to find a position of node next to source node 1. Nodes 2 and 3 are eligible for the position, which can be easily found according to adjacent relations among nodes. The value of the gene is index of selected node in successive sets. So we can get the next node obviously and put it into the path. Because the value in locus 1 is 3, there are two adjacent nodes in successive set of node 1. We can get the next node position of path by calculating (3 **mod** 2); it equals 1, so the next node index can be got from first node in successive sets, here it is 2, and put it into the path. Then we form the set of nodes available for next position and select the one with the current gene value of gene locus. Repeat these steps until we obtain a complete path ($1 \rightarrow 2 \rightarrow 4 \rightarrow 7 \rightarrow 8 \rightarrow 10$).

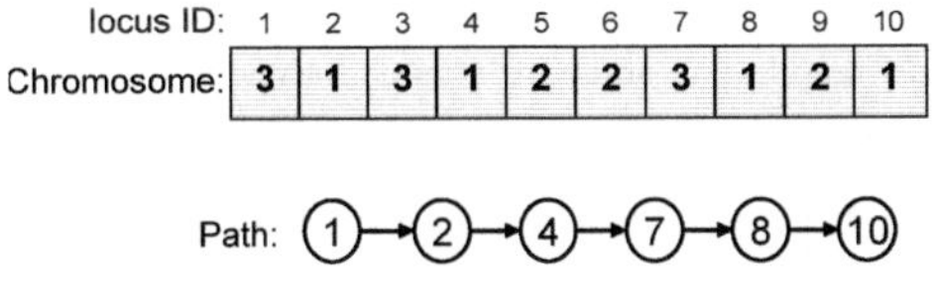

Locus ID i	$\lvert Suc_i[\cdot]\rvert$	$Suc_i[\cdot]$
1	2	{2, 3}
2	2	{4, 5}
3	3	{2, 5, 6}
4	2	{7, 8}
5	2	{4, 6}
6	2	{7, 9}
7	2	{8, 9}
8	2	{9, 10}
9	1	{10}
10	0	{Ø}

Fig. 9.59 An example of generated chromosome and its decoded path

Fig. 9.60 Chromosome encoding procedure

```
procedure 1: Improved Fixed-length Encoding
input: number of nodes n,
        array of successive nodes number Suc[]
output: k-th initial chromosome v_k[]
begin
   select max length assign to j in Suc[];
   for i =1 to n
      v_k[i] ← random[1, j];
   output v_k ;
end
```

Based on the improved fixed-length chromosome encoding method given in Fig. 9.60, we present the decoding process in Fig. 9.61. Using this procedure, a chromosome can be decode a path by a chromosome. The trace table of decoding procedure is shown in Table 9.40.

```
procedure 2: Improved Fixed-length Decoding
input: chromosome v[], no. of nodes n, origin ID O,
destination ID D, Suc[][]
output: path P[]
begin
   P[1] := O;
   for i := 2 to n              // initialize path with zero
      P[i] := 0;
   for i := 1 to n-1
      id := P[i];
      index := v[i] % |Suc[id][]|;
      if index = 0 then   index := |Suc[id][]|;
      P[i+1] := Suc[id][index];    // add current node into path
      if j <= i then break;
      if P[i+1] = D then break;    // find the destination node
   output P[];
end
```

Fig. 9.61 Chromosome decoding procedure

Table 9.40 Trace table of decoding procedure

iteration	node id j	$Suc[j]$	$\lvert Suci[\cdot]\rvert$	$Path[\,]$
1	1	{2, 3}	2	1
2	2	{4, 5}	2	1-2
3	4	{7, 8}	2	1-2-4
4	7	{8, 9}	2	1-2-4-7
5	8	{9, 10}	2	1-2-4-7-8
6	10	{Ø}	0	1-2-4-7-8-10

9.4.3.2 Genetic Operators

Weight Mapping Crossover: The next step is to generate a second generation population of solutions from those selected through genetic operators: crossover (also called recombination), and mutation. Weight mapping crossover (WMX) is combined that can be viewed as one of a one-point crossover for permutation representation.
Insertion Mutation: In this kind of mutation, a gene is randomly selected and inserted in a position, which is determined randomly.

9.4.3.3 Mixed Roulette Wheel Selection

To maintain the diversity of the population and Pareto set and enhance the local research ability of genetic operator, we should consider how to satisfy the driver's preference. The road map provided by Japan Digital Road Map Association (JDRMA) includes some road characteristics such as speed limits, link lengths, road classes, number of lanes, road width, slope condition, and the existence of signals. Usually, the route with higher road class will be considered when the car navigation system selects the route. So, here we propose a Mixed Roulette Wheel Selection approach to achieve this objective.

We can use factor value of each link in one route P to calculate the road class factor of P (as shown in Table 9.41), defining $g(P)$ as

$$g(P) = \sum_{(i,j)\in P} r_{ij}x_{ij}$$

After calculating the factor value of P, we should sort the chromosome in population and offspring by this value. When selection operator processing, roulette wheel selection procedure may choose the same chromosome. We select the chromosome with the higher $g(P)$ value instead of the duplicated one. In this way we can obtain the diversity character and let the search process migrate to the route with the highest road class.

Table 9.41 Road class of link classified by JDRMA

Road class of Link (i,j)	Code	r_{ij}
National highway	1	8
Express way	2	7
Common national road	3	6
Principal local road	4	5
Prefectural road	5	4
Common principal road	6	3
Municipal road	7	2
Others	9	1
Not investigation road	0	0

The whole procedure of Mixed Roulette Wheel Selection is shown in Fig. 9.62.

procedure 4: Mixed Roulette Wheel Selection
input: population $P(t\text{-}1)$, $C(t\text{-}1)$
output: population $P(t)$, $C(t)$
begin
step 1: Calculate the total fitness for the population and offspring
step 2: Calculate selection probability p_k for each chromosome v_k
step 3: Calculate cumulative probability q_k for each chromosome v_k
step 4: Generate a random number r from the range [0, 1]
step 5: **if** $r \le q_1$ **then**
select the first chromosome v_1;
else if v_k is unselected **then**
select the kth chromosome v_k $(2 \le k \le popSize)$ that $q_{k\text{-}1} < r \le q_k$ and mark v_k as selected.
Total selected number of chromosome assigned to *selectedNum*.
step 6: Select *popSize-selectedNum* chromosome with biggest class factor value from population and offspring which marked unselected.
output $P(t)$, $C(t)$;
end

Fig. 9.62 Mixed roulette wheel selection

9.4.3.4 Dynamic Route Selection Method

In a real driving environment, after selecting a route by the route select method, the traffic condition will change as time passes. So, dynamic route selection with VICS data should be considered, the procedure shown in Fig. 9.63.

```
procedure 6: Dynamic Route Selection with VICS data
input: real road map, VICS data,
output: the best compromised route Path[]
begin
    select a best compromised route Path[] by Priority-based GA for Multiobjective
        Route Selection Problem;
    driving on the selected route Path[];
    while (not reach destination node) do
        receive VICS data;
        if congestion occurred in Path[] then
            select a best compromised route Path[] by multiobjective GA for Route selection;
    end
    output Path[];
end
```

Fig. 9.63 Dynamic route selection with VICS data

9.4.4 Numerical Experiment

In order to make a large number of solutions in Pareto-optimal set S^*, first we calculate the solution sets with special GA parameter settings and much longer computation time by each approach which are used in comparison experiments, then combine these solution sets to calculate the reference set S^*. Furthermore, we will combine small but reasonable GA parameter settings for comparison experiments, thus ensuring the effectiveness of the reference set S^*. A simulation was carried out using the road map shown in Fig. 9.64. The database of this area is provided by Japan Digital Road Map Association.

In this section, the performance of m-moGA is compared with other evolutionary algorithms; one is NSGA-II [69], the other is a proposed method using the common Roulette Wheel Selection method, named moGA, and the set S is defined as evaluated solution set. We used these parameter to find the reference solution set S^*: population size, *popSize* = 300; crossover probability, p_C =0.90; mutation probability, p_M = 0.90; immigration rate; Stopping criteria: *maxGen* = 100000. We chose nondominated solutions as reference solutions from the test problem. We show the obtained reference solution sets for the 410-node / 678-arc test problem in Fig. 9.66. The obtained reference solutions for our test problems are summarized in Table

9.42, where the range width of the ith objective over the reference solution set S^* is defined as:

$$W_{f_i}(S*) = \max\{f_i(r) \mid r \in S*\} - \min\{f_i(r) \mid r \in S*\}$$

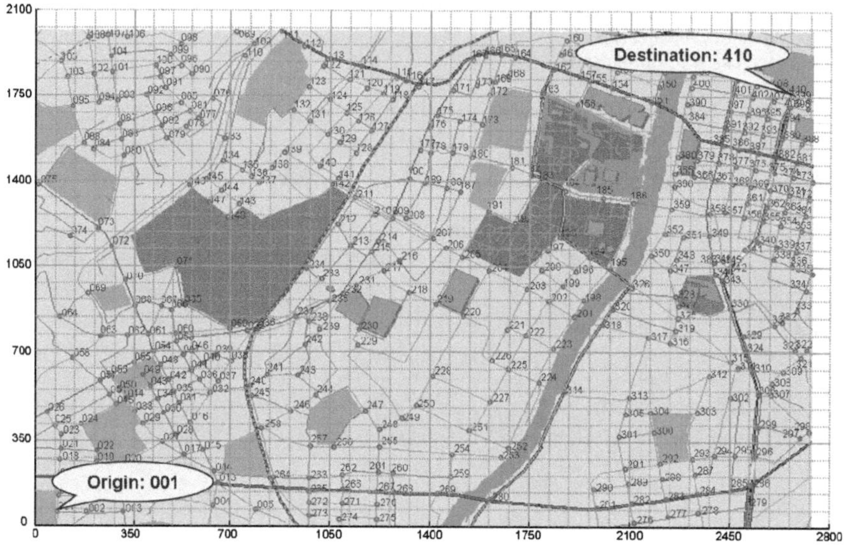

Fig. 9.64 The real road map near Kokura station in Kitakyushu city

Table 9.42 Obtained reference solutions

Test problem(# of nodes / # of arcs)	# of obtained solutions $\lvert S\rvert$	Range width $W_f(S^*)$		
		$f_1(r)$	$f_2(r)$	$f_3(r)$
410/678	157	366	1970	11.3

We compared these three algorithm through computational experiments on the 410-node / 678-arc test problems under the same GA parameter settings: population size, *popSize* =20; crossover probability, $p_C = 0.70$; mutation probability, $p_M = 0.60$; stopping condition, *maxGen* = 5000. Each simulation was run 10 times. Here we used average distance $D1_R$(S) to find an average distance of the solutions of S from S^*; its definition is as follows:

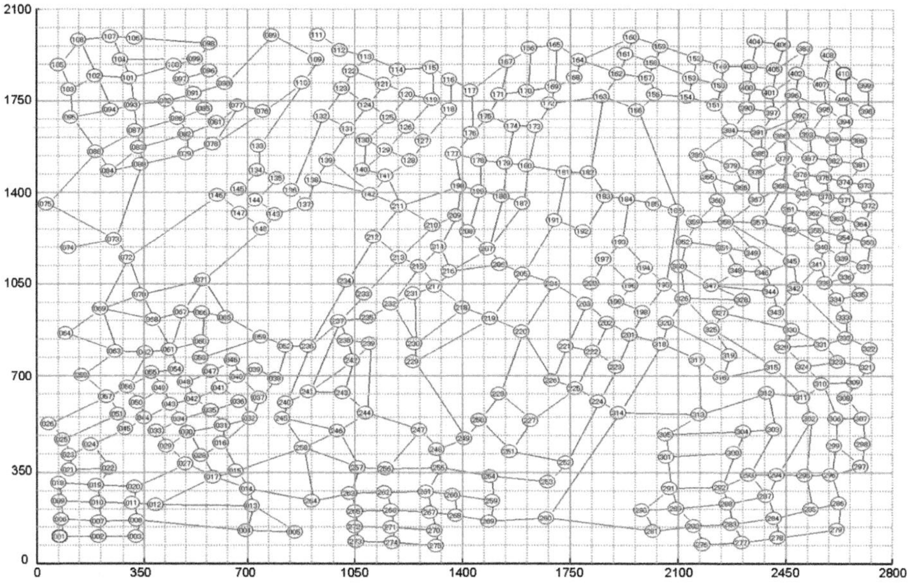

Fig. 9.65 The network model of road map near Kokura station in Kitakyushu city

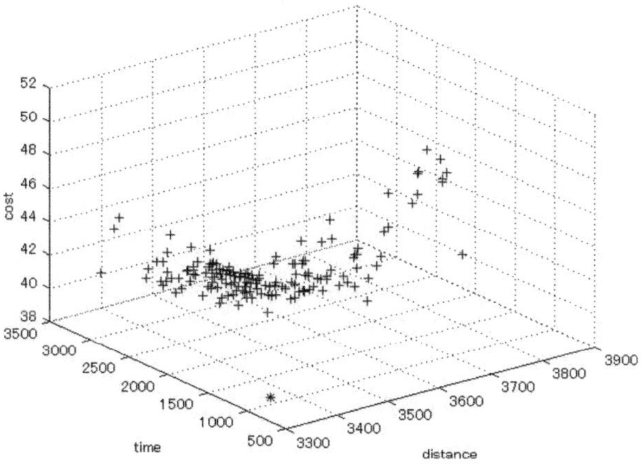

Fig. 9.66 Reference solutions

$$D1_R(S_J) = \frac{1}{\mid S* \mid} \sum_{r \in S*} \min\left\{d_{rx} \mid x \in S_j\right\}$$

where d_x is the distance between a current solution $x \in S$ and a reference solution r in the three-dimensional normalized objective space. $q = 1, 2, 3$ is defined as objective index:

$$d_{rx} = \sqrt{(f_1(r) - f_1(x))^2 + (f_2(r) - f_2(x))^2 + (f_3(r) - f_3(x))^2}$$

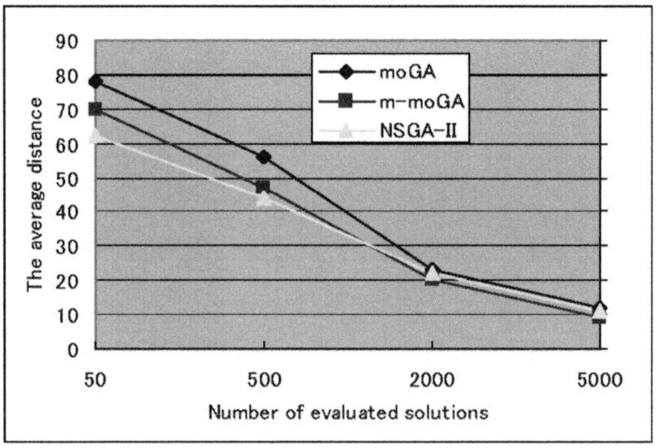

Fig. 9.67 Number of evaluated solutions

The solution set S is better if the value of $D1_R(S)$ is smaller. This measure explicitly computes the closeness of a solution set S from the set S^*. As depicted in Fig. 9.67 the best results of $D1_R(S)$ are obtained by our m-moGA. The Pareto result of experiment is shown in Fig. 9.68; from this figure, the ideal point and best solution point are labeled.

We propose a factor weight method in our algorithm–Factor weight of each objective for optimal route choice in Pareto solution set. Here we use w_{L}, w_{T}, w_{C} as factor weight for route length objective, time objective, cost objective respectively, and w_{L}+w_{T}+ w_{C}=1. All factor weight should be assigned by user as his preference and input in a car navigation device before departure. Here we proposed a weighted distance method to decide which route is the best compromised route in Pareto set:

$$d_{rl} = \sqrt{w_{\mathrm{L}}\left(\frac{f_1(x) - f_1(l)}{z_1^{max} - f_1(l)}\right)^2 + w_{\mathrm{T}}\left(\frac{f_2(x) - f_2(l)}{z_2^{max} - f_2(l)}\right)^2 + w_{\mathrm{C}}\left(\frac{f_3(x) - f_3(l)}{z_3^{max} - f_3(l)}\right)^2}$$

where here, x is current solution and l is ideal solution in the three-dimensional normalized objective space. Using this method we can get the compromised best solution. The best solution of different methods are listed in Table 9.43

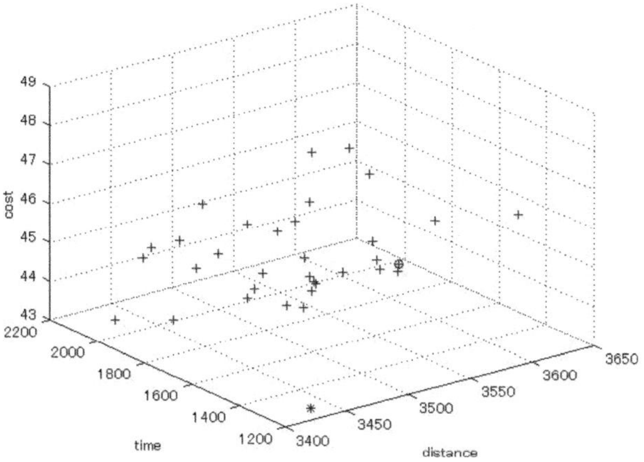

Fig. 9.68 Result from the m-moGA approach

Table 9.43 Comparison of best solution in different methods. ($w_L = w_T = w_C = 1/3$)

	m-moGA	NSGA-II	moGA
Length (m) $f_1(x)$	3524	3524	3830
Time (sec.) $f_2(x)$	1366.2	1459.6	1475.6
Cost (yen) $f_3(x)$	45.7	45.5	46.3

9.5 Summary

Everywhere we look in our daily lives, many problems bear resemblance to a network problem, and we want to learn how to model application settings as mathematical objects known as network design models and to study various ways (algorithms) to solve the resulting models. In this chapter, the network model of airline fleet assignment problems, container terminal optimization, AGV dispatching problem and car navigation routing problem were introduced.

First, the fleet assignment problem deals with assigning aircraft types, each having a different capacity, to the scheduled flights, based on equipment capabilities and availabilities, operational costs, and potential revenues. We introduced GA approaches to solve two basic fleet assignment problems, considering connection networks and time-space networks. Second, we introduced quay crane scheduling in the berth allocation planning problem, we introduced the problem as a network model with considering the arriving time, berth location and number of QCs associated with berth *etc.*, and presented a multiobjective GA to solve the problem.

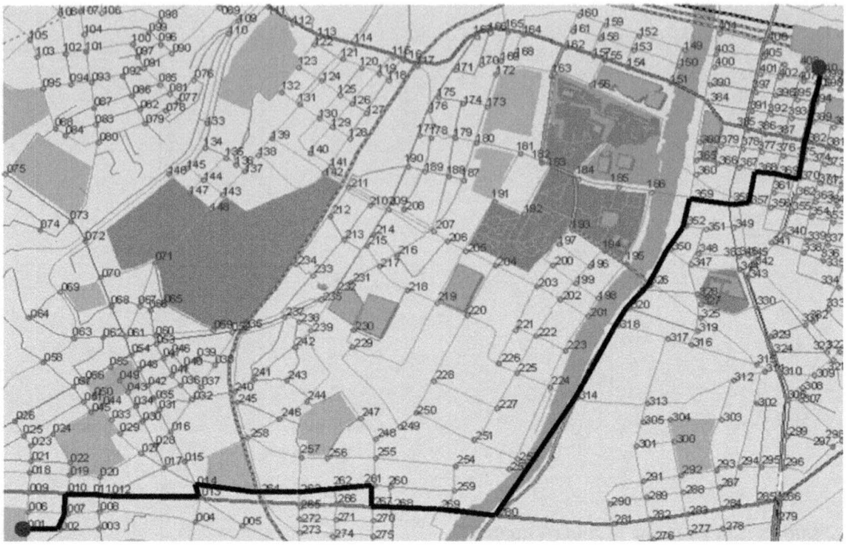

Fig. 9.69 The selected route

Third, we introduced a network model for the simultaneous scheduling and routing of AGVs in a flexible manufacturing system (FMS). We modeled an AGV system by using network structure. This network model of AGV dispatching has simplexes decision variables considering most AGV problem's constraints. For applying a GA approach to AGV dispatching problem that minimize the time required to complete all jobs (*i.e.*, makespan), a random key-based encoding method was presented. Numerical analyses for case study showed the effectiveness of proposed the approach.

Last, we introduced a routing model in a car navigation system. We analyzed the structure of car navigation systems, extracted the mathematical model from the multiobjective route selection problem in this field and combined with GA. According to the flexibility in dynamic routing, we developed improved fixed-length based multiobjective GA as an effective approach for finding the optimal route relative to the multi objective in car navigation systems.

References

1. Ahuj, R. K., Magnanti, T. L., & Orlin, J. B. (1993). *Network Flows*, New Jersey, Prentice Hall.
2. Abara, J. (1989). Applying integer linear programming to the fleet assignment problem, *Interfaces*, 19, 20–28.
3. Subramanian, R., Scheff, R. P., Quillinan, J. D., Wiper, D. S., & Marsten, R. E. (1994). Coldstart: Fleet assignment at Delta Air Lines, *Interfaces*, 24, 104–120.

4. Hane, C., Barnhart, C., Johnson, E. L., Marsten, R. E., Nemhauser, G. L., & Sigismondi, G. (1995). The fleet assignment problem: Solving a large-scale integer program, *Mathematical Programming*, 70, 211–232.
5. Rushmeier, R. A., & Kontogiorgis, S. A. (1997). Advances in the optimization of airline fleet assignment, *Transportation Science*, 31, 159–169.
6. Barnhart, C., Johnson, E. L., Nemhauser, G. L., Savelsbergh, M. W. P., & Vance, P. H. (1998). Branch-and-Price: column generation for solving huge integer programs, *Operations Research*, 46, 316–329.
7. Yan, S. & Tseng, C. H. (2002). A passenger demand model for airline flight scheduling & fleet routing, *Computer & Operations Research*, 29, 1559–1581.
8. Lohatepanont, M. & Barnhart, C. (2004). Airline schedule planning: Integrated models & algorithms for schedule design & fleet assignment, *Transportation Science*, 38, 19–32.
9. Desaulniers, G., Desrosiers, J., Dumas, Y., Solomon, M. M., & Soumis, F. (1997). Daily aircraft routing & scheduling, *Management Science*, 43, 841–855.
10. Rexing, B., Barnhartm, C., Kniker, T., Jarrah, A., & Krishnamurthy, N. (2000). Airline fleet assignment with time windows, *Transportation Science*, 34, 1–20.
11. Bandet, P. O. (1994). Armada-Hub: An adaption of AF fleet assignment model to take into account the hub structure in CDG, *Proceedings of AGIFORS 34th Annual Symposium*.
12. Desaulniers, G., Deserosiers, J., Solomon, M. M., & Soumis, F. (1994). Daily aircraft routing & scheduling, *GERAD technical report*, Montreal, Quebec, Canada.
13. Yan, S. & Young, H. F. (1996). A decision support framework for multi-fleet routing & multi-shop flight scheduling, *Transportation Research*, 30, 379–398.
14. Desrochers, M. & Soumis, F. (1988). A generalized permanent labeling algorithm for the shortest path problem with time windows, *Infor*, 26, 191–212.
15. Ioachim, I., Gelinas, S., Desrosiers, J., & Soumis, F. (1998). A dynamic programming algorithm for the shortest path problem with time windows & linear node costs, *Networks*, 31, 193–204.
16. Barnhart, C., Kniker, T. S., & Lohatepanont, M. (2002). Itinerary-based airline fleet assignment, *Transportation Science*, 36, 199–217.
17. Lee, L. H., Lee, C. U., & Tan, Y. P. (2007). A multi-objective genetic algorithm for robust flight scheduling using simulation, *European Journal of Operational Research*, 177, 1948–1968.
18. Yan, S, Tang, C. H., & Lee, M. C. (2007). A flight scheduling model for Taiwan airlines under market competitions, *Omega*, 35, 61–74.
19. Gu, Z., Johnson, L., Nemhauser, G.L. & Wang, Y. (1994). Some properties of the fleet assignment problem, *Operations Research Letters*, 15(2), 59-71.
20. Sherali, H.D., Bish. E.K., & Zhu. X. (2006). Airline fleet assignment concepts, models, and algorithms, *European Journal of Operational Research*, 172(1), 1-30.
21. Stojkovic, M., & Soumis, F. (2001). An Optimization Model for the Simultaneous Operational Flight and Pilot Scheduling Problem, *Management Science*, 47(9), 1290-1305.
22. Bélanger, N., Desaulniers, G., Soumis, F., Desrosiers, J. & Lavigne, J. (2006). Weekly airline fleet assignment with homogeneity, *Transportation Research Part B: Methodological*, 40(4), 306-318.
23. Bélanger, N., Desaulniers, G., Soumis, F. & Desrosiers, J. (2006). Periodic airline fleet assignment with time windows, spacing constraints, and time dependent revenues, *European Journal of Operational Research*, 175(3), 1754-1766.
24. Kim, K. H. & Kim, H. (1998). The optimal determination of the space requirement & the number of transfer cranes for import containers, *Computers & Industrial Engineering*, 35, 427–430.
25. Bose, J., Reiners, T., Steenken, D., & Voss, S. (2000). Vehicle dispatching at seaport container terminals using evolutionary algorithms, *Proceedings of Annual Hawaii International Conference on System Sciences*, 2, 1–10.
26. Zhang, C. Q., Wan, Y. W., Liu, J. Y., & Linn, R. J. (2002). Dynamic crane deployment in container storage yards, *Transporation Research*, Part B, 36, 537-555.

27. Chung, R. K., Li, C., & Lin, W. (2002). Interblock Crane Deployment in Container Terminals, *Transportation Science*, 36, 79–93,
28. Linn, R., Liu, J., Wan, Y., Zhang, C., & Murty, K. (2003). Rubber tired gantry crane deployment for container yard operation, *Computers & Industrial Engineering*, 45, 429–442.
29. Kim, K. Y. & Kim, K. H. (2003). Heuristic algorithms for routing yard-side equipment for minimizing loading times in container terminals, *Naval Research Logistics*, 50, 498–514.
30. Ng, W. C. & Mak, K. L. (2005). Yard crane scheduling in port container terminals, *Applied Mathematical Modeling*, 29, 263–276.
31. Ng, W. C. (2005). Crane scheduling in container yards with inter-crane interference, *European Journal of Operational Research*, 164, 64–78.
32. Kim, K. H. & Kim, K. Y. (1999). An optimal routing algorithm for a transfer crane in port container terminals, *Transportation Science*, 33, 17–33.
33. Kozan, E. & Preston, P. (1999). Genetic algorithms to schedule container transfers at multimodal terminals, *International Transactions in Operational Research*, 6, 311–329.
34. Preston, P. & Kozan, E. (2001). An approach to determine storage locations of containers at seaport terminals, *Computers & Operations Research*, 28, 983–995.
35. Kozan, E. & Preston, P. (2006). Mathematical modelling of container transfers & storage locations at seaport terminals, *OR Spectrum*, 28, 519–537.
36. Imai, A., Sasaki, K., Nishimura, E., & Papadimitriou, S. (2006). Multi-objective simultaneous stowage & load planning for a container ship with container rehandle in yard stacks, *European Journal of Operational Research*, 171, 373–389.
37. Kim, K. H. & Park, Y. (2004). A crane scheduling method for port container terminals, *European Journal of Operational Research*, 156, 752–768.
38. Han, M., Li, P., & Sun, J. (2006). The algorithm for berth scheduling problem by the hybrid optimization strategy GASA, *Proceedings of 9th International Conference on Control, Automation, Robotics & Vision*, 1–4.
39. Jung, S. H. & Kim, K. H. (2006). Load scheduling for multiple quay cranes in port container terminals, *Jorunal of Intellignet & Manufacturing*, 17, 479–492.
40. Zhou, P. F., Kang, H. G., & Lin, L. (2006). A dynamic berth allocation model based on stochastic consideration, *Proceedings of the Sixth World Congress on Intelligent Control & Automation*, 7297–7301.
41. Imai, A., Nishimura, E., Hattori, M., & Papadimitriou, S. (2007). Berth allocation at indented berths for mega-containerships, *European Journal of Operational Research*, 179, 579–593.
42. Hansen, P., Oguz, C., & Mladenovic, N. (2003). Variable neighborhood search for minimum cost berth allocation, *European Journal of Operational Research*, In Press.
43. Imai, A., Chen, H., Nishimura, E., & Papadimitriou, S. The simultaneous berth & quay crane allocation problem, *Transportation Research, PartE, Logistics & Transportation Review*, In Press.
44. Lee, D., Wang, H., & Miao, L. (2008). Quay crane scheduling with non-interference constraints in port container terminals, *Transportation Research Part E: Logistics & Transportation Review*, 44, 124–135.
45. Wen, S. Q. & Zhou, P. F. (2007). A container vehicle routing model with variable traveling time, *Proceedings of 2007 IEEE International Conference on Automation & Logistics*, 2243–2247.
46. Zeng, Q. C. & Yang, Z. Z. (2006). A bi-level programming model & its algorithm for scheduling at a container terminal, *Proceedings of International Conference on Management Science & Engineering*, 402–406.
47. Zhou, P. F., Kang, H. G., & Lin, L. (2006). A fuzzy model for scheduling equipments handling outbound container in terminal, *Proceedings of the Sixth World Congress on Intelligent Control & Automation*, 7267–7271.
48. Lau, H. Y. K. & Zhao, Y. Integrated scheduling of handling equipment at automated container terminals, *International Journal of Production Economics*, In Press.
49. Gen, M. & Cheng, R. (1997). *Genetic Algorithms & Engineering Design*, New York: John Wiley & Sons.

50. Gen, M. & Cheng, R. (2000). *Genetic Algorithms & Engineering Optimization*, New York: John Wiley & Sons.
51. Holland, J. (1992). *Adaptation in Natural & Artificial System*, University of Michigan Press, Ann Arbor, MI, 1975; MIT Press, Cambridge, MA.
52. Hwang, H., Moon, S., & Gen, M. (2002). *An integrated model for the design of end-of-aisle order picking system & the determination of unit load sizes of AGVs*, Computer & Industrial Engineering, 42, 294–258.
53. Kim, D. B. & Hwang, H. (2001). A dispatching algorithm for multiple-load AGVs using a fuzzy decision-making method in a job shop environment, *Engineering Optimization*, 33, 523–547.
54. Kim, S. H. & Hwang, H. (1999). An adaptive dispatching algorithm for automated guided vehicles based on an evolutionary process, *Internatioanl Journal of Production Economics*, 60–61, 465–472.
55. Kim, K., Yamazaki, G., Lin, L., & Gen, M. (2004). Network-based Hybrid Genetic Algorithm to the Scheduling in FMS environments, *Journal of Artificial Life & Robotics*, 8(1), 67–76.
56. Le-Anh, T. & Koster, D. (2006). A review of design & control of automated guided vehicle systems, *European Journal of Operational Research*, 171(11), 1–23.
57. Lin, J. K. (2004). *Study on Guide Path Design & Path Planning in Automated Guided Vehicle System*, PhD Thesis, Waseda University.
58. Moon, S. W. & Hwang, H. (1999). Determination of unit load sizes of AGV in multi-product multi-line assembly production systems, *International Journal of Production Research*, 37(15), 3565–2581.
59. Naso, D. & Turchiano, B. (2005). Multicriteria meta-heuristics for AGV dispatching control based on computational intelligence, *IEEE Transactions on System Man & Cybernetics*, part B, 35(2), 208–226.
60. Proth, J., Sauer, N., & Xie, X. (1997). Optimization of the number of transportation devices in a flexible manufacturing system using event graphs, *IEEE Transactions On Industrial Electronics*, 44(3), 298–306.
61. Qiu, L., Hsu, W., Huang, S., & Wang, H. (2002). Scheduling & routing algorithms for AGVs: a survey, *Internatioanl Journal of Production Research*, 40(3), 745–760.
62. Vis, Iris F. A. (2004). Survey of research in the design & control of automated guided vehicle systems, *European Journal of Operational Research*, 170(3), 677–709.
63. Yang, J. B. (2001). GA-Based Discrete Dynamic Programming Approach for Scheduling in FMS Environment, *IEEE Transactions on System Man & Cybernetics*, part B, 31(5), 824–835.
64. Gen, M., Cheng, R., & Oren, S. S. (2001). Network Design Techniques using Adapted Genetic Algorithms, *Advances in Engineering Software*, 32(9), 731–744.
65. Dijkstra, E. W. (1959). A note on two problems in connexion with graphs, *Numerische Mathematik*, 1(2), 269–271.
66. Hart, E. P., Nilsson, N. J., & Raphael, B. (1968). A formal basis for the heuristic determination of minimum cost paths, *IEEE Transactions on Systtem Science Cybernetics*, SSC-4(2), 100–107.
67. Chakrabory, B., Maeda, T., & Charkrabory, G. (2005). Multiobjective Route Selection for Car Navigation System using Genetic Algorithm, *Proceedings of IEEE Workshop on Soft Computing*, 190–195.
68. Kanoh, H. (2007). Dynamic route planning for car navigation systems using virus genetic algorithms, *International Journal of Knowledge-Based & Intelligent Engineering Systems*, 11(1), 65–78.
69. Deb, K., Pratap, A., Agarwal, S., & Meyarivan, T. (2002). A Fast & Elitist Multiobjective Genetic Algorithm: NSGA-II, *IEEE Transactions on Evolutionary Computation*, 6(2), 182–197.
70. Gen, M., Wen, F. & Ataka, S. (2007). Intelligent Approach to Car Navigation System for ITS in Japan, *Proceedings of International Conference on Computers, Communication & Systems*, 19–26.

Index

GPSR Compliance

The European Union's (EU) General Product Safety Regulation (GPSR) is a set of rules that requires consumer products to be safe and our obligations to ensure this.

If you have any concerns about our products, you can contact us on ProductSafety@springernature.com

In case Publisher is established outside the EU, the EU authorized representative is:

Springer Nature Customer Service Center GmbH
Europaplatz 3
69115 Heidelberg, Germany

Batch number: 10755036

Printed by Printforce, the Netherlands